The ASQ Certified Software Quality Engineer Handbook

The ASQ Certified Software Quality Engineer Handbook

Third Edition

Linda Westfall

ASQExcellence
Milwaukee, Wisconsin

Published by ASQExcellence, Milwaukee, WI
Produced and distributed by Quality Press, ASQ, Milwaukee, WI

Publisher's Cataloging-in-Publication Data

Names: Westfall, Linda, 1954–, author.
Title: The ASQ certified software quality engineer handbook, third edition / Linda Westfall.
Description: Includes bibliographical references and index. | Milwaukee, WI: ASQExcellence, 2024.
Identifiers: LCCN: 2024942717 | ISBN: 9781636941257 (hardcover) | 9781636941264 (pdf) | 9781636941271 (epub)
Subjects: LCSH Electronic data processing personnel—Certification. | Computer software—Examinations—Study guides. | Computer software—Quality control. | BISAC COMPUTERS / Software Development & Engineering / Quality Assurance & Testing | COMPUTERS / Computer Engineering | TECHNOLOGY & ENGINEERING / Quality Control | TECHNOLOGY & ENGINEERING / Systems Engineering
Classification: LCC QA76.3 .W466 2024 | DDC 004.092—dc23

ASQ and ASQExcellence advance individual, organizational, and community excellence worldwide through learning, quality improvement, and knowledge exchange.

Bookstores, wholesalers, schools, libraries, businesses, and organizations: Quality Press and ASQExcellence books are available at quantity discounts for bulk purchases for business, trade, or educational uses. For more information, please contact Quality Press at 800-248-1946 or books@asq.org.

To place orders or browse the selection of all Quality Press and ASQExcellence titles, visit our website at: http://www.asq.org/quality-press.

28 27 26 25 24 GP 6 5 4 3 2 1

Quality Press
600 N. Plankinton Ave.
Milwaukee, WI 53203-2914
Email: books@asq.org
Excellence Through Quality™

ASQE™
Inspire Excellence

For Robert Westfall, my husband, my partner, my best friend, and my playmate. Thank you for all of your support and patience while I wrote the three editions of this book and as I volunteered countless hours to Quality Press and other organizations over the years. Thank you for sharing your life with me, making me laugh out loud, cooking all of those fantastic meals, and sharing your passion for fireworks with me. Life with you continues to be a blast!!!

Table of Contents

Part III System and Software Engineering Processes

Part IV Project Management

Part V Software Metrics and Analysis

Part VI Software Verification and Validation

Part VII Software Configuration Management

List of Figures and Tables

Preface

Continuous improvement is a mantra implicit in the quality profession. So as software quality engineers, we should not be surprised that our own discipline has continued to evolve and change. By "practicing what we preach" in our field, adopting "lessons learned" from implementing software quality principles and practices, and proactively staying involved in managerial, procedural, and technological advances in software engineering and the quality arena, software quality engineers have learned to increase the value they add to the end software products.

One of the primary roles of a software quality engineer is to act as a management information source that keeps software quality as visible to software management as cost and schedule are when business plans and decisions need to be made. In order to fulfill this role, software quality engineers must continuously improve their skills and knowledge sets. The software quality profession has moved beyond the limits of using only testing or auditing as the primary tools of our trade. Software quality has emerged as a multifaceted discipline that requires us, as software quality engineers, to be able to understand and apply knowledge that encompasses:

- *Software quality management*: The processes and activities involved in setting the organization's strategic quality goals and objectives; establishing organizational, project, and product quality planning; and providing the oversight necessary to ensure the effectiveness and efficiency of the organization's quality management system. Software quality management provides leadership and establishes an integrated, cross-functional culture where producing high-quality software is "just the way we do things around here."
- *Software quality engineering*: The processes and activities needed to define, plan, and implement the quality management system for software-related processes, projects, and products. This includes defining, establishing, and continuously improving software-related systems, policies, processes, and work instructions that help prevent defects and build quality into the software.

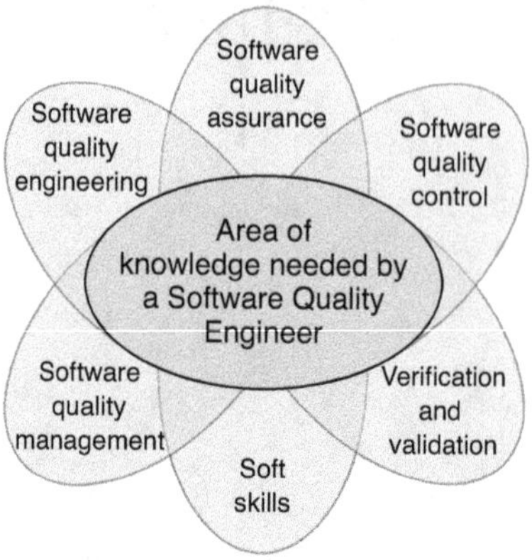

- *Software quality assurance*: The planned and systematic set of all actions and activities needed to provide adequate confidence that the:
  - Software work products conform to their standards of workmanship, and that quality is being built into the products
  - Organization's quality management system (or each individual process) is adequate to meet the organization's quality goals and objectives, is appropriately planned, documented, and improved, is being followed, and is effective and efficient.
- *Software quality control*: The planned and systematic set of all actions and activities needed to monitor and measure software projects, processes, and products to ensure that special causes have not introduced unwanted variation into those projects, processes, and products.
- *Software verification and validation*: The processes and activities used to ensure that software products meet their specified requirements and intended use. Verification and validation help ensure that the "software was built right" and the "right software was built."
- *Soft skills*: A software quality engineer also needs what are referred to as the "soft skills" to be effective in influencing others toward quality. Examples of "soft skills" include leadership, team building, facilitation, communication, motivation, conflict resolution, negotiation, and more.

The ASQ Certified Software Quality Engineer (CSQE) Body of Knowledge (BoK) is a comprehensive guide to the "common knowledge" software quality engineers should possess about these areas. To keep the CSQE BoK current with industry and practitioner needs, a modernized version of the CSQE BoK is released on a periodic basis. This handbook contains information and guidance that supports all of the topics of the 2023 version of the CSQE BoK upon which the CSQE exam is based. Armed with the knowledge presented in this handbook to complement the required years of actual work experience, qualified software quality practitioners may feel confident they have taken appropriate steps in preparation for the ASQ CSQE exam.

However, my goals for this handbook go well beyond it being a CSQE exam preparation guide. I designed this handbook not only to help software quality engineers but as a resource for software development practitioners, project managers, organizational managers, other quality practitioners, and other professionals who need to understand the aspects of software quality that impact their work. It can also be used to benchmark their (or their organization's) understanding and application of software quality principles and practices against what is considered a cross-industry "good practice" baseline. After all, by taking stock of our strengths and weaknesses, we can develop proactive strategies to leverage software quality as a competitive advantage.

New software quality engineers can use this handbook to gain an understanding of their chosen profession. Experienced software quality engineers can use this handbook as a reference source when performing their daily work. I also hope that trainers and educators will use this handbook to help propagate software quality engineering knowledge to future software practitioners and managers. Finally, this handbook strives to establish a common vocabulary that software quality engineers and others in their organizations can use to communicate about software and quality, thus increasing the professionalism of our industry and eliminating the wastes that can result from ambiguity and misunderstandings.

For me, personally, obtaining my CSQE certification, participating in the development of the ASQ CSQE program, and even the writing of this book were more about the journey than the destination. I have learned many lessons from my colleagues, clients, and students since I first became involved with the ASQ CSQE effort in 1992, as well as during my 45-plus-year career in software. I hope that you will find value in these "lessons learned" as they are embodied in this handbook. Best wishes for success in your software quality endeavors!

Linda Westfall
lwestfall@westfallteam.com

Acknowledgments

I would like to thank Louise Tamres and Eloise Henry, who helped review this third edition of my handbook.

I would also like to thank all the people who helped review the first and second editions of this handbook: Cathy Vogelsong, Brenda Fisk, Geree Streun, Watson Chan, Louise Tamres, Stuart Yarost, Robin Dudash, Darius Panahpour, Doug Hamilton, T. Scott Ankrum, Sue Carroll, Nancy Pasquan, Theresa Hunt, Tom Allen, Eloise Henry, Lynneth Lohse, Zigmund Bluvband, Dan Campo, Carolee Cosgrove Rigsbee, Margery Cox, Ruth Domingos, Scott Duncan, Eva Freund, Tom Gilchrist, Steven Hodlin, James Hutchins, Yvonne Kish, Matthew Maio, Patricia McQuaid, Vic Nanda, Ponmurugarajan Thiyagarajan, Bill Trest, and Rufus Turpin.

I would like to express my appreciation to the people at Quality Press, especially Erica Barse, for helping turn this book into reality.

Finally, I would like to thank all of the people who volunteered their time, energy, and knowledge to work with the ASQ and the Software Division to turn the Certified Software Quality Engineer (CSQE) exam into reality and who continue to support the ongoing body of knowledge and exam development activities.

Certified Software Quality Engineer (CSQE) Body of Knowledge

The topics in this Body of Knowledge include additional detail in the form of subtext explanations and the cognitive level at which the questions will be written. This information will provide useful guidance for the Examination Development Committee and the candidates preparing to take the exam. The subtext is not intended to limit the subject matter or be all-inclusive of what might be covered in an exam. It is intended to clarify the type of content to be included in the exam. The descriptor in parentheses at the end of each entry refers to the highest cognitive level at which the topic will be tested. A more comprehensive description of cognitive levels is provided at the end of this document.

I. General Knowledge (16 questions)

   A. Benefits of software quality engineering within the organization

      Describe the benefits that software quality engineering can have at the organizational level. (Understand)

   B. Ethical and Legal Compliance

      1. ASQ Code of Ethics for professional conduct

         Determine appropriate behavior in situations requiring ethical decisions such as identifying conflicts of interest and recognizing and resolving ethical issues. (Evaluate)

      2. Regulatory and legal issues

         Describe the importance of compliance to federal, national, international, and statutory regulations on software development (e.g., EU GDPR and China's Data Protection Laws). Determine the impact of issues such as copyright, trademark, intellectual property rights, product liability, and data privacy [e.g., Health Insurance Portability and Accountability Act (HIPAA) and US Data Protection Laws for personal identifying information (PII)]. (Understand)

   C. Standards and models

      Define and describe the ISO standards such as 9000 and 27000, NIST Special Publication 800-53, IEEE software standards, and the Capability Maturity Model Integration (CMMI) assessment models. (Understand)

D. Leadership skills

1. Organizational leadership

Use leadership tools and techniques (e.g. organizational change management, knowledge transfer, motivation, mentoring and coaching, and recognition). (Apply)

2. Facilitation skills

Use facilitation and conflict resolution skills and negotiation techniques to manage and resolve issues. Use meeting management tools to maximize meeting effectiveness. (Apply)

3. Communication skills

Identify and apply various communication methods in oral, written, and presentation formats. Apply various techniques for working in multi-cultural environments. Identify and describe the impact that culture, communications, and Diversity, Equity, and Inclusion (DEI) can have on an organization. (Apply)

E. Team management

Use various team management skills including assigning roles and responsibilities, identifying the classic stages of team development (forming, storming, norming, performing, and adjourning), and monitoring and responding to group dynamics. Work with diverse groups and in distributed work environments. Use techniques for working with virtual, in-person, and hybrid teams. (Apply)

II. Software Quality Management (27 questions)

A. Quality management system

1. Quality goals and objectives

Design software quality goals and objectives that are consistent with business objectives and regulations. Incorporate software quality goals and objectives into high level program and project plans. Develop and use documents and processes necessary to support software quality management systems. (Create)

2. Quality metrics and monitoring

Determine software quality goals and objectives for measuring and monitoring software including software product and process metrics, and analysis and reporting techniques. (Analyze)

3. Customers and other stakeholders

   Assess and evaluate the effect of various stakeholder group requirements (e.g., quality requirements for contractual obligations) on software projects and products. (Evaluate)

4. Outsourcing

   Determine the impact that outsourced services can have on organizational goals and objectives. Identify criteria for evaluating suppliers/vendors and subcontractors including governance and quality requirements for contractual obligations. (Analyze)

5. Business continuity, data protection, and data management

   Design plans for business continuity, disaster recovery, business documentation, change management, information security, and protection of sensitive and personal data. (Analyze)

B. Methodologies

1. Cost of Quality (COQ) and Return on Investment (ROI)

   Analyze cost of quality (COQ) categories (e.g., prevention, appraisal, internal failure, and external failure) and return on investment (ROI) metrics in relation to products and processes. (Analyze)

2. Process improvement

   Define and describe elements of benchmarking, lean processes, and the six sigma methodology. Use the Define, Measure, Act, Improve, Control (DMAIC) model and the Plan-Do-Check-Act (PDCA) model for process improvement. (Analyze)

3. Corrective and preventive action procedures

   Evaluate corrective and preventive action procedures and impact assessments related to software defects, process nonconformances, and other quality system deficiencies. (Evaluate)

4. Defect prevention

   Design and use defect prevention processes such as technical reviews, software tools and technology, and special training. (Evaluate)

5. Metrics and monitoring

   Apply metrics and monitoring techniques for adherence to software quality goals and objectives, process improvement, and defect prevention. (Apply)

C. Audits

1. Audit types

Define and distinguish between various audit types (e.g., process, compliance, supplier, system, internal, and external) and methods (e.g., virtual). (Understand)

2. Audit roles and responsibilities

Identify roles and responsibilities for audit participants including the client, lead auditor, audit team members, and auditee. (Understand)

3. Audit process

Define and describe the steps in conducting an audit, developing and delivering an audit report, and determining appropriate follow-up activities. (Apply)

III. System and Software Engineering Processes (32 questions)

A. Lifecycles and process models

1. Waterfall software development lifecycle

Apply the waterfall lifecycle and related process models and identify their benefits and when they are used. (Apply)

2. Incremental/iterative software development lifecycles

Apply the incremental and iterative lifecycles and related process models (e.g., rapid application development methodology) and identify their benefits and when they are used. (Apply)

3. Agile software development lifecycle

Apply the agile lifecycle and related process models (e.g., lean agile and SCRUM methodology) and identify their benefits and when they are used. (Apply)

4. DevOps

Employ DevOps to combine software development and IT operations to continuously release software. (Apply)

B. Systems architecture

Identify and describe various architectures (e.g., embedded systems, client-server, n-tier, web, wireless, messaging, and collaboration platforms) and analyze their impact on quality. (Analyze)

C. Cloud computing models and platforms

Describe cloud computing models such as Software as a Service (SaaS), Platform as a Service (PaaS), and Infrastructure as a Service (IaaS) and cloud computing platforms such as multi-tenant systems and distributed systems. (Understand)

D. Requirements engineering

1. Product requirements

   Define and describe various types of product and user requirements including but not limited to system, feature, function, non-functional, interface, integration, performance, globalization, and localization. (Understand)

2. Data/information requirements

   Define and describe various types of data and information requirements including data management, data integrity, and privacy by design. (Understand)

3. Quality requirements

   Define and describe various types of quality requirements, including reliability, usability. (Understand)

4. Compliance requirements

   Define and describe various types of regulatory, safety, and data retention requirements. (Understand)

5. Security requirements

   Define and describe various types of security requirements including data security, information security, cybersecurity, data privacy, and encryption requirements for data in transit and at rest. (Understand)

6. Derived requirements

   Apply derived requirements such as environmental requirements and user-defined security. (Apply)

7. Nonfunctional requirements

   Define and describe various types of nonfunctional requirements such as documentation, performance, and technical maintenance. (Understand)

8. Requirements elicitation methods

   Describe and use various requirements elicitation methods such as customer needs analysis, use cases, human factors studies, usability prototypes, Joint Application Development (JAD), and storyboards. (Apply)

9. Requirements evaluation

   Assess the completeness, consistency, correctness, and testability of requirements and determine their priority. (Evaluate)

E. Requirements management

1. Requirements change management

   Assess the impact that changes to requirements will have on software development processes for all types of life-cycle models. (Evaluate)

2. Bidirectional traceability

   Use various tools and techniques to ensure bidirectional traceability from requirements elicitation and analysis through design and testing. (Apply)

F. Software analysis, design, and development

1. Design methods

   Identify the steps used in software design and their functions. Define and distinguish between software design methods. (Understand)

2. Quality attributes and design

   Analyze the impact that quality-related elements (e.g., safety, security, reliability, usability, reusability, and maintainability) can have on software design. (Analyze)

3. Software reuse

   Define and distinguish between software reuse, reengineering, and reverse engineering and describe the impact these practices can have on software quality. (Understand)

4. Software development tools

   Analyze and select the appropriate development tools for modeling, code analysis, requirements management, and documentation. (Analyze)

G. Maintenance management

1. Maintenance types

   Describe the characteristics of corrective, adaptive, perfective, and preventive maintenance types. (Understand)

2. Maintenance strategy

   Describe various factors affecting the strategy for software maintenance including service-level agreements (SLAs), short- and long-term costs, maintenance releases, and product discontinuance and their impact on software quality. (Understand)

3. Customer feedback management

   Describe the importance of customer feedback management including quality of product support and post-delivery issues analysis and resolution. (Understand)

IV. Project Management (16 questions)

A. Planning, scheduling, and deployment

1. Project planning

   Use product acceptance, configurating acceptance, forecasts, resources, schedules, task, cost estimates, etc. to develop project plans. (Apply)

2. Work breakdown structure (WBS)

   Use Work Breakdown Structure (WBS) in scheduling and monitoring projects. (Apply)

3. Project deployment

   Use various tools including milestones, objectives achieved, and task duration to set goals and deploy projects. (Apply)

B. Tracking and controlling

1. Phase transition control

   Use various tools and techniques such as entry/exit criteria, quality gates, Gantt charts, and integrated master schedules to control phase transitions. (Apply)

2. Tracking methods

   Calculate project-related costs such as earned value, deliverables, and productivity and track the results against project baselines. (Apply)

3. Project reviews

   Use various types of project reviews such as phase-end, management, and retrospectives and post-project reviews to assess project performance and status, to review issues and risks, and to discover and capture lessons learned from the project. (Apply)

4. Program reviews

   Define and describe various methods for reviewing and assessing programs in terms of performance, technical accomplishments, and resource utilization. (Understand)

C. Risk management

1. Risk management methods

   Use risk management techniques (e.g., risk impact assessment, prevention, mitigation, and transfer) to evaluate project risks. (Evaluate)

2. Software security risks

   Evaluate risks specific to software security including deliberate attacks (e.g., hacking and sabotage), inherent defects that allow unauthorized access to data, and other security breaches. Plan appropriate responses to minimize their impact. Use threat analysis and modeling tools to identify and prevent potential security risks. (Evaluate)

3. Safety and hazard analysis

   Evaluate safety risks and hazards related to software development and implementation and determine appropriate steps to minimize their impact. (Evaluate)

V. Software Metrics and Analysis (21 questions)

A. Process and product measurement

1. Terminology

   Define and describe metric and measurement terms such as reliability, internal and external validity, explicit and derived measures, and variation. (Understand)

2. Software product metrics

   Choose appropriate metrics to assess various software attributes (e.g., size, complexity, the amount of test coverage needed, requirements volatility, and overall system performance). (Apply)

3. Software process metrics

   Measure the effectiveness and efficiency of software processes (e.g., Functional Verification Tests (FVT), cost, yield, customer impact, defect detection, defect containment, Total Defect Containment Effectiveness (TDCE), Defect Removal Efficiency (DRE), and process capability). (Apply)

4. Data integrity

   Describe the importance of data integrity from planning through collection and analysis. Apply various techniques to ensure data quality, accuracy, completeness, and timeliness. (Apply)

B. Analysis and Reporting Techniques

1. Metric reporting tools

   Use various metric representation tools, such as dashboards and stoplight charts to report results. (Apply)

2. Classic quality tools

   Describe the appropriate use of classic quality tools (e.g., flowcharts, Pareto charts, cause and effect diagrams, control charts, and histograms). (Apply)

3. Problem-solving tools

   Describe the appropriate use of problem-solving tools (e.g., affinity, tree, matrix, activity network, root cause analysis, and Data Flow Diagrams (DFD)). (Apply)

VI. Software Verification and Validation (32 questions)

A. Theory

1. V&V methods

   Use software verification and validation methods (e.g., static analysis, structural analysis, mathematical proof, simulation, and automation) and determine which tasks should be iterated because of modifications. (Apply)

2. Software product evaluation

   Use various evaluation methods on documentation, source code, etc. to determine whether user needs and project objectives have been satisfied. (Analyze)

B. Test planning and design

1. Test Strategies

   Select and analyze test strategies (e.g., test-driven design, good-enough, risk-based, time-box, top-down, bottom-up, black-box, white-box, simulation, automation, and continuous) for various situations such as Software as a Service (SaaS) and new product applications. (Analyze)

2. Test plans

   Develop and evaluate test plans and procedures such as system, acceptance, and validation to determine whether project objectives are being met and risks are appropriately mitigated. (Create)

3. Test designs

   Select and evaluate various test designs, including fault insertion, fault-error handling, equivalence class partitioning, boundary value. (Evaluate)

4. Software tests

   Evaluate and execute various tests including unit, functional, performance, integration, regression, usability, acceptance, certification, environmental load, stress, worst-case, perfective, exploratory, and system. (Evaluate)

5. Tests of external products

   Determine appropriate levels of testing for integrating supplier, third-party, and subcontractor components and products. (Apply)

6. Test coverage specifications

   Evaluate the adequacy of test specifications such as functions, states, data and time domains, interfaces, security, and configurations that include internationalization and platform variances. (Evaluate)

7. Code coverage techniques

   Use and identify various tools and techniques to facilitate code coverage analysis techniques such as branch coverage, condition, domain, and boundary. (Apply)

8. Test environments

   Select and use simulations, test libraries, drivers, stubs, harnesses, etc. and identify parameters to establish a controlled test environment. (Analyze)

9. Test tools

   Identify and use test utilities, diagnostics, automation, and test management tools. (Apply)

10. Test data management

    Ensure the integrity and security of test data through the use of configuration controls. (Apply)

C. Reviews and inspections

   Use desk checks, peer reviews, walk-throughs, inspections, etc. to identify defects. (Apply)

D. Test execution documents

   Evaluate and manage test execution documents such as test results, defect reporting and tracking records, test completion metrics, trouble reports, and input/output specifications. (Evaluate)

VII. Software Configuration Management (16 questions)

A. Configuration infrastructure

1. Configuration management team

Describe the roles and responsibilities of a configuration management group. (Understand) [NOTE: The roles and responsibilities of the configuration control board (CCB) are covered in area VII.C.2.]

2. Configuration management tools

Describe configuration management tools as they are used for managing libraries, build systems, and defect tracking systems. (Understand)

3. Library processes

Describe dynamic, static, and controlled library processes and related procedures such as check-in/check-out and merge changes. (Understand)

B. Configuration identification

1. Configuration items

Describe software configuration items (e.g., baselines, documentation, software code, and equipment) and identification methods (e.g., naming conventions and versioning schemes). (Understand)

2. Software builds and baselines

Describe the relationship between software builds and baselines and describe methods for controlling builds and baselines (e.g., automation and new versions). (Understand)

C. Configuration control and status accounting

1. Item change and version control

Describe processes for documentation control, item change tracking, and version control that are used to manage various configurations. Describe processes used to manage configuration item dependencies in software builds and versioning. (Understand)

2. Configuration Control Board (CCB)

Describe the roles, responsibilities and processes of the Configuration Control Board (CCB). (Understand) [NOTE: The roles and responsibilities of the configuration management team are covered in area VII.A.1.]

3. Concurrent development

   Describe the use of configuration management control principles in concurrent development processes. (Understand)

4. Status accounting

   Apply various processes for establishing, maintaining, and reporting the status of configuration items such as baselines, builds, and tools. (Apply)

D. Configuration Audits

Define and distinguish between functional and physical configuration audits and how they are used in relation to product specification. (Understand)

E. Product release and distribution

1. Product release

   Assess the effectiveness of product release processes: planning, scheduling, and defining hardware and software dependencies. (Evaluate)

2. Customer deliverables

   Assess the completeness of customer deliverables including packaged, hosted, and downloadable products; license keys and user documentation; and marketing and training materials. (Evaluate)

3. Archival processes

   Assess the effectiveness of source and release archival processes: backup planning and scheduling, data retrieval, archival of build environments, retention of historical records, and offsite storage. (Evaluate)

LEVELS OF COGNITION

Based on Bloom's Taxonomy—Revised (2001)

In addition to **content** specifics, the subtext for each topic in this BOK also indicates the intended **complexity level** of the test questions for that topic. These levels are based on "Levels of Cognition" (from Bloom's Taxonomy—Revised, 2001) and are presented below in rank order, from least complex to most complex.

Remember

Recall or recognize terms, definitions, facts, ideas, materials, patterns, sequences, methods, principles,

Understand

Read and understand descriptions, communications, reports, tables, diagrams, directions, regulations,

Apply

Know when and how to use ideas, procedures, methods, formulas, principles, theories,

Analyze

Break down information into its constituent parts and recognize their relationship to one another and how they are organized; identify sublevel factors or salient data from a complex scenario.

Evaluate

Make judgments about the value of proposed ideas, solutions by comparing the proposal to specific criteria or standards.

Create

Put parts or elements together in such a way as to reveal a pattern or structure not clearly there before; identify which data or information from a complex set is appropriate to examine further or from which supported conclusions can be drawn.

Part I

General Knowledge

Chapter 1

Benefits of Software Quality Engineering within the Organization

> Describe the benefits that software quality engineering can have at the organizational level. (Understand)
>
> **Body of Knowledge I.A**

QUALITY DEFINED

Because this is a book about software quality engineering, it seems appropriate to start with a definition of *quality*. However, the industry has not, and may never, come to a single definition of the term *quality*. For example, the Institute of Electrical and Electronics Engineers (IEEE), the International Organization for Standardization (ISO), and the International Electrotechnical Commission (IEC) released *Systems and Software Engineering—Vocabulary* (IEEE/ISO/IEC 2017) has the following set of definitions for quality:

1. The degree to which the system satisfies the stated and implies needs of its various stakeholders, and thus provides value
2. The ability of a product, service, system, component, or process to meet customer or user needs, expectations, or requirements
3. The degree to which a set of inherent characteristics fulfills requirements

Based on his studies of how quality is perceived in various domains (e.g., philosophy, economics, marketing, and operations management), Garvin (1984) concluded, "Quality is a complex and multifaceted concept." He describes quality from five different perspectives:

- *Manufacturing perspective*. This perspective matches how Crosby (1984) defines quality in terms of conformance to the specification. Crosby's point is that an organization does not want a variety of people, throughout the development of a product, making their own judgments about what the stakeholders need or want. A well-written specification is the cornerstone for creating a quality product. Even in agile development, where requirements specifications are not formally documented, without good user stories and the subsequent conversations with the stakeholders to determine their requirements, it is difficult to write high-quality software. For software, however, this perspective of quality is necessary but may not be sufficient since, according to Wiegers and Beatty (2013), errors made during the requirements stage account for 40 percent to 50 percent of all defects found in a software project.

From another viewpoint, this perspective refers to the ability *to manufacture (replicate)* a product to that specification over and over within accepted tolerances. Before an organization can adjust or improve its processes, that organization must let them run long enough to understand what is really being produced. Then the organization can design its specifications to reflect the real process capabilities and work to improve those capabilities by removing variance from those processes. While the primary focus of software quality is on design and development activities, this manufacturing and "variance removal" quality perspective reminds software organizations that the replication or download process cannot be completely ignored and must be verified as well.

- *User perspective.* Juran (1999) cites "fitness for use" as the appropriate measure for quality. Software practitioners can probably all relate stories of software products that conformed to their specifications but did not function adequately when deployed into operations. This perspective of quality considers both the viewpoints of the individual users and their context of use.

  For example, what a novice user might consider a quality user interface might drive a power user to distraction with pop-up help and warning messages that require responses. Also, what is a secure-enough interface for a software database used for personal information at home might be woefully inadequate in a business environment.

- *Product perspective.* Quality is tied to the inherent characteristics of the product. These characteristics are the *quality attributes,* also called the "ilities" of the software product. Examples include reliability, usability, accessibility, availability, flexibility, maintainability, portability, installability, adaptability, and so on. Of course, they do not all end in *-ility*. Correctness, fault tolerance, integrity, efficiency, security, and safety are also examples of quality attributes (see Chapter 12 for more information about quality attributes). The more the software has high levels of these characteristics, the higher its quality is considered to be. The ISO/IEC 25000 *Software Engineering—Software Product Quality Requirements and Evaluation* (SQuaRE) standards series provides a reference model and definitions for external and internal quality attributes and quality-in-use attributes (ISO/IEC 2014). This standards series also provides guidance for specifying requirements, planning and managing, measuring, and evaluating quality attributes.

- *Transcendental perspective.* Quality is something that can be recognized but not defined. As stated by Kan (2003), "to many people, quality is similar to what a federal judge once said about obscenity: 'I know it when I see it.'" This perspective of quality takes the viewpoint of the individual into consideration. What is "obscenity" to one person may be "art" to another. What one stakeholder considers good software quality may not be high enough quality for another stakeholder. In his writings, Tom Peters cites the customers' (stakeholders') reaction as the only appropriate measure for the quality of a product. This requires that product developers keep in touch with their stakeholders to make certain that the specifications accurately reflect those stakeholders' real (and possibly changing) needs and that value is being built into the software for those stakeholders.

- *Value-based perspective.* Quality is dependent on the amount a customer is willing to pay for it. This perspective leads to considerations of "good enough" software quality. Would people be willing to pay twice as much for the word processor for their home computer if it was twice as reliable? Would people be willing to pay an extra thousand dollars for a car if it meant more secure (less likely to be hacked) automated accident avoidance or self-parking systems? How much extra are people willing to pay for high-quality software in medical devices or high-quality software in airplane navigation systems? This value-added perspective can be expanded to include other stakeholders besides just customers. High-quality software has a positive benefit-to-cost ratio for one or more stakeholders.

BENEFITS OF SOFTWARE QUALITY ENGINEERING

Software quality engineering is the study and systematic application of scientific, technological, economic, social, and practical knowledge and empirically proven methods to the analysis and continuous improvement of all stages of the software life cycle to maximize the quality of software processes and practices, and the products they produce.

At its most basic, increasing the quality of the software means reducing the number of defects in that software and in the processes that are used to develop and maintain that software. Defects in the software can result from mistakes that occurred during the software development (either in-house development or development by a third party), release, and maintenance processes, or from defects in the processes themselves. These mistakes introduce defects into the software work products. Mistakes can also lead to missing, *ambiguous,* or incorrect requirement defects that result in the development of software that does not match the needs of its stakeholders. The most cost-effective way of handling a defect is to prevent it. In this case, software quality is accomplished through continuous process improvement, increasing staff knowledge and skill, and other defect prevention techniques that keep defects out of the software. Every defect that is prevented eliminates *rework* to correct that defect and also eliminates the effort associated with that rework.

If a defect does get interjected into the software, the more quickly the defect is identified and corrected, the less rework effort is typically required to correct it. Eliminating the waste of excessive rework allows organizations to use the saved effort to produce additional value-added software. In this case, software quality

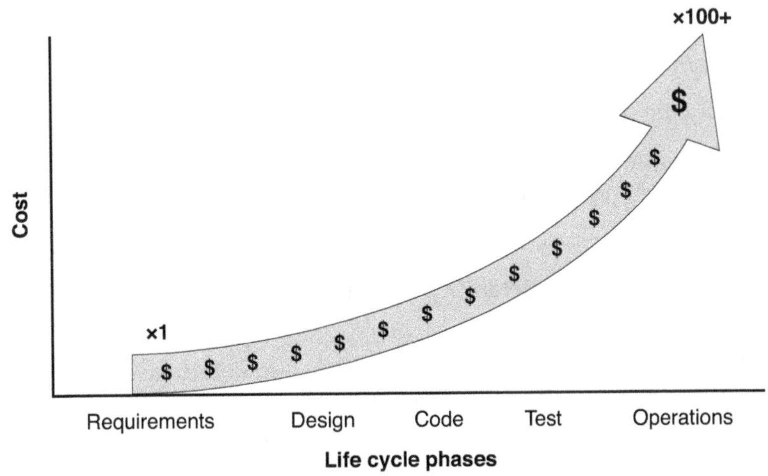

Figure 1.1 Cost of fixing defects.

is accomplished through techniques that improve defect detection and find the defects earlier.

For example, as illustrated in Figure 1.1, the cost to fix a defect will typically increase exponentially as it is found later and later in the life cycle. In fact, studies show that it can cost 100-plus times more to fix a requirement defect if it is not found until after the software is released into operations than it would have cost to fix that same defect if it was found in the requirements phase (Kan 2003; Pressman and Maxim 2015). The main point here is that software development involves a series of dependencies, where each subsequent activity potentially builds on and expands the products of previous activities. For example, when using traditional software development methods, a single requirement could result in four design elements that expand into seven *source code* modules. All of these have supporting documentation or tests. Therefore, if a defect in that requirement is not found until the design phase, costs increase because the requirement would have to be fixed, and the four design elements would have to be investigated and potentially fixed. If the defect is not found until the coding phase, the requirement would have to be fixed, and the four design elements and seven source code modules would have to be investigated and potentially fixed. By the coding phase, test cases (system, integration, and unit) and user documentation may also have been written based on the defective requirement, which would need to be investigated and potentially corrected. If the defect is not found until the testing phases, all of the work products based on that requirement would have to be investigated and potentially corrected, retested, and regression tested.

Agile methods attack these costs by significantly shortening the development cycle using shortened iterative or incremental development cycles, and through other techniques that shorten defect interjection to correction cycle times. At the end of each iteration, the goal is to have working software that can be demonstrated to the stakeholders, who provide feedback on the correctness and quality of the software.

Both studies and empirical evidence demonstrate that improving software quality benefits the development organization by:

- Reducing development and maintenance costs
- Decreasing process cycle times, which translate into shortened schedules and improvements in time-to-market
- Increasing the productivity, or velocity, of software practitioners
- Increasing the predictability of costs and schedules

Both defect *prevention* and defect detection help keep software defects from being delivered into operations. If fewer software defects are delivered, there is a higher probability of failure-free operations. Unlike hardware, software does not wear out with time. If a defect is not encountered during operations, the software performs reliably. Reliable software can increase the effectiveness and efficiency of work being done using that software. Reliable software reduces operational, failure, and maintenance costs to the software's stakeholders and thus reduces the overall *cost of ownership* of the software product to its customers and users. Reliable software also reduces software corrective maintenance costs to the organization that developed the software.

Taking a broader view, high-quality software is software that has been specified correctly and that meets its specification. If the software meets the stakeholders' needs and expectations and is value-added, it is more likely to be used instead of ending up as "shelfware." If the customers and users receive software that has fewer defects, is more reliable, and performs to their needs and expectations, then those customers and users will be more satisfied with the software. This is illustrated in Figure 1.2, which depicts Kano's model of the relationship between stakeholder satisfaction and quality. Kano talks about three types of quality, including the following:

- *Basic quality*. There is a basic level of quality that a stakeholder expects the product to have. This basic quality comes from satisfying the requirements that are assumed by the stakeholder to be part of the product and are typically not explicitly stated or requested during requirements elicitation activities. For example, people buying a new car expect that car to have four tires, a windshield, windshield wipers, and a steering wheel. They will not ask for these items when purchasing a new car; they just expect them to be there. This seems rather obvious for a car—people who build cars know what these basic requirements are and build them into their cars. But this is not necessarily true in software. A software developer may be writing telecommunications software one year and writing aerospace software the next. The developer knows software, but without a comprehensive knowledge of the customer's or user's business domain, it is easy to miss requirements that are assumed to be part of the product. Most experienced software developers can relate stories where the delivered product met its requirements but not its intended use because one of these assumed requirements was missed. Our stakeholders then hear the "you never told me that was a requirement" refrain while we argue that it is an

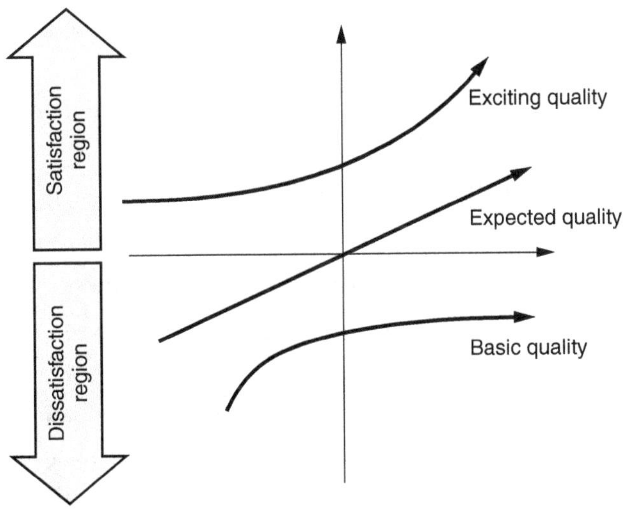

Figure 1.2 Kano model (Pyzdek 2000).

enhancement that they should be paying for, all the while they think we have just delivered a car without a windshield. This level of quality does not satisfy the stakeholders. (Note that the entire "basic quality" line is in the dissatisfaction region.) However, the absence of quality at this level will quickly increase a stakeholder's dissatisfaction.

- *Expected quality.* The "expected quality" line on the graph in Figure 1.2 represents satisfying those requirements that the stakeholder explicitly considers and requests. For example, a buyer will state preferences for the make, model, and options when shopping for a car. The stakeholder will be dissatisfied if this level of quality is not met. The stakeholder will be increasingly satisfied as this quality level increases when more and more of the stated requirements are satisfied by the software.
- *Exciting quality.* This quality level represents unrequested but innovative requirements. These are requirements that the stakeholders do not know they want, but will love when they see them. For example, when cup holders were first introduced in cars they were positively received. Note that the entire "exciting quality" line is in the satisfaction region. It should be remembered, however, that today's innovations are tomorrow's expectations. Now, in fact, most new car purchasers consider a cup holder to be part of the basic requirements for a car. However, care must always be taken to evaluate any innovative quality items to make sure that they are truly value-added to the stakeholders. This analysis should make certain that these innovations do not lead to "gold-plating" and overdone functionality that will lead to the software costing too much; taking too long to market; impacting software performance, safety, security, or other needed software attributes; and so on.

Increased stakeholder satisfaction can lead to the following benefits for the software development organization:

- Increased market share, and the ability to charge higher prices without affecting market share
- Increased profitability or return on investment
- Improvement to the organization's reputation in the industry
- Increased customer trust, loyalty, and repeat business
- Increased ability to prosper (or at least survive) even in bad economic times
- Decreased number of customer service requests and audits

An increase in the quality of the software can also increase the satisfaction of the software practitioners. For most software engineers, their favorite activity is not burning the midnight oil trying to debug critical problems reported from operations. Producing high-quality products makes the software practitioner's job easier and less frustrating. Practitioners can also take pride in what they are doing, thus enhancing their feelings of accomplishment and self-worth. Increases in employee satisfaction benefit the organization by increasing productivity/velocity and decreasing employee turnover.

Chapter 2

Ethical and Legal Compliance

The author of this book is not an ethicist or lawyer, nor has she received any legal training. All ethical and legal issues and risks that arise within an organization, on any software project, or with any contract should be referred to a lawyer or other legal professional. The following descriptions should be used as general information and summaries for no purpose other than as refresher materials for the ASQ Certified Software Quality Engineer (CSQE) exam.

1. ASQ CODE OF ETHICS FOR PROFESSIONAL CONDUCT

> Determine appropriate behavior in situations requiring ethical decisions such as identifying conflicts of interest and recognizing and resolving ethical issues. (Evaluate)
>
> **Body of Knowledge I.B.1**

ASQ requires its members and certification holders to comply with the ASQ Code of Ethics. This code of ethics establishes and communicates the behaviors that ASQ considers essential for quality professionals. It defines three fundamental principles and the underlying tenets that quality professionals should employ during their interactions with the public, employers, clients, and peers to confirm ethical behavior. A software quality engineer is expected to perform his or her duties professionally, including acting in an ethical manner when executing work activities and interfacing with other individuals and organizations. The ASQ Code of Ethics can serve as a guide for performing work in an ethical manner.

Quality professionals can also use the ASQ Code of Ethics as a model to help their organizations define acceptable and ethical behaviors. A code of ethics is important so that people in the organization know and understand what is expected of them when ethical issues arise.

ASQ Code of Ethics Fundamental Principles

ASQ requires its representatives to be honest and transparent. Avoid conflicts of interest and plagiarism. Do not harm others. Treat them with respect, dignity and fairness. Be professional and socially responsible. Advance the role and perception of the quality profession.

Expectations of a Quality Professional

A. **Act with Integrity and Honesty**

1. Strive to uphold and advance the integrity, honor, and dignity of the quality profession.
2. Be truthful and transparent in all professional interactions and activities.
3. Execute professional responsibilities and make decisions in an objective, factual, and fully informed manner.
4. Accurately represent and do not mislead others regarding professional qualifications, including education, titles, affiliations, and certifications.
5. Offer services, provide advice, and undertake assignments only in your areas of competence, expertise, and training.

B. **Demonstrate Responsibility, Respect, and Fairness**

1. Hold paramount the safety, health, and welfare of individuals, the public, and the environment.
2. Avoid conduct that unjustly harms or threatens the reputation of the Society, its members, or the quality profession.
3. Do not intentionally cause harm to others through words or deeds. Treat others fairly, courteously, with dignity, and without prejudice or discrimination.
4. Act and conduct business in a professional and socially responsible manner.
5. Allow diversity in the opinions and personal lives of others.

C. **Safeguard Proprietary Information and Avoid Conflicts of Interest**

1. Ensure the protection and integrity of confidential information.
2. Do not use confidential information for personal gain.
3. Fully disclose and avoid any real or perceived conflicts of interest that could reasonably impair objectivity or independence in the service of clients, customers, employers, or the Society.
4. Give credit where it is due.

5. Do not plagiarize. Do not use the intellectual property of others without permission. Document the permission as it is obtained.

(ASQ 2023a)

Conflicts of Interest

Most of the ASQ Code of Ethics principles and tenets are self-explanatory. However, conflicts of interest may need a bit more definition. A *conflict of interest* occurs when an individual in a position of trust, like a software quality engineer, has competing professional or personal interests that may make it difficult for that individual to perform his or her duties in an impartial or unbiased manner. A conflict of interest exists when a quality professional is in a position that can be exploited in some way for personal gain. A potential conflict of interest exists with any business connections, interests, or affiliations that might influence an individual's judgment or impair the equitable character of that individual's work. A potential conflict of interest can create the appearance of unethical behavior, even if no unethical behavior results from that conflict. Examples where conflicts of interest can occur include:

- Previous employment, regardless of the reason for separation, with the involved parties
- Multiple jobs or clients, where the interests of one job or client conflict with those of another
- Holding a significant amount of stocks or bonds whose value may be impacted by the individual's decisions or actions
- Previous or current close working relationship with one or more of the involved parties
- Auditing work performed by the auditor or processes documented by the auditor
- Desire to be hired by one or more of the involved parties
- Close friendship or family tie with one or more of the involved parties
- Offer of money, goods, or services like a bribe, kickback, or secret commission by one or more of the involved parties
- Acceptance of gifts from one or more of the involved parties

As stated in C.3 of the ASQ Code of Ethics, the best way to mitigate conflicts of interest is to "fully disclose and avoid any real or perceived conflicts of interest." Disclosing the actual or potential conflicts of interest to all parties allows informed decisions to be made about how to proceed.

2. REGULATORY AND LEGAL ISSUES

> Describe the importance of compliance to federal, national, international, and statutory regulations on software development (e.g., EU GDPR and China's Data Protection Laws). Determine the impact of issues such as copyright, trademark, intellectual property rights, product liability, and data privacy [e.g., Health Insurance Portability and Accountability Act (HIPAA) and US Data Protection Laws for personal identifying information (PII). (Understand)
>
> **Body of Knowledge I.B.2**

Compliance to Federal, National, International, and Statutory Regulations

Regulations are rules, laws, or instructions established by legislative or regulatory bodies, which set legal standards with which organizations must comply. Those bodies then verify compliance through audits, certifications, or other means. Typically, there are penalties for nonconformance to regulations, which can include fines, discontinuation of the ability to sell products, or even jail time for organizational officers. In some cases, regulators may have a licensing role where they issue a license before a new system can be used (Sommerville 2016).

Different countries establish regulations and laws in different areas to make certain that privately owned companies "follow certain standards to confirm that their products are safe and secure" (Sommerville 2016). For example, different governmental bodies in the United States establish and enforce regulations and rules for different industries. Examples of these governmental bodies in the United States include:

- Food and Drug Administration (FDA)—Medical device software, as well as software used to develop pharmaceuticals
- Federal Aviation Administration (FAA)—Airborne systems and equipment software
- Federal Communications Commission (FCC)—Telecommunications software
- Department of Energy (DOE)—Nuclear energy software
- Security and Exchange Commission (SEC)—Financial and banking software

Other examples of regulations are:

- Export regulations may restrict the types of software or software-intensive systems that can be exported to other countries.
- Data privacy and security laws like the European Union (EU) General Data Protection Regulations (GDPR), established in 2018, which "imposes obligations onto organizations anywhere, so long as they target or collect data related to people in the EU" (GDPR 2023). China also has data protection laws, including the Personal Information Protection Law (PIPL), established in 2021. The PIPL governs personal information processing activities carried out by entities or individuals within China. The Chinese Cybersecurity Law (CSL) and the Data Security Law (DSL) on cybersecurity and data protection also introduce new data protection requirements.

Contracts

Software development often involves contracts, either between software customers and the organization developing the software (contractor), or between the development organization and their vendors or suppliers (subcontractors). *Contracts* are legally binding agreements. To be legally binding, the promises made in the contract must be exchanged for appropriate consideration. In other words, there has to be something of value received by both parties. If a legal contract is breached, the parties to the contract can seek legal remedies. When selecting the type of contract to be used, consideration should be given to minimizing the risks for both parties and motivating the suppliers to perform optimally. There are many different contract types, including fixed-price, cost-reimbursement, incentive-based, time and materials, and indefinite-delivery contracts.

In contract law, the term *warranty* has a variety of meanings. Generally, a warranty is an agreement between the seller and buyer of a product that provides assurance that certain facts, conditions, or assertions are true. A software warranty is typically a promise by the organization developing the software regarding the quality of that software and the remedial actions that will be taken if the software does not perform as promised. There are two main types of warranties:

- An *expressed warranty* is a warranty that is directly stated (or "expressed") verbally or in writing. Typically, software express warranties are included in the software license, service level agreement, or contract. For example, an express warranty might be an agreement between the organization developing the software and the buyer that the developer will provide corrective maintenance for any defects identified in the software for a specified period of time.
- An *implied warranty* is an unstated promise that comes from the understanding of the buyer or the nature of the transaction when an express warranty does not exist. In the United States, the Uniform Commercial Code provides rules around implied warranties.

Intellectual Property Rights

Intellectual property is a legal area that includes inventions and ideas of the human mind, such as books, music, artwork, and software. Intellectual property rights give the creators of these works exclusive rights to their creations. Software intellectual property rights deal with legal issues involving the copyrighting, patenting, and licensing of software products, as well as trademarks. Intellectual property rights and the associated laws and regulations vary from country to country. A software practitioner should refer any questions or issues concerning intellectual property rights to the appropriate legal authority.

According to Futrell et al. (2002), "*Patents* protect ideas and are exclusive rights for novel inventions for a limited period of time." A patent owner has the right to exclude others from manufacturing, selling, or using products that are based on the patented idea or underlying concept for a specific period after the patent is issued. Kappos (2012), undersecretary of commerce for intellectual property and director of the U.S. Patent and Trademark Office, stated that "Patents are issued for process and apparatus, which are determined to be novel and non-obvious. Because many breathtaking software-implemented innovations power our modern world, at levels of efficiency and performance unthinkable even just a few years ago, patent protection is every bit as well-deserved for software-implemented innovation as for the innovations that enabled man to fly. . . . But it is equally important that patent protection be properly tailored in scope so that programmers can write code and engineers can design devices without fear of unfounded accusations of infringement."

Copyrights protect original written works, such as software, from being copied without permission. Owning a copyright on a software product means that the owner(s) are protected under the law from other people copying, distributing, or making adaptations to their software products without their permission. Unlike a patent, a copyright does not protect the underlying idea or concept but only protects the tangible expression of that idea or concept. In the United States, fair use is a defense against copyright infringement. "Fair use is a legal doctrine that promotes freedom of expression by permitting the unlicensed use of copyright-protected works in certain circumstances. Section 107 of the Copyright Act provides the statutory framework for determining whether something is a fair use and identifies certain types of uses—such as criticism, comment, news reporting, teaching, scholarship, and research—as examples of activities that may qualify as fair use" (copyright.gov/fair-use).

Most software in the United States is sold under a licensing agreement rather than just depending on copyrights for protection. A typical *software license* grants the license holder the right to either use or redistribute one or more copies of copyrighted software without breaking copyright law. *Proprietary software licenses* grant limited rights to the user to use one or more copies of the software while the ownership of that software remains with the software development organization. Typically, proprietary software licenses have a well-defined list of restrictions on what the users can do with the software. The user must accept the terms of the proprietary software license in order to use the software.

In contrast to proprietary software licenses, the ownership of the specific copy of the software is transferred to the user with an *open source license*. However, the copyright ownership remains with the original software developer. The user may use the software without accepting the open source license, or the user can optionally accept the license, in which case the user is typically granted additional privileges. *Copyleft* is a form of licensing that allows individuals the right to disseminate copies of, or even modified versions of, a software product (or other types of works). These licenses typically stipulate that any adapted or derived versions of the original software must also be made available through the same or similar licenses. Examples of copyleft licenses are the GNU General Public License and Share-alike. Organizations developing software should take great care when using open source and other publically available software to verify that they remain in compliance with the associated licensing agreements, especially if they intend for their software products to remain proprietary.

Copyrights and licenses are also important when commercial-off-the-shelf (COTS) or third-party developed software is being integrated with or built into another software product. Additional licensing requirements or royalty fees may be involved in this case. Care must also be taken when reusing software components with integrated COTS or third-party software to verify that any licensing agreements or royalties are carried forward with those reused components.

Trademarks are devices used to "brand" products or services and distinguish them from similar products or services in the marketplace. Examples of trademarks can include words, names, slogans, logos, symbols, and even sounds or colors. A *service mark* is similar to a trademark except that it identifies and distinguishes a service rather than a product. In the United States, the symbols "TM" for trademark and "SM" for service mark can be used "as soon as one intends to make a claim to the public as to the exclusive right to use a mark" (Vienneau and Johns 2008). However, there is a formal process for registering trademarks and service marks with the U.S. federal government if additional protection is desired. An example of a software trademark is the Capability Maturity Model Integration (CMMI®), where the ® symbol identifies it as a U.S. registered trademark.

Tort

Black's Law Dictionary (as cited in Futrell, Shafer, and Shafer 2002) defines a *tort* as "a wrongful act other than a breach of contract that injures another and for which the law imposes civil liabilities." There are five types of *tort lawsuits* that might be applied to software:

1. *Conversion*. Conversion is involved if the software was intentionally designed to steal from the customer or user or to destroy property. Conversion only involves tangible personal property. For example, conversion might apply if the software was intentionally designed to round financial transactions downward to the nearest penny and then deposit the fractions of a penny into someone's personal bank account.

2. *Negligence*. If the developer of a software system failed to take steps that a reasonable software developer would take, and because of this failure the software caused personal injury or property damage, that

developer might be liable to damages under tort law. There are four aspects to any negligence lawsuit:

- *Duty* means that there was an obligation to behave in a certain way (e.g., in a way that does not create an unreasonable risk of injury or property damage). For example, a private citizen has no duty to stop and render aid if he or she drives past an automobile accident with injuries. However, a doctor's license creates a legal duty to provide that aid. Another example would be that an organization developing a software product might create a duty based on provisions in their license or contract.
- *Negligent breach of duty* means that reasonable precautions were not taken to make sure that duty is satisfied. Negligence might occur if the software development organization did not use currently available "good software practices" when developing their software product.
- *Causation* means that the software had to cause injury or property damage.
- *Actual damages* had to result from the negligence (the award may include compensatory damages to compensate for the actual damages suffered, as well as punitive damages to punish the offending party).

3. *Strict product liability.* This type of tort lawsuit might be appropriate if the software caused injury or property damage because it is dangerously defective.
4. *Malpractice.* The software's author (or the program itself) provides unreasonably poor professional services—for example, if an accounting software application did not apply the generally accepted accounting principles (GAAP), or if an income tax package did not implement the tax laws.
5. *Fraud.* Fraud exists if the seller of the software knowingly misrepresents the capabilities of the product.

Data Privacy

Data privacy issues exist whenever personally identifiable information is collected, transmitted, or stored by the software. This information can include financial or credit information (including credit card information), medical records, personal identification information (e.g., social security numbers or mailing addresses), lifestyle information (religion, organizational or political affiliations, sexual preferences), and so on. Privacy laws around the world vary greatly, so the software development team must be aware of the laws that apply to their product and market. In the United States, data privacy laws include the following:

- The Health Insurance Portability and Accountability Act (HIPAA) is aimed at preventing a patient's health information from being disclosed without their permission.
- The U.S. data protection laws for personal identifying information (PII) are a set of hundreds of state and federal laws aimed at protecting personal data.

The privacy risk is that this information will be revealed to or obtained by unauthorized parties. One of the intents of software security measures is to prevent unauthorized access to private information. Data privacy also applies to encryption and how it applies to export, tort, and criminal laws. For example, based on the laws of individual countries, some encryption algorithms cannot be exported.

Compliance

Software practitioners must be aware of the regulatory and legal issues associated with the software they develop and verify compliance with any applicable laws or regulations to protect the public, their organizations, and themselves. When in doubt, software practitioners should seek appropriate advice from legal or regulatory experts.

Chapter 3
Standards and Models

> Define and describe the ISO standards such as 9000 and 27000, NIST Special Publication 800-53, IEEE software standards, and the Capability Maturity Model Integration (CMMI) assessment models. (Understand)
>
> **Body of Knowledge I.C**

A standard defines a disciplined, consistent approach to software development and other activities through the specification of rules, requirements, guidelines, or characteristics. Standards aim at promoting optimum community or organizational benefits and should be based on the combined results of science, technology, and practical experience.

A standard is used as a basis for comparison when specifying, developing, reviewing, auditing, assessing, or testing a system, process, or product. An organization or its personnel *comply* with standards when a comparison between what the standard says should be done matches what the organization or its personnel are actually doing. Products *conform* to standards when a comparison between what the standard requires matches the actual characteristics, content, or status of the product. A standard is usually specified by standard practice, or it is defined by a designated standards body (e.g., ISO, IEC, IEEE, OMG, and so on). A standard can specify requirements for an item or activity, including:

- *Size*: For example, an external interface standard might specify a communication packet size of 32 bytes
- *Content*: For example, the *IEEE Standard for Configuration Management in Systems and Software Engineering* (IEEE 2012) specifies what should be included in a Configuration Management Plan
- *Value*: For example, an external interface standard might specify values for error codes transmitted across that interface
- *Quality*: For example, modeling and coding standards provide workmanship standards for producing quality software work products, and the ISO 9001:2015 standard (ISO 2015) specifies the requirements for the effective implementation of a quality management system (QMS)

At the organizational level, standards make it easier for professionals to move between project and product teams within the organization, reducing the effort required for training. Using standards, the software developed by different groups within the organization is more consistent and uniform and, therefore, easier to integrate and reuse. The fact that everyone involved knows and understands the standard way of acquiring, developing, and maintaining the software products permits a uniform method for reviewing the product's status and the project.

At the industry level, standards can increase the professionalism of a discipline by providing access to good practices, as defined by experienced practitioners in the software industry. Many companies benchmark the ISO and IEEE standards as a basis for improving their processes and practices. Standards can also help introduce new technologies and methods into the software industry. For example, the Systems Modeling Language (SysML) Standards from the Object Management Group (OMG 2015) helped introduce a consistent methodology that can be used to specify, analyze, design, verify, and validate object-oriented requirements and designs for systems and systems-of-systems.

It should be noted that guidelines (guides) are different from standards. Both standards and guides are typically issued by some body of authority. However, standards define requirements, while *guidelines* define suggested practices, advice, methods, or procedures that are considered good practice but not mandatory.

A *model* is an abstract representation of an item or process from a particular point of view. A model expresses the essentials of some aspect of an item or process without giving unnecessary detail. The purpose of a model is to enable the people involved to think about, discuss, and understand these essential elements without getting sidetracked by excessive or complex details. Unlike standards, models are communication vehicles and not mandatory requirements. The Capability Maturity Model Integration (CMMI®) and life cycle models (waterfall, V, or spiral) are examples of models.

ISO 9000 Family of Standards

The "International Organization for Standardization (ISO) is a worldwide federation of national standards bodies (ISO member bodies)" (ISO 2015). ISO developed the 9000 family of standards to define good practice in the area of quality management and quality assurance. These standards define the basic, first level of a quality management system. Their implementation does not guarantee high quality.

Within the ISO 9000 family, the core standards include:

- *ISO 9000:2015 Quality management systems—Fundamentals and vocabulary*
- *ISO 9001:2015 Quality management systems—Requirements*
- *ISO 9004:2009 Managing for the sustained success of an organization—A quality management approach*
- *ISO 19011:2011 Guidelines for auditing management systems*

ISO also provides a set of supporting standards and guidelines to help organizations establish and improve their quality management systems, their processes, or their activities. A list of these supporting standards/guidelines is included in Annex B of the ISO 9001:2015 standard (ISO 2015).

The ISO 9000 family of standards is based on seven quality management principles applicable to any organization (e.g., software, manufacturing, or service including, ISO 2015; ISO 2015a):

- Customer focus
- Leadership
- Engagement of people
- Process approach
- Improvement
- Evidence-based decision making
- Relationship management

The ISO 9001:2015 standard defines the specific set of quality management system requirements that are used by registrars to audit and certify organizations. ISO 9001:2015 takes a process approach, which incorporates the Plan-Do-Check-Act (PDCA) cycle and risk-based thinking. Figure 3.1 illustrates a model of the major clauses of ISO 9001:2015.

Some industries have created industry-specific interpretations, or add-ons, to the ISO 9000 family of standards. This has been done to standardize the interpretation of ISO 9001:2015 for their industry and to add industry-specific additional requirements. These industry-specific interpretations also help make certain that auditors are trained in and understand the specific needs of those industries. Examples of industry-specific standards include:

- *ISO/IEC/IEEE 90003:2018 Software engineering—Guidelines for the application of ISO 9001:2015 to computer software*
- AS9100 for Aviation, Space and Defense (and AS9115 specific to deliverable software)
- ISO/TS 16949 for Automotive
- TL 9000 for Telecommunications
- ISO 13485 for Medical Devices and ISO 62304 for Medical Device Software
- ISO/TX 29001 for Petroleum, Petrochemical, and Natural Gas
- NQA 1 for Nuclear

ISO/IEC 27000 Family of Standards

ISO/IEC developed the 27000 family of standards to define good practice in the area of information security management, including information technology

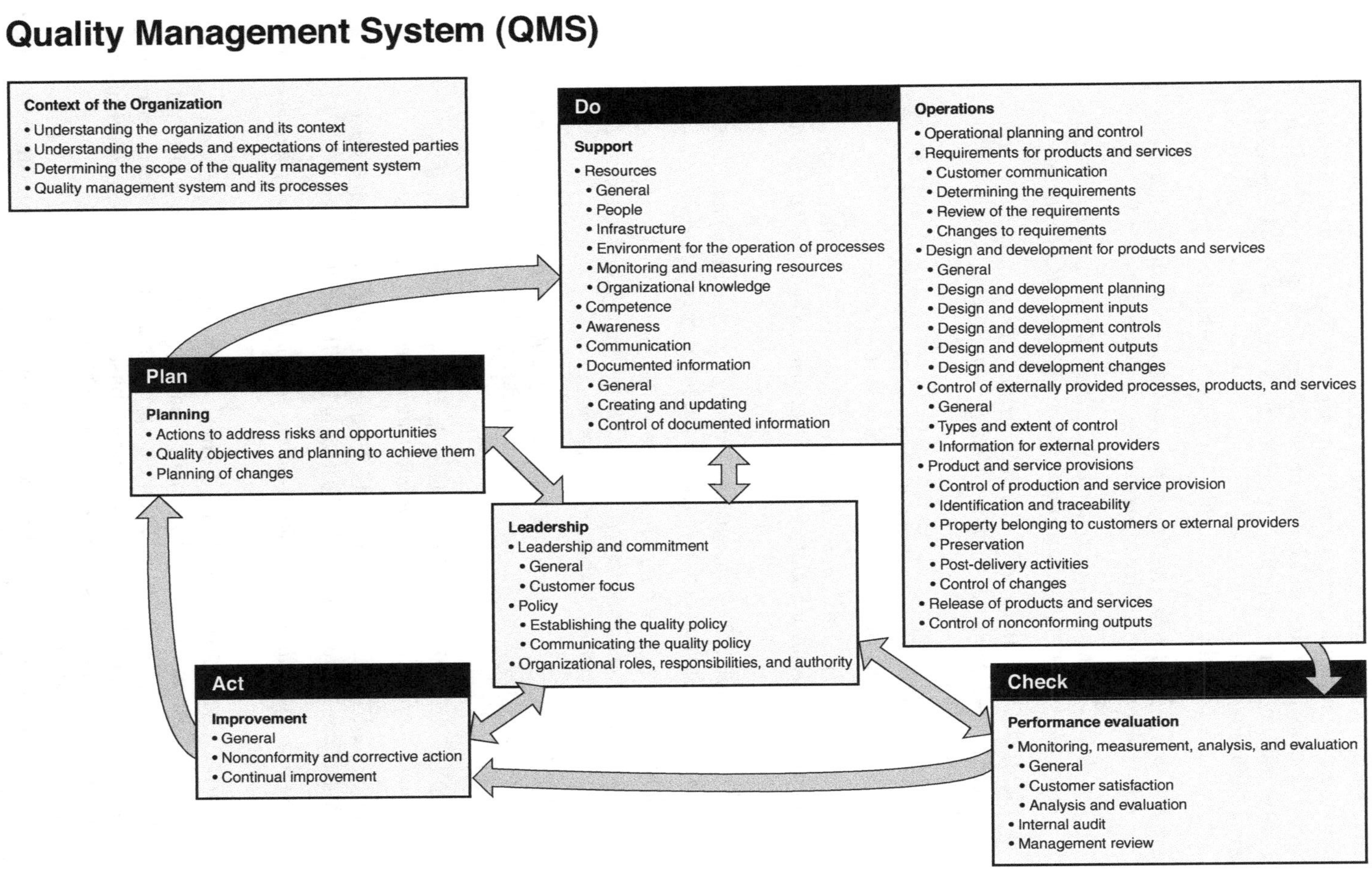

Figure 3.1 Major elements and interactions of an ISO 9001:2015 quality management system (ISO 2015).

(IT) security, cybersecurity, and privacy protection. This family of standards "provides the requirements, vocabulary, code of practice, and risk management techniques to implement and establish an effective IT security management system. It also provides guidance on auditing and certifying an information security management system" (from https://webstore.ansi.org/standards/iso/isoiec27000information).

Within the ISO/IEC 27000 family, the core standards include:

- *ISO/IEC 27000:2018—Information Technology—Security Techniques—Information Security Management Systems—Overview and Vocabulary*
- *ISO/IEC 27001:2022—Information Security, Cybersecurity and Privacy Protection—Information Security Management Systems—Requirements*
- *ISO/IEC 27002:2022—Information Security, Cybersecurity and Privacy Protection—Information Security Management Systems—Information Security Controls*
- *ISO/IEC 27003:2017—Information Technology—Security Techniques—Information Security Management Systems—Guidance*
- *ISO/IEC 27004:2016—Information Technology—Security Techniques—Information Security Management—Monitoring, Measurement, Analysis, and Evaluation*
- *ISO/IEC 27005:2022—Information Security, Cybersecurity and Privacy Protection—Guidance on Managing Information Security Risks*
- *ISO/IEC 27006:2015/Addendum 1:2020—Information Technology—Security Techniques—Requirements for Bodies Providing Audit and Certification of Information Security Management Systems*
- *ISO/IEC 27007:2020—Information Security, Cybersecurity and Privacy Protection—Guidelines for Information Security Management Systems Auditing*
- *ISO/IEC 27008:2019—Information Technology—Security Techniques—Guidelines for the Assessment of Information Security Controls*
- *ISO/IEC 27009:2020—Information Security, Cybersecurity and Privacy Protection—Sector-Specific Application of ISO/IEC 27001*
- *There are also additional standards in this family that provide industry-specific guidance.*

The ISO/IEC 27001:2022 standard defines the specific information security management system (ISMS) requirements that registrars use to audit and certify organizations. ISO/IEC 27001:2022 takes a process approach, which incorporates the Plan-Do-Check-Act (PDCA) cycle and risk-based thinking. Figure 3.2 illustrates a model of the major clauses of ISO/IEC 27001:2022.

The ISO/IEC 27001:2022 standard also includes an Annex A containing a list of 93 security controls that must be complied with to obtain registration. These controls are organized into the following four categories:

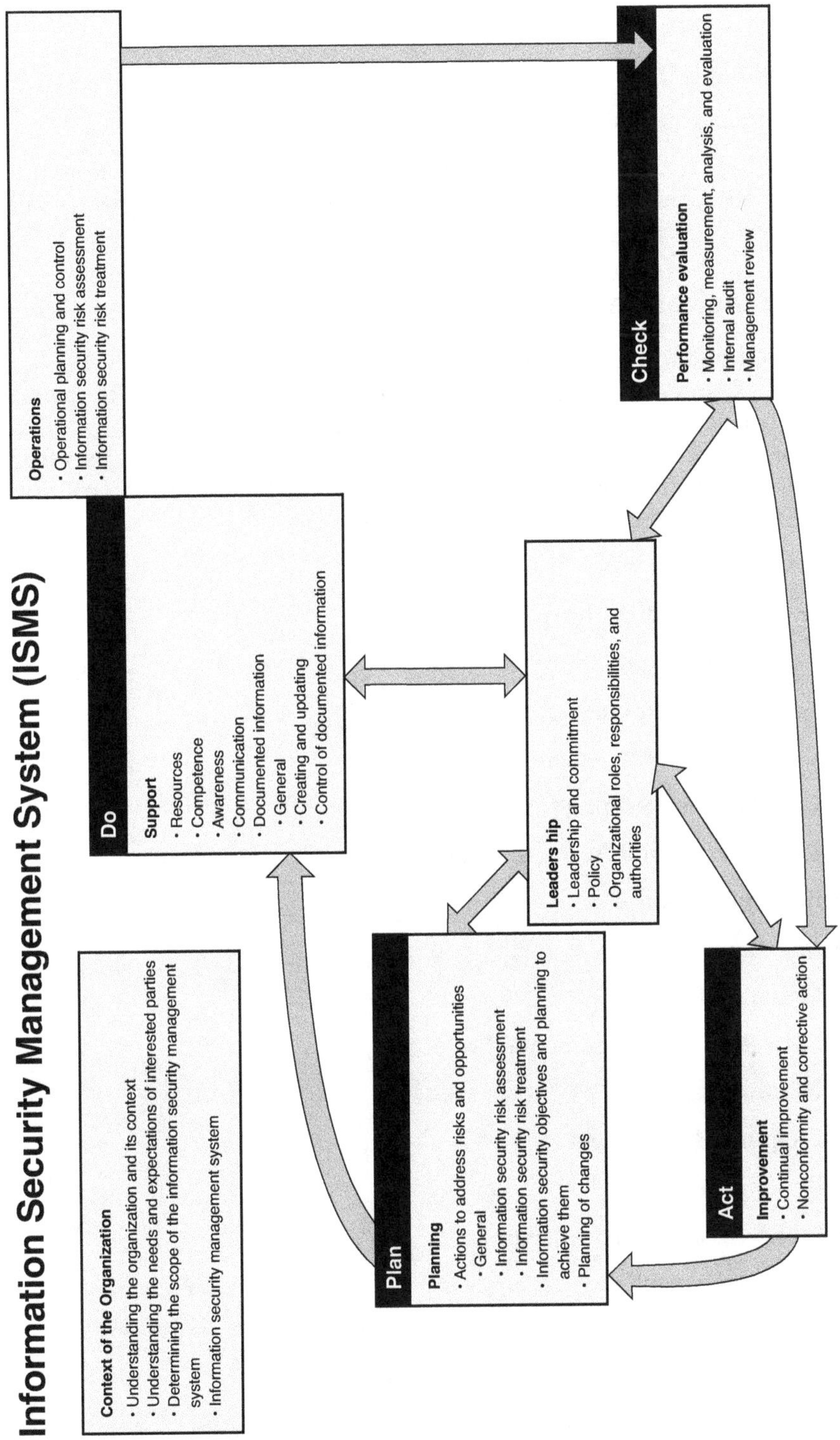

Figure 3.2 Major elements and interactions of an ISO 27001:2022 information management system (ISO/IEC 2022).

- Organizational controls
- People controls
- Physical controls
- Technological controls

ISO/IEC 27002:2022 includes guidance on implementing internationally recognized good practices when developing an ISMS incorporating these security controls.

NIST Special Publication 800-53

The National Institute of Standards and Technology (NIST) Special Publication 800-53:2020, *Security and Privacy Controls for Information Systems and Organizations,* describes a risk-based cyber security framework (CSF) that provides guidance on how organizations can prevent, detect, and respond to cyberattacks through the use of security and privacy controls. These security and privacy controls are organized into 20 families, including (NIST 2020):

- Access Control
- Awareness and Training
- Audit and Accountability
- Assessment, Authorization, and Monitoring
- Configuration Management
- Contingency Planning
- Identification and Authentication
- Incident Response
- Maintenance
- Media Protection
- Physical and Environmental Protection
- Planning
- Program Management
- Personal Security
- PII Processing and Transparency
- Risk Assessment
- System and Services Acquisition
- System and Communication Protection
- System and Information Integrity
- Supply Chain Risk Management

While private organizations can use NIST 800-53 as a guide for security, all U.S. federal government agencies and their contractors are required to comply with it.

IEEE Software Engineering Standards

The Software and Systems Engineering Standards Committee of the IEEE Computer Society develops and maintains a set of software and system engineering standards. This IEEE standards set is not always used verbatim, but these standards are used extensively as benchmarks, templates, and examples of industry good practices that organizations tailor to their own specific requirements. For organizations defining their software processes, these standards can provide guidance that minimizes time and effort. These standards can also serve as checklists that help verify that important items are not overlooked.

While ISO 9001:2015 and the Capability Maturity Model Integration (CMMI®) model provide road maps for what should occur in a good software quality engineering practice set, the IEEE software and systems engineering standards provide more detailed "how-to" information and guidance. As of the date of this publication, the current list of IEEE software and systems engineering standards includes the following standards that are closely related to the topics of the Certified Software Quality Engineer (CSQE) Body of Knowledge. See the IEEE software and systems engineering standards website for the latest versions of these and other IEEE software and systems engineering standards:

- 730-2014—*IEEE Standard for Software Quality Assurance Processes*
- 828-2012—*IEEE Standard for Configuration Management in Systems and Software Engineering*
- 829-2008—*IEEE Standard for Software and System Test Documentation*
- 982.1-2005—*Standard Dictionary of Measures of the Software Aspect of Dependability*
- 982.2-1988—*IEEE Guide for the Use of IEEE Standard Dictionary of Measures to Produce Reliable Software*
- 1008-1987 *IEEE Standard for Software Unit Testing*
- 1012-2016—*IEEE Standard for Systems, Software, and Hardware Verification and Validation*
- 1016-2009—*IEEE Standard for Information Technology—Systems Design—Software Design Descriptions*
- 1028-2008—*IEEE Standard for Software Reviews and Audits*
- 1042-1987—*IEEE Guide to Software Configuration Management*
- 1044-2009—*IEEE Standard Classification for Software Anomalies*
- 1061-1998—*IEEE Standard for a Software Quality Metrics Methodology*
- 1062-2015—*IEEE Recommended Practice for Software Acquisition*
- 1074.1-1995—*IEEE Guide for Developing Software Life Cycle Processes*

- 1228-1994—*IEEE Standard for Software Safety Plans*
- 1490-2003—*IEEE Guide Adoption of the Project Management Institute (PMI®) Standard–A Guide to the Project Management Body of Knowledge (PMBOK®) Guide*
- 1517-2010—*IEEE System and Software Life Cycle Processes—Reuse Processes*
- 1633-2016—*IEEE Recommended Practice on Software Reliability*
- 2675-2021—*IEEE Standard for DevOps: Building Reliable and Secure Systems Including Application Build, Package, and Deploy*
- 7000-2021—*IEEE Standard Model Process for Addressing Ethical Concerns During System Design*
- 7002-2022—*IEEE Standard for Data Privacy Process*
- 12207-2017—*IEEE/ISO/IEC Systems and Software Engineering—Software Life Cycle Processes*
- 14102-2010—*IEEE Standard for Adoption of ISO/IEC 14102:2008 Information Technology—Guidelines for the Evaluation and Selection of CASE Tools*
- 14471-2010—*IEEE/ISO/IEC Information Technology—Software Engineering—Guidelines for the Adoption of CASE Tools*
- 14764-2021—*IEEE/ISO/IEC Software Engineering—Software Life Cycle Processes—Maintenance*
- 15026-n—*IEEE/ISO/IEC 15026-n Systems and Software Engineering—Systems and Software Assurance*
  - *Part 1-2019: Concepts and Vocabulary*
  - *Part 2-2022: Assurance Case*
  - *Part 3-2015: System Integrity Levels*
  - *Part 4-2021: Assurance in the Life Cycle*
- 15288-2015—*IEEE/ISO/IEC Systems and Software Engineering—System Life Cycle Processes*
- 15289-2015—*IEEE/ISO/IEC—Systems and Software Engineering—Content of Life Cycle Information Products (Documentation)*
- 15939-2017—*IEEE/ISO/IEC—Systems and Software Engineering—Measurement Process*
- 16085-2020—*IEEE/ISO/IEC—Systems and Software Engineering—Software Life Cycle Processes—Risk Management*
- 16326-2019—*IEEE/ISO/IEC Systems and Software Engineering—Life Cycle Processes—Project Management*
- 20000-n—*IEEE Standard—Adoption of ISO/IEC 20000-n, Information Technology—Service Management*

- *Part 1-2013: Service Management System Requirements*
- *Part 2-2013: Guidance on the Application of Service Management Systems*

- 21839-2018—*IEEE/ISO/IEC Systems and Software Engineering—System of Systems (SoS) considerations in Life Cycle Stages of a System*
- 21840-2019—*IEEE/ISO/IEC Systems and Software Engineering—Guidelines for the Utilization of ISO/IED/IEEE 15288 in the Context of System of Systems (SoS)*
- 21841-2019—*IEEE/ISO/IEC Systems and Software Engineering—Taxonomy of System of Systems*
- 23026-2015—*IEEE/ISO/IEC Systems and Software Engineering—Engineering and Management of Websites for Systems, Software, and Services Information*
- 24748-n—*IEEE/ISO/IEC Systems and Software Engineering—Life Cycle Management*
    - *Part 1-2018: Guide for Life Cycle Management*
    - *Part 2-2018: Guide to the Application of ISO/IEC 15288 (System Life Cycle Processes)*
    - *Part 3-2020: Guide to the application of ISO/IEC 12207 (Software Life Cycle Processes)*
    - *Part 4-2016: Systems Engineering Planning*
    - *Part 5-2017: Software Development Planning*
- 24765-2010—*IEEE/ISO/IEC Systems and Software Engineering—Vocabulary*
- 24774-2021—*IEEE/ISO/IEC Systems and Software Engineering—Life Cycle Management—Specification for Process Description*
- 26511-2018—*IEEE/ISO/IEC Systems and Software Engineering—Requirements for Managers of Information for Users of Systems, Software, and Services*
- 26512-2017—*IEEE/ISO/IEC Systems and Software Engineering—Requirements for Acquirers and Suppliers of Information to Users*
- 26513-2017—*IEEE/ISO/IEC Systems and Software Engineering—Requirements for Testers and Reviewers of Information to Users*
- 26514-2021—*IEEE/ISO/IEC Systems and Software Engineering—Design and Development of Information for Users*
- 26515-2018—*IEEE/ISO/IEC Systems and Software Engineering—Developing Information for Users in an Agile Environment*
- 26531-2023—*IEEE/ISO/IEC Systems and Software Engineering—Content Management for Product Life Cycle, User and Service Management Information Users of Information to Users*

- 26702-2007—*IEEE/ISO/IEC Systems Engineering—Application and Management of the Systems Engineering Process*
- 29119-n—*IEEE/ISO/IEC Software and Systems Engineering—Software Testing*
    - *Part 1-2021: General Concepts*
    - *Part 2-2021: Test Process*
    - *Part 3-2021: Test Documentation*
    - *Part 4-2021: Test Techniques*
    - *Part 5-2016: Keyword Driven Testing*
- 29148-2018—*IEEE/ISO/IEC Systems and Software Engineering—Life Cycle Processes—Requirements Engineering*
- 41062-2019—*IEEE/ISO/IEC Software Engineering—Recommended Practice for Software Acquisition*
- 42010-2011—*IEEE/ISO/IEC Systems and Software Engineering—Architecture Description*
- 42020-2019—*IEEE/ISO/IEC Software, Systems and Enterprise Engineering—Architecture Processes*
- 42030-2019—*IEEE/ISO/IEC Software, Systems and Enterprise Engineering—Architecture Evaluation Framework*

It should be noted that in addition to the ISO 9000 and ISO 27000 family of standards, there are many ISO/IEC Information Technology, Software Engineering, and Systems and Software Engineering standards written/being written through ISO/IEC JTC 1/CS 7. Some of these standards have been adopted by IEEE, and others may potentially replace the current IEEE standards as listed.

Capability Maturity Model Integration (CMMI®)

The Software Engineering Institute (SEI) promotes the evolution of software engineering from an ad hoc, labor-intensive activity to a discipline that is well managed and supported by technology. As part of this work, the SEI established the CMMI® models (later supported by the CMMI® Institute and now owned by Information Systems Audit and Control Association [ISACA]), which are intended to communicate sets of good practices for use by organizations pursuing process improvement. The resulting CMMI® version 1.3 framework allowed for the generation of multiple CMMI® models for different disciplines:

- *Capability Maturity Model Integration (CMMI®) for Development (CMMI-DEV)* (CMMI 2010a)
- *Capability Maturity Model Integration (CMMI®) for Service (CMMI-SVC)* (CMMI 2010b)
- *Capability Maturity Model Integration (CMMI®) for Acquisition (CMMI-ACQ)* (CMMI 2010c)

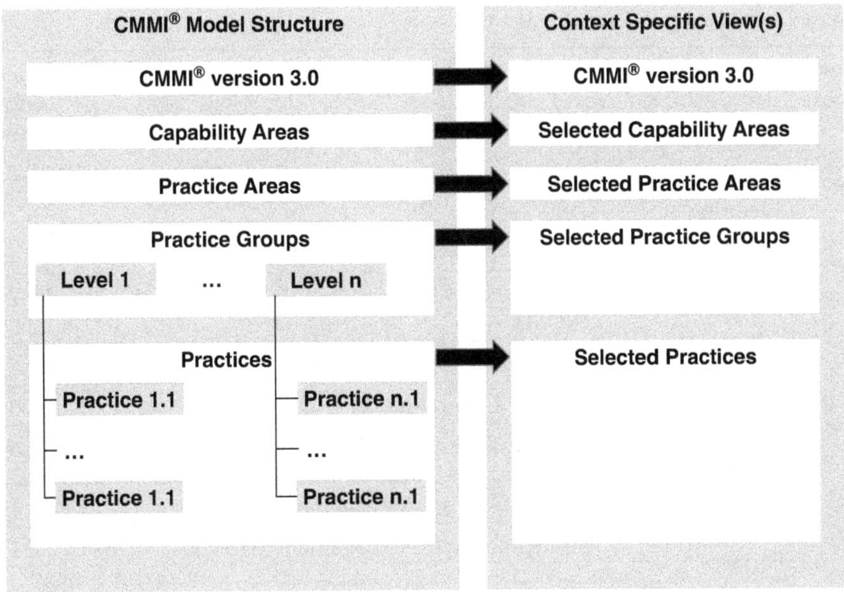

Figure 3.3 Structure of the CMMI® version 3.0 and associated views (based on CMMI 2023).

With version 2.0 of the CMMI®, these three models and the People Capability Maturity Model (CMU/SEI 2009) were merged into a single model. This version was a major change to both the structure and the content of the CMMI®. As of the writing of this book, the current version of the CMMI® model is version 3.0.

As illustrated in Figure 3.3, this single CMMI® version 3.0 model has multiple *views*, allowing users to select from predefined sets of the model's components or create custom views (combinations of capability areas, practice areas, or practice groups) that apply to different business environments. The predefined views for CMMI® version 3.0 include the following (CMMI 2023):

- CMMI® Development
- CMMI® Services
- CMMI® Data
- CMMI® Managing Security and Safety Capability Area
- CMMI® Planning and Managing Work Capability Area

These different views allow the users of the CMMI model to extract and use just the components of the model that are relevant to their specific context and domains. For example, the CMMI Development view would be the only view that extracted and showed the Product Integration and Technical Solution practice areas, while the CMMI Data view would be the only view that extracted and showed the Data Management and Data Quality practice area. Other core practice areas are included in multiple views.

Table 3.1 CMMI® version 3.0 categories and associated capability areas, practice areas, and practice groups (based on CMMI 2023).

Category	Capability Area	Practice Area	Practice Group (Level)				
			1	2	3	4	5
Doing	Ensuring Quality	Peer Reviews (PR)	1	4	1	-	-
		Process Quality Assurance (PQA)	1	4	1	-	-
		Requirements Development & Management (RDM)	1	5	7	-	-
		Verification & Validation (VV)	2	3	2	-	-
	Engineering & Developing Products	Product Integration (PI)	1	6	3	-	-
		Technical Solution (TS)	1	3	6	-	-
	Delivering & Managing Services	Service Delivery Management (SDM)	1	6	1	-	-
		Strategic Service Management (STSM)	1	3	1	-	-
	Selecting & Managing Suppliers	Supplier Agreement Management (SAM)	4	5	2	1	-
Managing	Planning & Managing Work	Estimating (EST)	1	3	2	-	-
		Planning (PLAN)	2	8	4	1	-
		Monitoring & Control (MC)	2	4	4	-	-
	Managing Business Resilience	Continuity (CONT)	1	3	3	-	-
		Incident Resolution & Prevention (IRP)	1	3	2	-	-
		Risk & Opportunity Management (RSK)	1	2	5	-	-
	Managing the Workforce	Enabling Virtual Work (EVW)	2	2	2	-	-
		Organizational Training (OT)	1	2	6	-	-
		Workforce Empowerment	1	3	3	-	-
Enabling	Supporting Implementation	Causal Analysis & Resolution (CAR)	1	2	5	2	1
		Configuration Management (CM)	1	6	-	-	-
		Decision Analysis & Resolution (DAR)	2	5	1	-	-
	Managing Data	Data Management (DM)	2	2	2	-	-
		Data Quality (DQ)	2	3	2	-	-
	Managing Security & Safety	Enabling Safety (ESAF)	2	3	3	-	-
		Enabling Security (ESEC)	2	4	3	-	-
		Managing Security Threats & Vulnerabilities (MST)	2	4	3	-	-
Improving	Improving Performance	Managing Performance & Measurement (MPM)	2	6	6	5	3
		Process Asset Development (PAD)	1	3	6	-	-
		Process Management (PCM)	3	2	6	1	-
	Sustaining Habit and Persistence	Governance (GOV)	1	4	2	1	
		Implementation Infrastructure (II)	1	2	3	1	-

Note: The numbers in the Practice Group columns indicate the number of practices for each capability level for each practice area.

As illustrated in Table 3.1, the CMMI® is organized into a tree structure of a hierarchy of related component types. These component types include:

- Categories
- Capability areas
- Practice areas
- Practice groups
- Practice

Within each practice area, the individual practices are organized into practice groups by capability level (levels 1, 2, 3, 4, 5). Different practice areas have different possible capability levels. For example, Configuration Management only has practices up to level-2 capability, while Causal Analysis & Resolution has practices up to level-5 maturity. This organization provides a road map for continuous improvement. For example, an organization would first implement the two level-1 practices for Verification and Validation. To continue to improve Verification and Validation, the organization would add the new functionality and sophistication of the three level-2 practices and then continue by adding the two level-3 practices.

Chapter 4

Leadership Skills

Defining *leadership,* like defining quality, can be a challenge. Example definitions from the literature include:

- "Leadership is the process of influencing others to willingly work toward common, shared goals" (unknown)
- "Leadership is the art of liberating people to do what is required of them in the most effective and humane way possible" (DePree 1989)
- "Leadership is the set of qualities that cause people to follow" (Loeb and Kindel 1999)
- "Leadership focuses on doing the right things, management focuses on doing things right" (Covey in Westcott 2006)

Effective leaders have many qualities and characteristics that help them influence others to follow them. These qualities and characteristics include:

- *Vision.* Leaders have the ability to create and nurture a vision, to communicate that vision, and inspire others to follow them toward the realization of that vision.
- *Courage.* Leaders do not balk at the sight of obstacles or fear failure. Leaders have the courage to act with confidence and take risks.
- *Decisiveness.* Leaders can make decisions and know when to act and when not to act.
- *Emotional intelligence.* Leaders can accurately recognize, understand, and manage their own emotions and those of others.
- *Emotional stamina.* Leaders can recover quickly from disappointment and bounce back from discouragement. Leaders have the ability to laugh and have a positive outlook.
- *Empathy.* Leaders have an appreciation for and understanding of the values and needs of others.
- *Self-confidence.* Leaders have an assuredness with which they meet the inherent challenges of leadership. Leaders have the ability to expect and accept criticism.

- *Accountability.* Leaders can accept responsibility for their actions and the actions of the team.
- *Credibility.* The leader's actions and words are trusted and believed.
- *Persistence.* Leaders can stick to a task or project until it is completed.
- *Dependability.* Leaders have the ability to meet commitments.
- *Stewardship.* Leaders can mentor, coach, guide, and develop team members, and effectively and efficiently utilize project resources.
- *Knowledge.* Leaders have a broad knowledge of the business domain, technical domain, management techniques, change management, and interpersonal skills. Leaders can apply that knowledge to aid their followers in performing the tasks at hand.
- *Ability to learn.* A leader is a perpetual student, learning about new systems, processes, techniques, methods, tools, and other changes to the business and technical domains in which they work.

1. ORGANIZATIONAL LEADERSHIP

> Use leadership tools and techniques (e.g., organizational change management, knowledge-transfer, motivation, mentoring and coaching, and recognition). (Apply)
>
> **Body of Knowledge I.D.1**

Organizational Change Management

Organizational change management is a mechanism that organizations use to grow and improve to stay competitive. There are two major types of change:

- *Incremental change* improves the existing systems, processes, or products to make them better. For example, the Model T has developed into the modern car through many incremental changes.
- *Evolutionary change* replaces the existing systems, processes, or products with better ones. For example, evolutionary change moved us from trains as the primary cross-country mode of transportation to automobiles and then to airplanes.

The steps to managing organizational change include:

Step 1: During the first step in the organizational change management process, the change leaders must identify and communicate the underlying need for change so that the people involved in and impacted by the change will buy into the need for change. This creates a sense of urgency to make the change.

Step 2: The next step is to initiate a change project and assemble a core team to lead the change effort. This step includes identifying the following:

- *Official change agent(s)*: An individual or team who is primarily responsible for planning and managing the change process.
- *Sponsor*: A senior leader/manager with the position and authority to legitimize and champion the change. This individual makes certain that adequate resources and staff are assigned to accomplish the change.
- *Advocates*: People who see the need for the change and will help sell it to the sponsor and the organization. Look to the "early adopters" for potential advocates.
- *Informal change agents*: People other than the official change agent who will help plan, manage, and implement the change process (based on Pyzdek 2001).

This core team establishes and communicates a *vision* of what the organization, systems, processes, or products will look like after the change is implemented. This vision is the destination on the road map to change. Continuous and active communication throughout the change process is essential to success. Change involves modifying the way people think and modifying the norms of the organization. "All change begins with the individual, at a personal level. Unless the individual is willing to change his behavior, no real change is possible" (Pyzdek 2001).

Step 3: In order to change, people must be empowered to change. That means providing them with the resources, tools, knowledge, skills, and motivation they need in order to do things in a different way. To accomplish this, the change effort should be treated as a project and planned and managed with the same rigor and oversight as a software development project. This planned project includes creating a series of small, short-term steps toward accomplishing the long-term goal (vision) with specific assigned responsibilities. Doing this has several advantages. First, the organization can see progress more easily and celebrate its successes. Second, the organization can learn from what has been accomplished and leverage these achievements into momentum toward future process improvement. Finally, defining responsibilities lets people know the boundaries and rules that define the scope of what they are personally allowed to change and when they need higher-level approval before changes are made. During this planning effort, forces that enable the change and make it more likely to occur should be identified and leveraged. Forces that act as barriers to the change should also be identified and proactive actions taken to remove or minimize the impacts of these barriers. Figure 4.1 illustrates a *force field analysis diagram* showing some of the forces that may drive an organization toward change and forces that may act as potential barriers to change.

Step 4: The final step is to institutionalize the change within the organization. This requires changing organizational norms that guide people's behavior so that the change becomes "just the way business is conducted." This may involve changing standards and processes, control systems, patterns of behavior, and reward systems.

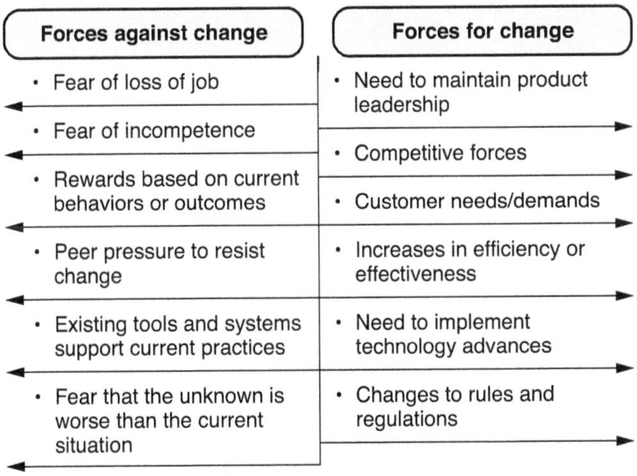

Figure 4.1 Change force field analysis.

The *Satir change model*, shown in Figure 4.2, illustrates the relationship between change and productivity. Within a certain level of variation, the existing way of doing things has a set level of productivity (old status quo). When change is first introduced, it causes a period of chaos in productivity where variation increases. Initially, some individuals are trying the new way and being successful, while others are not as successful or are resisting the change. Assuming that the change is positive, a learning curve occurs while people learn the new ways of doing things. During this learning curve, productivity generally increases, and the amount of variation in productivity starts decreasing. Finally, a new status quo is reached when the change is institutionalized and productivity levels out at a new higher level. Measures of the success of the change effort can be taken during the learning curve, but the full benefit of the change cannot be measured until institutionalization has occurred.

Motivation and Recognition

According to Blohowiak (1992), "if people believe that what they are doing has meaning—that it makes a contribution, that someone appreciates it—then they are motivated." *Motivation* requires two key elements:

1. For people to be motivated, they must have a clear understanding of what is expected of them
2. Meeting that expectation (or not meeting it) must be reinforced

According to Ryan and Deci (2000), there are two types of motivation. *Extrinsic motivation* is motivation that comes from a desire to obtain an external outcome or reward. The reinforcement for extrinsic motivation comes from the satisfaction of psychological or material needs by others through incentives or rewards. For example, external reinforcement might come from public recognition, a raise, a promotion, or a bigger office. *Intrinsic motivation* is a self-motivating process in

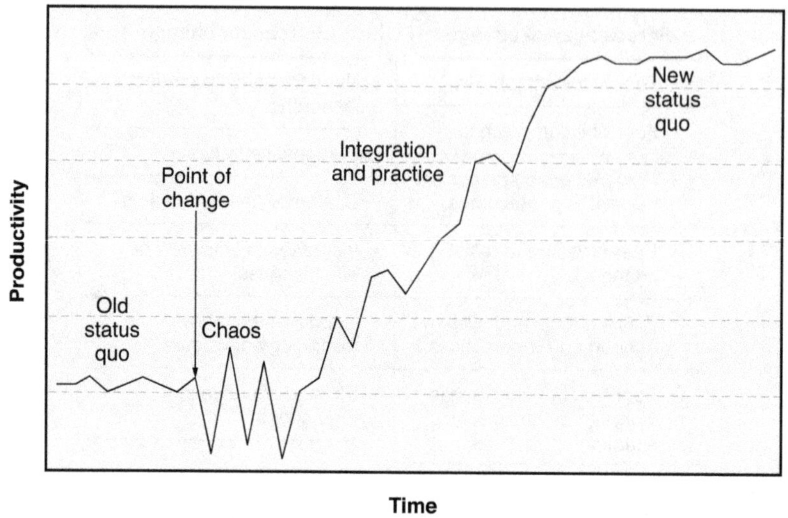

Figure 4.2 Satir change model (based on Weinberg 1997).

which the individual seeks out new things, experiences, or knowledge or takes on tasks because that individual enjoys or is interested in the activity. The individual obtains internal reinforcement by personally valuing characteristics of the situation itself. In intrinsic motivation, the internal reinforcement might come from gaining a sense of achievement or power, feeling creative, feeling a sense of belonging, having the satisfaction of making a contribution, or from self-actualization.

Westcott (2006) discusses three motivational theories specifically related to rewards:

- *Equity theory*: For rewards to be motivational, people have to believe that rewards (and punishments) are being equally distributed as deserved. People seek equity between the amount of effort (input) they put into a task and the rewards (output) they receive for doing it. For example, if everyone on the team gets the same reward no matter how much or little they contributed, then those rewards can actually demotivate team members.
- *Expectancy theory*: A person will be motivated to perform an activity based on their belief that:
  - Putting in the effort will actually lead to better results
  - The extra effort will be noticed, and that those better results will actually lead to personal rewards
  - Those personal rewards are valuable
- *Reinforcement theory*: People will be motivated to perform an activity based on their perception of a *trigger* (a signal) to initiate the behavior and the historic consequences of that behavior.

Different types of extrinsic and intrinsic forces motivate different people. Therefore, "one size does not fit all" when it comes to using recognition and rewards as reinforcement. Examples of different types of *recognition* and *rewards* include:

- Public praise and appreciation
- Thank-you letters or notes
- Compliments from important people while receiving the undivided attention of those people
- Gifts and other tokens of appreciation
- Conference or training opportunities
- Teaching or mentoring opportunities
- Special projects, or time for pet projects
- Time away from work
- More independence or autonomy
- Money, promotions, better office

As illustrated in Figure 4.3, Maslow defines a hierarchy of needs that prioritizes the types of rewards that motivate people. For example, if basic physiological needs for food, water, air, and shelter are not met, people will be motivated to fulfill those needs first. People will put themselves in danger or leave their social groups, if necessary, to meet those basic physiological needs. At the next level, people need to feel physically and economically safe. Once physiological and safety needs are met, people will be motivated by the need to belong, to be accepted by family and friends. Higher-level needs include self-esteem and self-actualization (the need to perform at one's best and self-improve).

Herzberg's *motivation-hygiene model*, also called the *two-factor model*, asserts that the factors that drive employee work satisfaction (*motivators*) are fundamentally different from the factors that cause employees to be dissatisfied at work (*hygiene factors*). For example, salary is a hygiene factor, which means that promises of salary increases will not typically satisfy or motive employees. However, if employees

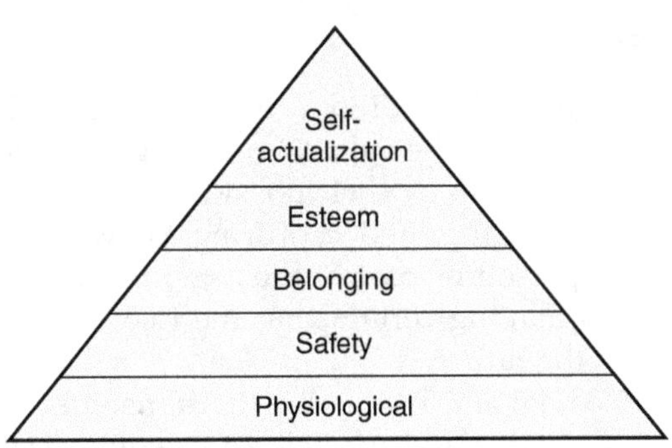

Figure 4.3 Maslow's hierarchy of needs.

Table 4.1 Herzberg's hygiene factors (based on Heller and Hindle 1998).

Hygiene Factors
Salary and benefits: Income, fringe benefits, bonuses, and vacation time
Working conditions: Work hours, overtime, infrastructure, office space, facilities, and equipment
Company policies: Formal and informal rules, restrictions, and regulations
Status: Rank, authority, and relationships with others
Job security: The degree of confidence the employee has regarding continued employment
Supervision and autonomy: The degree of control the employee has over the content and execution of their work
Office life: The level and type of job-related interpersonal relationships the employee has
Personal life: Work restrictions on the time the employee has with family, friends, and other interests

Table 4.2 Herzberg's motivation factors (based on Heller and Hindle 1998).

Motivation Factors
Achievement: Reaching or exceeding work and task objectives
Recognition: Acknowledgement of achievements by leaders, managers, and others
Job interest: The matching of the job to the employee's personal interests so that the job provides positive, satisfying pleasure
Responsibility: Opportunity to exercise authority and power, including leadership, risk-taking, decision-making, and self-direction
Advancement: Promotions, career progress, and increased rewards for achievement

feel that they are not being paid enough for the work they are doing or that their salary is inequitable compared to other people's salaries, those employees will become dissatisfied or demotivated. Table 4.1 lists basic hygiene factors, and Table 4.2 lists motivators.

Knowledge Transfer

According to Harrington (2006), "Today, more than ever, knowledge is the key to organizational success. . . . The value of most organizations is defined by their intellectual capital rather than by their physical assets." *Knowledge transfer* is the action of creating, organizing, and capturing knowledge and conveying that knowledge from one person or part of the organization to another. Knowledge transfer also involves making certain that the knowledge remains available to other stakeholders in the future.

Takeuchi and Nonaka (1995) describe the transfer of two different types of knowledge: tacit knowledge and explicit knowledge. *Tacit knowledge* is gained through experience rather than through formal reading or education. This knowledge is learned by "being there," seeing, participating, and experiencing.

From \ To	Tacit knowledge	Explicit knowledge
Tacit knowledge	Socialization	Externalization
Explicit knowledge	Internalization	Combination

Figure 4.4 Knowledge transfer (Takeuchi and Nonaka 1995).

Software development methods rely on the tacit knowledge gained through on-the-job experience. *Explicit knowledge* is tangible knowledge transmitted in a formal manner through various documents, formal training or education, video media, e-mail, artifacts, or books. Software development methods rely on the explicit knowledge gained through written policies, processes, plans, work instructions, and specifications. As illustrated in Figure 4.4, different mechanisms are needed when transferring between these different types of knowledge.

- Transferring tacit knowledge to tacit knowledge happens through socialization, the sharing of experience. For example, socialization occurs when an experienced member of the team explains coding standards to a new team member or when two people pass knowledge back and forth about a source code module during pair programming.
- Transferring tacit knowledge to explicit knowledge happens through externalization, for example, when a business analyst documents the stakeholders' requirements or when a project formally documents its lessons learned.
- Transferring from explicit knowledge to tacit knowledge happens through internalization, for example, when an individual reads a book on software testing and uses that knowledge to define a systematic set of test cases for a software system.
- Transferring from explicit knowledge to explicit knowledge happens through combination. *Combination* is a process of distilling multiple sources of explicit knowledge into a new source, for example, when an author of a book reads multiple books and other references on software quality engineering and then combines that information into their text.

The steps involved in knowledge transfer include:

- Identifying the knowledge needs
- Identifying the knowledge holders (within the organization or external to the organization)
- Motivating the knowledge holders to share their knowledge

- Selecting mechanisms and planning for knowledge transfer (mentoring, coaching, on-the-job-training, formal training/education, documentation, intranet, and so on)
- Implementing the knowledge transfer
- Analyzing the success of the knowledge transfer to determine if the knowledge was actually transferred
- Applying the new knowledge—receiver of the knowledge can use the knowledge for actionable benefit
- Creating a knowledge management system for long-term storage and propagation of knowledge
- Continuously improving the knowledge management system

Mentoring and Coaching

Mentoring is a relationship between two people where a more experienced or knowledgeable individual passes valuable skills, knowledge, and insights on to a less experienced individual in order to help the less experienced individual develop or improve their abilities, performance, knowledge, or other capabilities. A mentor counsels, advises, and directs based on their own personal and professional experience and expertise. Mentoring may be used to accomplish different goals—teaching toward a required outcome, transferring job skills, or improving job performance.

When mentoring to establish performance expectation, the mentor tells the individuals being mentored what to do and how to do it. The mentor describes the job in terms of major outcomes and how it aligns with higher-level goals and objectives. Measurable performance objectives and priorities are agreed to, and necessary skills, resources, and guidelines are mutually identified. The mentor reviews and verifies the understanding and commitments of the individuals being mentored. Dates are then set for periodic progress reviews. Mentoring is used to establish performance expectations when the mentored individuals:

- Are unclear about what is expected
- Need help sorting out priorities because they have taken on more than they can handle
- Are acting on inaccurate advice from coworkers about job procedures or standards
- Produce substandard work
- Are exceeding current expectations and need new challenges

When mentoring to transfer job skills, the mentor explains the why as well as the how of the job skill. The mentor shows the individuals being mentored how to do the work, and then lets the individuals perform the work under the mentor's guidance. Finally, the mentored individuals perform the work on their own while the mentor stays available to answer questions if necessary. Mentoring is used to transfer job skills when the mentored individuals:

- Need help learning to do a new task or assignment
- Seem confused about how to do a job
- Resist taking on or doing a job because they do not have the right skills
- Ask to take on an assignment for which they are not adequately skilled
- Are seeking career paths or advancements that require new skills

Mentoring to improve job performance involves the mentor describing the performance issues and the need for corrective action. The mentor seeks the opinions of the individuals being mentored about the root cause of the performance issues and ways in which performance can be improved. The mentor provides feedback on those opinions and then adds their own input. The mentor and mentored individuals jointly create and commit to an action plan to improve performance and set follow-up dates. The mentor expresses confidence and support in the mentored individual's ability to correct the performance issue. Mentoring is used to improve job performance when the mentored individuals:

- Are not meeting performance expectations
- Are meeting expectations but want to continue to improve
- Are meeting expectations, but the mentor believes that more effective or efficient ways to perform the work exist
- Are faced with an important career opportunity

Coaching, in contrast to mentoring, is about equipping individuals with the tools they need to find the answers for themselves. Coaching helps individuals and teams identify and articulate their professional challenges and goals and supports them in achieving those goals. Experienced coaches facilitate learning through the use of powerful questions that empower the people being coached to move themselves forward. The people being coached select the area of focus and direction—what they want to gain from the coaching. Coaching should be applied when the coachee(s):

- Are skilled but lack confidence or motivation
- Are working hard but not delivering the expected outcomes
- Need guidance in clarifying the way forward on a new project or task
- Would like to improve their leadership or interpersonal skills
- Want to unleash their own potential and learn how to better think through things for themselves
- Need to be empowered to take ownership
- Are dissatisfied with the path their career is taking, or their lack of career success

One popular framework for coaching is the GROW framework (Hancock 2024). The following list defines the elements of GROW and includes questions that a coach could use at each of the steps in the GROW framework:

- G for Goal—Bringing issues to the surface and identifying the real goal
  - What can you tell me about the problem?
  - What does the situation look like when the problem is resolved?
  - What would achieving the goal mean to them?
  - Where are you today in progress towards achieving this goal?
  - What tasks are you procrastinating on?
- R for Reality—Determining what is really going on
  - What are you experiencing?
  - What have you already tried?
  - What obstacles are in your way?
  - What resources do you have to help achieve this?
- O for Options—Exploring options or alternatives
  - What are your options?
  - What would you do if you knew you could not fail?
  - What have you seen work in the past?
- W for Will Do—Planning the path forward
  - What are your next steps?
  - What will you do?
  - What type of support do you need? From whom?

The goal of both coaching and mentoring is to improve the knowledge, skills, and abilities of individuals so that they can improve their performance, meet their objectives, and have successful careers. In addition, the collaborative nature of both mentoring and coaching develops interpersonal relationships that increase employee engagement, motivation, and job satisfaction.

The agile coach, also called a *Scrum master* in the Scrum methodology, applies a specific application of coaching techniques to the agile environment. The agile coach applies their coaching skills and strategies to help people, teams, and organizations create value-added software by applying agile methods (e.g., extreme programming, Scrum, lean, test-driven development, or some combination of these methods). The goal of an agile coach is to grow productive, self-managing agile individuals, teams, and organizations that can think for themselves rather than relying on management to dictate and direct agile practices. Figure 4.5 uses a *cause-and-effect diagram* to illustrate an overview of agile coaching.

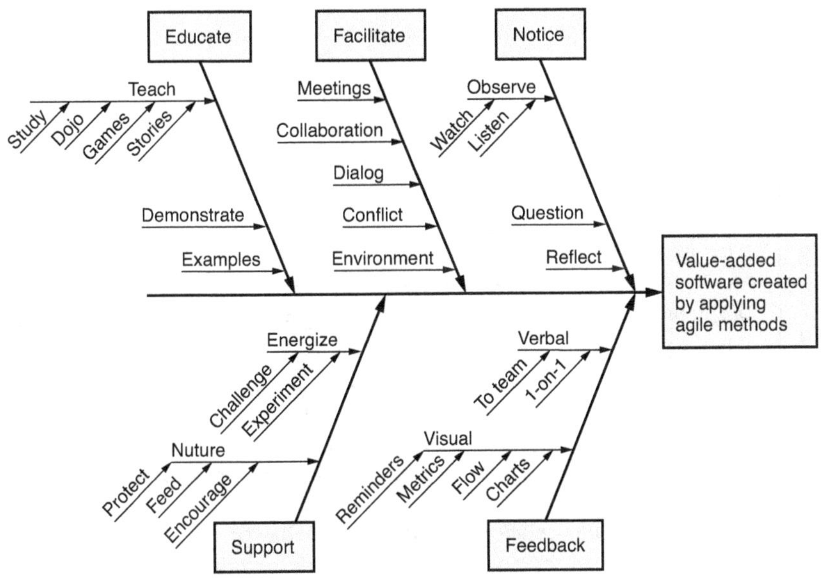

Figure 4.5 Cause-and-effect diagram—overview of agile coaching (based on Davies and Sedley 2009). (Note: A *dojo* is a place where training occurs.)

Situational Leadership

The concept behind *situational leadership* is that the style of leadership that is needed depends on the situation and the "readiness" of the followers to be led. As illustrated in Figure 4.6, the followers' needs and abilities dictate the amount of relationship/supportive behaviors and the amount of task/directive behavior that the leaders must use in a given situation, and thus the leadership style to be employed:

- Relationship/supportive behaviors include:
  - Two-way or multiway communications
  - Listening
  - Encouraging
  - Facilitating/coaching
  - Socioemotional support
- Task/directive behaviors include:
  - One-way communications from leader to follower
  - Spelling out of duties and responsibilities
  - Telling people what to do
  - Telling people how, when, where, and with whom to do it
  - Telling people who is to do it

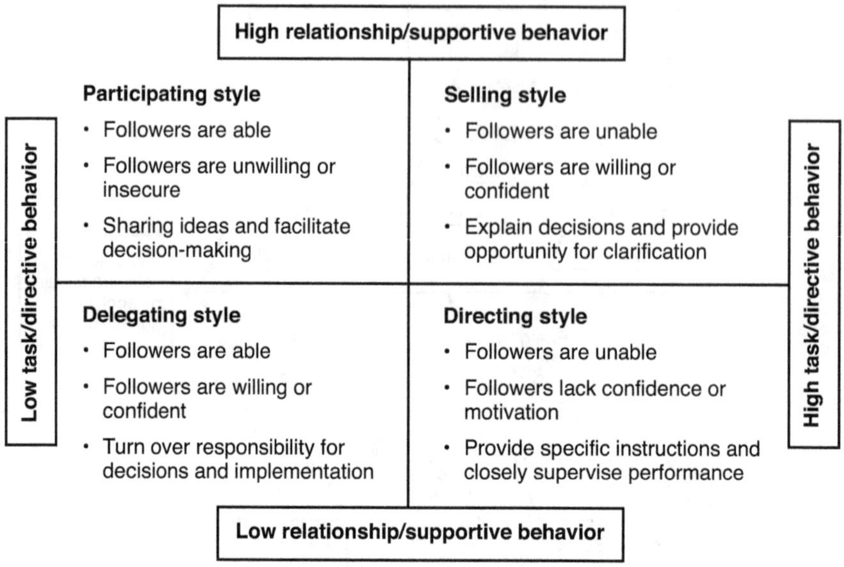

Figure 4.6 Situational leadership styles (based on Hersey 1984).

According to Hersey (1984):

- A *directing leadership style,* with high levels of task/directive behavior and low levels of relationship/supportive behaviors, should be used when the followers are unable to do the job and also lack the confidence to do it. For example, this directing style would be used for a new software practitioner, fresh from the university, who requires more supervision performing tasks like configuration management or requirements analysis, because they were not trained in these skills.
- In the *selling leadership style,* with high levels of both task/directive behaviors and relationship/supportive behaviors, decisions are explained and opportunities are provided to ask questions and clarify instructions. For example, the selling style would be used for a newly trained tester who is trying to apply new skills but who does not understand the need to follow the steps in the prescribed test development process.
- The *participating leadership style,* with low levels of task/directive behaviors and high levels of relationship/supportive behaviors, emphasizes the sharing of ideas and responsibilities and provides encouragement to the participants. For example, the participative style is appropriate when practitioners obtain the needed level of skills but are reluctant to take on total responsibility for the work because they feel unsure of themselves.
- Once practitioners are both able and willing to perform their work, the *delegating leadership style* may be most appropriate, where responsibility for decisions and implementation is turned over to the practitioners.

The self-organized teams used in agile projects are examples of the delegating style of leadership.

2. FACILITATION SKILLS

> Use facilitation and conflict resolution skills and negotiation techniques to manage and resolve issues. Use meeting management tools to maximize meeting effectiveness. (Apply)
>
> **Body of Knowledge I.D.2**

Facilitation

A *facilitator* is an individual who creates an environment in which a team can direct its own work. An effective facilitator structures the activities of the team and guides the team through its problem-solving activities. However, the team prioritizes issues, evaluates, and judges the alternatives; makes the decisions; and solves the problems. The facilitator helps stimulate discussion, makes sure everyone is participating, asks questions, and helps clarify key points. An effective facilitator helps the team reflect, expand, and summarize its discussions, options, alternatives, and decisions. The facilitator does not spend much time leading the meeting or talking. In fact, if the team is performing effectively, the facilitator simply sits back and lets the team members direct and control their own activities. However, the facilitator does watch closely in case problems arise that need attention. For example, the facilitator keeps the team focused on the processes and problems and not on people and personalities. The facilitator never presents himself or herself as a subject matter expert—the team members are the subject matter experts. The facilitator's expertise is the team tools, which aid in the team's ability to analyze issues, solve problems, or make decisions. For agile teams, the agile coach or Scrum master typically takes on the role of facilitator.

Conflict Management and Resolution

The traditional view of *conflict* is that it is caused by troublemakers and should be avoided or suppressed. The contemporary view of conflict is that it is inevitable and often desirable because it is necessary for creativity and innovation. If everything is good, there is no reason to do anything new or different. Problems cannot be fixed if the team is not aware that those problems exist. However, conflict must be appropriately managed.

Positive conflict, also called *constructive conflict*, can be beneficial. Sometimes it takes a conflict to bring problems to the forefront and make them visible. Discussing different points of view, and digging down into the issues help create collaborative solutions that take the needs of a diverse set of stakeholders into consideration. Conflict can create insight and innovation through combining multiple different solutions into a single, better-integrated solution. Managing positive conflict can help remove roadblocks to productivity by eliminating "brooding" and other negative feelings that sap motivation. Positive conflict creates incentives to challenge and change outmoded policies, methods, processes, and assignments. By making it all right to talk about and manage conflict instead of squelching it, people become aware of how behaviors and actions affect others so that they can correct or avoid similar behaviors in the future. Conflict resolution also increases interpersonal skills as people practice and become proficient at understanding the viewpoints of others in order to seek win–win outcomes. Creativity and innovation come from conflict. Discussion and problem-solving help people release stress. Finding solutions to problems and getting them resolved also increases morale. When managed correctly, positive conflict can present a fun challenge and a break from the normal routine, for example, in a rousing debate or challenging competition.

The facilitator should watch for cues that there is too much conformity and not enough positive conflict, including:

- Most issues are decided with little or no discussion
- Certain members of the team give opinions, others nod, and the decision is considered made
- Instant and silent "agreement" with the opinions of the leader or other influential members
- Controversial issues receive little or no open discussion—all sides of the controversy are not raised
- Rejection of minority opinions without consideration

The facilitator helps the team take corrective action if too much conformity, and not enough positive conflict, exists within the team. Steps to corrective action include:

- Openly rewarding divergent points of view
- Asking outright for other points of view
- Encouraging conflict and controversy arising from opposing ideas and opinions
- Making sure the team stays focused on products and processes and not on personalities
- Protecting or giving equal weight to minority opinions
- Asking for opinions from silent team members
- Making certain that power and influence are approximately equal
- Creating a devil's advocate role and valuing that role highly
- Having leaders hold their opinions until other team members have had an opportunity to be heard

On the other hand, managing conflict is a balancing act. Too much *negative conflict*, also called *destructive conflict*, can be damaging to the team synergy and inhibit the team's ability to accomplish its objectives. The facilitator should watch for signs that there is too much negative conflict. These include:

- Team members avoiding each other to the detriment of work progress
- Team members withholding important information or resources from each other, making it difficult to work effectively and efficiently
- Emotions running high, resulting in unproductive tension between team members
- Team members unwilling to work together
- Team members complaining that their work is being adversely affected by conflict between other team members

When faced with a negative conflict, it is easy to respond with emotion or instinct. Team members must learn to move to the logical, thinking part of their brains in order to handle conflicts productively. One way to recognize the potential for negative conflict before it escalates is to recognize "provocative" statements. Recognizing provocative statements can help people avoid using the same types of statements in their own interactions. Examples of these provocative statements include (Bernstein and Rozen 1990):

- "I just want to tell you my *real* feelings about . . ."
- "You always . . ."
- "You never . . ."
- "I checked with Joe and Mary, and we *all* feel . . ."
- "Why was I not consulted about . . ."
- "How come some people are allowed to . . . ?"
- "I thought I was in charge . . ."
- "In this company we do things a certain way . . ."
- "In the good old days . . ."

The facilitator helps the team take corrective action if too much negative conflict exists on the team. Steps to corrective action include:

- Promptly letting the people involved know how their conflict is affecting performance
- Setting up a joint problem-solving approach involving relevant stakeholders to resolve the conflict
- Asking the people involved to present their viewpoints objectively
- Searching for areas of agreement and common goals
- Getting agreement on the problem that needs to be solved

- Refocusing the team members on process and product issues, and away from personalities
- Having each person involved in the conflict generate possible solutions
- Getting commitment on what each person will do to solve the problem
- Summarizing and setting a follow-up date to make sure the conflict has been resolved

When there is too much negative conflict, the leader, facilitator, or conflicting individuals can select one of several conflict resolution strategies as outlined in Table 4.3.

Table 4.3 Conflict resolution strategies (based on Covey 1998).

Conflict Resolution Strategy	Benefits	Issues
I win–you lose: Solving the conflict through an authoritarian approach, where the person in authority gets their way and the others involved in the conflict do not.	• Expedient • Protects against exploitation	• Authoritarian approach • Sets up resentment • Temporary solution • Fosters competition, not cooperation • Solution at others' expense
I lose–you win: Solving the conflict through one party simply giving in or capitulating to the other.	• Preserves harmony • Avoids unhealthy competition • Settles unimportant issues in little time	• Abdicates your position • Give in or give up but harbor ill feelings • Repressed emotions • Allows exploitation • Self-esteem lowered
I lose–you lose: The conflict is not solved, and everyone loses.	• None	• Fosters need to "get back" or "get even" • Destructive
Compromise: Solving the conflict through each party getting part of what they want and giving up other things they want.	• Better than destructive bickering • Partly solves the problem • Some commitment felt	• Appeasement • Issues not really settled • Not motivating or developmental
I win–you win: Solving the conflict through seeking mutual benefit in the outcomes of the conflict.	• Solution is mutually beneficial and mutually satisfying • All parties feel good about the solution • Encourages cooperation and commitment • Permanent solution • Develops problem-solving ability	• Time consuming • Requires participation • Requires trust

Negotiation Techniques

Negotiation involves two or more parties, who each have something the other wants, reaching a mutually acceptable agreement through a process of bargaining. People do lots of negotiation in their business lives as well as their personal lives. For example, on the job:

- Salaries and benefits are negotiated between organizations and employees
- Requirements, prices, and schedules are negotiated between acquirers and suppliers
- Job assignments, priorities, and deliverables are negotiated between project managers and project team members and functional managers

One of the keys to successful negotiation is to realize that each party to the negotiation must give in order to receive. Each party must also gain something of value to them in compensation for any concessions that they make.

Core skills required for a successful negotiation include the ability to:

- Prepare well
- Define a range of objectives and prioritize them clearly
- Remain flexible
- Identify the other person's needs
- Separate people from the process
- Explore a wide range of options
- Focus on common interests
- Communicate (listening, questioning, verbal communication)

Figure 4.7 illustrates the basic steps in the formal negotiation process. During the *preparation step of negotiation*, objectives for the negotiation are established and prioritized. This includes establishing ideal objectives (if the negotiators could get everything they want, what it would look like), realistic objectives (what the negotiators expect to be able to get), and minimum objectives (the minimum the negotiators will accept without walking away). Preparing also includes doing research to obtain as much advance information as possible to help the negotiators understand the people they are negotiating with and their needs. Using their objectives and the gathered information, the negotiators can then plan and prepare a strategy for what they believe to be the possible paths the negotiation might take.

The actual negotiation session begins with the *proposing step of the negotiations*. The negotiators present uncontroversial, general points and stress the need for agreement from the outset. The negotiator making the initial proposal starts with their ideal objectives and leave room to maneuver and not give away everything up front. Each negotiator should wait for the other party to finish before responding with their own initial proposal. The negotiators should listen carefully to each other's proposals. They may be closer to agreement than they think.

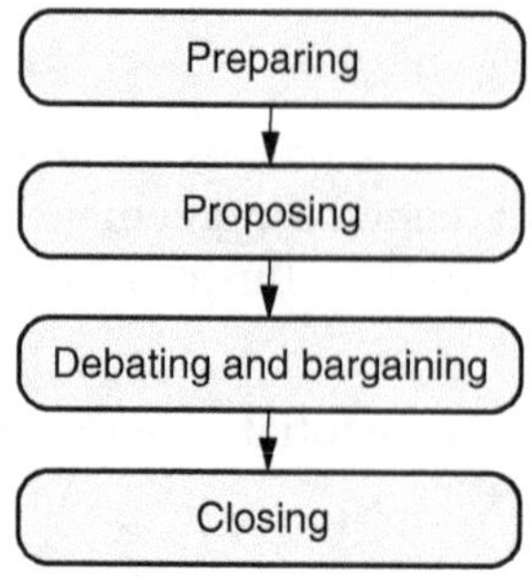

Figure 4.7 Formal negotiation process.

The *debating and bargaining step* then begins with asking clarifying questions as necessary—the more information the negotiators have the better job they can do in negotiations. It may also be a good idea to summarize the other party's proposal—"what I understand you to be asking for is . . ."—in order to verify understanding. The negotiators should look for similarities in their negotiating positions as starting points to build agreement. They then look for areas where they can counteroffer by trading off low-priority objectives that match up with the other party's higher priorities, in exchange for receiving their high-priority objectives. The negotiators should not concede one of their objectives without receiving something in return.

The *closing of the negotiation* involves creating a written record of the final agreement, including definitions of any words that may be ambiguous. If the negotiators cannot come to a mutually beneficial agreement, then everyone just walks away.

Meeting Management

For a meeting to be productive, the leader/facilitator should determine specific objectives for the meeting. The meeting objectives are basically the requirements statement for the meeting:

- Why is the team holding a meeting?
- What is the purpose?
- What is the expected outcome?

As a checkpoint, the leader/facilitator always evaluates and determines if there is a better way to accomplish these objectives other than holding a meeting.

Once the objectives are established, the leader/facilitator must plan how the team is going to accomplish those objectives during the meeting. This is the design stage for the meeting:

- Who needs to be at the meeting in order to accomplish the objectives?
- Who should not be at the meeting?
- What methods or tools are going to be used during the meeting?

An agenda is then created that outlines the meeting plan and assigns specific responsibilities for each section of the meeting. The leader/facilitator needs to make sure that the meeting logistics are handled in advance, as well. This includes:

- Deciding where and when to hold the meeting
- Making reservations for the meeting room
- Handling any special arrangements for equipment (flip charts, projectors)
- Making copies of materials needed in the meeting
- Ordering special supplies

Finally, the meeting objective and agenda need to be distributed to the meeting attendees. If there are any instructions for preparing for the meeting, or any materials that need to be reviewed before the meeting or brought to the meeting, these requirements need to be made clear to the attendees.

The leader/facilitator arrives early for the meeting in order to make sure that everything is set up as required and verifies that any equipment is in working order. Arriving early also gives the leader/facilitator the opportunity to select an appropriate position in the room from which to run the meeting. If the leader/facilitator wants to take control of the meeting and direct it, an authority position at the head of the table should be selected. If the leader/facilitator wants to join the group in an open discussion, a seat in the middle of the table may be more appropriate.

The leader/facilitator should establish a reputation as someone who starts meetings on time. Do not waste the team's time reviewing for latecomers. Ask the latecomers to stay after the meeting, and review with them after the other attendees have left the meeting. The objective is to get people in the habit of coming to meetings on time.

The meeting itself starts with reviewing and confirming the meeting's objectives and expected outcomes, followed by a review of the meeting's agenda. This will help focus the team on the meeting's goals and how those goals are going to be accomplished. It is also a good idea to review any ground rules for the meeting if problems have occurred in prior meetings. The leader/facilitator confirms with the participants that all required preparation has been accomplished. This may require postponing the meeting until a later date if key people do not attend or key preparation has not been accomplished.

During the meeting, the leader/facilitator helps the team stick to the agenda, and makes certain that everyone is participating. Using a "parking lot" to record off-topic ideas or items for later discussion can aid the team in staying focused on the meeting objectives and stick to the meeting agenda. A record (meeting minutes) is also kept to document important information, action items, and decisions.

At the end of the meeting, the list of action items is reviewed to verify that they have been adequately captured with an assigned owner and a target date for completion. The leader/facilitator helps the team summarize its accomplishments and evaluate the meeting. The next meeting is also planned as necessary. Productive meetings end on, or before, schedule. If all of the objectives have not been met, follow-up meetings can be scheduled.

After the meeting, the meeting record (meeting minutes) is formalized and distributed. As appropriate, that record is archived as a quality record. The leader/facilitator follows up with team members that are assigned action items to verify that those items are being addressed. The leader/facilitator then starts the meeting planning process over again for the next meeting.

3. COMMUNICATION SKILLS

> Identify and apply various communication methods in oral, written, and presentation formats. Apply various techniques for working in multicultural environments. Identify and describe the impact that culture, communications and Diversity, Equity, and Inclusion (DEI) can have on an organization. (Apply)
>
> **Body of Knowledge I.D.3**

Software practitioners spend a large portion of their time working with, and communicating with, other individuals. In fact, according to DeMarco and Lister (1999), software developers only spend 30 percent of their time working alone. They spend 50 percent of their time working with one other person and the last 20 percent of their time working with two or more other people. Strong *communication* skills are needed to make certain that these interactions are effective.

One-way communication occurs when the sender of a message communicates that message to the receiver without obtaining feedback. Examples of one-way communications include speeches and written documents. As illustrated in Figure 4.8, in one-way communications, the sender has a message they want to convey. This message is encoded using the sender's filters. These filters include the sender's:

- Paradigms about how things are (the way the sender sees the world)
- Experiences and vocabulary about the specific subject of the message
- Feelings and emotions about the other person, about the subject, and about life in general at the time the message is being encoded

The sender's encoded message is then interpreted through the receiver's filters. Because of these filters, what the receiver thinks the message is may be very different than what the sender actually intended to communicate.

Two-way communication occurs when the sender of a message communicates that message to the receiver and obtains feedback. Examples of two-way communications include discussions and interviews. Figure 4.9 illustrates a two-way communication model. The first half of the two-way communication model is identical to the one-way communication model. However, the two-way model includes the receiver providing feedback that is encoded through the receiver's

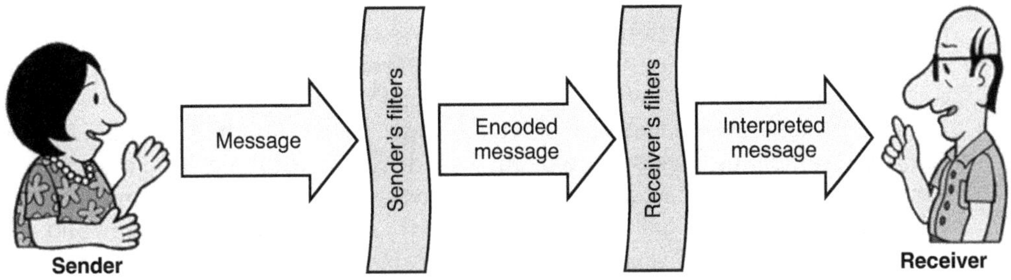

Figure 4.8 One-way communication model.

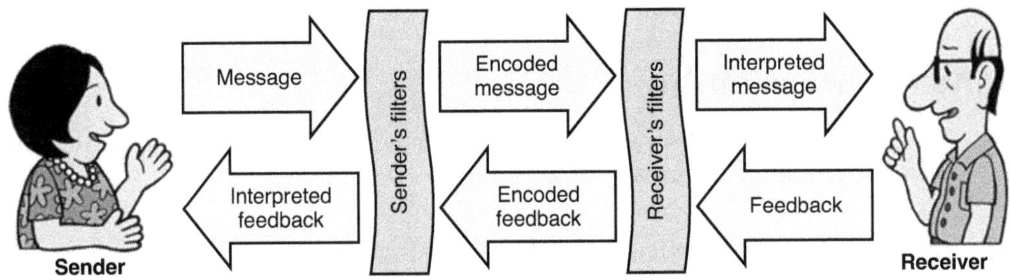

Figure 4.9 Two-way communication model.

filters and interpreted through the sender's filters (which includes their belief about what the content of the original message was).

Typical message problems include:

- The message was misinterpreted
- The message was considered unimportant and was ignored
- The message arrived too late to be effective
- The message ended up in the hands (or ears) of the wrong person

Does this sound familiar? The manager gives detailed instructions to an employee. The manager asks, "Do you have any questions?" and the employee responds, "No." The manager asks, "Do you understand what needs to be done?" and the employee responds, "Yes." Later, the manager finds out that the employee has done something completely different than the manager intended. Whose fault is it?—the answer is, the manager's. Lewis (1995) says, "the responsibility for communications rests with the communicator—not the other person." In this case, the manager did not obtain adequate feedback to verify that their instructions were correctly interpreted. The manager should have asked open-ended questions or had the employee restate or summarize the instructions in their own words to verify successful communications. If a message is misinterpreted, it may mean that the tone, word choice, medium, or timing (or any combination of these) used to convey the message were deficient. It may also mean that the sender's and receiver's filters were different enough to cause communications to break down. In this case, the sender may need to try a different approach to make certain that the message is received, as for example, having a third party act as an interpreter. An interpreter may be necessary even if both parties to the communications speak the same language but are using very different vocabularies. For example, a business analyst may act as an interpreter to facilitate communications between the customer, who is speaking in the vocabulary of the business domain, and the software engineer, who is speaking in the vocabulary of the technical domain.

Oral Communications

Table 4.4 lists examples of various types of oral communications and the advantages and disadvantages of each type.

Written Communications

Table 4.5 lists examples of various types of written communications and the advantages and disadvantages of each type.

Impacts of Communications on the Organization

Good communication is essential for the organization to produce high-quality software products through high-quality processes that are effectively and efficiently implemented during software projects. Examples of the impacts of poor communications on the organization include:

- Issues in communications with product stakeholders may result in missing, ambiguous, or incorrect requirements, potentially leading to software that does not meet stakeholder needs or its intended use, resulting in stakeholder dissatisfaction
- Failure to adequately communicate requirements to the software developers and testers may lead to misinterpretations of those requirements and their misalignment with designs, code, and tests, potentially resulting in excessive rework and field failures costing the organization time, money, and loss of reputation
- Issues in communicating project status may result in failure to take corrective actions when needed or unnecessary corrective actions, slipped schedules, cost overruns, or misalignment of stakeholder expectations with project objectives and results
- Issues in communicating software product or project risks may result in risks going unidentified or unmitigated
- Issues in communicating policies, standards or processes to the practitioners may result in their incomplete/inconsistent implementation, potentially leading to nonconformances, confusion about roles and responsibilities, or software product defects

Table 4.4 Oral communication techniques.

Informal Verbal Interchanges	
Advantages	**Disadvantages**
• Great way to see the action and keep in touch with what is happening • Immediate feedback on current problems • Can discuss wide range of topics in detail • Two-way communications	• Might be perceived as nosy if quality engineers are not genuine in their interest or if they visit too often • Can be disruptive • No permanent record • If unfocused, can waste time
Formal Meetings	
Advantages	**Disadvantages**
• Allows multiple people to express opinions and issues • Creates team synergy • Can discuss wide range of topics and issues in detail • Two-way communications	• Can degrade into time-wasters • Some people may feel uncomfortable offering opinions in public • All topics may not be relevant to all people, or needed people may be missing from meeting • Difficult when people are at remote locations (solution—teleconferencing)
Telephone Calls	
Advantages	**Disadvantages**
• Good for short, focused communications that need personal touch • Easy to reach people in remote locations • Two-way communications	• Phone tag • Disruptive to receiver • No permanent record • Loss of nonverbal communications
Voice Mail	
Advantages	**Disadvantages**
• Good for short, focused messages • Quick way to reach people in remote locations • Can be broadcast to multiple people • Receiver can listen to message at their convenience	• Messages can be lost without sender or receiver knowing there is a problem • Some people do not check their messages regularly • No permanent record • One-way communications
Formal Presentations	
Advantages	**Disadvantages**
• Great for presenting complex status reports or new training materials to audiences of various sizes • If done professionally, can have lasting positive impression on the audience • Limited two-way communications with questions and answers	• Requires considerable planning and skill to achieve a positive impression • Time-consuming • Poorly presented materials can negatively affect attitudes toward an otherwise successful project

(continued)

Table 4.4 (Continued)

Video Conferencing	
Advantages	**Disadvantages**
• Can be used for informal verbal exchanges, formal meetings, or formal presentations with the same advantages • Allows two-way communications with remote audiences	• Can be used for informal verbal exchanges, formal meetings, or formal presentations with the same disadvantages • Not the same sociological impact as face-to-face interactions • Significant work time differences can negatively affect team dynamics and performance

Table 4.5 Written communication techniques.

Formal Reports and Memos	
Advantages	**Disadvantages**
• Provide permanent record • Good for communicating updates, important procedures, or major changes	• Require precise wording to make certain that desired message is conveyed • Cannot be easily taken back if a problem or misinterpretation arises • One-way communications
E-mail, Texting, and Instant Messaging	
Advantages	**Disadvantages**
• May provide written records without generating paper • Allows the attachment of other documents • Quick way to reach people in remote and distributed locations • Find out if mail has been received (return receipt) • Messages can be forwarded to others	• Lose the "personal touch,"—tone of voice, nonverbal communication, and so on • Not good for sensitive issues • Some people do not read their messages regularly • Lack of privacy—can be forwarded easily to almost anyone • One-way communications
Handwritten, Short Notes	
Advantages	**Disadvantages**
• Good for kudos and thanks • Personal touch • Quick, simple, friendly, cheap	• Can get lost in the shuffle • Some people's handwriting is illegible • Difficult if people not in the same location • One-way communications

- Issues in communications between team members could lead to conflicts, lack of synergy, or duplication of effort
- Issues in communications between auditors and auditees could result in incorrect audit findings, missed nonconformances, or missed opportunities for improvement

Effective Listening

Nichols, who pioneered the study of listening, says that bad listening is the true cause of lost sales and lost customers, most accidents and production breakdowns, personality clashes and poor morale, bad communications, and misguided management (Toastmasters 1990). So how do people listen more effectively? Toastmasters (1990) recommends the following nine techniques (note that the first letter of each technique spells out the word "listening"):

1. *Like to listen*: The listeners must show a willingness to listen and enjoy what (and whom) they are listening to.
2. *Ignore distractions*: Listeners must learn to ignore distractions and visually and mentally focus on the speaker and what they are saying. Listeners need to listen with their eyes as well as their ears.
3. *Summarize*: Listeners can mentally summarize what the speaker is saying as they speak. Summarization techniques include:
    - Picking out main points, concepts, or key words.
    - Mental outlining or taking notes.
    - Comparing and contrasting what the speaker says with what the listeners already know.
    - Thinking ahead of the speaker and trying to predict their next point. If listeners are right, it will reinforce their memory of what was said. If the listener's prediction is incorrect, they will remember what was said because it was unexpected.
    - Numbering the points as listeners hear them.
4. *Tame emotions*: If listeners have negative feelings about what is being said or about the speaker, they can throw up mental barriers that make listening difficult. Listeners must control their mental attitude and listen objectively.
5. *Eliminate hasty judgments*: Listeners must work to ignore any biases they may have based on a person's appearance, accent, or even title. Listeners should listen with empathy by trying to place themselves in the speaker's shoes and attempt to listen from their point of view. Listeners need to remind themselves to hear the speaker out before making judgments.
6. *Never interrupt*. There are many ways to interrupt a speaker besides speaking up while they are talking:
    - Arguing mentally with the speaker.
    - Continuing to ponder something that was said while another point is being made.
    - Mentally questioning, at length, a statistic that was presented.

- Disagreeing as a point is being made.
- Not remaining open to reasons, arguments, and data.

7. *Inspire openness*: Listeners should look at the speaker. Listeners can also nod in agreement or smile when they hear something they like, look friendly, and react positively to the speaker. Listeners can also communicate their receptivity and be aware of their own body language.
8. *Need to listen*: Listeners can acknowledge that they need to listen. This makes listening a conscious decision.
9. *Generate conclusions*: As listeners listen, they can decide what it is they are going to accept from the speaker and what they will reject. Conclusions are different from hasty judgment because they are thought out and reasoned using logic.

Verbal listening involves listening for and analyzing word choice, and the direct and implied meanings of the words being chosen by the speaker. It also means listening for and analyzing tone of voice, emphasis, inflection, and the pacing of the communications. *Nonverbal listening* involves watching for and analyzing the eye contact, facial expressions, and body language of the speaker.

Interviews

Interviews are an excellent mechanism for obtaining detailed information from another person. For example, interviews are used in selecting candidates for open positions (job interviews), for obtaining data for an audit, in identifying risks, eliciting requirements, and in determining the level of stakeholder satisfaction.

Interviews are used to gather information from interviewees through their answers to questions. The intent of interview questions is to prompt the interviewee to do most of the talking. In order to accomplish this, the interviewer should ask open-ended, context-free interview questions. Table 4.6 includes examples of open-ended versus closed-ended questions. Open-ended questions include "what," "when," "where," "who," and "how" questions. Interviewers should be very careful when using "why"-type questions because they may sound accusatory

Table 4.6 Open-ended versus closed-ended questions—examples.

Open-Ended Questions	Closed-Ended Questions
What procedures do you follow when performing your work?	Do you follow the XYZ procedure when you do your work?
What reviews and approvals are involved in releasing your work products?	Do you perform a peer review on your work products? Does the software lead approve your work products?
How do you communicate the problems you encounter?	Do you record the problems you find in the defect-tracking tool?

Table 4.7 Context-free questions—examples.

Ask This	Not This
What documentation do you use when performing this task?	How do you use the statement of work template when performing this task?
How do you verify the quality, completeness, and consistency of your work products?	What steps do you use when conducting unit testing of your work product?
How do you track your project's progress?	How do you use Microsoft Project to track your project's progress?

when interpreted through the filters of the interviewee. The interviewer might be intending to ask a neutral question like, "Why are you doing it that way?" but based on the interviewee's past experiences and emotions, their interpretation of the message might come across as, "Why would anyone ever want to do it that way, you idiot!" To avoid these problems the interviewer might want to rephrase "why" questions, for example, as "what are the reasons behind . . . ?"

Context-free questions remove most of the context from the question itself and therefore do not put limitations on the scope of the answers the interviewee might give. Table 4.7 includes examples of context-free questions. Preparing interview questions in advance helps make certain that good, open-ended, context-free questions will be asked during the interview. Preparing the questions in advance also allows the interviewer to arrange the questions in a logical flow of topics rather than jumping around from topic to topic.

When asking an interview question, the interviewer should know what the "right," or expected, answer is. For example, if an audit interviewer is asking the question, "How do you communicate the problems you encounter?" that interviewer should know, based on the process requirements and procedure documentation, that problems are reported by entering them into a specific problem-reporting tool.

During the interview, the interviewer starts by introducing themself and explaining the purpose and process of the interview. The interviewer should put the interviewee at ease, and then ask questions and actively listen to the responses provided by the interviewee. However, preparing questions in advance does not lock the interviewer into using just those questions. Some prepared questions may need to be supplemented with follow-up questions to adequately cover an area, and some prepared questions may become unnecessary depending on the responses the interviewee gives to other questions. At the end of the interview, the interviewer handles any administrative matters and explains any next steps. The interviewer should also thank the interviewee for their time. After the interview, the interviewer should take time to further document the interview. Notes taken during the interview may be sketchy and need to be expanded on while the interview is still fresh in the interviewer's mind. The interviewer takes any follow-up actions that are necessary as a result of conducting the interview.

Working in Multicultural Environments

Working closely together on teams or one-on-one with others brings up social issues including personality types and personal preferences, cultural differences and diversity issues, and even political correctness concerns. The globalization of the software industry has expanded our horizons, but it also brings even more differences in culture, norms, expectations, and even languages into the workforce. The increasing utilization of cross-functional, team-based software development using agile techniques escalates the need to deal with these human differences and cultural issues. This shift to more awareness of cultural and sociological issues may be new or even uncomfortable to software practitioners who are used to working as individual contributors, spending long hours alone with their computers.

Communication filter differences, as well as other people issues, can become more pronounced in a multicultural setting. For example, words, phrases, and hand gestures that are commonly used in the United States may have derogatory meanings in other countries. Extra care must be taken to make sure that communications are effective and that customs are respected.

Cultural competence is the ability of individuals and organizations to work effectively with people from diverse ethnic, cultural, religious, and geographic groups. This competency requires awareness of, and respect for, cultural differences, beliefs, and biases, as well as the knowledge and skills to successfully interact in cross-cultural situations. Proactive steps should be taken to become culturally aware of the differences and expectations of other people during interactions in the workplace. This awareness, coupled with a willingness to learn about other people's needs, cultures, and preferences, will help avoid misunderstandings and allow people to work together effectively within a diverse work environment. Mutual respect and understanding are essential in handling any issues as they arise.

Diversity, Equity, and Inclusion (DEI)

Diversity, equity, and inclusion (DEI) is a framework of organizational policies, programs, practices, and behaviors that help ensure that all employees are treated fairly, can fully participate, and have access to resources, opportunities, and chances for successful outcomes. Many countries worldwide have laws and regulations that focus on eliminating prejudice and discrimination in the workforce by making it illegal.

Diversity is "the combination of the differences that shape our perspectives, approaches and overall world view" (Nekah 2022). As leaders in the workforce, we must recognize, respect, and value diversity. There are four types of diversity (based on Nekah 2022):

- *Internal diversity* includes characteristics like our race/ethnicity, gender, sexual orientation, physical and mental abilities, national origin, age, and so on.
- *External diversity* includes education, religion, beliefs, socioeconomic status, citizenship, relationship/marital status, personal interest, and so on.
- *Organizational diversity* includes type of employment (manager vs. employee, exempt vs. hourly), pay rate, profession (analyst, developer,

tester, quality assurance engineer, technician), work location (in-house vs. remote, manufacturing floor vs. office), and so on.

- The other three types influence *worldview diversity*, which is constantly changing. This type of diversity includes political beliefs, affiliations, morals, ideas, perspectives, ethics, values, and so on.

Equity promotes fairness and impartiality by providing equal treatment and access to opportunities, advancements, and resources. Equity is different than equality. With equality, all employees get precisely the same things. Equity considers each person's unique needs and characteristics and focuses on outcomes by providing employees with what they need to be successful. For example, equality would be giving everyone identical 10-speed racing bikes. Equity would give people different bikes depending on what they need and are trying to accomplish (e.g., a racing bike for racers, a mountain bike for off-roaders, a tricycle for young children, a specially equipped bike for people with disabilities, and so on). Equity acknowledges that diversity also brings different circumstances and starting points and compensates for those differences.

Inclusion is creating a work environment and culture that appreciates diversity. Inclusion focuses on utilizing the skills, talents, and perspectives of diverse employees to benefit the organization and those employees. People are welcomed, accepted, respected, empowered, and meaningfully engaged in an inclusive work culture. People feel like they are heard and contributing.

Chapter 5

Team Management

> Use various team management skills, including assigning roles and responsibilities, identifying the classic stages of team development (forming, storming, norming, performing, and adjourning), and monitoring and responding to group dynamics. Work with diverse groups and in distributed work environments. Use techniques for working with virtual, in-person, and hybrid teams. (Apply)
>
> **Body of Knowledge I.E**

Teams are an essential part of any quality system. Management teams establish, implement, and monitor the quality management system. Other teams are formed to improve the performance of interdependent processes and tasks. Team assignments are made to integrate complementary skills and knowledge and to create synergy, where the strength of the team as a whole is greater than the sum of its individual members. The team approach stimulates innovation and creativity by creating an environment that encourages people to try new approaches. Newer team members benefit from the mentoring of more experienced members while also contributing different approaches or ideas to an established team. Teamwork promotes trade-offs in problem-solving and helps individual members accept the challenges of change.

Teams also have weaknesses:

- Team goals can become misaligned with the goals of the organization.
- When teams and their decisions are not accepted and supported by management, the efforts of the team are wasted.
- Time is needed to build and maintain high-performance teams.
- Decision-making is typically slower using a team because of the time it takes to make a decision by consensus.
- Teams can also create negative synergy, where time is wasted simply because of the effort it takes to work with other people or

because teams get distracted with off-task activities, including social interactions.

- Teams can be impacted by "group think," in which the desire for harmony or conformity in the team results in a lack of diversity of thought, shutting down of alternative points of view, and dysfunctional decision making.

There are many different types of teams at various levels in the organization with software process and product quality roles. The *quality council,* which may be called by many different names, is comprised of high-level executive management charged with setting the strategic quality direction and establishing the quality policies for the organization. This team is responsible for establishing, implementing, and monitoring the quality management system at the organizational level to make certain it is *effective.*

Under the quality council, multiple *quality management teams* are responsible for implementing the quality management system and quality policies for individual departments, projects, or teams within the organization.

Cross-functional teams are established to deal with quality areas or issues that cross organizational boundaries. For example, a cross-functional team charged with system reliability may include members from hardware, software, manufacturing, technical support, and quality. Their role is to take a system-wide perspective of the area or issue and to verify that one department or team does not optimize their part of the process to the detriment of overall process performance.

Quality action teams, also called *quality circles, quality improvement teams,* or *Six Sigma teams,* are typically established to improve a targeted process or product. They analyze and prioritize improvement ideas, plan improvement actions, verify the implementation of those plans, and evaluate the results of that implementation. They then repeat the process for the next-highest-priority improvement idea.

Unlike a quality action team, which is established with a goal of continuous improvement of a process or set of processes, a *tiger team,* also called a *SWAT team* or *short-term team,* is established to deal with a specific quality issue. After that issue is satisfactorily resolved, the tiger team is usually disbanded.

An *engineering process group (EPG)* is a team of process experts that act as consultants to help document and implement standardized processes across the organization. The EPG then aids individual departments, projects, or teams in tailoring those standardized processes to their individual needs. The EPG also acts as a clearinghouse for identified lessons learned and best practices and for propagating them for the benefit of the entire organization.

While the EPG is a set of process experts, the experts do not own the processes. The processes are owned by the *process owner teams,* which include representatives of the individuals who implement those processes as part of their daily work. For example, the system testers own the system testing process, while the coders own the coding process and associated coding standards. In large organizations, where many individuals share each process, process owner teams may be formed for each major process to act as "users" groups or process change control boards. One of the roles of these process owner teams is to identify and share lessons learned and improvement ideas. Another role may be to evaluate the impact of proposed process changes from quality action teams, EPGs, or other groups or individuals. The process owner teams then act as the final authority to accept, defer, or reject the proposed changes to the organization's standardized processes and verify the implementation of the changes that are accepted.

System/software development and operations teams, including agile teams, are tasked with planning, building, operating, supporting, and maintaining high-quality products by implementing the quality management system policies, processes, and work instructions.

Team Roles and Responsibilities

Different team roles have specific responsibilities associated with them. The *team champion* is usually a senior member of management who selects and defines the team's mission, scope, and goals, setting the vision and chartering the team. The champion is responsible for establishing the team, selecting the team leader or facilitator, and working with them to identify team members. The champion reviews the team's progress, provides ongoing support and direction, represents the team to upper management, and runs interference for the team with the rest of the organization and other stakeholders. The team escalates issues to its champion if the team itself cannot resolve those issues. The champion maintains overall executive-level responsibility for the success of the team's efforts.

The team champion may also act as the team sponsor, or another member of management may sponsor the team. The *team sponsor* provides ongoing funding and other necessary resources to the team.

The *team leader* is the person responsible for managing the team. This includes:

- Focusing the team on its objectives and monitoring team progress toward accomplishing those objectives
- Calling, arranging, and chairing team meetings
- Handling or assigning administrative details
- Directing the team, including making assignments and taking follow-up action as required
- Managing and directing the utilization of team resources
- Overseeing the preparation and presentation of team reports and presentations
- Representing the team to the rest of the organization, including interacting with the team champion and sponsor

Self-directing teams, for example, agile teams, may not have an individual designated as the team leader, or the leader role may rotate to different team members as appropriate.

The *team facilitator* has experience working with teams and can guide the team in their work and the use of team tools. The facilitator keeps the team running smoothly and makes sure that all team members have an opportunity to participate and express their ideas. The facilitator also handles nonproductive behaviors and other issues of team dynamics, including helping the team resolve negative conflict. In agile processes, the team facilitator role is typically filled by the Scrum master or the agile coach.

The *team members* are responsible for working together to accomplish the objectives of the team. This includes:

- Actively participating in team activities
- Offering ideas, alternatives, and suggestions
- Actively listening to other team members and leveraging their inputs to create synergistic solutions and improvements
- Completing assigned tasks and action items on schedule
- Eliciting information from the groups or organizational units they represent, adequately representing those groups/units during team activities and discussions, and communicating team decisions and other information back to those groups/units

The *recorder*, also called the *scribe*, is responsible for generating, publishing, and maintaining minutes and other records from team meetings, including tracking action items and team decisions. The recorder may be a temporary or rotating position within the team.

Stages of Team Development

Teams typically go through a set of predictable stages of development as they move from formation through the execution of their assigned responsibilities, as illustrated in Figure 5.1. These stages can vary in duration and intensity depending on the makeup and personalities of the team membership, the organizational culture and environment, and the effectiveness and skills of the team leader or facilitator. Sometimes a team can move through forming, storming, and norming into performing in just a meeting or two. Other teams may take a much longer time to become a cohesive unit and move to high performance. If enough problems exist, a team may never make it out of the storming stage and therefore never accomplish their mission. A reorganization of the team or other issues can also cause the team to regress back one or more stages and then have to progress back through the team development stages. Team reorganization occurs when one or more team members leave the team, new members are added to the team, or major changes in work assignments occur.

The *forming* stage occurs when a new team is first created or when an existing team encounters a major reorganization. The forming stage is a period where the team members start to get to know each other and understand each other's values, priorities, work preferences, and abilities. The team leader and facilitator work with a team in the forming stage to clarify the mission and scope of the team and to collaborate in the definition of specific objectives, tasks, roles, and responsibilities needed to achieve the team's mission. The facilitator also works to establish an environment of openness and trust between team members and to train team members in the tools and methods that the team will use.

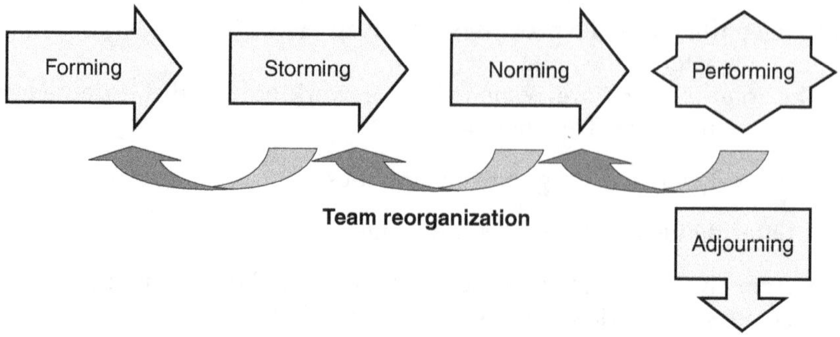

Figure 5.1 Stages of team development.

The forming stage is typically followed by the *storming* stage. During the storming stage, team members become aware that they are personally going to have to change in order to work together effectively, and there is a period of resistance to that change. During this stage, team members vacillate between working toward individual goals and working as part of the team. This stage is marked with arguments, frustration, anxiety, and the testing of the authority of the team leaders. During this stage, the team leader may need to take a primary role in guiding the team and reinforcing their mission. The facilitator understands that tension and negative conflict are normal during this stage and helps the team learn to identify and manage the resulting issues by creating an atmosphere where team members can safely express their feelings. The facilitator also makes certain that all of the team members are participating and have a voice in defining how the team will accomplish its mission.

The team then progresses to a *norming* stage where the team members work out their differences, come together, and start acting as a cohesive unit that cooperates to perform their work. The team establishes agreed-to methods of operation and ground rules (norms, team etiquette) for conducting business. During the norming stage, the team leader starts sharing more responsibility with the team and allows the team to begin to self-organize. The facilitator supports team members as they take on more individual responsibility for conducting team meetings and start identifying, analyzing, and solving their own problems. "The key to this stage is to build the team's confidence in their ability to resolve differences without anyone feeling left out or discounted" (Scholtes et al. 2003).

Truly effective teams then move into the *performing* (high performance) stage, where everyone knows how to work together effectively and efficiently. The team leader shifts from an authority position to more of a coaching role. The facilitator's main role during the performing stage is monitoring team interactions for issues that may require intervention. The facilitator continues to mentor the team by recommending tools and providing training as necessary.

When a team completes one assignment, they may move on to the next assignment or the next priority. However, at some point, the team may complete all of its work and enter an *adjourning* phase where the team accomplishes its mission, shuts down its efforts, and disbands. It is important to disband when the work is complete and not keep the team meeting just in case a need arises.

Group Dynamics

Teams are collections of individuals, each with their own style, personality, culture, work preferences, and personal goals. These differences can cause group dynamics problems to occur that threaten the cohesion and synergy of the team. Table 5.1 lists examples of these problems and their potential solutions.

Working with Diverse Groups

Teams need diversity to perform effectively. They need leaders and followers. They need people who focus on strategies and those who emphasize tactics. They need people with different skills and knowledge (e.g., domain knowledge, technical knowledge, tool skills, interpersonal skills, analysts, programmers, testers, and technical writers). Teams also need people with different personality types (socializers, timekeepers, persuaders, big-picture people, and people good at taking care of all the details). Without diversity, teams may not have enough of the positive conflict necessary to be creative and innovative.

On the other hand, in today's multicultural, global economy, diversity also brings the potential for misunderstandings and miscommunication that can lead to negative conflict and other issues. For example, in Asia a business card represents the person and should be treated with the utmost respect. A business card is presented and accepted with both hands, it is read and acknowledged when presented, and it should never be written on or stuck in a hip pocket. Humor varies significantly between cultures. Praise and criticism may be handled differently in different cultures. Even colors have different meanings. While unnecessary conflict may be avoided by focusing the team on products and processes instead of people, teams cannot ignore the people issues. Team leaders, facilitators, and members must have the courage to proactively identify and address people issues in order to avoid the potential "land mines" caused by these differences. With more and more diversity comes the need to learn about, understand, and be continuously aware and accepting of differences in the language, personality traits, and culture of team members and other stakeholders in order to take advantage of the benefits of diversity.

Working in Distributed Work Environments

In the new world order of telecommuting, remote work, and the globalization of the workforce, software practitioners and teams often no longer reside in a single work location. They may be distributed around the city, around the country, and even around the world. According to Gutierrez (2015), "highly skilled and talented distributed workers know that they can work just as effectively from anywhere while also attaining a better—worklife balance." Kvarme (2013) states that "a study of global IT teams revealed that key factors for successful distributed teams included a set of common goals, open dialogue, attention to building interpersonal relationships, and making sure that someone was responsible as a facilitator to support collaboration and communication."

When people need to work together in distributed work environments and face-to-face interactions are limited by distance, virtual teams can form by using technology to support communications and collaboration. These technologies can include:

- Virtual meeting tools that allow teams to nearly match the advantages of meeting face-to-face: being able to see body language and facial expressions linked with pacing and tone of voice; providing a mechanism for interactive, two-way communication; and providing bonuses like screen sharing, which is a very useful tool for team members to share information
- Online task and project management tools that provide distributed visibility with dynamic project and task creation, prioritization, assignment, and tracking
- Group chat tools that allow teams to have synchronous, interactive text conversations (on both public and private channels that include historic archives), which can be more beneficial than individual, asynchronous emails
- Shared intranet areas can help team communication when the team is spread across multiple time zones; however, large work time differences can negatively affect team dynamics and performance

Working with Virtual, In-Person, and Hybrid Teams

With virtual, in-person, and hybrid teams, leaders and facilitators need to nurture the team's culture and build trust between team members. This helps confirm that team members share a common vision while reinforcing a feeling of belonging. Techniques for doing this include:

- Building shared goals, with open debate about those goals and ongoing goal reinforcement
- Actively facilitating win–win solutions
- Minimizing the power differences between team members
- Spending time allowing team members to get to know each other personally by chatting before or after online or in-person meetings or creating opportunities for other social interaction
- Providing an area within the team space where team members can share their personal interests, funny stories, news, ideas, and innovations
- Using team retrospectives and reflections to determine what is working and what needs to be improved
- Creating joint learning opportunities where team members or guests present webinars, lessons learned, and other topics of interest, or listen to externally offered webinars, and by holding team discussion of the topics after the presentations

Table 5.1 Team problems and potential solutions.

Problem	Potential Solution
Problems starting or ending a task or activity	• Review mission statement • Review the action plan • Review established entry/exit criteria • Brainstorm potential next steps
Dominating participant	• Utilize structured participation technique (e.g., the *nominal group technique*) • Proactively ask for opinions from other team members • Use a round-robin technique by going around the team and letting each person take a turn talking • Talk to the dominating person outside the meeting to discuss the behavior
Reluctant participants	• Utilize structured participation technique or round-robin technique • Divide agenda into tasks with individual or small-group assignments and reports (some people are shy about participating in larger groups) • Proactively seek out the opinions of reluctant members • Talk to the reluctant person outside the meeting to discuss the behavior

Participant attempting to use position or authority to direct team decisions	• Utilize structured participation technique
	• Talk to authority figures outside the meeting and ask them to:
	  • Hold their opinions until other team members have had an opportunity to be heard
	  • Use questions when stating their views (e.g., "what does the team think about . . .")

Using opinions rather than facts	• Stress the importance of basing decisions on facts as part of the team norms • Ask if there are supporting data, or who could collect the data • Ask how the team can confirm or verify the facts
Rushing to solutions or completion	• Provide an opportunity for everyone's views to be heard • Provide the team with constructive feedback on this behavior • Utilize structured problem-solving techniques • Openly ask for alternatives or other opinions • Create a "devil's advocate" role
Digression and tangents	• Use a written agenda and stick to it • Restate the topic being discussed and redirect the conversation back to the appropriate topic • Review mission statement, objectives, or action plan • Use a "parking lot" to capture ideas or issues and postpone their discussion to a later time
Feuding team members	• Refocus group on process and products, not on people and personalities • Develop or restate ground rules • Utilize conflict resolution techniques • Adjourn meeting and talk to offending parties outside the meetings
Too much conformity	• Openly reward divergent points of view • Ask outright for other points of view • Encourage positive conflict and controversy arising from opposing ideas and opinions • Protect or give equal weight to minority opinions • Create a "devil's advocate" role

- Holding virtual or in-person meetings with other teams, including management, on a regular basis so the team feels connected with the bigger organization and understands the larger vision, goals, and objectives of the organization

Distributed and hybrid teams need to select and tailor development practices that fit into their virtual context, or they need to tailor existing practices to meet the needs of their distributed team.

Part II

Software Quality Management

Chapter 6
Quality Management System

The primary purpose of a *quality management system (QMS)* is to implement an organization's chosen quality strategy. The QMS accomplishes this by focusing on areas that are critical to successfully achieving quality goals and objectives, providing high-quality products and services, and satisfying stakeholders. A QMS is the aggregate of the organization's quality-related organizational structure, policies, processes, work instructions, plans, supporting tools, and infrastructure. The QMS provides the standards, guidance, and techniques needed to plan, implement, maintain, and continuously improve the activities and actions related to quality management, quality planning, quality engineering, quality assurance, quality control, and verification and validation.

1. QUALITY GOALS AND OBJECTIVES

> Design software quality goals and objectives that are consistent with business objectives and regulations. Incorporate software quality goals and objectives into high level program and project plans. Develop and use documents and processes necessary to support software quality management systems. (Create)
>
> **Body of Knowledge II.A.1**

Quality Goals and Objectives

The primary goal of establishing a QMS is to institutionalize quality-related activities into every aspect of the organization and its key business practices. This starts at the highest levels of management, where *quality goals* or targets are established and "integrated into the strategic plans of the organization" (Furterer and Woods 2021). These quality goals should align with the organizational business goals and objectives, the mission of the organization, applicable laws and regulations, contractual obligations, and the needs of its stakeholders.

The following examples of organizational quality goals are based on the ISO 9000 *Quality Management Principles* (ISO 2015a):

- Meet customer requirements and strive to exceed customer expectations by continuously focusing on customer value
- Establish unity of purpose and direction, and create conditions in which people are engaged in achieving the organization's quality objectives
- Make sure that people at all levels throughout the organization are competent, empowered, and engaged
- Establish and maintain a coherent system of interrelated processes that are understood, implemented, and managed to effectively and efficiently achieve consistent and predictable results
- Maintain an ongoing focus on improvement
- Make informed, fact-based decisions by using analysis and the evaluation of data and information
- Actively manage relationships with key, relevant stakeholders

Quality objectives are then planned to achieve these quality goals. These objectives should be *SMART* (specific, measurable, achievable, realistic, and time-framed). As part of the organizational-level quality plans, "quality objectives are set to specify what specific actions will be taken, within a given time period, to achieve a measurable quality outcome" (Furterer and Woods 2021). Examples of organizational software quality objectives include:

- The organization will be formally assessed and achieve level 3 maturity on the CMMI® version 3.0, by the end of the second quarter of 2024
- Software developers will be cross-trained so that 90 percent of software source code modules under active development or maintenance are supported by at least two developers by the end of the second quarter of 2025
- The percentage of development dollars being spent on rework activities will be reduced to 15 percent or less per project by the end of the fourth quarter of 2024
- The percentage of test cases that have been automated in each product's regression test suite will be increased to 50 percent or more by the end of the third quarter of 2024
- Stakeholder satisfaction levels will be increased to 90 percent or more by the end of 2024, as measured by customer satisfaction survey responses scored in the top two boxes
- The number of defects reported post-release will be decreased by 20 percent or more for each consecutive new functional release of the same product until the total defect containment effectiveness measure for the product reaches at least 95 percent

These organizational-level quality objectives should be propagated into lower-level division, team, and individual objectives, and into the program, project, product, and process objectives that support them. Specific responsibilities for the quality objectives must be assigned at all levels of the organization. These objectives need to be communicated so that individual employees are aware of their responsibilities and how their work impacts progress toward achieving the objectives.

Quality Planning

At the organizational level, *quality planning* needs to be incorporated into the annual operating plans for the organization. These plans should include defined quality goals and objectives for the coming fiscal year. As illustrated in Figure 6.1, organizational quality planning should also align with, and implement, the organization's business objectives.

At the project level, quality plans should include specifics for how the project intends to implement the organization's QMS. Project quality plans must include strategies and tactics for how the project intends to implement the organizational-level planning, including quality goals and objectives propagated down to that project. As illustrated in Figure 6.1, the project quality plans may be separate documents referenced by the project plans or included as sections in the project plans. The project quality plans must also align with other project plans.

Quality Management System (QMS) Documentation

The *QMS documentation hierarchy* defines the organization's strategy and tactics for achieving its quality goals and objectives. Figure 6.2 illustrates the different levels and types of documentation in a QMS documentation hierarchy.

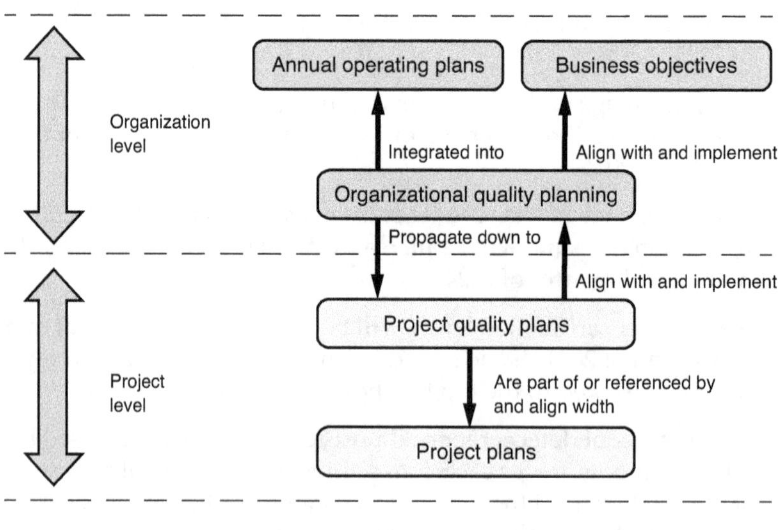

Figure 6.1 Quality planning hierarchy.

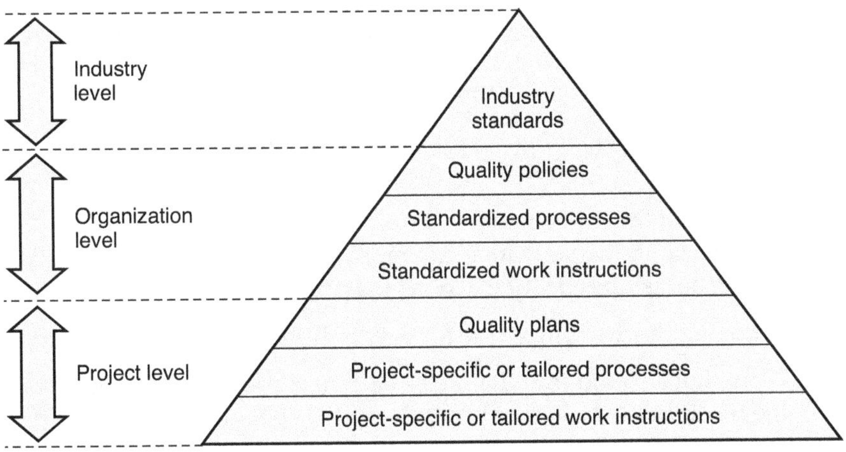

Figure 6.2 QMS documentation hierarchy.

Industry Standards

At the highest level, the organization's QMS is typically based on a framework of one or more industry standards or models. There are a variety of quality management systems and software industry standards and models that can be used (e.g., ISO 9001 and associated industry-specific interpretations or add-ons to ISO 9001, the IEEE software standards, or the CMMI® model described in Chapter 3). These industry standards and models represent "good practices" as defined by industry experts. Using one or more of these standards allows an organization to leverage this expertise and avoid wasting time by "reinventing the wheel." However, these defined industry good practices may need to be selected, adopted, and adapted based on the specific circumstances, needs, and context of the organization to turn them into "best practices" for that organization.

Quality Policies

At the organizational level, *quality policies* are formally established by senior management to define the overall direction and principles to be followed when making decisions and performing activities that impact quality. Upper-level management usually establishes quality policies to communicate the intent of the QMS and its objectives.

Standardized Processes

Standardized processes define the mechanics of what is required to implement the QMS activities at the organizational level. A *process* is a definable, repeatable, measurable sequence of tasks used to produce a quality product. ISO 9001 requirements state that "the organization shall establish, implement, maintain, and continually improve a quality management system, including the processes needed and their interactions" (ISO 2015).

By standardizing and documenting the definitions of the software processes, an organization can describe and communicate what usually works best. This standardization can help the organization by:

- Making certain that important steps in the processes are not forgotten or omitted
- Facilitating the propagation of lessons learned into organizational learning so the organization can repeat its successes project after project and stop repeating actions that lead to problems
- Reducing the learning curve for new or transferred staff members
- Providing a sound foundation for each new project while also allowing the flexibility to tailor the processes to the specific needs of that project

Defined and documented standardized processes provide the structured basis to create the metrics needed to understand process capabilities and analyze process results. They are necessary for consistent training, management review, and tool support. They provide the basis for organizational learning and continual process improvement.

There are many different methods for mapping processes. One simple way of documenting process definitions is to use the *ETVX (Entry criteria, Tasks, Verification steps, eXit criteria) method,* illustrated in Table 6.1.

- *Entry criteria*: The specific, measurable conditions that must be met before the process can be started.
- *Tasks*: The individual steps or activities that must be performed to implement the process and create the resulting product. The task definitions should also include responsibility assignments (roles) for each step.
- *Verification*: Steps that describe the mechanisms used to verify that the tasks are performed as required and that the deliverables meet required quality levels. Typical verification steps include reviews, tests, or sign-offs.
- *Exit criteria*: The specific, measurable conditions that must be met before the process can be completed.

Examples of entry and exit criteria include:

- Other processes or activities that must be satisfactorily completed
- Plans or documents that must be in place or must be updated
- Reviews, tests, or other evaluations that must be satisfactorily completed or approvals that must be obtained
- Specific measured values that must be obtained
- Staff with appropriate levels of expertise that must be available to perform the process
- Other resources that must be available and ready for use during the process

Table 6.1 ETVX method–example.

System Test Execution Process		
Entry Criteria	**Tasks**	**Exit Criteria**
• System Test Plan approved and under CM control • System Test Cases and System Test Procedures approved and under CM control • User documentation and software installation procedures approved and under CM control • Software build promoted to system test state • System test lab ready and available for use • Integration Test successfully complete • Testing staff available with the appropriate levels of expertise	T1. Execute System Tests and Report Anomalies: The system tester executes a selected set of test cases for each system test load. If the system test is suspended and restarted, these selected test cases will include cases to test all corrections and to regression test the software as appropriate. Any anomalies identified during system test are reported by the tester in accordance with the anomaly reporting process. T2. Debug and Correct Defects: The owner(s) (e.g., software development, technical publications) of each work product that is suspected to have caused the anomaly debugs that work product and corrects any identified defect(s) in accordance with the Problem Resolution process. T3. Build and Freeze Next Revision of System Test Work Products: Configuration management builds any updated revision of the software product(s) (e.g., software load, user manual, and installation instructions) that include the identified corrected components in accordance with the Software Build Process. T4. Write System Test Report: At the end of the final cycle of system test execution, the tester writes a System Test Report that includes a summary of the results from all of the system test cycles. T5. Promote Work Products: After the final test report is approved, all of the system test work products are promoted to the acceptance test status in accordance with the Configuration Management Promotion Process.	• System test completion criteria are met (as specified in the system test plan). • System Test Report is approved by test management. • Final system test software work products are promoted to beta test status.
	Verification: V1. Conduct Periodic Test Status Reviews: System test status review meetings are held on a periodic basis (as specified in the system test plan) during system test. If at any time it is determined that the suspension criteria (as specified in the system test plan) are met, system test execution is halted until the *resumption criteria* (as specified in the system test plan) are met and new revisions of the software work products are built and frozen. These meetings are also used to determine when system test is complete based on the system test completion criteria (as specified in the system test plan). V2. Peer Review Test Report: the testers peer review the system test report in accordance with the Peer Review process. V3. Review and Approve Test Report: Test management reviews and approves the final test report for distribution.	

One expansion on the ETVX diagram is the *EITVOX (Entry criteria, Inputs, Tasks, Verification steps, Outputs, eXit criteria) diagram* that adds the definitions of the inputs and outputs to the traditional ETVX diagram. The CMMI® for Development also expands on the ETVX method and states that "a defined process clearly states the purpose, inputs, entry criteria, activities, roles, measures, verification steps, outputs, and exit criteria" (CMMI 2010a).

The process's *purpose* is a statement of the value-added reason for the process. The purpose defines what the organization is attempting to accomplish by executing the steps in that process. For example, the purpose of a software system test execution process might be to verify the software system against the approved requirements, validate it against its intended use and identify product defects before the product is released.

Inputs are the tangible, physical artifacts that are input into and utilized during the process. These inputs may be work products created as part of other processes or items purchased or supplied by sources external to the organization (e.g., by customers or subcontractors). Example inputs to the software system test execution process might include the:

- System test cases and procedures (from the software system test design process)
- Software build (promoted from the software integration test execution process)
- User manual (from the user manual documentation process).

The distinction between inputs and entry criteria can sometimes appear confusing. For example, are test cases an input into the process, or is the fact that the test cases have been reviewed, approved, and placed under configuration management one of the entry criteria into the process? Some process definitions combine inputs and entry criteria into a single section to remove this confusion.

A *process flow diagram* can make that process easier to understand by showing the relationships between the various tasks, verification steps, and deliverables and by illustrating who (what role) is responsible for each task or verification step. The first step in creating a process flow diagram is to define the various roles of the process. The roles are the individuals or groups responsible for the steps that are part of the process. Their roles are listed in the "swim lanes" along the left side of a process flow diagram, as illustrated in Figure 6.3. These roles should be labeled in generic terms (e.g., project manager, developer, tester, test manager, SQE, SCM librarian) rather than using specific organizational titles or the actual names of individuals or groups. This is a key concept of good process definition, and the benefits of doing this include:

- Keeping the organization from being forced to update the process definition every time there is reorganization, someone gets a promotion or title change, or there is staff turnover.
- Allowing individual projects of various sizes to use the processes without the need for tailoring by assigning individuals or teams to these generic roles. For example, on very large projects, an entire team

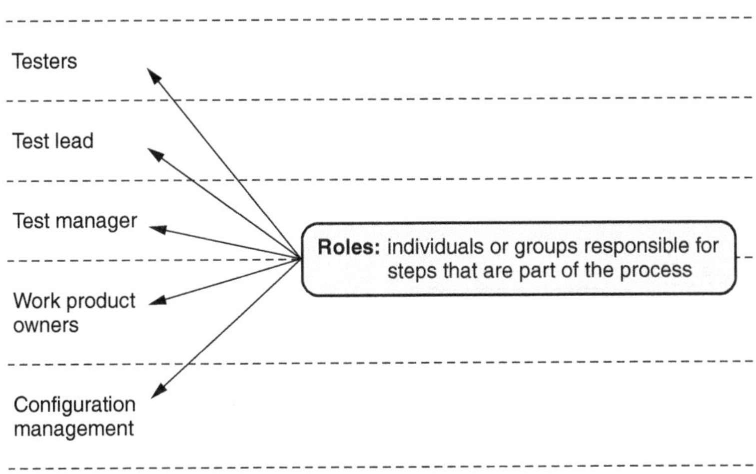

Figure 6.3 Roles in the swim lanes of a process flow diagram—example.

might be assigned to one of the process roles, while on tiny projects a single individual might be assigned several roles. Another example might be to use a generic role like *Approval Authority*, which could be assigned to a Configuration Control Board (CCB), a review team, or an individual, depending on the work product being approved.

- Saving time and money by avoiding non-value-added rework and unnecessary tailoring
- Making configuration management easier
- Enabling reuse

A flowchart can then be drawn across the swim lanes. The flowchart in Figure 6.4 illustrates the various tasks, verification steps, decisions, and deliverables of the software system test execution process. This example process starts at step T1: Assign Test Cases and follows the process flow through to the end at step T7: Promote Work Products. This method makes it easy for an individual assigned to a role to read across their "swim lane" and identify their responsibilities. If more than one role is responsible for a step, the box for that step simply spans multiple swim lanes (e.g., verification step V1 in Figure 6.4 and its associated internal decisions are the responsibility of the testers, test management, and project manager, and verification step V3 in Figure 6.4 is the responsibility of both the test manager and the project manager). Process tasks or verification steps may also call lower-level processes (e.g., task T2: Execute Test Cases Process and verification step V2: Peer Review Process in Figure 6.4 are calls to lower-level processes as denoted by the double bar on the left and right of the boxes). In this example, normal process flows are denoted with solid arrows, and process flows that occur only under specific conditions are denoted with dashed arrows. Optionally, the process flow diagram can also show the deliverables of the processes (e.g., deliverable D2: System Test Report in Figure 6.4). This example, using flow diagrams and flow charts to

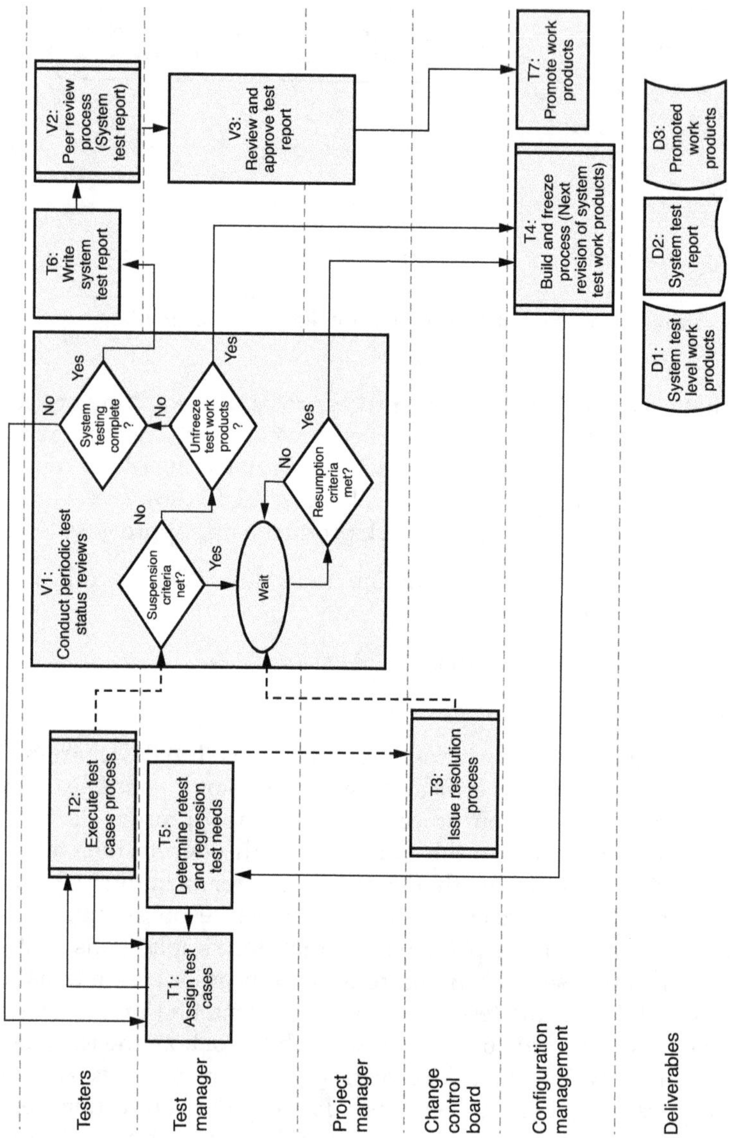

Figure 6.4 Process flow diagram—example.

illustrate process flow, is just one way to represent a process graphically. There are many different models and techniques that can be used when defining a process.

For each task or verification step defined in the process flow diagram, a textual description should be included in the process definition that describes the task or verification step in detail. These descriptions may include:

- Pointers to the detailed work instructions that describe how to accomplish the task or verification step.
- Additional descriptions of specific responsibilities. For example, "The test manager is responsible for conducting periodic test status reviews with the review participants, including testers, the test lead, work product owners, and configuration management."
- Required levels of expertise (or pointers to the descriptions of required levels of expertise) that must be possessed by those responsible for the task or verification step. For example, "Testers must be proficient on the XYZ testing tools set and the ABC simulator."
- Pointers to standards (e.g., formal inspection process standards or work product workmanship standards).
- Pointers to standardized templates (e.g., document, report, or meeting agenda templates) for creating the outputs of the task or verification step.
- Other resources (e.g., tools or hardware) that should be used in the task or verification step.

The expanded process definition might also include descriptions of (or pointers to the descriptions of) specific metrics collected or used, either as part of the process or to measure the quality of its products.

The *outputs* from the process include both deliverables and quality records. *Deliverables* are tangible, physical objects or specific, measurable accomplishments that are the outcomes of the tasks or verification steps. *Quality records,* also called *documented information* (ISO 2015), are outputs that provide evidence that the appropriate activities took place and that the execution of those activities met the required standards. Examples of quality records include:

- Minutes from meetings (e.g., reviews, configuration control boards, audits)
- Reports (e.g., technical review reports, managerial review reports, test reports, audit reports, status reports, metric reports)
- Change requests, action item lists, corrective actions, and so on
- Completed checklists
- Logs (e.g., test logs, engineering notebooks, issue logs, and risk logs)
- Formal sign-off and approval pages
- Completed forms (e.g., purchase orders, customer questionnaires, and approval forms)

As with inputs and entry criteria, the distinction between outputs and exit criteria can sometimes appear confusing. Some process definitions combine outputs and exit criteria into a single section to remove this confusion.

Process Architecture

Before each individual process is defined and documented, the *process architecture* should be designed to define the ordering of the individual processes, their interactions and interdependencies, the work product flow between the processes, and their interfaces with external processes. Figure 6.5 illustrates an example of a high-level process architecture flow diagram. Note that the System Test process defined in detail in Figure 6.4 is just one box in this higher-level architecture. There may be several levels of process architecture depending on the organization need's and the complexity of that organization's processes. For example, the Software Development Iteration box in Figure 6.4 might have its own lower-level architecture that defines lower-level processes including software requirements, architecture, component design, coding, unit testing, integration testing, and so on. The highest-level process architecture is the life cycle model (see Chapter 9 for more information on life cycle models).

Standardized Work Instructions

Going back to the QMS documentation hierarchy illustrated in Figure 6.2, standardized processes define the "what to do" requirements. However, the "how to do it" details should be left to lower-level *standardized work instructions*.

Guidelines are one type of work instruction. For example, the Execute Test Case process might have a task for "opening a problem report in the problem reporting database." By leaving this generic, the same process document can be used no matter which problem-reporting tool a project selects. To support this process, there may be several guideline-type work instructions, each providing details on how to open a problem report in a different tool, which fields to complete, and descriptions of the data that should be entered into each field. This example demonstrates how guidelines can document variations in a process without requiring the creation of multiple new processes to handle these variations.

Templates and *forms* are another type of work instruction. For example, the Plan Project process typically has a task for *documenting the project plans*. To support this process, several template-type work instructions may provide documentation guidance for each of several different project types (e.g., large projects, small projects, maintenance projects, agile projects). This is again an example of how templates can document variations in a process without requiring the creation of multiple new processes to handle these variations. Examples of forms might include purchasing forms, problem-reporting forms, forms for reporting audit nonconformances, or corrective actions plan forms.

Templates and forms allow practitioners to concentrate on content rather than format. Templates and forms act as an outline with instructions on what to document in each section, subsection, or field, thus preventing omission of needed information. Templates and forms can be hardcopy, electronic, or implemented using tools. For example, many requirements management tools allow the entry

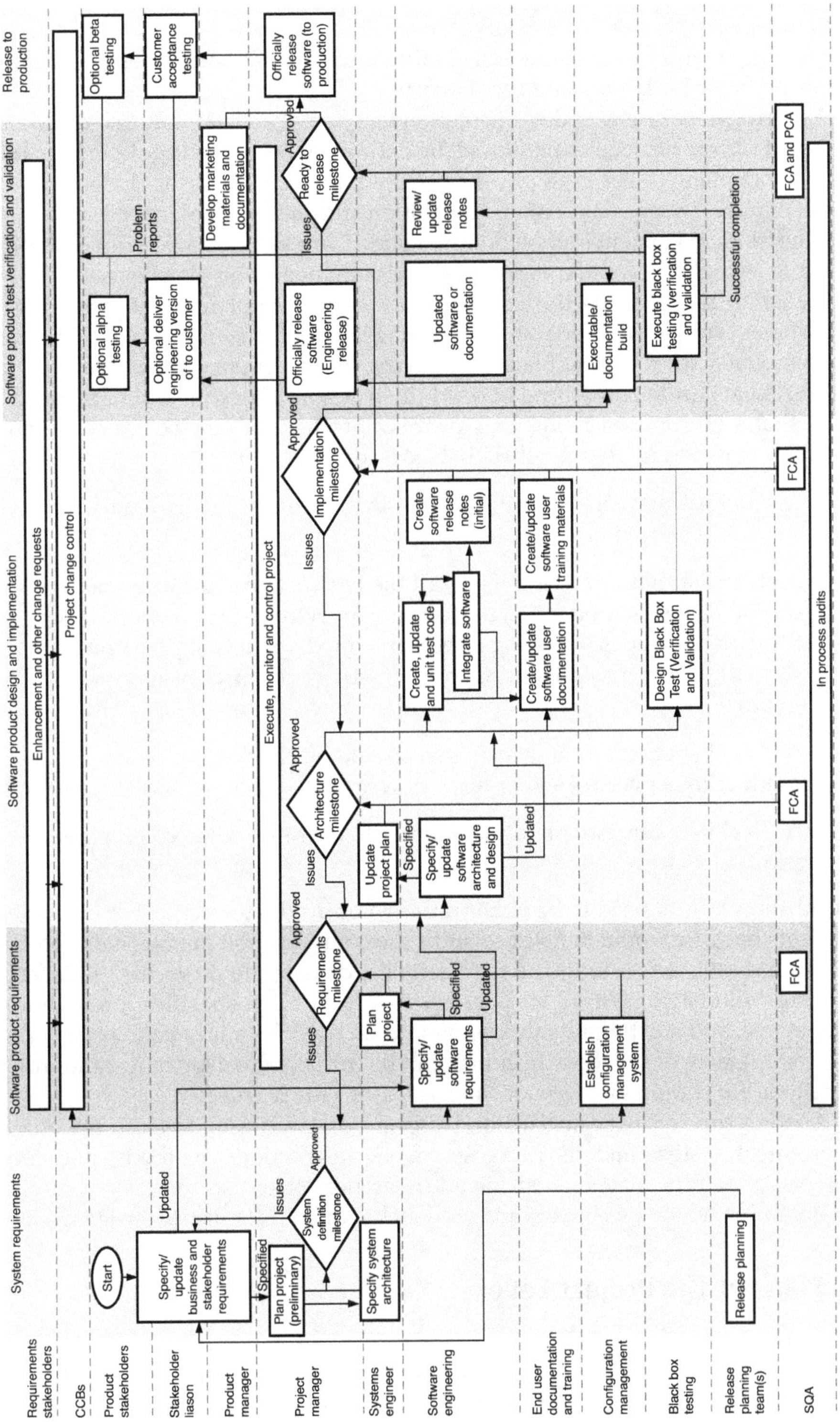

Figure 6.5 High-level process architecture—example.

of requirements information that can then be published using pre-established or customizable templates. Other tools allow for the automated completion of forms. One of the advantages of using tools is that tools usually include error handling that can enhance accuracy and consistency.

Utilizing a hierarchy of documentation templates can help reduce redundancy of content across the documentation set by defining where each type of information will be documented. For example, should each different planning document (e.g., project plan, software quality plan, verification and validation plan, and so on) have its own risk management section defining specific risks associated with each plan, or should a single, separate comprehensive risk management plan be used instead? With separate risk sections, the templates for each type of plan would include a risk management section. If a separate risk management plan is used, it would have its own template with the other plans referencing that risk management plan.

Checklists are a third type of work instruction. *Checklists* are tools used to verify that all important items for a process, product, or service are considered. Examples of different uses for checklists include:

- A checklist that includes all the steps in an activity so that none of the steps are omitted or done out of order
- A checklist that includes a list of all the factors and attributes that should be considered when performing an activity (e.g., a checklist of the different types of testing to be performed, including performance, security, safety, usability, load/volume/stress, resource usage, and so on)
- A checklist of common defects that should be checked for when conducting a peer review or test
- A checklist of questions to ask or areas to examine when conducting an audit

Checklists typically need to be either process- or product-specific. For example, a different checklist would be used to audit the configuration management process than would be used to audit the requirements development process. A different checklist would also be used for peer reviewing a design specification than for a code review, and a different checklist would be used for a Java code review than for a code review of C code. Checklists can be initially created from standards or guidelines but should evolve over time as lessons are learned.

Another type of work instruction is *training materials* used to teach the quality management system and its processes, work instructions, methods, and tools. Examples of training materials can include reference books, user manuals, computer-based training, videos, webinars, and student handbooks for instructor-led training.

Quality Plans at the Project Level

Project-level quality plans define the specifics of how a project intends to implement the organization's QMS to meet the quality goals and objectives of the organization and of that project. The software quality plans can be defined in a stand-alone *Software Quality Assurance Plan* document or incorporated into the project plans. Other quality plans, including *verification and validation plans, configuration management plans, supplier management plans, risk management plans,* and so on, may also be incorporated as sub-plans in either the Project Management Plan or other documents, or they may be stand-alone planning documents. Agile projects should do quality planning, but there may be little or no formal documentation of those plans ("just enough documentation"). For agile projects, quality planning may be documented as tasks or items in the iteration backlog (or Scrum backlog). In other words, the format of the plans is not what is important. The important thing is that quality planning takes place and is documented to the level of rigor appropriate to the needs and risk level of the project.

According to the *IEEE Standard for Software Quality Assurance Processes* (IEEE 2014), a project-level *Software Quality Assurance Plan* (SQAP) should include the following:

- A statement of the purpose and scope of the SQAP, definitions for all relevant terms and acronyms, and the identification of applicable references.
- An overview of the SQAP that addresses:
    - The organization of quality practitioners and their independence, including quality-related roles and responsibilities, and relationships to other stakeholder groups (project management, development, organizational quality management, suppliers, and so on) and the degree of independence of the quality-related roles to facilitate objectivity
    - Identification of software product risks, including safety, security, and other product risk (as distinguished from project risks)
    - Description of tools used to perform SQA tasks
    - Identification of the specific quality-related standards, practices, conventions, and metrics (e.g., documentation standards, modeling and coding standards, and test standards)
    - Estimates of effort, assignment of people and other resources, and a schedule of SQA milestones and planned SQA activities, tasks, and outcomes
- The definition of specific activities, outcomes, and tasks for product assurance, including the evaluation of:
    - Plans for conformance
    - Product for conformance and acceptability
    - Product life cycle support for conformance
    - Whether measurements objectively demonstrate the quality of the products
- The definition of specific activities, outcomes, and tasks for process assurance, including the evaluation of:

  - Life cycle processes for conformance
  - Environments for conformance
  - Supplier processes for conformance
  - Whether measurements objectively demonstrate the quality of the processes
  - Staff skills and knowledge
- The definition of specific activities, outcomes, and tasks for other SQA considerations, including:
  - Contract review
  - Quality measurement
  - Waivers and deviations
  - Tasks iteration
  - Communication strategies
  - Nonconformance process
- The identification of records and reports associated with SQA including activities and tasks related to analyzing, identifying, filing, maintaining, and disposing of those records and reports and making them available as required

Project-Specific or Tailored Processes

While the standardized processes define what "usually works best," they do not always match the exact needs of a specific project (program or product). One way of handling this situation is to tailor the standardized processes. *Process tailoring* alters or adapts a process to a specific use and allows standardized processes to be implemented appropriately for the project's needs. Tailoring can elaborate the process description to provide additional details to fit the purpose or context of a specific project. Tailoring can remove steps from a process that are unnecessary for a specific project. Tailoring may also include scaling the breadth and depth of the content or steps of a process. For example, on a smaller-than-usual project, the standardized processes may be tailored to remove unnecessary reviews or approvals because the same person is assigned multiple roles. On projects that are larger than typical, additional steps may need to be added to the standardized processes to allow for additional communication channels or levels of management. Process tailoring for a project may be defined as part of the project's plans or in separate process tailoring documentation.

Another way to handle the process requirements of a specific project is to create new, project-specific processes. For example, a joint venture project with another organization may require processes that are significantly different enough from the standard organizational processes that tailoring would be laborious and confusing. In this case, writing project-specific processes could be the best solution.

Project-Specific or Tailored Work Instructions

As with process documentation, based on the project's needs there may be a need to tailor standard work instructions. For example, audit checklists might need to be tailored to include any tailoring done to the processes being audited.

The project may also need to create project-specific work instructions. For example, a joint venture project might use a completely different problem-reporting tool than is typically used by the organization. In this case, project-specific guidelines on how to open a problem report using the new tool would be created.

Program-Level and Product-Level Documentation

Many software development or maintenance initiatives are run as projects. However, depending on the context of the development or maintenance effort, it may be more appropriate to implement quality plans and specific or tailored processes or work instructions at the program or product level instead of at the project level. Examples of when program-level or product-level documentation might be appropriate include:

- If multiple projects are being coordinated as a single program, and the same plans, processes, and work instructions need to be shared by all of those projects
- If incremental development is being used, and the same plans, processes, and work instructions need to be shared across each increment
- If iterative or agile development is being used, and the same plans, processes, and work instructions need to be shared across the entire product, iteration after iteration

2. QUALITY METRICS AND MONITORING

> Determine software quality goals and objectives for measuring and monitoring software including software product and process metrics, and analysis and reporting techniques. (Analyze)
>
> **Body of Knowledge II.A.2**

Metrics Defined

The term *metrics* means different things to different people. When someone buys a book or picks up an article on software metrics, the topic can vary from project cost and effort prediction and modeling to defect tracking and *root cause analysis* to a specific *test coverage* metric, to computer performance modeling, or even to the application of statistical process control charts to software.

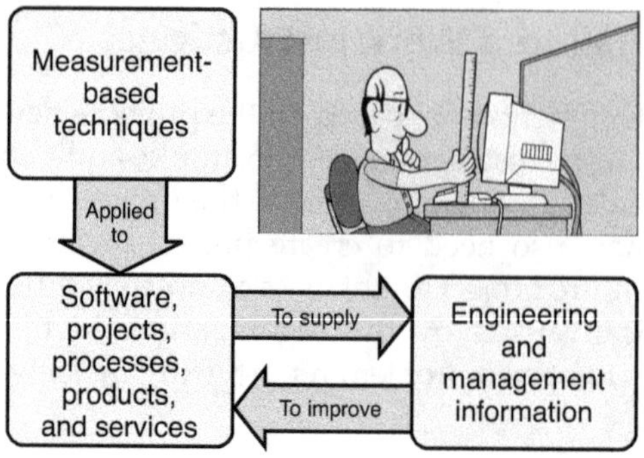

Figure 6.6 Metrics defined.

Goodman (2004) defines software metrics as "The continuous application of measurement-based techniques to the software development process and its products to supply meaningful and timely management information, together with the use of those techniques to improve that process and its products."

As illustrated in Figure 6.6, Goodman's definition can be expanded to include software projects and services. Examples of software services include:

- Responding with fixes to problems reported from operations
- Providing training courses for the software
- Installing the software
- Providing the users with technical assistance

Goodman's definition can also be expanded to include engineering information as well as management information. In fact, measurement is a key required element to move the software from a craft to an engineering discipline.

Software metrics are standardized ways of measuring the attributes of software projects, processes, products, and services to provide the information needed to improve those projects, processes, products, and services. The same metrics can then be used to monitor the impacts of those improvements, thus providing the feedback loop required for continual improvement.

Measurement Defined

The use of measurement is common. People use measurements in everyday life to weigh themselves in the morning or when they glance at the clock or the odometer in their car. Measurements are used extensively in most areas of operations and manufacturing to estimate costs, calibrate equipment, and monitor inventories. Science and engineering disciplines depend on the rigor that measurements provide. But what does measurement mean?

According to Fenton and Pfleeger (1997), "*measurement* is the process by which numbers or symbols are assigned to attributes of entities in the real world in such

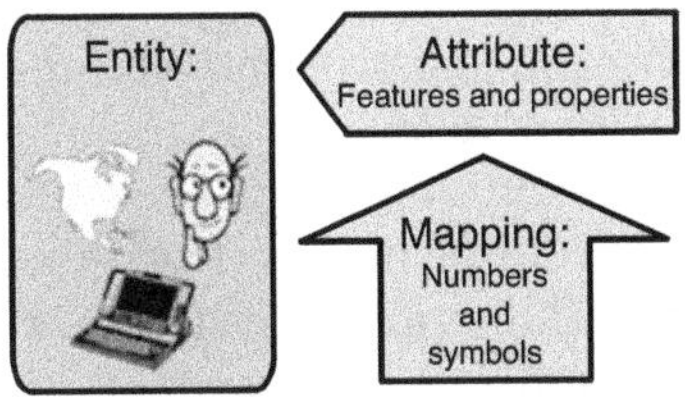

Figure 6.7 Measurement defined.

a way as to describe them according to clearly defined rules." Figure 6.7 illustrates this definition of measurement.

Entities are nouns, for example, a person, place, thing, event, or time period. An attribute is a feature or property of an entity. To measure, the entity being measured must first be determined. For example, a car could be selected as the entity. Once the entity is selected, the attributes of that entity that need to be described must be chosen. According to IEEE (1998d), an *attribute* is a measurable physical or abstract property of an entity. Attributes of a car include its speed, its fuel efficiency, and the air pressure in its tires.

Software product entities are the outputs of software development, operations, and maintenance processes. These include all of the artifacts that are produced. Examples of software product entities include:

- Requirements documentation
- Software design specifications and models (entity diagrams, data flow diagrams)
- Code (source, object, and executable)
- Test plans, specifications, cases, procedures, automation scripts, test data, and test reports
- Project plans, budgets, schedules, and status reports
- Customer call reports, change requests, and problem reports
- Quality records and metrics

Examples of attributes associated with software product entities include size, complexity, number of defects, test coverage, volatility, reliability, availability, performance, usability maintainability, effectiveness, efficiency, stability, reusability, and many others.

Finally, a *mapping system,* also called the *measurement method* or *counting criteria,* must be defined and accepted. It is meaningless to say that the car's speed is 65 or its tire pressure is 32 unless people know that they are talking about miles per hour or pounds per square inch. So what is a mapping system?

In ancient times there were no real standard measurements. This lack caused a certain level of havoc with commerce. Was it better to buy cloth from merchant A or B? What were their prices per length? England solved this problem by standardizing the "yard" as the distance between King Henry I's nose and his fingertips. The "inch" was the distance between the first and second knuckle of the king's finger, and the "foot" was literally the length of his foot.

To a certain extent, the software industry is still in those ancient times. As software practitioners try to implement software metrics, they quickly discover that very few standardized mapping systems exist for their measurements. No standard mapping system has been widely accepted, even for a seemingly simple metric such as the *severity* of a software defect. Examples from different organizations include:

- Outage, service-affecting, non-service-affecting
- Critical, major, minor
- C1, C2, S1, S2, NS
- 1, 2, 3, 4 (with some organizations using 1 as the highest severity and other organizations using 1 as the lowest)

An important element of a successful metrics program is the selection, definition, and consistent use of mapping systems for selected metrics. The software industry as a whole may not be able to solve this problem, but each organization must solve it to have a successful metrics program.

Quality Metrics and Monitoring

One of the organizational quality goals defined in ISO 9000 *Quality Management Principles* (ISO 2015a) is to "make informed, fact-based decisions by using analysis and the evaluation of data and information." Software metrics must be selected and defined to provide those data and information. Therefore, software metrics programs are designed to provide the specific information necessary to manage software projects and improve software products, processes, and services. Organizational, project, and task goals are determined in advance, and metrics are selected based on those goals. These metrics can determine the organization's effectiveness in meeting those goals. The foundation of this approach is aimed at making practitioners ask not so much, "What should I measure?" but "Why am I measuring?" or "What business needs does the organization wish its measurement initiative to address?" (Goodman 1993)

Measuring software is a powerful way to track progress toward an organization's goals. As Grady (Grady 1992) states, "Without such measures for managing software, it is difficult for any organization to understand whether it is successful, and it is difficult to resist frequent strategy changes." Appropriately selected metrics can help both management and engineers maintain their focus on their goals.

Metric Customers

With all of the possible software entities and attributes, it is easy to see that a huge number of possible metrics could be implemented. So how does an organization or team decide which metrics to use? The first step is to identify the customer. The customer of the metrics is the person or team who will be making decisions or taking action based on the metrics. The customer is the person who needs the information supplied by the metrics.

Step 1—Identify the metric customer: The first step in Westfall's *12 Steps to Useful Software Metrics* (12-Step) process (Westfall 2023) is to identify the customers for each metric. The customer of the metric is the person (or people) who will be making decisions or taking action based upon the metric—the person or people who need the information supplied by the metric.

There are many different types of customers for a metrics program. This diversity adds complexity to the program because each customer may have different information requirements. Customers may include:

- ***Functional management***: These people are interested in applying greater control to the software development process, reducing risk, and maximizing return on investment.
- ***Project management***: These people are interested in being able to accurately predict and control project size, effort, resources, budgets, and schedules. They are interested in controlling the projects they are in charge of and communicating facts to their management.
- ***Software engineers/programmers***: The people who do software development are interested in making informed decisions about their work and work products. These people are responsible for collecting a significant amount of the data required for the metrics program.
- ***Test managers/testers***: The people responsible for verification and validation activities are interested in finding as many new defects as possible in the time allocated to testing and obtaining confidence that the software works as specified. These people are also responsible for collecting a significant amount of the required data.
- ***Specialists***: Individuals performing specialized functions (e.g., marketing, software quality assurance, process engineering, software configuration management, audits and assessments, customer technical assistance) are interested in quantitative information upon which they can base their decisions, findings, and recommendations.
- ***Customers/users***: The people who purchase and use the software are interested in the on-time delivery of high-quality software products and reducing the overall ownership cost.
- If a metric does not have a customer, it should not be produced. Metrics are expensive to collect, report, and analyze, so if no one is using a metric, producing it is a waste of time and money.

The customers' information requirements should always drive the metrics program. Otherwise, we may end up with an information product without a market and a metrics program that wastes time and money. Recognizing potential customers and involving those customers early in the metric selection and definition effort increases the chances of success. Remember, software metrics don't solve problems. People solve problems. Software metrics act as indicators and provide

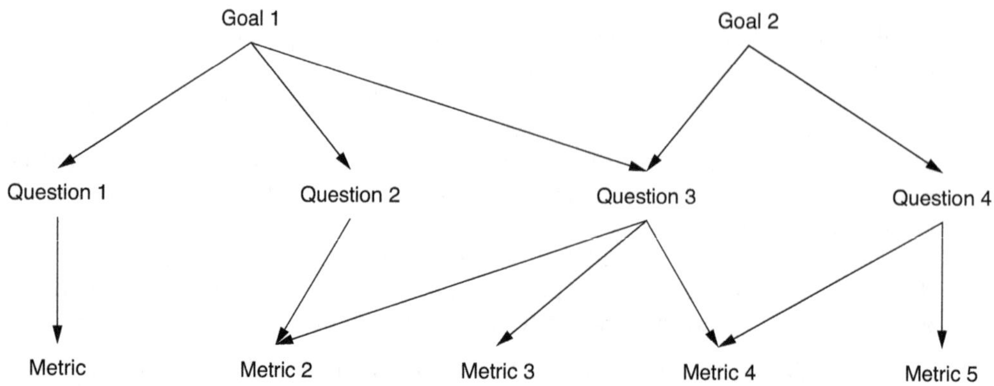

Figure 6.8 Goal/question/metric paradigm.

information so people can make more informed decisions and intelligent, fact-based choices.

Selecting Metrics

Basili and his colleagues defined a *goal/question/metric paradigm*, which provides an excellent mechanism for defining a goal-based measurement program (Grady 1992). The goal/question/metric paradigm is illustrated in Figure 6.8. This goal/question/metric paradigm is used in the next three steps of the 12-Step process (Westfall 2023) to select appropriate metrics based on the metrics customer's goals.

Step 2—Target the goals: The second step in the 12-step process (Westfall 2023) is to start implementing the goal/question/metric paradigm by selecting one or more measurable goals for the customer of the metrics:

- At the organizational level, these are typically high-level strategic goals. For example, QMS goals (as discussed earlier in this chapter) include being the low-cost provider, maintaining a high level of customer satisfaction, or meeting projected revenue or profit margin targets.
- At the project level, the goals emphasize project management and control issues or project-level requirements and objectives. These goals typically reflect the project success factors such as on-time delivery, finishing the project within budget or on schedule, delivering software with the required level of quality or performance, or effectively and efficiently utilizing people and other resources.
- At the specific task level, goals emphasize task success factors. Many times these are expressed in terms of satisfying the entry and exit criteria for the task.

Step 3—Ask questions: The third step in the 12-step process (Westfall 2023) is to determine the questions that need to be answered to determine whether each goal is being met or if progress is being made in the right direction. For example, if the goal is to maintain a high level of customer satisfaction, questions might include:

- What is our current level of customer satisfaction?
- What attributes of our products and services are most important to our customers?
- How do we compare with our competition?
- How do problems with our software affect our customers?

Step 4—Select the metric: Finally, in the fourth step, metrics are selected that provide the information needed to answer each question. Each selected metric now has a clear objective—to answer one or more of the questions that need to be answered to determine if we are moving toward our goals or meeting our goals.

When selecting metrics, be practical, realistic, and pragmatic. Metrics customers are turned off by metrics that they see as too theoretical. They need information that they can interpret and utilize easily. Avoid the "ivory tower" perspective that is completely removed from the existing software engineering environment (currently available data, processes being used, and tools). Customers will also be turned off by metrics requiring additional work to collect new or extensive data. Start with what is possible within the current process. Once a few successes are achieved, the metrics customers will be open to more radical ideas—and may even come up with a few metric ideas of their own.

It should also be noted that an organization's information needs will change over time. The metrics that are needed during requirements activities will be different than the metrics needed during testing or once the software is being used in operations. Some metrics are collected and reported periodically (daily, weekly, monthly), others are needed on an event-driven basis (when an activity or phase is started or stopped), and others are used only once (during a specific study or investigation).

An individual metric performs one of four functions. Metrics can help us ***understand*** more about our software projects, products, processes, and services. Metrics can be used to ***evaluate*** our software projects, products, processes, and services against established standards and goals. Metrics can provide the information we need to ***control*** projects and processes used to produce our software. Metrics can be used to ***predict*** attributes of software entities in the future. (based on Humphrey 1989). A comprehensive metric program includes metrics that perform all of these functions.

Figure 6.9 illustrates a requirement statement template that formally defines a selected metric in terms of one of these four functions, the attribute of the entity being measured, and the goal associated with the metric. Figure 6.9 also includes an example of the use of this template for the "percentage of known defects corrected" metric.

Having a clearly defined and documented requirements statement for each metric has the following benefits:

- Provides a rigor and discipline that helps ensure a well-defined metric based on customer goals

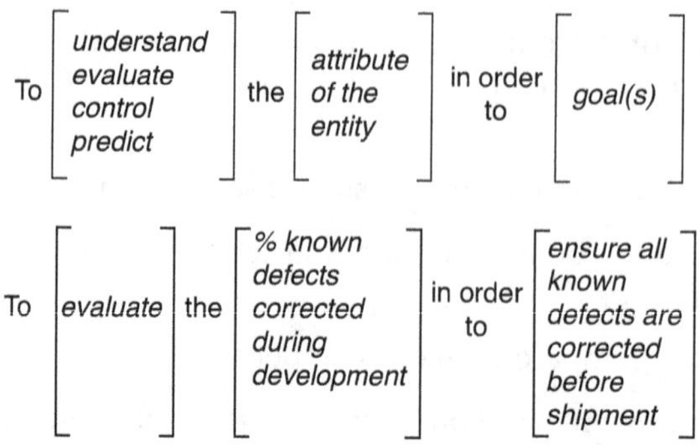

Figure 6.9 Metric requirement statement template with example.

- Eliminates misunderstandings about how the metric is intended to be used
- Communicates the need for the metric, which can help in obtaining resources to implement the data collection and reporting mechanisms
- Provides the basis for the design of the metric

Designing Metrics

Step 5—Standardize the definitions: The fifth step in the 12-step process (Westfall 2023) is to agree to standard definitions for the entities and their measured attributes. When practitioners use terms like *defect, problem report, size,* and even *project,* other people may interpret these words in their context with meanings that differ from the intended definition. These interpretation differences increase when more ambiguous terms like quality, maintainability, and user-friendliness are used. Additionally, individuals may use different terms to mean the same thing. For example, the terms *defect report, problem report, incident report, fault report,* or *customer call report* may be used by various organizations to mean the same thing. Unfortunately, they may also refer to different entities. One external customer may use a *customer call report* to refer to their complaint and a *problem report* as the description of the defect in the software. In contrast, another customer may use a *problem report* for the initial complaint. Differing interpretations of terminology may be one of the most significant barriers to correctly interpreting the measurement information.

Unfortunately, there is little standardization in the software industry of the definitions for most software attributes. Everyone has an opinion, and the debate will probably continue for many years. An organization's metrics program cannot wait that long. The suggested approach is to adopt standard definitions within the organization and then apply them consistently. The organization can use those definitions within the industry as a foundation to get started. For example, definitions from the *IEEE/ISO/IEC 24765 Systems and Software Engineering—Vocabulary* (IEEE/ISO/IEC 2017) or those found in software engineering and metrics literature. Pick the definitions that match the organizational objectives or use industry definitions as a basis for tailoring the definition to match the organization's needs.

Step 6—Choose the measurement function: In simple terms, the measurement function defines how to calculate the metric. Some metrics, called base measures, direct measures, or metric primitives, are measured directly, and their measurement function typically consists of a single variable. Examples of base measures include the

- Number of lines of code reviewed during an inspection
- Minutes spent preparing for an inspection meeting
- Number of defects found during the inspection

More complex metrics, called derived measures, are calculated using mathematical combinations (e.g., equations or algorithms) of base measures or other derived measures. An example of a derived measure would be the inspection's preparation rate, calculated as the number of lines of code reviewed divided by the number of preparation hours.

Many measurement functions include an element of simplification. This simplification is both their strength and their weakness. When practitioners create their measurement function, they need to be pragmatic. If they try to include all elements that affect the attribute or characterize the entity, their function can become so complicated that it's useless. Being pragmatic means not trying to create the most comprehensive function. It means picking the aspects that are the most important. Remember that the function can always be modified to include additional levels of detail in the future. Ask the questions:

- Does the function provide more information than we have now?
- Is the information of practical benefit?
- Does it tell us what we want to know?

There are two methods for selecting a measurement function: use an existing function or create a new one. In many cases, there is no need to "reinvent the wheel." Many software metrics functions exist that have been used successfully by other organizations. These are documented in the current literature and in proprietary products that can be purchased. With research, practitioners can utilize these functions with little or no adaptation to match their environment.

The second method is to create a new function. The best advice here is to talk to the people responsible for the product or resource or who are involved in the process. They are the experts. They know what factors are significant.

To illustrate the selection of a function, consider a metric for the duration of unplanned system outages. Suppose a software system installed at a single site is being measured; a simple function such as minutes of outage per calendar month may be sufficient. However, suppose different software releases installed on varying numbers of sites are being measured. That might require selecting a function such as minutes of outage per 100 operation months. Finally, if the focus is on customer impact, a function based on minutes of outage per site per year might be selected.

Step 7—Define the measurement methods: The seventh step is breaking down the function into its lowest-level base measures and defining the measurement method used to assign a value to each base measure. The measurement method defines the mapping system that assigns numbers or symbols to the attributes.

Some measurement methods are established using standardized units of measure (e.g., inches, feet, pounds, liters, dollars, hours, and days). Other measurement methods are based on counting criteria, simple counts of items with specific characteristics. For example, if the metric is the monthly problem report arrival rate, we could count all problem reports in the database with an open date that month. However, if we wanted defect counts instead of problem report counts, we might exclude all the reports that didn't result from a product defect (e.g., works as designed, user error, or withdrawn). Other rules may also be used for the measurement method. For example, the rules an organization uses for assigning severity to a problem report. These rules might include judging how much of the software's functionally is impacted (e.g., more than 50%, <=50% but >20%, <= 20%) or the duration of that impact (for more than 60 seconds, <=60 seconds but < 15 seconds, <= 15 seconds) or other criteria.

The importance of the need for defining a measurement method can be illustrated by considering the lines-of-code metric. The lines-of-code measure is one of the most used and most often misused of all of the software metrics. The problems, variations, and anomalies of using lines of code are well documented (Jones 1986), and there is no industry-accepted standard for counting lines of code. Therefore, if an organization is going to use a metric based on lines of code, a specific measurement method must be defined. This method must be known and understood by the metrics customers who receive the metric. Without this, invalid interpretations of the information are almost inevitable.

The base measures and their measurement methods define the first level of data to be collected to implement the metric. The function of minutes of system outage per site per year can be used to illustrate this. One of the base measures for this model is the number of sites. At first, counting this base measure seems simple. However, when the dynamics of adding new sites or installing new software on existing sites is considered, the counting criteria become more complex. Is the number of sites on the last day of the period used, or is some average number of sites during the period calculated? Either way, the date the system was installed on the site must be collected as data. In addition, if different software releases need to be compared, the releases installed on each site and when each was installed will need to be collected as data.

Step 8—Define the decision criteria: According to the 15939-2017—*IEEE/ISO/IEC—Systems and Software Engineering—Measurement Process* standard, decision criteria are the "thresholds, targets, or patterns used to determine the need for action or further investigation, or to describe the level of confidence in a given result" (IEEE/ISO/IEC-17). In other words, the decision criteria provide the guidance that will help the metrics customers interpret the measurement results.

Returning to Humphrey's four reasons to measure (i.e., control, evaluate, understand and predict (Humphrey 1989), control metrics monitor the software projects, processes, products, and services and identify areas where corrective or management action may be required. Metrics used for control act as "red flags" to warn metrics customers when things are not as expected or planned. If all is going well, the decision should be "everything is fine; no action is required." The decision criteria for control metrics usually take the following forms (Westfall 2023):

- *Threshold*: A value that, when crossed, indicates that an out-of-control condition may exist. For example, if a project manager planned the project assuming a 15 percent staff turnover, that manager could track the actual staff turnover against a 15 percent threshold.
- *Variances*: The difference between two values. When that delta exceeds a specific value, it indicates an out-of-control condition may exist. For example, a project manager could track the variance between the project's budget and actual expenditures.
- *Control limits*: Specific upper and lower boundary values used to indicate that an out-of-control condition might exist, for example, the upper and lower control limits in a statistical process control chart.

Evaluate metrics are used to examine and analyze measurement information as part of a decision-making processes. For evaluate metrics, the decision criteria help define what is good. For example, suppose an organization wants to make a 15 percent return on investment (ROI) for a project. In that case, the decision criteria could indicate that the benefit-to-cost ratio must be at least 1.15, or the project shouldn't be initiated. Testers might also establish decision criteria for their exit from software qualification testing. If decision criteria are not met, then testing should continue. Examples of these software qualification test decision criteria could include:

- At least X percent of all planned test cases must be executed, and at least Y percent of those must have passed.
- Zero non-closed critical problems can exist, no more than X major nonclosed problems can exist, all of which have workarounds, and no more than X minor nonclosed problems can exist. (We use the term *nonclosed* instead of open because of the possible ambiguity of the word *open*. For example, developers may consider a problem no longer "open" when they have corrected it, but it may not have been integrated into the product and retested).
- The arrival rate of new problem reports must be decreasing towards zero with no new critical problems reported in the last X number of days.

For metrics used to understand and predict, it is typically the "level of confidence in a given result" portion of the IEEE/ISO/IEC 15939 definition that applies. How confident is the metrics customer that the understanding they gained from the metric reflects reality? How confident is the metrics customer that the prediction reflects what the actual values will be in the future? One method to use is to calculate statistical confidence intervals. However, the metrics customer can also obtain a more subjective confidence level by considering factors including:

- The completeness of the data used. For example, if the metrics customer is trying to understand the amount of effort expended on software development, does the organization's effort reporting system include all the time spent, including unpaid overtime?

- Are the data used subjective or objective? Have human or other types of bias crept into the data?
- What is the integrity and accuracy of the data? For example, consider time card data. Are people completing their time cards as they complete tasks, or are they waiting until the end of the week and then estimating how much time they spend on various projects?
- How stable is the project, product, process, or service being measured? For example, suppose the project is using a new process or a new programming language; the metrics customer may not be very confident in a prediction based on historical data, or the organization may not have any relevant historical data upon which to base that prediction.
- What is the variation within the data set? For example, looking at the distribution in a data set with slight variance, the metrics customer can be reasonably confident that the data set's mean (average) accurately represents that data. However, if the data set has a large amount of variance or a bimodal distribution, it will decrease the confidence level that the mean accurately represents the data.

Step 9: The ninth step is to decide how to report the metric. This step includes defining the report format, data extraction and reporting cycle, reporting mechanisms, distribution, and availability.

The report format defines what the information product looks like. Is the information product a table with multiple measurement values? Is the new measure added as the latest value in a trend chart that tracks values for the metric over multiple periods? Should that trend chart be a bar, line, or area graph? Is it better to compare values using stacked bars or a pie chart? Do the tables and graphs stand alone, or is there detailed analysis text included with the report? Are goals, control values, or other decision criteria included in the report? (See Chapter 20 for more information on analysis and reporting techniques.)

The data extraction cycle defines how often data snapshots are required and when they will be available for use for calculating the measurement.

The reporting cycle defines how often the report is generated and when it is due for distribution. For example, root cause analysis measurements may be triggered by some event, like the completion of a phase or iteration in the software development process. Other metrics like the defect arrival rate may be extracted and reported daily during testing or extracted monthly and reported quarterly after the product is released to the field.

The reporting mechanism outlines the way that the metric is delivered (i.e., hard copy report, email, online electronic data).

Defining the distribution involves determining who receives regular copies of the report or access to the metric.

The availability of the metrics defines any restrictions on access to the metric (i.e., need to know, internal use only) and the approval mechanism for additions and deletions to access or standard distribution.

Step 10—Select the additional qualifiers: The 10th step is determining the additional metric qualifiers. A good metric is a generic metric. That means the metric is valid for an entire hierarchy of additional qualifiers. For example, we can talk about the duration of unplanned outages for an entire product line, an individual product, or a specific release of that product. We could look at outages by customer or business segment. Alternatively, we could look at them by type or cause.

The additional qualifiers provide the demographic information needed for these various views of the metric. The main reason additional qualifiers need to be defined as part of the metrics design is that they determine the second level of data collection requirements. Not only is the metric primitive data required, but data also have to exist to allow the distinction between these additional qualifiers.

Implementing Metrics

Step 11—Collect the data: The question "what data to collect?" was answered in steps 7 and 10 of the 12 steps. The answer is to collect all the data required to provide the metrics primitives and the additional qualifiers. The 11th step defines the following:

- Who collects each data item?
- When and how is each data item collected?
- Where is each data items store?
- What format is each data item store (e.g., number, character set, date, and so on)?

(See Chapter 19 for more information on data collection and integrity.)

Step 12—Consider the human factors: No discussion on selecting, designing, and implementing software metrics would be complete without looking at how measurements affect people and people affect measurements. Whether a metric is ultimately valuable for an organization depends upon the people's attitudes in collecting the data and calculating, reporting, and using the metric. The simple act of measuring will affect the behavior of the individuals being measured. When something is being measured, it is automatically assumed to have importance. People want to look good; therefore, they want the measures to look good.

When creating a metric, a decision must be made about what behaviors the metric should encourage. For example, a team uses a metric to focus on reducing technical debt. The organization must also consider what negative behaviors might result from using or misusing the metric—for example, management using the metrics to punish individuals or teams. Another example would be people trying to manipulate the measurement results. One of the author's favorite measurement quotes is, "Don't underestimate the intelligence of your engineers. For any one metric you come up with, they will find at least two ways to beat it" (source unknown).

This concern does not mean that organizations give up on a metrics program because people might misuse the metrics. It means they have to be conscious of potential problems and proactively plan how to deal with them. As the last step in the 12-step process, the organization determines the positive behaviors that they want as a result of implementing the metric and the possible negative behaviors that might result from its implementation. They can then create an action plan on how they will emphasize the positives and eliminate or at least minimize the negatives.

A good way to deal with these human factors issues in working with metrics is to follow some basic rules:

- *Don't measure individuals*: The state of the art in software metrics is not up to this yet. Individual productivity measures are the classic example of this mistake. Remember, our best people are typically given the most challenging work, and they are expected to mentor others in the group as well. If individual productivity is measured in lines of code per hour, these people may concentrate on their work to the detriment of the team and the project. Even worse, they may come up with unique ways of programming the same function in many extra lines of code. Focus on processes and products, not people.
- *Never use metrics as a "stick"*: The first time metrics are used against an individual or a group is the last time the data will be valid.
- *Don't ignore the data*: A sure way to kill a metrics program is to ignore the data when making decisions. "Support your people when their reports are backed by data useful to the organization" (Grady 1992). If the goals the organization establishes and communicates don't agree with its actions, then the people in the organization will perform based on the behavior, not the goals.
- *Never use only one metric*: Software is complex and multifaceted. A metrics program must reflect that complexity. A balance must be maintained between cost, product (including functionality and quality), and schedule attributes to meet the customer's needs. Focusing on any single metric can cause the measured attribute to be improve at the expense of other attributes, resulting in an anorexic software development process.
- *Select metrics based on goals*: Metrics act as a big spotlight focusing attention on the measured area. By aligning the metrics with organizational goals, people's attention is focused on the things that are important.
- *Provide feedback*: Providing regular feedback to the data collectors about the data they help collect has several benefits:
  - It helps maintain focus on the need to collect the data. When the data collectors see the data being used, they are more likely to consider data collection important.
  - If data collectors are kept informed about precisely how the data are used, they are less likely to become suspicious or fearful that they are being used against them.
  - By involving data collectors in data analysis and process improvement efforts, the organization benefits from the data collectors' unique knowledge and experience.

- Feedback on data collection problems and data integrity issues helps educate the data collectors responsible for data collection. The benefit can be more accurate, consistent, and timely data.

- *Obtain "buy-in"*: To have buy-in to both the goals and the metrics in a measurement program, data collectors need to feel ownership. Participating in the definition of the metrics will enhance this feeling of ownership. In addition, the people who work with a process daily will have intimate knowledge of that process. This participation gives them a valuable perspective on how the process can best be measured to ensure accuracy and validity and a better interpretation of the measured result to maximize usefulness.

A metrics program based on an organization's goals will help communicate, measure progress toward, and eventually attain those goals. People will work to accomplish what they believe to be important to management. Well-designed metrics with documented objectives can help an organization obtain the information it needs to continue improving its software project, products, processes, and services while focusing on essential goals. A practical, systematic, start-to-finish method of selecting, designing, and implementing software metrics is valuable in ensuring the consistent collection, reporting, and use of measurement information. For more on metrics and monitoring, see:

- Chapter 7: Methodologies (for Quality Management)
- Chapter 16: Planning, Scheduling, and Deployment
- Chapter 17: Tracking and Controlling
- Chapter 18: Risk Management
- Chapter 19: Process and Product Metrics
- Chapter 20: Analysis and Reporting Techniques

3. CUSTOMERS AND OTHER STAKEHOLDERS

> Assess and evaluate the effect of various stakeholder groups requirements (e.g., quality requirements for contractual obligations) on software projects and products. (Evaluate)
>
> **Body of Knowledge II.A.3**

Stakeholders are individuals, groups, or organizations who affect or are affected by the decisions, activities, or outcomes of a product, project, or process and therefore have some level of influence over the requirements for that product, project, or process. Stakeholders can be at any level. They do not have to have a management title or any official level of authority. An individual contributor or user may be an important stakeholder if they can influence the requirements, acceptance, quality perceptions, or other factors of the product, projects, or process, either as individuals or as a collective group.

All stakeholders "have particular issues, priorities, and concerns, and it is important that an organization understand each of these stakeholder groups and the issues most important to them" (Furterer and Woods 2021). Therefore, it is essential that an organization identify its product, process, and project stakeholders and assess and evaluate their motivations, needs, and requirements.

Product Stakeholders

As illustrated in Figure 6.10, there are three main categories of *product stakeholders*: the suppliers of the software, the acquirers of the software, and other stakeholders.

The *acquirer-type stakeholders* can be divided into two major groups. First, there are the *customers* who select, request, purchase, or pay for the software in order to meet their business objectives. The organization may have contractual obligations to its customers. The second group is the *users* who use the software directly or indirectly by receiving reports, outputs, or other information generated by the software. Another type of users is the *unfriendly users* that the software should be "unfriendly" to by preventing them from accomplishing their objectives. For a *gas station pay at the pump* system, examples of unfriendly users include hackers, credit/debit card thieves, counterfeiters, people with expired or over-the-limit credit/debit cards, and people trying to gain unauthorized access to the software's data or functionality. There may also be many different types of direct users.

There are novice users, occasional users, and power users of a software product. There may also be direct users with different levels of knowledge or skill, different roles or objectives, different access privileges, or different motivations.

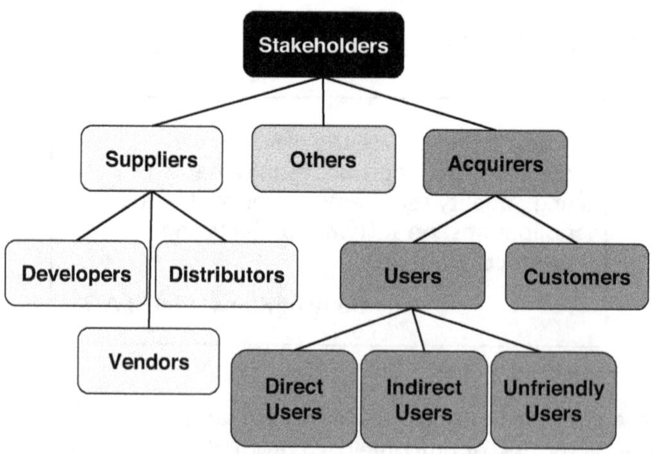

Figure 6.10 Categories of product stakeholders.

For example, users of a "pay at the pump" system include the gas station manager, the attendant, car owners purchasing gas, 18-wheeler drivers purchasing diesel, illiterate people, people who do not speak English, and so on. Each of these stakeholder groups may place different requirements on the software.

The *supplier stakeholders* of the software are divided into two groups. The *developers and vendors* are the individuals and groups that are part of the organization that develops or maintains the software. Developer or vendor stakeholders include project/technical manager, requirements analysts, designers, coders, testers, software quality engineers, configuration management specialists, marketing, sales, manufacturing, hardware development, legal, technical support personnel, or other groups internal to the supplier's organization. The *vendors,* also called subcontractors are third-party developers of the software. The *distributors* are the individuals and groups that distribute the software. Examples of distributor stakeholders include the office supply store that sells a word processor software package, or a construction firm that installs an energy management and building automation system in the high-rise they are building.

Other stakeholders are those who do not belong to either the supplier or acquirer groups but still may have some level of influence over the software, including:

- Lawmakers or regulatory agencies that create laws, regulations, and standards that impact the software product
- Organizations that create industry standards or guidelines or define industry good practices for the software product
- Other groups or individuals that are affected by the actions or decisions of the product's acquirers or suppliers
- Even society at large can have a vested interest in the software

Project Stakeholders

The Project Management Institute (PMI) *Project Management Body of Knowledge Guide* (PMBOK Guide) defines project stakeholders as "persons and organizations . . . that are actively involved in the project, or whose interests may be positively or negatively affected by the execution or completion of the project" (PMI 2013). Project stakeholders include:

- Individuals funding, initiating, or championing the project
- Project managers and members of the project team
- Individuals supporting the project who are not on the project team (e.g., lawyers, regulatory specialists, auditors, and purchasing)
- Stakeholders of the products produced by the project
- Individuals from other projects/programs that must interface or coordinate with the project
- Project/program management office

Process Stakeholders

Process stakeholders are individuals who affect or are affected by a software process or its outcome. Process stakeholders include:

- Individuals directly involved in the planning, management, execution, tracking, or control of the process activities or steps
- Individuals defining, documenting, and improving the process
- Process sponsors responsible for funding the process activities
- Individuals championing the process
- Stakeholders of the products produced by that process
- Stakeholders from other processes that must interface or coordinate with the process
- Individuals responsible for auditing or assessing the process

Stakeholder Management

Figure 6.11 shows the four stakeholder management processes defined in the PMBOK and the flow of information internally, between those four processes, and externally with other project management processes. These four processes can also be generalized and applied to product and process stakeholders.

The purpose of the *identify stakeholders* process is to identify the individuals, groups, or organizations who affect, or are affected by, the project's decisions, activities, or outcomes. This process also analyzes and documents the project-related requirements, contractual obligations, and motives of those stakeholders.

The purpose of the *plan stakeholder management* process is to plan the appropriate strategy and tactics to effectively interact with the identified stakeholders throughout the project. This plan includes identifying stakeholder communication mechanisms and stakeholder participation plans.

The purpose of the *manage stakeholder engagement* process is to effectively communicate and work with stakeholders throughout the project life cycle to meet their expectations and requirements, manage issues, and foster appropriate stakeholder project activities. According to Dudash (2015), "the project manager must manage stakeholder expectations by listening to current business needs, addressing any yet unstated stakeholder requirements, and adjusting project deliverables to address those needs."

The purpose of the *control stakeholder management engagement* process is to monitor stakeholder communications, interactions, and relationships and to take corrective action when necessary.

Benefits of Identifying and Involving Relevant Stakeholders

There are many benefits to identifying and involving relevant stakeholders in decisions about software products, processes, and projects. Identifying and considering the needs of all the different stakeholders can help prevent

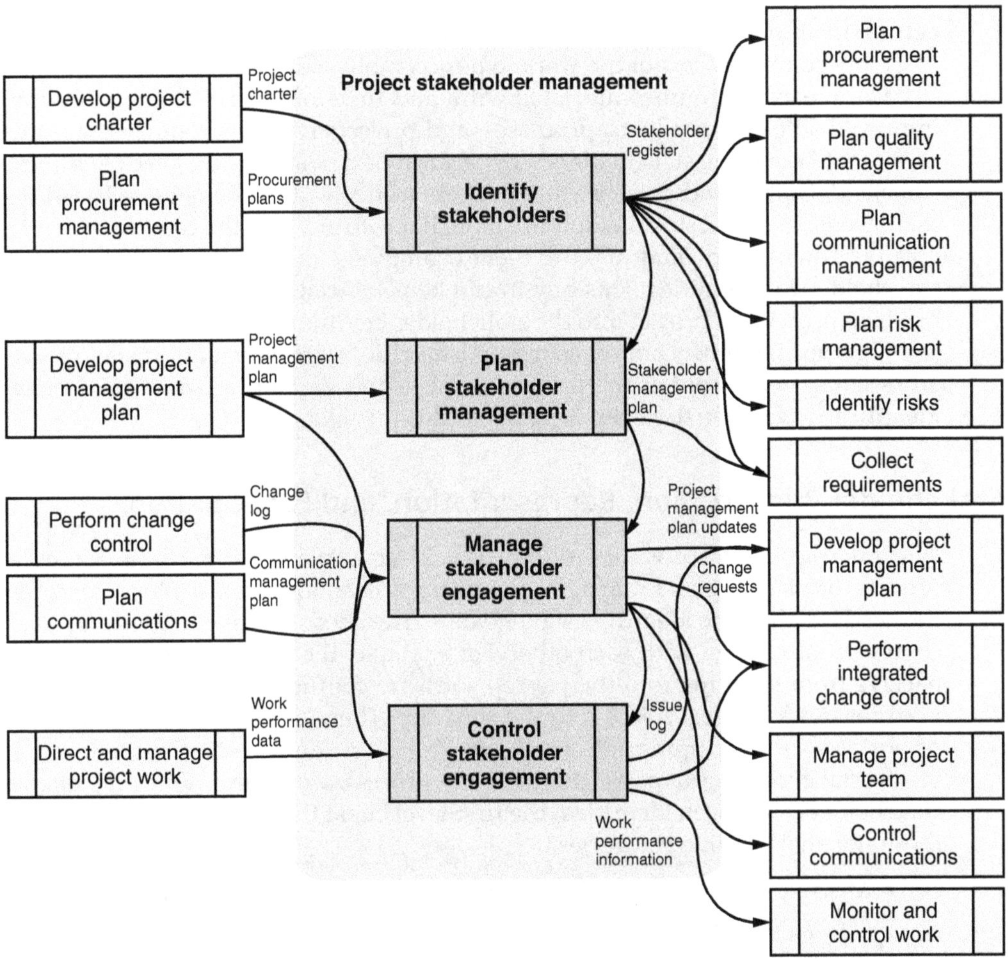

Figure 6.11 PMBOK project stakeholder management processes and their interrelationships (based on PMI 2013).

requirements from being overlooked. For example, suppose a project is creating a payroll system and they do not consider charities as one of the stakeholders. They might not include the requirements for the software to withhold, handle, track, and report charitable payroll deductions. Getting stakeholders involved in requirements engineering eliminates two of the most ineffective requirements elicitation techniques: clairvoyance and telepathy (based on Wiegers and Beatty 2013).

Software practitioners can never know as much about a stakeholder's work as that stakeholder knows. Identifying and involving key stakeholders provides access to the stakeholders' experience base and domain knowledge. The practitioner's job is to analyze, expand on, synthesize, resolve conflicts between, and combine the inputs from all the stakeholders into an organized set of product requirements, process definitions, or project plans. Identifying the different stakeholders and getting them involved also brings different perspectives to the table that can aid in a more complete view of the work to be accomplished.

Many people are uncomfortable with, and therefore resist, change. New or improved software products, processes, and projects typically require changing the way at least some stakeholders will perform part or all of their jobs. Obtaining stakeholder input and participation gets them involved in the change based on their needs. Involved stakeholders are more likely to buy into the completed work, creating "ownership" and making them champions for the change within their stakeholder communities. This buy-in can be beneficial in transitioning the new product, process, or project into the stakeholder environments.

"Perhaps the most common single mistake in development efforts is to leave an essential person out of the process" (Gause and Weinberg 1989). Stakeholder identification and participation help avoid this mistake.

Stakeholder Prioritization, Representation, and Participation

When developing a software product, designing, implementing, or improving a process, or planning and managing a project, it is almost impossible to consider the needs of all of the identified stakeholders. The needs of stakeholders may also contradict or conflict with each other. For example, the need to keep unfriendly hackers from breaking into the payroll software conflicts with the accountant's need for quick and easy access to the software. Therefore, stakeholders must be prioritized so that appropriate trade-offs can be made based on the relevance of the stakeholder groups to the product, process, or project. Once all of the stakeholders have been identified, the first-level cut at this prioritization is to sort those stakeholders into categories:

- *Must include*: These stakeholders must be included in the product, project, or process activities.
- *Like to include*: These stakeholders will only be included in the product, project, or process activities if schedules and costs allow.
- *Not directly involved*: These stakeholders will not be directly included in the product, project, or process activities.

Criteria that should be considered when sorting stakeholders into these groups include the stakeholder's:

- Power/authority
- Interest in the product/project/process
- Influence with other stakeholders
- Impact or ability to effect changes.

It should be noted that just because a stakeholder is not directly included in the product, project, or process activities, it does not mean the stakeholder's needs and requirements are ignored. It just means that those needs or requirements must be identified through means other than direct involvement.

For each stakeholder group that is included, a decision must be made about who will represent that stakeholder group in the product, process, or project activities. There are three main choices:

- *Representative*: Select a stakeholder champion to represent the group of stakeholders. For example, if there are multiple testers, the lead tester might be selected to represent this stakeholder group. The lead tester would then participate in the activities and be responsible for gathering inputs from other testers and managing communication with them. For example, for agile projects using Scrum, the product owner is the representative for all the acquirer-type stakeholders.
- *Sample*: Sampling may be appropriate for large stakeholder groups or for groups where direct access is limited for some reason. In this case, it would be necessary to devise a sampling plan for obtaining inputs from a representative set of stakeholders in that particular group. For example, if the company has several thousand employees, it may decide to take a sample set of employees to interview about their needs for the new accounting system.
- *Exhaustive*: If the stakeholder group is small or critical enough to success, it may be necessary to obtain input from every stakeholder group member. For example, if the software product has a small set of customers, it might be important to obtain input from each of those customers.

The second decision is to determine when and how each included stakeholder will participate in the activities. Will they participate throughout the entire product life cycle, process, or project, or only at specific times? What activities are they going to participate in? For example, a stakeholder's input may be gathered during the definition of the process, but that stakeholder might not be involved in the peer review of that definition or be part of the ongoing process change approval authority. For product requirements development, one key stakeholder might just be interviewed during requirements elicitation activities, while another key stakeholder is considered part of the requirements development team and participate in multiple requirements elicitation, analysis, specification, verification, and management activities.

The final decision is to establish a second level of priority for each included stakeholder group based on their relative importance to the success of the software product, process, or project. As conflicts arise between the needs of various stakeholders, priorities help determine whose voice to listen to.

Stakeholder Needs and Motivations

Individual stakeholders have different needs and motives that must be identified and understood for a product, process, or project to be successful. One method for accomplishing this is to meet with each stakeholder representative individually to encourage open communication and sharing of both positive and negative perspectives of the product, process, or project. However, even if a stakeholder is not included in product, project, or process activities, that stakeholder's needs or motivations may still need to be identified and considered.

Stakeholders have commonly acknowledged business needs that they can easily identify. However, stakeholders may also have unacknowledged motives (hidden agendas) that might drive the direction of their influence on the effort. Both the acknowledged business needs and these less obvious motives need to be uncovered and understood to effectively plan the product, process, or project. Table 6.2 illustrates an example of business needs and motives for the development of a payroll software package.

4. OUTSOURCING

> Determine the impact that outsourced services can have on organizational goals and objectives. Identify criteria for evaluating suppliers/vendors and subcontractors including governance and quality requirements for contractual obligations. (Analyze)
>
> **Body of Knowledge II.A.4**

In today's fast-paced software industry, where customers demand more extensive, more complex software products with high quality, lower costs, and faster times to market, software organizations are forced to seek whatever competitive advantages they can. *Outsourcing* the development or maintenance of all or part of a software product to a supplier outside the organization can sometimes provide an advantage. Outsourcing can be accomplished in a number of ways:

- Onsite, where the supplier is co-located with the acquirer
- *Offsite,* where the supplier is located in the same city, state, or country as the acquirer
- *Offshore,* where the supplier is located in a different country than the acquirer

Depending on the circumstances, outsourcing can provide several different competitive advantages, including:

- Reducing staff by taking advantage of less expensive but equally competent outsourced personnel.
- Reducing operating costs through reduced overhead and facilities costs in other parts of the country or other countries. For example, costs per square foot of office space are lower in Plano, Texas, than in Manhattan, New York.

Table 6.2 Business needs and motives—payroll software system example.

Key Stakeholder	Business Needs and Motives
Accounting	• Convenient mechanisms for capturing time worked, vacations, and so on, for each employee • Track, reconcile, manage, and report payroll • Eliminate the need to "chase" employees and supervisors for completed/approved time sheets each pay period • Keep jobs (do not automate people out of employment)
HR	• Automate and track employee-elected deductions, contributions, savings, and so on • Eliminate the tedious task of dealing with paper forms from their workload
Employees	• Convenient mechanisms for reporting time worked, vacations, and so on • Detailed statement of earnings, deductions, contributions, savings, and so on • On-time delivery of accurate paychecks and automatic deposit of payroll checks to banks • Eliminate different, complicated (sometimes hard-to-obtain) forms for requesting/changing deductions, contributions, insurance benefits, savings plans, and so on
IRS, Social Security, and state tax offices	• Automate wage garnishment, reporting, and funds transfers for taxes owed • Collect taxes owed on or before due date • Eliminate labor-intensive paperwork
Insurance companies	• Automate premium payments through payroll deduction • Sell more policies and options to employees
Charities	• Establish and automate long-term contributions through payroll deduction
Employee's bank	• Automate transfer of payroll into employee bank accounts • Get payroll check into employee bank account sooner • Eliminate labor-intensive paperwork and need for human interaction with paycheck
Company's bank	• Automate transfer of payroll into employee bank accounts • Automate transfer of withheld taxes from company account • Keep the company's money in their bank accounts as long as possible • Eliminate labor-intensive paperwork and need for human interaction with paycheck

- Providing access to additional skilled people, and other resources, without the time and expense required to hire and train individuals or to purchase nonhuman resources.
- Reducing time to market by providing more staff and resources with reduced lead time and thus allowing more work to be accomplished

on a shorter schedule. Remember that these additions must be appropriately planned and integrated into the project. As Fredrick Brooks (1995) notes in *The Mythical Man-Month*, "adding manpower to a late software project makes it later."

- Improving quality by working with suppliers with more expertise in a particular area or more mature processes than are available within the organization. For example, a telecommunications company might outsource its relational database development to an organization with more expertise in that area.
- Taking advantage of innovations available from the supplier.
- Allowing the organization to focus on core competencies while obtaining the expertise needed to create software outside those competencies; for example, a biomedical company might focus on developing biomedical software and outsource its payroll or inventory control software.
- Mitigating or sharing risks by transferring some of the risks to the supplier.

While many potential outsourcing benefits exist, outsourcing can also be risky. "Mismanagement, the inability to articulate stakeholder needs, poor requirements definition, inadequate supplier selection, and contracting processes, insufficient technology selection procedures, and uncontrolled requirements changes are factors that contribute to project failure" (CMMI 2007).

Acquisition Process

Acquisition is the process of obtaining software. That acquisition may be accomplished through in-house development, outsourcing, or a combination of these methods. Outsourcing can include the purchase of commercial off-the-shelf (COTS) software, the development of custom-built software by a third-party supplier, or both. Figure 6.12 illustrates the basic steps in the software acquisition process.

Having a defined software acquisition management process also facilitates the propagation of lessons learned from one acquisition project to the next so we can repeat our successes and stop repeating actions that lead to problems.

Step 1: Initiate and Plan the Acquisition

The *initiate and plan the acquisition* step begins when the idea or need is established. In this first step of the acquisition process:

- Acquisition stakeholders are identified, and needs and motivations are determined.
- The business needs for the software are described: The *business need* defines the "why" behind the software acquisition. A business need could be a problem that needs to be addressed or an opportunity the business wants to take advantage of through acquiring software. The business need also includes the description of any assumptions on

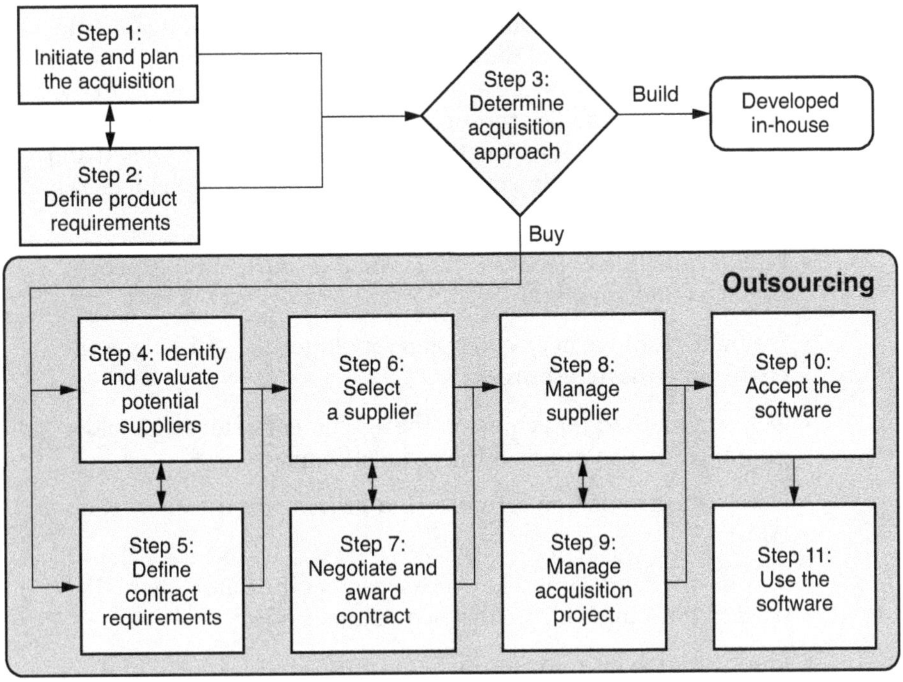

Figure 6.12 Acquisition process.

which the project is based and any constraints on the project, including factors such as schedule, budget, resources, software to be reused, technology to be employed, and interfaces to other products.

- The acquisition project is chartered by management: There are always limited resources within a company. Therefore, cost/benefit and risk analysis must be performed to determine if funding and performing the potential acquisition project is the correct choice for the organization. If it is, the acquisition project is chartered.
- Key acquisition roles are assigned, including:
  - *Acquisition sponsor*: The role responsible for providing organizational influence to help justify and sell the acquisition within the acquirer's organization
  - *Acquisition manager*: This role acts as the project manager for the acquisition project
  - *Members of the acquisition team*: The role of individual acquisition team members is to represent their stakeholder organization adequately and to perform assigned tasks
  - *Representatives*: From the customer or user community and other key stakeholders working with the acquisition team to define the business needs and software requirements

- Acquisition planning is done to define the strategies and tactics to be employed throughout the acquisition project's life cycle. The time spent up-front to define the acquisition strategy will pay off in the long run by ensuring stability throughout the acquisition process and the life of the software. However, these plans will also be updated and progressively elaborated as the acquisition project is executed. The acquisition planning process should:
  - Link acquisition objectives and tasks to resources (time, people, funds, and technology)
  - Estimate resource needs and plan for obtaining, utilizing, and controlling those resources
  - Define a method for achieving the agreement of all stakeholders to guarantee the adoption of the acquisition plan
  - Define the acquisition activities and provide for the integration of the effort
  - Specify the reviews and key measurement indicators that will be used to track and control that acquisition project
  - Identify, analyze, mitigate, and control risks

Step 2: Define Product Requirements

The *define product requirements* step in the acquisition process defines the software product itself. This definition includes refining the list of business needs into defined, business-level and stakeholder-level requirements. This step also defines the scope of the acquisition. The desired product must be adequately analyzed, and its business objectives, stakeholder functional requirements, business rules, and quality attributes determined and documented. The acquisition team should prioritize the requirements and separate needs from wants. Successful acquisition projects are dependent on clearly defining the required product.

Step 3: Determine Acquisition Approach

During the *determine acquisition approach* step, the acquisition team must determine the mechanism for acquiring the software. There are three basic choices:

- *Develop the software in-house*: In this case, the acquisition project becomes a software development project and is not considered outsourcing (see Chapters 3, 4, and 6 for more information about software development)
- *Outsource the software*: Outsourcing the software acquisition can be accomplished by acquiring one of the following:
  - Commercial-off-the-shelf (COTS) software is ready-made software that can be acquired and used by the organization. COTS can include:

    - Commercially available software that is used as is (e.g., word processers or support tools like compliers, *test automation* tools, or analyzers)
    - *Modifiable-off-the-shelf (MOTS) software* that may be adapted or customized to meet individual requirements (e.g., database tools that allow customization of fields and reports)
    - *Government-off-the-shelf (GOTS) software* that is government-supplied ready-to-use software products
  - Open source software
  - Customized software developed by an outsourced organization (supplier/ vendor)
- *A combination of these acquisition methods*: The organization may elect to use a combination of in-house developed and outsourced software. They may develop most of the software in-house but subcontract one or more functions to suppliers with specialized skills. For example, they might subcontract the GUI interface or security functions to a supplier with specific expertise. Another option would be to integrate COTS software that performs a specific function into a customized in-house or outsourced software product. For example, a COTS database manager could be integrated into an inventory tracking software product that is being developed in-house.

Step 4: Identify and Evaluate Potential Suppliers

During the *identify and evaluate potential suppliers* step, the acquisition team performs a market search to identify possible suppliers. Data collected during the market search can be used as feedback to reassess the original product definition and to determine whether modification to that definition will result in greater overall value in terms of cost, performance, availability, reliability, and other attributes. The market analysis should also evaluate maintenance and support, test results, and user satisfaction analyses. Using the information from the market search, the acquisition team narrows the list of all available suppliers down to the few potential suppliers that best match the business needs and product definition. This targets the evaluation and keeps evaluation costs to a minimum. The acquisition team then evaluates these selected potential suppliers by examining their capabilities, quality systems, and products (e.g., through *supplier qualification audits* or the evaluation of demonstration copies of the software). Another method for accomplishing this is to start with a preferred supplier list of prequalified suppliers. The depth required for this analysis on the needs and characteristics of the acquisition. However, care should be taken to obtain enough information to compare the qualifications of each potential supplier to make an informed decision.

A formal *request for proposal* (RFP) process is typically used for larger projects involving customized software to be developed by the supplier. It is a very formal process where specific proposal requirements and questions are outlined by the acquirer in the RFP and responded to by the supplier in a written proposal. Advertising an upcoming RFP may identify unknown suppliers as well as encourage suppliers to offer technological input and business advice. Potential suppliers may be able to suggest new technologies and capabilities not previously known or considered by the acquirer. An RFP should include quantifiable, measurable, and testable tasks to be performed by the supplier, specifications, and standards used for the project, acquirer-furnished equipment, information, or software to be used, and requirements for the products and services to be produced (GSAM 2000).

Other mechanisms for obtaining information and evaluating potential COTS suppliers include:

- *Supplier demonstration*: For COTS software, holding supplier demonstrations provides an excellent opportunity to see the product firsthand and ask questions. Actually seeing the features of different software packages in a live demonstration provides a context for comparison. Perhaps the basic functionality is comparable, but one product has a more intuitive user interface. The level of product knowledge and confidence of the supplier representative, and their willingness to answer tough questions, can be an indicator of the future relationship. Another indicator of future support is whether the supplier presents a "canned" demonstration, or has spent the time to customize its demo to specifically address the acquirer's business needs and requirements.
- *Evaluation copies*: For many COTS products, evaluation copies are available to demonstrate the software functions and capabilities and eliciting user buy-in.
- *References*: Evaluating references and past product performance can provide information that can be used to evaluate the functionality and quality of the software.

Other mechanisms for obtaining information and evaluating potential suppliers of custom software include:

- *Past performance*: Evaluating references and past project performance can provide helpful information. A supplier with a consistent history of successfully providing software is more likely to perform effectively in the future. Past performance is a strong indicator of whether the supplier has the capability and ability to complete delivery within schedule, on budget, and with the required level of functionality and quality. Previous experience with similar products is also a credible indicator of the likelihood that a supplier can successfully perform in the future.
- *Prequalification audits*: Information from the audits of potential suppliers can be used to determine if they have the capacity and capability to produce products of the required quality, integrity, safety, or security levels.

- *Prototypes*: For custom-built software, prototyping can be useful in determining if the supplier understands the requirements or as proof of concept.

Step 5: Define Contract Requirements

The *define contract requirements* step of the acquisition process is typically necessary only for suppliers who will be developing custom-built software. The type of contract or supplier agreement is selected, and the contents of the desired contract or agreement are defined. The contents of a contract or agreement are based on the original requirements, and the products and capabilities that are identified as potential suppliers are evaluated. Trade-offs between cost, schedule, and scope will also impact the contents of the contract or agreement. Steps 4 and 5 of the acquisition process are thus done iteratively, as they impact each other.

Step 6: Select a Supplier

During the *select a supplier* step of the acquisition process, the results of the supplier evaluation are judged against established selection criteria. Risks associated with each supplier are identified and analyzed, and a *cost/benefit analysis* is conducted. Based on this information, the final supplier of the outsourced package is selected.

One or more supplier evaluation scorecards can be created to summarize all the evaluation criteria information and scores for individual cost, schedule, product, and process attributes. Care should be taken that all information is gathered and the suppliers are scored in a way that eliminates variances in the scoring. This evaluation is best performed in a group-scoring meeting where participants have all of the gathered information available to assist in their scoring decisions. Using these forms, and evaluating scores as a group, can assist in maintaining momentum, avoiding personal bias, and ensuring consistency. Different techniques can be used for assigning a score to each attribute or criterion. For example:

- The *calculated scoring method*: Scores are assigned to each attribute based on predefined formulas, as illustrated in Figure 6.13.
- A *weighted scoring method*: A numerical weight is assigned to each of the evaluation criteria, and that weight is multiplied by the grade given to each prospective supplier to obtain a score for the individual criterion. Individual scores are then summed to give a total score, as illustrated in Figure 6.14.
- The *indexed scoring method*: Specific predefined criteria are used to select each score, as illustrated in Figure 6.15.

Once the primary supplier candidate has been selected, a supplier qualification audit may be used as a final in-depth evaluation of that supplier's quality system, and capability to produce the required software, before final selection.

Attribute	Max score	Supplier 1	Supplier 2	Supplier 3
Ability to deliver by date needed	10	10	7	8
Purchase price/licensing costs	10	7	5	10
Licensing restrictions	5	5	4	5
Operating costs	15	12	15	5
Maintenance costs	10	5	10	7
Process capability	10	10	8	5
Product functionality matches needs	20	18	16	8
Product quality	20	20	15	15
Ease of integration with existing systems	5	3	5	3
Ease of integration with our business processes	10	10	7	10
Ability to customize product	5	5	4	5
Technical support	5	5	3	2
Training availability	10	10	5	5
Total score	135	120	104	88

Ability to deliver by date needed = 10 points minus one point for each week past needed date

Product functionality meets needs = (# requirements met / Total requirements) × 20

Figure 6.13 Supplier scorecard—calculated scoring method example.

Attribute	Max score	Supplier 1	Supplier 2	Supplier 3
Ability to deliver by date needed	10	10	7	8
Purchase price/licensing costs	10	7	5	10
Licensing restrictions	5	5	4	5
Operating costs	15	12	15	5
Maintenance costs	10	5	10	7
Process capability	10	10	8	5
Product functionality matches needs	20	18	16	8
Product quality	20	20	15	15
Ease of integration with existing systems	5	3	5	3
Ease of integration with our business processes	10	10	7	10
Ability to customize product	5	5	4	5
Technical support	5	5	3	2
Training availability	10	10	5	5
Total score	135	120	104	88

Process capability	Grade	Weight	Score
Software quality	1	×1	1
Project management	1	×2	2
Configuration management	1	×1	1
Requirements management	0	×1	0
Systems and software engineering	1	×2	2
Verification and validation	1	×2	2
Risk management	0	×1	0
		Total score:	8

Grade: 1 = Meets or exceeds requirements
0 = Does not meet requirements

Figure 6.14 Supplier scorecard—weighted scoring method example.

Step 7: Negotiate and Award Contract

Once the supplier has been selected, the contract or supplier agreement terms are negotiated and the contract or agreement is awarded. Now is the time to do the final negotiation with the preferred supplier. Depending on the type of acquisition, this step may be as simple as issuing a purchase order for COTS software or as complex as negotiating a formal legal contract. When a formal contract or agreement is needed, a well-written contract or agreement minimizes the probability of misunderstandings. It is a significant contributor to a congenial relationship between the acquirer and the supplier. Experience has shown that

Attribute	Max score	Supplier 1	Supplier 2	Supplier 3
Ability to deliver by date needed	10	10	7	8
Purchase price/licensing costs	10	7	5	10
Licensing restrictions	5	5	4	5
Operating costs	15	12	15	5
Maintenance costs	10	5	10	7
Process capability	10	10	8	5
Product functionality matches needs	20	18	16	8
Product quality	20	20	15	15
Ease of integration with existing systems	5	3	5	3
Ease of integration with our business processes	10	10	7	10
Ability to customize product	5	5	4	5
Technical support	5	5	3	2
Training availability	10	10	5	5
Total score	135	120	104	88

Ease of integration with existing systems =

5: Seamless integration expected

4: Minimal work required for integration (< 1 week)

3: Average work required for integration (1–3 weeks)

2: Major work will be required for integration (> 3 weeks)

1: Existing systems will have to be replaced to accommodate integration

0: Integration expected to fail

Figure 6.15 Supplier scorecard—indexed scoring method example.

when the contract or agreement is unambiguous and clearly defines the duties and responsibilities of each party, the animosity that arises from quibbling over performance obligations can usually be avoided. A legal opinion should always be sought when dealing with formal contracts and is recommended for most other agreements. The contract should be customized to consider both the acquirer's and supplier's strengths and weaknesses.

If the negotiation results in an inability to reach an agreement with the supplier, another supplier may need to be selected. In this case, steps 6 and 7 of the acquisition process are done iteratively.

Step 8: Manage Supplier

The *manage supplier* step includes monitoring the supplier's performance throughout the execution of the acquisition project to verify successful completion of the contract or agreement. Again, the level of rigor in this step depends on the type of acquisition. It may be as simple as verifying that the order of COTS software arrived and is installed correctly without problems, or it may be as complex as managing the supplier over a multiyear development project.

For acquisitions that involve customized software development, ongoing formal evaluations and metrics should be used to monitor the supplier's progress against baselined budget, schedule, and quality standards and to manage the risks associated with the acquisition. This monitoring may include ongoing supplier audits, joint project or product reviews, and partnering for joint process improvement actions. The goal is to provide the acquirer with enough visibility into the supplier's work activities to have confidence that the contractual or agreement obligations and product requirements are being met or to identify issues that need corrective action. "When the supplier's performance, processes, or products fail to satisfy established criteria as outlined in the supplier agreement, the acquirer may take corrective action" (CMMI 2010c).

For an acquisition project of any length, change will occur. Therefore, this step also involves managing and maintaining the requirements throughout the execution of the acquisition project. Requirements management mechanisms must be implemented to manage and control changes to those requirements and to confirm that approved changes are integrated into the acquisition plans, software products, and activities (see Chapter 13 for more information on requirements management). This step can also involve managing any changes to the contract or agreement during the project.

Step 9: Manage Acquisition Project

The acquisition project should be managed like any other project throughout the acquisition process. This management includes executing the project, ongoing monitoring, taking corrective actions, tracking corrective actions to closure if the variation between actuals and plans becomes unacceptable, controlling change, and progressively elaborating project plans as more information is obtained (or replanning, as necessary).

When the final software product includes in-house software integrated with software from one or more suppliers, establishing adequate communications between the acquirer and their suppliers is necessary. An effective mechanism for this is to form an *integrated product team* (IPT). An IPT, as illustrated in Figure 6.16, is a cross-functional team formed for the specific purpose of delivering an integrated product. An IPT is composed of representatives from all appropriate organizations (acquirer and suppliers) and functional disciplines. IPT team members, working with a team leader, plan and implement a successful program (interrelated set of acquirer and supplier development projects), identify

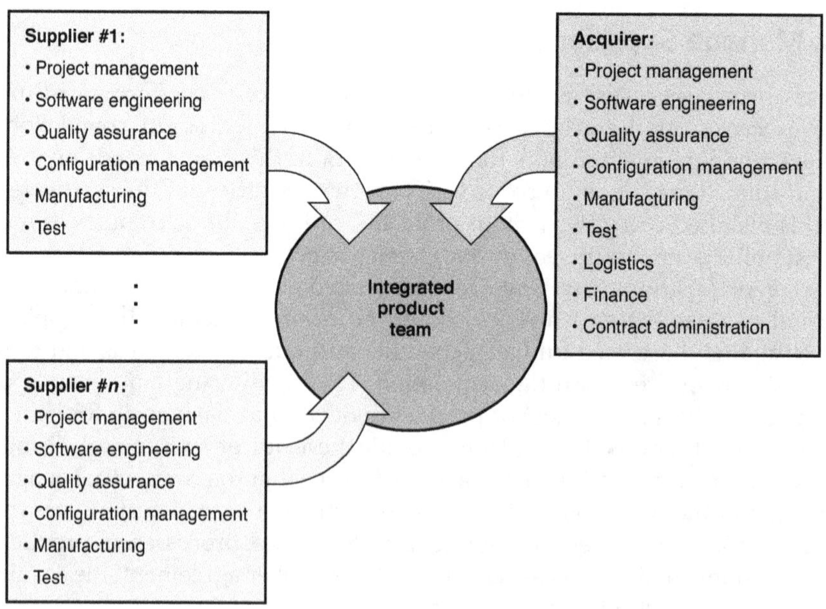

Figure 6.16 Integrated product team.

and resolve cross-functional issues, and make sound and timely decisions. IPT members should have complementary skills and be committed to a common purpose, performance objectives, and approach for which they hold themselves mutually accountable.

Step 10: Accept the Product

The supplier should perform verification and validation (V&V) activities throughout the software development process. For custom-built software, the acquirer should monitor these activities as part of the supplier management process. As part of product *acceptance,* the acquirer should also conduct V&V activities on the delivered products. Example V&V activities may include alpha, beta, or acceptance testing or functional and physical configuration audits. These V&V activities should be conducted against negotiated *acceptance criteria* included in the contract or agreement to verify that the delivered products meet all agreed-to requirements. The supplier should correct any identified product problems.

The accepted software will either be used as delivered or incorporated into the acquirer's software as one of its components. If the software is being integrated into a more extensive product, the acquirer should also perform integration and system testing on that larger product.

As with any other project, an acquisition project will end and must be closed. This closure includes transitioning the product into operations and its support into maintenance mode. If the acquiring company has not purchased the rights to the software source code, that code may be transitioned into escrow, so that if the supplier goes out of business or no longer supports the product, the acquirer is not substantially harmed. The acquirer can obtain the source code from escrow and still support its products.

This step also includes the completion and closing of the contract or agreement and any other related activities.

Step 11: Use the Software

The supplier's role typically does not end when the software is accepted and put into operations. The supplier continues to support the software with technical support, corrective releases to fix identified problems (corrective maintenance) or feature releases to add new or updated functionality (perfective, preventive, or adaptive maintenance). Service level agreements need to be negotiated either as part of the initial supplier agreement or as a supplementary agreement to establish the supplier's obligations for this ongoing support.

Preferred Supplier Relationships

One way to minimize the effort needed to select and monitor suppliers is to create preferred supplier relationships. Potential or current suppliers are evaluated and rated against standardized criteria, such as the effectiveness of their quality management systems and past performance. Suppliers who meet or exceed the standards are identified and recognized as preferred suppliers entitled to additional benefits. Those benefits may include selection preference, reduced oversight, reduced verification and validation activities, and additional business opportunities.

Preferred suppliers may also be considered for strategic partnerships across multiple acquisition projects. A strategic partnership is a corporate strategy for teaming up with one or more suppliers that have complementary resources to achieve a mutual business objective. Benefits of preferred suppliers and strategic partnerships can include:

- Leveraging of competencies
- Reducing costs and improving service
- Freeing-up in-house resources
- Reducing the number of suppliers to manage
- Sharing of best practices and process improvement initiatives to provide mutual benefit

5. BUSINESS CONTINUITY, DATA PROTECTION, AND DATA MANAGEMENT

> Design plans for business continuity, disaster recovery, business documentation, change management, information security, and protection of sensitive and personal data. (Analyze)
>
> **Body of Knowledge II.A.5**

Business Continuity

Business continuity is "the capability of the organization to continue delivery of products or services at acceptable predefined levels following a disruptive incident" (ISO 2012). Business continuity is about planning, establishing, implementing, operating, monitoring, reviewing, maintaining, and continually improving a documented management system to:

- Protect against disruptive incidents by reducing the probability that disruptive incidents will occur
- Prepare for disruptive incidents if they do occur by creating response and recovery strategies and plans, so the business will be ready to take appropriate action and thus minimize potential damage

This is done regardless of whether the disruptive incident is a negative issue that is part of "business as usual" or a major disaster. According to the Business Continuity

Institute, "business continuity is about building and improving resilience in your business. *Resilience* is widely defined as the ability of an organization to absorb, respond to and recover from disruptions" (BCI 2018).

The management of business continuity is fundamentally part of risk management. However, business continuity also includes aspects of information security, information technology management, compliance, and governance. The Business Continuity Institute defines a business continuity management life cycle which includes six good practices (based on BCI 2018):

- Two management practices:
  - *Policy and program management*: Defining an organizational policy for business continuity and strategies and tactics for how that policy will be implemented, controlled, and evaluated using business continuity program management
  - *Embedding business continuity*: The institutionalization of business continuity practices into the ongoing, daily business activities and the organizational culture so that everyone in the organization is aware of and participating in business continuity
- Four technical practices:
  - *Analysis*: Analysis and assessment of the organization and its key business functions to identify potentially disruptive incidents, their probability of occurrence, and the impact on the organization's ability to continue essential business functions if those disruptive incidents do occur
  - *Design*: Identify and select appropriate strategies and tactics to respond to, and recover from, disruptive incidents if they do occur to minimize any stoppage of, or hindrance to, the organization's essential business functions
  - *Implementation*: Create and document a business continuity plan that includes instructions and procedures that enable the organization to respond to, and recover from, disruptive incidents with minimal stoppage or hindrance to the organization's essential business functions
  - *Validation*: Evaluate and confirm that the business continuity management program meets the business continuity policy objectives, and that the organization's business continuity plan will fulfill its purpose

Disaster recovery is a subset of business continuity that focuses on the organization's ability to recover or continue the operations of vital information technology and technical systems that support critical business functions following a disaster (natural or human-induced).

Data Protection and Security

Data security verifies that the right people can collect, access, utilize, and update data correctly, while also detecting, preventing, or recovering from inappropriate security attacks on the data. Examples of security attacks include attempts to:

- Gain unauthorized access to the data (e.g., accessing the data through a back door, or through spoofing, hacking, or *tunneling*)
- Compromise the data's integrity, availability, or confidentiality (e.g., through viruses, worms, ransomware, or denial-of-service attacks)
- Inappropriately collect, falsify, or destroy data (e.g., unauthorized or inadvertent disclosure of confidential information, breaching data encryption, or data contamination)

Security attacks can come from the outside, through hackers, stolen identification or passwords, or others trying to exploit security vulnerabilities in the data or databases. Security attacks can also come from the inside through sabotage, malicious code (e.g., *time bombs,* logic bombs, or Trojan horses), or back doors created by practitioners. Data security focuses on four main objectives:

- Maintaining access control (passwords, firewalls, multilevel privileges)
- Preserving the integrity of the data
- Providing backup and recovery, if security mechanisms should fail and result in the corruption or loss of databases or data
- Providing an audit trail of data access and use to provide the information needed for accountability

Backing up data is the single most important thing an organization can do to protect the data from loss and allow for quick recovery from data corruption or disasters. (See Chapter 29 for more information on backups.)

Data Management

In essence, the term *data management* practically defines itself—that is, it is simply the management of data. Data management is also called *information management, enterprise information management, enterprise data management, data resource management, information resource management,* and *information asset management.* Data management activities include:

- Planning for data management
- Coordinating/controlling data collection and storage activities
- Coordinating data maintenance activities
- Coordinating/controlling data dissemination activities to responsively and economically acquire, access, and distribute data
- Coordinating data archival and disposal activities

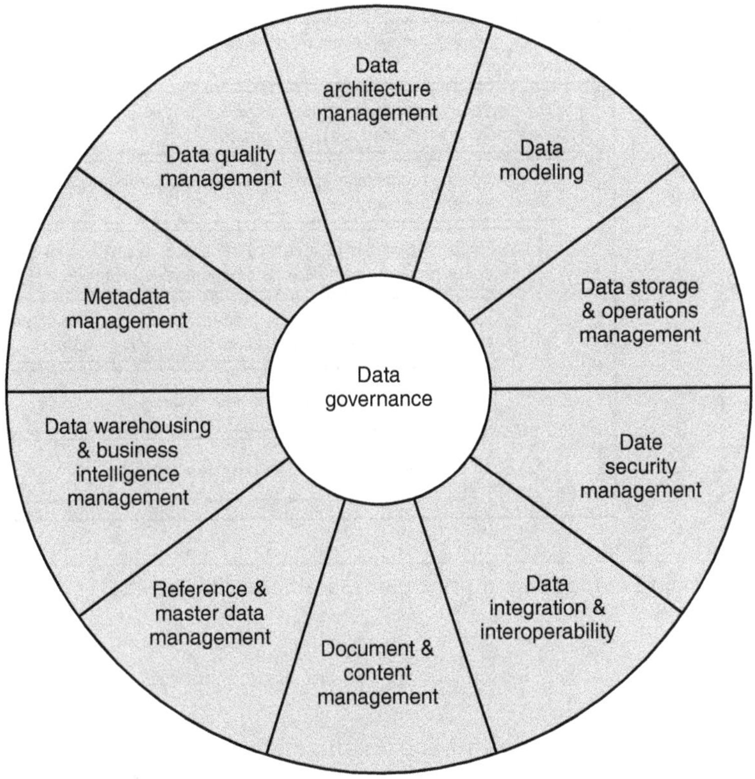

Figure 6.17 Functions of data management (DAMA 2017).

Inputs into the data management process include data products to be managed from data suppliers and data requests from the organization, projects, people, and other processes. The outputs of the data management process are stored in protected data repositories, and data and information products are delivered to the organization, projects and people, and other processes. The functions of data management, as defined by the Data Management Association's (DAMA's) *Data Management Body of Knowledge* (DAMA 2017), are illustrated in Figure 6.17.

The GEIA-859B *Data Management* standard (GEIA 2022) "is intended to articulate contemporary data management principles and methods that are broadly applicable to management of electronic and non-electronic data in both the commercial and government sectors." This standard defines the eight data management principles illustrated in Figure 6.18.

Data Management—Governance

Data governance is a centralizing and integrating activity of data management. Data governance involves data management planning and the monitoring and controlling of data assets. Data governance activities include:

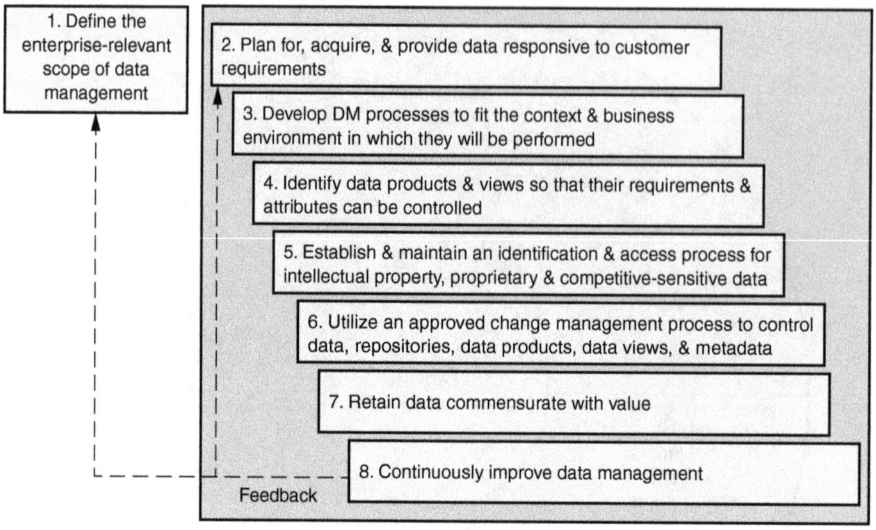

Figure 6.18 Eight data management principles (based on GEIA 2022).

- *Planning:*
  - *Strategy*: Understanding the organization's strategic data requirements and developing, defining, and executing a data strategy to meet those requirements
  - *Organization and roles*: Defining and establishing organizations and roles involved in performing and managing each of the data management and governance activities
  - *Policies and standards*: Establishing, documenting, implementing, and maintaining data policies, standards, and procedures
  - *Projects and services*: Initiating, planning, and sponsoring data management projects and services, including reviewing and approving data architectures and estimating data asset values and associated costs
- *Monitoring and controlling:*
  - *Oversight*: Supervising both the data professional organizations and staff, managing data management projects and services, and coordinating data governance activities
  - *Issues*: Tracking data management issues to resolution
  - *Change management*: Requesting, documenting, conducting impact analysis, reviewing, approving, and tracking requested changes to controlled data, data records, data architectures, and information products to resolution

- *Assurance*: Monitoring and verifying regulatory compliance, and conformance to data policies, standards, procedures, and architectures
- *Championship*: Communicating and promoting the value of data assets

Data Management—Planning

Data management planning establishes and communicates the processes for all data management within the organization or within the program or project, depending on the level of data management addressed by the planning. As with other planning activities, the rigor and amount of documentation required for data management planning may vary based on the needs of the organization and the risk involved. Data management planning considers plans for:

- Data management stakeholder identification
- The collection or receipt of data, including data flow within the organization, program, or project and from external stakeholders (e.g., customers or suppliers)
- Unique data identification, including versioning
- Data management roles and responsibility/authority assignments
- Data quality assurance
- Storing, transforming, transmitting, and providing access to the data
- Tools used in data management
- Data change management
- Training related to data management

Data Management—Data Collection and Storage

Coordinating and controlling data collection and storage activities start with determining the *data requirements*. This activity includes requirements for identifying *metadata*, which are the data about the data. Data identification characterizes the data and data products, to confirm adequacy, uniqueness, and consistency. It safeguards data interoperability among team members so everyone is collecting, interpreting, and using the data in a consistent manner. Mechanisms are established to assign identifying information to distinguish similar or related data products from each other (e.g., the configuration control board [CCB] name and meeting date might be used to distinguish one set of CCB minutes from another).

The data management architecture is then designed. The *data management architecture* is a set of integrated specifications defining the information needs, data models used to design the data solutions, architectures for databases, data integration, data warehouses, business information, and metadata. The data management architecture defines formal names, data definitions, data structures, data integrity rules, and robust data documentation for each data item required to fulfill those information needs. The data management architecture helps establish a common vocabulary for the important program or project entities and the data attributes about those entities.

Once the architecture is established, solution components must be designed and implemented, including databases and other data structures, data capture, access and usage components, and user interface components.

If the right data are not collected correctly, then the objectives of the data management program cannot be accomplished. Data analysis is pointless without good data. Therefore, establishing good data collection processes is the cornerstone of any successful data management program. The goals are that data collection should be:

- *Objective*: The same person collects the data the same way every time.
- *Unambiguous*: Two different people collecting the same data for the same item will collect the data the same way.
- *Convenient*: Data collection must be simple enough not to disrupt the working patterns of the individual collecting the data. This means that data collection must become part of the process and not an extra step performed outside of the workflow.
- *Accessible*: In order for data to be useful and used, easy access to the data is required. This typically means that even if the data are collected manually on forms, they must ultimately be entered into a database.
- *Timely*: Data collection and reporting must be timely.

Data Management—Data Maintenance

Data maintenance activities include:

- Safeguarding the performance, reliability, integrity, and security of the databases through tuning, monitoring, access control, error reporting, corrective action, and change control
- Providing mechanisms that support data availability requirements
- Implementing and controlling the database environments
- Installing and administering data technology
- Supporting data technology usage and related issues
- Performing technical data integrity checks on collected data, monitoring stored data to confirm compliance with content and format requirements, and identifying errors in, or corruption of, stored data
- Handling data technology licenses
- Verifying the existence of a current backup copy for emergency restoration
- Replenishing current storage media if the life expectancy of the data is longer than that of the media or migrating the data to modern storage media

If problem areas or opportunities for improvement are identified, data maintenance activities include implementing appropriate corrective or *preventive actions*. Examples of data management corrective or preventive actions include:

- Correcting data anomalies
- Improving data collection or reporting mechanisms
- Correcting or improving data management processes to make them more efficient or effective
- Making improvements to data management tools or infrastructure, including:
    - Updating to a new version of a tool
    - Migrating to a better tool
    - Purchasing faster data storage devices
    - Adding more capacity to the networks to increase the availability

Data maintenance activities also include implementing approved change control processes used to control changes to:

- Data and metadata
- Data requirements, data management architecture, data component designs, and implementations
- Data products and views
- Database environment and its configuration
- Data management processes
- Data management tools and infrastructure

Data Management—Data Dissemination

Coordinating and controlling data dissemination activities to provide data to authorized parties includes:

- Creating, implementing, and maintaining documentation that provides data access instructions
- Receiving, evaluating, and responding to requests for data and information
- Confirming that electronic access rules are followed before allowing access to the database and before data and information are released or transferred to the requester

As illustrated in Figure 6.19, when data are used to produce a formal information product, they usually go through a process where the data supplier creates and delivers the information product, which then goes through verification and validation before it is formally accepted. Data and information products can be

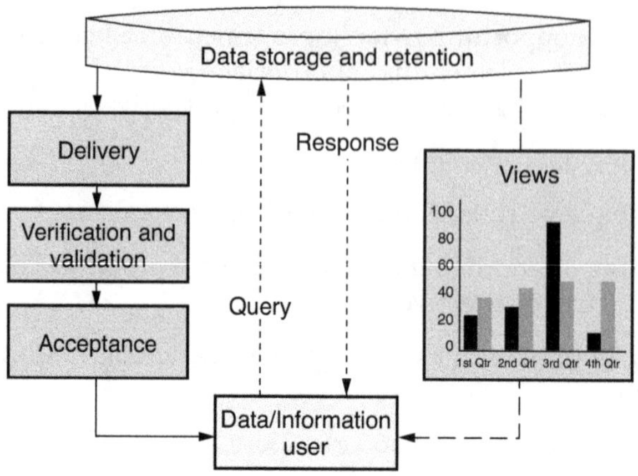

Figure 6.19 Data distribution.

less formally distributed through direct access query and response pairs. For example, a user might use a structured query language (SQL) to request a report of all defects reported against a specified product during a specified time period, and the response would be a list of those defects. Many databases have *predefined reports,* called *views,* which allow the users to directly produce their own reports by selecting that view and entering specific parameters. For example, a project management tool might have an earned value reporting view that can be selected for a currently active project.

Data Management—Data Archiving and Disposition

Coordinating data archival and disposal activities include:

- Determining when data are no longer actively being used (but still have value to the organization) so they can be archived (note that there may be laws and regulations that specify record retention periods and that therefore impact data archival and disposition practices)
- Disposing of data when data are no longer of value

Chapter 7

Methodologies (for Quality Management)

An organization's quality management system (QMS) focuses on the strategies and tactics necessary to allow that organization to successfully achieve its quality goals and objectives and provide high-quality products and services. As part of the QMS, methods and techniques need to exist that:

- Monitor the effectiveness of those strategies and tactics (cost of quality and return on investment)
- Continually improve those strategies and tactics (process improvement models)
- Deal with problems identified during the implementation and performance of those strategies and tactics (correction, corrective/preventive action)
- Prevent problems from occurring during the implementation and performance of those strategies and tactics (defect prevention)

1. COST OF QUALITY (COQ) AND RETURN ON INVESTMENT (ROI)

> Analyze cost of quality (COQ) categories (e.g., prevention, appraisal, internal failure, and external failure) and return on investment (ROI) metrics in relation to products and processes. (Analyze)
>
> **Body of Knowledge II.B.1**

Cost of Quality (COQ)

Cost of quality, also called *cost of poor quality,* is a technique organizations use to attach a dollar figure to the costs of producing or not producing high-quality products and services. In other words, the cost of quality is the cost of preventing, finding, and correcting defects (nonconformances to the requirements or intended use). The costs of quality represent the money that would not need to be spent if the products could be developed or the services could be provided perfectly the first time, every time. According to Krasner (1998), "cost of software quality is an accounting technique that is useful to enable our understanding of the economic trade-offs involved in delivering good quality software." The costs of building the product or providing the service perfectly the first time do not count as costs of quality. Therefore, the costs of performing software development and perfective and adaptive maintenance activities do not count as costs of quality, including:

- Requirements elicitation and specification
- Architectural and detailed design
- Coding
- Creating the initial build and subsequent builds to implement additional requirements
- Shipping and installing the initial release and subsequent feature releases of a product into operations

There are four major cost of quality categories: prevention, appraisal, internal failure costs, and external failure costs. The total cost of quality is the sum spent on these four categories.

- *Prevention costs of quality* is the total cost of all activities used to prevent poor quality and defects from getting into products or services. Examples of prevention cost of quality include the costs of:
  - Quality training and education
  - Quality planning
  - Supplier qualification and supplier quality planning
  - Process capability evaluations, including quality system audits
  - Quality improvement activities, including quality improvement team meetings and activities
  - Process definition and process improvement
- *Appraisal costs of quality* is the total cost of analyzing the products or services to identify any defects that make it into those products and services. Examples of appraisal cost of quality include the costs of:
  - Peer reviews and other technical reviews focused on defect detection of new or enhanced software
  - Testing of new or enhanced software
  - Analysis, review, and testing tools, databases, and test beds
  - Qualification of supplier's products, including software tools
  - Process, product, and service audits
  - Other verification and validation (V&V) activities
  - Measuring product quality

- *Internal failure costs of quality* is the total cost of handling and correcting failures found internally before the product or service was delivered to the customers or users. Examples of internal failure cost of quality include the costs of:
    - Scrap—the costs of software that was created but never used
    - Recording failure reports and tracking them to resolution
    - Debugging the failure to identify the defect
    - Correcting the defect
    - Rebuilding the software to include the correction
    - Re–peer reviewing the product or service after the correction is made
    - Testing the correction and regression testing other parts of the product or service
    - Other V&V activities performed on the correction
- *External failure costs of quality* is the total cost of handling and correcting failures found after the product or service has been made available externally to the customers or users. Examples of external failure costs of quality include many of the correction costs that were included in the internal failure costs of quality:
    - Recording failure reports and tracking them to resolution
    - Debugging the failure to identify the defect
    - Correcting the defect
    - Rebuilding the software to include the correction
    - Re–peer reviewing the product or service after the correction is made
    - Testing the correction and regression testing other parts of the product or service
    - Other V&V activities performed on the correction

    In addition, external failure costs of quality include the costs of the failure's occurrence in operations. Examples of these costs include:

    - Warranties, service level agreements, performance penalties, and litigation
    - Losses incurred by the customers, users, and other stakeholders because of lost productivity or revenues due to product or service downtime
    - Product recalls
    - Corrective releases and installation of those corrective releases

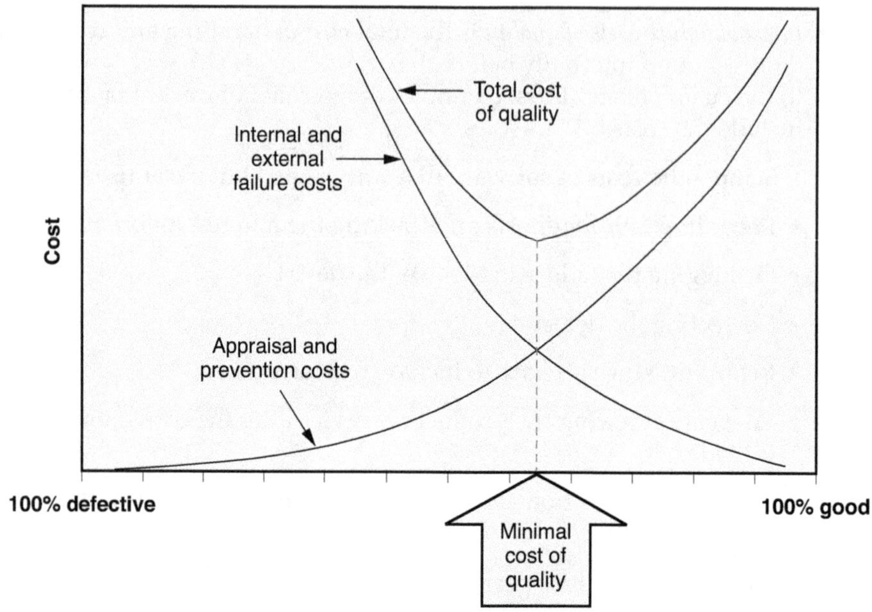

Figure 7.1 Classic model of optimal cost of quality balance (based on Campanella 1990).

- Technical support services, including help desks and field service
- Loss of reputation or goodwill
- Customer dissatisfaction and lost sales

In order to reduce the costs of internal and external failures, an organization must typically spend more on prevention and appraisal. As illustrated in Figure 7.1, the classic view of cost of quality states that there is, theoretically, an optimal balance where the total cost of quality is at its lowest point. However, this point may be very hard to determine because many of the external failure costs, such as the cost of stakeholder dissatisfaction or lost sales, can be extremely hard to measure or predict.

Figure 7.2 illustrates a more modern model of the optimal cost of quality. This view reflects the growing empirical evidence that process improvement activities and prevention techniques are subject to increasing cost-effectiveness. This evidence indicates that near-perfection can be reached for a finite cost (Campanella 1990). For example, Krasner (1998) quotes a study of 15 projects over three years at Raytheon Electronic Systems (RES) as they implemented the Capability Maturity Model (CMM). At maturity level 1, the total cost of software quality ranged from 55 to 67 percent of the total development costs. As maturity level 3 was reached, the total cost of software quality dropped to an average of 40 percent of the total development costs. After three years, the total cost of software quality had dropped to approximately 15 percent of the total development costs, with a significant portion of that being the prevention cost of quality.

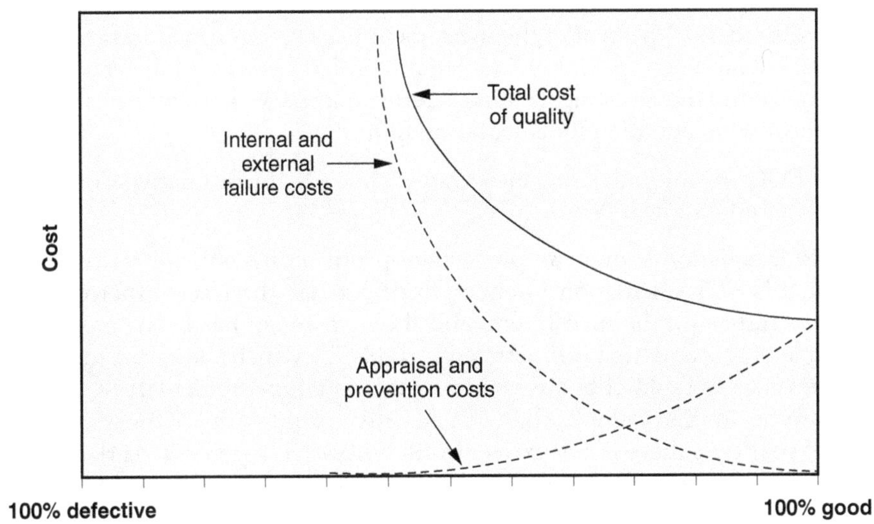

Figure 7.2 Modern model of optimal cost of quality (based on Campanella 1990).

Cost of quality information can be collected for the current implementation of a project or process, and then compared with historic, baselined, or benchmarked values or trends over time and considered with other quality data to:

- Identify process improvement opportunities by identifying areas of inefficiency, ineffectiveness, and waste.
- Evaluate the impacts of process improvement activities. Provide information for future risk-based trade-off decisions between costs and product integrity requirements.

Return on Investment (ROI)

Return on investment is a financial performance measure, used to evaluate the benefit of an investment or to compare the benefits of different investments. "ROI has become popular in the last few decades as a general purpose metric for evaluating capital acquisitions, projects, programs, and initiatives, as well as traditional financial investments in stock shares or the use of venture capital" (business-case-analysis.com).

There are two primary equations used to calculate ROI. The first one is

- ROI = (Cumulative Income/Cumulative Cost) × 100% (Westcott 2006)

This ROI equation shows the percentage of the investment returned to date. When this ROI calculation is equal to or greater than 100 percent, there has been a positive return on the investment and the investment has been "paid back." An example of the use of this equation is illustrated in Table 7.1. At the end of year one, in this example, only 12.5 percent of the investment has been returned. At the end of year two, 40.3 percent of the investment has been returned. At the end of year three, 131.0 percent of the investment has been returned indicating that the investment has been "paid back" with a profit of 31 percent. At the end of year four, 235.7 percent of the investment has been returned, or 2.35 times return on investment.

The second equation for calculating ROI is:

- ROI = ((Cumulative Income − Cumulative Cost)/Cumulative Cost) × 100% (based on Juran 1999)

This ROI equation shows the percentage profit on investment returned to date. When this ROI calculation is equal to or greater than zero, there has been a positive return on the investment, and the investment has been "paid back." An example of this equation is illustrated in Table 7.2, which shows the identical costs and incomes as Table 7.1 but uses this second equation to calculate ROI. At the end of year one, in this example, there is a negative profitability of 87.5 percent. At the end of year two, there is a negative profitability of 59.7 percent. At the end of year three, the investment has been paid back, and there is a positive profitability of 31.0 percent. At the end of year four, there is a positive profitability of 135.7 percent.

It should be noted that ROI is a simple method for evaluating an investment or comparing multiple investments. Neither of the two ROI equations takes into consideration the net present value (NPR) of an amount received in the future. In other words, ROI ignores the time value of money. In fact, time is not taken into consideration at all in ROI. For example, which of the following is the better investment?

- Investment A has a cumulative net cash flow of $75,000.00.
- Investment B has a cumulative net cash flow of $60,000.00. Both required a total cost of $50,000.00.

Table 7.1 ROI as benefit to the investor.

Year	Cost	Income	Net Cash Flow	Cumulative Cost	Cumulative Income	ROI	Cumulative Net Cash Flow
1	$(20,000.00)	$2,500.00	$(17,500.00)	$(20,000.00)	$2,500.00	12.5%	$(17,500.00)
2	$(14,000.00)	$11,200.00	$(2,800.00)	$(34,000.00)	$13,700.00	40.3%	$(20,300.00)
3	$(5,000.00)	$37,400.00	$32,400.00	$(39,000.00)	$51,100.00	131.0%	$12,100.00
4	$(5,000.00)	$52,600.00	$47,600.00	$(44,000.00)	$103,700.00	235.7%	$59,700.00

Table 7.2 ROI as percentage profit.

Year	Cost	Income	Net Cash Flow	Cumulative Cost	Cumulative Income	ROI	Cumulative Net Cash Flow
1	$(20,000.00)	$2,500.00	$(17,500.00)	$(20,000.00)	$2,500.00	(87.5)%	$(17,500.00)
2	$(14,000.00)	$11,200.00	$(2,800.00)	$(34,000.00)	$13,700.00	(59.7)%	$(20,300.00)
3	$(5,000.00)	$37,400.00	$32,400.00	$(39,000.00)	$51,100.00	31.0%	$12,100.00
4	$(5,000.00)	$52,600.00	$47,600.00	$(44,000.00)	$103,700.00	135.7%	$59,700.00

If the ROI is calculated as ((Income − Cost) / Cost)×100%), investment A has an ROI of 50 percent, and investment B has an ROI of 20 percent. From just this information, investment A looks like the better investment. However, consider the time factor. Investment A took five years to see the 50 percent ROI, so it returned an average of 10 percent yearly. Investment B took only one year to see the 20 percent ROI. Now which is the better investment?

Another issue is that there are no established standards for how organizations measure income and cost as inputs into ROI equations. For example, items like overhead, infrastructure, training, and ongoing technical support might or might not be counted as costs. Some organizations may consider only revenues as income. In contrast, others may assign a monetary value to factors like improved reputation in the marketplace or stakeholder goodwill and include them as income. "This flexibility, then, reveals another limitation of using ROI, as ROI calculations can be easily manipulated to suit the user's purposes, and the results can be expressed in many different ways" (Investopedia.com 2024).

Care must be taken when using ROI as a financial performance measure to ensure that the same equation is used, that incomes and costs are measured consistently, and that similar time intervals are used. This care is especially needed when making comparisons between investments.

2. PROCESS IMPROVEMENT

> Define and describe elements of benchmarking, lean processes, and the six sigma methodology. Use the Define, Measure, Act, Improve, Control (DMAIC) model and the Plan–Do–Check–Act (PDCA) model for process improvement. (Analyze)
>
> **Body of Knowledge II.B.2**

Benchmarking

Benchmarking is the process used by an organization to identify, understand, adapt, and adopt outstanding practices and processes from others, anywhere in the world, to help that organization improve the performance of its processes, projects, products, or services. Benchmarking can provide management with the assurance that quality and improvement goals and objectives are aligned with the good practices of other teams or organizations. At the same time, benchmarking helps make certain that those goals and objectives are attainable because others have attained them. Using benchmarking can help an organization "think outside the box" and can result in breakthrough, evolutionary improvements. Figure 7.3 illustrates the steps in the benchmarking process.

Step 1: The first step is to determine what to benchmark, that is, which process, project, product, or service the organization wants to analyze and improve. This step involves assessing the effectiveness, efficiencies, strengths, and weaknesses of

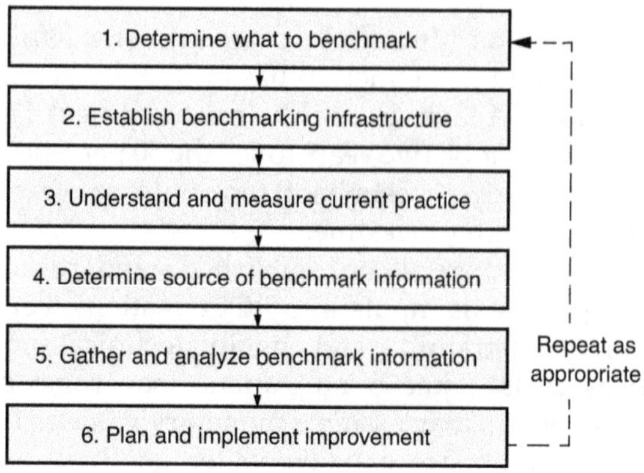

Figure 7.3 Steps in the benchmarking process.

the organization's current practices, identifying areas that require improvement, prioritizing those areas, and selecting the area to benchmark first. "Examples of how to select what to benchmark include systems, processes, or practices that:

- Incur the highest costs
- Have a major impact on stakeholder satisfaction, quality, or cycle time
- Strategically impact the business
- Have the potential of high impact on competitive position in the marketplace
- Present the most significant area for improvement
- Have the highest probability of support and resources if selected for improvement" (Westcott 2006)

Step 2: The second step in the benchmarking process is establishing the infrastructure to do the benchmarking study. This step includes identifying a sponsor to provide necessary resources and champion the benchmarking activities within the organization. This step also includes identifying the members of the benchmarking team who will perform the benchmarking activities. Members of this team should include individuals who are knowledgeable and involved in the area being benchmarked and individuals who are familiar with benchmarking practices.

Step 3: In order to make an accurate comparison, the benchmarking team must obtain a thorough, in-depth understanding of the current practices in the selected area. Key performance factors for the benchmarked area are identified, and the current values of those factors are measured. Current practices for the selected area are studied, mapped as necessary, and analyzed.

Step 4: Determine the source of benchmarking good practice information. Notice that the information-gathering steps of benchmarking focus on good practice. There are many good practices in the software industry. Good practices become best practices when adopted and adapted to meet the exact requirements, culture, infrastructure, systems, and products of the organization performing the benchmarking. During this fourth step, a search and analysis is performed to determine the good practice leaders in the selected area of study. Several choices can be considered, including:

- *Internal benchmarking*: Looks at other teams, projects, functional areas, or departments within the organization for good practice information
- *Competitive benchmarking*: Looks at direct competitors, locally or internationally, for good practice information. This information may be harder to obtain than internal information, but industry standards, trade journals, competitor's marketing materials, and other sources can provide valuable data
- *Functional benchmarking*: Looks at other organizations performing the same functions or practices, outside the industry. For example, an information technology (IT) team might look for good practices in other IT organizations in other industries. IEEE, ISO, IEC, other standards, and the CMMI® model are likely sources of information, in addition to talking directly to individual organizations.
- *Generic benchmarking*: Looks outside the box. For example, an organization that wants to improve:
  - On-time delivery practices might look to FedEx
  - Just-in-time, lean inventory practices might look to Wal-Mart or Toyota
  - Its product's graphical user interface (GUI) might look to Google or Amazon's web pages

  It does not matter if the organization is not in these fields.

Step 5: Benchmarking good practices information is gathered and analyzed. There are many mechanisms for performing this step, including site visits to targeted benchmark organizations, partnerships where the benchmark organization provides coaching and mentoring, research studies of industry standards or literature, evaluations of good practice databases, Internet searches, attending trade shows, hiring consultants, stakeholder surveys, and other activities. The objective of this step is to:

- Collect information and data on the performance of the identified benchmark leader or on good practices
- Evaluate and compare the organization's current practices with the benchmark information and data
- Identify performance gaps between the organization's current practices and the benchmark information and data to identify areas for potential improvement and lessons learned

This comparison is used to determine where the benchmark is better and by how much. The analysis then determines why the benchmark is better. What specific practices, actions, or methods result in superior performance?

Step 6: For benchmarking to be useful, the lessons learned from the good practice analysis must be used to actually improve the organization's current practices. To complete the final step in the benchmarking process:

- Obtain management buy-in and acceptance of the findings from the benchmarking study
- Incorporate the benchmarked findings into business analysis and decision-making
- Create a plan of specific actions and assignments to adapt (tailor) and adopt the identified good practices, and turn them into best practices for the organization by filling the performance gaps
- Pilot those improvement actions, and measure the results against the initial values of identified key factors (identified in step 3) to monitor the effectiveness of the improvement activities
- If the piloting was successful, propagate those improvements throughout the organization; for unsuccessful pilots or propagations, appropriate corrective action must be taken

Lessons learned from the benchmarking activities can be leveraged to improve future benchmarking activities. The prioritized list created in the first step of the benchmarking process can be used to consider improvements in other areas, and of course, this list should be updated as additional information is obtained over time. Benchmarking should be a continuous process that looks at current performance and also continues to monitor key performance indicators into the future as industry practices change and improve.

Plan–Do–Check–Act (PDCA) Model

Many different models describe the steps to process improvement. One of the simplest models is the classic *Plan–Do–Check–Act* (PDCA) *model,* also called the *Deming circle* or the *Shewhart cycle.* Figure 7.4 illustrates the PDCA model, which includes the following steps:

- The *plan* step includes studying the current state of the practice and determining what opportunities or needs exist for process improvements. Priorities are established, and one or more process improvements are selected for implementation. For each selected improvement, a plan is created to define the specific objectives, tasks or activities, assignments, resources, budgets, and schedules needed to implement that improvement.
- The *do* step implements the plan. This step includes:
  - Identifying and involving relevant stakeholders
  - Identifying root causes of problems or nonconformances, investigating alternative solutions, and selecting a solution

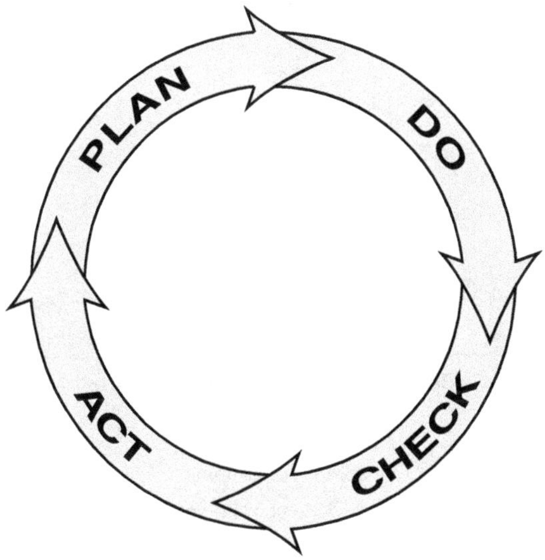

Figure 7.4 Plan–do–check–act model.

- Implementing and testing the selected solution by creating or updating systems, policies, standards, processes, work instructions, products or services, or metrics as needed
- Developing or providing training as needed
- Communicating about the change and its status to stakeholders

- The *check* step, also called the *study* step (*plan–do–study–act* model), monitors or measures the outcomes of the improvement and analyzes the resulting process after the plan is implemented to determine if the objectives were met, if the expected improvements actually occurred, and if any new problems were created. In other words, did the plan and its implementation work?
- During the *act* step, the knowledge gained during the check step is acted upon. If the plan and implementation worked, action is taken, and controls are put in place to institutionalize the process improvement throughout the organization. Then the cycle is restarted by repeating the plan step for the next improvement. If the plan and implementation did not result in the desired improvement or other problems were created, the act step identifies the root causes of the resulting issues and determines the needed corrective actions. In this case, the cycle is started over by repeating the plan step to plan the implementation of those corrective actions. The act step can also involve abandoning the change.

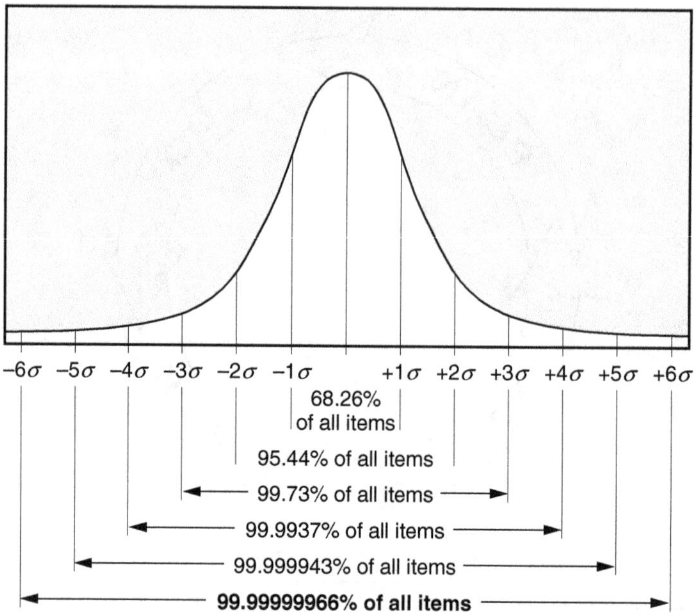

Figure 7.5 Standard deviation versus area under a normal distribution curve.

Six Sigma

The Greek letter *sigma* (σ) is the statistical symbol for standard deviation. As illustrated in Figure 7.5, assuming a normal distribution, plus or minus six standard deviations from the mean (average) would include 99.99999966 percent of all items in the sample. This statistic leads to the concept of a *six sigma measure of quality,* which is a near-perfection goal of no more than 3.4 defects per million opportunities.

More broadly than that, Six Sigma is a:

- Philosophy that "views all work as processes that can be defined, measured, analyzed, improved and controlled" (Kubial and Benbow 2009)
- Fact-based, data-driven methodology for eliminating defects in processes, through focusing on understanding stakeholder needs, prevention, continual improvement of processes, and a reduction in the amount of variation in those processes through the implementation of a rigorous step-by-step approach (DMAIC and DMADV, as illustrated in Figure 7.6)
- Business management strategy that has evolved into "a comprehensive and flexible system for achieving, sustaining, and maximizing business success" (Pande et al. 2000)
- Collection of qualitative and quantitative techniques and tools that organizations and individual practitioners can use to drive process improvement

DMAIC Model

The Six Sigma *DMAIC model* (Define, Measure, Analyze, Improve, Control) is used to improve existing processes that are not performing at the required level, as measured against *critical to x (CTx)* requirements (where x is a critical customer or stakeholder requirement including quality, cost, process, safety, delivery, and so on), through incremental improvement.

- The *define* step in the DMAIC model identifies the stakeholders, defines the problem, determines the requirements, and sets goals for process improvement consistent with stakeholder needs (*"voice of the customer"*) and the organization's strategy. An improvement project is chartered and planned. A team is formed that is committed to the improvement project and provided with management support (a champion) and the required resources.
- During the *measure* step in the DMAIC model, the current process is mapped (if a process map does not already exist). The CTx characteristics of the process being improved are determined. Metrics to measure those CTx characteristics are selected, designed, and agreed upon. A data collection plan is defined, and data are collected from the current process to determine the baselines and levels of variation for each selected metric. This information is used to determine the current process capability and define the current process's benchmark performance levels.
- During the *analyze* step in the DMAIC model, statistical tools are used to analyze the data from the measure step to fully understand the influences that each input variable has on the process and its resulting outputs. Gap analysis is performed to identify the differences between the current performance of the process and the required performance. Based on these evaluations, the root cause(s) of the problem or variation in the process are determined and validated. The objective of the analyze step is to understand the process well enough that it is possible to identify alternative improvement actions during the improve step.
- During the *improve* step in the DMAIC model, alternative approaches (improvement actions) are considered to solve the problem or reduce the process variations. The team then assesses each alternative's costs and benefits, impacts, and risks and performs trade-off studies. The team comes to a consensus on the best approach and creates a plan to implement the improvements. The plan contains the appropriate actions needed to meet the stakeholders' requirements. Appropriate approvals for the implementation plan are obtained. A pilot is conducted to test the solution, and measures against the CTx requirements from that pilot are collected and analyzed. If the pilot is successful, the solution is propagated to the entire organization, and measures against the CTx requirements are again collected and

analyzed. If the pilot is unsuccessful, appropriate DMAIC steps are repeated as necessary.

- During the *control* step in the DMAIC model, the newly improved process is standardized and institutionalized. Controls are put in place to make certain that the improvement gains are sustained into the future. This step includes selecting, defining, and implementing key metrics to monitor the process or product to identify future "out of control" conditions. The team develops a strategy for project handoff to the process owners. This strategy includes propagating lessons learned and creating documented procedures, training materials, and other mechanisms necessary to guarantee ongoing maintenance of the improvement solution. The current Six Sigma project is closed, and the team identifies the next steps for future process improvement opportunities.

DMADV Model

The Six Sigma *DMADV model* (Define, Measure, Analyze, Design, Verify), also known as *Design for Six Sigma (DFSS)*, is used to define new processes and products at Six Sigma quality levels. The DMADV model is also used when the existing process or product has been optimized but still does not meet required quality levels. In other words, the DMADV model is used when evolutionary change (radical redesign), instead of incremental change, is needed. Figure 7.6 illustrates the differences between the DMAIC and DMADV models.

- During the *define* step in the DMADV model, the goals of the design activity are determined based on stakeholder needs and aligned with the organizational strategy. This step mirrors the define step of the DMAIC model.
- During the *measure* step in the DMADV model, CTx characteristics of the new product or process are determined. Metrics to measure those CTx characteristics are then selected, designed, and agreed upon. A data collection plan is defined for each selected metric.
- During the *analyze* step in the DMADV model, alternative approaches to designing the new product or process are considered. The team then assesses each alternative's costs and benefits, impacts, and risks, and performs trade-off studies. The team comes to a consensus on the best approach.
- During the *design* step in the DMADV model, high-level and detailed designs are developed, implemented, and optimized. Plans are also developed to verify the design.
- During the *verify* step in the DMADV model, the new process or product is verified to make sure it meets the stakeholder requirements. This step may include simulations, reviews, pilots, tests, or other V&V activities. The new process or product design is then implemented into operations. The team develops a strategy for project handoff to the

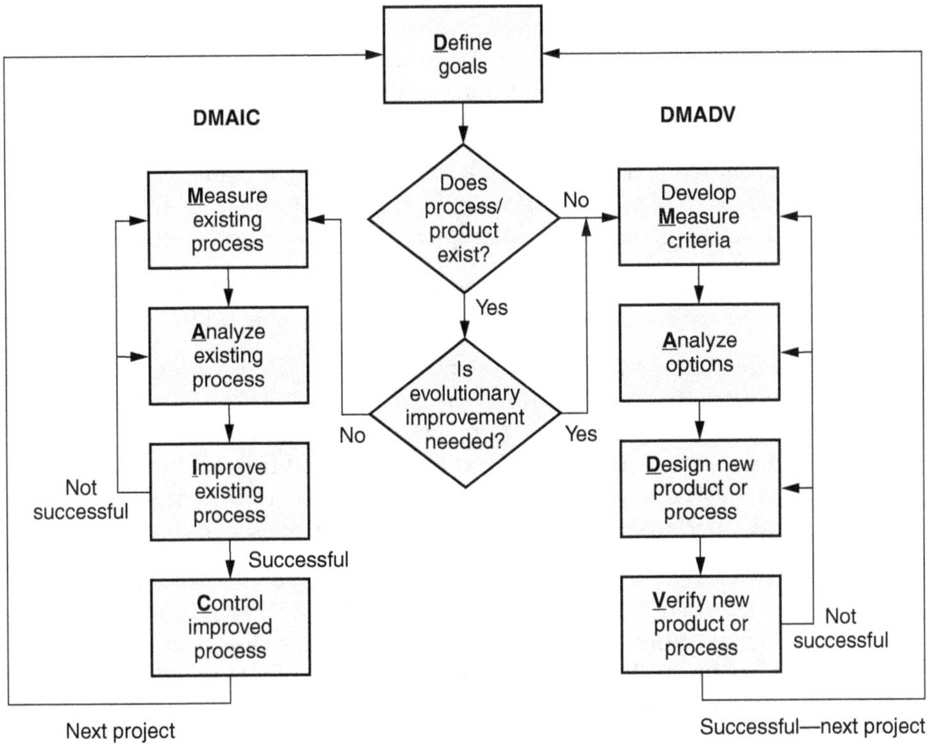

Figure 7.6 DMAIC versus DMADV Six Sigma models.

process owners. The current Six Sigma project is closed, and the team identifies the next steps for future projects.

Lean Techniques

While *lean principles* originated in the continuous improvement of manufacturing processes, these lean techniques have now been applied to software development. The Poppendiecks adapted the seven lean principles to software as follows (Poppendieck and Poppendieck 2007):

- Eliminate waste
- Build quality in
- Create knowledge
- Defer commitment
- Deliver fast
- Respect people
- Optimize the whole

Waste is anything that does not add value or gets in the way of adding value, as perceived by the stakeholders. Organizations must evaluate the software development and maintenance timelines and reduce those timelines by removing non-value-added wastes. Examples of waste in software development include:

- *Incomplete work*: If work is left in various stages of being partially completed (but not fully complete), it can result in waste. If a task is worth starting, it should be completed before moving on to other work.
- *Extra processes*: The process steps should be optimized to eliminate any unnecessary work, bureaucracy, or extra non-value-added activities.
- *Extra features or code*: A fundamental software quality engineering principle is that "gold plating" should be avoided. That is, avoid adding extra features or "nice to have" functionality not in the current development plan or cycle. Everything extra adds both direct and hidden costs due to increased testing, complexity, difficulty in making changes, potential failure points, and even obsolescence.
- *Task switching*: Switching back and forth between tasks wastes the time and energy it takes to put one task away, start another task, and get back into flow. Belonging to multiple teams causes productivity losses due to task switching and other forms of interruption (DeMarco 2001).
- *Waiting*: Any step in a process that results in delays or causes personnel at the next step to wait should be reevaluated. Anything interfering with progress is a waste since it delays the early realization of value by the stakeholders.
- *Unnecessary motion* (e.g., to find answers or information): Additional non-value-added steps or unnecessary approval cycles interrupt concentration and cause extra effort and waste. It is important to determine how much effort is required to learn enough useful information to move ahead with a project.
- *Defects*: Another fundamental software quality engineering principle is that rework to correct defects is a waste. The goal is to prevent as many defects as possible. If defects get into the product, the goal is to find those defects as early as possible, when they are the least expensive to correct.

To eliminate waste, one technique to use is *value stream mapping*, which traces a product from raw materials to use. The current value stream of a product or service is mapped by identifying all of the inputs, steps, and information flows required to develop and deliver that product or service. This current value stream is analyzed to determine areas where wastes can be eliminated. Value stream maps are then drawn for the optimized process, and that new process is implemented.

Building quality in requires that quality be built into the software rather than trying to test it later (which never works). This means that developers focus on:

- *Perceived integrity*: Involves how the whole experience with a product affects a stakeholder both now and as time passes
- *Conceptual integrity*: Comes from system concepts working together as a smooth, cohesive whole
- *Refactoring*: Involves adopting the attitude that internal structure will require continual improvement as the system evolves
- *Testing*: Becomes even more critical after a product goes into operations because 50 percent or more of product change occurs when the product is in operations; a healthy test suite helps maintain product integrity and helps document the system

Knowledge is created and learning is amplified through providing feedback mechanisms. For example, developing a large product using short iterations allows multiple feedback cycles as iterations are reviewed with the customers, users, and other stakeholders. The continuous use of metrics and reflections or retrospective reviews throughout the project and process implementation creates additional opportunities for feedback. The team should be taught to use the scientific method to establish hypotheses, conduct rapid experiments, and implement the best alternatives.

Deferring commitment means making irrevocable decisions as late as possible. This delay helps address the difficulties resulting from making those decisions when uncertainty is present. Of course, "first and foremost, we should try to make most decisions reversible, so they can be made and then easily changed" (Poppendieck and Poppendieck 2007). Making decisions at the "last responsible moment" means delaying decisions until the point when failing to make a decision would eliminate an important alternative and cause decisions to be made by default. As the process progresses, gathering as much information as possible allows better, more informed decisions to be made.

Deliver fast means getting the product to the customer or user as quickly as possible. The sooner the product is delivered, the sooner feedback from the customers, users, or other stakeholders can be obtained. Fast delivery also means the stakeholders have less time to change their minds before delivery, which can help eliminate the waste of rework caused by requirements volatility.

Organizations and leaders must *respect people* to improve systems. This respect means creating teams of engaged, thinking, technically competent people who are motivated to design their own work rather than just waiting for others to order them to do things. This respect also requires strong, entrepreneurial leaders who provide people with a clear, compelling, achievable purpose; give those people access to the stakeholders; and allow them to make their own commitments. Leaders provide the vision and resources to the team and help them when needed without taking over.

Optimize the whole is all about systems thinking. Winning is not about being ahead at every stage (optimizing or measuring every task). It is possible to optimize an individual process and actually sub-optimize the entire system. All decisions and changes are made with consideration of their impacts on the entire system, as well as their alignment with organizational goals and critical customer or stakeholder requirements.

3. CORRECTIVE AND PREVENTIVE ACTION PROCEDURES

> Evaluate corrective and preventive action procedures and impact assessments related to software defects, process nonconformances, and other quality system deficiencies. (Evaluate)
>
> **Body of Knowledge II.B.3**

Corrective and preventive actions are those actions taken to eliminate the root cause of a problem or potential problem (nonconformance, noncompliance, defect, or other issue) in order to prevent its future recurrence or ensure the occurrence of a positive outcome. Unlike corrective action, which is intended to eliminate future repetition of problems that have already occurred, *preventive actions* are taken to prevent problems that have not yet occurred. For example, an organization's supplier, Acme, experienced a problem because they were not made aware of the organization's changing requirements. As a result, the organization established a supplier liaison whose responsibility it is to communicate all future requirements changes to Acme in a timely manner. This is corrective action, because the problem had already occurred. However, the organization also established supplier liaisons for all of its other key vendors. This is preventive action because those suppliers had not yet experienced any problems.

The Plan–Do–Check–Act, Six Sigma, and lean improvement models discussed previously in this chapter are examples of models that can be used for corrective or preventive action, as well as process improvement. The CMMI® specifically addresses corrective and preventive actions in its Causal Analysis and Resolution practice area (CMMI 2023).

Corrective and preventive action begins once the problem has been reported. However, corrective and preventive action is more than just taking the remedial action (correction) necessary to fix the problem. Corrective and preventive action also involves:

- Researching and analyzing one or more problems to determine their root causes
- Exploring alternatives and developing a plan to eliminate those root causes
- Implementing that plan to improve the software products or processes to prevent the recurrence of the problem
- Verifying the implementation
- Verifying the effectiveness of the improvement
- Analyzing the effects of the problem on the products produced while the problem was occurring and determining whether these products can be used "as is," if they need to be reworked/corrected,

or potentially even be recalled if they have already been released into operations

- Closure, including the retention of quality records for future analysis and management review

As illustrated in the example in Figure 7.7, the corrective and preventive action process starts with identifying a problem. Problems in current products, processes, or systems can be identified through various sources. For example, they can be identified:

- As nonconformances or other negative observations during audits
- Through suggestion systems
- By quality action teams
- Through lessons learned during project, process, or system implementation
- Through the performance of root cause analysis on one or more product problems
- Through the identification of unstable trends or out-of-control states using metrics
- Through product verification and validation activities (e.g., peer/technical reviews or testing)

The second step in the corrective and preventive action process is to assign a champion and sponsor for the action and assemble an action planning team. This team determines if remedial action is necessary to stop the problem from affecting the quality of the organization's products and services until a longer-term solution can be implemented. If so, a remedial action team is assigned to make the necessary corrections. For example, suppose the source code module being produced does not meet the coding standard. A remedial action might be for team leads to review all newly written or modified code against the coding standard prior to it being baselined. While this is not a permanent solution and may cause a bottleneck in the process, it helps prevent any more occurrences until a long-term solution can be reached. If remedial action is needed, it is implemented as appropriate. For a software product problem, remedial action is typically the correction of the immediate underlying defect(s) that is the direct cause of the problem. Remedial action has an immediate impact on the quality of the software process or product by eliminating the specific, identified problem. Typical remedial actions include:

- Fixing the product defect, testing the correction, and regression testing the product
- If the product problem was reported from operations:
  - Including the fix in a corrective release of the product
  - Including the fix in the next planned feature release of the product
- Rewriting or correcting the process or product documentation

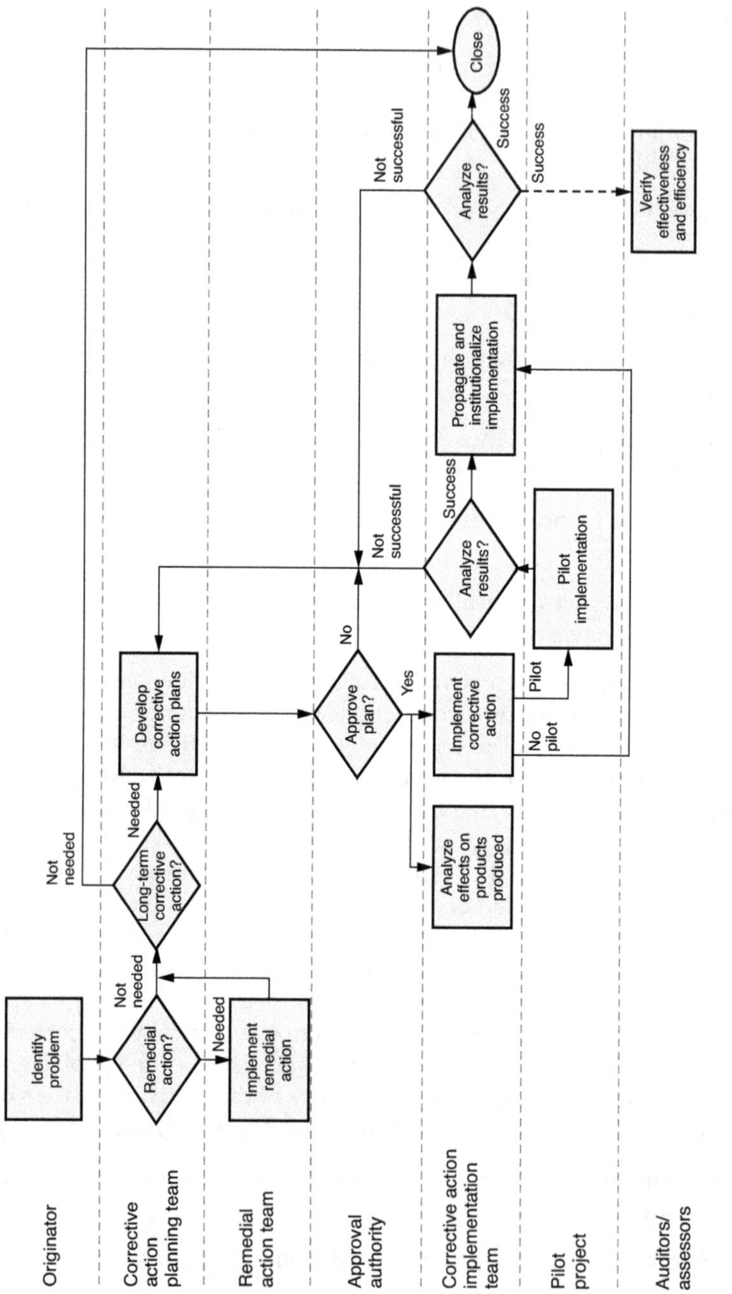

Figure 7.7 Corrective and preventive action process—example.

- Rewriting or correcting the process or product training materials
- Training or retraining employees or other stakeholders

All corrective and preventive actions should be commensurate with the risk and impact of the problem. Therefore, it may be determined that it is riskier to fix the problem than to leave it uncorrected, in which case the problem is resolved with no action taken, and the corrective and preventive action process is closed.

The third step in the corrective and preventive action process is initiating a long-term corrective or preventive action. If the problem is determined to be an isolated incident with minimal impact, remedial action may be all that is required. However, this may simply eliminate the symptoms of the problem and allow the problem to recur in the future. For multiple occurrences of a problem, a set of problems, or problems with higher impact, more extensive analysis is required to implement long-term corrective and preventive actions to reduce the possibility of additional occurrences of the problem in the future. If the action planning team determines that long-term action is needed, that team researches the problem. It identifies its root cause through statistical techniques, data-gathering metrics, analysis tools, or other means. For the coding standard example, the root cause might be that:

- The coding standard is out of date because a newer coding language is now in use
- New hires have not been trained on the coding standard
- Management has not enforced the coding standard, and therefore the engineers consider it optional

Once the root cause is identified, the action planning team develops alternative approaches to solving the identified root cause based on their research. That team analyzes each alternative solution's cost and benefits, risks, and impacts and performs trade-off studies to come to a consensus on the best approach. The action plan addresses improvements to the control systems that will avoid the potential recurrence of the problem. For example, if the action planning team determines that the root cause is lack of training, not only must the current staff be trained (corrective action), but controls must be put in place to make certain that all future staff members (new hires, transfers, outsourced personnel) receive the appropriate training in the coding standard as well (preventive action).

The output of this step is one or more corrective and preventive action plans, developed by the action planning team, that:

- Define specific actions to be taken
- Assign the individual responsible for performing each action
- Assign an individual responsible for verifying that each action is performed
- Estimate the effort and cost for each action
- Determine due dates for completion of each action
- Select mechanisms or measures to determine if desired results are achieved

The approval authority then reviews and approves those action plans, as defined in the quality management system processes. For example, if changes are recommended to the organization's quality management system (QMS), approval for those changes may need to come from senior management. If individual organizational-level processes or work instructions need to be changed, the approval body may be an *engineering process group* (EPG) or *process change control board* (PCCB) made up of process owners. On the other hand, if the action plans recommend training, the managers of the individuals needing that training may need to approve those plans. For product changes, the approval may come from a Configuration Control Board (CCB) or other change authority, as defined in the configuration management processes. The lead auditor may also need to review the corrective and preventive actions that result from nonconformances found during an audit.

During this plan approval step, any affected stakeholders should be informed of the plans and allowed to provide input into the impact analysis and approval. The approval/disapproval decisions from the approval authority are also communicated to impacted stakeholders. If the action plans are not approved, the action planning team determines what appropriate future actions to take (replanning).

In the implement corrective action step, the implementation team executes the action plan(s). Depending on the actions to be taken, this implementation team may or may not include members from the original action planning team. If appropriate, the implementation of the action plans may also include a pilot project to test whether the action plan works in actual application. If a pilot is held, the implementation team analyzes the results of that pilot to determine the success of the implementation. If a pilot is not needed, or if it is successful, the implementation is propagated to the entire organization and institutionalized. Results of that propagation are analyzed, and any issues are reported to the corrective action planning team. If the implementation or pilot did not correct the problem or resulted in new problems, then these results may be sent back to the action planning team to determine what appropriate future actions (replanning) to take.

The action implementation team must also analyze the problem's effects on past products:

- Are there any similar software products that may have similar product problems?

- Does the organization need to take any action to correct the products or services created while a process problem existed?

For example, assuming that training was the root cause of not following the coding standard, all of the modules written by the untrained coders need to be reviewed against the coding standard to understand the extent of the problem. Then a decision must be made about whether to correct those modules or accept them with a waiver.

Successful propagation and institutionalization may close the problem, depending on the source of the original problem (e.g., if it was a process improvement suggestion or problem identified by the owners of the process). However, at some point in the future, auditors, assessors, or other individuals should perform an evaluation of the completeness, effectiveness, and efficiency of that implementation to verify its continued success. If the corrective and preventive action resulted from a nonconformance identified during an audit, the lead auditor would have to be

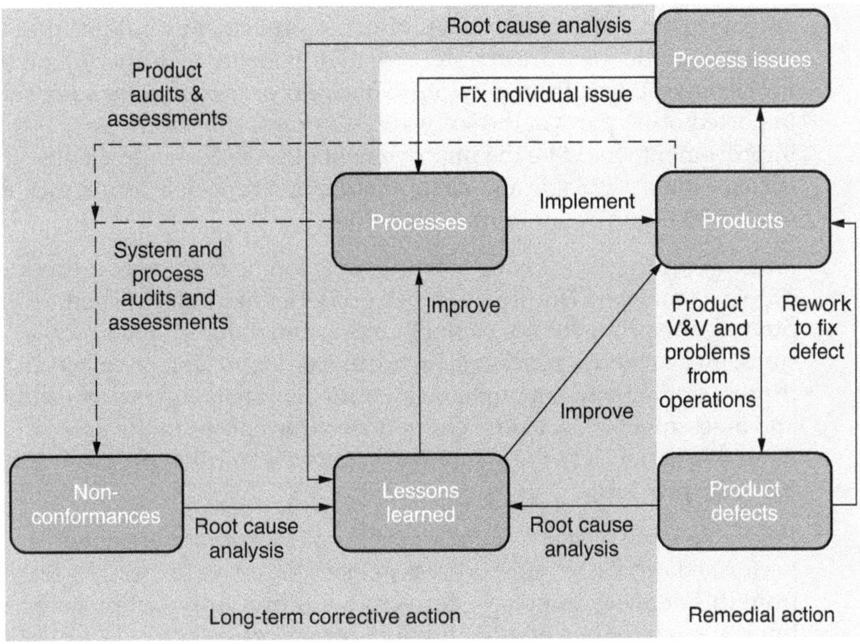

Figure 7.8 Remedial action (fix/correct) versus long-term corrective and preventive action for problems.

provided evidence that the nonconformance was resolved and agree with that resolution before the corrective action is closed.

So where does corrective and preventive action fit in with the other software processes? As illustrated in Figure 7.8, every organization has formal or informal software processes that are implemented to produce software products. These software products can include both the final products that are used by the customers or users (e.g., executable software, user documentation) and interim products that are used internally by the organization (e.g., requirements, designs, source/object code, plans, test cases). Product and process corrective and preventive action fit in as follows:

- *Product fix/correction*: During software development, various V&V processes are used to provide confidence in the quality of those products and to identify any defects in those products. Once the product has been released into operations, additional problems may be reported, including software operational failures, usability issues, stakeholder complaints, and other issues. Analysis of these field-reported problems may uncover additional software defects. Remedial corrective action activities take the form of rework to fix/correct these identified defects in the software products.
- *Product corrective and preventive action*: Root cause analysis can be performed on one or more identified product defects. Based on the lessons learned from this analysis, one or more products or processes

are improved. An example of a product improvement would be the redesign of the software architecture to provide more decoupling if the problems resulting from changes to one part of the software adversely impacted other parts of the software. An example of a process improvement would be the improvement of unit testing processes because too many logic and data initialization type defects are escaping into later testing cycles or into operations.

- *Process fix/correction*: During software development, various process problems may be identified as those processes are implemented to produce the products. For example, extra, non-value-added process steps, areas where a process is ineffective or inefficient, or defects in the process documentation might be identified. System and process audits and assessments may also identify nonconformances in the processes. Remedial action activities fix or correct these identified problems in the software processes.
- *Process corrective and preventive action*: Root cause analysis can be performed on one or more process defects. Based on the lessons learned from this analysis, long-term corrective and preventive action may be implemented for one or more products or processes.

4. DEFECT PREVENTION

> Design and use defect prevention processes such as technical reviews, software tools and technology, and special training. (Evaluate)
>
> **Body of Knowledge II.B.4**

Defect preventive action is proactive in nature. Establishing standardized processes based on industry good practice, as defined in standards such as ISO 9001 and the IEEE software engineering standards or in models such as CMMI®, can help prevent defects by propagating known good practices throughout the organization (see Chapters 3 and 6). A standardized process approach strives to control and improve organizational results, including product quality and process efficiencies and effectiveness. Creating an organizational culture that focuses on employee involvement, creating stakeholder value, and continuous improvement at all levels of the organization can help make sure that people not only do their work but think about how they do their work and how to do it better.

Techniques such as risk management, failure modes effects analysis (FMEA), statistical process control, data analysis (e.g., data trending in a problematic direction), and audits/assessments can be used to identify potential problems and address them before those problems occur. Once a potential problem is identified, the defect preventive process is similar to the corrective and preventive action process discussed previously. The primary difference is that instead of identifying and analyzing the root cause of an existing problem, the action team researches potential causes of a potential problem. Industry standards, process improvement models, good practices, and industry studies all stress the benefits of preventing defects over detecting and correcting them.

Since software development is knowledge work, one of the most effective forms of preventive action is to provide practitioners with the knowledge and skills they need to perform their work with fewer mistakes. If people do not have the knowledge and skill to catch their own mistakes, those mistakes can lead to defects in the work products. Training and on-the-job mentoring or coaching are very effective techniques for propagating necessary knowledge and skills. This includes training or mentoring/coaching in the following areas:

- *Customer/user's business domain*: So that requirements are less likely to be missed, defined incorrectly, or misinterpreted after they are specified
- *Design techniques and modeling standards*: So that defects are less likely to be inserted into the architecture and design
- *Coding language, coding standards, and naming conventions*: So that defects are less likely to be inserted into the source code module
- *Tool set*: So that misuse of tools and techniques do not cause the inadvertent insertion of defects
- *Quality management system, processes, and work instructions*: So that practitioners know what to do and how to do it, resulting in fewer mistakes

Technical reviews, including peer reviews and inspections, are used to identify defects, but they can also be used as a mechanism for promoting the common understanding of the work product under review. For example:

- If designers and testers participate in the peer review of requirements, they may obtain a more complete understanding of the requirements, preventing misinterpretations from propagating into future design, code, and test problems and rework.
- If a defect is identified and discussed in a peer review, the attendees of that peer review are less likely to make similar mistakes when developing similar work products.
- Inexperienced practitioners, who attend the technical reviews of expert practitioners' work products, also have the benefit of seeing what high-quality work products look like and can use that reviewed work product as an example or template when developing their work products later.

Other examples of problem-prevention techniques include:

- Incremental and iterative development methods, agile methods, and prototyping all create feedback loops that help prevent mistakes by verifying a consistent and complete understanding of the stakeholder requirements.
- Good design techniques, including decoupling, cohesion, and information hiding, help prevent future defects when the software is modified and maintained.

- Good V&V practices throughout the life cycle prevent defects from propagating forward into other work products. For example, a single defect found during a requirements review could have potentially become multiple design, code, and test case defects if it had not been found and corrected.
- Software tools and technologies also help prevent problems. For example, modern build tools help prevent problems by automatically initializing memory to zero so that missing variable initialization code does not cause product problems. Security analysis tools can identify security holes in the code and so on.
- Mistake-proofing processes using well-defined, repeatable processes, standardized templates, and checklists can prevent problems. When practitioners understand what to do, when to do it, how to do it, and who is supposed to do it, they are less likely to make mistakes. Checklists can prevent problems that might arise from missed steps, activities, or items for consideration.

Defect prevention also comes from benchmarking and identifying good practices within the organization and in other organizations and adapting and adopting them into improved practices within the organization. Holding *lessons learned sessions,* also called *post-project reviews, retrospectives,* or *reflections,* can also identify potential improvements that can prevent defects. These techniques allow the organization and its teams to learn lessons from the problems encountered by others without having to make mistakes and solve the problems themselves.

5. METRICS AND MONITORING

> Apply metrics and monitoring techniques for adherence to software quality goals and objectives, process improvement, and defect prevention. (Apply)
>
> **Body of Knowledge II.B.5**

The *IEEE/ISO/IEC 15939 Software Engineering—Software Measurement Process* "defines a software measurement process applicable to all software-related engineering and management disciplines." As illustrated in Figure 7.9, this process is described through a model that "defines the activities of the measurement process that are required to adequately specify what measurement information is required, the measures and analysis results are to be applied, and how to determine if the analysis results are valid" (IEEE/ISO/IEC2017b).

Evaluating an organization's success in meeting its software quality goals and objectives, implementing process improvement implementation (including corrective and preventive actions) and preventing defects, and verifying its continued success over time requires an organization to establish a commitment to a metrics and monitoring program.

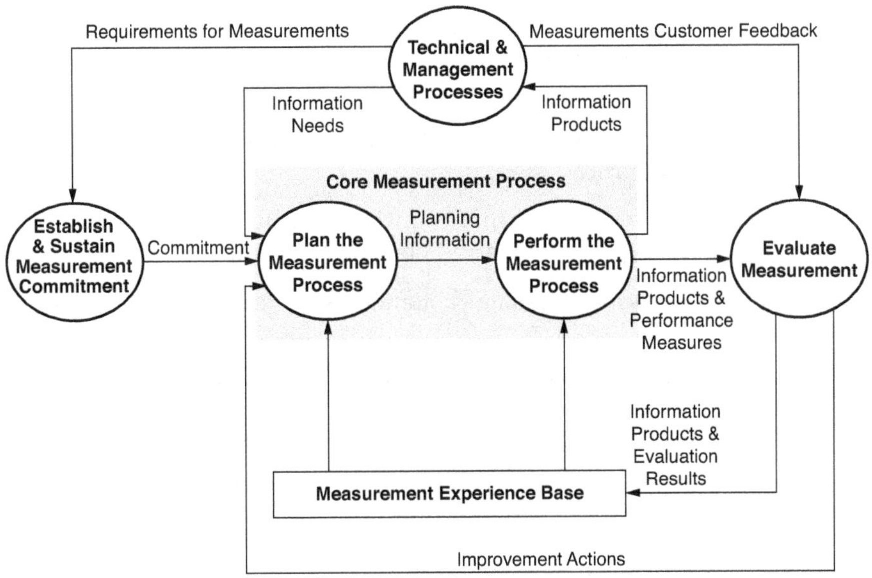

Figure 7.9 Software measurement process (based on IEEE/ISO/IEC-17b).

The core to the measurement process includes planning the measurement process by determining of the CTx characteristics associated with the organization's goals, objectives, process improvement, and products (as discussed earlier in this chapter). Metrics to measure and monitor those CTx characteristics must be selected, designed, agreed upon, and implemented as part of the QMS (as discussed in Chapter 6).

This planning information must then be applied to the organization's project, processes, products, and services by performing the measurement process. This performance includes the collecting and storing of data, transforming that data into information, and creating, distributing the information products and performance measures to the metrics customers. These information products are analyzed and used to make fact-based decisions, even if the decision is "everything is fine, no action is necessary."

The software measurements process should also be monitored and improved like all of the other software processes. This is done by evaluating the measurement, and based on the results of that analysis, the process continues by:

- Feeding the measurement customer's feedback into the technical and management processes to:
  - Improve the organization's projects, processes, products, and services
  - Determine the requirements for additional measurement, which sustains the measurement commitment to the metrics program
  - Feed additional information needs into planning or improving the existing metrics process and metrics by iterating the plan measurement process step

- Storing the information products and evaluation results in the measurement experience base to be used to improve the planning and performing of the measurement process
- Feeding improvement actions into the iteration of the plan measurement process step

Examples of CTx characteristics that might be evaluated as part of monitoring and measuring software quality goals and objectives include:

- Positive trends in increasing customer, user, and stakeholder perceived value
- Increases in employee satisfaction and engagement
- Positive impacts on team knowledge, skills, or abilities
- Increases in process efficiencies and effectiveness (reduced costs, improved cycle times, reduced resource requirements, improved product quality)
- Improvements in process capability and maturity scores
- Decreases in non-conformances, noncompliances, or defects found in audits, assessments, reviews, and verification and validation activities

Examples of CTx characteristics that might be measured and monitored as part of process improvement include:

- Reductions in total cost of quality, cost of development, or overall cost of ownership
- Reductions in development or delivery schedules or cycle times
- Positive impacts on product functionality, quality, performance, reliability, maintainability, technical leadership, or other quality attributes
- Reductions in the number of problem reports that were returned to the originator because not enough information was included
- Reductions in the number of problem reports that were not actual defects in the product (e.g., operator errors, works as designed, and could not duplicate)
- Positive impacts on team knowledge, skills, or abilities
- Positive impacts on stakeholder satisfaction or success

Examples of CTx characteristics that might be measured and monitored as part of product improvements include:

- Positive impacts on product functionality, quality, performance, reliability, maintainability, technical leadership, or other quality attributes
- Reductions in technical debt

- Decreases in the fix-on-fix ratio, where future V&V activities or operations determine that all or part of the problem was not corrected
- Decreases in the number of identical problems being identified later in other products or other versions of the same product
- Reductions in the number of unauthorized changes being made to baselined configuration items

Examples of CTx characteristics that might be measured and monitored as part of evaluating the efficiency and effectiveness of the corrective and preventive action process itself include:

- Decreases in the cycle times for the entire process or individual steps in the process so problems are being corrected in a timely manner
- Removal of bottlenecks or excessive wait times in the process
- Removal of wastes in the process
- Reductions in the cost to correct problems
- Reductions in the reoccurrence of problems that have been previously corrected
- Risk priority, impact, exposure, timeframe, and risk reduction leverage

Examples of CTx characteristics that might be measured and monitored as part of defect preventive action include:

- Decreases in total cost of quality, cost of development, or overall cost of ownership
- Reductions in rework costs
- Increased first-pass yields
- Reductions in defect density
- Positive trends in phase containment and defect removal effectiveness
- Reductions in the number of reported post-release problem
- Positive impacts on team knowledge, skills, or abilities
- Positive trends in the increase of automation
- Positive impacts on stakeholder satisfaction, including employee satisfaction

See Chapters 6, 17, 18, 19, and 20 for more about some of these metrics and other metric examples.

Chapter 8
Audits

INTRODUCTION TO AUDITS

The ASQ *Audit Division* (Coleman 2020) defines an *audit* as "a systematic, independent, and documented process for obtaining audit evidence and evaluating it objectively to determine the extent to which the audit criteria are fulfilled." IEEE (2008) includes a similar definition of an audit as "an independent examination of a software product, software process, or set of software processes performed by a third party to assess compliance with specifications, standards, contractual agreements, or other criteria." According to ISO (2015), internal audits are a required part of a quality management system. The IEEE *Standard for Software Reviews and Audits* (IEEE 2008) and ISO 19001 *Guidelines for Auditing Management Systems* (ISO 2018) include process information for the execution of audits.

Software audits are planned activities—there are no surprise software audits. Software audits are not trying to "catch" anyone at their worst. The people involved in a software audit should be aware of the audit's scope, objectives, schedule, and their roles and responsibilities during the audit. Audits are conducted by individuals who are independent of the area being audited. This independence helps make sure that an objective evaluation is conducted. Software audits have documented plans and reports. In addition, corrective action plans are documented, as required, for any nonconformances discovered during the audit. Software audits evaluate some aspect of a software system, process, project, product, or supplier and provide management with information from which fact-based decisions can be made. Software audits use a set of agreed-to criteria as the requirements to conduct these evaluations.

Audit Objectives

Software audits should be value-added activities conducted to provide information to management based on an evaluation of whether the:

- Organization's standards, processes, systems, and plans are adequate to enable the organization to meet its policies, requirements, goals, and objectives.
- Organization complies with those documented standards, processes, systems, and plans during the execution of its work activities.

- Organization's standards, processes, systems, and plans and their implementation are effective. In other words, are the policies, requirements, goals, and objectives actually being met?
- Resources, including human and other nonhuman resources, are being efficiently and effectively utilized.
- Products conform to their required specifications and workmanship standards, and products are actually fit for use by their intended audience.

Software audits also help identify areas needing improvement and identify best practices within the organization that should be propagated to other areas. In fact, identifying and propagating best practices is one of the primary objectives of audits in more mature organizations where few or no nonconformances exist.

Audit Program

An overarching internal *audit program* should be planned as part of an organization's quality management system to make certain that required audits are performed regularly and audits of critical functions are performed frequently. The audit program also makes certain that only trained, qualified, and independent auditors perform audits and that audit processes are standardized and continually improved. Audits within the audit program should be scheduled to closely align with the organization's strategies, stakeholders' requirements, senior management's concerns, and organizational risks. Individual audits should be scheduled within the audit program to minimize the inconvenience for the auditee. For example, an internal audit of a project should not be scheduled to coincide with a major release when everyone on the project is scrambling to take care of the final details. A better time to schedule such an audit might be while that release is in development or after the major release push is over.

Audit Considerations

Audit findings must be based on facts. However, since humans are identifying and interpreting those facts, the audit staff must be as *independent* as possible to enhance objectivity. *Objectivity* is the absence of any bias that will influence the audit results. While total independence on the part of an internal auditor is impossible (at some level of the organization, everyone reports to the same management), the goal is still to maintain enough independence so that the auditor can objectively evaluate and analyze the evidence and produce unbiased audit results.

Conducting an audit consumes resources, including money and time. Auditors typically need to be able to spend time with and talk to software managers and individual contributors to perform an effective audit. This communication can take time away from developing and maintaining software, which is the primary mission of a software organization. Care should be taken to confirm that the value of the information received from the audit is worth more than the expected cost. In other words, audits should be value-added activities. This means that the audit information must have a client (customer), someone who will use the information to make decisions, even if that decision is "everything is great—no action is necessary."

An audit may result in one or more findings. This result means that more resources will be needed to perform root cause analysis of any identified nonconformance and make corrective action plans and implement improvement actions or propagate best practices. Things will improve, but in the meantime, more resources will be consumed.

The mere act of conducting an audit establishes expectations in the participants' minds that management will act on the information and improve things. If the audit findings are not adequately followed up on, the entire audit process may be perceived as a waste of time and energy. Unless everything is perfect, value-added audits should result in actions and follow-through.

Based on past experience with pure conformance-type audits, people may perceive audits in a negative light as "policing actions." This perception may lead to reluctance to cooperate with the audit or suspicions that the audit results will be used against the participants. Care should be taken by the auditors, and the auditor's and auditee's management, to focus the audits on process and product improvement or systematic issues and not on individuals.

1. AUDIT TYPES

> Define and distinguish between various audit types (e.g., process, compliance, supplier, system, internal, and external) and methods (e.g., virtual). (Understand)
>
> **Body of Knowledge II.C.1**

Audits are typed by who conducts the audit (internal versus external audits or first-, second-, and third-party audits) or by what is being audited (systems, process, product, project, or supplier audits). There are also special-purpose audits, including follow-up audits and desk audits.

Internal First-Party Audits

An *internal audit,* or *first-party audit,* is an audit that an organization performs on itself. As illustrated in Figure 8.1, in a first-party audit, the people conducting the audit (auditors), the people being audited (auditees), and the client (the person or organization that requested the audit) are all members of the same organization. The audit criteria for an internal audit can come from inside the organization (e.g., the organization's quality management system, individual processes, or plans). The audit criteria can also come from outside the organization. For example, the standards or

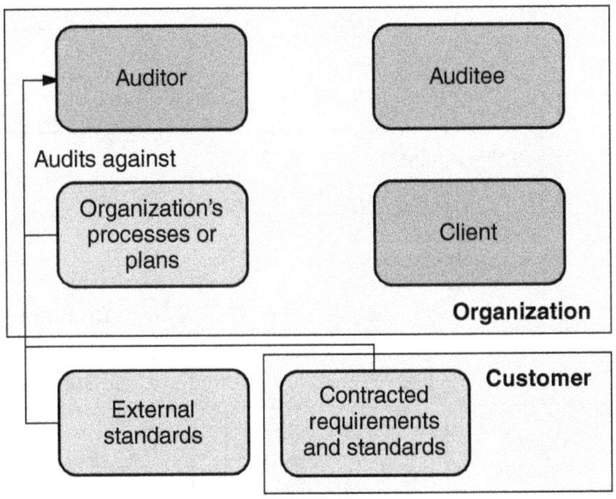

Figure 8.1 Internal first-party audit.

requirements of the organization's customers or external industry standards can be used as audit criteria.

For organizations and teams using agile methods, internal audits may continue to be used based on the risk levels, requirements imposed by industry standards (like ISO 9001, or its industry-specific counterparts like NQA-1, AS 9100, and so on), regulatory requirements, or contractual obligations. "Agile or not, a team ultimately has to meet legal and essential organizational needs, and audits help to ensure this" (Ambler and Lines 2012). However, in lower-risk industries with less rigorous requirements, agile teams may substitute self-monitoring through review, retrospectives, or reflections to fulfill the software quality assurance objectives of some or many of their internal audits.

In some cases, an organization can hire or outsource its internal audits to an external firm or team of consultants. This outsourcing is still considered an internal, first-party audit because the hired auditors are temporary employees.

External Second-Party Audits (Supplier Audits)

A *second-party audit* or *supplier audit* is an external audit where a customer, or an organization contracted by a customer, performs an audit on its supplier. As illustrated in Figure 8.2, in a second-party audit, the auditor and clients are in the customer's organization, and the auditee is in the supplier's organization. The audit criteria for an external second-party audit can come from inside the supplier's organization. For example, the supplier's systems, processes, or plans can be used as audit criteria. The audit criteria can also come from the customer's organization, including the customer's standards, contracts, and requirements. In a second-party audit, the supplier can also be audited to external standards if the contract or agreement with the supplier requires adherence to those standards.

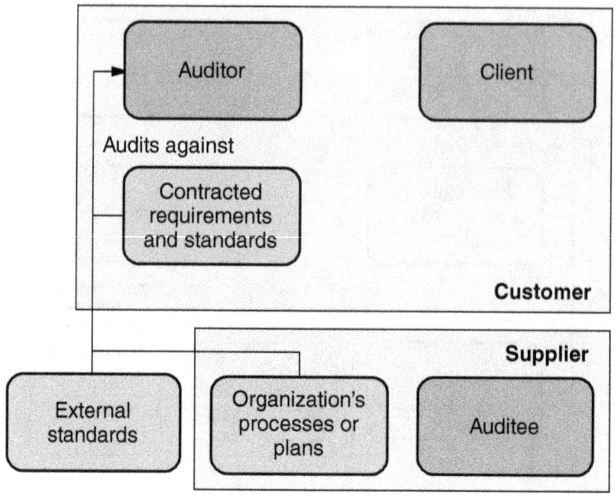

Figure 8.2 External second-party audit.

An example of a second-party audit would be a supplier qualification audit, where a customer audits a potential supplier prior to awarding a contract to verify that the supplier has the capability and capacity necessary to produce products of the required quality level. Another example would be a supplier surveillance audit, where a customer performs audits of its suppliers as part of ongoing supplier monitoring activities during the execution of a contract or agreement. Of course, the requirements for these audits should be documented in the contracts or other agreements with the supplier.

External Third-Party Audits

A *third-party audit* is performed on an auditee by an external auditor other than their customer. In a third-party audit, the client may be the organization being audited or a third-party organization. As illustrated in Figure 8.3, the audit criteria for a third-party audit may be the audited organization's internal systems, processes, or plans, or external standards may be used as the audit criteria.

An example of a third-party audit would be government regulators auditing an organization to verify that all required external regulations (standards) are being met. Another example would be a registrar conducting an audit prior to granting certification to ISO 9001. Even though the registrar's fees are paid by the organization being audited, they are considered an independent third party because they report to an accreditation board.

System Audits

System audits are conducted on management systems that evaluate all of the policies, processes, work instructions, supporting plans, activities, training, and other components of those systems. A system audit can be considered a mega-

Figure 8.3 External third-party audit.

process audit because it looks at all of the processes in the system. For example, a quality management system audit evaluates an organization's existing quality management system's adequacy, conformance, and effectiveness. Other systems that might be audited include an organization's environmental system, security system, or safety system.

Process Audits

A *process audit* evaluates a single process (or small set of processes) to determine if the process is defined, deployed, and adequate to meet the required quality objectives. A process audit also evaluates whether the process is being implemented correctly with due diligence and to determine if the process works effectively and efficiently. A process audit covers only a specific portion of an entire system. An example of a process audit would be an organization auditing its requirements specification inspection process to evaluate conformance to its documented peer review process, and the effectiveness of that process. Another example might be auditing the software configuration management process or just one of its subprocesses, such as the configuration identification process.

Product Audits

A *product audit* looks at a product and evaluates conformance to the product specifications, performance requirements, workmanship standards, or the stakeholders' requirements. A product audit may look at the results of software peer review, testing, or other verification and validation (V&V) activities, review V&V or other product quality records, or include a sampling of repeated V&V activities (re–peer review or retest). An example of a product audit would be an organization auditing a sampling of source code modules to determine the level of conformance to internal coding standards. Another example would be to audit a finished subsystem to evaluate its compliance with its allocated functional requirements.

Project Audits

A *project audit* examines the processes and activities used to initiate, plan, execute, track, control, and close a project. It evaluates the project's conformance to documented instructions or standards. A project audit also evaluates the project's effectiveness and efficiency in meeting the project's intended goals and objectives, and the adequacy and effectiveness of the project's controls.

Follow-Up Audits

Any audit may produce nonconformances or noncompliances that require corrective action. When those corrective actions are completed, a *follow-up audit* is one of the mechanisms that can be used to verify the completeness and effectiveness of the implementation of those corrective actions. A follow-up audit can be combined with the next scheduled audit of the area to minimize time and expense.

Desk Audits

A *desk audit* or *document audit* is limited to evaluating the organization's documentation, including quality records. These audits can be conducted at the auditor's desk since no on-site visit is required, where people are interviewed and activities are observed. A desk audit might also be conducted by mail, where questionnaires are sent to the auditee, and the auditee provides the required objective evidence for the auditor to review without the need for an on-site visit.

Virtual Audits

A *virtual audit* is conducted remotely using a virtual environment that includes online meeting tools for interviews, digital documents and records, online cameras for witnessing activities, and other online communication tools. There may be challenges with performing virtual audits, including:

- The availability of hardcopy-only documents or quality records
- Lack of face-to-face, in-person communications (e.g., loss of rapport between the auditor and the auditee or the inability to monitor nonverbal language)
- Proprietary information that cannot be taken off-site or outside controlled areas

2. AUDIT ROLES AND RESPONSIBILITIES

> Identify roles and responsibilities for audit participants including the client, lead auditor, audit team members, and auditee. (Understand)
>
> **Body of Knowledge II.C.2**

Client

The *client,* also called the *initiator* or *customer* of the audit, is the person or organization requesting the audit. The client determines the need for the audit and provides the authority to initiate the audit. The client:

- Defines the purpose (objectives) and scope of the audit
- Determines the audit criteria
- Determines the type of audit to be conducted
- Selects the auditing organization

The client is the primary customer of the audit report and determines its distribution. The client also acts as the final arbitrator for any audit-related issues that cannot be handled at a lower level.

Auditor Management

The *auditor management* is the management of the auditing organization. The auditor management is responsible for assigning a lead auditor to each audit and working with that lead auditor to select additional members for the audit team. The auditor management makes sure that the selected lead auditor and other auditors have appropriate training, knowledge, skill level, and independence. The auditor management provides the funding and other resources required to plan, prepare for, execute, report, follow up on, and manage the audit.

In the case of internal audits, the auditor management may also be responsible for establishing an effective audit program, including supporting procedures, processes, and tools and for evaluating the performance of that audit program and of the individual auditors. The auditor management may also plan the audit program, which includes setting audit priorities and defining the organization's audit program schedule. An organization may have a full-time team of internal auditors or a group of part-time, trained auditors who perform audits as just one of their work activities. Either way, while performing their auditing duties, the internal auditors should report to a manager within the organization with enough authority to:

- Promote the independence of auditors and audits and maintain the operational freedom necessary to conduct audits
- Make certain that the quality audit program has a broad-enough focus to include all of the:
  - Elements of the quality management system
  - Functional areas within the organization that impact quality
  - Organization's products and services
- Provide adequate resources to conduct the audits and train the auditors
- Authorize access to records, personnel, and physical properties relevant to the performance of audits

- Make certain that the management of the audited organization adequately considers and addresses the audit findings
- Make certain that the management of the audited organization takes the appropriate actions to correct the root cause of any problems or nonconformances found during the audit

Lead Auditor

The *lead auditor,* also called the *audit team lead,* is designated to manage the individual audit and its audit team and is responsible for the overall conduct and success of that audit. In addition to fulfilling the responsibilities of an auditor, the lead auditor:

- Plans the audit and documents the audit plan, including assigning responsibilities to audit team members and working with the auditee management to plan the schedule and logistics for the audit
- Manages the audit team, including:
  - Assisting in the selection of the team
  - Verifying that the auditors have the skills, knowledge, and information they need to conduct the audit successfully
  - Making adjustments to team assignments if necessary
  - Mentoring and evaluating the auditors on the team
  - Providing feedback on their performance
- Makes decisions about how the audit will be conducted, including controlling conflict and handling any problematic situations or issues that may arise during the audit
- Acts as a focal point for communications with the auditee management, including negotiating with the auditee management over issues that arise during the audit or disagreements about the content of the audit report
- Coordinates the activities and logistics of the audit, including conducting the opening meeting, daily audit team and feedback meetings, and closing meetings
- Reviews the findings and observations found during the document review and audit execution, and prepares and distributes the audit report
- Coordinates the review of the corrective action plans
- Verifies the implementation of the corrective action plans

Auditors

The *auditors* are the individuals conducting the audit. For more extensive audits, there may be an audit team made up of a group of auditors. For small audits, the lead auditor may be the only auditor. The auditors are responsible for:

- Preparing for the audit by:
  - Understanding the purpose and scope of the audit
  - Reviewing the audit plan and audit criteria (requirements)
  - Creating checklists, interview questions, and other audit tools
  - Performing the documentation review
- Gathering objective evidence during audit execution
- Evaluating evidence against audit criteria to determine audit findings
- Attending opening, closing, and daily audit team meetings
- Reporting audit findings to the lead auditor as input to the audit report
- Maintaining confidentiality and professionalism

Auditee Management

The *auditee management* is the manager of the organization being audited, who has been assigned to coordinate with the lead auditor on matters related to the audit. The auditee management is responsible for:

- Working with the lead auditor to plan the schedule and logistics for the audit
- Informing employees of the audit and its purpose and scope
- Providing the auditors access to auditee people, facilities, and resources
- Attending the opening, daily feedback, and closing meetings of the audit
- Providing all appropriate information requested by the auditors
- Providing a liaison or escort to the auditors as needed
- Responding to findings with corrective action plans and working with the lead auditor to resolve any issues that arise during the review of the corrective action plans
- Making sure that corrective actions are implemented
- Providing the lead auditor with evidence of successful implementation of corrective action

Auditee

The *auditees* are the individuals being interviewed or observed during the execution of the audit. The auditees are responsible for providing appropriate and accurate answers to the auditors' questions and for providing all appropriate information requested by the auditors.

Escort

For external audits, for any audit where the auditors are unfamiliar with the auditees' organization or physical location, or for any audit in areas with special safety or security concerns, escorts from the auditee organization may be assigned to each auditor or group of auditors. The *escort* accompanies the auditor during objective-evidence gathering and serves as a liaison between the audit team, the auditees, and the auditee management. The escort is responsible for:

- Introducing auditee personnel to the auditor
- Providing clarifying information as necessary, including the interpretation of terminology or acronyms
- Providing or requesting supplies, records, and so on needed by the auditor
- Acting as a guide for the auditor
- Making certain that the auditor complies with company rules and safety and security protocols

3. AUDIT PROCESS

> Define and describe the steps in conducting an audit, developing and delivering an audit report, and determining appropriate follow-up activities. (Apply)
>
> **Body of Knowledge II.C.3**

The basics steps in the audit process are illustrated in Figure 8.4 and include:

- *Initiate the audit*: A software audit starts with the formal audit initiation by the client of that audit.
- *Plan the audit*: The audit must then be planned, and that plan must be documented and communicated to the audit stakeholders.
- *Prepare for the audit*: During the audit preparation step, the lead auditor and other auditors prepare for the audit by gathering, organizing, and analyzing as much information as possible. By the time the auditors conduct the execution step of the audit, they should be familiar with the auditee's organization, the audit criteria, and the systems, processes, procedures, or projects being audited. During this step, the audit inputs are evaluated against the audit criteria. Audit checklists, interview questions, sampling plans, and other tools are also prepared during this step.
- *Audit execution*: The audit execution, also called the audit performance or audit implementation, is the fieldwork of the audit. It is the

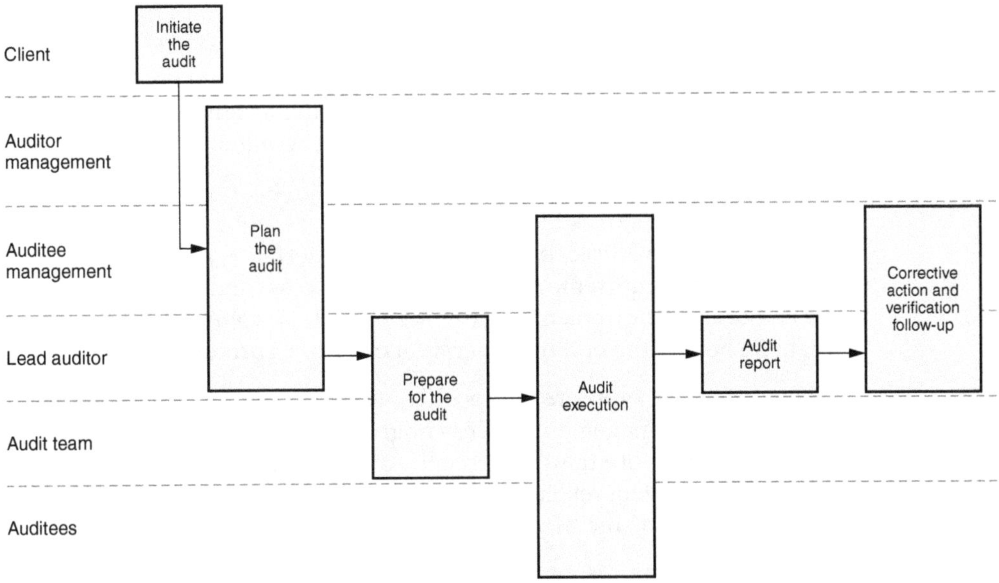

Figure 8.4 Audit process.

data- and information-gathering portion of the audit. During the execution step, the audit plans are implemented, and the audit tools are utilized to gather objective evidence, which is then analyzed to produce the audit results. During audits that include an on-site visit to the auditee or virtual audits, an opening meeting, daily feedback meetings, and a closing meeting are part of the audit execution step. During a desk audit, the audit execution step is limited to reviewing documents, quality records, and other information available from the auditor's desk.

- *Audit report*: During this step, the audit results are formally documented in the audit report, and all of the quality records for the audit are collected and appropriately stored.
- *Corrective action and verification follow-up*: The auditee management is responsible for creating and implementing corrective actions for any nonconformances or noncompliances identified during the audit. The auditee management may also choose to implement process improvement and best-practice propagation actions. Either verification follow-up by the lead auditor based on information supplied by auditee management or a follow-up audit is used to confirm the effective implementation of those corrective actions.

Audit Initiation

The frequency of audits varies based on an organization's goals, objectives, risks, and contractual requirements. Audits may be initiated when:

- Defined criteria in the audit program establish the need for the audit:
  - Every element of the quality management system should be periodically scheduled for an audit to make sure a comprehensive evaluation occurs. Critical elements may be scheduled for audits more frequently than less critical elements.
  - When major milestones in the software development project are completed. For example, if a project uses a waterfall-type life cycle, an audit of the requirements specification process might be held at the end of the requirements phase. If a project uses agile, an audit might be held at the end of an iteration or prior to product release.
  - When measured values reach specified targets. For example, an audit of the project management process might be held if the project goes over budget by more than 20 percent. Another example would be to audit the problem resolution process if more than 10 major field-reported problems remain unfixed for more than 30 days.
  - When a significant discrepancy or problem is identified. An audit may be conducted to determine what factors led up to the discrepancy or problem or to identify improvements that will prevent similar issues from occurring in the future.
- A client requests an audit. For example, an internal manager requests an audit of their area, or an external customer requests an audit based on contractual or agreement requirements.
- Regulatory or governmental audit requirements must be met. For example, a regulatory agency may require a specific type of audit be conducted annually or when a major milestone is met.
- Contractual audit requirements must be met. For example, a supplier may be required to conduct internal audits of their processes and systems and report the results to their customers, or the customer may require second-party audits where they audit the supplier's processes and systems.
- A major organizational change has occurred and the new management is seeking information about the organization's current status.

During the initiation steps, the client selects the auditing organization; defines the purpose, scope, and objectives of the audit; and designates the audit criteria. This initiation includes defining why the audit is being conducted and what organizations, processes, projects, or products will be audited. All required inputs to the audit must also be available before the audit can start.

The *audit purpose* statement acts as a mission statement for the audit. It should define the purpose (reason) and major objectives of the audit. Examples of audit purpose statements include:

- *Internal first-party audit example*: To confirm continued, compliant implementation of the peer review process, to evaluate its effectiveness, and to identify improvement opportunities at the Northeast site

- *External second-party audit example*: To evaluate Acme subcontractor's software development processes to determine if they should be qualified as a preferred supplier for the graphical user interface (GUI) designs
- *External third-party audit example*: To perform a gap analysis of the software quality management system to determine the readiness for an ISO 9001 certification audit

The *audit scope* defines the boundaries of the audit by identifying the exact items, groups, locations, or activities to be examined. While the client of the audit is responsible for defining the purpose and scope of the audit, for internal audits, it many times falls to the lead auditor to formally document the purpose and scope statement as part of the audit plan. The scope of the audit is also used to focus the auditors' attention on what they are supposed to be evaluating. Auditors should not actively seek to identify problems outside the scope of the audit. However, if problems outside the scope are identified, they should be reported to the auditee management. If severe or systemic problems could have a major impact on the quality of the software products or services or that could have legal ramifications if not corrected, they should also be reported to the client as a negative observation in the audit report.

The documents that will be evaluated during the audit preparation should be made available at this time as input to the audit. This availability allows the auditors to perform a documentation review before the execution of the audit. For example, suppose an internal audit is being conducted for a project using the organization's software configuration management (SCM) process as the audit criteria, audit-related input documents might include the project's SCM plans and project-specific/tailored SCM processes or work instructions. Note that the level of these input documents depends on the scope of the audit. For example, while the project-level audit in the foregoing example looked at project-level documentation as inputs, for a system-level audit, the input documents would include system-level documentation (project-level documentation might be sampled during the audit execution but would not be requested as input documentation). Quality records are typically not considered inputs into the audit process. However, they are sampled, as appropriate, as objective evidence during audit execution.

Other inputs to the audit may include background information about the auditee's organization and information from prior audits. Background information can help the auditor understand the context in which the audit is taking place. This background can include information about the business strategies and objectives, product information, domain information, and information about the industry (e.g., competitive factors, and regulatory environment). If prior audits have been conducted for the organization, the system, or processes being audited, then the audit reports, corrective action plans, and follow-up information from those audits are also inputs to the current audit.

Audit criteria provide the objective requirements against which conformance and compliance are evaluated. Examples of audit criteria include:

- Written organizational quality policies
- Documented objectives (e.g., budgets, programs, and contracts)

- Customer or organizational quality specifications, standards, processes, or plans
- Product requirements or workmanship standards
- Governmental or regulatory requirements
- Industry standards

Audit Planning

During the planning step, the lead auditor selects the appropriate strategy for the audit. Audit strategies include:

- *Element method audit strategy*: This strategy organizes the audit into manageable tasks based on the various elements of the audit requirements. For example, an audit against the ISO 9001 standard might be divided into elements based on the various requirements, such as the context of the organization, planning, support, operations, performance evaluation, improvement, and leadership. The auditors then examine each element across all organizations or locations being audited.
- *Departmental audit strategy*: This strategy organizes the audit into manageable tasks based on the organization's departmental or functional structure or by physical location. For example, that same ISO 9001 audit could also be structured by departments such as project management, systems engineering, software development, software testing, software quality assurance, software configuration management, and so on. The auditors then examine all of the elements of the standard for each department.
- *Discovery method audit strategy*: In this strategy, also called *random auditing* or *exploratory auditing*, the auditors go into an area and investigate what is currently happening in that area. For example, if the auditors visit a software engineering group and that group is currently conducting a peer review, then the auditors evaluate that peer review. If, instead, they are doing unit testing or writing a requirements specification, then that is what the auditors evaluate.

  While discovery method auditing does examine current work practices and finds just-in-time issues, it does not lend itself to systematic auditing or complete audit coverage.

Based on the strategy selected, the lead auditor estimates effort and staffing needs for the audit. The lead auditor negotiates with the auditor management, who provides the audit's funding, resources, and staffing. The lead auditor then coordinates with the auditee management to establish the detailed audit schedule, makes arrangements to obtain audit-related input documents and information about the auditee organization, and coordinates the audit logistics.

The lead auditor documents the audit plan to formally communicate the audit details to the auditor and auditee management and the audit team. The audit plan is the primary communication vehicle to inform audit stakeholders of the impending audit. According to IEEE (2008), the audit plan describes the:

- Purpose and scope of the audit
- Audited organization (location[s] and management), including business units to be audited
- Software systems, processes, products, projects, or suppliers to be audited
- Audit criteria
- Auditor responsibilities
- Examination activities (document reviews, interviews, activities being observed, records being examined, and so on)
- Resource requirements
- Audit schedule
- Requirements for confidentiality
- Checklists and other tools
- Audit report formats
- Audit report distribution
- Required follow-up activities

The audit plan is a living document and should be updated and progressively elaborated as information is gathered during the audit. According to IEEE (2008), the client should review and approve the audit plans and changes to those plans.

Audit Preparation

Auditors prepare for the audit by studying the input documentation prior to the audit execution. The purpose of this document review is to evaluate the audit input documents against the audit criteria in order to assess them for compliance, completeness, consistency, and effectiveness. The subsequent execution step will then determine if the system, processes, and plans defined in those documents have been implemented as documented and are functioning effectively and efficiently. During the preparation step, the auditors use information from the input documents and audit criteria to prepare checklists, objective evidence-gathering plans, interview questions, and other tools for use during the audit. Standardized checklists, interview questions, or other tools may be used but should be tailored to the needs of each specific audit.

"An audit checklist is the primary tool for organizing an audit" (Coleman 2020). *Checklists* are lists of yes/no questions that correspond to the audit criteria. Each audit criterion (requirement) can be translated into one or more checklist items. Checklists are used to bring structure to the audit and to confirm complete coverage within the audit scope. Each item in the checklist should be precise, measurable (it must be possible to gather objective evidence to determine if that item is being met), and factual. The purpose of a checklist is to guide the gathering of information during the execution step of the audit. Checklists help the auditor remember what to look for and observe in a particular area. The information recorded on the checklists can then be analyzed and used to substantiate audit findings and conclusions. However, checklists are simply tools and should not be used to limit or restrict the execution of the audit. Checklists can be modified or even abandoned during the audit execution as appropriate.

For many types of audits, standardized checklists can be created (e.g., see the standardized checklists for functional and physical configuration audits in Chapter 28 of this handbook). The advantages of using standardized checklists include the following:

- Guiding audit team members, the auditees, and management to help achieve consistency, uniformity, completeness, and continuity of the application of the audit process
- Reducing the effort involved in preparing for each audit
- Providing a basis for analyzing lessons learned from previous audits, which can then be incorporated into those standardized checklists to improve future audits, as part of continual process improvement
- Providing a set of training aids to promote common understanding

Prior to each audit, these standardized checklists should be reviewed and updated, as necessary, to make sure that they reflect any changes made in the standards, policies, processes, or plans since the last audit was conducted. These generic checklists should also be supplemented and tailored to the exact circumstances of each audit. For example, if the corrective actions against prior audit findings are being verified with the current audit, specific checklist items for those actions may be added to the checklist. Another example might be the auditing of small projects, where specific optional processes or process steps do not apply and are tailored. Only processes required for that specific project should be included in its audit checklist. "Optional" or "not applicable" processes should either be identified as such or removed from the checklist.

The auditors should prepare a plan for objective evidence gathering for each checklist item. This plan will be used during the audit execution to gather evidence to determine whether the checklist item has been met. If the auditors decide to use interviewing as the objective evidence-gathering mechanism for one or more checklist items, interview questions should also be prepared in advance. Interview questions should be open-ended to prompt the auditee to do most of the talking during the interview and context-free to not limit the scope of the discussion or signal the expected answer through the question wording.

In addition to preparing for the audit as an auditor, the lead auditor also prepares the audit team by making certain that all of the auditors have received orientation and any necessary training required by the audit. The lead auditor continues to work with the auditee management to keep communication lines open, answer questions, and handle ongoing logistical issues.

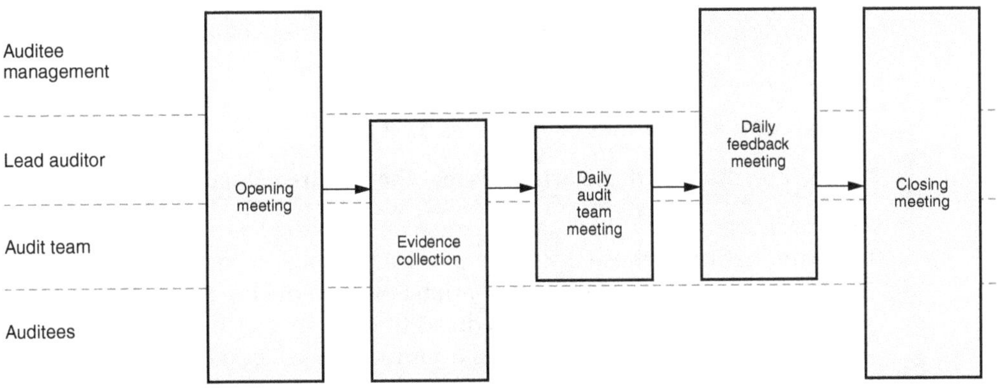

Figure 8.5 Audit execution process.

Audit Execution

The audit execution step is the actual fieldwork of the audit. The audit execution is the data- and information-gathering (objective evidence–gathering) portion of the audit. As illustrated in Figure 8.5, the execution step of the audit starts with an *opening meeting* between the audit team and the auditee organization. The lead auditor conducts this meeting, and at least one member of the auditee management should attend the opening meeting. However, additional members of the auditee management and other auditees and interested parties may attend at the discretion of the auditee management. The objectives of the opening meeting include:

- Introducing the audit team
- Reviewing the conduct of the audit
- Reviewing audit schedule and logistics
- Making sure the auditee understands what to expect from the audit

Any known issues, problems, or adjustments that need to be made to the audit schedule or logistics should be handled at this meeting through consensus between the auditee management and the audit team. For example, if interviewees are unavailable at the scheduled time, the audit schedule can be adjusted. If there are any disagreements, required clarifications, or questions from the auditees, these should also be addressed at this time.

Most of the time and effort of the audit execution step is spent gathering objective evidence. "The job of the auditors is to collect factual information, analyze and evaluate it in terms of the specified requirements, draw conclusions from this comparison, and report results to management" (Arter 1994). *Objective evidence* is information that can be proven true based on facts obtained through observation, document review, measurement, test, or other means. One of the reasons that the independence of the auditor is so essential to the audit is that it helps make sure that the observed or documented evidence is uninfluenced by prejudice, emotion, or bias. Techniques for gathering objective evidence include:

- *Examining quality records*: Quality records provide the evidence that:
  - The products conform to the requirements
  - Appropriate activities or processes took place
  - The execution of those activities met the required standards
- *Reviewing documents*: If the documents were supplied as audit inputs, an examination of those documents was accomplished during the preparation step. Based on information obtained during the audit execution, it may be necessary to do additional reviews of those documents supplied as input or to examine other documents requested during the audit execution.
- *Witnessing an event or process*: An auditor can witness or observe work-in-progress events taking place, or a process being implemented, to see if what is actually happening meets the audit criteria or the documentation. This involves witnessing how work is actually being performed by watching or being present without participating actively in the event or process. The auditor observes to determine that:
  - The product was made or activity performed according to documented procedures or work instructions
  - Activities were performed by the designated responsible person
  - The proper equipment, environment, and tools were used
  - Participants were familiar with policies and procedures, and they knew their roles and responsibilities (based on Russell 2013)
- *Finding patterns*: The auditor can examine data or software metrics and look for patterns or trends. For example, the auditor could review problem report data to look for error-prone components or common root causes.
- *Interviewing auditees*: The auditor can interview auditees to obtain information. However, a single person saying something is true does not make it objective evidence. The auditor must seek corroboration through more than one auditee saying the same thing or through quality records, documentation, observation, or other methods.
- *Examining physical properties*: The technique of examining physical properties for objective evidence is used most often in product-type audits. While this technique is used widely in manufacturing, hardware, and some service industry audits, where physical products or outputs are more prevalent, it is less applicable to software-type audits. It could, however, still be used to examine the properties of items such as screen displays, compliance with coding standards, or media labeling.

When auditing an organization or team that uses agile methods, there may not be as many quality records or documents available as when auditing traditional software development methods. Therefore, the auditors may have to depend more heavily on other objective evidence-gathering techniques.

There is never enough time to examine all possible objective evidence during an audit. For example, there may be hundreds of source code modules or thousands of sets of meeting minutes (from peer reviews, status meetings, change control board meetings, and so on) or other quality records. Audits are always based on sampling from all of the available objective evidence. Random samples should be taken whenever possible, so that results can be generalized to the entire population.

Tracing is an example of one approach for gathering objective evidence. Tracing follows the chronological progress of an item as it is being processed. Tracing can begin at the beginning, middle, or end of the process. Then an action is chosen, and work products are sampled from that action. For example, the auditor could start in the middle of the software development process and select the action of coding a module. The auditor could then randomly sample specific source code modules from the list of modules that have been coded. The auditor gathers information about the sampled items in five areas:

- *Labor*: Information about who coded the module(s), the experience or qualifications of the coder, how much effort was expended, or what the estimated effort was for each module
- *Equipment*: Information about the development platform and tools
- *Methods*: Information about coding standards, code inspection processes, naming conventions, or source control processes used
- *Environment*: Information about the quality of the environment where the coders work (e.g., noise level, interruptions)
- *Measurements*: Information about what measurements were collected as part of the coding process

The auditor can then trace the sampled modules forward into integration test, system test, and release, looking at items such as the integration methods, change control, and problem reports against the modules. The auditor can also choose to trace the modules backward and ask to see the associated detailed design for each module, the higher-level design that specified each module, or the tracing of each module back to its associated requirements. The benefit of tracing is that if a *random sampling* of items can be traced forward or backward through the process, there is a high confidence level that all of the other items can be traced.

Each day of the audit, the audit team should get together in a *daily audit team meeting* to share gathered objective evidence, tentative conclusions, and problems. Based on the progress made during audit execution, these meetings may also be used to replan the audit activities. For example, the team may decide that additional data need to be gathered to clarify incomplete or contradictory results, or delays in one area of the audit can result in the shifting of assigned activities between auditors. Even if only a single auditor is performing the audit, that auditor should take time each day to consolidate information and plan further activities as appropriate. During the last day of the audit, the daily team meeting is used to create the draft audit report that will be presented during the closing meeting.

Utilizing the objective evidence obtained during the execution phase allows audit conclusions and results to be fact-based.

After the audit team meeting each day, a *daily feedback meeting* (also called a *daily briefing*) is held to update the auditee management on the information collected so far and on any potential nonconformances, problems, or areas of concern. The goal here is "no surprises." These meetings also allow the auditee management to provide additional objective evidence to clarify misconceptions and prevent incorrect information from being included in the audit report.

Holding a *closing meeting* completes the audit execution step. The auditee management should attend this meeting with members of the audit team and other interested stakeholders, including auditees. The lead auditor conducts the closing meeting by starting with a summary of the audit and then presenting the audit results in detail. An open discussion about the results follows this presentation. The lead auditor or audit team members responds to any questions and clarifies information as necessary. This discussion also gives the auditee management a final opportunity to "point out any mistakes with respect to the facts that have been collected" (Juran 1999). During the closing meeting, the lead auditor also explains the requirements for the corrective action response to any nonconformances or noncompliances identified and the associated follow-up activities. The objective of the closing meeting is to confirm that the auditee management clearly understands the audit results so they can start working on corrective actions.

Audit Reporting

During the reporting step, the audit results are formally documented in the audit report. The *audit report* contains (based on Russell 2013):

- Introduction:
  - Purpose and scope of the audit
  - Auditee, client, and auditing organization
  - Audit team members and lead auditor
  - Audit dates and locations
  - Standards used for the audit
  - A statement that qualifies the results due to the sample taken
  - A discussion that addresses any confidentiality issues associated with the audit
  - Auditee personnel involved in the audit (optional)
  - Audit report distribution
- A summary of the audit results, including:
  - A synopsis of the findings
  - The audit team's evaluations of the extent of the auditee's compliance to the audit criteria

  - The audit team's assessment of the auditee's ability to achieve their objectives based on the current systems, processes, or plans and their actual implementation
  - The effectiveness and efficiency of the system or processes (first-party system or process audits only)
- The detailed audit findings, including:
  - A *major nonconformance* is a significant breakdown of a system, control, or process; a nonfulfillment of specified requirements or audit criteria; or a number of minor nonconformances related to the same requirement summarized into a single major nonconformance.
  - A *minor nonconformance* is an isolated or single observed lapse in an audit criterion (requirement) or part of an audit criterion not being met. Nonconformances or noncompliances are always within the scope of the audit.
  - An *observation* is anything else that was observed by the auditors that is important enough to be communicated to the client and that is not a nonconformance or noncompliance. *Positive observations* contribute to the product, system, process, or project quality. *Negative observations* detract from the product, system, process, or project quality and if not addressed may result in future issues. Any observed potential nonconformance outside the scope of the audit is also reported as a negative observation.
  - *Process improvement opportunities* are identified areas where proactive steps can be taken to prevent future issues or improve the effectiveness or efficiency of a system, process, or product.
  - *Best practices* are identified areas where the auditee's practices are worthy of being brought forth as examples that other teams, groups, or projects within the organization could use as benchmarks to improve their practices.
- Mechanisms for recording corrective actions and follow-up activities

Audit Corrective Action and Verification Follow-up

For each nonconformance or noncompliance in the audit report, the auditee management needs to make sure that corrective actions are planned and implemented. The auditee management may also decide that corrective actions need to be planned and implemented for negative observations and process improvement opportunities. The auditee management may either do this planning and implementation themselves or delegate those activities to others. The corrective action plan should include specific actions with schedules and responsibility assignments for each action. If actions have already been taken to correct the nonconformance or noncompliance or observation, these should also be documented in the corrective action plans. If the auditee management determines that corrective action is not appropriate after additional review of the audit findings, they should document the reasons why no action is necessary in their response to the lead auditor.

Upon receipt of the auditee management's corrective action response, the lead auditor coordinates the review of the corrective action plans. This may be accomplished by the lead auditor reviewing the corrective action plans, by the audit team reviewing those plans, or, if necessary, by using a technical expert to review those plans. This review can help prevent the auditee's organization from wasting time and resources implementing corrective actions that will not eliminate the nonconformance or noncompliance or will not solve the problem.

If the corrective action plans, or justifications for why no corrective actions were needed, are acceptable, the lead auditor informs the auditee. If one or more of the proposed corrective actions or justifications are unacceptable, the lead auditor contacts the auditee management and resolves the issue. If the lead auditor and auditee management cannot resolve the issue, it can be escalated to the client.

As the auditees complete the implementation of each corrective action, they inform the lead auditor. The lead auditor then coordinates the verification of each implemented corrective action:

- Through written communications with the auditee
- By reviewing the revised documentation or new quality records
- By conducting a follow-up audit or by re-auditing against the associated findings during a future audit

If the corrective action is completed before the end of the audit, the follow-up on that corrective action's implementation may be completed as part of that same audit, and the audit report can include both the reporting of the nonconformance or noncompliance and the fact that a corrective action was implemented and verified.

The lead auditor confirms that the actions taken to verify the corrective action are documented. This documentation becomes part of the quality records for the audit.

Each corrective action is closed as its successful implementation is verified. Each nonconformance or noncompliance is closed when all of its associated corrective actions are verified and closed. The audit is closed when all nonconformances or noncompliances for that audit are closed.

Audits, like any other process, should be auditable. This means that quality records related to the audit must be collected and retained. Based on the needs of the auditing organization, the required quality records for an audit should be included in the audit process definition. Examples of quality records maintained for an audit might include:

- The audit plan
- Opening and closing meeting minutes, including attendance lists
- The audit report
- Corrective action, follow-up, and closeout documentation for each nonconformance or noncompliance
- Optionally, completed audit working papers, such as checklists and auditor notes

Part III

System and Software Engineering Processes

Chapter 9
Life Cycles and Process Models

Software development life cycles (SDLC) and *process models* are high-level representations of the software development process. These models define the stages (phases or iterations) through which software development moves, and the activities performed in each of those stages. Each SDLC model represents one software project, iteration, or increment, from conception until that version of the product is completed or released.

Software product life cycle models represent the software product's life, from conception until the product is retired from use. For a successful software product, the product life cycle usually encompasses multiple passes through the software development life cycle or process model.

Life cycle and process models provide the framework (highest-level process architecture) for the detailed definitions of the individual software development processes. Defined models act as a road map for development teams to make sure that all necessary steps are implemented during software development to produce a high-quality product. Following predefined models and their associated processes helps confirm that critical steps are included to improve quality. These models depict interrelationships between major milestones, baselines, and project deliverables. These models can help project managers and teams plan activities and track progress by breaking the development effort into stages, each with a defined set of activities, including stage transition reviews. These models also provide a consistent framework for:

- Gathering lessons learned
- Implementing process improvements
- Comparing the results across multiple projects
- Establishing metrics
- Providing useful organizational process assets that can be used to better estimate and plan future development, maintenance, and acquisition projects

These models continue to be refined over the years through experience and research. Organizations can learn from these refinements and then adopt and adapt their own life cycle and process models to match their needs, business domain, culture, and approach to software development.

1. WATERFALL SOFTWARE DEVELOPMENT LIFE CYCLES

> Apply the waterfall life cycle and related process models and identify their benefits and when they are used. (Apply)
>
> **Body of Knowledge III.A.1**

Waterfall Model

The *waterfall model* is based on the concept that software development can be thought of as a simple sequence of phases. Each phase proceeds from start to finish before the next phase is started. The work products produced in one phase in the waterfall model are typically the inputs into the subsequent phases. The premise of the waterfall model is that a project can be planned before it is started and that it will progress in a reasonably orderly manner through development. In the waterfall model, a well-defined set of requirements is specified before the design begins, the design is complete before coding begins, and the product is tested after it is built.

Figure 9.1 illustrates an example of the waterfall model. This example shows six phases, but on an actual project, the number of phases depends upon the needs of that project and the organization conducting the project. Some waterfall models include only the downward arrows, depicting the flow of activities downward through the development project. This example also includes upward arrows to indicate that some iteration is allowed in the development activities.

Many critics of the waterfall model say that it is antiquated and does not reflect what truly goes on in software development. It should be remembered, however, that a model is an abstract representation of an item or process from a particular point of view. A model is a simplification meant to express the essentials of some aspect of the item or process without giving unnecessary detail. The purpose of a

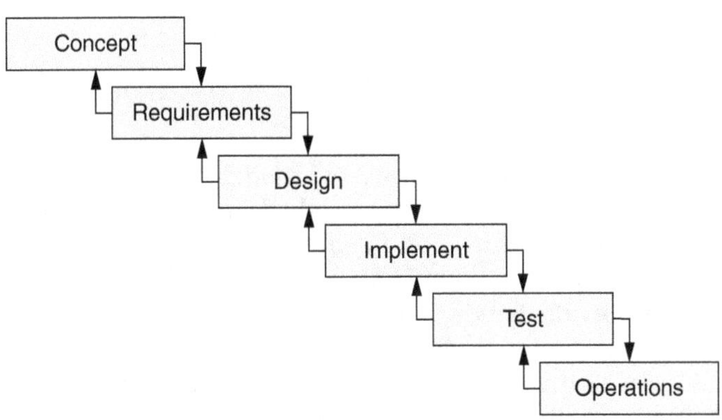

Figure 9.1 Waterfall model—example.

model is to enable the people involved to think about and discuss these essential elements without getting sidetracked by all of the detail. In this respect, the waterfall model is still a useful tool because it is easy to understand and discuss. Even the most unsophisticated stakeholders can understand the basic concepts of the waterfall model. The waterfall model removes almost all of the detail, including most of the iterations that typically occur during development. However, many of those details may be accounted for in lower-level process definitions under the model.

The waterfall model was the first model to define a disciplined approach to software development. It has been thoroughly studied, is easy to understand, and is well defined. One of the strengths of the waterfall model is that it correlates directly to the deliverables of software development, which can aid in project management activities. Many tools exist to support the waterfall model.

A significant weakness of the waterfall model is that most, if not all, requirements must be known at the beginning of development. The waterfall model does not readily accommodate change. The software product is also not available for use until the project is complete or nearly complete. Hence, the model includes no provisions for intermediate feedback from stakeholders. However, prototypes may be used to provide early feedback during waterfall-based development.

The waterfall model is most appropriate for projects where the requirements are well known and expected to be stable, and the development process is expected to progress in an orderly, disciplined manner. For example, the waterfall model might be appropriate for a project to port an existing product to a new platform, environment, or language where the constraints are well known. Other examples might include enhancement projects to update the software to adhere to new government regulations or automating a report currently being implemented manually. The waterfall life cycle model may also be appropriate for projects with only a few experienced programmers and many junior programmers or new development projects in very mature domains where yet another software product, just like the last one, is being built. The waterfall model would typically not be appropriate for projects where the requirements are fuzzy or expected to have high volatility or projects that are not expected to progress in a linear manner. For large, high-risk projects, a more risk-based approach, such as the spiral model or incremental development, may be more appropriate than the waterfall model.

V-Model

The *V-model* is a variation on the waterfall model that highlights the relationship between the testing phases and the products produced in the early life cycle phases, as illustrated in Figure 9.2. In this example, acceptance testing evaluates the software against the stakeholders' needs, as defined in the concept phase, and *system testing* evaluates the software against the product-level requirements specified during the requirements phase, and so on. Test planning and design can start as soon as the corresponding early development phase is significantly completed. For example, once a significant number of stakeholder requirements (concept, in the example) are defined, acceptance test planning and design can be initiated. As soon as a significant number of product-level requirements are defined, system test planning and design can be started.

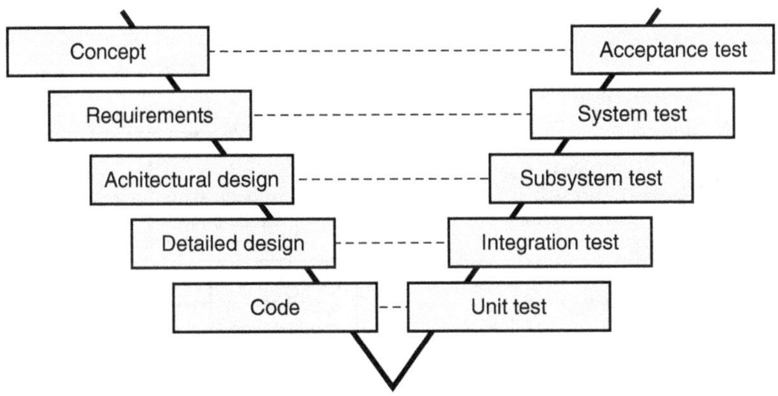

Figure 9.2 V-model—example.

W-Model

The *W-model* is another variation of the waterfall model, as illustrated in Figure 9.3. The W-model has two paths (or crossing Vs), each representing the life cycle for a separate organization or independent team during development. The first path represents the development organization or team responsible for developing the requirements, design, and code. The second path represents the independent verification and validation (V&V) organization responsible for independently analyzing, reviewing, and testing the work products of each phase of the software life cycle.

2. INCREMENTAL AND ITERATIVE SOFTWARE DEVELOPMENT LIFE CYCLES

> Apply the incremental and iterative life cycles and related process models (e.g., rapid application development methodology) and identify their benefits and when they are used. (Apply)
>
> **Body of Knowledge III.A.2**

Spiral Model

The *spiral model,* initially defined by Boehm (1988), is a risk-based process model that expands on previous models with details, including exploring alternatives, prototyping, and planning. As illustrated in Figure 9.4, the spiral model subdivides each software development cycle into four quadrants. The life cycle activities are then incorporated into these four quadrants in a spiral fashion, starting in the

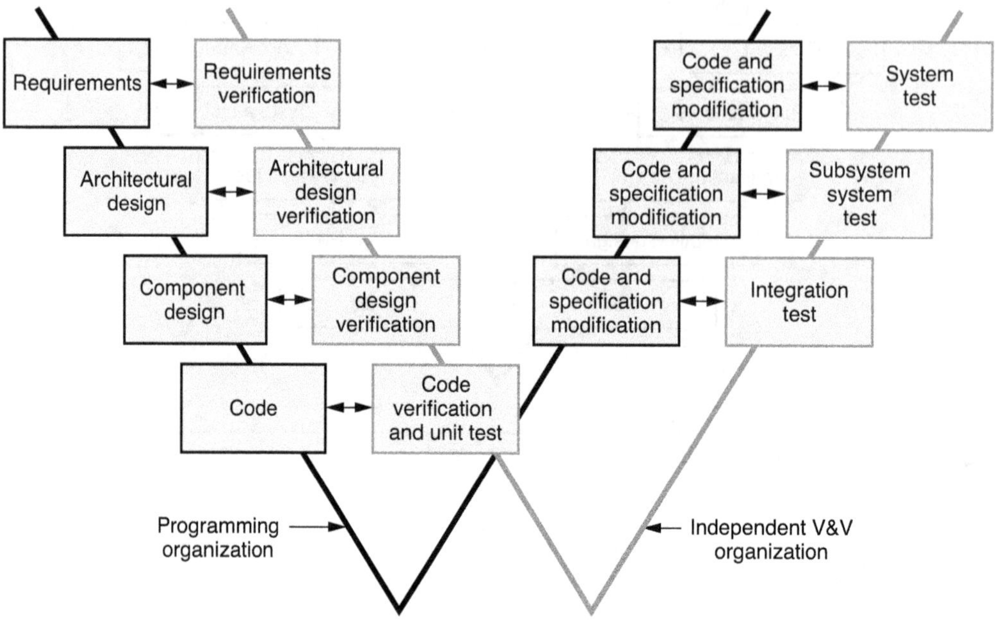

Figure 9.3 W-model—example.

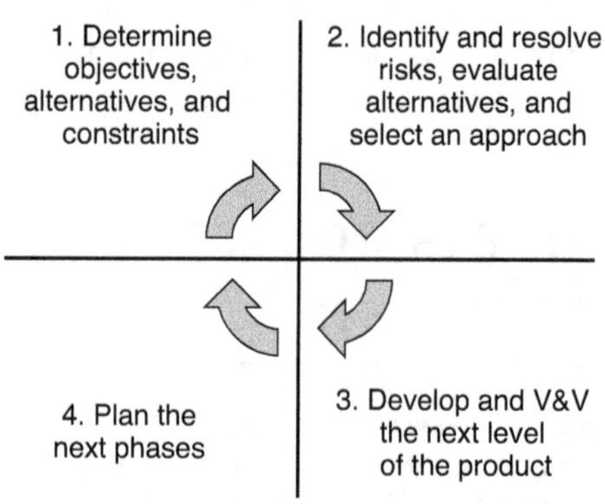

Figure 9.4 Spiral model steps.

middle with the concept phase, as illustrated in the example in Figure 9.5. The four quadrants are:

1. *Determine objectives, alternatives, and constraints*: The development team determines the objectives cycle's, alternatives, and constraints. For example, for the concept phase, this might include exploring buy-versus-build options or determining which business-level or stakeholder-level requirements should be addressed during the project.

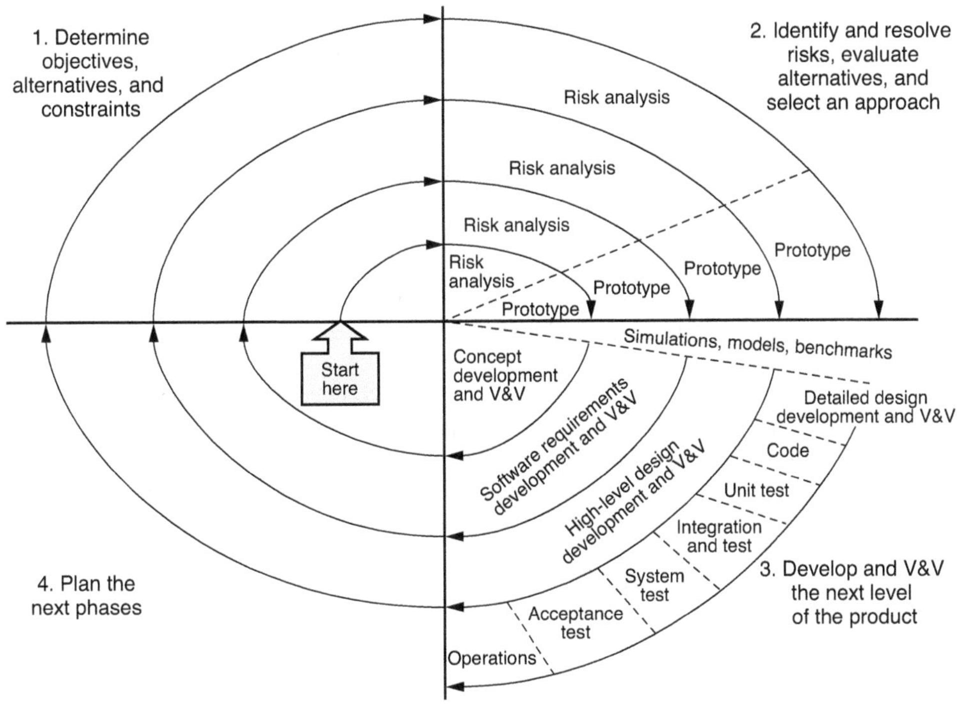

Figure 9.5 Spiral model—example.

2. *Identify and resolve risks, evaluate alternatives, and select an approach*: Alternatives are evaluated, risks are identified and analyzed, and prototyping may be done to select the approach that is the most suitable for this cycle of the project. For example, for the concept phase, this might include *risk identification* and analysis for the buy-versus-build option, a cost/benefit analysis of the alternatives, and prototyping to evaluate the choices to determine if building the software in-house is the appropriate decision.

3. *Develop and V&V the next level of the product*: Develop the software products for that phase and perform V&V activities for those products. If appropriate, this may also include creating simulations, models, or benchmarks. For example, for the concept phase, this might consist of eliciting, analyzing, and specifying the business-level and stakeholder-level requirements and performing V&V activities to validate those requirements.

4. *Plan the next phase*: During these first three activities, decisions are made, and additional information is obtained that will impact the project plans. Therefore, in the fourth activity of the spiral model, the project plans are progressively elaborated with further details for subsequent cycles. This fourth step typically ends with a review or other cycle transition activities. For example, this might include using

the information and decisions from the first three activities of the concept phase to update the project plans for the requirements phase and other subsequent phases.

These four quadrants are then repeated for the requirements and the high-level (architectural) design cycle. With the completion of the first two activities of the detailed design cycle most of the major decisions for the project have been made, and the spiral model finishes the third activity, which includes detailed design specification and V&V, coding and code V&V, unit test, integration and test, system test, acceptance test, and operations. As with all of the other software development life cycle models, the spiral model can be tailored to the needs of the project and organization.

The spiral model shifts emphasis from product development to risk analysis and avoidance. The spiral model focuses on exploring options early by obtaining feedback through prototyping to verify that the right product is being built appropriately. The spiral model also includes mechanisms for handling change through the iteration of any given activity or cycle as many times as necessary before moving on to the next cycle. The spiral model incorporates solid project management techniques by emphasizing project plan updates as more information is obtained.

Boehm et al. (2000) list six characteristics common to successful applications of the spiral model:

1. "Concurrent rather than sequential determination of artifacts
2. Considerations in each spiral cycle of the main spiral elements:
    - Critical-stakeholder objective and constraints
    - Product and process alternatives
    - Risk identification and resolution
    - Stakeholder review
    - Commitment to proceed
3. Using risk considerations to determine the level of effort to be devoted to each activity within each spiral cycle
4. Using risk considerations to determine the degree of detail of each artifact produced in each spiral cycle
5. Managing stakeholder life cycle commitments with three anchor point milestones:
    - Life Cycle Objectives (LCO): The stakeholders agree that the technical and management approach to this project are adequately defined before this milestone is passed
    - Life Cycle Architecture (LCA): The stakeholders agree that the selected solution will adequately meet their needs before this milestone is passed

- Initial Operational Capability (IOC): The stakeholders agree that the product is ready to launch, including site preparation and the readiness of the users, operations, support staff, and maintainers

6. Emphasis on activities and artifacts for system and life cycle rather than for software and initial development"

The spiral model is complex and is not well understood or easily grasped by some stakeholders. Therefore, stakeholders may find it challenging to communicate and use. The spiral model requires a high level of risk management skills and analysis techniques, which makes it more people-dependent than other models. This also means the spiral model requires a strong, skilled project manager. One of the spiral model's weaknesses is the fact that it does not correlate as well to the needs of software development done under contract (e.g., mapping to controls, checkpoints, and intermediate deliveries) as the waterfall-based models.

The spiral model is appropriate for projects where the software development approach is nonlinear or contains multiple alternative approaches that must be explored. For smaller, low-risk, straightforward projects, the extra risk management, analysis, and planning steps may add unnecessary cost and effort. Because of the extensive flexibility and freedom built into the spiral model, fixed-cost or fixed-schedule projects may not be good candidates for using this model. The spiral model may also not be appropriate for projects with less-experienced staff.

Iterative Model

The *iterative model* is a software development model where steps or activities are repeated multiple times. This may be done to add more and more detail to the requirements, design, code, or tests, or it may be done to implement small pieces of new functionality, one after another. There are many different iterative models. For example, the spiral model discussed previously can be implemented as an iterative model, and the test-driven development and feature-driven development methods described in the following Agile Software Development Life Cycles subsection are also iterative models. Figure 9.6 illustrates a generalized example of the iterative development model demonstrating the principle of flow. In this example, a development cycle starts with short-term, focused planning, which defines the functionality to be implemented in that iteration. Each cycle consists of designing, writing tests, developing code, executing tests, and integrating successfully developed code into the "baseline" for that cycle and then more designing, writing tests, and so on, cycling through the loop. This continues with feedback (from other developers, the customer, users or other stakeholders, and the software itself) occurring at all points until the functionality is completed within the time frame allocated to the increment. That increment's output is then released to stakeholders who may try it out and put it into operations or just provide feedback for the next development cycle (iteration).

The benefits of the iterative models include needing to know only some of the requirements up front. Well-defined requirements can be implemented in the software, while fuzzier requirements are being investigated or stabilized. New requirements can be added to future iterations as they are identified. Breaking high-risk products or projects into smaller pieces can also help reduce risks.

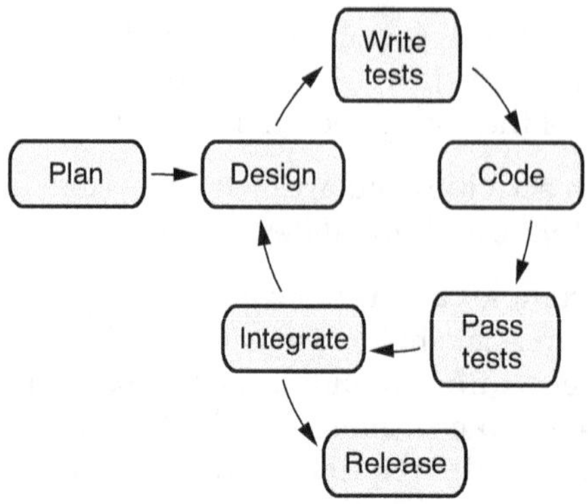

Figure 9.6 Iterative model—example.

The continuous feedback loops built into the iterative models provide cycles of learning that help eliminate the propagation of similar errors into other parts of the product and allow quality to be built into the product with more confidence.

Incremental Development

Incremental development is the process of constructing increasingly larger subsets of the software's requirements through the use of multiple passes through software development. In incremental development, after the requirements have been determined, they are prioritized and allocated to planned increments, each of which is one pass through software development, as illustrated in Figure 9.7. Each subsequent version or release is usable but only has part of the functionality (except the last delivery, which includes all of the requirements). Each increment can have its own software development life cycle model (e.g., waterfall, V, iterative). There is no requirement to use the same model for each increment. For example, the first two increments could use the waterfall model, and the next increment could switch to an iterative model.

Incremental development can be done sequentially, where one increment is completed before the next increment starts, as illustrated in increments 1 and 2 in Figure 9.8. The increments may also be done in parallel, as illustrated by increments 2 and 3 in Figure 9.8. For example, once the development team has completed the coding process and turned the software over to the test team for increment 2, they can start development of increment 3.

One of the strengths of incremental development is that building smaller subsets is less risky than building one large system. This allows customers and users to receive or evaluate early versions of product, which contain their highest priority operational needs. This provides opportunities for validation and feedback of the delivered software much earlier than traditional waterfall-based projects. Incremental development also accommodates change very well. If new or changed requirements are discovered during the development of an increment, they can be allocated to future increments without disrupting the schedules and plans for the current development cycle. However, if the new or

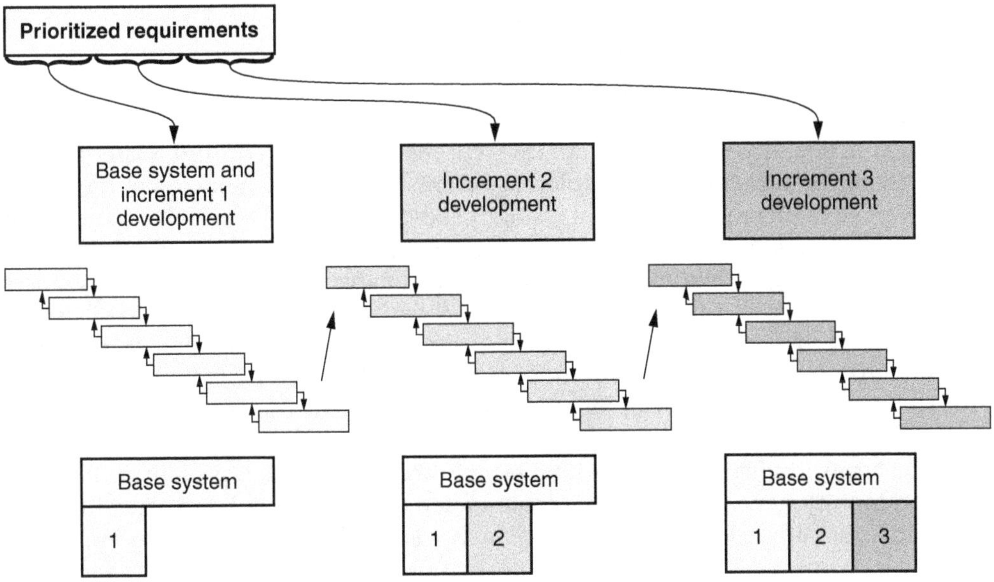

Figure 9.7 Incremental development process—example.

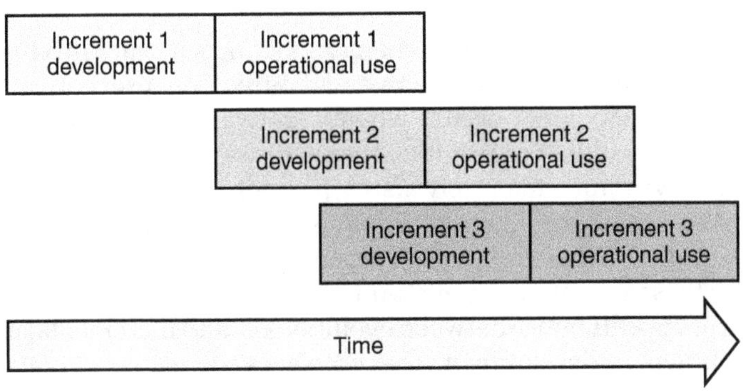

Figure 9.8 Incremental development over time—example.

changed requirements are of high enough priority that they must be added to the current release, then either unimplemented requirements with lower priorities can be slipped to the next increment, or updated schedules and other plans may need to be negotiated.

One weakness of the incremental development model is that most of the requirements must still be known up front. Depending on the system and the development methodology the project is using, additional work may be needed to create an architecture that can support the entire system. That architecture must be open enough to accept each new increment of functionality as it is implemented. Each released increment must be a complete working system, even though it does not include all the intended functionality. Therefore, incremental development is sensitive to how the content of a specific increment is selected if it is intended for release into operations. For example, it would not be feasible to release an inventory control system increment that allowed the checking-in of inventory if the functionality of checking-out inventory is not scheduled until a later release. Incremental development also places portions of the product under configuration control earlier, requiring formal change procedures, with the associated increased overhead, to start earlier. Finally, one of the benefits of incremental development is the ability to get high-priority software functionality into the hands of customers or users faster and thus get their feedback earlier. However, this may also be a weakness if the software is of poor quality. In other words, the development organization may be so busy fixing defects in the last increment that they have no time to work on new development for the next increment. Incremental development does not require an iterative product life cycle, but they are often used together. The terms *iterative development* and *iteration* are being used, especially in the agile community, generally to mean a combination of an incremental product life cycle and an iterative development life cycle. For example, as illustrated in Figure 9.9, the extreme programming principle of flow combines iterative development cycles with incremental product development where the outputs of one iteration become the inputs into the next, and so on.

Evolutionary Development

Evolutionary development happens when an existing software product in operation is updated to implement perfective, adaptive, or preventive maintenance. Evolutionary development occurs when a software product is successful. If the customers or users like the product, they will want to continue using it, but the operational environment is rarely static. Technologies change, stakeholder needs and priorities change, business domains change, standards and regulations change, and so on. Therefore, smart organizations plan their software strategies to consider these potential changes and plan for evolutionary development.

The primary difference between evolutionary and incremental development is that in evolutionary development, the complete system with all of its requirements has been in operational use for some time, as illustrated in Figure 9.10. Like incremental development, evolutionary development builds a series of successively different software versions. However, with evolutionary development, all of the original requirements are built into the first evolution of the software released into operations. Incremental development techniques could be used to build any of the software evolutions using the legacy software as input. Each evolution can use the same or a software development model different from its predecessor evolutions, as illustrated in Figure 9.11.

One of the strengths of the evolutionary model is that it focuses on the long-term success of the software as it changes and adapts to its stakeholders' needs and other changes that occur over time. Only the requirements for the current evolution are known, but this approach provides opportunities for customer/user validation and feedback as releases are delivered. Product evolutions can be sold to fund further development and provide profit to the organization. In fact, these evolutions are many times "cash cows" for the development organization, as they

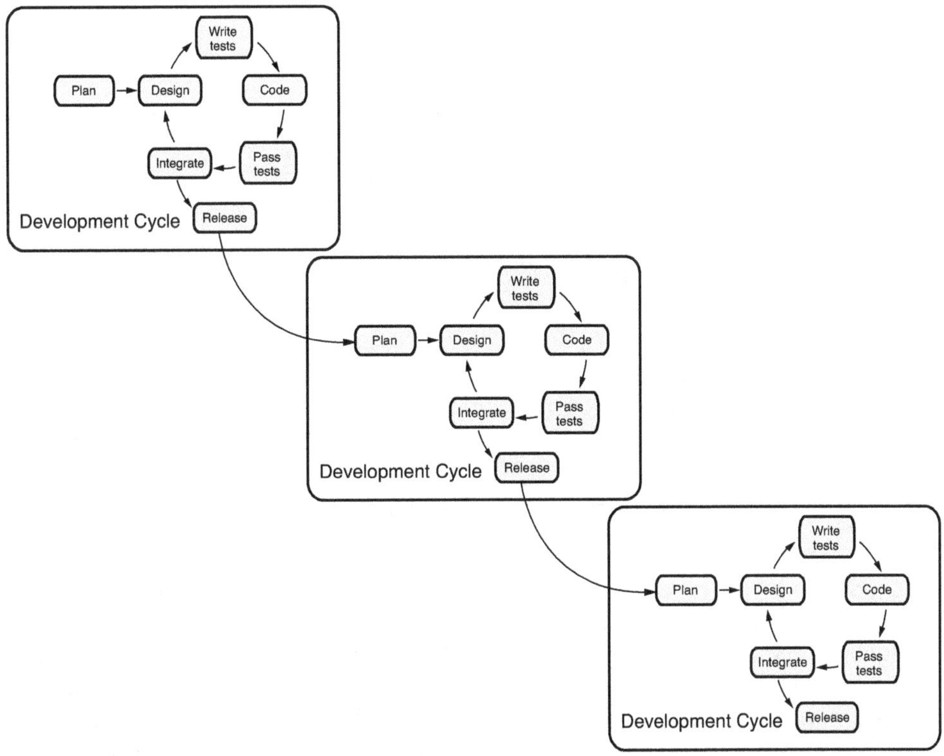

Figure 9.9 Combination of iterative and incremental models—example.

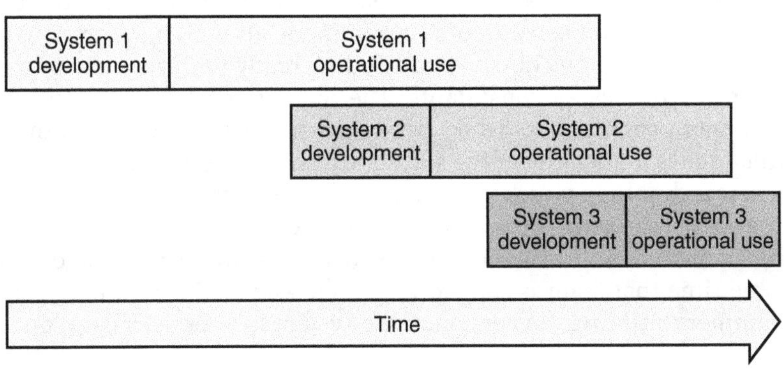

Figure 9.10 Evolutionary development over time—example.

receive funding for new and updated software without the investment necessary to create an entirely new system from scratch.

One weakness of the evolutionary development model may be the software's success. The longer the span of time between evolutions, or the more evolutions there are, the higher the probability that knowledgeable people have moved on

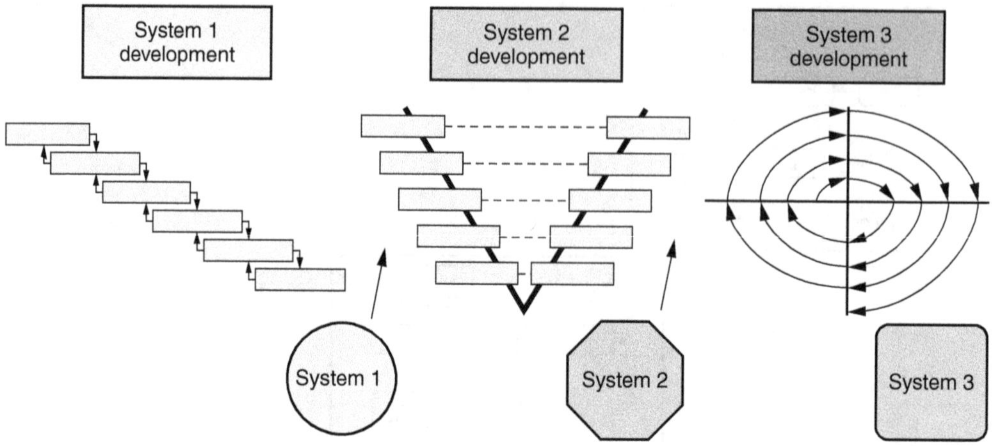

Figure 9.11 Evolutionary development process—example.

to other projects or organizations. Effort, expense, and other project attributes can be very difficult to estimate for long-term support and future development needs for requirements that the organization will not even know about for months or even years. Evolutionary development will only be successful with strong customer/user involvement and input. There is also the risk (and potential benefit) of never-ending evolutions. It is not unusual for software built decades ago to still be operational. At some point, it may become difficult to find staff with the appropriate skill sets, or the hardware and other technologies, to support these very old software systems. For example, the author knows of a company that is still supporting an old legacy payroll system written in COBOL in the 1960s. Their product support team consists of programmers, all with ages ranging from their late 50s to mid-80s, many of whom are getting ready for retirement or have retired and been hired back as consultants. The management of this company cannot find younger programmers who want to learn COBOL. Of course, the answer to this dilemma is to reengineer the software into a modern programming language using new technology, which is the significant investment that management is now making so that the software will be supportable into the future. They are currently paying two development teams, one of which is supporting the old system. At the same time that team is transferring their knowledge to a new generation of programmers building the version of the system in a modern language.

3. AGILE SOFTWARE DEVELOPMENT LIFE CYCLES

> Apply the agile life cycle and related process models (e.g., lean, agile, and Scrum methodology) and identify their benefits and when they are used. (Apply)
>
> **Body of Knowledge III.A.3**

Agile Methods

There are many agile methods and frameworks, including Scrum, extreme programming (XP), test-driven development (TDD), *feature-driven development (FDD)*, crystal, lean, Kanban, and others. Many of these methods focus on different aspects of software development and are complementary. The term *agile* or *agile development* is typically used generically to describe one of these methods, or the merging of several of these methods, adopted and adapted by an organization to meet its needs and the needs of its team(s), projects, and stakeholders.

The *Agile Manifesto*, illustrated in Figure 9.12, was developed at a meeting in Snowbird, Utah, in 2001. This was the first meeting of what later became known as the *Agile Alliance*. It is important to emphasize the phrase "there is value in the items on the right" since some criticisms of agile methods make it sound like such methods reject processes, plans, contracts, and documentation. The word "over" is key. The Manifesto does not say "instead of," which many critics of agile methods seem to imply.

The Agile Alliance (www.agilealliance.org) has also defined the following 12 principles to support the Agile Manifesto:

1. "Our highest priority is to satisfy the customer through early and continuous delivery of valuable software.
2. Welcome changing requirements, even late in development. Agile processes harness change for the customer's competitive advantage.
3. Deliver working software frequently, from a couple of weeks to a couple of months, with a preference to the shorter timescale.
4. Business people and developers must work together daily throughout the project.
5. Build projects around motivated individuals. Give them the environment and support they need, and trust them to get the job done.

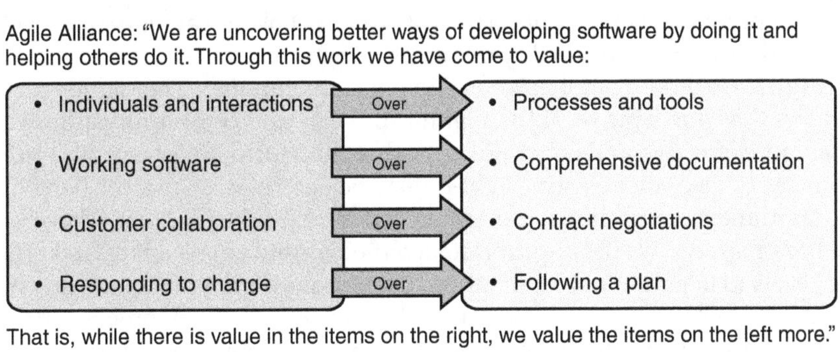

Figure 9.12 Agile Manifesto.

6. The most efficient and effective method of conveying information to and within a development team is face-to-face conversation.
7. Working software is the primary measure of progress.
8. Agile processes promote sustainable development. The sponsors, developers, and customers or users should be able to maintain a constant pace indefinitely.
9. Continuous attention to technical excellence and good design enhances agility.
10. Simplicity, the art of maximizing the amount of work not done, is essential.
11. The best architectures, requirements, and designs emerge from self-organizing teams.
12. At regular intervals, the team reflects on how to become more effective, then tunes and adjusts its behavior accordingly."

Plan-driven software development methodologies focus on defining the "rules" for how procedures and methods, people, tools, and equipment interact, as illustrated in the triangle on the left in Figure 9.13 from the Capability Maturity Model Integration for Development (CMMI 2010a). In plan-driven methodologies, there is typically an emphasis on technology solutions to process issues.

The triangle on the right in Figure 9.13 shows the relationships emphasized in most agile methods. The primary focus is on communication between the stakeholders, developer, and product because the behavior of the three toward one another matters significantly. Developers must communicate with stakeholders to understand their needs and what is value-added to those stakeholders. The developers then create the product and communicate with that product through reviews and testing to verify that the product matches its requirements. The product is communicated to the stakeholders through frequent prototypes and demonstrations of the as-built product so the stakeholders can validate that the product meets its intended use and works as needed. The stakeholders then communicate any issues and additional needs to the developers, and the feedback continues. This frequent, high-quality communication provides a continuous feedback loop to confirm that the right product is being built correctly to catch problems earlier in the development cycle and to improve the quality of the resulting product.

Agile methods emphasize sociological solutions. The agile community believes that "focusing on skills, communication, and community allows a project to be more effective and more agile." Rather than technology-oriented rules, agile software development uses "light-but-sufficient rules of project behavior" and "human and communication-oriented rules" (Cockburn 2002). These solutions are then paired with the appropriate technologies to enable increased efficiencies (e.g., tools to facilitate test automation, continuous deployment, and so on).

Agile methods are appropriate for software development efforts with "uncertain requirements combined with unpredictable implementation risk." If the development work "depends on knowledge creation and collaboration," agile methods may be more appropriate than plan-driven methods (James 2012).

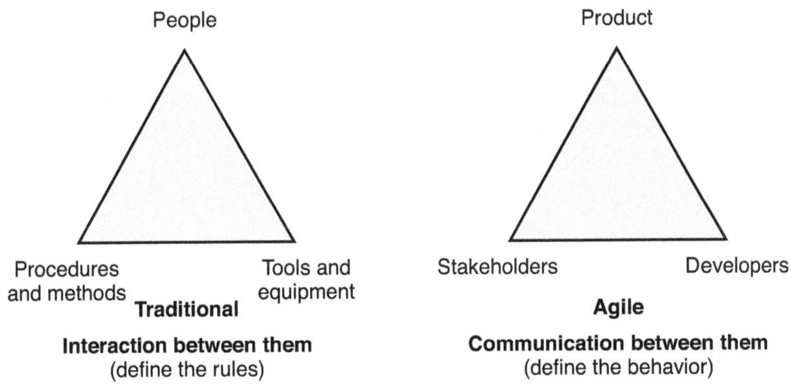

Figure 9.13 Methodology triangles.

Scrum

The term *scrum* originally derives from a strategy in the game of rugby where it denotes "getting an out-of-play ball back into the game" with teamwork (Schwaber and Beedle 2002). Scrum is a software development framework that defines a structure of roles, meetings, work products, and rules. Scrum uses a short (e.g., two weeks or four weeks) incremental development cycle called a *sprint*. The goal of each sprint is to develop a completed increment of deliverable product functionality by the end of the sprint.

Scrum defines three specific roles with responsibilities:

- The *product owner* is responsible for establishing the overall product vision and for the software development effort's return on investment (ROI). Other product owner responsibilities include:
  - Acquiring initial and ongoing funding for product development
  - Representing the other stakeholders by acting as a final arbitrator for requirement questions, decisions, and priorities
  - Managing, controlling, and making the product backlog visible so that each iteration addresses the most valuable requirements remaining in the product backlog
  - Accepting or rejecting the completed version of the software from each sprint and deciding when the product will be delivered
  - Planning the long-term release strategy
  - Deciding when to terminate development
- The *Scrum team*, also called the *Scrum development team*, is a self-managing, self-organizing, collaborative team with the authority to make decisions and take the actions necessary to achieve the goals of each sprint. The Scrum team is cross-functional and diverse. The team

must include members with all of the skills needed to implement the software—requirements, design, code, technical publications, white-box and black-box testing, quality, configuration management, domain or regulatory expertise, and so on. Besides these activities, the Scrum team is also involved in:

- Estimation
- Creating the sprint backlog
- Helping the product owner review and prioritize the product backlog
- Identifying impediments to the development effort

The ideal Scrum team size is from five to nine members. For larger-scale development efforts that require more people to be involved, the Scrum methodology suggests forming multiple Scrum teams and using a *Scrum of Scrums*, made up of representatives from each of the individual Scrum teams, to allow coordination between the teams.

- The *Scrum master* is responsible for making sure that the development effort is conducted according to the practices, values, and rules of Scrum and that it progresses as planned. The Scrum master interacts with the Scrum team, the product owner, and other stakeholders to:
  - Teach the Scrum process and act as an agile coach
  - Protect the Scrum team from outside distractions and interference
  - Help acquire resources and remove impediments
  - Help adopt, adapt, and continuously improve the Scrum process to meet the needs of the Scrum teams and the organization
  - Facilitate the sprint planning meetings, daily Scrum meetings, sprint review meetings, and sprint retrospective meetings
  - Facilitates gathering of team agreements, including agreements on how they will do their work
  - Capture empirical data for tracking progress and determining Scrum team velocity (the team's overall ability to deliver work) during each sprint

In Scrum, the Scrum team is self-managing; management is in charge of final decision-making and establishes the charters, standards, and conventions the project must follow. Management also participates in the setting of goals, objectives, and requirements for the project. Management is involved in selecting the product owner, gauging progress, and reducing the product backlog.

The customer, users, and other stakeholders participate in the tasks related to identifying and refining product backlog items for the system being developed or enhanced. The customers and users receive the delivered software and provide feedback to the Scrum team. Customers may also be involved in selecting the product owner.

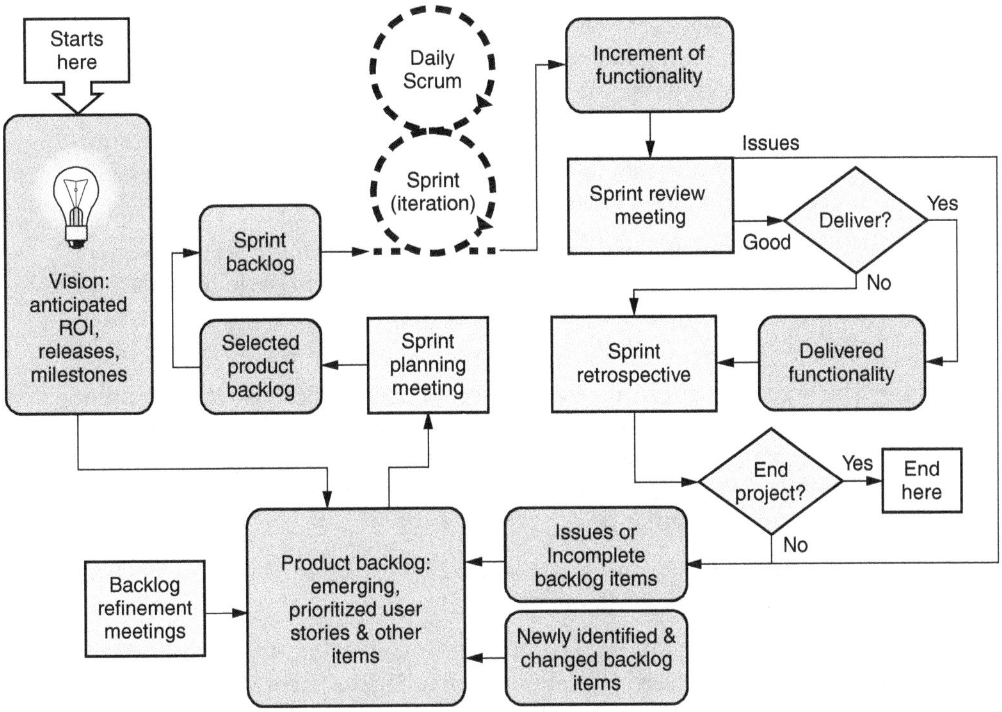

Figure 9.14 Scrum process.

The *product backlog* comprises a prioritized and continuously updated list of business and technical requirements for the system being built or enhanced, often stated as user stories (see Chapter 12 for a discussion of user stories). The product backlog items can include desired features, functions, nonfunctional requirements, bug fixes, defects, requested enhancements, and technology upgrades. The product backlog can also include process improvements.

As illustrated in Figure 9.14, the Scrum process starts with a vision, including anticipated ROI, releases, and milestones. The initial product backlog contains a list of prioritized product requirements. Over time, additional requirements emerge and are prioritized and added to the product backlog. Items on the product backlog may also be changed or updated over time as additional information is obtained or as stakeholder needs change.

Each sprint starts with a sprint planning meeting facilitated by the Scrum master. This meeting is actually two meetings held back-to-back, but many organizations combine these two meetings into a single session. During the first part of the sprint planning meeting, the Scrum team and Scrum master meet with the product owner to:

- Identify a sprint "goal," which is a short description of the overall work to be achieved during the sprint. This goal is used to focus the effort and tie the sprint together conceptually.

- Reprioritize the items remaining in the product backlog based on new information, or any items added, or returned to the product backlog since the previous sprint.
- Obtain answers from the product owner to any questions the Scrum team has about the items in the product backlog, including questions on content, purpose, meaning, and intention.
- Select as many items from the product backlog as the Scrum team believes they can turn into a completed increment of deliverable product functionality by the end of the sprint—creating the *selected product backlog.*
- The Scrum team estimates the selected user stories, using story points or effort. The team compares the estimated total to their performance during past sprints as a way of measuring if they are over- or under-committing the work in the sprint. (Estimation and velocity as discussed in Chapter 16.)
- The Scrum team commits to the product owner that it will do its best to deliver these selected items.

During the second part of the sprint planning meeting, the sprint activities are planned by translating each selected product backlog item into a preliminary set of tasks needed to implement that item. Additional tasks may also be added to conduct the sprint as required. The team may not know all the tasks up front, and additional tasks may be added to the list as the sprint progresses.

During the sprint, the Scrum team turns the selected product backlog into an increment of functionality through the execution of the tasks that translate the sprint backlog items into one or more software products (software build, user documentation, and so on). During the sprint, the Scrum team also holds brief daily meetings called *Scrums* or *daily standup meetings*. Each Scrum team member answers three questions during these daily Scrums:

- What have I worked on since the last Scrum meeting?
- What will I work on before the next Scrum meeting?
- What impediments (issues or problems) do I have?

The output of each sprint is "a potentially shippable product increment" (Schwaber and Cohn 2007), which is working software (tested and complete according to the agreed-to definition of "done") of high enough quality to be shipped, even though it may not have all the functionality desired to actually be shipped. When the Scrum team delivers the new functionality at the end of the sprint, the process includes:

- A *sprint review meeting* with the product owner and any other interested stakeholders to demonstrate the new functionality and obtain feedback. A decision is made on whether to release the product (or this decision may be made during release management planning as discussed in Chapter 29)

- A *sprint retrospective* where the Scrum master and Scrum team evaluate and assess the execution of the last sprint and adjust the team's processes and working agreements as needed to improve for the next sprint

At the end of each Scrum, the product owner decides whether or not there is anything left on the product backlog that has enough value to continue to the next sprint. New requirements may be identified during the sprint or as output from the sprint review meeting based on stakeholder feedback or other stakeholder needs. In addition, there may be incomplete sprint backlog items from the sprint that just ended. These items are prioritized and added back into the product backlog. The next sprint planning meeting is held to start the process again, repeating the cycle until all features are implemented or the product owner decides to terminate the development effort.

The *backlog refinement meeting* is not one of the official Scrum meetings. However, many organizations find these meetings useful to prepare the items in the product backlog for the next sprint planning meeting by:

- Breaking down large stories/functions or *epics* (a *story* that is too large to implement in a single sprint) into smaller stories/functions
- More clearly defining poorly understood or ambiguous stories/ functions
- Having additional conversations with stakeholders to define acceptance criteria for the stories
- Reprioritizing added or refined items

Extreme Programming

Extreme programming (XP) is an agile, iterative development methodology that takes the ideas of agile development "to the extreme." XP includes test-driven development (TDD) techniques where developers first write a test that the current software cannot pass because they have not yet written the corresponding code and then write the code that will pass that test. Using refactoring, the developers then improve the quality of that code, eliminating any duplication, complexity, or awkward structure. This establishes what Beck (2005) calls a natural and efficient "rhythm" of test–code–refactor, test–code–refactor, and so on, as illustrated in Figure 9.15.

XP is based on the following values, many of which are shared across the agile community:

- *Communication:* XP emphasizes communication. According to Beck (2005), "What matters most in team software development is communications."
- *Simplicity:* To avoid commitment to vaguely understood needs and structures, XP asks that the design be kept to the most essential thing that can provide the functionality needed within the timeframe of the next delivery or iteration. A straightforward, widely communicated, and understood software structure means that there will be less for people

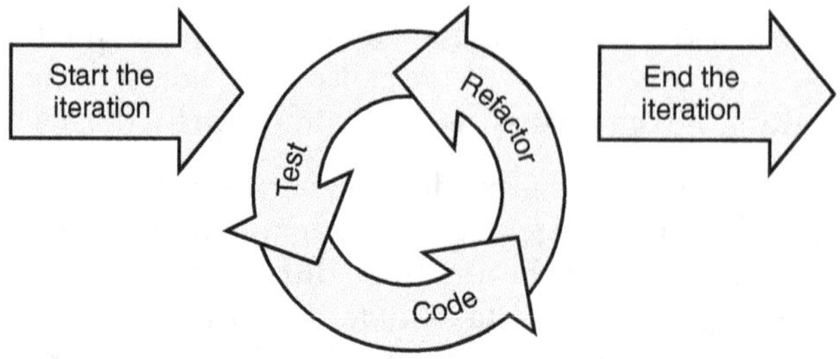

Figure 9.15 Test–code–refactor rhythm.

to have to "communicate" about in a formal fashion. This helps remove ambiguity that may muddy effective communications. XP coined the acronym *YAGNI*, "You ain't gonna need it," to emphasize not adding structure and design that current (known) functionality does not or will not need. If you think you may need it, wait until the future to add it (based on Succi and Marchesi 2001). However, YAGNI does not mean that you cannot have complex software. YAGNI means that the software should be as simple as is possible and still meet the customer's need.

- *Feedback:* XP practitioners believe that because things will change, it is important to shorten the feedback cycles from weeks or months to minutes or hours. XP teams continuously work to obtain and utilize as much feedback as they can.
- *Courage:* Courage will sometimes manifest itself as a person's willingness to do or try something when an obstacle surfaces, to make mistakes, learn from them, discard what is not working, and start over. At other times, courage can mean waiting, listening, and learning until a better, simpler solution presents itself. XP practitioners must have the courage to speak the truth, admit what is really there, and deal with it.
- *Respect:* A team cannot be as effective as it should or can be if anyone's contributions are not valued by the other members. The team must also be highly motivated to succeed and enthusiastic about the work environment. To accomplish this, team members must respect each other and the project to build trust and collective responsibility.

XP promotes the following principles, again many of which are shared across the agile community:

- *Economics*: XP methods focus on economics by:
  - *Deferring costs*: Encouraging projects to wait "until the last *responsible* moment" before making design/development commitments
  - *Earning payments*: Committing to delivering valuable, working software as soon as possible and frequently thereafter means that the customer will derive the advantages of the software, and the development organization will receive payment sooner

- *Software reuse*: Creating software that is intended for reuse means that its value can be realized over and over instead of just once
- *Business alignment*: Emphasizing customer interaction and feedback creates software products that are more likely to have business value, align with business goals and objectives, and address the needs of the business

- *Mutual benefit*: Mutual benefit means focusing on activities and products that deliver value to the stakeholders now, the developers now, and both the stakeholders and developers later.
- *Accepted responsibility*: XP team members "signing up" for implementation tasks. Once the list of needs for the next iteration is defined and prioritized and broken down into the tasks required to implement those features, the developers self-assign themselves to those tasks.
- *Redundancy*: This means using redundant activities with similar goals as long as they add value. For example, testing after pair programming is redundant because both activities are aimed at removing defects—however, in practice, testing typically finds additional defects that were not uncovered in development, so testing is value-added.
- *Self similarity*: Looking for patterns that work and repeating them elsewhere. Looking for patterns that help solve today's problems using yesterday's solutions. Unless it is the only choice, do not go to a unique solution for management or process issues or for requirements, design, or code solutions.
- *Quality*: Quality must be built into the software. To quote Beck (2005):
  - "Pushing quality higher often results in faster delivery; while lowering quality standards often results in later, less predictable delivery."
  - "There is no apparent limit to the benefits of quality, only limits in our ability to understand how to achieve quality."
  - "Quality isn't a purely economic factor. People need to do work that they are proud of."
- *Flow*: XP considers software development as a continuous flow from one iteration to the next rather than a set of discrete phases.
- *Failure*: Failure is not waste—it allows us to learn from our mistakes. Trial and error may be the cheapest way to find out what works.
  - *Opportunity*: Learn to see problems as opportunities for:
  - Improving the way we are doing things
  - Creating better software
  - Building interpersonal relationships

  - Expanding our personal knowledge and skill sets
  - Team and organizational learning
- *Reflection*: Improvement comes from learning about the causes of problems (or potential ones) and finding solutions to correct (or avoid) problems. XP asks the team to continuously consider what they are doing and why. This provides input and feedback that allows immediate reuse of good ideas and quick elimination of problems. We should not hide our mistakes. We should expose them and learn from them.
- *Humanity*: XP increases the focus on human issues so that people and teams can work closely together. This focus may be new or even uncomfortable to those who are used to working as individual contributors.
- *Improvement*: Both agile and plan-driven methods agree that continuous improvement is the best way to effectively and efficiently achieve higher-quality, value-added software.
- *Baby steps*: XP emphasizes making changes in small steps. Teams should look for the least that can be done that moves in the right direction, then prioritize changes and implement them incrementally, reducing both overhead and risk.
- *Diversity*: Having diversity on the team can generate positive conflict (new ideas, different opinions, alternative approaches).

To implement its values and principles into the daily work of software development, XP includes a set of primary and corollary practices. XP primary practices include:

- *Team co-location*: Software should be developed in a workspace large enough for the whole team, including the customers, to work together in a single space.
- *The whole team*: The entire cross-functional, self-directed XP team works as a unit to accomplish its goals when implementing XP projects.
- *Informative workspaces*: XP teams use the walls of their workspace to communicate important, active information. At a glance, people should be able to get a feeling of how the project is progressing and identify current or potential problems simply by looking at the information posted around the workspace.
- *Energized work*: XP emphasizes working only as many hours as practitioners can be productive at a sustainable pace (eight-hour days, 40-hour weeks, and not coming to work sick). XP also encourages a reasonable amount of playtime at work to rejuvenate individuals and stimulate creativity.
- *Stories*: Developers create the requirements for the system by asking their stakeholders to tell stories about who will use the system, how it will be used, and why. These stories are documented and become the

first level of prioritized requirements fleshed out in more detail only when selected for implementation.

- *Quarterly cycle*: On a quarterly cycle, plan to (Beck 2005):
    - "Identify bottlenecks—especially those controlled outside the team
    - Focus on the big picture—where the project fits within the organization
    - Initiate repairs
    - Plan the theme(s) for the quarter
    - Pick a quarter's worth of stories to address those themes"
- *Weekly cycle*: The team's job is to write test cases and get them running in the next five days. Weekly planning meetings are held to:
    - Review how actual progress matches expected progress
    - Select customer stories to implement that week
    - Further redefine the requirements and acceptance criteria for each story as needed
    - Estimate the duration of the effort needed to implement each story
    - Divide stories into tasks
    - Have each team member self-select tasks they accept responsibility for performing
- *Slack*: Ideally, the team will finish all the committed work in just the amount of time required, but leaving some slack in the weekly schedule allows for making up for any unforeseen events during the weekly cycle.
- *Pair programming*: Pair programming involves two people, one constantly reviewing what the other is developing. (See Chapter 23 for more details on pair programming.)
- *Test-driven development (TDD)*: TDD advocates writing the test cases that the code must pass before writing the code.
- *Incremental design*: Agile methods in general, and specifically XP, do not write what some call big design up front (BDUF). The design emerges over time through gradual, persistent effort devoted to reflecting on what already exists in the system and what is needed next.
- *Continuous integration*: XP practices integrate each piece of changed (or new) software code into the system continuously, creating and testing a new build within a few hours of a code change. This is the opposite end of the spectrum from the traditional waterfall life cycle behavior where all (or most) of the software code is written before integration starts.

- *10-minute build*: To integrate as frequently as desired in an XP project, it must be possible to build the system and run tests very quickly. The goal represented by a 10-minute build is to make this possible.

XP corollary practices include:

- *Team continuity*: High-performance teams should be kept together over time.
- *Real customer involvement*: Martin (2003) says the customer is "the person or group who defines and prioritizes features." The customer needs to be in close contact with the development team (if not team member) and take responsibility for the software by participating actively in its definition and validation (writing test cases and possibly even executing test cases).
- *Negotiate scope*: The XP philosophy says that if something has to "slip" within a given increment, it should be the scope (functionality) rather than manipulating cost, schedule, or quality. Functionality missed because of scope reduction is moved to the next increment.
- *Single code base*: As noted earlier in discussing continuous integration, maintaining a single code baseline is very desirable. However, branches can occur briefly and then be brought back together through frequent integrations, done at least daily.
- *Shared code*: After the team has a sense of collective responsibility, anyone on the team can improve any part of the system at any time (Beck 2005). This requires adoption of team-wide coding standards. "If a change needs to be made, the person who is in the best situation to make it (the developer who sees the immediate need for the change) can make that change" (Astels, Miller, and Novak 2002).
- *Code and tests*: XP's "minimalist" approach to documentation suggests that the code and tests should be the basis for any other documentation, not the reverse. For good or bad, the product is what the software does, so the software code and its tests are the final authority on documenting the product's capabilities.
- *Incremental deployment*: By deploying software incrementally, early in the project, and often after that, the project demonstrates tangible and working results regularly. This allows the customer to provide helpful feedback just as frequently, so that little time passes between work done and work reflected on. Review of small increments allows for frequent adjustment of project direction and quick identification of defects. Planning for each small increment is also more accurate.
- *Daily deployment*: The XP ideal is to get working software into the customer's hands every day. Except in minor cases, this is probably not practical for various reasons. However, it is important to realize that the longer software is not being actively used, the more likely development may diverge from what the customer ultimately needs

(even if it matches the initial direction the customer thought made sense).

Lean

Scrum or eXtreme Programming (XP) are methodologies that have elements of a mindset (values and principles) but they also have practices and processes. Lean, however, is purely a mindset. There are no Lean practices. "Lean teams look at the way they're working today, and figure out where they are habitually running into problems. Then they make changes to improve the way they work" (Stellman and Greene 2017).

According to Jenny Stuart, "lean's most significant contribution was encouraging and enabling people to continuously improve the way in which they work" (Stuart 2017).

Lean principles are discussed in Chapter 7 of this handbook.

Kanban

Kanban is a merging of two Japanese words, "kan" which means visual and "ban" which means sign or card. Kanban is an agile system that uses tokens as a mechanism for managing the flow of a process. The core practices of Kanban include:

- *Visualizing workflow*: Kanban teams use Kanban boards to visualize the workflow as whole systems.
- *Limiting work-in-progress (WIP)*: Once the Kanban team understands its workflow, it can start setting and adjusting work-in-progress limits for each step in that workflow.
- *Managing flow*: The Kanban team can start managing their process workflow by understanding how quickly work moves through the process steps. The team can use WIP limits and a pull system to manage that workflow.
- *Making process policies explicit*: A Kanban team defines and continuously evaluates and changes as necessary specific policies to guide them with their work. These policies define what has to be true for that step to be considered satisfactorily completed (its exit criteria).
- *Implementing feedback loops*: "Feedback loops are all about testing all of the policies and improvements the team is implementing to measure their effects and make sure they are working" (Stellman and Greene 2018). Data should be collected and measurements are used to provide fact-based feedback.
- *Improving collaboratively and evolving experimentally*: The Kanban culture encourages reflection and continuous improvement of the workflow, the resulting work products, and the process policies. A Kanban team collaboratively works together on continuous improvement by using the scientific method (hypothesize and experiment to validate the hypothesis).

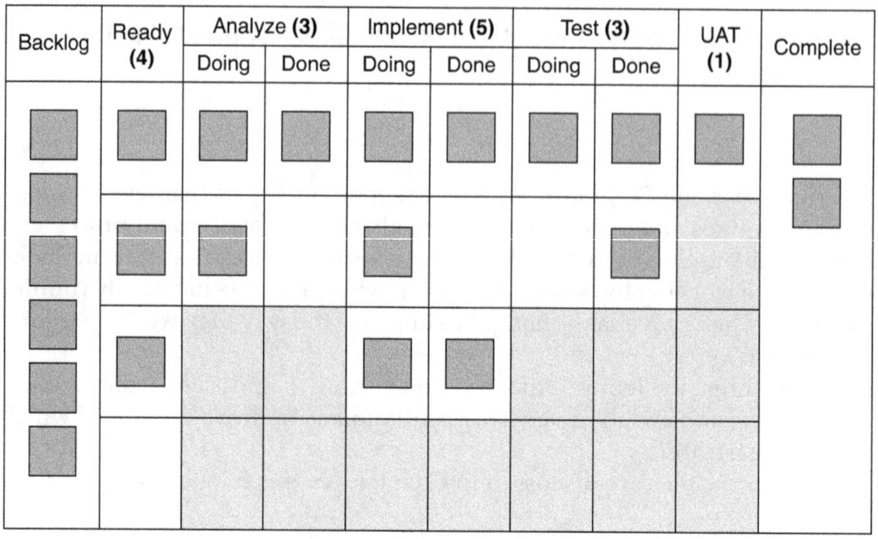

Figure 9.16 Kanban board—example.

As illustrated in the example in Figure 9.16, Kanban boards are used to visualize:

- Work items and their status
- The flow of those work items through the steps in the process

Like with Scrum, the Kanban board backlog contains user needs or other work items that have not yet been selected to begin, based on their priority. However, rather than just having ready, in-work, and done columns like a Scrum task board, a Kanban board has columns for every major step in the process flow. These process steps may be different for different types of work. Kanban boards will be set up differently for each type of work. For example, a software development Kanban board may have steps that include analyze, implement, test, and user acceptance test (UAT) like the Kanban board in Figure 9.16. As another example, a problem report correction board might have process steps including impact analysis, CCB approval, correction, and retest or regression test. A team should never just use someone else's Kanban board. Each team should use their work flows and the steps in their processes and set up one or more boards to reflect the way they do business.

For each step in the process, the doing column shows what tasks are currently being worked. The done column shows tasks that are available to pull forward into the next step in the process.

By limiting the work in progress at each step, it optimizes the amount of work being done by the system without causing bottlenecks or multitasking. The fastest way to do two things is to do one completely first and then move on to do the second thing. As Stuart says, we need to "Stop starting and start finishing" (Stuart 2017). The goal is to create a constant movement or flow through the system and minimize total cycle time through the workflow.

WIP limits are set based on the capacity of the people doing the work. Ideally, each person should only be doing one item at a time. A team should consider setting their initial WIP limits to 1.5–2 times the number of people working on that step. For example, if two people are doing analysis, 1.5 times 2 would set the work limit to 3. If only one person is working on a step and they are working half-time, the WIP limit might be set to 1. WIP limits apply to the combination of the items in work and the done items for each process step, so that work does not stack up as partially done work.

Kanban metrics are discussed in Chapter 19 of this handbook.

4. DEVOPS

> Employ DevOps to combine software development and IT operations to continuously release software. (Apply)
>
> **Body of Knowledge III.A.4**

"DevOps is the emerging professional movement that advocates a collaborative working relationship between Development and IT Operations, resulting in the fast flow of planned work (e.g., high deploy rates) while simultaneously increasing the reliability, stability, resilience, and security of the production environment" (Kim et al. 2023).

Historically, agile methods have focused on the development of software. This can result in problems if development releases software more and more rapidly but traditional information technology (IT) operations have not changed their methods to keep up with rapid-fire software releases and deployments into operations.

With its success, DevOps has been extended to integrate other groups that influence the technical software-value streams into the DevOps team, including security (DevSecOps), quality (DevQualOps), product management, marketing and business management, thus breaking down the silos between these various groups. This extension allows a single team to manage the entire software product life cycle and all of its practices from initial planning, through development, build, and testing, to integration, release, deployment, and monitoring, to looping back into planning the next cycle.

The goals of DevOps are to:

- Increase interpersonal communications by breaking down the silos and increasing feedback loops
- Shorten software product cycle times to frequently deliver enhancements, fixes, and other updates that provide value to the stakeholders (e.g., up to hundreds or even thousands of releases per day) without impacting the stability and reliability of operations
- Quickly detect and correct failures and other issues when they occur and reduce technical debt

- Apply lean management principles to software development
- Implement automation where possible to remove the need for repetitive and tedious manual operations that can be error-prone
- Establish measurements that supply useful information to manage software products and processes
- Rigorously improve software products and processes
- Create "a culture that fosters two things: continual experimentation, which requires taking risks and learning from success and failure; and understanding that repetition and practice is the prerequisite to mastery" (Kim et al. 2023).

The flow of the DevOps practices is typically represented as an infinity sign, as illustrated in Figure 9.17 because the practices loop back and are continuously repeated. DevOps practices include:

- *Plan*: The development team meets with the product owners and other stakeholders to determine their goals and create a plan for implementing those goals.
- *Implement*: The development team works with the product owners and other stakeholders to refine their business needs. The development team then designs and codes the system and its software to meet those needs. Distributed version control tools help ensure that each developer is always working on the most current version of the code.
- *Continuous build*: As soon as new or modified software is checked into the control libraries, automation is used to create a new build that includes those updates. Tools feed back any issues found during the build to the appropriate developer(s).

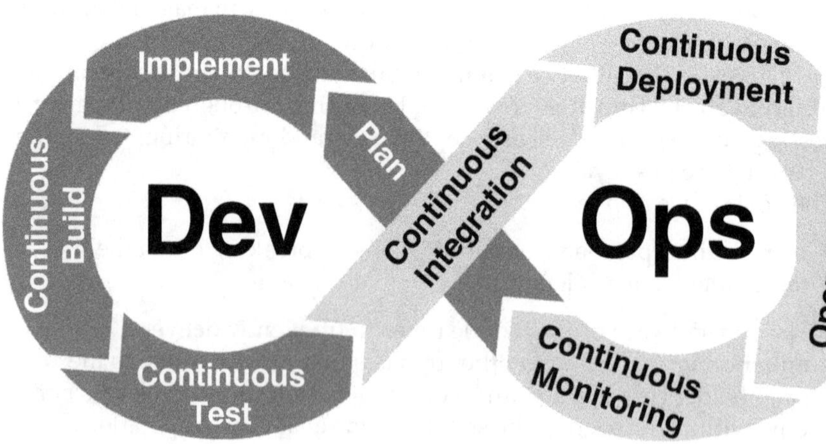

Figure 9.17 DevOps practices.

- *Continuous test*: Each new build is then tested using automated testing techniques and tools to ensure software quality. Any identified issues are again fed back to the appropriate developer(s).
- *Continuous integration*: When testing is complete, the new or updated software is automatically integrated (packaged) into the existing codebase in a production-like staging environment.
- *Continuous deployment*: Once the software is staged, it can automatically be deployed into the operational environment (push system), or it can be deployed on demand from production (pull system).
- *Operations*: The software is used and supported in the production environment(s), and those production environments are controlled and managed.
- *Continuous monitoring*: Monitoring is performed in operations to provide information about the quality and health of the software. Automated tools can be used during this activity to monitor the software and its platform to check the performance of the system. The resulting information (e.g., field failures, improvement opportunities, and issue) are fed into the next planning activity, and the cycle is restarted for the next updates.

Chapter 10
Systems Architecture

> Identify and describe various architectures (e.g., embedded systems, client–server, n-tier, web, wireless, messaging, and collaborative platforms) and analyze their impact on quality. (Analyze)
>
> **Body of Knowledge III.B**

A *system architecture* defines the system-level components (including subsystems), interfaces between those components, and the allocation of system-level requirements to those components and interfaces. The architectural design of a system may also consider timing and bandwidth of those interfaces. System-level components can be software, data, hardware, and humans (who implement processes and manual operations) and may also include "processes (e.g., processes for providing services to users), procedures (e.g., operator instructions), facilities, materials, and naturally occurring entities" (ISO/IEC 2011).

Good systems architectures address higher-level concepts and abstractions for the system. (Lower-level details are dealt with during the *component design*, also called *detailed design*, activities, which define the internals of each component at the hardware, software, data, or process engineering level.) Stakeholders of the systems architecture should be able to:

- Understand what the system does
- Understand how the system works
- Understand how the system relates to its environment
- Work on one piece of the system, independent of other pieces
- Extend the system
- Reuse one or more components or parts of the system to build another system

"Architecture is what remains when you cannot take away any more things and still understand the system and explain how it works" (Kruchten 2000).

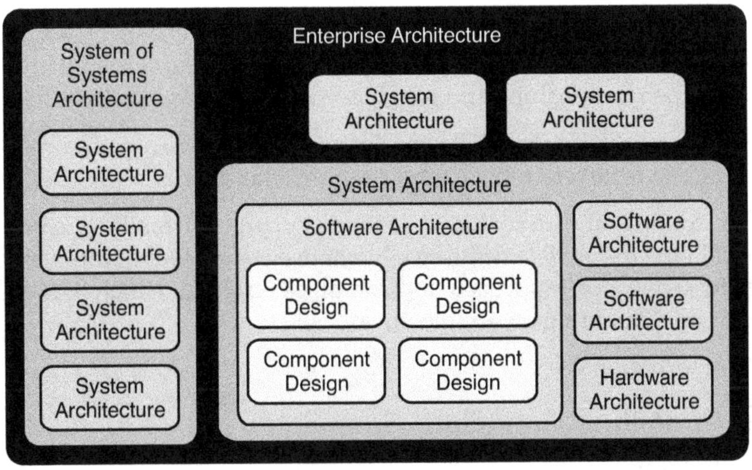

Figure 10.1 Levels of architecture and design—example.

Levels of Architecture and Design

As illustrated in Figure 10.1, there can be a hierarchy of architectures, which may include the following levels:

- *Enterprise architecture*: The enterprise architecture is the structure of organization-level components (e.g., processes, systems, personnel, teams, organizational sub-units, and so on), and the relationships and interfaces between them. If software is part of the architecture used by the enterprise, then both the enterprise and its system(s) will have an impact on the design of that software.

- *System architecture*: For software that is part of a larger system, design may also include the architectural design of the system, which will be discussed in this chapter. As part of the system architecture, the system-level requirements are allocated to various constituent components (subsystems or packages, including hardware, software, database, and manual operation subsystems within the system, and so on). The major design activity at this level is to determine how to best segregate the system into its components, and to define the communications, interfaces, and interactions between those subsystems.

  This step may be skipped for software development projects that only include a single smaller software application, and not a complex software system or hardware/software system.

- *System of systems*: A system may also be part of a larger system of systems. For example, an organization's financial system may be a system of systems, including a time-accounting system, payroll system, accounts payable system, accounts receivable system, and so on. A car is a system of systems, including an antilock brake system, fuel injection system, airbag system, and so on. The system of systems architecture defines the systems that make up the system of systems and the relationships and interfaces between those systems.

- *Software architecture and design*: Software architecture and component design are covered in Chapter 14 of this handbook.

Another example of different levels of architectural is from *The Common Approach to Federal Enterprise Architecture* (see whitehouse.gov/sites/default/files/omb/assets/egov_docs/common_approach_to_federal_ea.pdf2012), which lists eight levels of scope for implementing an architecture:

1. International
2. National
3. Federal
4. Sector
5. Agency
6. Segment
7. System
8. Application

Embedded System

An *embedded system* is a computer system with a dedicated function embedded as part of a larger system, for example, a software/hardware braking system or fuel injection system that is a component of a larger system (car, airplane). If the software is installed into a hardware component, it is often referred to as *firmware* and is stored in read-only memory (ROM) in a flash memory chip or other device.

Embedded systems can vary widely in capability. Many embedded systems are limited to the performance of one or a limited number of dedicated functions or tasks, as opposed to a personal computer or mainframe. Digital watches, microwaves, mobile phones, traffic lights, and handheld calculators are just a few examples of embedded systems. A system of systems like a modern automobile or aircraft has multiple embedded systems. Medical devices and weapons systems are examples of embedded systems that are complex and multifunctional. Embedded systems can range from having no user interface to a complex graphical user interface similar to a personal computer. Many embedded systems are also restricted by timing, size, power utilization, reliability, and other constraints imposed by the system's environment.

n-Tier

In an *n-tier architecture,* also called a *multitier architecture,* the system is logically decomposed into two or more layers (levels), each with independent processing capability, creating a modularized approach to the architecture. The "n" in an n-tier architecture implies any number, such as two-tier, four-tier, five-tier, and so on.

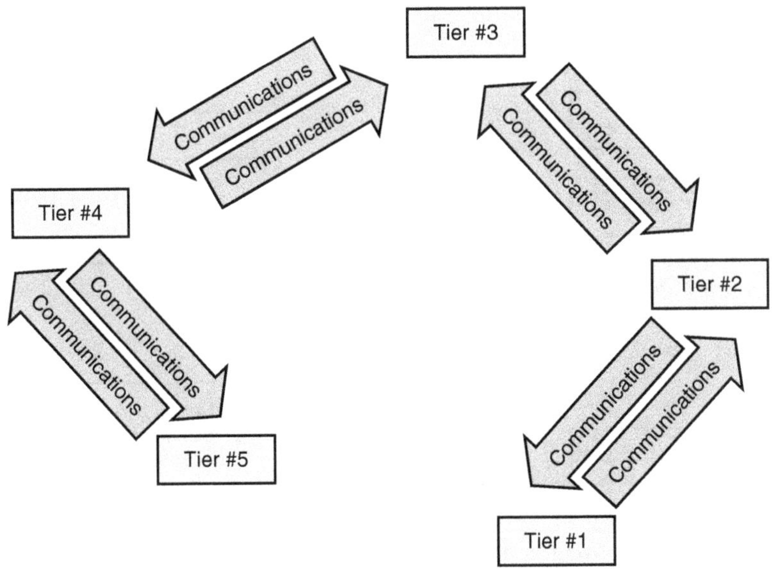

Figure 10.2 Five-tier architecture—example.

Well-defined interfaces and information-hiding (abstraction) characterize a tiered or layered architectural style. As illustrated in Figure 10.2, each tier interfaces with the tiers just above and below it and performs part of the whole system's functionality. If communication is severed between two tiers, partial independent functionality within each tier or between nonsevered tiers is still possible. For example, the telecommunications network is a classic tiered architecture. If the private branch exchange (PBX) inside a company loses connections with the local switch, employees cannot make outside phone calls but may still be able to call other employees using the same PBX. If, however, the local switch has a good connection to the PBX but loses connection with the class 5 switch, employees can still call each other. They may also have limited outside calling capacity with others sharing the same local switch.

The tiers provide abstraction because a tier does not need to know how the operations of the layers above or below it are implemented. In a tiered architecture, layers can be added, upgraded, or replaced independently without impacting other parts of the system.

Client/Server

Client/server architectures are a specific type of two-tier architecture, as illustrated in Figure 10.3. For example, many computer networks within businesses are based on client–server architectures. Each computer or device on the network is a client or a server. Central *server* computers or devices on a network provide services, manage network resources (e.g., the sharing of software licenses), and centralize data.

The *client* is the requester of services and typically runs on a decentralized access terminal, computer, or workstation. These clients download or access the

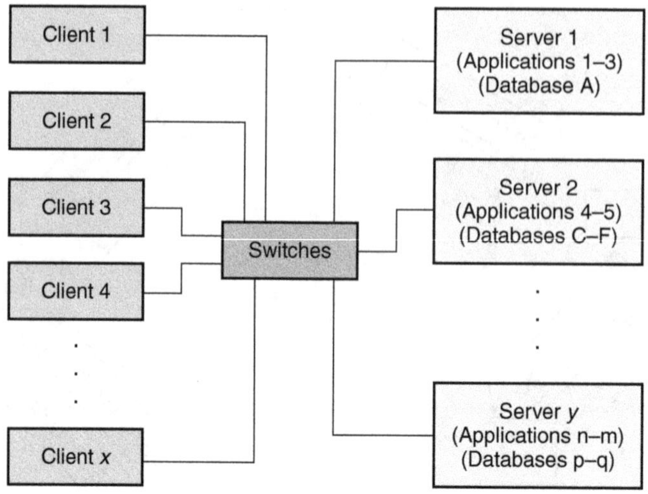

Figure 10.3 Client–server architecture—example.

centralized data, share the network resources, and utilize the centralized processing capability available from the server. A *thin client,* also called a *slim* or *lean client,* primarily focuses on the user interface and on sending and receiving inputs or outputs from the server. A thin client depends on the server for most of its processing activities. Typically, the only software installed on a thin client is the user interface, networked operating system, and possibly a limited number of frequently used applications. In contrast, a *fat client,* also referred to as a *rich* or *thick client,* typically has many applications installed and performs most of the processing functionality locally, depending on the server chiefly for shared data. A thick client can run much more independently. In fact, a thick client may be able to run without connectivity to the server for some time, requiring only periodic connection for data refreshing and synchronization with the rest of the system and its users.

Advantages of a client–server architecture include the:

- Decentralized distribution of roles and responsibilities across the network
- Sharing of resources and data
- Simplified system management (e.g., single-point propagation of new or updated software and centralized backups)
- Clients that can be disconnected from the server for job site remote entry or modification of data, coupled with the ability to refresh centralized, shared databases on the servers from distributed clients
- Ability to have increased security protection at the server level
- Potentially lower costs, especially if software licenses can be economically shared or if lower-cost equipment can be utilized (e.g., using thin clients)

Disadvantages can include:

- Degraded performance or resource unavailability under high network traffic conditions
- Limitations on the level of processing available to the client if the connection with the server is severed
- Potential issues with backing up organizational assets that are not uploaded to the servers if the clients are disconnected at the time of backup

Peer-to-Peer Architecture

A *peer-to-peer architecture* is a distributed applications architecture that partitions tasks or workloads between equally privileged, interconnected nodes (peers). As illustrated in Figure 10.4, these peers are interconnected and share resources across a network without using one or more centralized administrative systems (servers in the client-server architecture). Each peer can function as a client and a server to other peers on the network.

Web

There are three major types of web architectures, as illustrated in Figure 10.5, including:

- The *internet* is a global network connecting millions of computers. The Internet is characterized by *business-to-consumer (B to C)* architectures, where internet pages are created by businesses, organizations, or individuals to provide open access to consumers. This creates a community that consumers will revisit because of dynamic content-delivery capabilities.

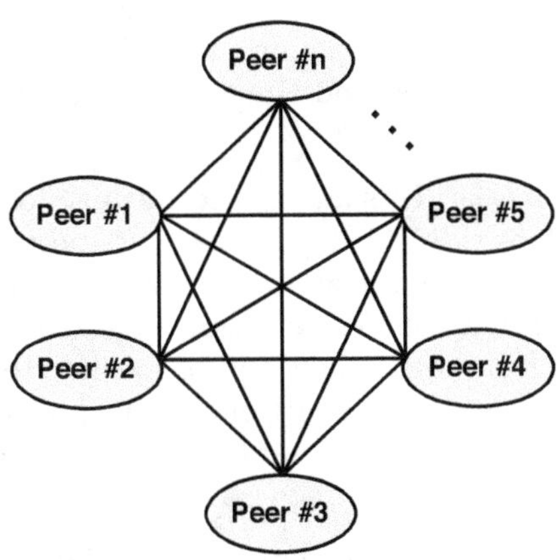

Figure 10.4 Peer-to-peer architecture—example.

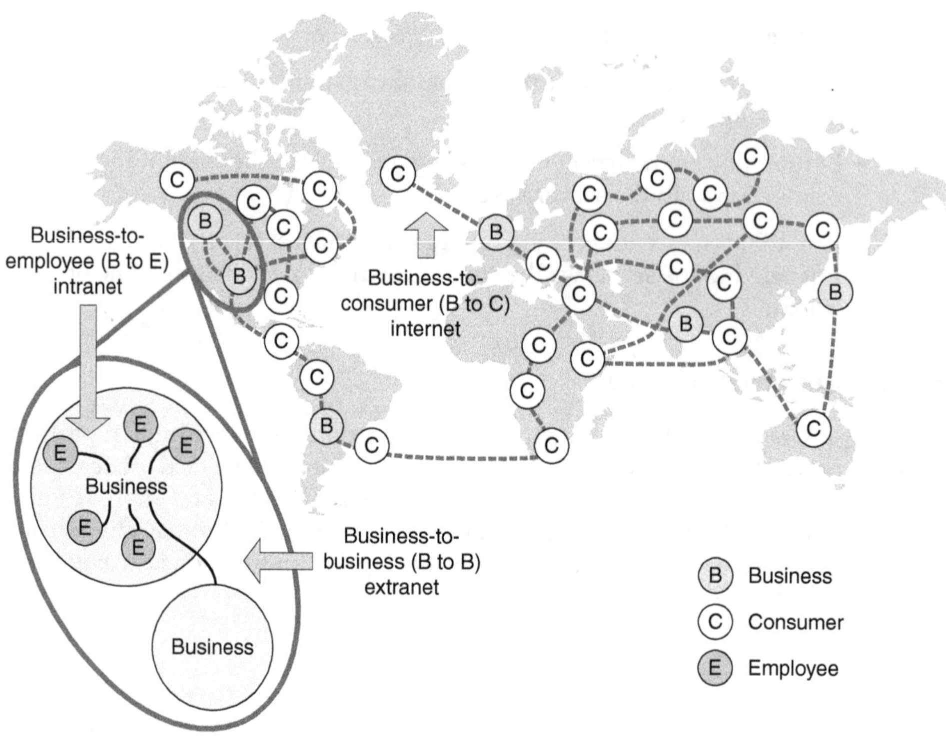

Figure 10.5 Web architecture—example.

The *TCP/IP network architecture* is made up of four layers that interact to make the internet work, including the following:

- *Application layer*: This layer provides services that can be used by applications on the host to communicate data to other applications on the same or different hosts. It defines how the internet services operate and how those services can be used by the applications. For example, these services include FTP services for file transfer, HTTP services for web pages, and Simple Mail Transfer Protocol (SMTP) for e-mail.
- *Transportation layer*: This layer performs host-to-host communications. On the sender's side, this layer breaks up messages that an application wants to transfer into packets and attempts to deliver those packets. On the receiver's side, this layer retrieves the information from the packets and passes it to the appropriate application. This is done using one of two protocols, Transmission Control Protocol (TCP), or User Datagram Protocol (UCP).
- *Internet layer*: This layer transports Internet Protocol (IP) packets from the originating host across network boundaries to the destination host. For outgoing packets, this layer is responsible for the functions

of addressing, routing, and congestion control for Internet Protocol (IP) packets. It transmits the packet to the next-hop node by passing to the appropriate network interface layer. For incoming packets, this layer retrieves the packets and sends them to the transportation layer.

- *Network interface layer*: This layer, also called the *network access layer*, is responsible for packet forwarding through the network, including routing through intermediate routers, to the appropriate destination node using its unique Internet Protocol (IP) address. Packets from the same message may take different routes. Some may never end up at their destination, in which case, the TCP protocol is used. TCP requests retransmission of lost or damaged packages. If the UDP protocol is used, the application must take care of any errors.

- An *intranet* resides behind the firewall of an organization and is only accessible to people within the organization. Intranets typically implement a *business-to-employee (B to E)* architecture, where intranet pages integrate and consolidate enterprise information. This allows employees to access and utilize existing information more effectively from a point of self-service access, facilitates the capturing and sharing of new information, reduces data overload, and makes certain that current, up-to-date information is readily available. There can be many portals within a single organization for different departments, projects, geographic locations, and so on.
- *Extranets* refer to intranets that have multiple levels of access, through various security protections, available to authorized outsiders. An extranet allows organizational partners to exchange information based on a *business-to-business (B to B or B2B)* architecture. Extranet pages provide business information and application functionality to external partners, including real-time access to important data. This can decrease the costs of partner support activities. Each portal has the ability to be customized and personalized to the needs of individual partners (e.g., customers, suppliers, and partners in joint ventures).

Wireless

In today's environment, many individuals need to take their computer processing power with them as they travel, without being tethered via cabling to a network. People need *wireless* access to the Internet, to stored information and data, and so on, as they travel from site to site. For example, a doctor or nurse might use a personal electronic device to access patient records from wherever that information is needed. Salespeople, consultants, law enforcement, students, and many others need to stay in contact while traveling, which can be accomplished through wireless access on their electronic devices. Wireless access is often provided to guests of a company or for use in conference rooms or manufacturing areas where wires might be a problem.

Messaging

Messaging system architectures are designed to accept messages from, or deliver messages to, other systems. An e-mail system or instant messaging (IM) are examples of messaging systems. In its simplest form, a messaging architecture:

- Accepts messages from external systems, determines the internal recipients, and routes those messages appropriately
- Accepts messages from internal sources, determines their destination systems, and routes them internally or externally as required

Collaboration Platforms

According to Wikipedia, "*Collaboration platforms* offer a set of software components and software services that enable individuals to find each other and the information they need, and to be able to communicate and work together to achieve common business goals." A collaboration platform helps individuals and teams work together regardless of how geographically dispersed they are. Examples of key components of a collaboration platform include:

- Messaging via e-mail, databases, or lists of contacts or customers, and coordinated calendars and scheduling tools
- Virtual meeting tools, including instant messages, web-based meetings, audio or video conferencing, and desktop sharing
- Information sharing via shared files and data refreshing, document repositories with search capabilities, and mechanisms for sharing ideas and notes
- Social computing tools such as blogs, wikis, chat rooms, and social media (e.g., Facebook, LinkedIn, and so on)

Chapter 11

Cloud Computing Models and Platforms

> Describe cloud computing models such as Software as a Service (SaaS), Platform as a Service (PaaS), and Infrastructure as a Service (IaaS) and cloud computing platforms such as multi-tenant systems and distributed systems. (Understand)
>
> **Body of Knowledge III.C**

According to the National Institute of Standards and Technology (NIST 2011), *cloud computing* "is a model for enabling convenient, on-demand network access to a shared pool of configurable computing resources (e.g., networks, servers, storage, applications, and services) that can be rapidly provisioned and released with minimal management effort or service provider interaction." The cloud provides infrastructure, platforms, and software resources "as a service" through cloud service providers (CSPs) over the internet to both individuals and organizations, typically on a cost per usage basis (e.g., per monthly subscription, amount of usage time, amount of storage used and so on).

NIST's definition of cloud computing identifies "five essential characteristics," which include the following (NIST 2011):

- *On-demand self-service*: The ability to register, provision, and use resources directly via an administration console (web access) or programmatically without any human interaction with the CSP.
- *Network access*: All capabilities, including access to computer resources, data, and applications, are available via the network using heterogeneous devices (e.g., laptops, desktops, mobile devices) and software such as web browsers.
- *Resource pooling*: The computer resources are shared across multiple tenants in a multitenant model. Resources are dynamically assigned and reassigned without consumers having to know or worry about resource constraints or the resource's physical location.

- *Rapid elasticity*: The ability to increase capacity rapidly by adding resources via self-service during peak demand and reducing capacity when the demand drops.
- *Metered service*: Usage is on a pay-per-use basis; that is, usage is continuously monitored on a per-second, per-minute, or per-hour basis or the quantity used, such as application programming interface (API) calls or gigabytes (GB) stored, and invoiced monthly. Consumers get invoiced for what they used, similar to a monthly electricity bill.

Traditional On-Premise Data Center

According to Jarvis et al. (2022), "A *data center* is a physical facility where companies store their data and run their applications. It is the central location from which an organization's computing power is delivered to customers and employees. This structure could be as big or as small as the needs of the organization that it supports. Also, the facility could comprise real or virtualized hardware or, more recently, cloud-based hardware resources."

In traditional on-premise data centers, the organization is responsible for identifying requirements (including capacity planning); purchasing, designing, building, operating, maintaining, and updating the data center; and sourcing, procuring, designing, building, and operating everything in the data center including its infrastructure, platform, and software applications. The organization must have competent personnel responsible for performing all of the data center processes.

Cloud Computing Models

There are four basic cloud computing models, including the following:

- Infrastructure as a service (IaaS)
- Platform as a service (PaaS)
- Software as a service (SaaS),
- Functions as a service (FaaS)

Infrastructure as a Service (IaaS)

Infrastructure as a service (IaaS) is a cloud computing model where the CSP owns, supplies, and manages the hardware and other infrastructure components rather than having those components on-premise at the organization. IaaS gives the organization the highest level of control over their information technology (IT) resources while providing the infrastructure traditionally provided by the on-premise data center. As illustrated in Table 11.1, this infrastructure includes:

- Computers
- Networking/firewalls

Table 11.1 Source of computing resources.

Resource	Traditional On-Premise	Cloud Computing Models		
		IaaS	PaaS	SaaS
Applications	User	User	User	CSP
Data	User	User	User	CSP
Developer tools	User	User	CSP	CSP
Middleware	User	User	CSP	CSP
Operating system	User	User	CSP	CSP
Virtualization	User	CSP	CSP	CSP
Servers	User	CSP	CSP	CSP
Storage	User	CSP	CSP	CSP
Networking/firewalls	User	CSP	CSP	CSP
Data center	User	CSP	CSP	CSP

- Storage
- Servers
- Virtualization

Examples of IaaS include:

- Amazon Web Services (AWS)
- Microsoft Azure
- Google Compute Engine (GCE)

Platform as a Service (PaaS)

Platform as a service (PaaS) is a cloud computing model where the CSP provides a cloud-based platform for software developers to develop, manage, and maintain their software applications throughout the entire life cycle. As illustrated in Table 11.1, the PaaS CSP provides the infrastructure (resources also provided by IaaS) and adds the operating system(s), middleware, and development tools resources.

With PaaS, companies can focus on their core competencies of application development, testing, deployment, and maintenance, without being distracted by the need to set up, configure, support, manage, and maintain the development environment.

Examples of PaaS include:

- AWS Elastic Beanstalk
- Windows Azure
- Google App Engine

Software as a Service (SaaS)

Software as a service (SaaS) is a cloud computing model where the CSP delivers a full application stack as a service, from underlying infrastructure to maintenance and updates to the provided application software. As illustrated in Table 11.1, the SaaS CSP provides the infrastructure and necessary platform (resources also provided by IaaS and PaaS) and adds the applications(s), and data resources. This allows the cloud customer to run the applications directly on the cloud. Those users can focus on using the SaaS software without worrying about how that software is maintained and updated over time. Examples of SaaS software include:

- Office 365
- Zoom, WebEx, GoToMeeting
- Drop Box
- Netflix and Hulu

Functions as a Service (FaaS)

According to Jarvis et al. (2022), "FaaS takes computing to the next level by allowing software developers to create and run a cloud function without worrying about the underlying computer infrastructure. CSPs manage the underlying infrastructure, resources, and capacity and provide APIs to the developer that can configure and manage their function."

Cloud Computing Platforms—Multitenant

The *cloud computing platforms,* also called *cloud deployment models,* define where computing resources are positioned and who has control over them.

In a *multitenant cloud* computing platform, also called the *public cloud,* multiple cloud customers share the usage of on-demand services and resources. A good analogy for a multitenant platform is an apartment building. In an apartment building, each tenant has access to all of the common public areas in the building (lobby, pool, laundry area), while they also each have their own private apartment with their own key where they control access. On a multitenant platform, common resources are shared by all the cloud customers. However, each cloud customer has their own private areas (e.g., their data storage) that only they have access to.

Cloud Computing Platforms—Private

Private cloud computing platforms are owned by and dedicated to a single individual or organization. These platforms are completely isolated from the public and have no shared resources with anyone outside the owner. Organizations use private networks for greater control and for increased security.

Cloud Computing Platforms—Hybrid

Hybrid cloud computing platforms are a combination of multitenant and private platforms. This allows individuals and organizations to leverage the advantages of multitenant services while maintaining their most sensitive data and information on a private cloud platform for security and compliance reasons. Utilization of a hybrid cloud platform can add integration complexities and increase costs.

Benefits of Cloud Computing

The benefits of utilizing cloud computing include the following:

- *Access*: The cloud customer has access to state-of-the-art infrastructures, platforms, and software applications without having to purchase, physically manage, maintain, and upgrade them.
- *Elasticity and scalability*: Resources can be added or removed quickly, on an on-demand basis, allowing the cloud customer to rapidly response to changing business conditions.
- *Cost*: Pay-per-use or subscription models reduce capital expenditures both up-front (instead of creating an on-premise data center) and on an ongoing basis (including support costs and the costs of upgrading to newer technologies over time).
- *Staffing*: Cloud computing offers reductions in the number of staff required to manage an on-premise data center.
- *Security and compliance*: The CSP assumes the responsibility for establishing and maintaining security and compliance, which minimizes the need to have this expertise in-house.
- *Availability*: Cloud computing can provide access to resources across multiple locations worldwide.
- *State-of-the-art technology*: CSPs typically stay on top of the latest technologies and upgrade their offerings with the latest innovations including staying up-to-date with software updates and the latest generation of computing hardware.
- *Redundancy*: Cloud computing can provide redundant infrastructures and platforms to increase reliability, availability, and business continuity and facilitate disaster recovery.
- *Efficiency*: With PaaS, organizations can develop new software and get it into production quickly without worrying about the underlying infrastructure and platform.
- *Data storage, backups, and recovery*: Cloud computing provides the ability to store, back up, and recover large volumes of data.
- *Access to development tools*: Cloud computing provides access to a wide range of development tools for building and testing applications that individuals or small to mid-size organizations might not be otherwise able to access.
- *Big data analytics*: Cloud computing offers almost unlimited resources for storing and analyzing large volumes of data.

- *Control*: IaaS customers maintain control over their platform, including their operating systems, middleware and tools, and applications and data.

Challenges of Cloud Computing

Challenges of utilizing cloud computing include the following:

- *Cost*: While cloud computing does not have the capital costs incurred by on-premise data centers, pay-per-use or subscription costs can get out of control if not managed appropriately.
- *Vendor dependence*: Cloud computing may result in becoming overly dependent on a single CSP and make it difficult to move to a different vendor.
- *Process changes*: Cloud customers may need to change their process and work instructions to accommodate working in an IaaS environment.
- *Support*: Based on the service level agreement, support for the infrastructure, platform, or software applications may be costly or limited.
- *Control*: On-premise resources give individuals and organizations more control than resources provided by a CSP, and this can limit customization.
- *Connectivity*: Since cloud computing is delivered via the internet, slow connections or loss of connectivity can have a major impact on resource availability and performance.

Chapter 12

Requirements Engineering

A *requirement* is a capability, attribute, or design constraint of the software that provides value to one or more stakeholders or is needed by one or more stakeholders. The requirements are what the customers, users, software product suppliers, and other relevant stakeholders must determine and agree upon before that software can be built. The requirements define the "what" of a software product:

- *What the software must do to add value for its stakeholders*: These functional requirements define the capabilities of the software product.
- *What the software must be to add value for its stakeholders*: These quality attributes or *product attribute* requirements (also called the "ilities") define the characteristics, properties, or qualities that the software product must possess. They define how well the product performs its functions.
- *What limitations exist on the developers/suppliers' choices when developing the software*: The design constraints define these limitations. Design constraints define "how" to develop the software rather than "what" needs to be developed. For example, design constraints might include requirements to use a specific programming language, algorithm, communications protocol, encryption technique, or input/output mechanism (e.g., push a button or sound an alarm).

The task of eliciting, analyzing, and specifying requirements is the most difficult part of software engineering. To quote Fredrick Brooks (1995), "The hardest part of building a software system is deciding precisely what to build. No other part of the conceptual work is as difficult as establishing the detailed technical requirements, including all of the interfaces to people, to machines, and to other software systems. No other part of the work so cripples the resulting system if done wrong. No other part is more difficult to rectify later." In other words, to quote Karl Wiegers (2004), "If you don't get the requirements right, it doesn't matter how well you do anything else."

Requirements Engineering

Requirements engineering is a disciplined, process-oriented approach to defining, documenting, validating, and maintaining the requirements throughout the product life cycle. Requirements engineering is made up of two major processes: requirements development, discussed in this chapter, and requirements management, discussed in Chapter 13.

Requirements Development

Requirements development encompasses all of the activities involved in eliciting, analyzing, specifying, and validating the requirements, as illustrated in Figure 12.1.

Requirements development is an iterative process. A development team should not expect to go through the steps in the process in a one-shot, linear fashion. For example, the requirements analyst may talk to one stakeholder and analyze what that stakeholder told them. The analyst may document their understanding of that part of the requirements and then go back to that stakeholder for clarification (validation). The analyst then talks to another stakeholder or holds a joint requirements workshop with several stakeholder representatives, then merges the newly acquired information into the requirements documentation. To further analyze the new requirements, the analyst builds a prototype that is shown to a focus group. Based on the input from the focus group, the analyst documents additional requirements in the specification and holds a requirements walk-through to validate that set of requirements. The analyst then moves on to eliciting the requirements for the next feature, and so on, with each step adding new requirements and further analyzing, specifying, and validating those requirements in relationship to the previously obtained requirements.

The *requirements elicitation* step includes all of the activities involved in identifying the requirement's stakeholders, selecting representatives from each key stakeholder group, and collecting information to determine the needs of each group.

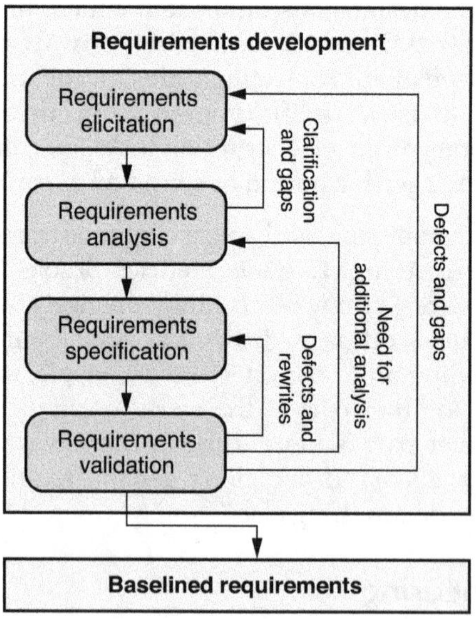

Figure 12.1 Requirements development processes.

The *requirements analysis* step includes taking those stakeholder needs and refining them with further levels of detail. It also includes representing the requirements in various forms (e.g., prototypes and models), establishing priorities, analyzing feasibility, looking for gaps that identify missing requirements, and identifying and resolving conflicts between various stakeholder needs. The knowledge gained in the analysis step may necessitate iterations with the elicitation step, as clarification is needed, conflicts between requirements are explored, or missing requirements are identified.

During the *requirements specification* step, the requirements are documented so that they can be communicated to all the product stakeholders. Requirements may be documented in many ways (e.g., in one or more specification documents, user stories, use cases, or as data entries in a requirements management tool).

The last step in the requirements development process is *requirements validation* to make certain that the requirements are well-written, complete, and will satisfy the stakeholder's needs. Validation may lead to the iteration of other steps in the requirements development process because of identified defects, gaps, additional information or analysis needs, clarification, or other issues.

Requirements Baseline

After one or more iterations through the requirements development process, part or all of the requirements are deemed "good enough" to baseline and become the basis for software planning, design, and development. A good saying to remember at this point is, "When is better the enemy of good enough?" The requirements will never be perfect—analysts can always do more and more refinement and gather more and more input (which may or may not improve the requirements). At some point, the law of diminishing returns starts to apply, where the additional information is not worth the additional effort to obtain it. Requirements baselining is a business decision that should be based on risk assessment. Are the requirements "good enough" to proceed with development? Remember that developers typically obtain valuable information as they develop the software, which can be fed back into requirements updates and enhancements at a future date. Are the requirements "good enough" that the primary customers of the requirements have enough information to proceed? Those primary customers include:

- The developers that are going to create the work products from the requirements
- The testers that are going to use the requirements to create test cases and procedures to verify and validate that the work products successfully implement those requirements
- The customer, users, and other stakeholders whose expectations are being managed based on those requirements

The requirements development process does not assume any specific software life cycle model. Requirements development may be incremental, as illustrated in Figure 12.2. But whether the project defines all of the requirements at once, a part of the requirements, or one requirement at a time, each requirement must

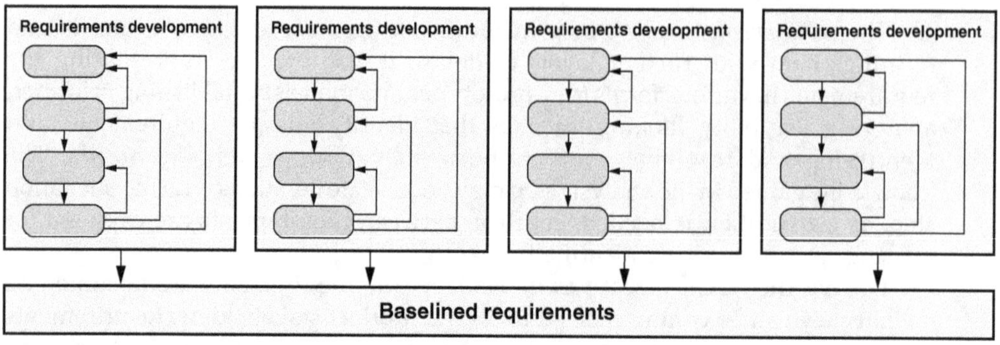

Figure 12.2 Incremental requirements development.

be determined and baselined before the project can implement that part of the software. (See Chapter 26 for more information on baselines.)

1. PRODUCT REQUIREMENTS

> Define and describe various types of product and user requirements, including but not limited to system, feature, function, nonfunctional, interface, integration, performance, globalization, and localization. (Understand)
>
> **Body of Knowledge III.D.1**

Most stakeholders talk about "the requirements" as if they were all the same thing. However, by recognizing that there are different levels and types of requirements, as illustrated in Figure 12.3, stakeholders gain a better understanding of what information is needed when defining the requirements.

Business Requirements

Business requirements, also called *business needs,* define the business problems to be solved or the business opportunities to be addressed by the software product. In general, the set of business requirements defines why the software product is being developed. Business requirements are typically stated in terms of high-level features and the objectives of the customer or organization requesting the development of the software, and they may also include the objectives of other stakeholders.

Stakeholder Functional Requirements

Stakeholder functional requirements (user functional requirements) look at the functionality, behaviors, or capability of the software product from the perspectives

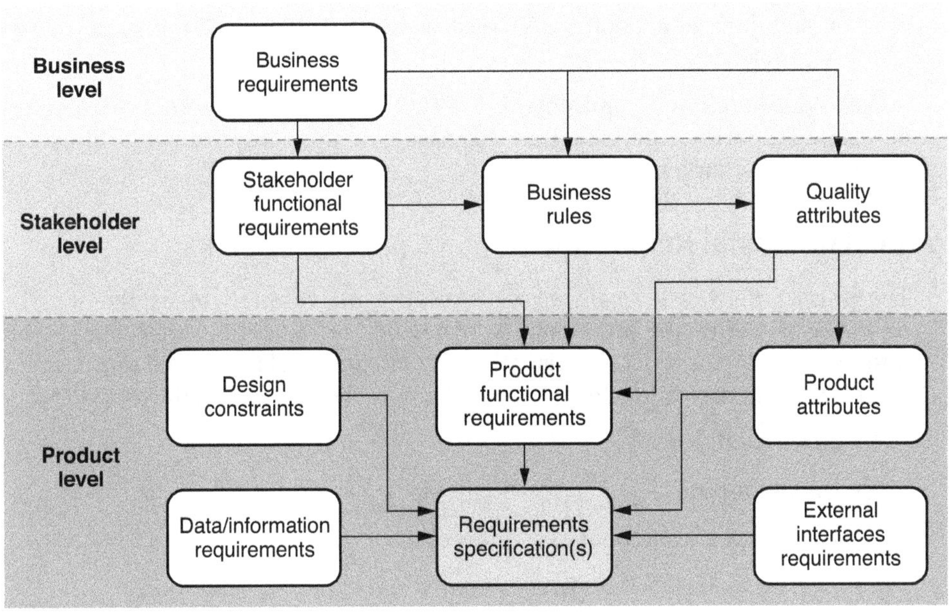

Figure 12.3 Levels and types of requirements.

of the various stakeholders of that product. These requirements define what the software has to do in order for the users, customers, developers, and other stakeholders to accomplish their objectives (or, in the case of "unfriendly" stakeholders like hackers or criminals, to keep them from accomplishing their objectives).

Multiple stakeholder-level functional requirements may be needed in order to fulfill a single business requirement, with each different stakeholder type generating different functional requirements. For example, if the business requirement is to *allow the gas station customer to pay for gas at the pump*, stakeholder functional requirements might include:

- Requirements for the *customer of the gas station*:
  - Input credit, debit, or ATM card
  - Enter security personal identification number (PIN)
  - Select the type of gas they want
  - Request a receipt at the pump
- Requirements for the *gas station attendant*:
  - Receive an alarm if the receipt printer is out of paper
  - Monitor ongoing gas purchase transactions
- Requirements for the *gas station manager*:
  - Set and change the price for each type of gas

- Get the credit, debit, or ATM card validated before allowing gas to be pumped

Stakeholder functional requirements are typically defined in terms of user stories or usage scenarios (use cases) as discussed in the "Requirements Elicitation" section of this chapter.

Product Functional Requirements

The *product functional requirements* (typically just referred to as the *functional requirements*) define the functionality, behaviors, or capabilities that must be built into the software product to satisfy the stakeholder-level functional requirements.

Functional requirements also describe how the software will respond to:

- Error conditions
- Invalid inputs
- Actions performed out of sequence
- Actions or inputs that do not occur when expected (time-outs)
- Inability to access or allocate resources
- Other faults

Multiple product functional requirements may be needed to fulfill a single stakeholder functional requirement. For example, the stakeholder functional requirement that the *customer of the gas station can input their credit, debit, or ATM card* might translate into multiple product functional requirements for the software to:

- Prompt the customer to input the card using the card reader
- Detect the card
- Determine if the card was incorrectly read and prompt the customer to re-input the card if necessary
- Parse the information from the magnetic strip on the card
- Prompt the customer to enter the PIN for debit or ATM cards or a ZIP Code for credit cards
- Accept the PIN or ZIP Code entry
- Report errors to the customer if the PIN or ZIP Code was entered incorrectly (too many or too few characters), or the PIN/ZIP Code was not entered (time-out)
- And so on

It is always easier for people writing requirements to start with an example or a template. Example templates for writing product functional requirements include:

- When <*condition/event*>, the <*subject/noun*> shall <*do what*>

- When *<condition/event>*, if *<status/state/condition>*, then the *<subject/noun>* shall *<do what>*
- When *<condition/event>*,
  - if *<status/state/condition>*, then the *<subject/noun>* shall *<do what>*
  - else, if *<status/state/condition>*, then the *<subject/noun>* shall *<do what>*
  - else, if *<status/state/condition>*, then the *<subject/noun>* shall *<do what>*
- When *<condition/event>*, the *<subject/noun>* shall:
  - *<do what>*
  - followed by *<do what>*
  - followed by *<do what>*

However, templates like these are just a starting point. Any requirement written using a template should be expanded on as necessary for completeness and wordsmithed into a clear, concise, unambiguous, well-written sentence.

In agile development, requirements are often only documented to the level of stakeholder requirements. These stakeholder requirements then act as a reminder to have further conversations with the stakeholders just before each requirement is implemented. If this is the case, product functional requirements may be expressed in:

- Acceptance criteria associated with the user stories
- Associated test cases
- The further refinement of the information in the use case (including additional detailed steps, business rules, and assumptions)

Functional requirements do not include design considerations (constraints or limitations) that restrict the choices that can be made when designing and implementing the software. The focus of functional requirements is on *what* needs to happen, not *how* it will happen. Functional requirements, and, in fact, all requirements, should be able to be verified from outside the software without looking into the design or code. There are exceptions, of course; if there truly are requirements restricting design and implementation choices, they should be included as design constraint–type requirements.

Interface Requirements

The *interface requirements* are a subset of nonfunctional requirements at the product level. They define the required information flow across shared interfaces to external entities outside the boundaries of the software product being developed. These interfaces include communications with hardware components, humans, other software applications, the operating system, file systems, and communication functions (such as email, web browsers, network protocols, and electronic forms).

At the business or stakeholder requirements level, a context diagram can be used to identify the external entities and the high-level inputs and outputs to each of those entities. Other requirements models can also be used to identify external entities to the software that may require *external interface requirements.* For example:

- In a use case diagram, the actors represent external entities
- In data flow diagrams, rectangles represent external entities

At the product level:

- For software that is part of a larger system, the interface requirements to entities external to the system may be defined in the system requirements and allocated to the software requirements. Interface requirements from the software to other system components may be defined in the system architecture.
- For stand-alone software that is not part of a larger system, the inputs and outputs between the software and the external entities are defined in the external interface requirements.

Examples of the types of information specified for interface requirements include:

- Data or messaging types, formats, or structures, including valid values or ranges
- Control values
- Interrupts, communication priorities, or interactions
- Data translations or bit mappings
- The use of standardized communications protocols
- User interface style guides for usability or product branding purposes
- Other nonfunctional requirements affecting the interfaces (e.g., transfer rates, response times, frequencies, capacity, security controls, handshaking)
- Reports, prompts, and the reporting of errors
- Devices supported

Performance Requirements

At the stakeholder level, performance requirements are a subset of nonfunctional, quality attribute requirements. They are typically described in qualitative terms by the customers, users, and other stakeholders. For example, a user might state that *all reports should be processed quickly.* The role of the requirements analyst is to work with that user to determine how quick "quickly" really is and translate the qualitative stakeholder requirement into a quantitative, measurable, and testable product-level product attribute requirement.

At the product level, *performance requirements* are a subset of nonfunctional requirements that define the software's expected or worst-case ability to perform work under various defined conditions. Specific, measurable ranges, limits, or values should be stated for each performance product attribute. A product-level requirement that states *When a report is requested, that report shall be processed within 30 seconds* is better than *processes quickly,* but it still leaves room for ambiguity. For example, the developer might implement the reports, and the tester verifies that it takes an average of 10 seconds, and no more than 15 seconds, to produce any of the reports on the test bed. However, when the users request reports in the field, it takes an average of 15 minutes for the reports to be printed and ready for pick up on the network printer because their network is under capacity and slow. This would result in very unhappy users who believe their requirement was not met.

The problem with this example is the ambiguity of the word *processed* and the fact that speed of printing the report on the user's network is out of the scope of the software product and therefore cannot be controlled by that software. Performance requirements should be specific in defining precisely what is being measured. When does the measurement start? When does it end? Performance requirements should make sure that all work between those points is within the scope and under the control of the software. Therefore, a better version of this example requirement would be, *When any report is requested, it will take no more than 30 seconds from the time the request is received until the completed report is posted to the network for printing.*

The following are examples of performance-type quality attributes and their associated product attribute templates:

- *Throughput*: A measure of the amount of work performed by a software system over a period of time (e.g., transactions per hour, jobs per day)

  Example throughput requirement template: <*acceptable range, average quantity or worst-case quantity*> of <*defined work units*> which can be successfully <*completed from where to where*> per <*time unit*> under <*defined conditions*>

- *Capacity*: A measure of the amount of entities, activities, actions, or events that can be concurrently connected to, or handled by, the software or system (e.g., the maximum number of concurrent users, the maximum number of records in a database, the maximum number of pumps that can all be pumping gas at the same time)

  Example capacity requirements template: <*acceptable range, average quantity or worst-case quantity*> of <*entities or components*> that can simultaneously <*be performing a task/action or be in a state*> under <*defined conditions*>

- *Response time*: A measure of the average or maximum amount of *calendar time* required to respond to a stimulus (e.g., user command, input from another software application or hardware signal)

  Example response time requirement template: <*acceptable time range, average time or worst-case time*> for <*defined response from where to where*> to a <*defined event/ stimulus*> under <*defined conditions*>.

The *<defined conditions>* part of the previous templates might be stated as "under normal operating conditions" or specify special conditions like peak load or specific system states (maintenance/test mode, emergency conditions). Requirements may also need to be specified for degrading performance under specific conditions. This may include defining acceptable performance criteria for when the software or system has been degraded (e.g., failed hardware or system overloads). Is it permissible for a system to become slower as loads increase? Can certain types of users be denied access or even dropped under certain conditions? For example, if a telephone switch is at maximum capacity, can normal user calls be restricted or dropped in favor of 911 calls or police or fire department calls?

It should be noted that generalized product attribute requirements are not specific to any individual product functional requirement. They apply across multiple product functional requirements and are typically specified separately from the functional requirements. However, if a quality attribute or product attribute applies to a single functional requirement, it is usually incorporated as part of that functional requirement and not specified separately. For example, the template for a product functional requirement/response time combination might be: When *<condition/event>*, the *<subject/ noun>* shall *<do what>* within *<acceptable time range, average time, or worst-case time>* as measured *<from where to where>*.

In agile development, performance (and other quality or product attributes) requirements may be:

- *Stated as an individual user story*: For example, *As the gas station manager, I can request any report, and it will take no more than 30 seconds from that request until the time the completed report is posted to the network for printing, in order to keep me from standing around waiting for a report.*
- *Stated as part of a functional user story*: For example, *The gas station attendant can poll the pump controller for the current transaction values and have them displayed within three seconds in order to process post-pay attendant transactions.*
- *Included as additional information in a use case*: For example, frequency of use, business rules, or assumptions.
- Expressed in the acceptance criteria associated with the user stories or in the associated test cases.

Globalization—Internationalization and Localization Requirements

Internationalization requirements are another subset of nonfunctional requirements that deal with developing software that can be adapted to target markets in different geographical locations around the world without the need to change the software. This includes adapting the software to different:

- Languages
- Regional and cultural differences
- Customs and standards

Localization is the actual process of customizing the internationalized software to a region or location through translation or adding location-specific components.

The term *globalization* is used to combine the concepts of internationalization and localization. Examples of globalization considerations include:

- *Character sets*: Verifying that the software uses the appropriate character set for the location. For example, ASCII (American Standard Code for Information Interchange) is appropriate for the United States but not for countries like Russia or China that have very different character sets. Capitalization of characters is applicable in some languages and not in others. Some Asian languages require a "double byte" character set due to the complexity of the characters.
- *Text layout direction*: Most European languages are written from left to right, while Hebrew and Arabic languages use right to left, and some Asian languages optionally use a vertical text direction.
- *Keyboards*: Keyboards may vary with locale, and the software must correctly interpret key codes according to the appropriate character set.
- *Text filters and string lengths*: The software may accept only certain characters in a certain field. The software must allow and display every character in the appropriate places. The number of characters in the text strings, required to display, input, or output a word, may vary by language. Translated language tends to expand in length. Other considerations may include how the software converts from upper to lower case characters or how it sorts alphabetically.
- *Operating systems*: Can the software handle variances in wildcard symbols, filename delimiters, and common operating system commands?
- *Hot keys*: Do any hot keys from the original language have their old effect even though they do not appear in the localized menu?
- *Garbled in translation*: If the software builds messages from fragments, how does that appear in the localized version?
- *Spelling, underlining, and hyphenation rules*: Spelling, underlining, and hyphenation rules are not the same across languages. Spelling rules can even vary across dialects of the same language. Does the software's spell checker work correctly for the location?
- *Paper size*: Paper sizes differ in different countries.
- *Culturally bound graphics*: For example, in some cultures it would be inappropriate to use photos of women with the head or face uncovered.
- *Data formatting*: Different countries have variations in:
  - Time and date display formats and calendar formats
  - Money formats and currency symbols

- Number formats and the numeric separators—Americans put a comma in numbers; other countries put a decimal point or a space
- Rulers, grids, and measurements (distance, volume, speed, and so on) must be in the correct unit of measure for the location.
- Address formats, including ZIP Codes (e.g., the United States uses all numbers, Canada uses numbers and letters, and Curaçao does not use ZIP Codes)

Design Constraints

The design constraints are a subset of product-level, nonfunctional requirements that define any restrictions imposed on the choices that the software supplier can make when designing and implementing the software. For example, there may be a requirement that specific user interface formats, screen designs, or report formats be used. Other design constraints might include requirements to use a specific programming language, operating system, hardware platform, encryption methodology, algorithm, database format, or communications protocol.

Features versus Functions

Stakeholders often talk about software in relation to its features. Marketing literature or product vision statements often list a set of new or enhanced features that a new software product or new release of an existing software product will include. However, there is really no consistently used definition of the term *feature* in the software literature. Wiegers (2013) defines a *feature* as "one or more logically related system capabilities that provide value to a user and are described by a set of functional requirements." The IEEE (IEEE 2008a) definition of a feature as "a distinguishing characteristic of a system item (includes both functional and non-functional attributes such as performance and reusability)" expands the definition beyond the capabilities of the software to include the software's quality attributes.

What does seem to be pervasive in the literature is that the term *feature* is used to represent the perspective of the software customers or users and other external stakeholders when discussing requirements. This perspective ignores the software's implementation and looks at the software as a black box. It focuses on the value-added scope of the software.

The term *function* takes the perspective of the software developers. Functions define the functionality, behaviors, or capabilities that the developers have to build into the software for it to implement the features needed by the external stakeholders.

System versus Software Requirements

When the software is part of a larger system that includes other components, the business- and stakeholder-level requirements feed into the *product-level system requirements*. The system architecture, as discussed in Chapter 10, then allocates sets of system requirements downward to the software, hardware, and manual operations components. In this case, the *product-level software requirements* are the requirements that have been allocated to one or more software components of the system.

The system architecture also defines *requirements* for the interfaces between software components of the system and other components of the system, and between software components of the system and components external to the system. The system architecture also defines the requirements for integrating the system components, including software components, back together to build the complete system.

2. DATA/INFORMATION REQUIREMENTS

> Define and describe various types of data and information requirements, including data management, data integrity and privacy by design. (Understand)
>
> **Body of Knowledge III.D.2**

Many software products provide value through the input, manipulation, storage, and output of data. *Data/information requirements* define the specific data items or data structures and information products that must be included as part of the software product to fulfill the needs of the stakeholders. For example, the pay-at-the-pump system would have data requirements for purchase transaction data, gas price data, gas pump status data, and so on. It would also have information requirements for receipts, periodic reports, tax collection reports, and so on.

Information requirements define how data are transformed into information that is useful to the stakeholders of the software. According to the *Data Management Body of Knowledge* (DAMA 2017), "information is data in context." The raw material of information is data. By adding the context, collected data starts to have meaning. For example, a data item stored as the number 14 does not by itself provide us with any usable information. But if we know its context—for instance, if it has a definition such as "the number of newly detected defects," a time frame such as "last week," and relevance such as "while system testing software product ABC"—that data item is converted to information. Information can be stored as additional data.

The requirements analyst must determine the stakeholder information needs, identify users of the information, as well as when the information will be needed. The analyst must then determine the data requirements to provide that needed information. The goal of data/information requirements is to make sure that the collection, integrity, and accessibility of data are adequate to provide accurate, timely, and consistent information. Data/information requirements should support the stakeholders' business, operational, production, support, and maintenance information needs, including information needed for legal, regulatory, tax, governance, management, historical, audit, and other purposes.

Defining Data/Information Items and Data Relationships

After the requirements analysts identify the high-level inputs and outputs of the system, those data/information items should be assigned names and definitions, added to a defined data items list, and used consistently throughout the requirements. As stakeholder- and product-level requirements are further defined and refined, additional data/information item names and definitions continue to be added to that list. New data items can come from additionally identified inputs and outputs (including details from information in forms or reports) or data that need to be manipulated or stored by the system. The requirements should only define the required business data. The internals of the data structures and database design should be left to the designers unless they are design constraints.

One way to identify data/information items is to look for the nouns used in other requirements. If those nouns are data or information items, they should be added to the defined data items list. Having a single list of data items can greatly reduce the ambiguity and misunderstandings caused by using different names to refer to the same data item or using the same name to mean different items.

Using a *data dictionary* is a simple way to specify the data items list. A data dictionary can include the following types of information for each data item in the software:

- Name
- Definition, meaning, or description
- If the item is a data structure, a list of the other data items included in the composition of that structure (which provides a single point of definition so this information is not buried in, or redundantly repeated in, the requirements)
- Data type
- Size or length
- Format
- Default values
- Allowable values
- Units of measure

During requirements development, the data dictionary should only include required data and information. Additional details can be added later by the designers and implementers. For example, the definitions of data types, size or length, format, and default/allowable values are typically left until later unless those information items are design constraint–type requirements.

The requirement analyst can also use data models, including data flow, entity relationship, and class diagrams, as a "visual representation of the data objects and collections the system will process and the relationships between them" (Wiegers 2013).

Data Management

As part of data management, each required data/information item should be associated with the functional requirements responsible for its generation and use. These basic data management functions are defined by the *CRUD acronym*:

- *Create*: Specify requirements for creating the data item, or instances of the data item, in the software
- *Read (reference, retrieve)*: Specify requirements for the reading, searching for, or querying of the data item, or instances of the data item, in the software
- *Update (modify)*: Specify requirements for updating the data item, or instances of the data item, in the software
- *Delete*: Specify requirements for the deleting of the data item, or instances of the data item, in the software

One variation on the CRUD acronym is CRUDL, which adds *listing* for specify requirements for the listing or reporting of the data item, or instances of the data item or information product like a report in the software.

Examples of other data management requirements include:

- *Data access*: Specify data access methods for direct real-time access, batch access or data as a service
- *Data refreshing*: Specify how data from clients or other remote devices will be refreshed/updated into the centralized data store
- *Data sharing*: Specify requirements for sharing data with other systems or for data interoperability
- *Performance*: Specify any performance requirements associated with the data
- *Metadata*: Specify any required metadata (data about the data), including audit trails of data item access

Data/Information Integrity

Withall (2007) states that *information integrity* "can be summed up by the *ACID properties* possessed by any reliable database, which are:

- *Atomic*: Any change is made either completely or not at all
- *Consistent*: Whichever way you look at the data, you get the same picture
- *Isolated*: No change is affected by any other changes that's in progress but not yet completed
- *Durable*: Any completed change stays there."

Examples of data/information integrity requirements include the following:

- *Data longevity*: Specify how long data items must be retained within the system, including any requirements for archiving data items, restoring data items from the archive, and for disposing of data items
- *Data mirroring, backup, restoring, and recovery*: Specify requirements for:
  - Mirroring the data (storing it in multiple data stores or on multiple devices)
  - Backing up the data to safeguard its durability even when storage media fails or is corrupted
  - Restoring the data from the backup
  - Recovering any changes that have been made since the last backup
- *Data disposing*: Specify the disposition of data (including temporary data, metadata, residual data such as deleted records, cached data, local copies, archives, interim backups)
- *Data verification*: Specify how:
  - The accuracy of input or stored data will be verified
  - Inaccurate or duplicate data will be handled or reported
  - Incomplete data entries will be handled or reported (e.g., from incomplete/inaccurate user entry or because of system failures or aborts that result in partial data entry)
- *Coordination of multiple data stores*: Specify how inconsistencies between the same data stored in multiple places or on multiple systems will be handled

Privacy By Design

"*Privacy by design* (PbD) means privacy is seamlessly integrated into products, services, and system designs by default. Protecting customer data becomes a guiding force in the user experience, taking the same level of importance as functionality" (Hiscock 2023).

The ISO 31700-1 Privacy by Design standard "defines clear rules for organization around how consumer's personal information is processed and how consumer privacy is addressed throughout the product life cycle (Thielova 2023).

PbD has seven main principles:

1. Companies should be proactive by identifying privacy risks and mitigating them through the use of processes and procedures rather than being reactive when privacy breaches occur.
2. Privacy by default means that the users of your software products automatically receive the highest data security level without worrying about their personal privacy settings. This includes data collection, use, retention, and disclosure limits and minimizing the amount of data collected.

3. Data privacy requirements must be built into the software design and implementation and verified/validated throughout the life cycle, not tacked on after the software is developed.
4. "A fatalistic attitude won't work with Privacy by Design. Those who argue trade-offs must be made with the user experience or with security protocols have a zero-sum attitude. Those who work to integrate privacy into every design element seamlessly take a positive-sum approach . . . where privacy is increasingly a market mover, not just an issue of legal compliance" (Hiscock 2023).
5. The data are secure and protected throughout their entire life, from the point when they are collected until they are removed from the software and destroyed.
6. Privacy policies and procedures should be communicated to your users. "Users should never be in the dark about how you deal with their data. Transparency leads to trust, and this journey is made possible by clear documentation and communication" (Thielova 2023).
7. The privacy of user data should be the primary concern through the use of user-centric approach to data privacy.

3. QUALITY REQUIREMENTS

> Define and describe various types of quality requirements, including reliability and usability. (Understand)
>
> **Body of Knowledge III.D.3**

Quality Attributes

Stakeholder-level *quality attributes* are characteristics that define the product's quality from the stakeholders' perspective. Quality attributes typically define "how well" the software must perform to meet the quality expectations of the stakeholders.

A quality attribute may translate into product functional requirements for the software that specify what functionality must exist to meet that attribute. For example, an ease-of-learning usability requirement might translate into the functional requirement of having the system display pop-up help when the user hovers the cursor over an icon. A quality attribute may also translate into nonfunctional, product attribute requirements. For example, an ease-of-use usability requirement might translate into product attribute requirements for response time to user commands or report requests.

The *ISO/IEC 25000:2005 Software Engineering—Software Product Quality Requirements and Evaluation (SQuaRE)—Guide to SQuaRE* (ISO/IEC 2014, 2015) standards series provides a reference model and definitions for external and internal quality attributes and quality-in-use attributes. This standards series also guides planning, managing, measuring, evaluating, and specifying requirements for quality attributes. (See Chapter 14 for information about design and quality attributes.)

Reliability Requirements

Reliability requirements define the quality characteristics of the software associated with its ability to satisfactorily and correctly perform its intended functions over time and under various conditions. Examples of reliability-type quality attributes and associated product attribute templates include:

- *Reliability*: The extent to which the software can perform its functions without failure for a specified period of time under specified conditions

  Example product-level reliability requirements template: <*mean time or average time*> for a <*defined product or component*> to experience <*defined failure type*> under <*defined conditions*> (based on Gilb 2005)

- *Availability*: The extent to which the software or service is available for use when needed, which includes reliability and the ability or time required to recover from a failure if one occurs

  Example product-level availability requirements templates:

  <*Percentage*> of <*defined time period*> a <*defined product or product component*> is <*available*> for its defined <*tasks*>

  A <*defined product or product component*> will experience <*an average or a maximum amount of time*> of <*defined failure type*> during a <*time period*> under <*defined conditions*>

Usability Requirements

Usability requirements define the quality characteristics of the software associated with ease of learning or ease of use for a particular type of direct user. What may be user-friendly for a novice or occasional user might drive a power user crazy with extra details or actions. Conversely, what may be user-friendly to a power user might make a system too complex or unintuitive for a novice or occasional user. Examples of usability type quality attributes and associated product attribute templates include:

- *Usability*: The ease with which a user can operate or learn to operate software. Example product-level usability requirements templates (based on Gilb 2005):

  Ease of learning: <*average or maximum amount effort*> required for a <*user type or role*> to become proficient in performing a <*task or action*> under <*defined conditions*>

  Productivity: <*quantity*> of <*defined tasks or actions*> that can be performed by a <*user type or role*> using the product per <*time unit*> under <*defined conditions*>

  Response time: <*mean time or maximum time*> of a <*defined response*> to a <*defined user stimulus*>

  Error rate: <*quantity*> of user <*error type*> mistakes made by a <*user type or role*> per <*time unit*> under <*defined conditions*>

Likeability: <*quantity or percentage*> of <*user type or role*> that report <*liking*> the product

Other Quality Attributes

Examples of other quality attributes that might be desirable to customer and user type stakeholders include:

- *Accessibility*: The degree to which the software is available to as many people as possible (the ability to access). Accessibility is often focused on people with special needs or disabilities.
- *Accuracy*: The extent to which the software provides precision in calculations and outputs.
- *Efficiency*: The extent to which the software can perform its functions while utilizing minimal amounts of computing resources (e.g., memory or disk space).
- *Flexibility*: The ease with which the software can be configured, reconfigured, or customized by the user.
- *Interoperability*: The degree to which the software functions correctly and shares resources with other software applications or hardware operating in the same environment.
- *Robustness*, also called *fault tolerance* or *error tolerance*: The extent to which the software can handle invalid inputs or other faults from interfacing entities (e.g., hardware, other software applications, the operating system, or data files) without failure.
- *Scalability*: The ability of the software to be extended to accommodate more interactions, hardware connections, or users.

Remembering that the developers, vendors, and distributors of the software are also stakeholders, examples of other quality attributes that might be desirable to supplier-type stakeholders include:

- *Installability*: The ease with which the software product can be installed on the target platform
- *Maintainability*: The ease with which the software or one of its components can be modified or changed after it has been released to operations
- *Portability*: The effort required to migrate the software to a different platform or environment or the ability of the software to run well on multiple computer configurations (hardware, operating systems, browsers, and so on)
- *Reusability*: The extent to which one or more components of a software product can be reused when developing other software products
- *Supportability*: The ease with which the technical support staff can isolate and resolve software issues reported by the users or other stakeholder

4. COMPLIANCE REQUIREMENTS

> Define and describe various types of regulatory, safety, and data retention requirements. (Understand)
>
> **Body of Knowledge III.D.4**

Business Rules Including Regulatory Requirements

Business rules are stakeholder-level requirements that define how the stakeholders do business. The software product must adhere to these rules to perform appropriately within the user's domain. Business rules may be propagated into stakeholder functional requirements, quality attributes, or product-level requirements, including product functional requirements, product attributes, and even design constraints. As opposed to the business-level requirements, business rules are:

- Government regulations, laws, and guidelines
- Industry standards and guidelines
- Organizational/corporate policies, standards, and practices

Examples of business rules that the pay-at-the-pump software might need to implement include:

- Laws or regulatory requirements governing pumping gas (e.g., environmental protection or weights and measures regulations)
- Types of credit or debit cards taken (e.g., the software may only accept certain types of credit or debit cards or may not accept gas station credit cards from competing gas companies)
- Rules about what can be charged to a credit card (e.g., a maximum limit of $100 per transaction or no more than two transactions per card per day)

The cost of litigation can be a significant risk for software suppliers. Lawyers or regulatory experts should be consulted to determine which laws and regulations are applicable, and to confirm accurate interpretation of those laws and regulations. Requirements analysts must be "aware of the laws that apply to your kind of product, and write requirements to ensure that the product complies with these laws" (Robertson and Robertson 2013).

Safety Requirements

Safety-critical software is software that:

- Can result in an accident if it:
  - Inadvertently responds to stimuli

    - Fails to respond when required
    - Responds out of sequence
    - Responds in combination with other responses
- Is intended to prevent an accident
- Is intended to mitigate the results of an accident
- Is intended to recover from the results of an accident

An *accident* is anything that injures or kills a human being, damages or destroys property, or has an adverse impact on the environment or society. Examples of product-level safety attribute templates include:

- The system shall *<prevent, detect, report, react to>* a *<type of failure>* such that *<type of accident>* is prevented under *<defined conditions>*
- The system shall not cause more than *<average or maximum number>* of *<type of safety incidents>* per *<time interval or number of events>* under *<defined conditions>*
- The system shall not allow *<a safety risk>* to exceed *<specified level or variance>* under *<defined conditions>*
- The system shall react to *<type of accident or hazardous condition>* by performing *<type of action(s)>*
- The system shall *<prevent, detect, react to>* *<type of accident>* *<percentage of the time>*
- The system shall not allow *<this potentially hazardous event or action to happen>* unless *<this state, action, or event>* has already occurred

Data Retention Requirements

Many government regulations around the world address data retention requirements to protect data privacy. For example, the European Union (EU) General Data Protection Regulations (GDPR 2023) and Personal Information Protection Law (PIPL) discussed in Chapter 2.

5. SECURITY REQUIREMENTS

> Define and describe various types of security requirements including data security, information security, cyber security, data privacy, and encryption requirements for data in transit and at rest. (Understand)
>
> **Body of Knowledge III.D.5**

Software security is the ability to protect data and functionality against unauthorized access and to withstand malicious or inadvertent interference with its operations. Software security focuses on the probability that an attack on the software will occur and be detected, prevented, or recovered from. Examples of security attacks include attempts to:

- Gain unauthorized access to the software's data or functionality (e.g., accessing the software through a back door or spoofing, *spyware*, hacking, or tunneling)
- Compromise the software's integrity, availability, or confidentiality (e.g., through viruses, worms, or denial of service attacks)
- Inappropriately collect, falsify, or destroy data (e.g., unauthorized or inadvertent disclosure of private or personal data)

Types of security requirements include:

- *Cyber security*: Specify any protections of the software "from theft or damage to the hardware, the software, and the information on them, as well as from disruptions or misdirection of the services they provide. This includes controlling physical access to the hardware, as well as protecting against harm that may come via network access, data, and code injection, and due to malpractice by operators, whether intentional, accidental, or due to them being tricked into deviating from secure procedures" (Wikipedia.org 2016).
- *Data and information security*: Specify which sets of users have access/authority (or do not have access/authority) to read, write, modify, transfer, or delete specific data or information items and at what times and under what conditions, including any associated laws, regulations, standards, or contractual obligations.
- *Data privacy and confidentiality*: Specify requirements related to:
  - Safeguarding privacy and confidentiality of personal data, and defining an individual's access and rights to their personal data, including any associated laws, regulations, or standards
  - Restricting access to, or use of, secret information or data, intellectual property, proprietary information, or competition-sensitive data

Security attacks can come from outside the organization (e.g., from hackers, stolen identification or password, or other people trying to take advantage of security vulnerabilities in the software). Security attacks can also come from inside the organization through acts of sabotage, malicious code (time bombs, logic bombs, or Trojan horses), or back doors created by software practitioners. Software security requirements focus on four main objectives:

- Maintaining access control:
  - Limiting access to the system through passwords, firewalls, authentication, and physical security (access to facilities and equipment)

    - Limiting what the users can do once they are in the system through multilevel access control, privileges, security levels, and so on
    - Limiting exposure through access timeouts, service limitation, and so on
- Preserving the integrity and functionality of the software and the integrity and confidentiality of its data, through virus detection, encryption, handshaking on data communication, checksums, and so on
- Providing backup and recovery if security mechanisms should fail and result in corruption of the software or data
- Providing an audit trail by maintaining records of software access and use to provide the information needed for accountability

Examples of product-level security attribute templates include:

- *Attack handling*: <*probability*> that a <*defined product or product component*> shall <*handle (detect, prevent, recover from)*> a <*defined attack*> under <*defined conditions*>
- *Access control*:

  The systems shall <*lock, place out of service*> a <*external access*> after <*number*> unsuccessful attempts to <*login, handshake, authenticate*> within <*defined time period*>

  Access to <*specified data, assets, or functions*> is limited to users with <*defined roles, security levels, privileges, or authentications*>

  The system shall <*time out, log out, lock*> access to <*data, asset, function*> after <*defined time period*> of inactivity
- *Audit trail*: The system shall log all attempts to <*access secure data, or perform specific function*> by users having insufficient privileges

6. DERIVED REQUIREMENTS

> Apply derived requirements such as environmental requirements and user defined security. (Apply)
>
> **Body of Knowledge III.D.6**

According to the IEEE/ISO/IEC (2017) Systems and Software Engineering Vocabulary Standard, a *derived requirement* is:

1. A lower-level requirement that is determined to be necessary for a top-level requirement to be met.
2. A requirement that is not explicitly stated in customer requirements, but is inferred from contextual requirements (such as applicable

standards, laws, policies, common practices, and management decisions) or from requirements needed to specify a product or service component

Based on this definition, the product-level functional and product attribute requirements are derived during requirements analysis from the stakeholder-level requirements. Other derived requirements are introduced by solution decisions made during the system/software architecture and component design activities.

Environmental requirements define the required configuration(s) of hardware, software (operating system, browsers, databases, and so on), and other environmental factors in which the system or software operates.

Security requirements were discussed earlier in this chapter.

7. NONFUNCTIONAL REQUIREMENTS

> Define and describe various types of nonfunctional requirements such as documentation, performance, and technical maintenance. (Understand)
>
> **Body of Knowledge III.D.7**

Nonfunctional requirements are a category of requirements that includes all of the requirements that are not directly related to the functionality, behaviors, or capabilities of the software. Many times, nonfunctional requirements relate to multiple functional requirements. For example, an accuracy requirement that states, "All monetary calculations will be rounded to the nearest penny," would further define requirements for all functional requirements that deal with monetary calculations. Nonfunctional requirements include:

- Interface requirements
- Quality and product attributes
- Design constraints
- Data/information requirements

Documentation requirements are a subset of nonfunctional requirements that specify the documents that are delivered as part of the software product set. These documents may include:

- User manuals
- Operator manuals
- Help files
- Release notes, also called version description documents

- Training materials
- Specifications or planning documents
- Quality records
- Metrics that are periodically delivered during operations
- And so on

Performance requirements were discussed previously in this chapter. *Technical maintenance requirements* are typically defined in a service level agreement (SLA) or contract with the customers. (See Chapter 15 for more information on maintenance.)

8. REQUIREMENTS ELICITATION METHODS

> Describe and use various requirements elicitation methods, such as customer needs analysis, use cases, human factors studies, usability prototypes, Joint Application Development (JAD), and storyboards. (Apply)
>
> **Body of Knowledge III.D.8**

Requirements elicitation is the data and information gathering step in the requirements development process. The first step in requirements elicitation is to define the vision, scope, and limitations for the newly developed or updated software product. The *product vision* defines how the new or updated software product bridges the gap between the current state and the desired future state needed to take advantage of a business opportunity or solve a business problem. The *product scope* defines what will be included in the product, thus defining the boundaries of the product. As requirements analysts elicit and analyze potential business-, stakeholder-, and product-level requirements, they should always judge those requirements against the product's scope:

- If the requirements are within scope, they can be considered for inclusion in the software.
- If requirements are out of scope, they should be rejected unless they are important enough that the scope of the project should be adjusted to include them.

The *product limitations* document items that were considered for inclusion in the software but for whatever reason a decision was made not to include them. It is essential to document the scope and limitations because they help establish and maintain stakeholder expectations for the product. The scope and limitations documentation can also help in dealing with requirements volatility and gold-plating issues.

Stakeholder (Customer) Needs Analysis

Once the vision, scope, and limitations are defined, they are used to identify potential software product stakeholders. From the list of potential stakeholders, a requirements engineering stakeholder participation plan (as discussed in Chapter 6) is developed to define decisions about:

- Which stakeholders will participate in the requirements elicitation and other requirements engineering activities
- How each of those stakeholders will participate in those activities
- Who will represent them

Direct Two-Way Communications

Probably the most effective way of eliciting information to perform customer needs analysis is through direct two-way communications with the stakeholders. Direct communication with stakeholders has a major advantage over many other techniques because it is a two-way communication technique with a feedback loop. This feedback allows the requirements analyst to ask questions, request examples, obtain additional information, and clarify terminology or ambiguities. Feedback can also be nonverbal like facial expressions or body language. Direct two-way communications with the stakeholders can be accomplished through:

- Interviews
- Focus groups
- Facilitated requirements workshops
- The evaluation of prototypes

Interviews

"One of the most important and most straightforward requirements elicitation techniques is the interview, a simple, direct technique that can be used in virtually every situation" (Leffingwell and Widrig 2000). Interviews are an excellent mechanism for obtaining detailed information from another person. Before the interview, the requirements analyst must decide on the scope of the interview. For a software product of any size, there are often a huge number of topics that could be covered in an interview. If the analyst tries to cover everything simultaneously, the interview may be unfocused, and the information obtained may be superficial.

Focus Groups

Focus groups are small groups of selected stakeholders representing the "typical" type(s) of stakeholders in a single stakeholder group from whom input is needed. Lauesen (2002) recommends that a focus group include 6 to 18 people. The group should also be fairly homogeneous. For example, the gas purchaser stakeholder group might need to be divided into several homogeneous focus groups, separating family car drivers from drivers of 18-wheelers from fleet drivers of taxis and military vehicles.

Focus groups can be particularly valuable if the software product has one or more large or very diverse stakeholder groups. For example, the stakeholder group of gas station customers for the pay-at-the-pump system is very large and diverse. If a stakeholder group doesn't have any obvious candidates for a single stakeholder champion, the requirements analyst might want to consider using one or more focus groups to elicit requirements from this stakeholder type.

The focus group participants are brought together in a meeting to discuss one or more aspects of the requirements for the software product. While focus group members typically don't have any decision-making authority, their input into the requirements elicitation process can be valuable. Multiple focus group sessions using different samples from the same stakeholder group may also be helpful in confirming good coverage. For example, one focus group of family car drivers might focus on the user interface, while another group might concentrate on a frequent buyer program.

Open the focus group meeting with a presentation of the topic or theme of the meeting. It may also be beneficial to conduct an icebreaker activity to give the focus group members time to get to know each other and feel comfortable. The intent is to get the focus group members to interact and discuss the software product. For example, the facilitator might have the focus group members discuss what they like about the existing system, or the group could brainstorm a list of bad experiences with a similar product in the same kind of work domain. Another approach is to have the group discuss a future vision for the product and what they wish it would be like. The focus group could also be shown a prototype and asked what they like or dislike about it.

The goal of a focus group session is to record as many ideas as possible in a short amount of time. During the information-gathering session, the facilitator makes certain that everyone's ideas are heard and that those ideas are not critiqued, criticized, or rejected. Recording the ideas can be done directly, as part of the focus group meeting, or by an observer who is not actively participating in the meeting (e.g., someone observing the group through a two-way mirror).

After ideas are gathered, it may be useful to have focus group members select their top-priority ideas. For example, ask each participant to make a list of their top 10 choices. These lists then become inputs the requirements analysis team uses to help identify product requirements and their priorities.

Facilitated Requirements Workshops—Joint Applications Development

A *facilitated requirements workshop,* also called *joint application development (JAD)* or *joint application design,* is a process that brings together cross-functional groups of stakeholders to produce specific requirements work products (e.g., use cases, user stories, storyboards and so on). These workshops differ from focus groups in the group's make-up and the group's goal. A focus group is a homogenous group of stakeholders that meets to gather opinions and ideas. A facilitated requirements workshop is a heterogeneous team that meets with the goal of making decisions and creating one or more requirements deliverables. For example, a workshop might create a list of business rules, flesh out the details of a set of use cases, or create and analyze a requirements model.

By bringing together a diverse team of representatives from various stakeholder groups (e.g., customers, users, developers, testers, and other stakeholders), facilitated requirements workshops can help streamline communications and identify and resolve requirements issues between impacted parties. This process can reduce the time it takes to elicit the requirements and produce higher-quality requirements work products. These workshops focus on collaboration and team-building, which promote a sense of joint ownership of the deliverables and resulting product(s). Team synergy can also reduce the risk of missing requirements.

A trained, neutral facilitator is used during these workshops to help create productive participant interactions. The workshop team members are stakeholder representatives with decision-making authority and subject matter expertise. These team members are responsible for the content and quality of the requirement deliverables produced during the workshop. Each workshop should also have a recorder responsible for keeping a written record of each team meeting. This includes recording decisions and issues and any action items assigned during the session.

Facilitated requirements workshops can require a lot of work outside the workshop meetings themselves. The facilitator must prepare for the workshop by working with team members to determine the specific scope and deliverables for each meeting and planning the activities and tools that will be used. The facilitator must custom-design each meeting for the specific team based on the needs of the specific product or deliverables being produced. The workshop also typically requires pre-work on the part of some or all of the participants. For example, participants may brainstorm ideas with the team they represent to bring to the meeting or create a "straw man" version of a deliverable for the other members to review.

These workshops are an iterative process of discovery and creativity that requires the interaction and participation of all the team members in "serious play." Serious play "means playing with models as a means of innovation, invention, and collaboration." It also "means having fun at the meetings as a means of enhancing the group's productivity, energy, and interrelationships" (Gottesdiener 2002).

Prototypes

A *prototype* is a demonstration version of the software product. It can be used as a "straw man" to show to the stakeholders to help elicit and analyze requirements and determine if the requirements analyst's understanding of what was heard during the elicitation process was correct. Demonstrating prototypes to various stakeholders can also elicit new information or requirements that might have been missed in the initial discussions and validate the requirements. Prototypes can be particularly useful if:

- The customers, users, or other stakeholders don't really know, or have doubts about, what they want
- The developers have doubts about whether or not they understand what the stakeholders are telling them
- The feasibility of the requirement is questionable

There are various approaches to creating prototypes:

- *Paper*: Low-fidelity prototypes where, for example, the screens or reports are sketched out on paper
- *Wizard of Oz*: For example, interactive mock-ups, where a person behind the scenes or tool (e.g., Microsoft PowerPoint using hyperlinks) responds to user inputs with the subsequent display or output
- *Automation*: High-fidelity prototypes that resemble a functioning system. The risk here is that the stakeholder viewing the prototype may think the product is close to being finished and not understand why product release is not scheduled in the near future

Prototypes can be constructed as either throwaway prototypes or evolutionary prototypes:

- For *throwaway prototypes*, development creates the prototype, uses it for requirements elicitation or analysis, and then discards it. Typically, throwaway prototypes are "quick and dirty" development efforts focusing on the software product parts of interest. Extra work is not expended on error checking or code comments.
- On the other hand, *evolutionary prototypes* intended to become part of the final software product are typically built using a more rigorous development approach. More time is spent creating a high-quality prototype to evolve into a robust, reliable software product.

Prototypes can also be constructed as horizontal or vertical prototypes:

- A good analogy for a *horizontal prototype*, also called a mock-up, is the false front scenery on a movie set. Horizontal prototypes look and feel like the real system from the outside perspective, but under the surface, it's all smoke and mirrors. There is not much real code backing them up. Horizontal prototypes can be valuable when analyzing the software's user interface or performing usability evaluations.
- *Vertical prototypes*, also called *proof of concept prototypes*, are a single slice through the product that focuses on proving the feasibility of one specific aspect of the software product during requirements evaluation. For example, developers could prototype just the charge card reading and verification process to make sure that timing constraints can be met or that the communication protocols with the credit card clearinghouse are understood.

Prototypes can, therefore, be valuable tools to help with the following:

- *Elicit requirements information*: For example, by showing a user-interface prototype to a focus group and asking the participants what they like and dislike about that interface
- *Analyze the requirements*: For example, using a proof-of-concept prototype to analyze the feasibility of the requirements

- *Validate the requirements*: For example, using a prototype of a report to make certain that the specified report requirements meet the intended use of the report

Other Customer Needs Elicitation Techniques

Examples of other requirements elicitation techniques include:

- *Observation of work in progress*: This includes site visits, where the users are observed performing their actual jobs. Users can also be observed in a simulated environment (e.g., using a prototype).
- *Questionnaires or surveys*: Sent to stakeholders to obtain their opinions or input, including stakeholder satisfaction surveys or *marketing surveys*.
- *Process flow analysis*: Can provide an understanding the flow of the current processes and how the software is intended to automate or support that process.
- *Analysis of competitors' products*: Care must be taken to avoid legal or regulatory concerns.
- *Reverse engineering existing products*: This may be required for maintenance on older software products where requirements are unavailable and must be rediscovered. Care must be taken to avoid legal, licensing, or contractual concerns.
- *Benchmarking and best practices*: Evaluation can be done of high-quality products. For example, when creating a user interface, benchmarking of good websites might be done to elicit information about good practices in usability.
- *Human factors studies*: These studies consider how the human users of a software system physically interact with that system and its environment. The purpose of doing human factors studies in relationship to requirements is to verify that the software system conforms to the abilities and capabilities of its human user. These studies consider ease of use, ease of learning, *ergonomics*, usage preferences (e.g., some people prefer to use keyboard entry rather than a mouse, or certain screen color combinations may be aesthetically displeasing), education and training levels, physical handicaps, languages, customs, and so on. The study of human factors can be critical where the potential exists for possible human errors to cause safety concerns—for example, where performing tasks out of sequence might cause unsafe conditions.
- *Document studies*: Sometimes the requirements have already been written in, or identified from, some other form of documentation. For example, there may already be a detailed specification for how the software must interface with an existing piece of hardware, or a manually created report, and the software just needs to duplicate its contents and layout. As appropriate, requirements elicitation is implemented through studies of existing documentation, including:

- Industry standards, laws, or regulations
- Product literature from the development organization or its competition
- Process documentation and work instructions from the customers or users
- Change requests, problem reports, or help desk reports
- Lessons learned from prior projects or products
- Reports, information products, and other deliverables from the existing system

Capturing Stakeholder (Customer) Needs

There are multiple techniques for capturing stakeholder needs, including user stories, use cases, and storyboards.

User Stories

A *user story* is an agile development mechanism used to capture a description of a stakeholder need in a way that acts as a reminder to have a future discussion about that need. The intent is to defer elaborating a stakeholder need into detailed requirements until just before each story is implemented—just-in-time requirements development. This process eliminates the waste of up-front effort spent elaborating requirements that, for whatever reason, are never implemented (e.g., priorities change, stakeholder needs change, or the project gets canceled).

User stories are typically only one sentence long and are written in the language of the stakeholder or business. They are usually created during face-to-face conversations between the developers and the other stakeholders of the software. A user story identifies the stakeholder, the stakeholder need, and the justification for that need. For example, "As the gas station owner, I need all credit card transactions validated by the credit card clearinghouse to minimize my financial exposure from stolen, expired, or over- the-limit cards." User stories may also include other information deemed valuable to act as a reminder of the stakeholder's need and prompt future discussion to refine and elaborate that need and define its acceptance criteria, for example, during the backlog refinement meeting or iteration planning meeting.

Use Cases

A *use case* is a scenario created to describe a usage thread for the system being implemented. The entire set of use cases describes how the system will be used and can provide the basis for functional *usage scenario testing*. While use cases came out of object-oriented analysis techniques, they can be effectively used for many different types of applications. Use cases could be created from the information received during an interview. The requirements analyst could then walk through the use case with the interviewee to obtain feedback that elicits more information and validates the use case. Use cases could also be used in a facilitated requirements workshop to elicit, analyze, specify, and validate stakeholder needs.

Step 1: The first step in defining use cases is to determine the scope (boundaries) of the system under consideration. Determining the scope identifies the different actors. *Actors* are entities outside the system under consideration that interact with that system. The scope of the system under consideration determines who the actors are and are not. For example, if the gas pump is considered part of the system, it is not an actor. However, if the gas pump is outside the scope of the system under consideration, it would be an actor.

Actors are different than users. A user may perform several actor roles when interacting with the system. For example, a single user of a word processing package could have roles including author, reviewer, and editor of a document. Each of these roles would be a separate actor. Actors can also be hardware devices, other software applications, or other systems that interact with the software system under consideration. For example, the credit card clearinghouse actor for the pay-at-the-pump product is another software application with which the software must interact to obtain authorization for credit or debit card purchases.

Step 2: The second step is identifying and creating a list of interactions between the actors and the system under consideration. These interactions may be initiated by either the actor or by the software product on behalf of the actor. A *use case diagram,* as illustrated in Figure 12.4, is one mechanism for documenting this list of interactions and showing the related actors and the use case interrelationship to each other.

A use case diagram starts with a box representing the software system's scope. Outside the box are stick figures representing the various actors that interact with the product. Each stick figure is labeled with the name of that actor. For example, for the pay-at-the-pump application some actors might be the gas station customer, owner, and attendant and the credit card clearinghouse. The ovals inside the box represent the interactions or use cases. Use cases are typically named using a verb and object (e.g., *purchase ticket, print a boarding pass,* or *add a user*). They may also have additional qualifiers (e.g., using adjectives as in *generate monthly tax reports,* or other qualifiers such as *purchasing gas at pump using a credit card*).

The arrows in a use case diagram show the interactions between the actor and the use cases. If the arrow goes from the actor to the use case, it indicates a *primary actor* that either initiates that use case or for which that use case is initiated (e.g., the system might initiate automated nightly reports for the gas station owner). An arrow from a use case to an actor indicates a *secondary actor.* Secondary actors are also involved in the use case but are not the initiator of the use case. For example, in Figure 12.4, the primary actor for the *purchase gas at pump with credit card* use case is the customer who initiates the transaction with the software. Secondary actors include the credit card clearinghouse and the attendant.

Use case diagrams can also indicate the relationships between use cases:

- An *<<include>> use case* separates a behavior that is similar across multiple use cases into a separate use case in order to remove redundancy and complexity.
- An *<<extend>> use case* adds behavior to the base use case at a specific *extension point* under specific conditions. For example, *the purchase gas at pump with credit card* use case can be extended at a specific point with the *purchase car wash at pump* use case.

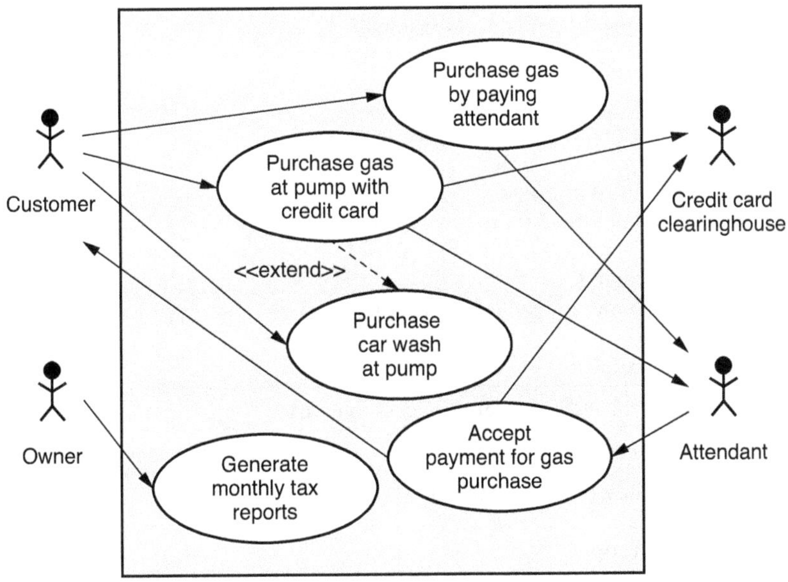

Figure 12.4 Use case diagram—example.

Of course, the use case diagram in Figure 12.4 has been simplified for illustration purposes and does not show all possible actors or use cases. In this example, other use cases might include purchase gas at pump with debit card, generate weekly reports, change gas price, change tax rates, and so on.

Step 3: Each use case is then developed to describe the details of each start-to-finish interaction between the actor and the system. A use case defines:

- *Primary actor*: The actor that initiates the use case
- *Secondary actors*: Other actors that interact with the use case
- *Preconditions*: Specific, measurable conditions that must be met before the use case can be initiated (that is, entry criteria)
- *Post-conditions*: Specific, measurable conditions that must be met before the use case is considered complete (that is, exit criteria)
- *Main success scenario*: Also called the *happy path*, the normal or typical sequence that results in a successful interaction with the actor(s)
- *Alternative scenarios*: Other alternate sequences are variations that still result in a successful completion of the task (that is, satisfaction of the post-conditions)
- *Exception scenarios*: Other exception sequences are variations that result in an unsuccessful completion of the task (that is, post-conditions are not satisfied)

There are many different ways to document a use case. Table 12.1 illustrates the *two-column use case method*, where actor actions are documented in one column

Table 12.1 Two-column use case method—example.

Use case scenario: Accept post payment for gas
Primary actor: Attendant
Secondary actors: Customer, credit card clearinghouse
Preconditions:

- Gas has been successfully pumped
- Customer has arrived at attendant to pay for gas

Post-conditions:

- Payment has been received
- Customer was able to pay for gas with their preferred payment type (cash, credit card, debit card)
- Customer received a receipt

Main Success Scenario	
Actor Actions	**System Responses**
1. Attendant greets customer and asks which pump was used 2. Customer identifies pump 3. Attendant polls system for pump information 5. Attendant confirms price with customer and acknowledges price with system 7. Attendant asks if there are other items to be purchased and customer responds "no" 8. Attendant asks for payment type and customer pays in cash 9. Attendant enters cash tendered into cash register 11. Attendant provides change and receipt to customer 12. Attendant ends transaction	4. System reports gallons pumped and total price 6. System accepts price and displays price on cash register display 10. System calculates and displays change and prints receipt 13. System saves transaction information and resets pump
Alternative scenarios:	
Actor actions	**System responses**
5a1. Customer identified *wrong pump* and changes pump identification 5a2. Attendant cancels previous pump 5a4. Return to step 3	5a3. System resets to "no pump identified"
7a1. Customer purchases other items 7a2. Attendant enters price for each item in cash register 7a4. Return to step 8	7a3. System accepts price and displays item names, prices, and running total price on cash register
Exception scenarios:	
Actor actions	**System responses**
4a2. Attendant manually checks pump display and enters amount into cash register 4a3. Return to step 7	4a1. System cannot communicate with pump

Table 12.1 (Continued)

Exception scenarios:	
Actor actions	**System responses**
8a1. Attendant asks for payment type and customer swipes invalid credit card type (not accepted type of card) 8a4. Return to step 8 8b1. Attendant asks for payment type and customer swipes invalid credit card (expired, reported stolen, or over limit) correctly (after one or more tries) 8b4. Credit card clearinghouse disapproves transaction 8b6. Return to step 8	8a2. System reads and parses magnetic strip 8a3. System displays error 8b2. System reads and parses magnetic strip 8b3. System establishes communications with credit card clearinghouse and transmits merchant information, credit card information, and transaction amount 8b5. System displays disapproval

and system responses are documented in the second column. Again, this example has been simplified. Other alternative scenarios, not shown in this example, might include *purchasing gas with a valid credit card or debit card, entering an incorrect PIN one or more times before entering it correctly,* and so on. Other exception scenarios, not shown in this example, might include *the system not being able to communicate with the pump display, the system not being able to print a receipt, the system not being able to read the card's magnetic strip, the customer entering the PIN wrong three times,* and so on.

Other methods for documenting use cases include the *one-column use case method,* where all the steps are listed in a single column, and creating diagrams of the use where each step is a bubble and arrows show the relationships between the actors and the steps.

The level of detail used to document the use case depends on the needs of the project at the specific time. Typically, use cases are documented at a very low level of detail initially, and then more details are added as necessary. For example, early in development the use case in Table 12.1 might have just included the following steps (one-column method):

1. Attendant polls pump for gas amount and price
2. Customer pays attendant
3. Attendant provides receipt and change, if any

Other information that can be defined for each use case includes:

- A unique use case identifier and the use case name
- The use case creation history (who created the use case and its creation date)
- The use case modification history (who modified it and the modification dates)
- Use case description
- Use case priority

Figure 12.5 Storyboard—example.

- Frequency of use
- Related business rules
- Related assumptions

Storyboards

Similar to the panels in a comic strip, a *storyboard* graphically illustrates who the characters are in a story and describes the order of what happens to those characters and how it happens. For example, Figure 12.5 shows a storyboard for the main success scenario from the use case in Table 12.1. Storyboards are mainly used in requirements engineering to describe human user interfaces. Pictorial sequences are often easier for stakeholders to visualize and interpret than written steps. They can aid in understanding and provide a basis for discussion about what needs to happen and how.

9. REQUIREMENTS EVALUATION

> Assess the completeness, consistency, correctness and testability of requirements and determine their priority. (Evaluate)
>
> **Body of Knowledge III.D.9**

Requirements Analysis

Requirements analysis takes the information gathered during requirements elicitation and evaluates it, looking for gaps that need to be filled through more requirements elicitation and conflicts that need to be resolved to have a complete, consistent set of stakeholder requirements. Requirements analysis also refines multiple views of the requirements to derive detailed product-level requirements. Other aspects of requirements analysis activities include evaluating technical feasibility and testability and analyzing requirement and failure mode risk.

Models are usually used to aid in the requirements analysis process and can be particularly beneficial because models:

- Present different summarized views of requirements
- Help decompose business- and stakeholder-level requirements into product-level requirements
- Aid in communications—"A picture is worth 1,000 words"
- Act as a transition between requirements and design
- Aid in the identification of:
    - Requirements defects
    - Missing requirements
    - Non-value-added requirements

Traditional software models, including data flow diagrams, entity relationship diagrams, and state transition diagrams and tables, come from structured analysis techniques. These models have been around for decades, but even though they are older than their object-oriented counterparts, they can still be very useful in analyzing requirements.

Object-oriented techniques bring several additional models to the requirements analysis tool belt. One advantage of these models is that they have been standardized by the Object Management Group (OMG). Object-oriented models include the use case diagrams and use cases discussed previously in this chapter, as well as class, sequence, and activity diagrams.

Other requirements analysis models come from outside the software-specific arena. For example:

- *Process flow diagrams*: Used for defining processes as part of a quality management system; can be used to describe processes that are being automated using software (see Chapter 6 for a discussion of process flow diagrams)
- *Decision trees*: Can be used to analyze system decisions as a basis for requirements analysis (see Chapter 22 for a discussion of decision trees)
- *Event/response tables* can be used to help analyze error-handling requirements

Data Flow Diagrams

A *data flow diagram* (DFD) is a graphical representation of how data flows through and is transformed by the software system. Figure 12.6 illustrates the Yourdon/DeMarco symbols used in a data flow diagram. In a requirements-type DFD, the circles represent a process or set of functions that must be performed (unlike a design-type DFD, where a circle represents specific software components or modules).

At the highest (most general) level, a data flow diagram can be used to define a *context diagram* of the system, where a single circle represents the entire software system, and the external entities are represented by rectangles, with arrows depicting the data flows between the system and those external entities. Data flow diagrams can also be used to decompose and describe the internal workings of the software in progressively more detail. For example, in the DFD illustrated in

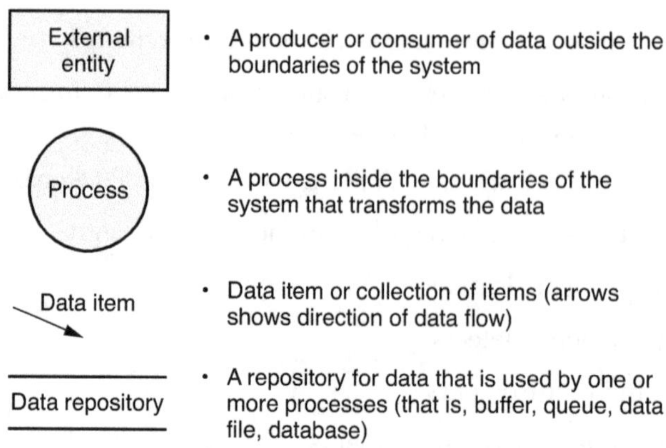

Figure 12.6 Data flow diagram (DFD)—Yourdon/DeMarco symbols.

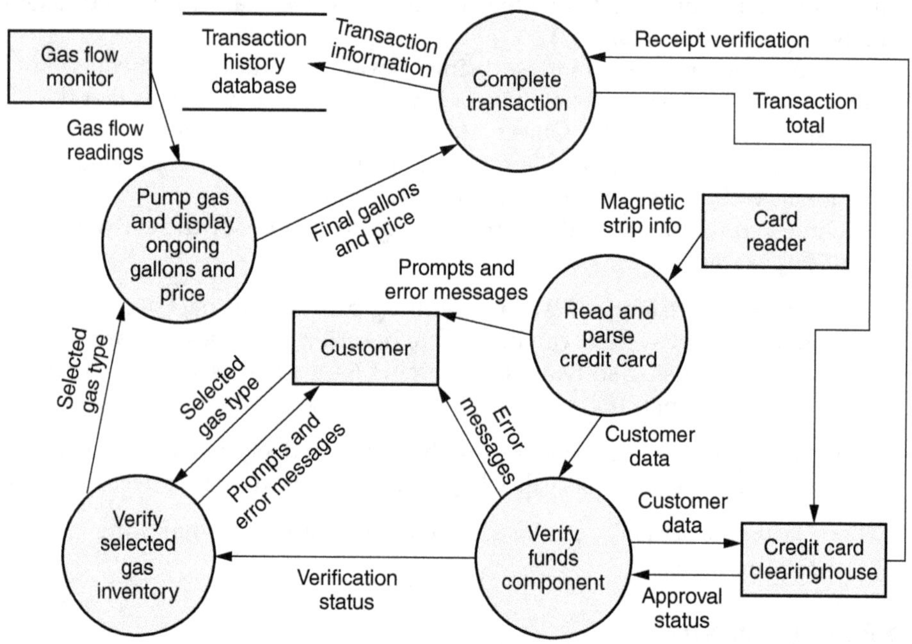

Figure 12.7 Data flow diagram (DFD)—example.

Figure 12.7, a circle represents the *pump gas and display ongoing gallons and price* process. This circle could be further broken down into a lower-level data flow diagram that describes more detail about the internals of that process, and so on.

Entity Relationship Diagrams

An *entity relationship diagram* (ERD) graphically represents the relationships between the system's data items (entities). The primary components of an ERD include:

- *Data items*: Represented by a labeled rectangle
- *Relationships*: Represented by labeled lines connecting the data items
- *Indicators for cardinality and modality*: Represented by symbols or numbers at the ends of the relationship lines

The example ERD illustrated in Figure 12.8a has five entities (*gas station, gas pump, type of gas, transaction, receipt*). The labeled lines show the relationships (*the gas station contains gas pumps, gas pumps pump different types of gas, gas pumps process transactions,* and *receipts record transactions*).

Cardinality is the specification of the number of occurrences of one data item that can be related to the number of occurrences of another item. As illustrated in Figure 12.8b, the outside set of symbols (on the relationship lines) represent the cardinality, with the line symbol indicating one relationship and a "crow foot" symbol indicating many relationships. Two entities can be related as follows:

- *One-to-one:* One occurrence of object A can be related to one, and only one, occurrence of the other object B. For example, each receipt records a single transaction, and each transaction is recorded on a single receipt.
- *One-to-many:* One occurrence of object A can be related to one or many occurrences of the other object B. For example, a gas station contains one or more gas pumps, but each gas pump is located at only one gas station.
- *Many-to-many:* Each occurrence of object A can be related to one or many occurrences of object B, and each occurrence of object B can be related to one or many occurrences of object A. For example, each gas pump can pump multiple types of gas, and each type of gas can be pumped from multiple gas pumps.

Modality indicates whether or not the data item is required. As illustrated in Figure 12.8c, the inside set of symbols (on the relationship lines) represent the modality. The modality relationship symbol is a circle if the relationship is optional. For example, not every transaction will have a printed receipt. The modality is a straight line if the relationship is mandatory. For example, every receipt must have an associated transaction.

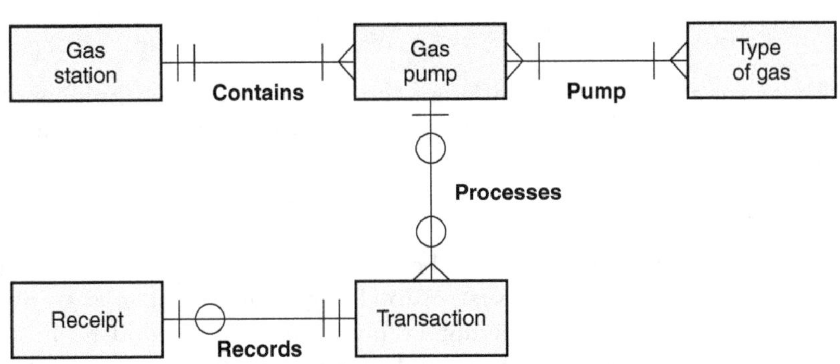

Figure 12.8a Entity relationship diagram (ERD)—example.

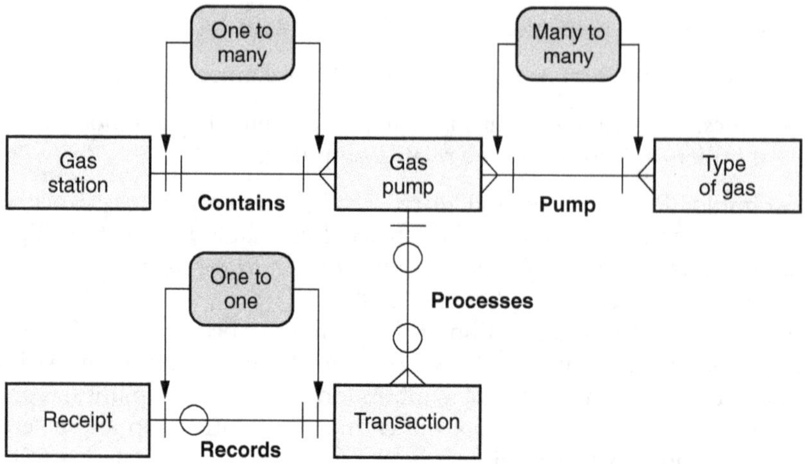

Figure 12.8b Entity relationship diagram (ERD) cardinality—example.

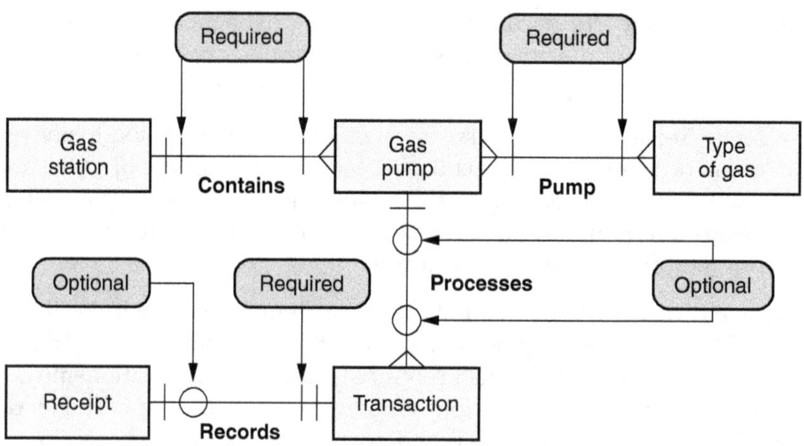

Figure 12.8c Entity relationship diagram (ERD) modality—example.

There are actually many different types of symbols and numbers that can be used to identify the cardinality and modality in an ERD, as illustrated in Figure 12.8d.

State Transition

State transition analysis evaluates the behavior of a system by analyzing its states and the events that cause the system to change states. A *state* is an observable mode of behavior. In the following examples, a printer could be in a *wait for print command, print receipt, error-jammed,* or *error-empty state.* State transition can be illustrated using a *state transition diagram,* as shown in Figure 12.9, or as *state transition table,* as shown in Table 12.2.

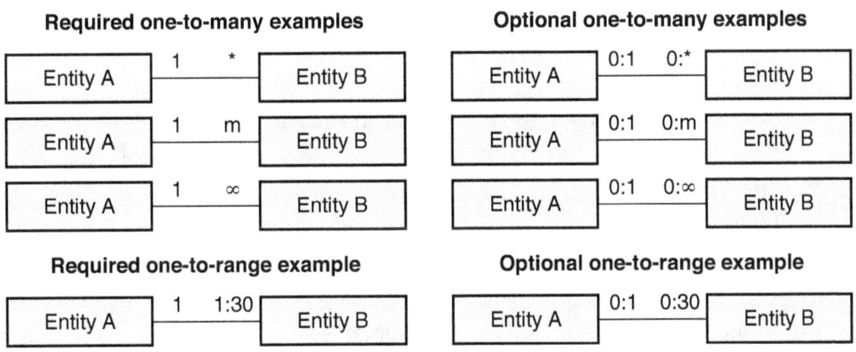

Figure 12.8d Other cardinality and modality symbols—example.

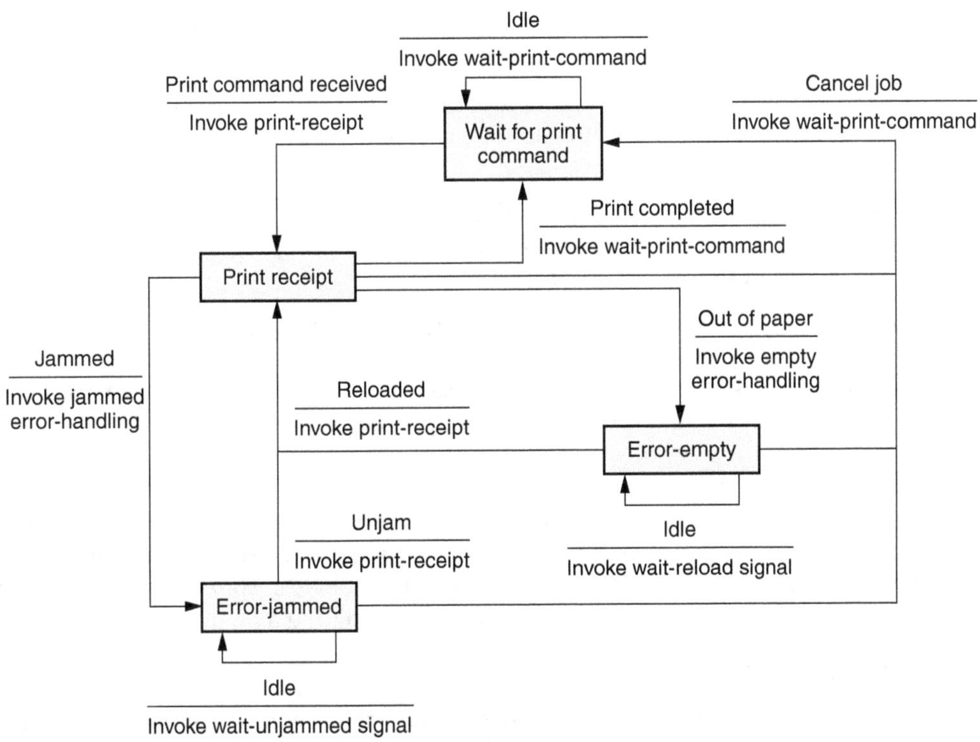

Figure 12.9 State transition diagram—example.

Class Diagrams

A *class diagram* is an object-oriented model of the system's static structure that shows the system's classes and their relationships. *Classes* are the physical or conceptual entities that make up the product, which, as illustrated in the example in Figure 12.10, are drawn in a class diagram as a rectangle with three partitions:

Table 12.2 State transition table—example.

		To State			
		Wait for Print Command	**Print Receipt**	**Error: Jammed**	**Error: Empty**
From State	**Wait for Print Command**	Idle	Print command received	—	—
	Print Receipt	Print completed or cancel job	—	Jammed	Out of paper
	Error: Jammed	Cancel job	Unjammed	Idle	—
	Error: Empty	Cancel job	Reloaded	—	Idle

- *Name*: Includes the class name that uniquely identifies the class and other general properties of the class (e.g., the gas pump class or the transaction class).
- *Attributes*: A list of the class's properties and data. For example, the gas pump might have a status (idle, in use, out of service) or values like "total gas pumped today."
- *Operations*: A list of functions that the class performs. For example, the gas pump might perform operations like pumping gas, displaying gallons pumped, or printing a receipt.

The lines between the classes in a class diagram show the relationships between the classes. The same cardinality and modality symbols used for ERDs are also used in class diagrams (see Figure 12.8d). For example, the *gas pump* class, illustrated in Figure 12.10, has a one-to-many relationship with the *transaction* class.

Sequence Diagrams

A *sequence diagram*, also called an *interaction diagram*, is an object-oriented model that records in detail how objects interact over time to perform a task by describing the sequences of messages that pass between those objects. The example illustrated in Figure 12.11 shows a sequence diagram for validating a credit card.

Activity Diagrams

An *activity diagram* is an object-oriented model that depicts a dynamic, activity-oriented view of the software product's functions and describes the workflow, or flow of control, through the activity. As illustrated in Figure 12.12, each rounded-corner box represents an action, and arrows show the sequencing of these actions.

Event/Response Table

Event/response tables list the various external events that can occur and define how the software product should respond to those events based on the given state when the events occurred. Events are initiated by users (actors), including hardware devices like sensors, buttons, and so on, or by time (e.g., the end of the day, the elapse of a certain amount of time). Table 12.3 shows an example of a partial event/response table.

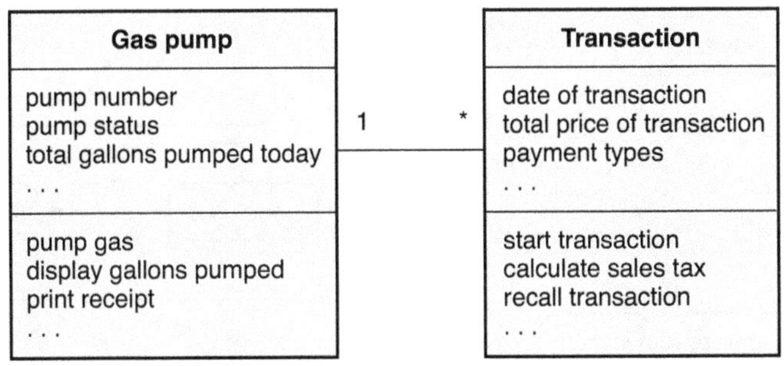

Figure 12.10 Class diagram—example.

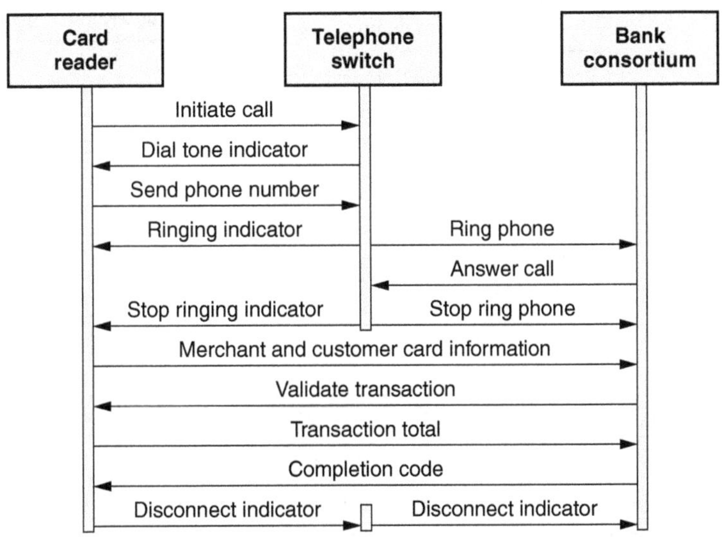

Figure 12.11 Sequence diagram—example.

Requirements Specification

The *requirements specification* can take many forms or be captured in one or more specification documents. For example, all of the requirements information may be documented in a single *software requirements specification (SRS)* document. On other projects, the requirements may be specified in multiple documents. For example:

- Business requirements may be documented in a *business requirements document (BRD), marketing requirements document (MRD), or project vision and scope document, or as part of a concept of operations document.* IEEE provides a Guide for Information Technology—System Definition—Concept of Operations (ConOps) Document (IEEE 1998b).

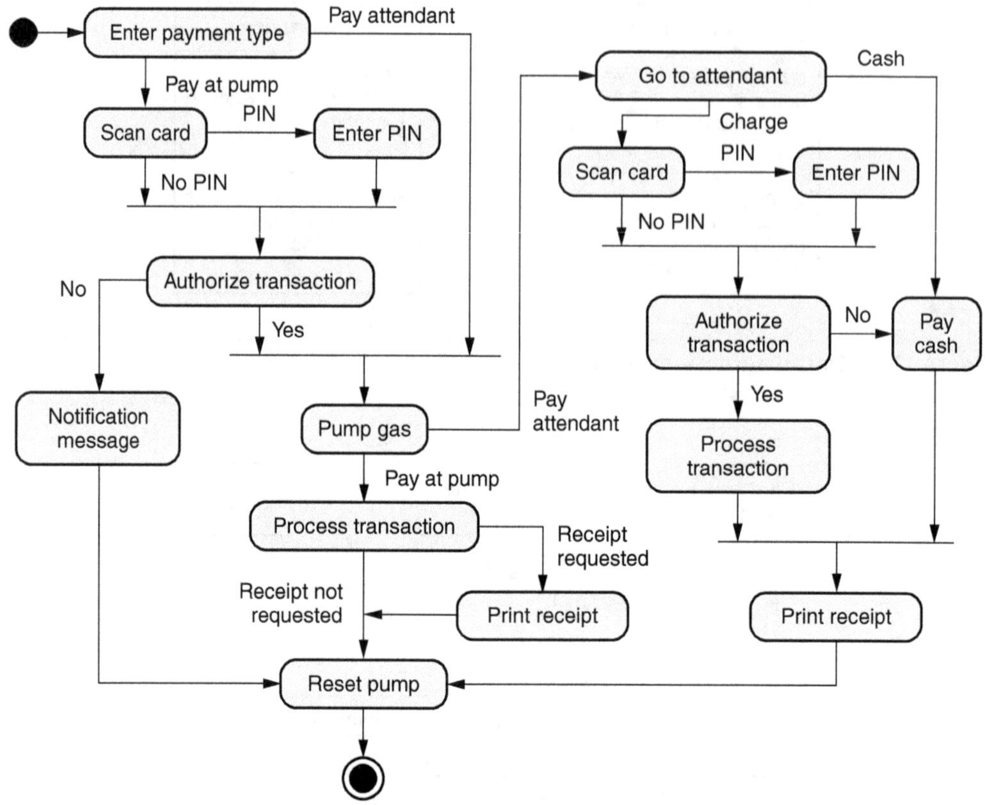

Figure 12.12 Activity diagram—example.

- Stakeholder requirements may be documented in a set of use cases, user stories, a *user requirements specification (URS)* document, or as part of a *concept of operations document.*
- The system requirements may be documented in a *system requirements specification.*
- The software functional and nonfunctional requirements, and other product-level requirements, may be documented in a *software requirements specification (SRS).* IEEE provides Recommended Practice for Software Requirements Specifications (IEEE 1998c).
- External interfaces may be included in the SRS or in separate *external interface requirements* documents.

In fact, there may not even be a document as such. Another way of specifying requirements is to document them in a requirements tool or database. In its simplest form, the requirements specification may be a list of "to dos," such as:

- Items in a Scrum product backlog or sprint backlog
- Elements in a spreadsheet

Table 12.3 Event/response table–example.

Event	Pump State	Response
Card inserted into card reader	Pump out of service	Not applicable (N/A)
	Idle	Start new transaction—prompt to remove card
	Card already scanned and transaction started—prompting for inputs	Transaction in progress—cancel previous transaction (y or n) message If n—Continue with previous card If y—Cancel previous transaction and start a new transaction with the new card
	Card being verified with credit card clearinghouse	Transaction in progress—cancel previous transaction (y or n) message If n—Continue with previous card If y—Cancel previous transaction (including cancel messages to clearinghouse) and start a new transaction with the new card
	Gas being pumped	Error message
	Gas pumping complete—awaiting transaction completion	Error message
Cancel button pressed	Pump out of service	N/A
	Idle	N/A
	Transaction started—prompting for inputs	Cancel transaction and return to idle
	Card being verified with credit card clearinghouse	Cancel transaction (including cancel messages to clearinghouse) and return to idle
	Gas being pumped	Stop pumping Transaction in progress—cancel transaction (y or n) message If n—enable pump for continued pumping If y—gas pumping is complete and move to transaction completion
	Gas pumping complete—awaiting transaction completion	Error message
End of day	System active	Process end-of-day reporting
	System out of service	Need mechanism to call up and print previous end-of-day reports

- Documented on sticky notes or index cards displayed on the project's war room wall

Requirements Validation

Requirements validation techniques make certain that the requirements, as specified, will meet the intended needs of the stakeholders. As requirements are elicited, analyzed, and specified, they should be continuously evaluated and validated to confirm their completeness, consistency, correctness, and testability. This can be done through facilitated requirements workshops, peer reviews, prototypes, agile backlog refinement meetings, iteration or sprint planning meetings, the writing of test cases, or discussions with stakeholders. If the requirements are not correct, it does not matter how well the rest of the project is performed. The project team can do a perfect job of building the wrong product. Requirements development is an iterative elicitation, analysis, specification, and validation process. At each step along the way, as intermediate work products are created, those products can be validated as part of the ongoing requirements activities.

Requirements Peer Reviews

As illustrated in Figure 12.13, *peer reviews* may be conducted throughout the requirements development process to validate intermediate requirements work products or sections of the requirements specification.

Once the requirements specification is complete, that specification should be evaluated and peer reviewed as a whole to validate that it is:

- *Complete*: Includes all of the requirements (functional, quality/product attributes, external interfaces, data, and design constraint type requirements) that must be satisfied by the software to meet the needs of the stakeholders and the software's intended use.
- *Consistent*: The requirements are externally consistent and do not conflict with higher-level requirements, including business- or stakeholder-level requirements, requirements in external interface specifications or system requirements specifications, or external standards, regulations, laws, or other business rules.

  The requirements should also be internally consistent so conflicts do not exist between requirements that result in the requirements contradicting each other. Consistent notations and symbols should be used, as well as consistent modeling techniques. Consistent terminology is also used in the requirements to minimize ambiguity:

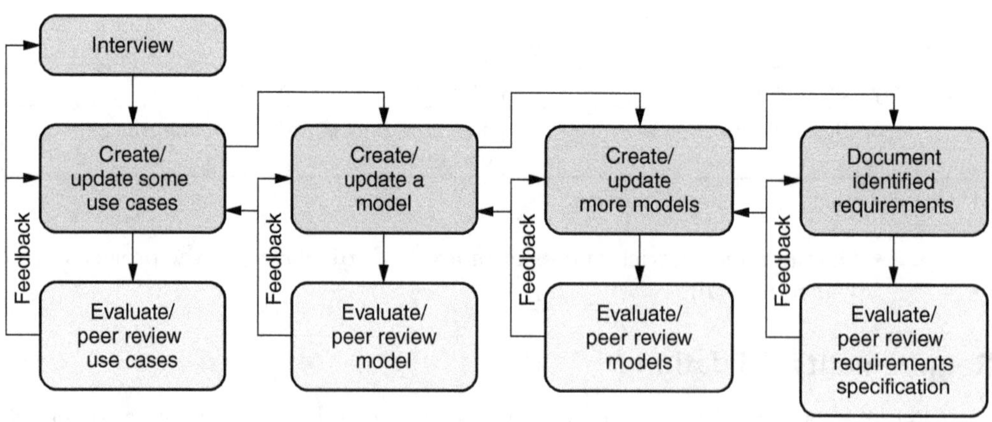

Figure 12.13 Iterative requirements evaluation.

- A word has the same meaning every time it's used. For example, the word *customer* should not refer to the person buying gas in one requirement and the person acquiring the software in another requirement.
- Two different words aren't used to mean the same thing. For example, the person who drives up and purchases gas should not be called the *driver* in one requirement and the *customer* or *purchaser* somewhere else.

- *Modifiable*: Each requirement is stated in only one place and referenced elsewhere, so that if changes are needed it is less likely that internal inconsistencies result.

  To the extent possible, each requirement should be able to be changed without excessive impact on other requirements. One way to avoid ambiguity is to remove all pronouns from the requirements specifications and replaced them with specific nouns. For example, a requirement initially reads, *The controller board displays an alarm when the metered temperature exceeds 300 degrees Fahrenheit, and it sends a reset signal to the burner unit. This requirement is later modified to read, The controller board and operator console display alarms when the metered temperature exceeds 300 degrees Fahrenheit, and it sends a reset signal to the burner unit.* There is now ambiguity as to whether the controller board or operator console sends the reset signal. By replacing the word *it* in the original requirement with its noun *the controller board,* the modified requirements would read, *The controller board and operator console display alarms when the metered temperature exceeds 300 degrees Fahrenheit, and the controller board sends a reset signal to the burner unit,* and the ambiguity does not occur when the modification is implemented.

- *Compliant* with the applicable standards, laws, and regulations.

In addition to evaluating the requirements specification documentation as a whole, each individual requirement should also be evaluated to validate that it is:

- *Clear and unambiguous*: Each requirement statement should have one, and only one, interpretation. Each requirement should be specified in a coherent, easy-to-understand manner. One trick to identifying ambiguous requirements is to search for words that end in *"ly"* (e.g., *quickly, efficiently, completely, user-friendly*) and *"ize"* (e.g., *optimize, minimize, maximize, standardize*). Not every *"ly"* or *"ize"* word will result in an ambiguous requirement—but many of them may, so they should be evaluated. Other keywords such as *flexible, accommodate, safe, easy, sufficient, improve,* and *reduce* may also add to ambiguity.

- *Concise*: Each requirement should be stated in short, specific, action-oriented language. Stating requirements in active, not passive, voice can also make them more concise and readable. The requirements analysts need to forget all of the flowery language they learned in college English composition class and learn the principles of good engineering writing. The requirements should minimize excessive or redundant information.

- *Finite*: A requirement should not be stated in an open-ended manner. For example, words such as *all, always, every, usually, sometimes, throughout,* and *etc.* should be avoided in requirements statements. Phrases such as *including but not limited to* should also be avoided.
- *Measurable*: Specific, measurable limits or values should be stated for each requirement, as appropriate. Words such as *fast, quick, large,* or *small* should be replaced with specific measurement values such as *within three seconds* or *within a maximum of not more than 3 millimeters.*
- *Feasible*: The requirement is able to be implemented using available technologies, techniques, tools, resources, and personnel within the specified cost and schedule constraints.
- *Testable*: There exists a reasonably cost-effective way to determine that the software satisfies the requirement. A requirement is considered testable if an objective and feasible test can be designed to determine whether the software meets the requirement.
- *Traceable*: Each requirement should be traceable back to its source (e.g., system-level requirements, standards, business rules, stakeholder needs, or *enhancement requests*). It should also be specified in a manner that allows traceability forward into design, implementation, and tests.
- *Value-added*: Each requirement is relevant to the defined scope of the product and adds value for at least one of the stakeholders.

One of the main benefits of involving multiple people in a peer review of the requirements specification documents results from the fact that different reviewers have different perspectives of the product and, therefore, find different types of defects. The reviewers should be selected to maximize this benefit by choosing peer reviewers that include:

- Other *requirements analysts* because they know the workmanship standards and best practices for the requirements specification. They are also most familiar with the common defects in requirements work products.
- The *designers* and *testers* have a strong vested interest in the quality of the requirements because their work depends on the requirements being correct. They are also excellent candidates for analyzing issues such as understandability, feasibility, and testability.
- *Customers/users* (or their representatives, including marketing) bring an essential perspective to the requirements inspection because they are the best candidates for identifying missing requirements and failures to address customer or user needs.

- *Specialists* may also be called on when special expertise can add to the effectiveness of the inspection. Specialists might include experts on efficiency, security, safety, graphical user interfaces, hardware, interfacing products, and so on.

Writing Test Cases

Writing test cases should begin early in the life cycle. Since functional testing only requires knowledge of the requirements and not of the internals of the software product, testers can start writing test cases against the requirements as soon as the requirements are defined. The advantages of writing these test cases early in the life cycle include the fact that writing test cases helps:

- Evaluate the requirements and confirm their quality
- Uncover defects in the requirements

Writing test cases early may result in some rework of the test cases if the requirements change, but the cost of that rework will be more than offset by the savings resulting from finding and fixing more requirements defects earlier.

Prioritizing Requirements

Prioritizing requirements can help the project team:

- Understand what's important to their stakeholders
- Work on the most important requirements first
- Balance project scope against project schedule, cost and staff constraints, and quality goals
- Trade off new high-priority requirements against lower-priority requirements that can be deferred
- Provide the highest possible value at the lowest possible cost
- Deliver the most valuable functionality sooner—saving less valuable functionality for later releases of the product

During the first pass at *requirements prioritization,* the requirements can simply be prioritized into three major groups:

- *High*: The high-priority requirements are the requirements that absolutely must be included in the next increment or release of the software product, or it will not be considered successful. If a high-priority requirement is not complete, would the project postpone the release of the software product? If the answer is no, then the requirement is not a high priority.
- *Medium*: The medium-priority requirements are the ones the stakeholders would like to have included in the next increment or release of the software product, but these requirements can wait until the following increment or release if necessary. The goal for medium-

priority requirements is to get as many of them implemented as possible in the next increment or release.

- *Low*: The low-priority requirements will be deferred to a later increment or release. Depending on business objectives, budgets, schedules, and other factors, the project team may never get around to some of these requirements.

During the second pass at requirements prioritization, the focus is only on the medium-priority requirements from the first pass. The high-priority requirements must be implemented, and the low-priority requirements will not be implemented. There is a possibility that the project will only have the resources to do some of the medium-priority requirements. Therefore, the project team wants to make sure that they implement the most important ones. To accomplish this, the project team needs to rank the medium-priority requirements. There are several ways to accomplish this, but according to Wiegers (2003), one of the most rigorous is to consider four factors:

- *Benefit*: The first factor considers the benefits that the stakeholder will enjoy if that requirement is present. For example, if the requirement to allow gas station customers to *pay for their gas purchases in cash at the pump* is included in the product, then those customers receive the benefit of saving the extra time it takes to go pay the attendant. The gas station owners will receive the benefit of a competitive advantage if other gas stations don't offer this service.
- *Penalty*: The second factor considers the penalties that would occur if the requirements were not included. The value here lies in avoiding the penalties by implementing the requirements. There might be legal penalties if a law or regulation was not met by the system, for example, if the system did not handle the taxes on gas purchases correctly. There might also be injuries or property damage if the system were unsafe.
- *Cost*: The third factor to consider is the relative cost of implementing one requirement over another. All other factors being equal, less-costly requirements are prioritized higher than more-costly ones.
- *Risk*: The fourth factor to consider is the technical risk of implementing the requirement. Technical risk is the probability that the project team

Table 12.4 Requirements prioritization matrix.

	Criteria and Weights				
	Benefit (.40)	**Penalty (.20)**	**Cost (.15)**	**Risk (.25)**	**Total**
Requirement 1	4	2	3	1	2.70
Requirement 2	3	4	1	4	3.15
Requirement 3	1	2	2	2	1.60
Requirement 4	2	3	4	2	2.50

will not be able to get the requirements correctly implemented on the first attempt and incur costs for rework if they do not. All other factors being equal, lower-risk requirements are prioritized higher than higher-risk ones.

These four factors can be used to create a *prioritization matrix* for the medium-priority requirements, as illustrated in Table 12.4. This example shows a simple weighted prioritization matrix where each requirement is ranked on a scale of one to five in terms of its impact on each of the four factors. Each score is multiplied by the relative weight for that factor, and the sum of the weighted scores for each requirement is used as a ranking.

Chapter 13

Requirements Management

Requirements management starts with getting stakeholder buy-in to the baselined requirements. (See Chapters 12 and 26 for discussions of baselines.) *Requirements management* encompasses all of the activities involved in:

- Controlling the baselined requirements
- Keeping requirements consistent with plans and other work products
- Tracking the status of the requirements as the project progresses through the software development process to make sure that the requirements are actually built into and verified and validated in subsequent work products
- Requesting changes to the baselined requirements, performing impact analysis for the requested changes, approving or disapproving those changes, and implementing the approved changes
- Negotiating new commitments based on the impact of the approved changes, as necessary

Requirements management activities are closely linked with requirements development activities. Both are part of requirements engineering as illustrated in Figure 13.1. If changes to the requirements are approved, the requirements development process must be repeated to elicit information about those changes, analyze the new or changed requirements, specify the new or changed requirements, and validate those changes and those requirements and any other requirements that the changes might impact. A new requirements baseline is then created.

The requirements process does not exist in a vacuum. Various organizational and external factors influence these activities, as illustrated in Figure 13.2. The context in which software requirements engineering is performed influences its practices and activities. Therefore, the tools and techniques used in software engineering should be tailored to the needs of the participants, the projects, and the organization.

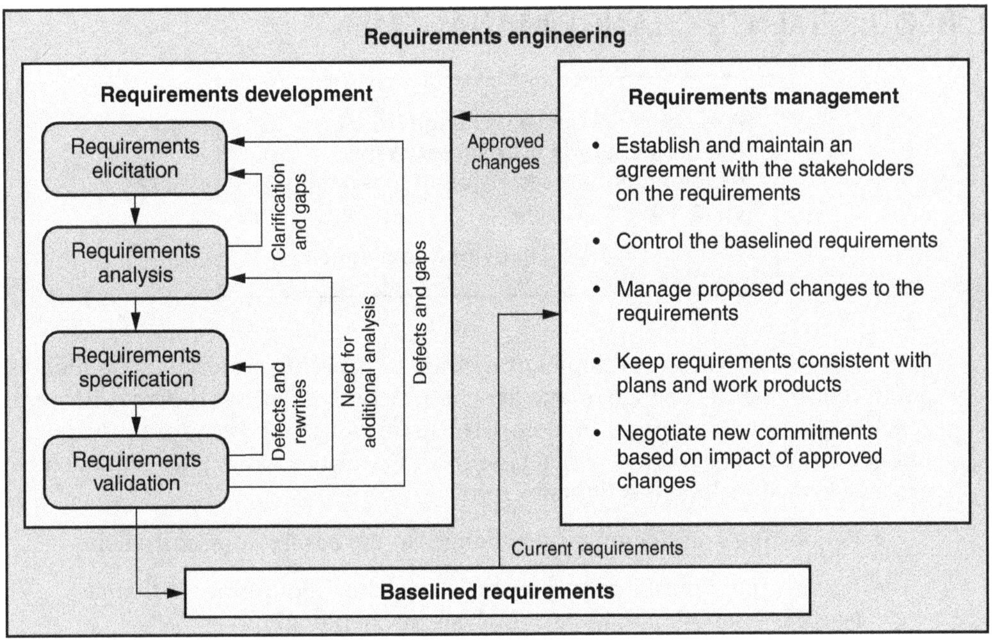

Figure 13.1 Requirements engineering process.

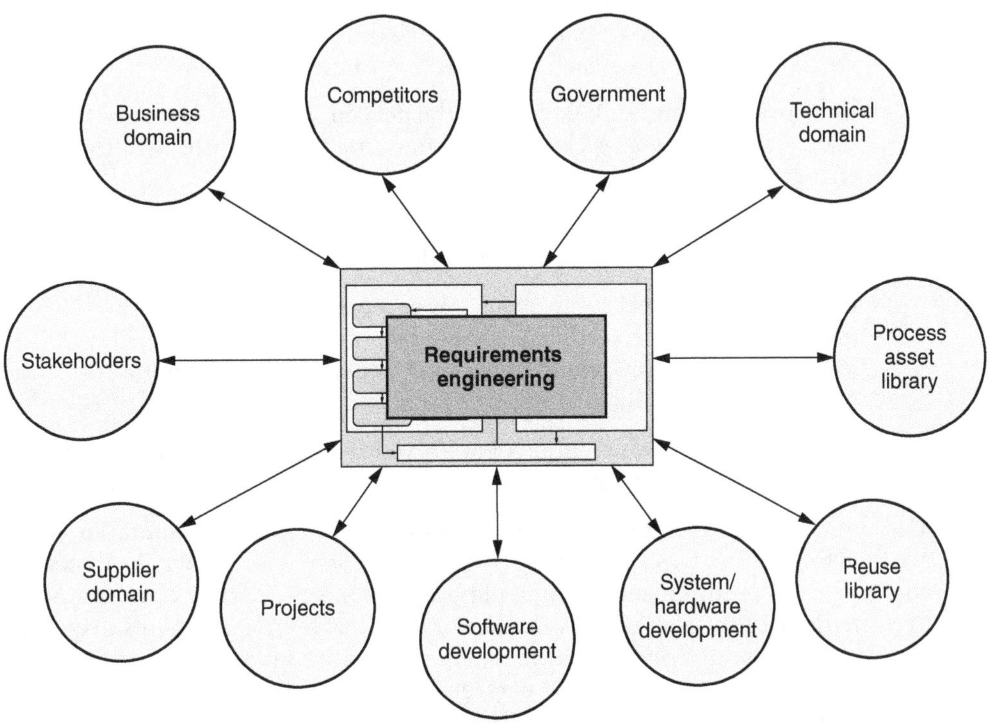

Figure 13.2 Requirements engineering organizational context.

1. REQUIREMENTS CHANGE MANAGEMENT

> Assess the impact that changes to requirements will have on software development processes for all types of life cycle models. (Evaluate)
>
> **Body of Knowledge III.E.1**

In traditional software development methods, including waterfall-type models, good requirements change management practices identify the requirements specification(s) as configuration items. Those configuration items are then placed under formal change control when baselined. Requirements change management processes must include mechanisms for:

- Requesting and documenting changes to the baselined requirements
- Performing impact analysis on each requested requirements change (see the discussion of impact analysis in Chapter 27)
- Informing affected stakeholders of each requested requirements change and soliciting their input to the impact analysis
- Having the appropriate authority, typically referred to as a configuration control board (CCB), make decisions on accepting, deferring, or rejecting each requested requirements change
- Informing affected stakeholders of the decision to accept, defer, or reject the requirements change and obtaining their commitment to the change if it is accepted
- Tracking requested requirements changes from submission through final disposition (rejection or completion of the change)
- If a requested requirements change is accepted, all affected work products and the project plans should be updated accordingly (and re-baselined as appropriate) so that they remain consistent with the revised requirements
- Verifying the implementation of approved requirements changes in all impacted software work products

The updated requirements specifications and other updated configuration items are then re-baselined, as appropriate. Version control practices track the history of changes to the requirements document baselines. Metrics should also be utilized to track the stability, also called volatility or churn, of the requirements over time.

Requirements change management is simpler in agile projects. Since each iteration only lasts a few weeks, it is rarely considered necessary to interrupt the iteration with requirements changes. Therefore, requirements changes are handled by posting the changes as new stories in the product backlog or by changing existing stories in the product backlog that have not yet been implemented. New or changed stories are prioritized with other product backlog stories and are handled in future iterations. However, if the requested requirements change is considered so critical that it requires interrupting the current iteration:

- The requirements changes are still handled by posting the changes as new stories in the product backlog or by changing existing stories in the product backlog that have not yet been implemented
- The current iteration is terminated, and all unfinished work is returned to the product backlog
- A new iteration is started, with the critical changes being prioritized to the top of the product backlog and therefore picked up to be implemented in the new iteration

If iterative, incremental, or evolutionary life cycles are used, each new set of requirements (new requirements baseline) constitutes a change to the requirements of the existing software. Requirements change management processes, including impact analysis, should be performed for the new or updated requirements in each baseline to make certain that the new requirements do not conflict with the functionality and quality or product attributes that have already been built into the existing software.

2. BIDIRECTIONAL TRACEABILITY

> Use various tools and techniques to ensure bidirectional traceability from requirements elicitation and analysis through design and testing. (Apply)
>
> **Body of Knowledge III.E.2**

The IEEE/ISO/IEC standard *Systems and Software Engineering—Vocabulary* (IEEE/ISO/IEC 2017) defines *traceability* as "the degree to which a relationship can be established between two or more products of the development process, especially products having a predecessor–successor or master–subordinate relationship to one another."

Traceability tracks the relationship between each unique product-level requirement and its source. For example, a product requirement might trace backward through a stakeholder functional requirement to a business-level requirement, a stakeholder request, a business rule, an external interface specification, an industry standard, law, or regulation or to some other source.

Traceability is also used to track the relationship forward between each unique product-level requirement and the work products to which that requirement is allocated. For example, a single product requirement might trace to one or more architectural elements, component design elements; objects or classes; source code modules; unit, integration, system, alpha, beta, acceptance tests; user or operational documentation topics; training materials; and so on that implement that requirement.

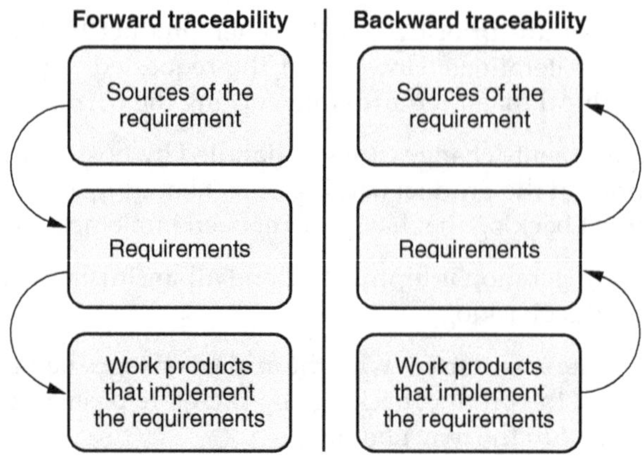

Figure 13.3 Bidirectional (forward and backward) traceability.

Good traceability practices allow for *bidirectional traceability,* meaning that the traceability chains can be traced in both the forward and backward directions, as illustrated in Figure 13.3.

Forward traceability looks at:

- Tracing the requirements sources to their resulting product-level requirement(s) to verify the completeness of the system and software requirements specification.
- Tracing each unique product-level requirement forward into the architecture, designs, source code modules, tests, documentation, processes, training materials, and other work products that implement that requirement. The objective is to make certain that each requirement is implemented in the software products and thoroughly tested.
- Forward traceability verifies the proper direction of the evolving product (that the project is building the right product) and indicates completeness of the subsequent implementation. For example, if a business rule cannot be traced forward to one or more product-level requirements, then the product requirements specification is incomplete and the resulting product may not meet the needs of the business.

  If a product requirement cannot be traced forward to its associated architectural design elements, then the architectural design is not complete, and so on.

If, on the other hand, there are changes in the business environment (e.g., a business rule change or a standard change), and if good forward traceability has been maintained, that change can be traced forward to the associated stakeholder- or project-level requirements and all of the work products that are impacted by that change. This greatly reduces the effort required to do a thorough impact analysis when there are changes. It also reduces the risk that any affected work product is forgotten, resulting in an incomplete implementation of the change.

Backward traceability, also called *reverse traceability*, looks at:

- Tracing each unique work product (e.g., design element, object or class, source code module, test, document, training material, and so on) back to its associated requirement. Backward traceability helps verify that the requirements have been kept consistent with the subsequent work products as they are being built based on those requirements.
- Tracing each requirement back to its source(s)—for example, business or stakeholder needs, regulations, laws, standards, and so on.

Backward traceability helps make sure that the evolving software products remain on the correct track regarding the original or evolving requirements (that the project is building the product correctly). The objective is to make sure that the scope of the product is not expanding through the addition of features or functionality not specified in the requirements ("gold-plating" or "creeping elegance"). If there is a change needed in the implementation—or if the developers come up with a creative, new technical solution—that change or solution should be traced backward to the requirements, and the business needs to make certain that it is within the scope of the desired product. If there is a work product element that does not trace back to the requirements, one of two things is true:

- The first possibility is that a requirement is missing because the work product element is needed. In this case, traceability has helped identify the missing requirement and can also be used to evaluate the impacts of adding that requirement to project plans and other work products (forward traceability again).
- The second possibility is that gold-plating is going on—something has been added that should not be part of the product. Gold-plating is a high-risk activity because project plans have not allocated time or resources to the work, and the existence of that part of the product may not be well communicated to other project personnel (e.g., the tester does not test it, or it is not included in user documentation).

Another benefit of backward traceability comes when a defect is identified in one of the work products. For example, if a source code module has a defect, then traceability can be used to help determine the root cause of that defect. Is it just a code defect, or does it trace back to a defect in the design or requirements? If it is a design or requirements defect, forward traceability can then be used to determine what other work products might be impacted by the defect? (Figures 27.5 and 27.6 illustrate the use of backward and forward traceability in defect impact analysis.)

Many modern requirements management tools and some agile tools include traceability mechanisms as part of their functionality that allow the generation of a traceability matrix directly from the tool. These tools support linking the requirements and their source and subsequent work products. Any automation is desirable since manually creating and maintaining a traceability matrix is extremely time-consuming.

The classic manual way to perform traceability is by constructing a traceability matrix. As illustrated in Table 13.1, a *traceability matrix* summarizes, in matrix form, a trace from the original identified requirements sources to their associated

Table 13.1 Traceability matrix–example.

Requirement source	Product requirements	Architectural design section #	Component design section #	Code module	Unit test case #	System test case #	User manual
Business rule 1	R00120 Credit card types	4.1 Parse magnetic strip	4.1.1 Read card type	Read_Card _Type.c Read_Card _Type.h	UT 4.1.032 UT 4.1.033 UT 4.1.038 UT 4.1.043	ST 120.020 ST 120.022	Section 12
			4.1.2 Verify card type	Ver_Card _Type.c Ver_Card _Type.h Ver_Card _Types.dat	UT 4.2.012 UT 4.2.013 UT 4.2.016 UT 4.2.031 UT 4.2.045	ST 120.035 ST 120.036 ST 120.037 ST 120.038	Section 12
Use case 132 step 6	R00230 Read gas flow	7.2.2 Gas flow meter Interface	7.2.2 Read gas flow indicator	Read_gas_flow.c	UT 7.2.043 UT 7.2.044	ST 230.002 ST 230.003	Section 21.1.2
	R00231 Calculate gas price	7.3 Calculate gas price	7.3 Calculate gas price	Cal_gas_ Price.c	UT 7.3.005 UT 7.3.006 UT 7.3.007	ST 231.001 ST 231.002 ST 231.003	Section 21.1.3

product requirements and then on to other work product elements. To construct a traceability matrix, each requirement, each requirement's source, and each work product element must have a unique *identifier* that can be used as a reference in the matrix. The requirements matrix has the advantage of being a single repository for documenting both forward and backward traceability across all of the work products. While maintaining a manual traceability matrix is a labor-intensive activity, the benefits that come from higher visibility, reduced impact analysis effort, and decreased risk can make it a value-added activity.

Another mechanism for implementing traceability is using *trace tags*. Again, each requirement, each requirement source, and each work product element must have a unique identifier. In trace tagging, however, those identifiers are used as tags in the subsequent work products to identify backward tracing to the predecessor document. As illustrated in Figure 13.4, for example, the software design specification (SDS) includes tags that identify the requirements implemented by each uniquely identified design element, and the unit test specification (UTS) includes trace tags that trace back to the design elements that each test case verifies. This tagging propagates through the work product set with source code modules that include trace tags back to design elements, integration test cases with tags back to architecture elements, and system test cases with trace tags back to requirements, as appropriate, depending on the hierarchy of work products used for the software. Trace tags have the advantage of being part of the work products, so a separate matrix does not have to be maintained. However, while backward tracing is easy with trace tags, forward tracing is very difficult using this mechanism.

All traceability implementation techniques require the commitment of a cross-functional team of participants to create and maintain the linkages between the requirements, their source, and their allocation to subsequent work products. The requirements analyst must initiate requirements traceability and document the original tracing of the product requirements to their source. As system and software architects create the architectural design, those practitioners add their information to the traceability documentation. Developers doing component

SRS R00104 The system shall cancel the transaction if at any time prior to the actual dispensing of gasoline, the cardholder requests cancellation.

SDS

SDS identifier	SRS tag	Component name	Component description	Type	Etc.
7.01.032	R00104	Cancel_Transaction	Cancel transaction when the customer presses cancel button	Module	

UTS

Test case #	SDS identifier	Test case name	Inputs	Expected results	Etc.
23476	7.01.032	Cancel_Before_PIN_Entry			
23477	7.01.032	Cancel_After_Invalid_PIN_Entry			
23478	7.01.032	Cancel_After_Start_Pump_Gas			

Figure 13.4 Trace tagging—example.

design, code, and unit testing must add additional traceability information for the elements they create, as do the integration, system, and alpha, beta, and acceptance testers. For small projects, some of these roles may not exist or may be done by a single practitioner, which limits the number of different people working with the traceability information. For larger projects, where traceability information comes from many different practitioners, it may be necessary to have someone who coordinates, documents, and does periodic audits of the traceability information from all its various sources, to verify completeness and consistency across all traced elements.

Chapter 14

Software Analysis, Design, and Development

During the software analysis, design, and development activities, the software developers must change their thinking from "what needs to be done" to "how it will be done." The design is the bridge between the requirements and the final, executable software product or implemented software-intensive system. Design can be defined as the:

- Process of defining the architecture, components, interfaces, data, and other characteristics of a software system or component to satisfy specified software requirements (IEEE/ISO/IEC 2017)
- Process of conceiving, inventing, or contriving a scheme for turning software requirements into an operational program

"The product's functionality, quality attributes, and constraints drive its architecture design" (Wiegers and Beatty 2013). According to Pressman and Maxim (2015), "Design is the place where creativity rules—where customer requirements, business needs, and technical considerations all come together in the formulation of a product or system." Software analysis and design is typically done at two major levels, as illustrated in Figure 14.1:

- Creating the software architecture
- Defining the software component design of the internal workings and defining the structure of each identified software component

The software architecture and component designs that are the outputs of software analysis and design become the inputs into the software development, as illustrated in Figure 14.2. Development activities translate these designs into actual software products. This translation includes the source code with comments and documentation (e.g., help files, user manuals, operations manuals, installation instructions, or the *version description document,* also called *release notes*).

Software analysis, design, and development, like the software requirements process, are iterative processes. The software product-level requirements are input into software analysis and design, which may uncover the need for further refinement of the business-, stakeholder- and product-level requirements. Development may uncover the need for further refinement of the software architecture or component designs or even the requirements. If an incremental, iterative, or agile life cycle is utilized, the continuous development, refinement,

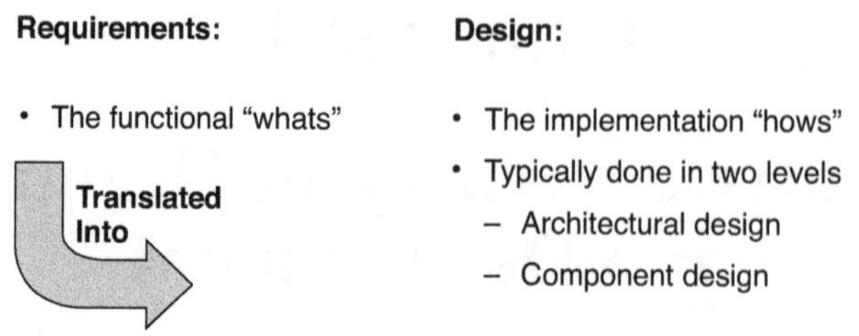

Figure 14.1 Activities of analysis and design.

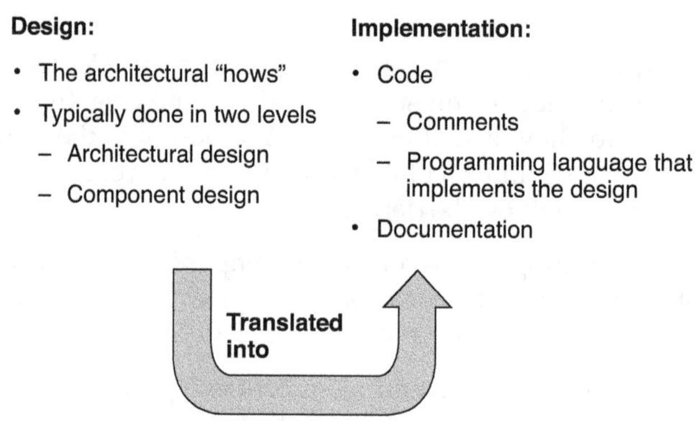

Figure 14.2 Activities of implementation.

and elaboration of the software architecture and component designs are repeated with each new increment.

1. DESIGN METHODS

> Identify the steps used in software design and their functions. Define and distinguish between software design methods. (Understand)
>
> **Body of Knowledge III.F.1**

Primary Goal of Analysis and Design—Manage Complexity

According to McConnell (2009), "Design is about managing complexity." Modern software and software-intensive systems are typically too large and complex for any one person to understand the entire software application as a single piece." Therefore, the primary goal of software analysis and design is to manage that complexity. A primary mechanism for managing this complexity is to decompose the software into components that compartmentalize the software information. Software requirements are allocated to those components so that they can be developed as independently from each other as possible. The relationships and interdependencies between the components are determined so that each component can be segregated and worked on independently. This decomposition also facilitates iterative or incremental development and the reusability of the software components.

Other Goals of Analysis and Design

Other goals of software analysis and design include:

- *Developing alternative solutions*: There are many possible solutions for software implementation. One of the goals of analysis and design is to select the "best" solution based on engineering trade-off analysis. If the same requirements are given to multiple designers, the result will be a variety of different design solutions. Their solutions will vary because of differences in engineering; trade-off decisions about functionality, performance, fault-tolerance, security, safety, and efficiency; and the technical methods selected to implement the requirements.

  Designing is all about analyzing, balancing, and selecting between these engineering trade-offs. Therefore, a large part of the design process explores alternative solutions and determines which alternative "best fits" the stakeholder needs, including cost constraints, schedule constraints, and risks, as well as the requirements.

- *Providing all of the information needed to code, integrate, and test the software*: The design also acts as a bridge between the stakeholder and product-level requirements, and the information needed to code, integrate, and test the software.

  The design documents the decisions about "how to" progressively elaborate the requirements into the solutions that will be used to develop and verify the software.

- *Making sure that quality attributes are built into the software*: According to Wiegers and Beatty (2013), "Shortcomings in design lead to products that are difficult to maintain and extend, and that do not satisfy the customer's performance, usability, and reliability objectives." Part of design is verifying that the quality attribute requirements, as well as functional requirements, are designed and built into the software.

- *Improving software maintainability and identifying opportunities for reuse*: By designing the software with components that are as independent from each other as possible, changes to one part of the software have minimal impact on other components. This independence increases the maintainability of the software. It also allows individual components to be segregated from one software product and more easily reused in another. Not only can components be built for reuse in other software products, but a good design considers preexisting components from other software products, as well as what commercial off-the-shelf (COTS) software can be used in the design for the current product.

Design Concepts—Abstraction

Abstraction is used in design to help manage software complexity. It does this by emphasizing the essential attributes of the software at a specific level while suppressing the implementation details. Modeling is a widely used mechanism for providing abstraction of a software product at various levels of precision and detail. For example, as illustrated in the model on the left in Figure 14.3, when designing a database, an entity relationship diagram (or class diagram) could be used to show the interrelationships between the database records while hiding the details of the internals of the records.

Abstraction can also help manage complexity by postponing decisions about the implementation details until later levels of the design or implementation. For example, as illustrated in the model on the right in Figure 14.3, the designer knows that detailed control flow decisions have to be made. For example, decisions need to be made about "what type of gas to pump" or "what the detailed steps are in authenticating a transaction." However, those detailed implementation decisions and associated error-handling implementation decisions can be abstracted in the architecture and left to detailed component design.

The architecture of the software is always an abstraction. Details are left to component and module design. At the highest level of abstraction, the architecture defines the major subsystems of the system (including hardware and software subsystems), allocates requirements to those subsystems, and defines their interfaces. Each software subsystem can then be defined at a lower level of abstraction by defining the components within that subsystem and their interactions. The next level of detail might still include abstraction by defining

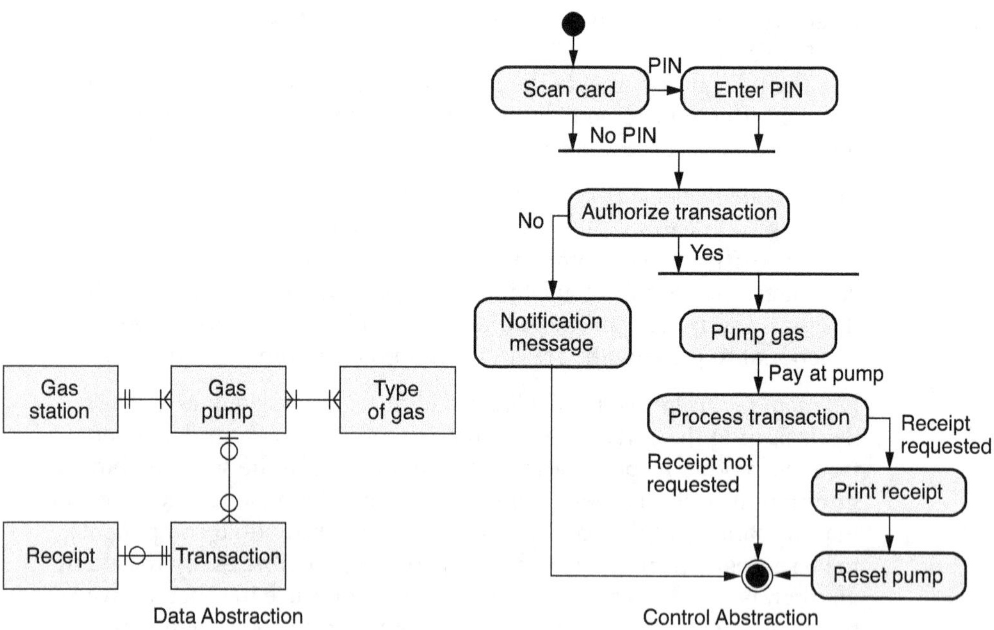

Figure 14.3 Models used for abstraction—examples.

the modules within each component and their interactions, while still hiding the internal implementation details of each module.

Design Concepts—Modularity

Modularity is the logical partitioning of the system, or software, into components or modules that can be treated as *black boxes* that can hide information. Inputs go in, and outputs come out, but the internal processes inside are unknown black boxes to other parts of the software.

The design process defines the interfaces and functionality of each component or module in such a way that, for any given input, the output can be accurately predicted without any knowledge of the internal workings of that component or module. Modularity is one of the primary mechanisms for implementing the architecture.

Design Concepts—Information Hiding

In *information hiding*, each component/module is designed in such a way as to "hide" its design decisions to keep them independent from other parts of the software. These hidden secrets might be implementation decisions, data structures, algorithms, specific data values, or anything else that is likely to change.

One of the important design decisions is to determine what information needs to be hidden and what information should be shared with the rest of the software. The designer starts with a list of design decisions and determines which ones are difficult, risky, unknown, or likely to change. These decisions are then assigned to a component or module to hide their implementation from other software components.

For example, assume that the design decision is that the maximum number of gas pumps for the software is 30. Instead of hard coding this information throughout the software, a single data item is created, MAX_PUMPS, and assigned a value of 30. The MAX_PUMPS data item is then used throughout the rest of the software, which must be designed and implemented so that changes to the MAX_PUMPS value will not impact those parts of the software. If the maximum number of gas pumps ever changes, then only one single data item value is changed and the rest of the software is not impacted by the change. However, if the MAX_PUMPS value ever changes, appropriate regression testing should be performed to verify that other parts of the software still function correctly.

As another example, assume that a new software product needs to calculate the sales tax. One way of doing this would be to have a variable TAX_RATE, and then whenever you need to calculate the sales tax, you could just multiply the TOTAL_PRICE by TAX_ RATE to calculate the taxes. This method may work for states that charge a single rate for every item purchased, but what about states like Texas, where some items are taxable and some are not or where there are sales tax–free days for school clothes and supplies twice a year? If the software product is internationalized, do all countries calculate taxes the same way? The way sales taxes are calculated is a design decision that should be hidden. The designer creates a single data item for the statement "TAXES=calculate_taxes (TOTAL_PRICE)" everywhere in the software that sale taxes are calculated. Then, if the tax calculation algorithm changes or becomes more complex in the future, that change would require the modification of only one source code module.

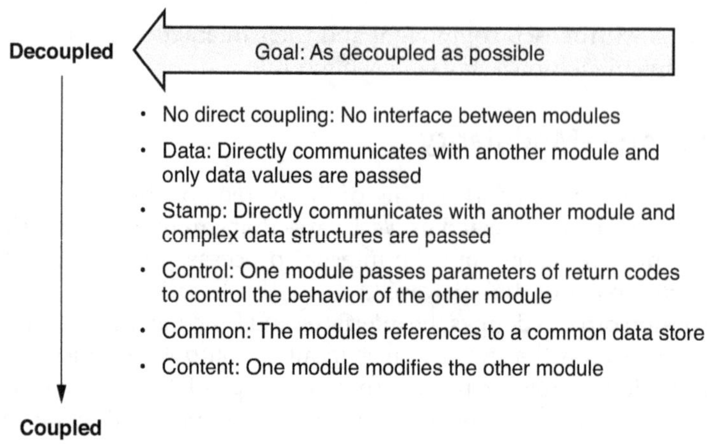

Figure 14.4 Levels of coupling—example (based on Pfleeger and Atlee 2010).

"Information hiding is useful at all levels of design, from the use of named constants instead of literals, to creation of data types, to class design, routine design, and subsystem design" (McConnell 2009).

Design Concepts—Coupling

Coupling measures the interconnectivity between modules in a software structure, as illustrated in Figure 14.4. The more coupled a component or module is, the larger the impact is if it needs to be changed, and the more regression testing is needed to verify that the changes to the coupled component or module do not adversely impact other unchanged parts of the software.

Coupling cannot be avoided because the software components or modules must function together and interact in complex systems. The goal is to keep coupling as loose as possible by making design decisions that minimize connections among different parts of the software. Good abstractions, information hiding, and modularity facilitate minimal coupling.

Design Concepts—Cohesion

As illustrated in Figure 14.5, *cohesion* measures the extent to which a component in the software design performs a single task or function. The goal is strong cohesion so that components or modules grouped together work well together to achieve a common goal. During design, requirements should be allocated to the modularized components or modules in such a way as to provide separation of concerns. This separation takes the concept of modularity one step further to localize any changes to the requirements to one or a few components or modules.

Software Analysis and Design Process

Software analysis and design is a disciplined, process-oriented approach to the definition of software architecture and component design, illustrated in

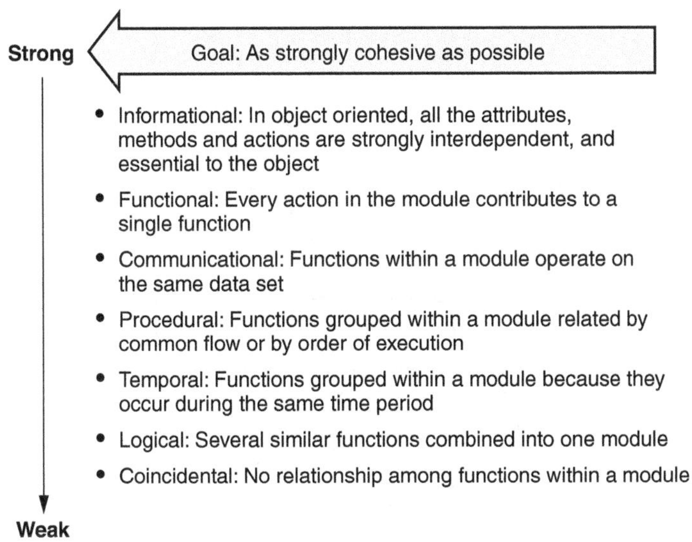

Figure 14.5 Levels of cohesion—example (based on Pfleeger and Atlee 2010).

Figure 14.6. The software analysis and design process encompasses all of the activities involved in:

- Analyzing alternative solutions and selecting the "best" solution for implementing the requirements
- Defining the software architecture for that solution
- Verifying that architecture
- Designing the internals of each component or module from that architecture
- Verifying the designs of those components

As discussed earlier, many possible solutions to implementing the software may exist. Therefore, the design process starts with identifying and analyzing alternative solutions. The most appropriate solution is then selected for the software based on engineering trade-offs.

Horst Rittel and Melvin Webber defined a "wicked" problem as one that could be clearly defined only by solving it, or by solving part of it. Software design is almost always a wicked problem (McConnell 2009). The completed software architecture or component design should look well organized, straightforward, easy to understand, and clean. However, the design process itself is "wicked" and sloppy. It is infested with false starts, blind alleys, mistakes, and potentially unfeasible solutions. In fact, these false starts, blind alleys, and mistakes are one of the points of doing design. The goal is to explore many options and possibilities during the design process, when issues are much cheaper to resolve than they will be later, during more expensive code development.

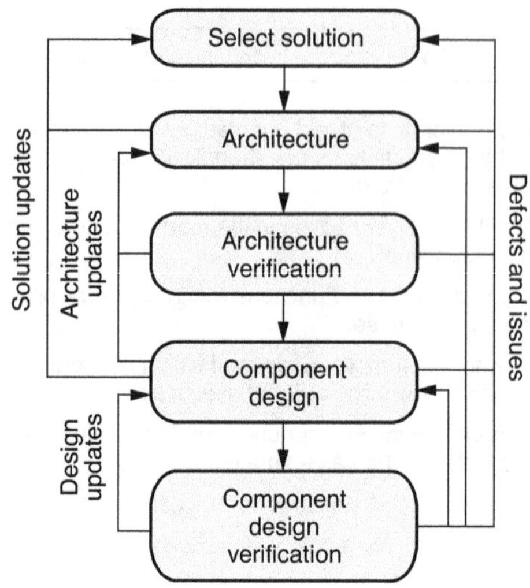

Figure 14.6 Analysis and design process.

Software analysis and design is an iterative process. Do not expect to go through the steps in the process in a one-shot, linear fashion. Every step may identify defects or issues in the previous step(s), or gaps that require the solution, architecture, or component design, or even the requirements, to be updated or refined.

Software Architecture

The first major level of software analysis and design is *software architectural design,* also known as the *preliminary, high-level,* or *top-level design.* At this level, analysis and design techniques are used to:

- Specify the underlying structure of the software and how that structure is partitioned into the various software components (subsystems, programs, source code modules, databases, and other data elements)
- Specify the interactions or interfaces between those software components
- Define the high-level design for all of the interfaces between the software components and other external components outside the software, including interfaces to:
    - Components within the system (hardware and human interfaces)
    - Entities outside the system
- Allocate the software requirements to the components of the system in a way that allows each component to be developed independently and then integrated back into a complete product and document this allocation through traceability.

In the next step, verification and validation activities are performed on the architectural design. This step may be done iteratively with the architectural design step, as parts of the architecture are completed. According to IEEE/ISO/IEC 12207 (2017a), this is done to evaluate:

- Traceability to and consistency with the software requirements
- Internal consistency between the components
- Appropriateness of design methods and standards
- Feasibility of creating high-quality component designs that will be consistent with the specified architecture
- Feasibility of operation and maintenance

Software Architecture Views

An *architectural structure* is a set of actual elements of the software (or system) held together by a relationship. An *architecture view* is a representation (abstraction) of a set of one or more of these architecture structures from a specific perspective or area of interest relevant to one or more stakeholders. "Perhaps the most important concept associated with software architecture documentation is that of the view" (Clements et al. 2011).

One of the primary purposes of a view is to manage the complexity of a system. This is accomplished by considering only a subset of that system's structures and their interrelationship at a time. That is, the architecture views divide up the complex, multidimensional software or system into a specific set of useful and manageable abstractions of that software or system from different points of view.

What the relevant views are for any given system will depend on the goals of the stakeholders for that system. Using the human body as an example, if the goal is to determine the cause of a patient's chest pains, the circulatory and digestive system views are probably much more valuable that the skeletal view. However, if the patient has a broken bone, the skeletal view may be the most relevant view. The same is true with software systems. Different views support different purposes, stakeholder concerns, and uses. Different views emphasize different functional and non-functional requirements.

Historically, one of the first documented examples of a specific set of views is Kruchten's *4+1 View Model*, as illustrated in Figure 14.7. The 4+1 View Model includes the following views (Kruchten 1995):

- *Logical view*: The logical view is the object-oriented decomposition of the system. It defines what services the system provides to the user and other stakeholders (functional requirements).
- *Process view*: The process view captures the concurrency and synchronization aspects of the design.
- *Physical view*: The physical view describes the mappings of the software onto the hardware, and reflects its distributed aspects, and is of particular use to system engineers.

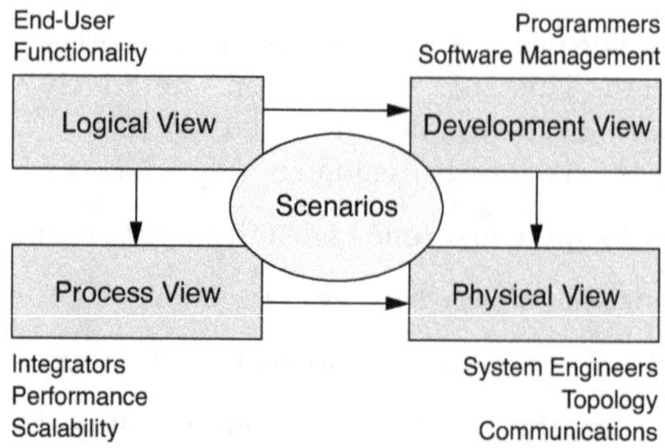

Figure 14.7 Kruchten's 4+1 View Model—architecture view example (Kruchten 1995).

- *Development view*: The development view describes the static organization of the software in its development environment. This view focuses on the software from the developers' perspective.
- *Scenarios*: The fifth view includes scenarios, which are typically described in a few selected use cases that tie the other four types of views together.

Other sets of views that have been proposed include:

- Conceptual, module interconnection, execution, and code views (Soni 1995)
- Functional, information, concurrency, development, deployment, and operational views (Rozanski and Woods 2005)
- Customer, application, functional, conceptual, and realization views (Philips research in Clements et al. 2011)

The modern approach is to create views that best serve the concerns of the stakeholders associated with the system under consideration (IEEE/ISO/IEC 2011). These views are then implemented using various models of the software architecture (see the discussion of various models in Chapter 12).

Software Component Design

The second major activity in software analysis and design is *software component design*, also called *detailed design*. The component design defines the internals of each software component in the architecture. Component design refines the architectural design down to individual software modules that can be coded. This includes the internal design of each database or other data element and the design of the details of each external interface. The purpose of the software component design is to provide a design that:

- Fully defines the structure and capabilities of the software's components
- Is traceable to, and fully implements, the requirements and software architecture

- Optimizes important product attributes requirements
- Allows for reuse, and the use of COTS components
- Can be verified against the requirements and software architecture
- Is sufficiently detailed to permit coding, testing, and maintenance of the software components

As part of this step, verification and validation activities are performed on each component design element. According to IEEE/ISO/IEC 12207 (2017a), this is done to evaluate:

- Traceability of the source code modules to the requirements and architecture
- External consistency with the software architecture
- Internal consistency between the components
- Appropriateness of design methods and standards
- Feasibility of testing
- Feasibility of operation and maintenance

As with the software architectural design, the primary goal of component design is to manage complexity.

- To minimize software source code component complexity, the component design should:
  - Define component interfaces to allow their internal workings to be hidden from other parts of the system or software
  - Avoid deep inheritance hierarchies
  - Avoid using global data
  - Avoid deep nesting of loops and conditionals
  - Design well-structured code (e.g., with single entry and exit points and no "go to" statements)
  - Define a consistent approach to error handling
  - Keep the components and modules small
  - Use design patterns
  - Use standardized naming conventions, including using clear, self-explanatory variable names
- To minimize interface complexity, the interface design should:
  - Pass parameters, not structures
  - Minimize the number of parameters passed
  - Pass only the parameters needed
  - Use standardized communication protocols

Structured Analysis and Design (SAD)

According to Yourdon and Constantine (1979), *structured analysis and design* (SAD) is "a collection of guidelines for distinguishing between good designs and bad designs, and a collection of techniques, strategies, and heuristics that generally lead to good design." SAD is a document-based approach to rigorous design, as opposed to the minimal-documentation approach of the agile philosophies. SAD techniques include:

- *Data modeling*: The data elements of the system or software and their relationships are analyzed, modeled, and documented (e.g., with entity relationship diagrams)
- *Data flow modeling*: The movement of data around the system or software and the associated data transformations are analyzed, modeled, and documented (e.g., with data flow diagrams)
- *Entity behavior modeling*: The events that impact data entities, or the sequences of those events, are analyzed, modeled, and documented (e.g., with control flow diagrams or state diagrams)

Object-Oriented Analysis and Design (OOAD)

The idea behind *object-oriented analysis and design* is to design the software around the elements that are least likely to change over time. For example, 50 years ago, gas stations had customers, attendants, owners, gas pumps, and transactions just like they do today. However, consider how the interactions between those entities have changed in those 50 years. In object-oriented analysis and design, the objects represent specific entities in the software system. Objects are self-contained with both data associated with them and associated procedures to manipulate the data. There have been lengthy discussions in the industry over what the word "object" really means. An equation for recognizing an object-oriented approach follows:

> Objects (encapsulations of attributes and exclusive services; abstractions of entities in the problem space, with some number of instances in the problem space) = Classification + Inheritance + Communication with messages (based on Coad and Yourdon 1990).

In object-oriented analysis and design, objects belong to classes. A *class* is a category (classification) of objects, or object types that defines all of the common properties of the different objects that belong to it. Classes are the building blocks of object-oriented software. For example, the object Cathy belongs to the class of ASQ members, and so does the object Tom. Because of this, they share certain attributes (data structures) from that class. For example, they both have a membership number, a membership type, divisions/sections they belong to, ASQ certifications they hold, and so on. They also share procedures (called operations) they can perform. For example, they can pay their dues, join another division or section, subscribe to an ASQ journal, recertify, and so on. The object Susan, who is not a member of ASQ, does not share these attributes or operations.

Inheritance is a relationship between classes that allows the definition of a new class based on the definition of an existing class. For example, if the class Gasoline already exists, three new classes can be defined (Regular, Super, Premium), to inherit all of the attributes defined for the parent class Gasoline without duplicating those attributes in the three new classes. The new classes can also execute all of the operations they inherit from their parent Gasoline class.

Developers can also use inheritance to combine common elements from several existing classes to create a new common parent class. For example, common attributes and operations from the two classes *Male* and *Female* can be extracted and placed into a new class, *Person*, thus allowing redundancy to be removed from the software.

Multiple levels of inheritance can exist. For example, the class *Female* inherits attributes and operations from the class *Person*, which inherits from the class *Mammals*, which inherits from the class *Living organisms*. The class *Female* possesses all of the attributes and can perform all of the operations defined in *Person*, *Mammals*, and *Living organisms*. The class of *Female* also possesses the unique attributes and operations defined within the class *Female*. The *operations* are actions that can be applied to an object to obtain a certain effect.

Operations fall into several categories:

- *Accessor* operations give information about an object such as the value of some attribute, or general state information. This kind of operation does not change the object on which the operation is being performed.
- *Modifier* operations modify the state of an object by changing one or more attributes to new values.
- *Constructor* operations are used to create a new object, including the initialization of the new instance when it comes into existence.
- *Destructor* operations are used to perform any processing needed just prior to the end of an object's lifetime.

Constructors and destructors are different from accessors and modifiers in that they are invoked implicitly as a result of the birth and death of objects.

A *method* is a piece of software code that implements an operation.

Another aspect of object-oriented analysis and design is polymorphism. *Polymorphism* means that the sender of a stimulus (or message) does not need to know the receiving object's class. The receiving object can belong to an arbitrary class. For example, if a person's friend can be either the Male class or the Female class, polymorphism exists because "friend" is used to refer to either of the two classes. The object that receives the stimulus determines its interpretation. If the receiver's class is known in advance, then polymorphism is not needed, but if the receiver can be of varying classes within limits, then the polymorphism characteristic must specify those limits. An example of this is a method used to draw graphics. If three dimensions are passed in the message, the method Draw creates a triangle; if four dimensions are passed, then Draw creates a rectangle, five a pentagon, and so on. In this case, there are multiple methods with the same name but they each accept a different number of variables in the method call. Generally, the selection of the variant is determined at run time (dynamic binding) by the compiler, based on, for example, the type or number of arguments passed (Petchiny 1998).

All information in an object-oriented system is stored within its objects. The only way to affect an object (to manipulate its attributes) is when the object is ordered to perform operations. The attributes and operations are encapsulated in the object. Objects support the concept of information hiding. That is, they hide their internal structure from their surroundings. Encapsulation means that all that is seen of an object is its interface, namely the operations that can be performed on that object. This reduces complexity. Since it is impossible to become involved in the object's internal structure, users can use them only according to their specifications, which define what operations they perform, not how they perform them.

2. QUALITY ATTRIBUTES AND DESIGN

> Analyze the impact that quality-related elements (e.g., safety, security, reliability, usability, reusability, and maintainability) can have on software design. (Analyze)
>
> **Body of Knowledge III.F.2**

According to Bass et al. (2013), "whether a system will be able to exhibit its desired (or required) quality attributes is substantially determined by its architecture." The ability of the system to implement its product attribute requirements, as derived from quality attributes, is facilitated by good architectural choices that focus on those requirements. However, while this is absolutely necessary, it is not sufficient. Product attribute requirements must be considered and well specified during the requirements engineering activities or they will not necessarily even be a focus of concern during the architecture activities. Product attribute requirements must also be built into the systems during component design and implementation.

Safety

For safety-critical systems and other mission-critical requirements, software analysis and design activities typically involve *failure mode and effects analysis* (FMEA), also called *failure mode effects and criticality analysis* (FMECA), to anticipate ways that the software might fail and anticipate the potential safety related effects (possible injury or death of people, damage or destruction of property, or harm to the environment or society), and the criticality, of those effects. Design tactics that help eliminate, minimize, or recover from safety-related effects include:

- *Increasing reliability/availability*: This is done by designing a system to eliminate, isolate, minimize, or recover from software or system failures and their effects. Since many safety issues involve the inability of the software to function correctly under failure conditions, the more reliable and available the software, the fewer failure conditions exist.

- *Increasing security*: The software can only be safe if it is secure. If someone can break into the software, that person can potentially make the software perform in an unsafe manner.
- *Adding interlocks*: This includes interlocks that enforce sequences of events in order to enhance safety or verify that safe conditions exist before an action is taken. For example, a microwave turns off when the door is opened, or the car brake must be depressed before the car can be put in gear.
- *Adding fail-safes*: If a failure occurs, the system defaults to the safest possible state.
- *Decoupling*: To prevent a failure in one area of the software from cascading into other parts of the software.

Security

While some security vulnerabilities are intentional in nature, such as malicious code, many of them are accidental in nature, caused by security holes. A *security hole* is an unintended function, mode, or state, caused by design flaws or coding defects, which a security attacker can exploit to result in a security breach.

Designing for security involves designing software that resists security attacks, detects those attacks when they happen, and can recover from successful attacks. Design tactics for resisting attacks include:

- Limitations on who can access the system (e.g., user authentication and firewalls)
- Limitations on what a user can do once they are in the system (e.g., access control privileges and security levels)
- Mechanisms for maintaining data integrity and confidentiality (e.g., encryption, handshaking on data communication, and checksums)
- Limitations on exposure (e.g., access timeouts and service limitations)

To detect attacks, intrusion detection software must be designed into the system. Recovering from security attacks includes restoration of service, recovering from any damage (e.g., backups of critical data and virus analyzers), and maintaining audit trails that aid in the identification of security vulnerabilities.

According to the Open Web Application Security Project (OWASP), proven security principles for software applications include (owasp.org 2013):

- *Applying defense in depth*: This is accomplished through the use of multilayered security mechanisms. If an attack breaches one level of security, another level of security might still be successful in identifying and repelling that attack.

- *Defining a positive security model*: That identifies what is allowed and rejects everything else. For example, when an input is filtered, screen for valid characters rather than attempting to screen for invalid characters. Some of the invalids might be missed and cause security vulnerabilities.
- *Failing securely when handling errors or failures*: For example, any failures encountered during an authorization operation should result in an "unauthorized" return code.
- *Executing each operation with the least amount of privileges required*: For example, if a function only needs to read information from a database, then open that database file with read-only privileges.
- *Not depending on obscurity or secrecy*: Obscurity and secrecy are ineffective ways to implement security. Somehow hackers always seem to manage to find what is hidden.
- *Keeping the security simple*: The more complex the software, the greater the opportunity for issues in the security software itself, and the more difficult it is to verify the security.
- *Collect audit trail data*: Detecting intrusions requires a log (audit trail) of security-relevant events, regular monitoring of that log, and proper responses to detected intrusions. Without proper intrusion detection, the attacker is given unlimited time to perfect or take advantage of their attack.
- *Not trusting the infrastructure or external services*: Perform error and security checking of any inputs that come from outside the software.
- *Establishing security defaults that emphasize security*: If necessary, allow the software's users to tailor the software to reduce the security levels, but default to the highest-level security requirements. For example, the default might be to log the user out after five minutes of keyboard inactivity, but allow the user to customize this delay to a longer interval.

Reliability and Availability

Software reliability and availability problems occur when defects in the software are encountered during the use of that software in operations, resulting in software failure. Therefore, good requirements, design, and coding practices should be used to prevent as many defects as possible from being introduced into software products. High software complexity is one of the major contributors to software reliability and availability issues. Designing software to increase cohesion and make individual software components as decoupled as possible can help increase the reliability and availability of that software. Good verification and validation practices should also be employed to identify and remove as many defects as possible from software products under development before their release into operations.

Software can also be designed to be more robust and fault tolerant, and therefore more reliable and available. Software components that process inputs and outputs should include appropriate error checking and error handling. This includes inputs and outputs to hardware and databases, as well as those to users. Fault-detection mechanisms can include periodic ping/echo or heartbeat tactics. In a *ping/echo tactic,* "one component issues a ping and expects to receive back an echo within a predefined time, from the component under scrutiny" (Bass et al. 2013). In a *heartbeat tactic,* one component periodically sends out a heartbeat signal and another component detects it. If either the ping/echo or heartbeat tactic fails, the non-responding component is assumed to have a problem.

To increase availability, mechanisms need to be designed and built into the software to recover from faults that do occur. One mechanism for accomplishing this is to create distributed or redundant architectures without any single point of failure. A less expensive alternative might be to have spares that can be activated if failures occur. If an external component is suspected of having a problem, for example, that component can be placed out of service. If an event or transaction does not occur in the expected time frame it can be timed out or retried. Retries may also be used to attempt to recover from certain types of failures (e.g., failure to allocate memory or buffers). Other recovery mechanisms include:

- Logging of the fault, or notification of the fault (e.g., to users or other components) as soon as it occurs
- Switching to a degraded mode of operations with less capacity or functionality
- Returning the system to a known state or shutting down the system

Usability

Usability involves the ease with which a user can operate or learn to operate the software. Characteristics of a usable design include:

- *Accessibility*: The degree to which the software is available to as many people as possible (the ability to access). Accessibility is often focused on people with different abilities. Does the product support people with the widest range of characteristics and capabilities, including visual, auditory, physical speech, cognitive language, learning and neurological (http://www.w3.org/TR/WCAG21/).
- *Responsiveness*: How quickly does the system react to user inputs and commands?
- *Efficiency*: Can users do what they want in a minimum amount of steps and time?
- *Comprehensibility*:
    - Do users understand the product structure and its user documentation?

  - Is the interface easy to remember and use even with infrequent use?
  - Is the user interface intuitive to use?
  - Does the system have consistent, logical user interfaces?
  - Does the software use vocabulary/terminology that is familiar to the typical user?
  - Are the prompts and error messages easy to interpret?
- *Flexibility*: The ease with which the software can be configured, reconfigured, or customized by the user.
- *Aesthetics*: Are the screens, reports, and other user interfaces pleasing to the senses?
- *Ease of learning* and *ease of use*: How intuitive is the software to learn and use? Is there adequate help? Is documentation easily searchable? Does the software align with the user's training, education level, technical, or domain expertise?

Examples of designing software for ease of use include designs:

- With continuous feedback to the users to keep them informed about what is going on
- Where the order of processing or data entry maps to the order in which users perform tasks
- That minimize the number of keystrokes or pages that the user needs to visit to perform a task or enter data
- With ways to abort, exit, or undo unwanted operations
- With mechanisms that allow users to tailor or customize their interactions with the software
- With error messages that clearly communicate issues, errors, or invalid inputs

Performance

Performance (capacity, throughput, response times, and so on) is all about time. When events occur in the system (e.g., interrupts, transactions, messages, or requests from users, the hardware, or other systems), the software or one of its components must react in a timely manner. Performance modeling during architecture activities is important in determining whether the software will meet its performance requirements. If performance is a concern, the architecture must take into consideration:

- Time-based events and associated behaviors, including:
  - Their usage of shared resources
  - Their processing and blocked time

- The frequency and volume of:
  - Intercomponent and intermodule communication
  - Network communication
  - Data access

Architecture tactics for handling performance concerns include:

- Managing event rates, for example, with event queues
- Prioritizing events
- Exploiting concurrency, for example, with parallelism
- Adding more resources
- Increasing the efficiency of existing resources
- Reducing processing overhead

Concurrency is implemented by decomposing work into cooperating or synchronizing processes that run in parallel. For example, concurrency occurs:

- Whenever the software creates a new thread (an independent sequence of control)
- When the software is executing on more than one processor
- When the software is utilizing other mechanisms that allow the software to multitask

Interoperability

Interoperability is the degree to which the software functions appropriately while sharing computer resources or data with other software applications or hardware operating on the same platform or in the same environment. Interoperability always has a context, including with whom or with what the system must be interoperable and under what circumstances. Types of interoperability include the following:

- Exchanging information via interfaces
- System of systems: In a system of systems, the systems may:
  - Directly interact and be subordinate to the system of systems
  - Operate in parallel with the system of systems but retain their own independence
  - Work together to address shared interests but without centralized management
  - Not be known to each other

Interoperability architecture tactics include:

- Adoption of standards (e.g., specific protocol, communication, or technology standards)
- Special components to route messages or coordinate interactions among systems (e.g., enterprise service bus or orchestration server)
- Utilization of a common operating system that manages shared resources (e.g., memory, buffers, or bandwidth)

Reusability

In order to be *reusable,* the software or software components should be designed to be independent of both the hardware and the software operating system. Reusable software designs are low in complexity, modular, cohesive, decoupled, and consistent. To reuse a component, the engineers must be able to understand what it does. This requires that the component be clearly documented and that documentation be kept up to date as the component is modified.

Maintainability and Modifiability

Software that is cohesive, decoupled, logically structured, and low in complexity is more *maintainable* and *modifiable*. Consistency of structure, language, design and coding standards, naming conventions, communication protocols, interfaces, and so on adds to the maintainability of the software. Well-written code comments and choosing the correct programming language to match the application type are also essential to maintainability.

Think of the maintenance engineer as one of the primary stakeholders of the architecture. For many systems, more than 80 percent of associated engineering efforts and costs are spent performing maintenance. That means that most systems that people work on are legacy systems. Many software developers spend their entire careers doing maintenance and never work on new development.

"A study of great designers found that one attribute they had in common was their ability to anticipate change" (Glass 1995). The goal of good architectural design is to isolate changeable items so that the impacts of the change will be limited to a single component or module. A change to one part of the architecture should not affect other pieces of the architecture. Isolation of components and information hiding minimizes the impact of changes on the integrity of the architecture and the system. Design tactics for creating a maintainable or modifiable architecture include:

- Identify items that are likely to change. If the requirements do not include information about items that are likely to change, then hold discussions with the stakeholders (including maintenance engineers) to determine these items. Items to consider are changes to the technical, legal, social, business, or customer aspects of the system or software.
- Separate items that are likely to change. Use modularity and information hiding to separate each changeable item into its component

or module (e.g., class or source code modules) or into a component or module with other volatile items that are likely to change at the same time.

- Isolate items that seem likely to change. Design the component or module interfaces to be insensitive to the potential changes. Design the interfaces so that changes are limited to the internals of the component or module and the change is hidden from other components or modules.

3. SOFTWARE REUSE

> Define and distinguish between software reuse, reengineering, and reverse engineering, and describe the impact these practices can have on software quality. (Understand)
>
> **Body of Knowledge III.F.3**

Reuse

Reuse occurs when one or more components of an existing software system or component are reused when developing a new software system or component. Reused software is used "as is" without modification. If even one minor change is made to the component, it is not considered a reused component because a second copy must be maintained separately. Reusability can be applied to any level in the structural hierarchy of the software. Requirements, architectural designs, component designs, source code modules, test cases, and test procedures can all be reused. Dunn (1990) states that "the higher the structural level of the reusable component, the more effectively the concept of reusability has been executed." The reuse of a software component has several benefits, including improved quality, reliability, productivity, and cost gains.

Software components that have withstood the test of time or been improved over time through removing latent defects have obvious quality advantages. By reusing these components in other systems, quality improvements can be realized. However, defects in reused components impact multiple products and therefore multiple stakeholders. Components that are intended for reuse should undergo rigorous development and verification and validation (V&V).

Typically, when software components are reused, less time needs to be spent creating the plans, designs, source code modules, documents, and data. Therefore, the same level of functionality can be delivered to the user for less effort, thus improving productivity. However, if the candidate for reuse is not low complexity, unambiguous, modular, and well documented, it may take considerable time to analyze and understand it in order to be able to reuse it. This can negatively impact potential productivity gains.

The cost benefits of reuse are calculated by subtracting the sum of the costs associated with reuse from the sum of the costs of producing the component from scratch. Pressman and Maxim (2015) lists the following costs associated with reuse:

- Domain analysis and modeling
- Domain architecture development
- Increased documentation to facilitate reuse
- Support and enhancement of reuse component
- Royalties and licenses for externally acquired components
- Creation or acquisition and operation of a reuse repository
- Training of personnel in design and construction for reuse

In order to make it easy to reuse a source code module, nothing that is changeable should be "hard coded" in the source code—for example, a set of reports is created that outputs to a printer. These reports include the customer's name in the title. Rather than "hard coding" the customer's name in the source code for each report, the source code reads the name from a file or global variable. This allows the report set to be reused for another customer without major rewrites.

In order to reuse a component, the engineer must be able to access it easily. It must be easier to reuse an existing component than to construct the component from scratch. One mechanism to allow easier access to reuse components is to create a reuse library of qualified components. Components in the reuse library must be under strict configuration control to prevent unauthorized changes and to make sure of the communication of authorized changes to all impacted parties.

Reengineering

Reengineering is the process of starting with a legacy software product and creating a new version of that product by redesigning or rebuilding it. The value of reengineering includes improved maintainability, performance, reliability, and so on. Reengineering may also be done to move to more modern or advanced technologies, languages, environments, and so on. A typical candidate for reengineering is a software product that continues to be useful but that is getting "old" because:

- It breaks too often
- It takes too long to repair
- When it is repaired, something else always seems to break
- It no longer represents the newest technology or can no longer be supported because the software language, operating system, or hardware platform are antiquated
- It can no longer meet updated, modern performance or security requirements

In reengineering, the existing legacy system often defines the requirements for the new system ("build a system just like the existing one except . . ."). In fact, the legacy system itself may be the only documentation of business processes or rules involved in the automated activities.

Reverse Engineering

In *reverse engineering,* the software requirements, design, interface information, scripts, tests, other software work products, and even the source code are re-created from the as-built software. Reasons for reverse engineering include:

- The lack of good configuration management practices when the software product was originally developed, resulting in the loss of needed work products
- Requirements, design, or other documentation were either never created or were not kept up to date and consistent as the software was modified over time

4. SOFTWARE DEVELOPMENT TOOLS

> Analyze and select the appropriate development tools for modeling, code analysis, requirements management, and documentation. (Analyze)
>
> **Body of Knowledge III.F.4**

There are many tools available to provide automated or semi-automated support for the various activities of software development.

Software requirements tools include:

- *Requirements templates*: Used to provide a template for creating a requirements specification
- *Requirements management tools*: Used to document, trace, and manage change to individual requirements
- *Modeling tools*: Used to create graphical representations of sets of requirements to aid in requirements analysis (these models can also be used to illustrate the requirements specification)
- *Prototyping tools*: Used to create trial versions of the requirements to help elicit, analyze and validate the requirements
- *Requirements analyzers*: Used to identify potential issues with the requirements (e.g., highlighting ambiguous wording, nonfinite requirements, or nonmeasurable requirements)

- *Traceability tools*:
  - Used during development to document the relationship between the various requirements levels and between the requirements and other work products of the development process
  - Used throughout the software product life cycle to identify the potential impacts of changes

These tools are used to aid in the development of the requirements through automation of requirements modeling and prototyping activities, and through documentation of the requirements. Software requirements tools can range from a simple template for a requirements document or use case to sophisticated requirements management tools. More sophisticated requirements tools can be integrated with other tools in the software development platform to provide mechanisms for automated bidirectional traceability of the requirements with predecessor work products.

Software design tools include:

- *Architectural design tools*: Used to model the overall software structure including defining the component of the architecture and their interfaces
- *Modeling tools*: Used to create graphical representations of different views of the architectural and component designs and aid in design analysis (these models can also be used to illustrate the design specifications)
- *Design optimization tools*: Used to identify areas of the design that can be improved or optimized
- *Prototyping tools*: Used to create trial versions of the design to help compare different design alternatives and validate the design or its feasibility
- *Design verification tools*: Used to aid in design reviews and to specify and detect violations in design rules and constraints

Software implementation tools include tools that are used when translating the requirements and design into machine-readable source code and its associated user documentation. These tools include:

- *Code generators*: Used to automate the creation of the source code from the design
- *Source code editors*: Used to manually write the source code
- *Static code analyzers*: Used to detect defects in the source code and identify areas of the code for improvement
- *Automated code reviewers*: Used to identify source code defects or check the source code against a predefined set of rules, best practices coding standards, or naming conventions

- *Security analyzers*: Used to detect source code security vulnerabilities
- *Debuggers*: Used to aid in the identification and correction of coding defects

Other software development tools include:

- Collaborative development tools
- Network tools
- Content management tools for web applications
- Reverse engineering tools
- Operating systems
- Word processors
- Spreadsheets
- Libraries
- Simulators
- Databases
- And all the other tools that can be part of the development environment

Chapter 15

Maintenance Management

Successful software products tend to have very long life spans when measured from initial release to final retirement. For example, the initial release of Microsoft Word was in 1983, so that software product has been in use for more than 30 years with no retirement in sight. However, according to Sommerville (2016), "During their lifetime, operational software systems have to change if they are to remain useful." The need to change existing software products may result from the following:

- The need for additional or changed functionality
- Incremental or evolutionary development
- Defects in the work products
- Changes to the business climate or business needs
- Changes to user requirements, industry standards, or regulations
- Changes to the hardware or target platform
- Technological advancements
- Competitive motivations and threats
- The need for more reliability, usability, security, or safety in the software

Therefore, changes are inevitable if the software is to remain technically current and competitive in the marketplace, retain its value, and continue to meet the stakeholders' needs.

Software maintenance is the process of making changes to software products after they have been released into operations. In some cases, these software products may be *legacy software*, a term used to mean older software written using antiquated methods, programming languages, or technologies. Legacy software refers to software that is out of date and that needs to be reengineered or replaced but that is still in use in operations.

Software maintenance adapts, optimizes, enhances, or corrects defects in a released software product while still preserving that product's quality and integrity. In other words, the objective of software maintenance is to change part of a released software product without breaking another part of that product. Maintenance process activities include:

- Maintenance process implementation
- Problem and modification analysis
- Modification implementation
- Maintenance release review and acceptance
- Emergency maintenance
- Migration
- Software retirement

Since agile development processes are based on a continuous flow of software development supported by an emerging product backlog, there is no real distinction between initial development and ongoing post-delivery maintenance. All iterations after the software's initial release could be considered maintenance iterations.

1. MAINTENANCE TYPES

> Describe the characteristics of corrective, adaptive, perfective, and preventive maintenance types. (Understand)
>
> **Body of Knowledge III.G.1**

Corrective Maintenance

Even with the best state-of-the-practice software quality assurance and control techniques, it is likely that the users of the software will encounter at least a few software failures once the software products are delivered to operations. The IEEE *Standard for Software Engineering—Software Life Cycle Processes—Maintenance* (IEEE 2006) defines *corrective maintenance* as "the reactive modification of a software product performed after delivery to correct discovered problems." Corrective maintenance is performed to repair the defects that cause operational failures.

Perfective Maintenance

As the software is used, the users, customers, or other external stakeholders may discover requirements they need to add to the software product. Marketing or engineering may also come up with new requirements to add to the software to continue innovating the software, to remain technically current, or to meet competitive threats. The ISO/IEC/IEEE *Systems and Software Engineering—Vocabulary* (IEEE/ISO/IEC 2017) defines *perfective maintenance* as "improvements in the software's performance or functionality, for example, in response to user suggestions or requests." Perfective maintenance is performed to add new requirements to the software (to "perfect" the product). Perfective maintenance is the normal enhancement of successful, working software products over time so that the software retains its value to the stakeholders.

Adaptive Maintenance

The IEEE (2006) defines *adaptive maintenance* as the "modification of a software product, performed after delivery, to keep a software product usable in a changed environment." As time progresses, a software product's operational environment will likely change.

Examples of changes to the external environment in which that software must function include:

- External interface definitions may change (e.g., communications, protocols, and external data file structures)
- Government regulations or other business rules may change (e.g., tax code or regulatory reporting requirements)
- Hardware and software that interface with the product may change (e.g., new versions of the operating system, new or updated peripherals or other hardware, new or updated interfacing software applications, additional or modified fields or records in external databases)

Preventive Maintenance

The IEEE (2006) defines *preventive maintenance* as "modification of a software product after delivery to detect and correct latent faults in the software product before they manifest as failures."

Preventive maintenance is a proactive approach to prevent problems with the software before they occur. For example, additional error handling and fault tolerance might be added to a safety-critical system, or an enhanced encryption methodology might be added to improve software security. Self-diagnostic testing might also be added to the software product.

As software products change over time they can deteriorate, especially if good design techniques were not employed when the software was initially developed. It may become necessary to modify an existing, released software product by refactoring it to eliminate duplication, complexity, or awkward structure or to reengineer it to make it easier to maintain going forward. These changes are another type of preventive maintenance.

2. MAINTENANCE STRATEGY

> Describe various factors affecting the strategy for software maintenance including service-level agreements (SLAs), short- and long-term costs, maintenance releases, and product discontinuance and their impact on software quality. (Understand)
>
> **Body of Knowledge III.G.2**

Maintenance Process Implementation

Maintenance process implementation includes all of the activities involved in the maintenance part of the quality management system (QMS). These activities include establishing and maintaining policies, standards, processes, and work instructions related to software maintenance. They also include:

- Performing maintenance planning for each software release or product. This planning defines the specifics for implementing the maintenance portion of the QMS for a software release or product. This planning also includes *tailoring* or defining specific maintenance processes or work instructions needed for that software release or product.
- Transitioning control of the software product from the developers to the maintainers if a separate group performs the maintenance activities.
- Negotiating service contracts or service level agreements (SLAs) with the acquirers of the software.
- Defining and implementing mechanisms for receiving, documenting, and tracking problem reports and enhancement requests to resolution.
- Coordinating with configuration management to manage changes to existing software products or product components.

Planning for these maintenance process implementation activities should occur early in the software development life cycle, well before the software product's release into operations. Typically, these activities are part of contract negotiations with the acquirer or are a part of the project planning activities.

A *service level agreement* (SLA) is a formally negotiated agreement between two parties (typically, the software's maintainer and the acquirer) that defines the agreed-upon level of service to be provided. An SLA defines both the services and measures of service support attributes, including availability, performance, responsiveness, timeliness, and completeness of that service. For example, an SLA might include requirements for the availability of help desk services ("technical help desk support is available on a 24/7 basis" or "80 percent of all help desk calls are answered within 30 seconds") or time limits on resolution of problem reports ("95 percent of all major defects are resolved within 30 days"). The SLA may also include service fee agreements, warranties, penalties for violations, escalation procedures, or defined responsibilities for both parties.

Problem and Modification Analysis

The maintainers perform problem and modification analysis after the software has been released to operations, which includes all activities related to evaluating problems or requested enhancements, determining their impacts, making resolution decisions, and developing solutions to approved changes. For problem reports, the evaluation process may require replication of the problem to debug it and identify its root cause, or the actual software defect(s) involved. This impact analysis and resolution decision is done through the configuration control process.

The effort involved in problem and modification analysis may dramatically increase if the released software is being changed:

- Is part of a system of systems because adaptive maintenance may be required on other interfacing systems
- Has one or more components that need to be changed, and those components are reused in multiple software products or multiple supported versions of the same product because corrective or adaptive maintenance may be required on those other software products or releases
- Has hardware or software dependencies because adaptive maintenance may be required on the dependent hardware or software

Modification Implementation

Once the appropriate authority has approved one or more changes, the maintainer updates the impacted software products, or product components, to incorporate those changes. These activities involve all of the software development activities (requirements, analysis and design, implementation, verification and validation) necessary to implement the changes. This effort may be as simple as creating a service pact for a corrective release. However, if approved changes are significant enough, their implementation may involve initiating a new software development project (e.g., evolutionary development).

Maintenance Release Review and Acceptance

The maintenance release review and acceptance includes verification and validation (V&V), acceptance, release, and distribution activities of the changed software products. V&V activities ensure that the changes are correct and that other defects were not interjected into the software when those changes were made. These activities also ensure that the resulting software meets the required integrity, quality, and performance levels.

Emergency Maintenance

While complete and rigorous software maintenance process activities are desirable, there may be conditions under which emergency maintenance activities must be implemented. For example, urgent changes might be needed if:

- A critical software failure occurs that disrupts normal operations or causes a significant safety or security concern
- Unanticipated changes occur in the software's environment that disrupt normal operations or cause a significant safety or security concern

Software maintenance processes must include mechanisms for performing emergency changes to the software code or executable without ensuring that other associated work products (e.g., requirements, designs, and documentation) are also updated. However, these processes must include mechanisms for recovering from the emergency state by:

- Making necessary changes to all the related products to keep them consistent with the software code or executable, including updating requirements, design, documentation, and so on
- Verifying the long-term integrity and consistency of all of the software work products

Migration

Migration consists of maintenance processes specific to porting the software and its associated data to another environment (e.g., a move to a new hardware platform, operating system, or database structure). In addition to many of the maintenance activities described previously, migration may also include:

- The development of migration tools
- Conversion of the software or data
- Mechanisms for handling migration over time, including parallel operations in both the old and new environments for some time
- Support for the old environment into the future

Software Retirement and Product Discontinuance

Software retirement is when the maintenance support of a particular software release is terminated or when the maintenance support of an entire software product is terminated. Issues related to retirement include:

- How long is the software product actively supported?
- How many past releases of the software are actively supported?
- Is retired software removed from its operational environment, and, if so, how is this accomplished?
- What mechanisms are used for archiving retired software and its related records?

Agreements about these issues should be included in negotiations between the software developers and acquirers early in the life of the initial software product and renegotiated as necessary with each subsequent software release to manage expectations on both sides.

3. CUSTOMER FEEDBACK MANAGEMENT

Describe the importance of customer feedback management including quality of product support, and post-delivery issues analysis and resolution. (Understand)

Body of Knowledge III.G.3

Customer Feedback

Customer feedback is the data from customers, users, and other external stakeholders about their experiences and perceptions when interacting with an organization or its products. For example, these data can take the form of:

- Customer satisfaction survey results
- Transaction-specific feedback
- Loyalty and repeat business analytics
- Word-of-mouth-generated business or sales data
- External stakeholder complaints and their resolution
- Post-delivery issues or problem analysis and their resolutions
- External stakeholder ideas, suggestions, or product and service enhancement requests

An organization should have specific business objectives in mind before actively collecting customer feedback information, to focus the customer feedback collection process. Information obtained from analyzing feedback should help answer specific questions about those business objectives and allow the organization to make informed decisions about the actions needed to move toward achieving those business objectives.

Customer Feedback Management

How customers, users, and other external stakeholders feel about an organization and its products significantly impacts that organization's business. Poor customer perceptions and experiences will impact customer loyalty and repeat business. There are many choices in today's software marketplace. If stakeholders do not perceive a software product to be user-friendly, reliable, high quality, safe, or secure, those stakeholders will go elsewhere. If stakeholders have a negative experience when interacting with an organization or its software product, they will go elsewhere—and they will tell their friends.

Customer feedback management is the process of gathering, organizing, analyzing, reporting, and making customer feedback actionable. In other words, customer feedback management provides an organized way to take the "voice of the customer" and turn it into:

- New and innovative products that are value-added to those stakeholders
- Enhancements and improvements to existing products that are also likely to improve the stakeholder perception of those products
- New or improved stakeholder services (e.g., technical support, user training, sales, and order management) or stakeholder interaction and communication mechanisms that will make those services more effective and efficient and improve stakeholders' satisfaction

Customer feedback management is a stakeholder-centric, proactive approach that allows external stakeholders to become indirectly involved in product and service development and improvement. LaMalfa (2010) lists the following nine ways to help an organization succeed at customer feedback management:

1. Have well-defined goals and objectives when collecting customer feedback
2. Get executive buy-in and internal support for a customer feedback program
3. Develop a formal "voice of the customer" program
4. Collect and manage customer feedback in a centralized system
5. Be a customer advocate throughout the feedback process
6. Communicate and share customer feedback with others
7. Collect feedback in real time, and on an ongoing basis
8. Integrate customer feedback into the business
9. Tie customer feedback programs to business outcomes

Tools, including web applications and portals, exist that are designed to help automate the implementation of customer feedback management.

Part IV

Project Management

Chapter 16

Planning, Scheduling, and Deployment

According to the Project Management Institute (PMI) (PMI 2013), a *project* is "a temporary endeavor undertaken to create a unique product, service, or result." A project has a specific purpose or goal to be accomplished. That goal is to create a unique outcome in the form of a new or updated product, service, or result.

The fact that a project is *temporary* means that a project is a set of activities performed a single time and not repeated. Projects differ in this respect from the organization's repeated, ongoing operational work. The goal of a project is typically to add something new to, change something in, or provide information to ongoing operations. Temporary also means that a project has a definite start and end point and lasts a specific, finite amount of time. However, temporary does not imply short. Projects may last only a few hours, or they may be multiple years long. A *successful project* ends when its objectives have been accomplished. It may also be terminated before the objectives are met, if funding is cut, the business needs for the product or service are eliminated, or if it becomes clear that the project cannot meet some or all of its objectives. Examples of software projects include:

- Development of a new software product
- Addition of new features to an existing software product
- Implementation of a corrective maintenance release for an existing software product

Nonproject activities include any ongoing or repetitive activity or activities that are part of the operations of the organization using the software. These are sometimes referred to as level-of-effort work activities. Examples of nonproject activities include:

- Production of monthly reports on field-released software defects
- Ongoing technical support or help desk activities for existing software packages

A software project requires human resources with various skills, along with nonhuman resources, to accomplish its objectives. A project is constrained by the limits on the resources available to that project. Human resources include requirements analysts, designers, programmers, testers, managers, and specialists (e.g., software quality engineers, auditors, configuration management specialists, security and safety specialists, and so on). Nonhuman resources include:

- Computers
- Support software and tools (e.g., compilers, build scripts, code analyzers, test drivers)
- Facilities (e.g., office space, test beds)
- Supplies (e.g., paper, USB flash drive)

Every software project has a customer or intended user for its created product, service, or result. The customer or user may be external to the organization implementing the project (e.g., software produced for sale, under a contract, or to a mass market), or the customer or user may be internal to the organization (e.g., management information systems and internally developed software tools).

Project Processes

As illustrated in Figure 16.1, project management is a system of processes for initiating, planning, executing, tracking and controlling, and closing the project. These project management activities are not one-time events but form a feedback loop that continues throughout the life of the project.

Project initiation includes deciding whether or not to start the project based on needs or a business case analysis:

- Is the project valuable to the organization?
- Does the project align with organizational strategies and goals?
- Is it feasible with current resources and technologies?
- Do the risks associated with this project outweigh its potential opportunities?

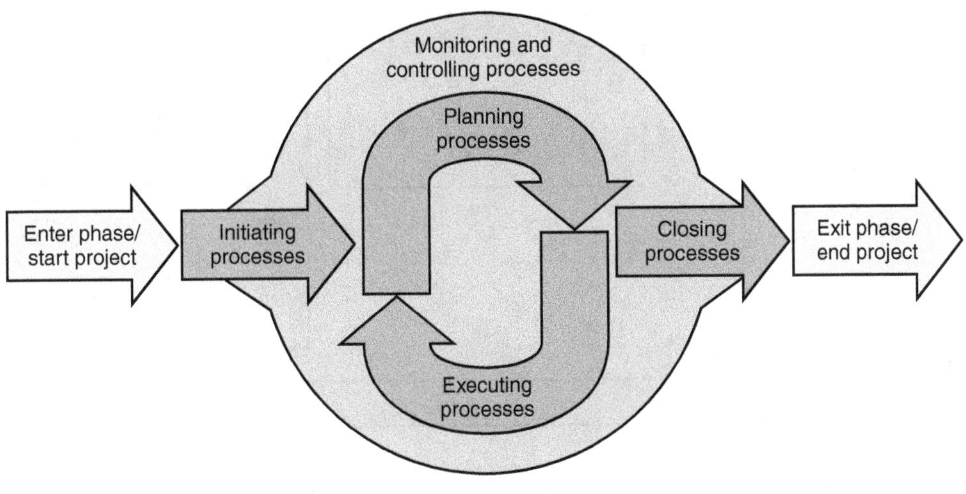

Figure 16.1 Project management process. (Source: PMI 2013. Reprinted with permission.)

During *project initiation*, the initial scope of the project is defined and resources are committed to the project. The Project Management Institute (PMI 2013) defines two processes in its project-initiating process group: *develop project charter* and *identify stakeholders*. The major output of the initiation processes is the project charter and a list of project stakeholders.

Project planning establishes the strategic and tactical plans for the project, including activities, processes, schedules, budgets, staffing, and resources. Planning also occurs throughout the project, as additional information is obtained and fed back as progressive elaboration into the project plans.

Project execution includes the activities and actions required to implement the project's strategic and tactical plans, which are performed in order to meet the project's objectives and create the project's deliverables.

Project monitoring and controlling, also called *tracking and control*, monitors the actual results of the project execution against the established plans. These activities also control significant deviation from the expected plans to keep the project in line with its objectives, and manage changes to the project's scope and objectives.

Project closure involves shutting down the project when it is completed (or terminated before completion). The Project Management Institute defines two processes in its project closing process group: *close project* and *phase and close procurements* (PMI 2013).

Not only should these project processes be performed at the project level, but, as illustrated in Figure 16.2, they should also be repeated as part of each phase

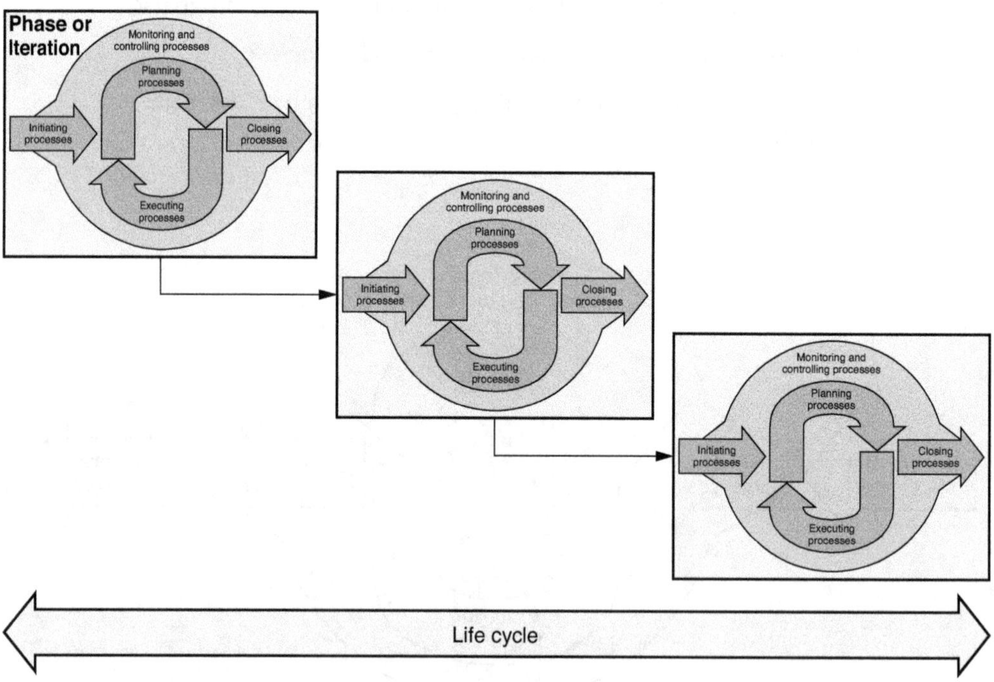

Figure 16.2 Life cycle phase project management processes. (Source: PMI 2013. Reprinted with permission.)

of the project. Initiation activities should be repeated at the beginning of each phase of the software development life cycle to make sure that the project is still viable and should continue or be terminated. Planning should be revisited at each phase to add progressive elaboration to the plans based on the information gathered by executing the previous phase. The activities of each phase are then executed and the phase is closed at completion. Throughout this process, the activities of the phase are monitored and controlled.

In Scrum (an agile software development framework), the initial sprint (iteration) is initiated when a decision is made to start the first Scrum. Planning is done at the beginning of each sprint during the sprint planning meeting. Plans are progressively elaborated, as needed, during the execution of the sprint. Execution is the work that is accomplished each day of the sprint. Monitoring is done through continuous communications and during daily team meetings where actual velocity to predicted velocity is tracked and impediments are discussed (velocity is discussed later in this chapter). If controlling actions are needed, they are taken by the Scrum team with the aid of the Scrum master. Project closure occurs when the sprint is completed, the new working increment of functionality is demonstrated, and a sprint retrospective is performed. Depending on the Scrum implementation, a decision is made at the end of each sprint or each release (which may include a set of multiple sprints) as to whether another sprint or release is needed. If so, the next sprint is initiated.

Project Drivers

Each project has three major drivers: cost, schedule, and scope. As illustrated in Figure 16.3, a good analogy of project management is the juggling of three bowling balls, with each ball representing one of the three drivers. If the project manager or project team pay too much attention to any one bowling ball, they are likely to drop one of the others and smash the foot of the project. The *cost driver* represents the cost of all of the resources that go into the project. The *schedule driver* represents the calendar time it takes to complete the project. The *scope driver* represents the quantity of work done during the project to produce outputs with the required functionality and quality. For software, the scope driver is a combination of the:

- Amount of software functionality
- Required quality level of that functionality (quality attributes)
- Level of required process rigor to be employed during the project

Each project will have one primary driver that is the most important to the project. Cost may be the primary driver on a project with a fixed-price contract or where profitability is essential to success. Software that is being built to target a specific market window may have schedule as its primary driver. For example, one company was working on a project where a communication system was being developed for military war games. If the software was not delivered on schedule in time for the military exercise, the government did not want it and would not pay for it. Scope may be the primary driver in a project to develop software that impacts the health or safety of people. However, for each project, all of the required activities must be performed at the required level of rigor to verify that a complete,

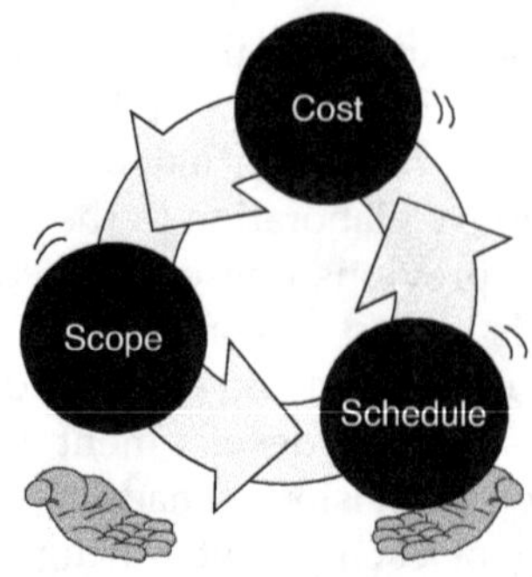

Figure 16.3 Cost–schedule–scope trilogy.

quality product is built. Part of the job of a software project manager is to determine the primary driver and maintain the focus on that driver without losing sight of the other two (dropping either or both of the other two bowling balls).

An old project management saying states that a project can fix any two of these elements but not all three. That means that given the project's current capability levels:

- If the project schedule needs to be shortened, it will require either spending more or reducing the scope of the work being done
- If a larger scope of work is required, because of the need for more functionality or a higher level of software quality, the project will take longer or cost more
- If cost reductions are necessary, the schedule or the scope will be impacted

The good news is that empirical evidence from the software industry shows that improving process maturity and implementing effective and efficient quality management systems allows a win–win–win across this trilogy—software projects can create increased scope (build in more functionality with higher quality), in less time, at a reduced cost.

A project is considered successful if the stakeholders' project objectives are completed on schedule, within budget, at the desired quality and performance level, without reducing the scope of the project, while effectively and efficiently utilizing people and other project resources. The goal of project management is to plan for and execute a successful project.

For Scrum projects, the cost and schedule drivers are fixed. The cost driver is fixed by the size of the Scrum team, typically seven people, plus or minus two, times the duration of the sprint (iteration). The schedule driver is fixed by the duration of the sprint, typically one to four weeks (30 days), and the number of planned sprints. Therefore, only the scope of each sprint is negotiated during the sprint planning meeting, when stories are selected and committed to by the Scrum team. The quality level for the product being produced is fixed by the team's "definition of done," which usually includes factors such as (based on James 2016):

- All code needed to implement the selected product backlog items has been created or updated through pair programming, or it has been peer reviewed
- Refactoring has been done to improve poorly designed or poorly written code, to remove redundant code, or to improve the code's quality, including its performance
- All documentation has been updated
- Manual and exploratory testing of new functionality has been executed and successfully completed
- All planned regression testing has been executed and successfully completed
- Any automated tests for new functionality have been created, reviewed, executed, and passed (e.g., tests that will be used in future regression testing)
- Checkout and build processes are fully reproducible

1. PROJECT PLANNING

> Use product acceptance, configuration acceptance, forecasts, resources, schedules, task and cost estimates, etc., to develop project plans. (Apply)
>
> **Body of Knowledge IV.A.1**

As illustrated in Figure 16.4, project planning provides the road map (project plan) for the project journey. It defines the reason for the trip (mission statement), the destination (project objectives), the vehicle (processes and methods), the people taking the trip (staff and other stakeholders), the expected cost of the trip (budget), the expected duration of the trip (schedule), and the stops along the way (milestones).

Project planning includes all of the activities necessary to:

- Select the appropriate life cycle and processes
- Define the various subsidiary plans
- Create estimates of scope, cost, staff, resources, and schedule
- Define the work breakdown structure (WBS) and activity network
- Identify, analyze, and mitigate risks
- Integrate the results into the overall software project plans for the project

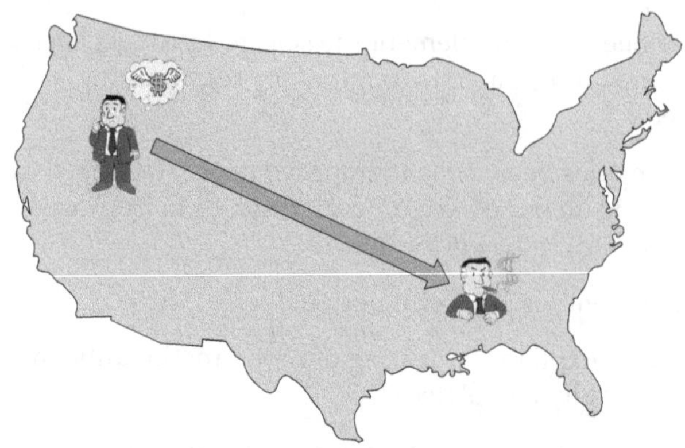

Figure 16.4 Project planning is the road map for the project journey.

As illustrated in Figure 16.5, the inputs into project planning include the project charter, environmental factors, and organizational process assets (these three inputs are discussed in more detail later in this chapter). The outputs of project planning are the software project plans (the details of the software project plan are also discussed later in this chapter).

Project planning is an iterative process, as illustrated by the arrows going both ways in Figure 16.5. Generic high-level activities can be planned in advance, but many activities cannot be planned to any level of detail until predecessor activities are completed. During the early phases of a project, the project team must usually settle for high-level general plans. As the project progresses, more and more detail can be added to these plans, as illustrated in Figure 16.6. For example, during the initial planning phase there may be an estimate that five programmers will be needed to code, inspect, and unit test 60 modules over a period of four months. By the end of the architectural design phase, the project may have identified:

- The specific 57 modules to be coded, inspected, and unit tested
- The order in which they will be implemented
- A detailed schedule for that implementation over a 17-week period
- Specific assignments for five programmers responsible for doing the coding of specific source code modules

Planning a project requires the integration of many different activities. Figure 16.7 illustrates the various processes in the Project Management Institute's planning process group, and how they interact and are integrated into developing the project management plan. The numbers preceding each process name in this figure indicate the section in the PMBOK Guide (PMI 2013) that describes that process.

Inputs into the Project Planning Processes

As illustrated at the top of Figure 16.5, the primary inputs into the project planning process are the project charter, project objectives, and organizational environmental factors and process assets.

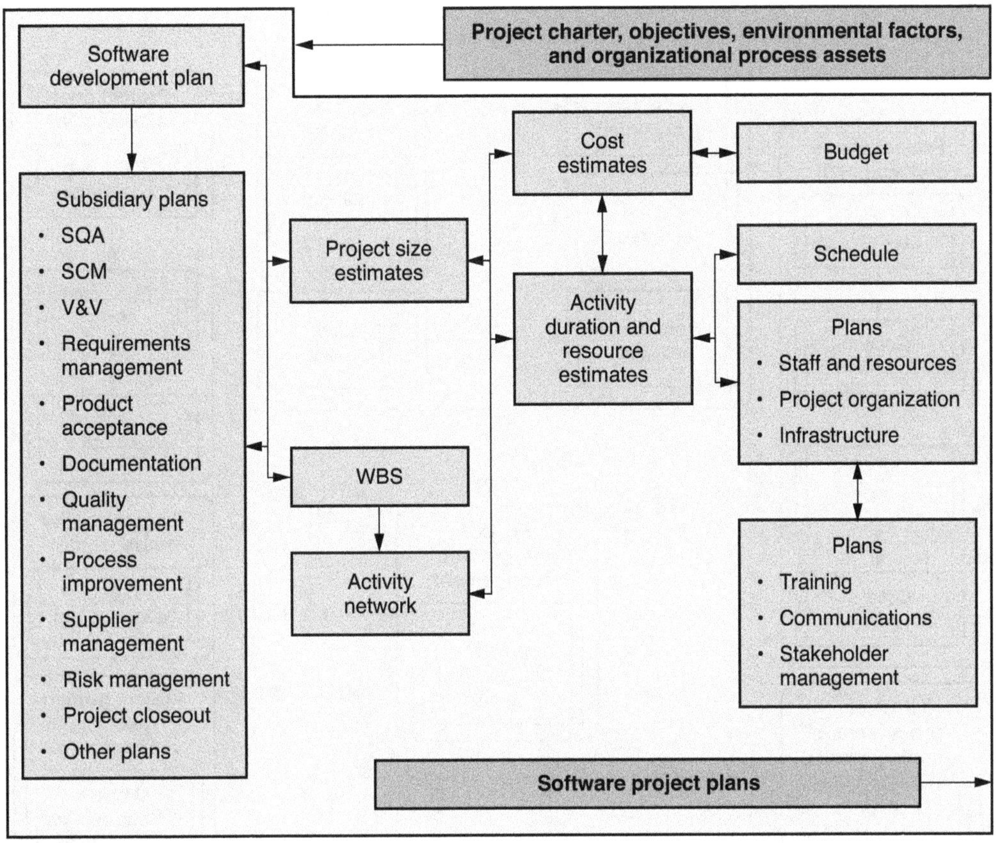

Figure 16.5 Project planning.

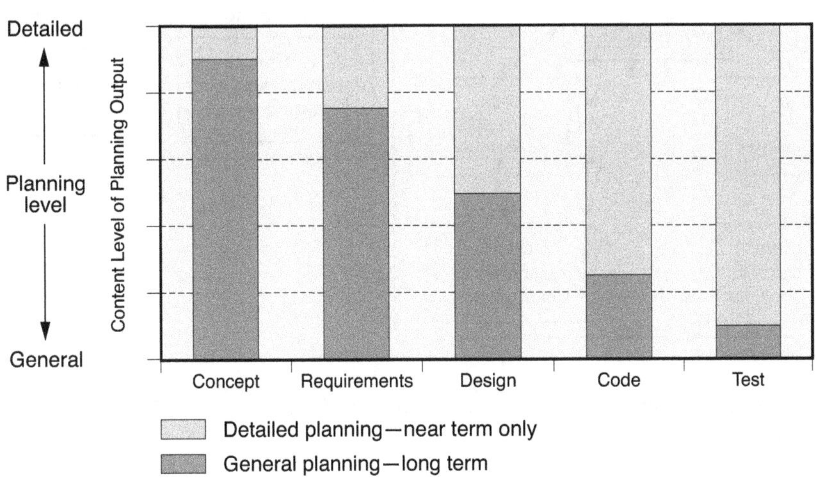

Figure 16.6 Long-term versus near-term planning.

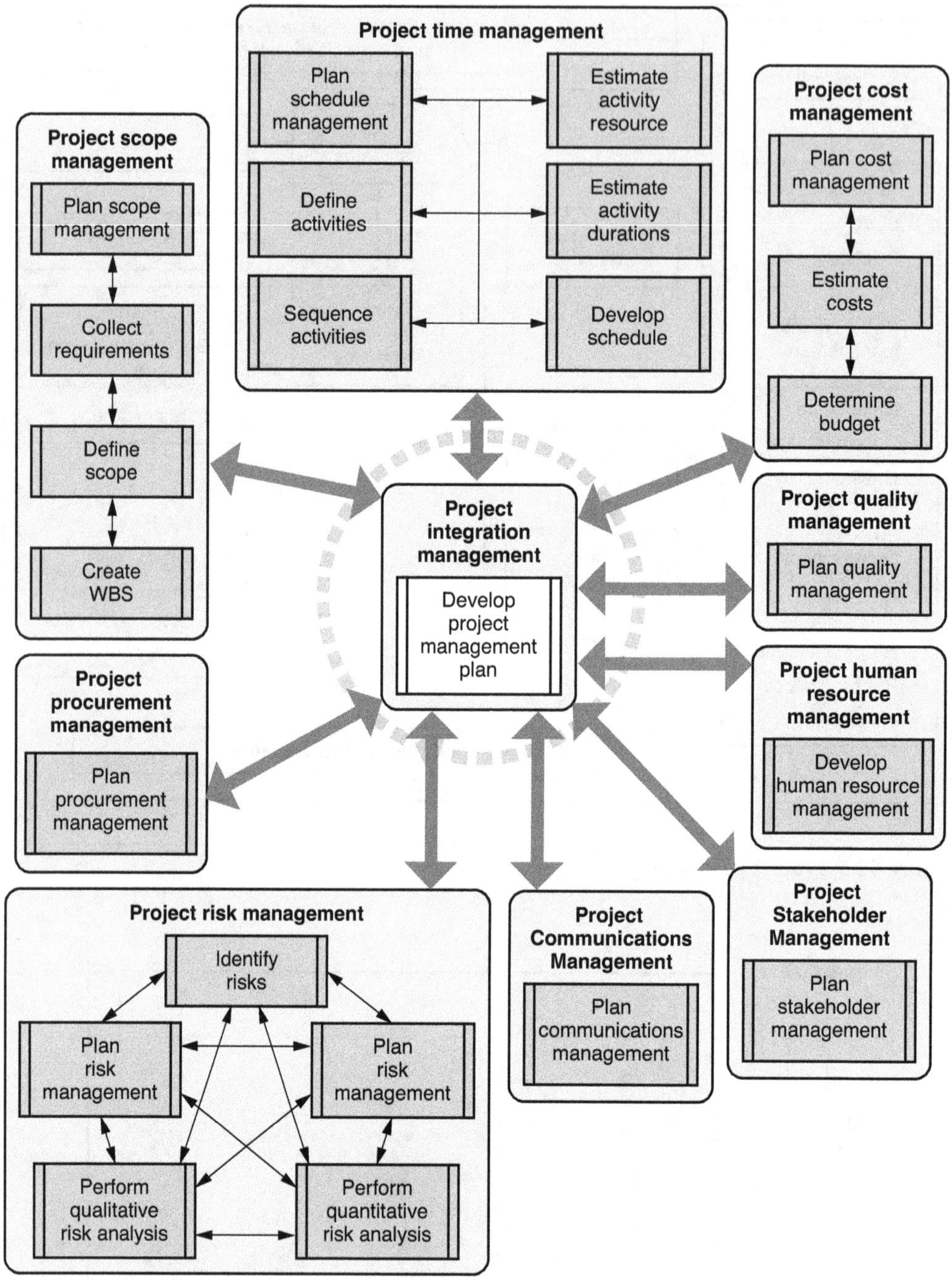

Figure 16.7 PMI project planning process group. (Source: PMI 20013. Reprinted with permission.)

Project Charter

The *project charter* is the document that formally authorizes the project and is created during project initiation. The contents of a typical project charter include:

- A justification for the project
- The initial scope and limitations statement of the business needs
- The project objectives
- A definition of the products, services, or results that are expected as the outcome of the project
- A list of the project stakeholders
- A list of major project risks, if any

Project Objectives

Project objectives define the specifics of what success looks like for the project from the perspective of its stakeholders. What exactly the project is trying to accomplish for its stakeholders from a business-level perspective. Each project objective should be SMART:

- Specific and understandable, by being stated in such a way that other people will know what the project is trying to achieve
- Measurable and verifiable, so that the project team knows when they have achieved it
- Attainable and aligned with higher-level organizational objectives
- Relevant and realistic, so that it can be achieved within the time, cost, and resource restraints of the project
- Time-framed, with specific due dates for accomplishment

There may be many different types of stakeholder objectives for the project. Examples of project objectives include:

- Business objectives:
  - Capture a market share of X percent within Y months
  - Achieve an X percent increase in profitability or return on investment within Y months
- Cost objectives:
  - Complete the project under a cost of $X
  - Reduce the cost of ownership of the new ABC system over X months (or years) by Y percent
- Schedule, milestone, or cycle time objectives:
  - Deliver the project or complete a major milestone by X date

- Achieve an X percent reduction in the cycle time for producing work product Y or for performing activity Z
- Product technical objectives:
  - Add ABC new functionality to an existing product
  - Increase the software's capacity so that it can handle X number of transactions in an hour
- Product quality and reliability objectives:
  - Keep the total post-release software outage levels below X minutes per site per year
  - Reduce user-reported defects by X percent below prior product releases
- Organizational or personnel objectives: train X additional software engineers in XYZ techniques
- Contract and procurement objectives: qualify a new XYZ tool supplier
- Management system objectives: pilot the new process improvement ABC

Project objectives for Scrum projects are defined during the sprint planning meeting when sprint objectives are defined and committed to by the Scrum team with agreement from the product owner.

Environmental Factors and Process Assets

Organizational environmental factors include all of the elements that surround the project, including items such as:

- Organizational culture
- Infrastructure (e.g., existing office space, computer equipment, and labs)
- Tools (e.g., software development tools, project management tools, testing tools, configuration management tools)
- Experience and training levels of the staff assigned to the project

Organizational process assets are the artifacts considered "useful to those who are defining, implementing, and managing processes in the organization" (CMMI 2010a). These process assets represent organizational learning and knowledge, which can be tailored to meet the needs of each program or project. These process assets may include:

- The organization's quality management system (e.g., policies, standard processes, standard work instructions)
- Definitions of the software life cycles approved for use within the organization

- Guidelines and criteria for tailoring the organization's set of standard processes
- Metrics repository, including historic data and metrics from previous projects, processes, and work products
- Historic information (e.g., plans, reusable work products and components, and lessons learned from previous projects)

Project Plans

As illustrated in Figure 16.5, there are multiple plans included as part of the project planning process.

Software Development Plan

The first step in the project planning process is to the software development plan which includes:

- Selecting and defining the software life cycle or process model that will be used for the project. (See Chapter 9 for a discussion of life cycle and process models.)
- Selecting, defining, or tailoring the software processes and work instructions that will be used during the project.
- Selecting and specifying the methods, tools, and techniques that will be used on the project. These include development methodologies, programming languages, and other tools, techniques, or workmanship standards (e.g., coding or documentation standards) used to specify, design, build, test, integrate, document, deliver, modify, and maintain the project's work product and deliverables.

Subsidiary Plans

The next step in project planning process includes working with the appropriate stakeholders within the organization to develop the subsidiary plans. These subsidiary plans may be incorporated as sections into a single project plan document, or they may be separate stand-alone documents that are referenced by this primary project plan document. IEEE has standards that provide guidance for many of these plans, and many of them are discussed elsewhere in this book. Subsidiary plans include:

- *Software quality assurance (SQA) plan* specifies the plans, tools, methods, techniques, schedules, and responsibilities for assuring that the required quality levels are met for the software processes and work. This includes planning any reviews or audits that will be done as part of the project activities.
- *Software configuration management (SCM) plan* specifies the plans, tools, methods, techniques, schedules, and responsibilities for configuration identification, control, status accounting, and auditing, and for release management.

- *Verification and validation (V&V) plan* specifies the plans, scope, tools, methods, techniques, schedules, and responsibilities for the verification and validation activities of the project.
- *Requirements management plan* specifies plans, tools, methods, techniques, schedules, and responsibilities for measuring, reporting, and controlling change to the product requirements, and for performing requirements traceability.
- *Product acceptance plan* specifies the plans, tools, methods, techniques, schedules, and responsibilities for obtaining the customer's or user's approval of the external product deliverables, including the objective criteria for accepting each deliverable.
- *Documentation plan* specifies the list of documents to be prepared and the plans, tools, methods, techniques, schedules, and responsibilities for preparing, reviewing, baselining, releasing, and distributing those documents.
- *Quality management plan* specifies process improvement activities, including:
  - Periodically assessing the project's processes
  - Determining issues and areas for improvement
  - Managing and correcting issues and implementing those improvements
  - Any piloting of organizational-level process improvement that will be done as part of the project
- *Supplier management plan* specifies the criteria for selecting suppliers and a tailored or negotiated management plan for managing each supplier selected to contribute outsourced work products to the project.
- *Risk management plan* specifies the plans, tools, methods, techniques, schedules, and responsibilities for identifying, analyzing, planning mitigations, mitigating, tracking, and controlling project-related risks and risks associated with the product produced by the project.
- *Project closure plan* specifies the activities for the orderly closure of the project, including:
  - Delivering the project's products into operations
  - Transferring ownership of the project's products from development to maintenance and technical support
  - Archiving project files and quality records
  - Transitioning people and other project resources to future projects or other areas
  - Closing down project charge numbers with accounting or finance
  - Conducting post-project reviews

- Other plans include any special plans needed for the project. For example, projects that are producing safety-critical software might need safety plans, or projects where security is a major concern might have separate security plans.

Staff and Resource Plans

Staffing plans specify the number of staff required, by skill levels, when they are needed, and their source (internal transfer, new hire, contracted). *Resource plans* specify the number of nonhuman resources required by type, when they are needed, and source (internal transfer, purchased, rented). Challenges that the project manager faces when allocating staff and other nonhuman resources to a project include:

- Lack of sufficient resources for the optimum schedule.
- Allocation or leveling of multiple resource types. Software projects require highly skilled people, and a project may involve many different specialists (e.g., requirements analysts, designers, coders, testers, technical writers, quality specialist). If one specialist has slack time, it does not mean they can be used to cover the overload in another specialty. The same is true with other resources. Having an extra computer does not necessarily mean that it has the correct configuration to use it as a test bed.
- Allocation or leveling of resources across multiple projects.
- Productive work time is not eight hours per day, because team members spend time on overhead (e.g., staff meetings, e-mail, status reports, phone calls), on supporting other projects, and on product maintenance.
- A person working on more than one activity at a time (multitasking) can involve extra effort because of switching back and forth between activities. This can also be true for resources that are being used on more than one project or activity. For example, the software, database, and tool set on a test bed might have to be reloaded each time a switch is made.
- More than one person working on the same activity can involve extra effort because of additional time required to communicate.

Project Organization Plans

Project organization plans should include a description of the internal project team organization and structure, including internal interfaces (e.g., lines of authority, responsibility, and communication) within software development and supporting organizations. This description typically includes a project organizational chart.

During the project, the project team members may also need to interact with other stakeholders external to the project team. Examples of these stakeholders include the parent organization, customer or user organizations, subcontracting organizations, marketing, legal, technical support, other projects, and training personnel. The project organization plan should describe these interfaces and identify the project team members who will act as *liaisons* to those stakeholders. For example, the customer does not want 30 different team members calling with requirements questions. In this case, a requirements analyst would typically be assigned as the liaison to that external customer for requirements issues. The other project team members would take their questions to that analyst. The requirements analyst may already know the answer, which keeps the customer from being annoyed with redundant questions. Even if the analyst does not know the answer, he or she will have established communications channels with that customer to make obtaining the needed answer easier. The project manager might also be assigned to that same external customer as the liaison for issues related to project budgets, schedule, or status reporting.

The project organization plan also defines the roles and responsibilities of the project team members. In Chapter 6, when process definition was discussed, generic roles were assigned to the various process activities. The project organization plans define the specific project personnel assigned to those generic roles for the project. For example, the system test process definition might include roles for test manager, tester, test bed coordinator, configuration manager, and developer. For a very large project, Brenda might be the test manager with 50 individuals assigned as testers; the test bed coordinators may be Tom and Nancy; the configuration managers are Sue, Scott, and Doug; and 100 other individuals are assigned to the developer role. On a tiny project, Watson may be assigned the test manager, tester, and test bed coordinator roles, while Darius is assigned to the project manager, development manager, developer, and configuration manager roles. This allows the process documentation to remain very generic while minimizing the amount of tailoring that needs to be done on the project.

Infrastructure Plans

The project's infrastructure refers to the technical structures that support the project. The *infrastructure plan* specifies the plans for developing and maintaining this technical structure. Infrastructure examples include:

- Office space, conference rooms, break rooms, cafeterias, and restrooms, and their associated building maintenance, janitorial services, and so on
- Computer networks and their associated hardware, operating system, software, information technology (IT) support, and so on
- Telecommunications and other collaborative communications systems
- Administrative support, including secretarial, legal, human resources, and so on

Training Plans

Training plans specify the training needed to make sure that the necessary skill levels required by the project are available, as are the plans, tools, methods, techniques, schedules, and responsibilities for obtaining that training.

Communication Plans

The *communication plans* define the information needs of the various project stakeholders and the communication methods and techniques that will be used to fill those needs, including descriptions, responsibilities, and schedules for project review meetings and selected metrics.

Stakeholder Management Plans

Stakeholder management plans define the "strategies for effectively engaging stakeholders of the project, based on their needs, interests and impact on project success" (PMI 2013). In other words:

- Which stakeholders are going to take an active role in which project activities?
- Who will represent each stakeholder?
- When will they be involved? How long will they be involved?
- How will they be involved?

Project Estimation and Forecasting

The project plan should include the specification of the tools, methods, and techniques used to estimate, re-estimate, and forecast the project elements. It should also include responsibility assignments for performing these activities. These project elements include estimates of the project's size, effort, costs, calendar time (schedule), and required computer resources. They also include *forecasts* (estimates for specific future point in time) of staffing needs, skill levels, work rates and availability, resource needs and availability, and project risks and opportunities. As illustrated in Figure 16.8, these estimates are interrelated. The size of the project is estimated, and the work breakdown structure (WBS) is defined. (The WBS is discussed later in Section 2 this chapter.) Depending on the methods used, the size estimate or the WBS is used as a basis for estimating project effort and cost. Effort and cost are then allocated to the activities in the WBS, and used (along with the activity network) to create the calendar-time estimates

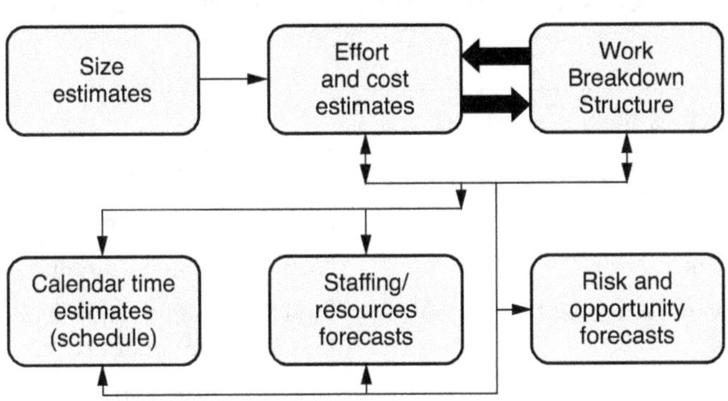

Figure 16.8 Project estimates and forecasts.

and staffing or resource forecasts. To do this, trade-offs are evaluated and choices are made based on the cost–schedule–scope trilogy.

Estimates for effort, cost, and calendar time, forecasts for staffing and resources, and the WBS all become inputs into forecasting risks and opportunities. Risk management activities may then feed back into additions or updates to the project's size, WBS, estimates and forecasts. (See Chapter 18 for a detailed discussion of risk management.)

One of the fundamental project estimations is the size of the software and other artifacts being produced by the project. These size estimates are often used as inputs to help estimate the effort and cost of the project. (Size metrics, including function points, lines of code, weighted requirements counts, story points, and so on, are discussed in Chapter 19.)

Effort is the amount of staff and engineering work time (staff-hours, staff-days, staff-months) it will take the people projected to be assigned to the project (or one of its activities) to complete that project (or activities), based on their normal work rate. *Work rates* can vary depending on the work environment and other factors. In an eight-hour workday, getting 80 percent (about 6.5 hours) on-task work from an individual contributor may be optimistic for many organizations. Where does the rest of the time go? It goes to phone calls, e-mails, bathroom breaks, conversations, sick time, interruptions, time spent task-swapping (putting one activity away and getting up to speed on the next one), and so on. Productivity and velocity are used to measure work rates. *Productivity* is measured in terms of the amount of work product produced per staff effort. As illustrated in Table 16.1, different metrics can be used to measure productivity, depending on the type of work. For Scrum projects, since effort is fixed (team size and iteration duration), velocity is used instead of productivity. *Velocity* is measured in terms of the amount of work completed per calendar time (days).

Calendar time is the duration of time that it will take to complete the project (or one of its activities) from start to finish. Calendar time is typically a function of effort and staff or resource loading, so trade-offs can be made between calendar time and staffing or resource estimates. For example, if the activity with the estimate of 40 hours of effort is assigned to a person working on that activity full time, it should take about 40 hours of duration (one work week). However, if that person can only work on the activity half time, that 40 hours of effort may take a little over two weeks to accomplish (with added time for activity switching). On

Table 16.1 Productivity metrics—examples.

Type of work	Productivity metric
Eliciting, analyzing, and specifying requirements	Number of requirements / amount of effort
Writing user documentation	Number of pages / amount of effort
Developing software component	Function points / amount of effort LOC / amount of effort
Testing	Test cases executed / amount of effort Test cases passed / amount of effort Defects found / amount of effort

the other hand, if two people are assigned to that activity full time, it may take a little over one-half week of calendar time (with added time for two people to work together and communicate). Of course, that assumes that more than one person working on the activity can shorten its calendar time. As the old adage goes, assigning nine women to the activity of having a baby will not result in producing a baby in one month rather than nine months.

Cost is the amount of money it will take to complete the project. Costs for the project include direct labor and overhead, travel, capital equipment, and other expenses such as supplies, rents, and subcontractor costs.

Critical computer resources should also be forecast for the project. Computer resources may be forecast for the development platform, the test platforms, or the target platform. Examples of critical computer resources include memory capacity, disk space, processor usage, and communications capacity.

There are two basic classes of estimation methods: expert judgment methods and *model-based estimation* methods. Since each technique has its own strengths and weaknesses, experts recommend the use of multiple estimation techniques. The resulting estimates are then compared and differences are evaluated to determine the final estimates that will be used for the project.

Estimates are not a one-time activity. The project should plan to re-estimate as additional information is obtained (e.g., at the completion of requirements, high-level design, and detailed design activities). Regular review of the project's progress, assumptions, and product requirements should be done to detect changes to, and violations of, the assumptions underlying the estimates. If variances between the actuals and estimates become too large, it may also signal the need to re-estimate.

Expert-Judgment Estimation Techniques

Expert-judgment techniques are manual techniques where individuals or teams make experience-based estimations or forecasts. Expert-judgment techniques typically start with the WBS and estimate effort for each activity. These estimates are then rolled up for an overall project-level effort estimate. Expert-judgment techniques include Delphi, wideband Delphi, planning poker, PERT, and other team techniques.

Delphi technique steps:

- *Step 1*: Coordinator presents each expert with project specifications and other relevant information
- *Step 2*: Several experts, in isolation from one another, use their personal experience and judgment to estimate the cost, size, or effort for each activity in the WBS
- *Step 3*: The coordinator collects the estimates from each expert and prepares a summary of the experts' responses and their rationales
- *Step 4*: Each expert receives this composite feedback and is asked to make new predictions based on that feedback
- *Step 5*: Repeat steps 1 through 4 until consensus is reached, that is, when the estimates are close enough to use their average as the estimate

Wideband Delphi is a variation of the Delphi technique (Futrell et al. 2002). The primary difference is that instead of the experts being in isolation, as in steps 2 and 4 of the Delphi method, the wideband Delphi technique includes estimation team meetings where the experts discuss their estimates and the reasons for them, considering them in the context of other estimates. These meetings typically reduce the time it takes to come to consensus.

Planning poker is an agile estimation technique where all of the members of the agile development team participate. Since agile teams are kept small, they will typically not exceed seven plus or minus two people. Each participant is given a deck of cards where each card has a single estimate written on it (e.g., the cards may have numbers from a *Fibonacci sequence* with the number on the next card being the sum of the numbers on the previous two cards: 1, 2, 3, 5, 8, 13, 21, 34, 55, 89, or the cards may be numbers in multiples of four or powers of two). For each user story being estimated, the facilitator (Scrum master or agile coach) presents a story, the team discusses the story, and the product owner answers questions. At the end of the discussion, each team member selects a card from the deck that represents his or her relative estimate for that story (either in story points or effort-hours). All cards are then simultaneously turned over to reveal the individual estimates. If the estimates vary, the team members with the highest and lowest estimates explain their estimates and the reasons behind them. Other group members can also join this discussion, and again the product owner answers questions. However, the facilitator actively works to minimize the time frame of the discussion. Each team member then reselects a card that represents their new estimate for the story based on this additional discussion, and those cards are again simultaneously revealed. This process of discussion and new estimations are repeated until the estimates converge enough to obtain consensus on an estimate for the story. Throughout this estimation process the facilitator records information from the discussion that might be helpful when the story is implemented and tested in the future.

Other variations of the expert judgment team techniques can also be used. For example:

- A *panel of experts* is convened, and each activity from the WBS is discussed. Based on this discussion and their own expert judgment, these experts estimate the effort for each activity. Mathematical averages are taken of the individual estimates to represent the team's estimate. This process is iterated until consensus is reached that the average is acceptable as an estimate. The main difference between this and wideband Delphi is that the estimation is not done privately and is compiled by a facilitator.
- In the *program evaluation and review technique* (PERT) method, three estimates are made for each activity in the WBS instead of one, including:
  - a = most optimistic estimate
  - m = most likely estimate
  - b = most pessimistic estimate

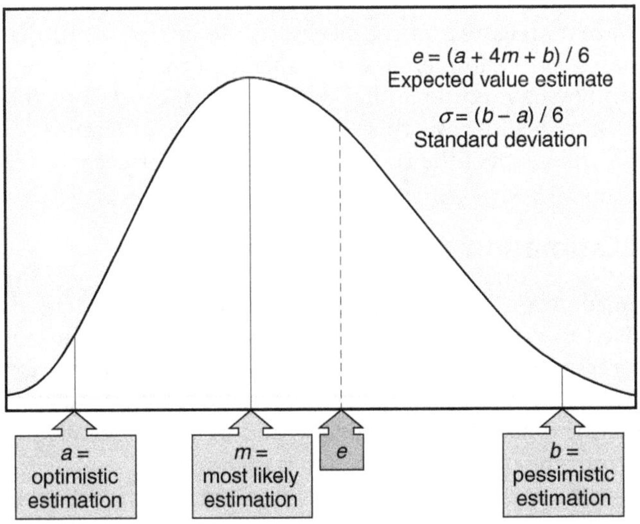

Figure 16.9 Program evaluation and review technique (PERT) method.

As illustrated in Figure 16.9, these three estimates are used to define a beta probability distribution for the estimate, and the final estimate (*e*) used for each work package is the expected value based on the estimates, which are calculated as $e = (a + 4m + b)/6$ with a standard deviation $\sigma = (b - a)/6$. Today the PERT method is typically replaced with more modern models, where the same three estimates are fed into a project estimation tool that uses Monte Carlo estimation techniques to sample values all along the curve to produce the final estimate.

The strengths of the expert-judgment techniques include:

- The fact that experts often understand differences between past experiences and the new techniques that will be used in the project being estimated.
- Experts can factor in special characteristics of the project such as personnel issues or political considerations. For example, the experts might know that the team for this project has extensive experience building similar software, which will reduce the time needed for design and coding. Another example might be a project for a customer that has been historically difficult to work with and therefore will require additional time to perform the requirements activities.

Weaknesses of the expert-judgment techniques include:

- The technique is no better than the expertise and objectivity of the experts. If the experts are lousy estimators, the result will be lousy estimates.
- The experts may be biased by a desire to please or a desire to win the project.

- Software managers and engineers tend to be very optimistic in their estimates. For example, they may include time to code, peer review, and unit test a module in their effort estimations but include no time in their estimates for rework, re-peer reviewing, or testing—because, of course, they expect the code will be written perfectly the first time.

Model-Based Estimation Techniques

A second major class of estimation techniques involves models (mathematical formulas) used to calculate the estimates for the project. These models typically start with an estimation of the software's size (e.g., lines of code or function points) and any other inputs related to the characteristics of the project. These models are based on historic data from the industry, and many of them can be tailored using actual data from the organization's own past projects.

The strengths of the model-based estimating techniques include:

- The fact that mathematical models eliminate most of the biases found in expert-judgment techniques
- Automated tools exist to support most models

Weaknesses of these techniques include:

- Counting function points requires skilled staff, and these individuals may not be available in some organizations.
- It may be difficult (if not impossible) to accurately estimate lines of code early in the project and lines of code are very dependent on the coding language selected.
- Using models with the default historic industry data may not accurately reflect the characteristics of the project being estimated. Tuning them to match the organization's or project's actual characteristics may require historic metrics data from similar projects, which many lower-maturity organizations simply do not have.

Examples of model-based estimation techniques include:

- The *constructive cost model—version 2* (COCOMO II) (Boehm et al. 2000) is an updated version of the classic COCOMO model documented by Barry Boehm in *Software Engineering Economics* (Boehm 1981). COCOMO II offers estimating capability at three levels of granularity. These three levels capture three stages of software development activity, and provide three levels of model *precision*: prototyping, early design, and post-architecture.
- The *software life cycle management* (SLIM) model developed by Putnam and Myers (2003), used to estimate effort, time, and cost. The SLIM model is based on the equations for the *Rayleigh curve* for the major project activities. The Rayleigh curve for the entire project is the sum of the separate curves for design and code, test and validation, maintenance, and management activities, as illustrated in Figure 16.10. The area under each Rayleigh curve corresponds to

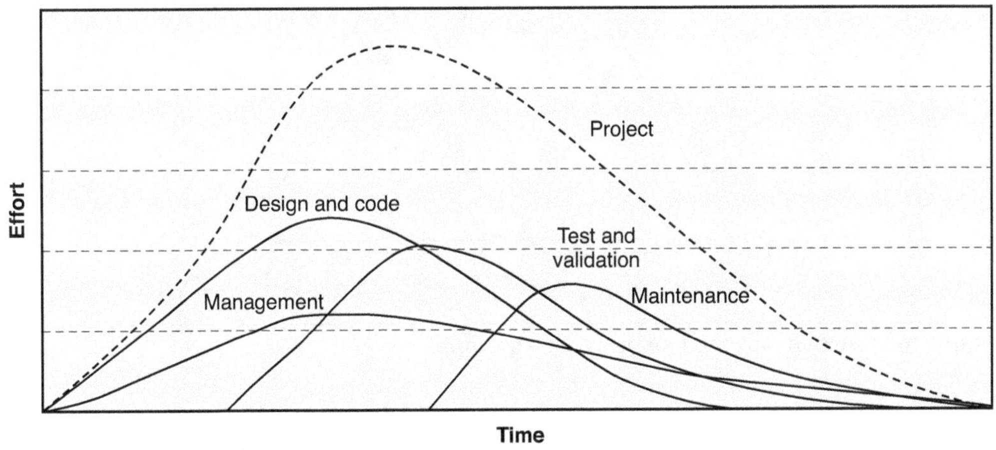

Figure 16.10 Rayleigh staffing curve.

cumulative effort for that activity. These curves for the individual activities are then added together to obtain a Rayleigh curve that represents the total effort for the project. The project curve does not include a requirements specification phase in the model. Rayleigh curves can also be used to estimate costs and defect arrival rates.

- The *functional point model* for software estimation turns function points into effort and cost estimates, using productivity and cost factors that are based on historic data from the organization, or from industry averages, by using the following formulas:
    - Effort = Adjusted Function Points × Productivity Factor (person-hours/function point)
    - Cost = Adjusted Function Points × Cost Factor ($/function point)
- Many organizations also build their own estimation models based on information from actual projects.

Activity Networks

Activities are the lowest-level branches of the WBS. When those activities are diagrammed together with their predecessor and successor relationships, an *activity network* is formed. Figure 16.11 and Figure 16.12 are examples of two different representations of the same set of activities and relationships:

- Figure 16.11 illustrates an example of an *activity on the line network*, where the lines represent the activities and the nodes represent events. An event is the beginning or ending point (in time) of an activity. In this example, activity A starts at event 1 and ends by event 2, activity C starts at event 2 and ends by event 3, while activity B also starts at event 1 but ends by event 3. The reason for using "by event 3" is because if the duration

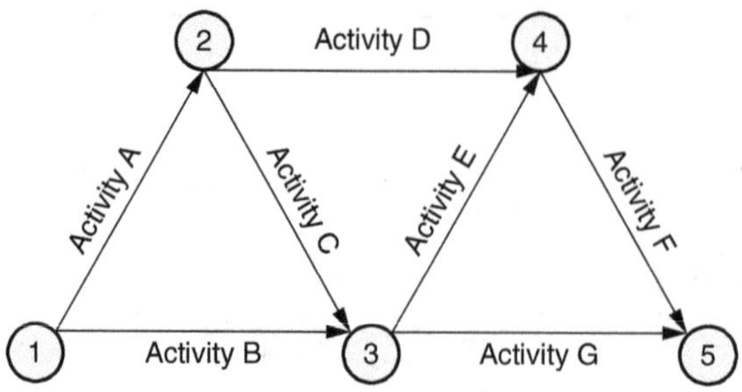

Figure 16.11 Activity on the line network—example.

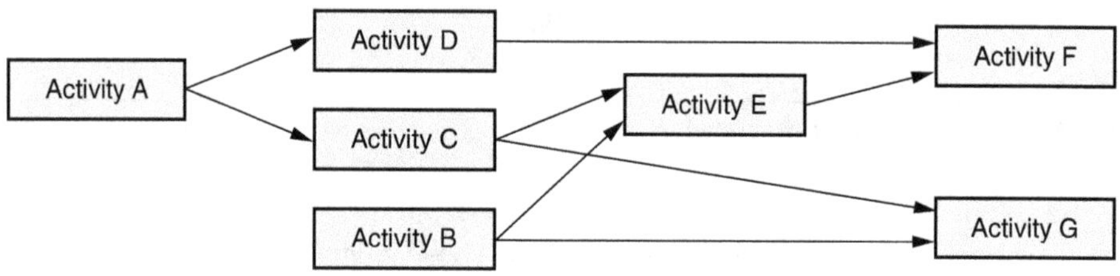

Figure 16.12 Activity on the node network—example.

of activity A plus activity C is longer than the duration of activity B, then activity B is said to have *slack* and can actually finish sooner but must finish before event 3, which starts activities E and G. Of course the opposite is true: if the duration of activity A plus activity C is shorter than the duration of activity B, then activities A plus C have *slack*.

- Figure 16.12 illustrates an example of an *activity on the node network*, where the nodes represent the activities and the lines simply represent their predecessor and successor relationships.

Figure 16.13 illustrates examples of the four different possible activity network relationships, using activity on the node–type diagrams:

- In a *finish-to-start relationship*, the first activity A must finish before the second activity B can start. A lag time may also be specified in this relationship. In this figure the lag time is five days. An example of this type of relationship might be that the design review (activity B) is held five days after the design document is completed and distributed (activity A) in order to give participants time to prepare for the review.
- In a *start-to-start relationship*, the first activity C must be started before the second activity D can start. Again, a lag time may be specified (10 days in this figure). An example of this type of relationship might be that the writing of the user manual (activity D) can start 10 days after the start of detailed design (activity C) to allow enough of the design to be defined so that enough information is available to start the user manual.

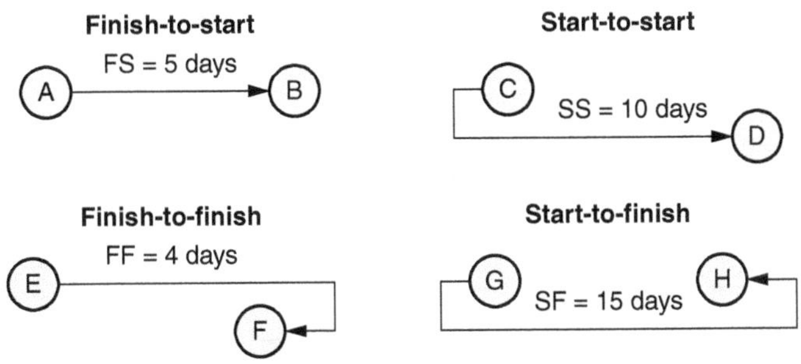

Figure 16.13 Types of activity network relationships.

- In a *finish-to-finish relationship,* the first activity E must be finished before the second activity F can finish. Again, a lag time may be specified (four days in this figure). An example of this type of relationship might be that the integration test (activity F) can run concurrently with the coding and unit test (activity E), but the integration test will not be completed until four days after the last source code module has passed the unit test.
- In a *start-to-finish relationship,* the finish of the second activity H must lag the start of the first activity G by the specified lag time (15 days in this example). There are not many practical examples of this type of relationship, and many project management tools do not even support it, but it is a possible relationship.

A common mistake that should be avoided when diagramming activity networks is diagramming activities in sequence when they could be done in parallel. This is typically done based on known people or resource dependencies. However, the activity network should only include natural dependencies based on logical flow of the activities. Staff and resource dependencies are handled later in project scheduling. Another common mistake occurs when an activity is shown as dependent on total completion of another activity, but it is actually only dependent on completion of part of that other activity.

Scheduling and Critical Path Analysis

The *critical path* is the longest path through the activity network, which defines the shortest amount of time in which the project can be completed (calendar time schedule). A *critical activity* is any activity in the activity network that must be completed by a certain time (event) and has no slack time. A critical activity is always on the critical path.

The first step in determining the critical path is to set the duration of each activity in the network to the effort estimate for that activity. Available staff and resources are then allocated across the activities and resource leveling is done. *Resource leveling* examines the staff and resource allocation, looking for areas where the use of staff or other resources is unbalanced or over-allocated, and resolves any identified issues. Resource leveling requires trade-offs in the cost–

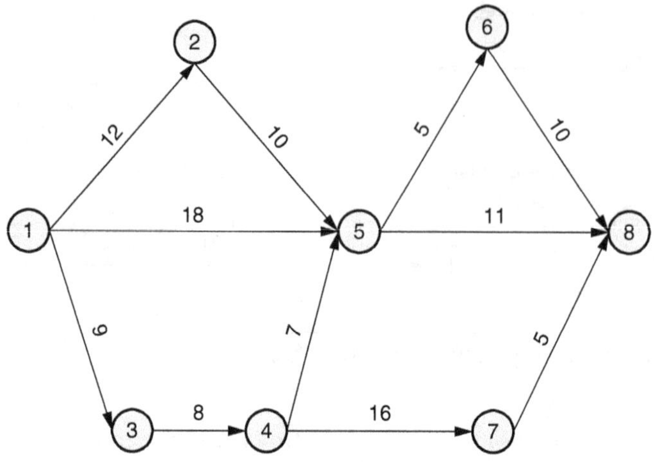

Figure 16.14 Activity on the line network—example.

schedule–scope trilogy. This step establishes the initial duration for each activity in the network, based on activity predecessor and successor relationships, efforts estimates, and staff and resource allocations.

The longest path (critical path) through the activity network is then determined. To illustrate this step, consider the "activity on the line" network example in Figure 16.14 and assume that the numbers on each line specify the estimated effort in days of the associated activity (e.g., the activity from event 1 to event 2 has a duration of 12 days). The duration of each path is then calculated by adding together the effort estimates for each activity on that path. In this example:

- The duration for the path through events 1-2-5-6-8 = 12 + 10 + 5 + 10 = 37
- The duration for the path through events 1-5-6-8 = 18 + 5 + 10 = 33
- The duration for the path through events 1-5-8 = 18 + 11 = 29
- The duration for the path through events 1-2-5-8 = 12 + 10 + 11 = 33
- The duration for the path through events 1-3-4-5-6-8 = 9 + 8 + 7 + 5 + 10 = 39
- The duration for the path through events 1-3-4-5-8 = 9 + 8 + 7 + 11 = 35
- The duration for the path through events 1-3-4-7-8 = 9 + 8 + 16 + 5 = 38

The longest path through this activity network is 39 days, so the path through events 1-3-4-5-6-8 is the critical path. The other paths through the activity network have *slack*, which is the difference between their duration and the duration of the critical path. In this example, the path through events 1-5-6-8 has a duration of 33 days and a slack of 39 – 33 = 6 days. That means that tasks along this path can be a total of 6 days late and the project will still be delivered on time.

To continue this example, assume this is a schedule-driven project and that the customer needs the software product being developed by this project in 35 days.

Since the original critical path was calculated at 39 days, the project must find a way to shorten the duration of one or more activities on the critical path. There are four ways to shorten the duration of an activity by making trade-off choices based on the cost–schedule–scope trilogy.

First, the human resources for the activity can be changed, either by adding more people or by assigning more efficient or experienced people to the activity. The trade-off here is a reduced schedule for an increased (personnel) cost. Each activity's duration was originally estimated based on the assumed skill level of the type of person that would be assigned to the activity. For example, if the activity that starts at event 5 and ends at event 6 was originally assigned to a junior-level engineer, a choice could be made to assign a senior-level engineer to that activity and re-estimate that activity's duration at four days instead of five days. Alternatively, the project manager might decide to assign two people to the activity that starts at event 6 and ends at event 8, which would result in the activity's duration being re-estimated at six days instead of ten days. The duration for the path through events 1-3-4-5-6-8 would then be reduced to 9+8+7+4+6=34 days.

Unfortunately, while the original critical path has now been reduced to less than the 35-day goal, it does not completely solve the problem. The path through events 1-3-4-7-8, which has the duration of 38 days, now becomes the new critical path for the project. Now that path has to be shortened to less than 35 days, and so on, until all paths through the program are less than 35 days long, except one, the resulting new critical path.

One note of caution: the project manager may be tempted to change the human resources factor by planning overtime into the project schedule. This is never a good idea. People can only work so many productive hours a day, and if overtime is scheduled into the project, people tend to burn out or make more mistakes because they are tired, both of which can have negative impacts on the project schedule in the long run. Also, if people are already working overtime, they have nothing left to give when additional effort is needed for corrective actions to bring a project back under control when problems occur.

A second way to shorten the duration of an activity is to reduce the objectives or specifications. In other words, do not do all of the work initially planned. The trade-off here is to reduce the schedule by reducing the scope. One method for doing this is to prioritize the requirements and shift one or more of the lowest-level requirements into a future incremental release, or simply remove them from the scope. Another method is to reduce the activities or processes. For example, hold fewer meetings or decide to run fewer test cases (or cycles of testing) than originally planned. A caution here is that not performing certain activities may result in the reduction of the quality of the product. This is a serious concern if it creates a risk that the product may no longer meet its required quality levels.

A third way to shorten the duration of an activity is to change the materials or technical resources being used for that activity. Sometimes resources other than human resources are the constraining factor on an activity. In this case adding more resources may help. For example, if test beds are a constraining resource, making additional test bed resources available to the project could reduce activity duration. Another option might be to assign testers to multiple shifts so test beds are more effectively utilized.

A fourth way to shorten the duration of an activity is to change the processes or methods of working. Process improvements specifically targeted at cycle time reductions can impact the duration of activities that use those processes. The word of caution here is to remember the learning curve. Making changes to the processes may actually negatively impact the cycle time for some period of time until everyone is up to speed on the new process.

When performing critical path analysis, if the project has more than one critical path, it increases the risk of the project slipping behind its schedule. Therefore, activities on selected multiple critical paths should be manipulated until only one critical path exists. Rules of thumb for selecting which of the multiple critical paths to reduce include selecting and reducing the path with the:

- Largest number of activities
- Least-skilled personnel
- Greatest technical risk or where there is questionable feasibility
- Risk factors that are outside the control of the project team
- Activities that would have the greatest cost impact if they slipped
- Activities that are new (no historical information used to estimate)
- Activities that have historically been problems
- Risky activities with no contingency (backup) plans
- Activities that are the most difficult

Budget

Initial *budgets* specify the plans for allocating the cost of the project across its various budget categories. During the execution of the project, actual costs for each category can be tracked against the budget to provide insight into when and how the project is spending its monetary resources. Examples of budget categories include:

- *Labor costs—staff and contractors* (engineering, management, support, administrative)
    - Wages, salaries, and benefits
    - Travel and relocation
    - Training (conferences, tuition, course fees)
- *Capital costs*
    - Computers and peripherals (host, test, and target computers)
    - Networks
    - Software engineering environment (tools, processes)
    - Physical facilities (office space and equipment, labs)

- *Other costs*
  - Subcontract/supplier cost
  - Supplies and materials
  - Rents

2. WORK BREAKDOWN STRUCTURE (WBS)

> Use work breakdown structure (WBS) in scheduling and monitoring projects. (Apply)
>
> **Body of Knowledge IV.A.2**

Work Breakdown Structure

A *work breakdown structure* (WBS) is a hierarchical decomposition of the project into subprojects and subtasks, down to individual activities. The goal is to break the project work down into manageable elements called *activities* that can be easily estimated, budgeted, scheduled, tracked, and controlled. The WBS becomes the basis for the effort, cost, and schedule estimates of the project. Just remember, "If an activity is not in our WBS, we will have to do it in zero time and for free" (Ould 1990). In other words, any activity that is not specified in the WBS will not be allocated effort, budget, staff, or resources. The benefits of creating a WBS include:

- Making sure that all work activities are identified and understood
- Helping to determine estimates and schedules for the project
- Modularizing the project into manageable pieces that can be tracked and controlled
- Helping quantify people and skills needs, as well as other resource needs for the project
- Aiding in communication of the work that needs to be accomplished
- Providing useful information for analyzing impacts of proposed changes

There are no hard-and-fast rules for how far to break the project down into activities when creating a WBS. The larger in size or more complex a project is, the more likely that it will require more breadth (substructures) and depth (levels) in the WBS. Projects that include suppliers, or that span multiple organizational entities, may require additional breadth and depth as well. The WBS should be broken down as far as possible based on current information, but only as far as necessary for understanding. This balance, between providing enough visibility without micromanaging the project, requires the application of good engineering judgment.

Activities in the WBS should have a cohesive time frame (no gaps for other activities) and should include only related work items. Each activity should produce one or more work products, and most of the steps should be the responsibility of the same team members. For example, the activity for coding module X might include the actual coding, peer review, and unit testing of that module. The author of the source code module is responsible for all of these processes, even though other individuals are involved in the peer review. A good rule of thumb is that an activity typically has a time frame of one to two people for 1 percent to 2 percent of total project time (about one to two weeks on a one- to two-year-long project).

One method used to represent a WBS is a tree diagram, as illustrated in the example in Figure 16.15. The top level of the tree diagram always contains a single item, the product being produced or the project being implemented. The second level contains the major components of that top-level item. The third level includes subcomponents of the second-level items. The fourth level includes subcomponents of the third-level items, and so on down the tree. An indented list or outline can also be used to represent a WBS. This is the way that the WBS is documented in most project management tools.

There are two basic types of WBS: the product type and the process type. Figure 16.15 illustrates an example of a *product-type WBS*, which partitions the project by breaking down the project's product, service, or result into smaller and smaller parts. At its lowest levels, the product is broken down into the activities that produce an individual component of the project's work products.

Figure 16.16 illustrates an example of a *process-type WBS*, which partitions the project into smaller and smaller processes. At its lowest levels, the process is broken down into individual activities. Utilizing a process-type WBS for software has the following advantages for software projects:

- Allows standardization of WBS templates, based on standardized process definitions, that can be tailored to each project—rather than starting from scratch with each new product
- Does not require locking in the project architecture early in the planning process
- Facilitates comparison between projects
- Provides a mechanism for defining processes and activities not directly linked to the creation of product components (e.g., project management, software quality assurance, software configuration management)

The two main WBS types can be combined into a *hybrid WBS*, as illustrated in Figure 16.17. Advantages of the hybrid-type WBS include all of the advantages of the process-type WBS, plus it links product components directly to the processes that create them. Using a hybrid-type WBS allows the project to start from a process-oriented work plan and then progressively elaborate that plan with additional detail once the software architectural information is available.

The Scrum equivalent of a WBS is the iteration task list. During the sprint planning meeting, the team plans the initial tasks (activities) that need to be done to implement each story. This task list defines the work that needs to be done to turn the sprint backlog items into a potentially shippable increment of functionality that has been successfully integrated with any previously completed increments of the software product. Additional tasks may be added to the list, as needed, for

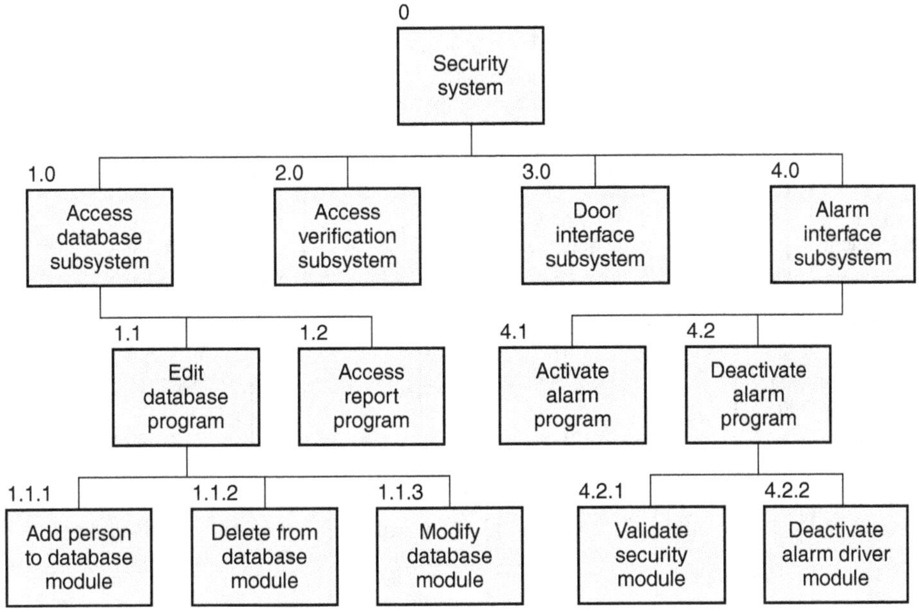

Figure 16.15 Product-type WBS—example.

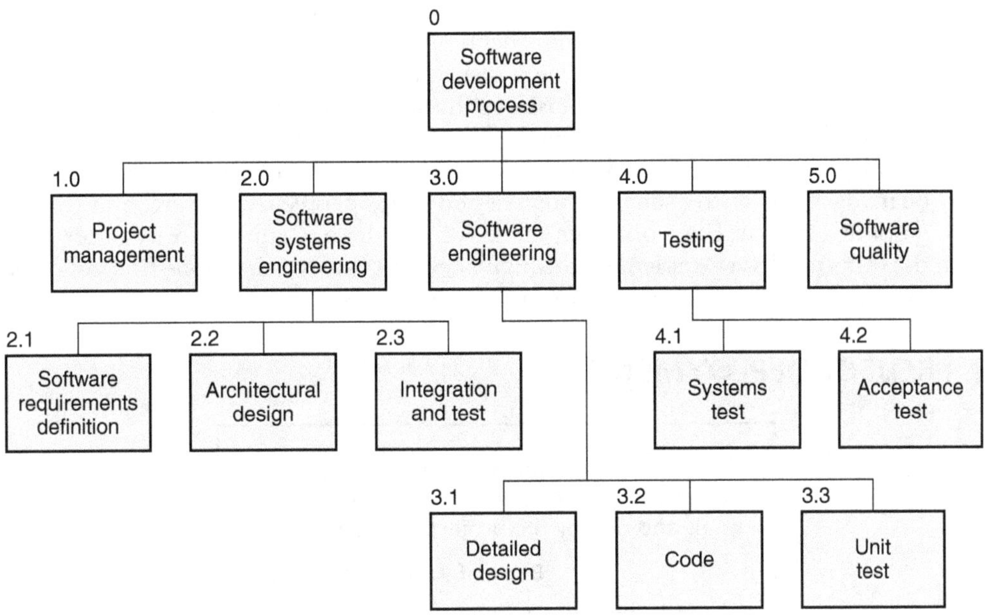

Figure 16.16 Process-type WBS—example.

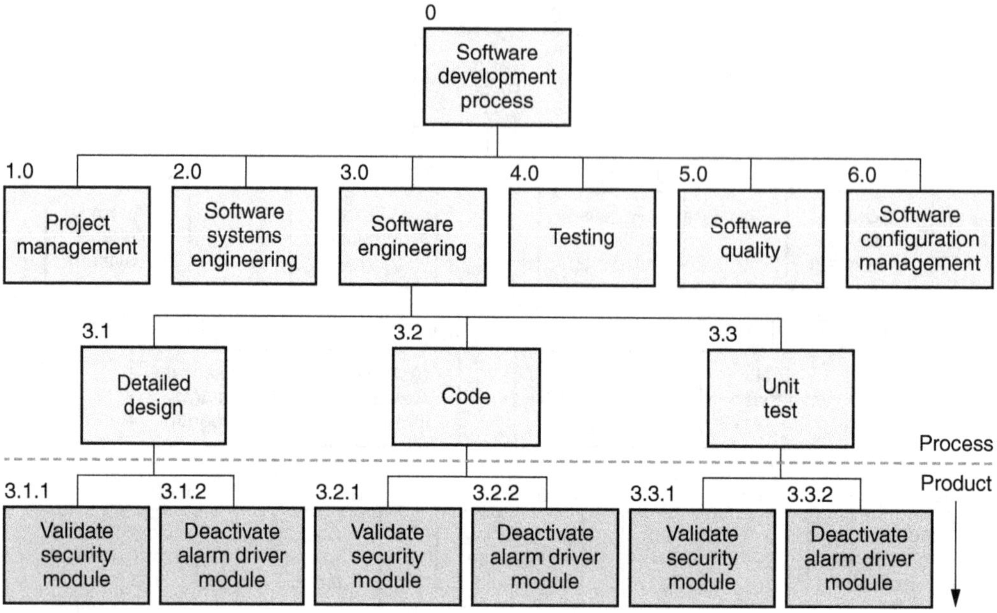

Figure 16.17 Hybrid WBS—example.

conducting the sprint. The team then estimates the amount of effort-hours each of these tasks will take. During execution of the sprint, this task list is added to (progressively elaborated) if new tasks are uncovered. Sprint team members self-assign themselves to these tasks as the sprint progresses. As a team member finishes one task, he or she selects the next task to be done (or pairs select the task if pair-programming is used for that task). This allows staff allocation decisions to be made at the last reasonable moment and keeps people from overcommitting to work in advance. The work from the task list is done in story priority order, with the tasks needed to implement the highest priority story being done first, followed by the next highest priority story, and so on.

3. PROJECT DEPLOYMENT

> Use various tools, including milestones, objectives achieved, and task duration to set goals and deploy the projects. (Apply)
>
> **Body of Knowledge IV.A.3**

Project deployment involves the actual execution of the project activities, including:

- Working toward achieving the project's objectives and milestones
- Implementing processes, methods, and standards

- Executing the planned actions and activities specified in the WBS
- Creating, controlling, verifying, and validating deliverables
- Expending effort and spending money
- Communicating with stakeholders and managing their expectations
- Staffing, training, and managing people
- Obtaining and using materials, equipment, facilities, and other resources
- Implementing risk plans and managing risks
- Selecting and managing suppliers
- Implementing approved changes

Most of the effort of a project is spent performing the executing process group activities. Figure 16.18 illustrates the various processes in the Project Management Institute's *executing process group* and how they interact and are integrated through the *directing and managing project execution process*. The numbers preceding each process name indicate the section in the *PMBOK Guide* (PMI 2013) that describes that process.

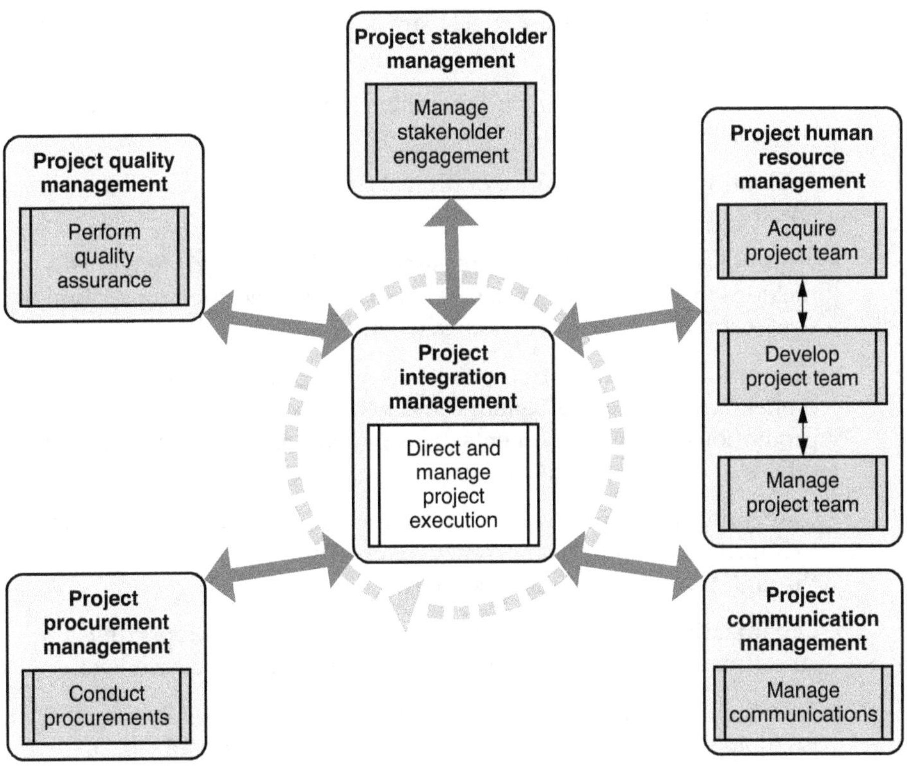

Figure 16.18 PMI executing process group. (Source: PMI 2013. Reprinted with permission.)

Project Milestones

A *milestone* is an event that represents a point in the project of specific significance. Throughout the deployment of the project, the project manager and project team members should focus project work efforts on accomplishing milestones on time. Individual activities may be finished ahead of schedule. Other activities may be on schedule or behind schedule. There will always be variation in accomplishing the work of the project. However, managing project execution to the milestone gives a summary view of how the project is doing against its overall schedule.

Meeting Project Objectives

Throughout the deployment of the project, the project manager and project team members should align strategies and tactics with, and focus project work efforts on, accomplishing the stakeholders' objectives for the project to verify that progress continues to be made toward their achievement. Once a project objective has been achieved, effort and other resources from the implementation of that objective can be refocused on other project objectives.

Acquire, Develop, and Manage Project Team

During project deployment, the project team is acquired, developed, and managed, including:

- Recruiting and selecting staff members to fill each position on the project.
- Evaluating the performance of project personnel and coaching to enhance their performance levels.
- Training staff members to increase their skills and knowledge. This includes the orientation of new team members and the propagation of new technologies and practices to existing team members.
- Compensation deals with the rewards system. This includes both monetary (salaries and benefits) and nonmonetary (recognition, promotions, plum assignments, trips or conferences, time off, better offices or equipment) mechanisms.
- Career development involves promoting the growth of team members to improve members' competency and readiness to move upward through the organizational hierarchy.
- Organizational design deals with the hierarchy of the reporting structure and the authority and responsibility given to each level.
- Team and culture development establishes the norms, cohesion, mythology, and membership requirements for the organization.

According to Curtis, the key people issues in software development are (CMU/SEI 2009):

- The overwhelming impact of individual differences
- The fact that knowledge is the raw material of software development—technology cannot make up for lack of knowledge
- Because of the increasing knowledge-intensiveness of software engineering—people have become our primary asset
- The importance of a work environment that supports a skilled, knowledgeable, competent work force
- The need to address people issues as part of an integrated organizational improvement process

Manage Communications

Effective communication with the project team and other stakeholders is essential to project success. Checklist items that could be used to monitor the effectiveness of communications with the project team and support staff include:

- The project objectives have been communicated, and project team and support staff members understand how their efforts contribute to accomplishing those objectives
- Project risks have been identified and communicated
- The people who will implement the project plans participated in their creation
- Project team and support staff members have access to up-to-date project plan documentation
- The project plan has been reviewed and committed to by project team and support staff members
- Project team members have been co-located to facilitate communications, or, where this is not possible, a mechanism is in place for virtual co-location
- Agreement exists that engineering and support staff managers inform project managers before reassigning project personnel to other nonproject work
- Roles and responsibilities of each project team and support staff member have been clearly defined and communicated
- Project team meetings are scheduled on a regular basis to review progress
- Transfer or termination of team members is coordinated with his or her replacement, or written instructions are left for successor
- Where cross-functional teams are required to complete tasks or activities, frequent coordination meetings are held to keep everyone informed

- Project team and support staff members are encouraged to provide early warnings of project risks and problems
- Consensus decisions are made when appropriate
- People are kept informed of decisions or changes that impact their work
- Upper management is aware of the project's status
- Managers interact routinely with project team and support staff members to provide assistance, listen to their problems, help solve problems, and understand the status of the project firsthand
- Metrics used to track the project are made available to project team and support staff members

Perform Quality Assurance

Perform quality assurance is the process of reviewing and auditing against the quality requirements, quality plans, standardized processes, workmanship standards, and quality control results to verify compliance. The goal is to make sure that quality is built into the software work product by assessing and improving development processes, verification and validation activities, and intermediate work products.

Conduct Procurements

During project deployment, the *conduct procurement* process includes:

- The identification and evaluation of prospective suppliers to perform outsourcing activities
- Defining contract requirements
- Selecting suppliers
- Negotiating and awarding contracts

Manage Stakeholder Engagements

The *manage stakeholder engagement* process includes:

- Communicating with the stakeholders throughout the project
- Engaging key stakeholders in the appropriate project activities as planned
- Managing stakeholder expectations
- Making sure stakeholder needs are met
- Addressing stakeholder issues as they are identified

Examples of checklist items that could be used to monitor the effectiveness of stakeholder engagement management include:

- The "real" project stakeholders, or models of the "real" stakeholders, have been identified and consulted or evaluated in order to define project requirements
- The stakeholders' objectives for the project have been communicated to all stakeholders
- The customer, users, and other external stakeholders are involved in the appropriate project activities, and coordination meetings are held, as appropriate
- Changes are understood and agreed to by all impacted stakeholders
- Senior management reviews all commitment changes, and new software project commitments, made to individuals and groups external to the organization, as appropriate and when changes occur
- External stakeholders are aware of the ongoing project status and are provided early warning of risks and problems that may impact them

Chapter 17
Tracking and Controlling

Tracking the project involves all of the activities required to monitor and review the project's ongoing activities, performance, and progress against the project plans and project performance baselines, to identify any significant variations, problems, issues, or risks. There are two primary methods for project tracking: meetings and metrics. Tracking activities include:

- Measuring actual progress against the project and organizational objectives
- Assessing adherence to methods and standards
- Evaluating actions and activities status against project plans
- Monitoring requirements volatility
- Measuring process and product quality
- Monitoring change requests and issues to resolution
- Measuring duration, effort, and cost actuals against projected estimates, schedules, and budgets
- Monitoring the utilization of materials, equipment, facilities, and people
- Tracking the status of risks and risk-handling activities
- Monitoring supplier performance against supplier agreements (contracts), plans, and requirements

Controlling the project involves all of the activities required to evaluate issues and problems to determine the appropriate corrective actions, identify preventive actions in anticipation of possible problems (risks), and manage those corrective and preventive actions to closure. Control activities include:

- Realigning with the project and organizational objectives
- Improving methods and standards
- Modifying or controlling actions and activities
- Controlling requirements volatility
- Controlling product quality
- Controlling change and issues

- Controlling costs and schedules or replanning them
- Controlling the utilization of people and resources (materials, equipment, facilities)
- Controlling risks
- Controlling communications and stakeholder engagements
- Coordinating with suppliers on any necessary corrective and preventive actions

Continuous tracking and controlling activities provide the project team and management with visibility into the project's health and early identification of any problems or risks that require investigation or action. Project tracking and control are used to identify differences between the plans and estimates of what should happen and what is actually happening as the project progresses. So why are actuals different than plans and estimates? Because of:

- *Bad estimates*: The software industry is notoriously bad about underestimating costs, effort, and schedules.
- *Forgotten tasks*: Necessary activities are sometimes not included in the work breakdown structure and are therefore not considered in the estimates. For example, if a six-month-long project has weekly project team meetings that last one hour with 20 people in attendance, that is 26 meetings × 20 people × 1 hour = 520 hours of effort, not including the time people spend preparing for these meetings and tracking and controlling issues and risks identified during these meetings. If the effort and schedule estimations do not include these meetings, the project is at least 520 hours behind before it starts.
- *Poorly executed tasks*: Tasks that require more rework or testing than estimated. For example, if a poor-quality product reaches system test, more defect correction and regression testing cycles may be required than were planned.
- *Natural and human disasters*: Fires, floods, snowstorms, illness, death, and other disasters may impact people's ability to even get to work in order to make progress on the project. These disasters can be positive as well as negative. For example, the lead architect on a project was getting married. He was useless for about six weeks before the wedding. In fact, all he wanted to do was sit and talk about the wedding with other team members, which impacted their ability to make progress as well. What was particularly amusing was that his future wife also worked on the project, and she was highly productive because she wanted all of her work done before the wedding.
- *Change*: There is an old military saying that "no battle plan ever survived the encounter with the actual enemy." Well, "no project plan ever survived the encounter with the actual project" either. Why? Because things change! Objectives change, requirements change, technology changes, people leave the project and new people come

on board, the economy changes, and the marketplace changes (e.g., an organization's competitors may announce a product, and the project may need to change to address this new competitive threat).

Since what happens on the project differs from what is planned, when are controlling actions necessary? The answer to this question goes back to the definition of project success. Corrective action is needed if the variance is significant enough to potentially cause the project to fail on any one or more of the five project success factors. In other words, if the variance between actuals and plans is large enough to impact the project's ability to:

- Deliver the product on time
- Deliver within budget
- Deliver with all of the required functionality
- Deliver at the needed quality and performance levels
- Effectively and efficiently utilizing people and resources

Remember the project plan that created the nice straight road map to the project's destination? Then reality happened. The path taken to the project's destination (the actual results) is very different from the original straight-line plan, as illustrated in Figure 17.1. Flexibility is key, because what happens on a project is different than what was estimated or planned. Project plans must be living documents that are designed to change as reality happens. *Project tracking* is the information system that provides the project team and management with the facts and information needed to make informed decisions about what needs to be changed. *Project control* provides the mechanisms to make those changes in an organized, intelligent manner, communicate them to the project stakeholders, and track them to closure.

Tracking and controlling a project requires the integration of many different process activities. Figure 17.2 illustrates the various processes in the Project Management Institute's *monitoring and controlling process group* and how they interact and are integrated.

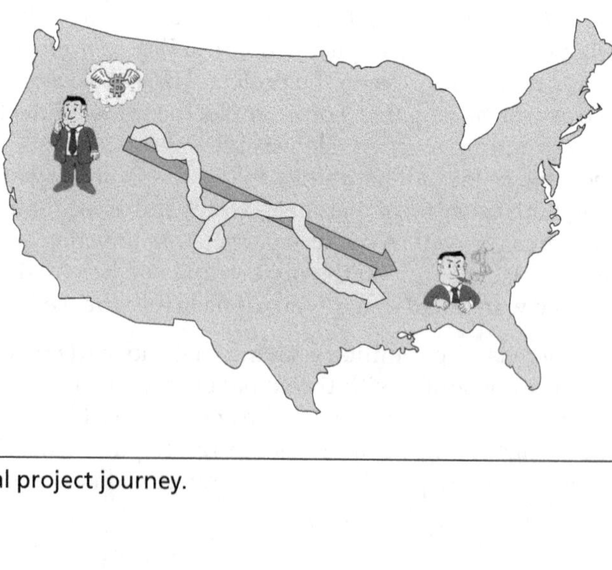

Figure 17.1 Actual project journey.

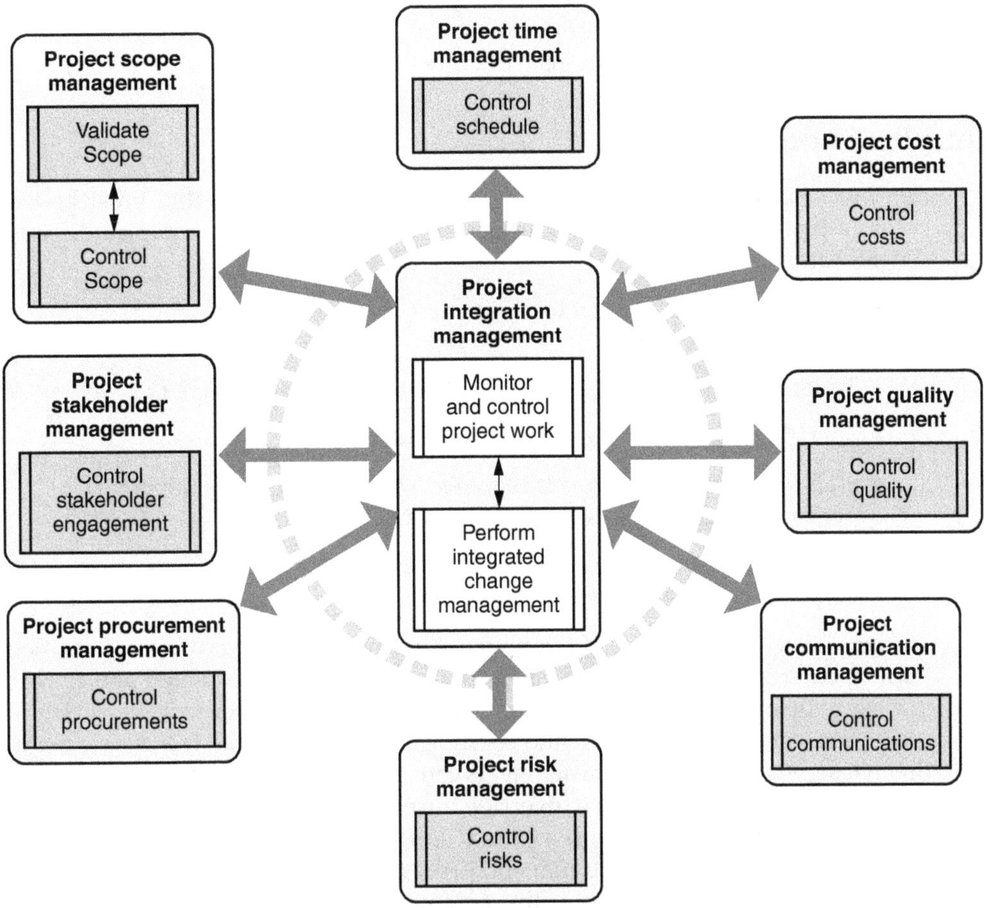

Figure 17.2 PMI monitoring and controlling process group. (Source: PMI 2013. Reprinted with permission.)

1. PHASE TRANSITION CONTROL

> Use various tools and techniques such as entry/exit criteria, quality gates, Gantt charts, and integrated master schedules to control phase transitions. (Apply)
>
> **Body of Knowledge IV.B.1**

Checkpoints can be used to confirm that the project remains in control as it transitions from one phase to another (or from one iteration, process, or activity to another). Tools and techniques, including entry/exit criteria, quality gates, Gantt charts, integrated master schedules, and budgets, can be used to monitor progress at these checkpoints. If the project is not on track, corrective actions should be implemented and managed to closure before the transition is considered complete.

Entry/Exit Criteria

Entry and exit criteria are the specific, measurable conditions that must be met before a phase, iteration, process, or activity can be started or completed. These criteria can include:

- Work products that must be completed, approved, or placed under configuration control
- Tasks or verification steps that must be satisfactorily completed
- Specific measured values that must be obtained
- Staff with appropriate levels of expertise that must be available to perform the phase, iteration, process, or activity
- Resources that must be available and ready for use during the phase, iteration, process, or activity

During the actual execution of the project, the entry and exit criteria are verified. If entry criteria are unmet, corrective actions should be implemented and tracked to closure before that phase, iteration, process, or activity is started. If exit criteria are unmet, corrective actions should be implemented and tracked to closure before that phase, iteration, process, or activity is considered complete. Of course, business decisions can always be made to override the entry and exit criteria. However, any exceptions are handled through a formal *deviation* (if the exception is temporary) or *waiver* (if the exception is permanent) process, with the appropriate approvals.

As discussed in Chapter 6, the entry and exit criteria for a process are documented as part of the definition of each process. However, the project plans may tailor these criteria to match the project's specific needs. The project plan may also define specific values for entry or exit criteria, as required by the project. For example, the standard system-test process might define one of its exit criteria as "The number of unresolved defects do not exceed the limits specified in the project plans." For a specific project, the project plans (quality, verification and validation plans or test plans) might then define the required values for this criterion as:

- No unresolved critical-severity defect can exist
- No more than 10 unresolved major defects can exist, all of which must have workarounds approved by the customer
- No more than 25 unresolved minor defects can exist

Quality Gates

A *quality gate* is a checkpoint or review that a work product must pass through to be acquired (baselined and placed under formal change control) or to transition to the next project phase, iteration, process, or activity. The project's configuration management plans define each configuration item's acquisition point quality gates.

The project's quality plans can define additional quality gates used throughout the project and the required acceptance criteria for each product passing through each quality gate. If the acceptance criteria for a given configuration item are not met at a quality gate, corrective actions should be implemented and tracked to closure before that product is considered to have transitioned through that quality gate. Again, exceptions can be handled through formal deviation or waiver processes with the appropriate approvals.

When the person responsible for a project activity indicates that they are done with that activity, and if that activity's work product(s) have a defined quality gate, a different person or team of people then conducts the checkpoint or review activities. The intent of the quality gate is to confirm that the work product meets its requirements and workmanship standards before it becomes the basis for future work. Quality gates might be as simple as the team leader checking that a source code module meets the coding standard before it is baselined or as complex as a formal phase-end review, as discussed subsequently. Peer reviews, product audits, and tests can also be quality gates. For example, once the peer review team has reviewed and accepted the system test report, it is considered to have passed its quality gate.

Gantt Charts

A commonly used project management tool is the Gantt chart. A *Gantt chart* is a bar diagram that shows the detailed status of the individual activities from the project's work breakdown structure (WBS) by graphing the schedule and duration for each activity. A Gantt chart also highlights milestones, which are typically depicted as diamonds. As illustrated in Figure 17.3, a *scheduling Gantt chart* shows the schedule for each activity with a bar that starts when the activity is scheduled to start and ends when the activity is scheduled to end.

Once the schedule is baselined, a second set of bars is added to the Gantt chart to represent the actual schedule. Figure 17.4 illustrates a *tracking Gantt chart* that shows the light gray planning bars from the baselined schedule and a second set of bars for the actuals. For example, the black tracking bar for activity A shows

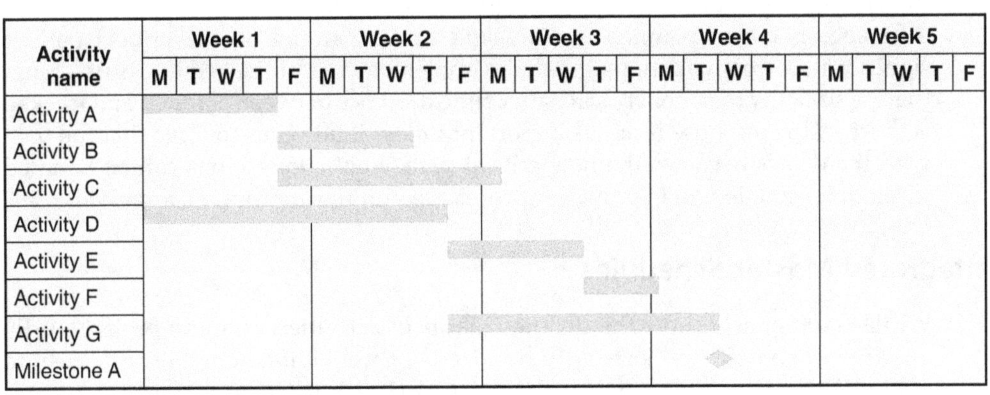

Figure 17.3 Scheduling Gantt chart—example.

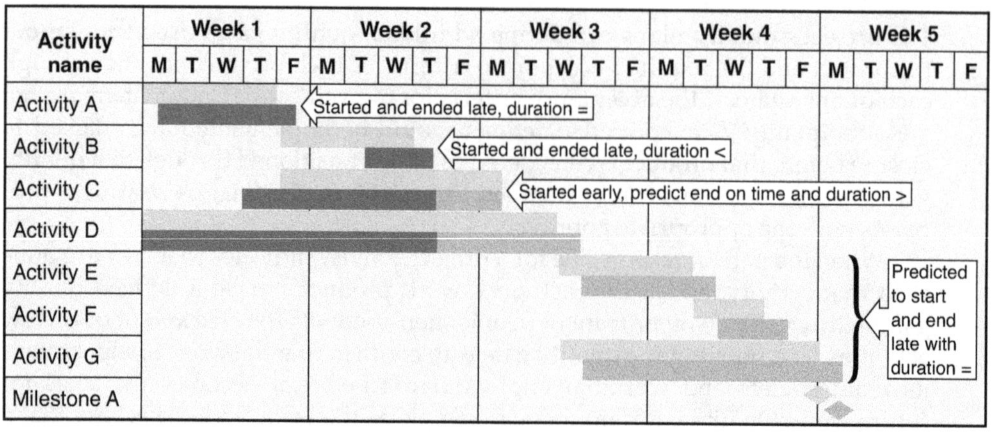

Figure 17.4 Tracking Gantt chart—example.

that it started late (the tracking bar starts to the right of the planning bar) and ended late (the tracking bar stops to the right of the planning bar) but had the same duration (the tracking and planning bars are the same length). Activity B started and ended late but had a shorter duration than planned (the tracking bar is shorter than the planning bar). The split tracking bar for activities C and D shows that these activities are only partly complete. The black part of the tracking bar shows when the activities started and their duration to date. The dark gray part of the tracking bar shows the predicted duration of the part of the activity yet to be completed and its predicted completion point. Based on this example, activity C started early, is about three-quarters complete, and is predicted to end on time. Activity D started on time, is about two-thirds complete, and is predicted to end late. The dark gray tracking bars for activities E through G show that they have not yet started but are predicted to start and end late and have the same duration as originally planned. Finally, based on the current status, milestone A is predicted to occur later than originally scheduled.

The most important concern when evaluating Gantt charts is monitoring the critical path because that path represents the project completion date. If the difference between planned and tracked activity status on the critical path is significant, corrective actions should be implemented and tracked to closure. Since there is slack in other, noncritical paths, the difference between planned and tracked activity status on those is not as important unless it becomes so large that the path with the delays becomes the new critical path. Again, exceptions can be handled through formal deviation or waiver processes with the appropriate approvals.

Integrated Master Schedules

While Gantt charts show the detailed status of activities from the project's WBS on large or complex projects with hundreds of individual activities, it is easy to get lost in these details. It may be beneficial to break these large or complex projects into subprojects, each with its own Gantt chart. If this is done, an *integrated master schedule* is used to summarize the status of these lower-level subprojects into a higher-level project view. Typically, tools are used to track individual project schedules. If integrated master schedules are used, many of those same tools have mechanisms for combining those individual schedules into the integrated master schedule.

Another use for integrated master schedules is with programs where multiple projects are coordinated together. In this case, the integrated master schedule provides a combined view of the high-level interrelationships and dependencies between the projects. A final use for integrated master schedules is when concurrent or incremental development is used to develop software. In this case, each increment can have its own detailed schedule, and the integrated master schedule is used to provide visibility into the interrelationships and dependencies between the increments.

If the difference between the planned and tracked master schedule is significant, corrective actions should be implemented and tracked to closure. Exceptions can be handled through formal deviation or waiver processes with the appropriate approvals.

Budgets

Budgets are tools for tracking project costs and expenditures, just as Gantt charts and integrated master schedules are tools for tracking project activities and schedules. The initial planning budget allocates the estimated project costs into cost categories and subcategories. Once the planning budget is baselined, actual costs to date are tracked against each budget category and subcategory. Budget reports can be generated and checked at phase, iteration, process, or activity transition points or at other designated times (e.g., as part of the input into project reviews). Corrective actions should be implemented and tracked to closure if actual costs deviate significantly from planned costs. Exceptions can be handled through formal deviation or waiver processes with the appropriate approvals.

Project Corrective Action

When a major project issue is identified (e.g., when there is a significant deviation from baselined budgets, baselined schedules, or required quality levels) corrective actions should be planned, implemented, and tracked to closure. When planning project corrective actions, there are two primary choices:

- *Return to plan,* also called *return to green* or *get to green*: The first choice is to return to the plan, that is, to take corrective action that realigns project actuals with the plan and continue with the implementation of the current plan.
- *Replan*: The second choice is to replan by making the appropriate adjustments to the project plans creating a new baselined plan.

Whenever possible, the project should select the first choice and return to the current plan because a project does not occur in a vacuum. For example:

- Project stakeholders may have made plans or commitments based on the current plan.

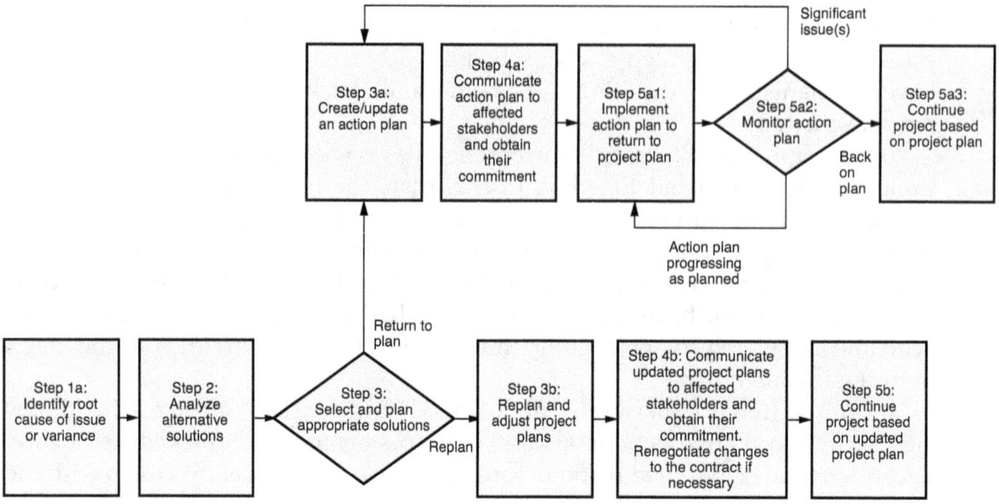

Figure 17.5 Steps in project corrective action process.

- Other projects may be dependent on this project releasing people or resources on schedule, and the success of those other projects may be impacted if those people or resources are not available.
- The project may be part of a coordinated program of several projects that will be impacted if this project is not successful.

The steps in the project corrective action process, as illustrated in Figure 17.5, include:

Step 1: The first step in the corrective action process is to identify the root cause of the issue or variance.

Step 2: Consider alternative approaches to solving or correcting that root cause. Trade-off analysis is performed to identify and analyze the risks, impacts, and costs and benefits for each approach. It is usually valuable to involve stakeholders affected by the issue in these discussions to obtain a broad perspective on the problem and the risks and impacts of its potential solutions.

Step 3: An appropriate alternative solution is selected and planned based on the trade-off analysis. If a return-to-plan approach is selected, an action plan is created that includes:

- Defining specific actions to be taken
- Assigning a person the responsibility for making sure that each action is performed
- Estimating effort, costs, staffing, and resources for each action
- Determining due dates for the completion of each action
- Selecting mechanisms or measures to determine if the desired effect is achieved

If a replan approach is taken, any and all of the project's plans may need to be adjusted, including:

- Adjusting the activities (add, change, or delete activities)
- Adjusting effort, cost, staffing, and resource estimates
- Changing activity duration estimates and schedules
- Changing staff and resource-acquisition plans or assignments
- Changing the budget
- Changing the project scope or product requirements
- Changing supplier agreements (contracts) or supplier management plans

When replanning is selected, baselines must be created for the new plans, and the original baselines should also be retained. This provides a broader, more factual information base for analysis during project retrospectives, post-project reviews, and other quality engineering efforts (e.g., studies of the project estimation accuracy as part of process improvement).

Step 4: The changed project plans, or corrective action plan, should be communicated to all affected stakeholders, and their commitment to the plan(s) is obtained. If replanning is required, and the changes result in the critical path extending past the current due date or significant changes to the budget or scope of the projects, this step may include negotiating changes to contracts with the customer or suppliers.

Step 5: In this step, the corrective action is implemented. If the replan option is selected, corrective action implementation means the project continues based on the updated plans. If the return-to-plan option is selected, the implementation step includes verifying that action plan activities are being performed and tracked to closure and that their outcomes have the desired impact on the project:

- Making sure that the corrective action plan stays on schedule
- Monitoring effort, cost, staffing, and resource estimates against actuals
- If significant problems are identified, repeating the corrective action steps as needed

2. TRACKING METHODS

> Calculate project-related costs, including earned value, deliverables, productivity, etc., and track the results against project baselines. (Apply)
>
> **Body of Knowledge IV.B.2**

Tracking Earned Value

According to the Project Management Institute (PMI 2005), "*Earned value management (EVM)* has proven itself to be one of the most effective performance measurements and feedback tools for managing projects." While many metrics exist that show the details of the project status, earned value metrics provide management with summary-level project metrics, giving insight into the project's overall performance of the project against its schedule and budget.

The steps involved in performing earned value tracking include:

- *Step 1*: Determining the critical resources to be tracked—typically, earned value is calculated in dollars and effort (staff-months or engineering-hours)
- *Step 2*: Allocating the resource budget to individual activities in the WBS
- *Step 3*: Calculating the planned value, earned value, and actual value metrics
- *Step 4*: Analyzing the earned value metric compared to the other metrics to date:
  - If earned value < actual value, then the project is over budget
  - If earned value < planned value, then the project is behind schedule

The basic earned value metrics include:

- *Planned value (PV)* = the budgeted cost of work scheduled (BCWS), which indicates the amount of resource (cost or effort) that was planned to be expended to do the work that was planned to be accomplished at the time the measurement was taken
- *Earned value (EV)* = the budgeted cost of work performed (BCWP), which indicates the amount of resource (cost or effort) that was planned to be expended to do the work that was actually completed at the time the measurement was taken
- *Actual value (AV)* = the actual cost of work performed (ACWP), which indicates the amount of resources (cost or effort) that was actually expended to do the work that was actually completed at the time the measurement was taken

The analysis earned value metrics include:

- *Cost variance (CV)* = BCWP – ACWP
  - If the CV is positive, the project is under budget
  - If the CV is negative, the project is over budget
- *Cost performance index (CPI)* = BCWP/ACWP
  - If the CPI is greater than one, the project is under budget
  - If the CPI is less than one, the project is over budget

- *Schedule variance (SV)* = BCWP – BCWS
  - If the SV is positive, the project is ahcad of schedule
  - If the SV is negative, the project is behind schedule
- *Schedule performance index (SPI)* = BCWP/BCWS
  - If the SPI is greater than one, the project is under budget
  - If the SPI is less than one, the project is over budget

There are two basic ways of calculating work performed. The first and simplest is to earn back the value only when the activity is complete. The second is to earn back a portion of the value based on the percentage of the activity completed by prorating the value earned back. For example, if an activity is scheduled to take 20 days and it is 50 percent done, then the budgeted cost of work performed would be 10 days, which can be compared to the actual number of days spent performing the activity to date. This second method is more complex because it requires intermediate reporting of progress-to-date and cost-to-date on uncompleted activities. Another issue with using the second method for software is that activities may be reported as 90 percent complete for the last 90 percent of the time spent performing them. For most software projects, earned value metrics are calculated using only work completed.

Table 17.1 includes information for an example of a small five-activity project. In this example, at the time earned value was measured:

- Activity A was scheduled to be completed, and it has been, but it took one more day to complete than budgeted
- Activity C was scheduled to be completed, and it has been, and it took the same amount of time to complete as budgeted
- Activity B was started, and two days were spent working on it, but an impediment occurred that stopped the work on Activity B
- So the practitioner(s) working on Activity B switched over and completed Activity E ahead of schedule, with Activity E taking two days less to complete than budgeted

Based on the information in Table 17.1, the calculation of the earned value metrics would be:

- The planned value (PV) is equal to the sum of the budgeted costs for the activities scheduled to be completed at the time of the measurement. Since activities A, B, and C were scheduled to be completed, BCWS = 3 + 12 + 9 = 24 days.
- The earned value (EV) is equal to the sum of the budgeted costs for the activities actually completed at the time of the measurement. Since activities A, C, and E are completed, BCWP = 3 + 9 + 8 = 20 days.
- The actual value (AV) is equal to the sum of the actual costs for the activities actually completed at the time of the measurement. Since activities A, C, and E are completed, ACWP = 4 + 9 + 6 = 19 days.

Table 17.1 Earned value–example.

	Activity A	Activitiy B	Activity C	Activity D	Activity E
Status	Done	Started	Done	Not started	Done
Schedule	Done	Done	Done	Not done	Not done
Budget	3 days	12 days	9 days	5 days	8 days
Actual	4 days	2 days	9 days	—	6 days

To analyze this example:

- The cost variance = BCWP − ACWP = 20 days—19 days = 1 day to complete the work that has been completed. In other words, 20 days were budgeted to do the work that it only took 19 days to do, so the project is one day under budget.
- The cost performance index (CPI) = BCWP/ACWP = 20/19 = 105%, or the project is getting 1.05 hours of work for every hour budgeted.
- The schedule variance = BCWP − BCWS = 20 days − 24 days = −4 days to complete the work that was scheduled to be completed. In other words, 24 days of work should have been done by now and only 20 days of work has been accomplished, so the project is four days behind schedule.
- The schedule performance index (SPI) = BCWP/BCWS = 20/24 = 83%, or the project is only getting .83 hours of scheduled work done for every hour worked.

The current values of the earned value metrics provide a snapshot of the current project status. Table 17.2 includes information that can be used to interpret the earned value metrics.

Trending the earned value metrics over time can provide even more valuable insight into the project. As illustrated in Figure 17.6, the planned value (BCWS) can be plotted out to the end of the project based on the current schedule. The earned value (BCWP) and actual value (ACWP) can then be plotted over time. Remember that the cost variance = BCWP − ACWP. As can be seen on this graph, not only is this project over budget and behind schedule at this time, but the trends also show that things are getting worse (the variances are increasing) as the project progresses. The earned value and actual value curves can also be extrapolated to determine when the project will be done and how much it will cost when completed based on current trends.

If the cost or schedule variance is considered significant, corrective actions should be implemented and tracked to closure. Exceptions can be handled through formal deviation or waiver processes with the appropriate approvals.

Tracking Deliverables

The ultimate goal of any project is to create releasable deliverables. Therefore, tracking the creation and verification of these deliverables against the baselined plans provides valuable information on the health of the project. Deliverables

Table 17.2 Interpreting earned value.

Relationships	Interpretation
ACWP=BCWP=BCWS	No variance—everything is going according to plan.
(BCWS=ACWP)<BCWP, ACWP<BCWS<BCWP or BCWS<ACWP<BCWP	Work is *ahead of schedule*, and costs are *under budget* for work that has been accomplished. Looks good but may lead to missed opportunities.
BCEW<(ACWP=BCWP)	Work is *ahead of schedule*, and costs are being maintained for work that has been accomplished.
ACWP<(BCWS=BCWP)	Work is on schedule, and costs are *under budget* for work that has been accomplished.
(ACWP=BCWP)<BCWS	Work is *behind schedule*, but costs are being maintained for work that has been accomplished.
(BCWS=BCWP)<ACWP	Work is on schedule, but costs are *over budget* for work that has been accomplished.
BCWS<BCWP<ACEP	Work is *ahead of schedule*, but costs are *over budget* for work that has been accomplished.
ACWP<BCWP<BCWS	Work is *behind schedule*, but costs are *under budget* for work that has been accomplished.
BCWP<(BCWS=ACWP), BCWP<BCWS<ACWP or BCWP<ACWP<BCWS	Work is *behind schedule*, and costs are *over budget* for work that has been accomplished.

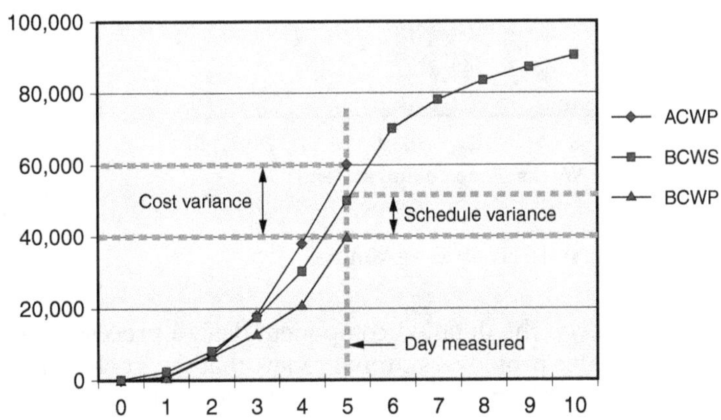

Figure 17.6 Interpreting earned value—example.

metrics help keep the product component of the project management trilogy of cost, schedule, and product visible to the project team and management.

The *design deliverable status metric* illustrated in Figure 17.7 can be used to track and control the number of design elements (modules, classes, data elements) from the architectural design that have completed the detailed component design and the number of detailed component designs that have been inspected against the

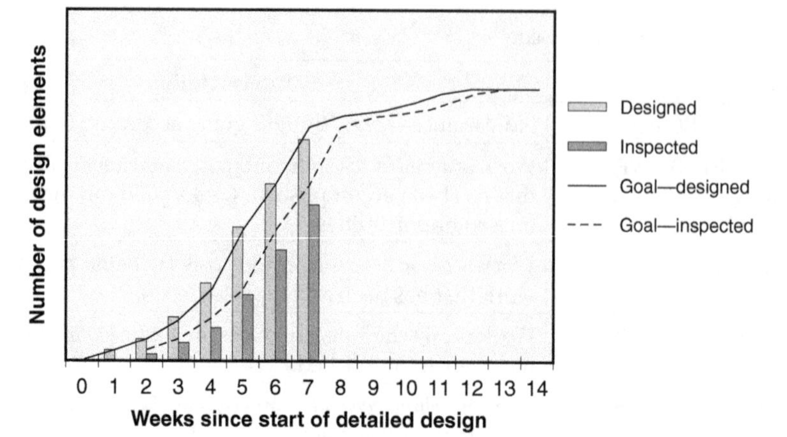

Figure 17.7 Design delivery status metric—example.

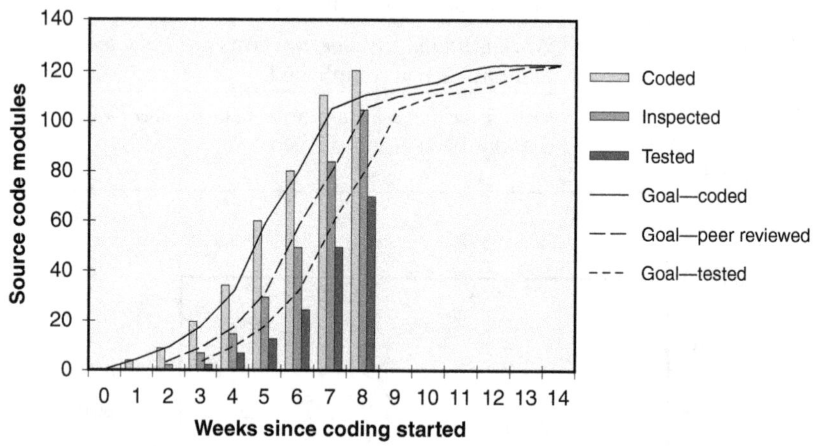

Figure 17.8 Code delivery status metric—example.

planned goals to keep the detailed component design process on schedule. This type of graph can also provide a summary view that can get lost in the details of a Gantt chart. For example, in this graph, it is easy to see that while the completion of the detailed component designs is on schedule, getting them peer reviewed is currently behind schedule at week seven.

As illustrated in Figure 17.8, similar graphs can be created for *code deliverable status metrics* that track the status of writing, peer reviewing, and unit testing code against the schedule to keep the code and unit test process on schedule. In this example, the project is ahead of schedule in getting source code modules coded and on schedule for peer reviewing them but falling behind the goal of getting the code unit tested at week eight. Depending on the project, tracking the completion status of other possible deliverables, including documentation, training materials, features, functions, product requirements, test case creation, or use cases (or use case steps), might also be useful. Each project decides what deliverables are appropriate to track for that project.

Tracking Productivity and Velocity

From a traditional project tracking and control perspective, a predicted productivity level is used when estimating project effort and schedule values. Agile methods use predicted velocity values to evaluate the amount of work that can be done during the next iteration. Typically, these predictions are based on historic productivity or velocity levels.

The project risk is that the actual productivity, or velocity, level will significantly differ from the predicted value. Therefore, as illustrated by the example in Figure 17.9, a metric can be used to track the variance between planned and actual productivity levels if this is considered a high-priority risk for the project. In agile iterations, velocity actuals versus predictions are typically tracked using *burn-up charts or burn-down charts*. As illustrated in Figure 17.10, a burn-up chart tracks progress by graphing the number of items completed, starting with zero, up to the total number of items to be done (goal). A burn-down chart begins with the total number of items that need to be completed and tracks progress by graphing the number of items left to complete down to having no work left to do.

Tracking Resources and Staffing

Resource utilization metrics are used to track and control the utilization of various resources during the software project's implementation to make sure that adequate resources exist to meet schedules and to monitor resource overruns that may affect the budget. Examples of these metrics might include:

- Number of personnel assigned to the project (actuals versus planned)
- Number of staff hours expended on the project (actuals versus planned)

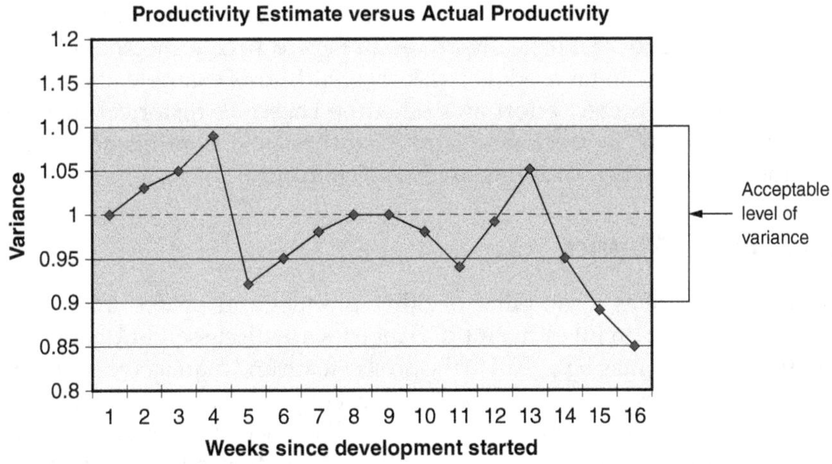

Figure 17.9 Productivity metrics—example.

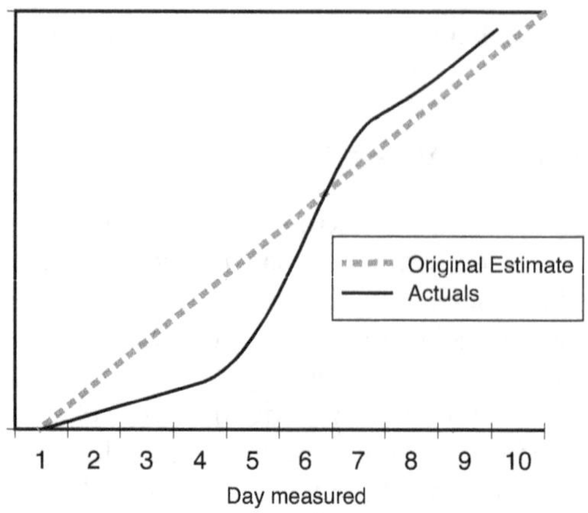

Figure 17.10 Velocity burn-up chart—example.

- Number of test bed hours (available/utilized versus planned)
- Number of square feet of office space (available/utilized versus planned)
- Number of dollars spent (actuals versus planned)

As illustrated in the example resource utilization metric in Figure 17.11, the actual resource utilization (solid line) can be tracked against the predicted resource utilization from the initial plan (dotted line). The planned resource utilization should be updated based on what is occurring on the project. The metric can also include the new current plan (dashed line).

Staff turnover is a complementary metric to monitor the number of people or staff hours available to the project. Because of the learning curve, even if an experienced engineer can be replaced immediately, productivity will be impacted until the replacement is up to speed on the processes and work products. In the example staff turnover metric illustrated in Figure 17.12, a threshold has been set so that the metric can be used as a risk trigger. In this example, the project manager planned the project effort and schedule based on historical turnover rates for similar projects. By tracking against that threshold, contingency plans can be implemented if the turnover rate exceeds 15 percent.

Selecting Project Metrics

Chapter 19 describes a sampling of other product and process metrics. Cost-of-quality and return on investment (ROI) metrics are discussed in Chapter 7. Any or all of these metrics may be useful to the project team and management in estimating, tracking, and controlling their project. As part of project planning, each project can use the goal–question–metric paradigm, as discussed in Chapter 19, or a similar process to select which metrics will be collected and tracked based on that project's goals and information needs. The chosen metrics should be documented in the project communication plans.

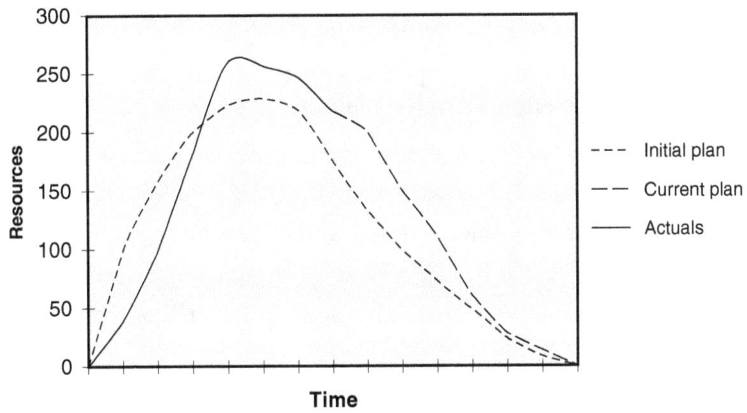

Figure 17.11 Resource utilization metrics—example.

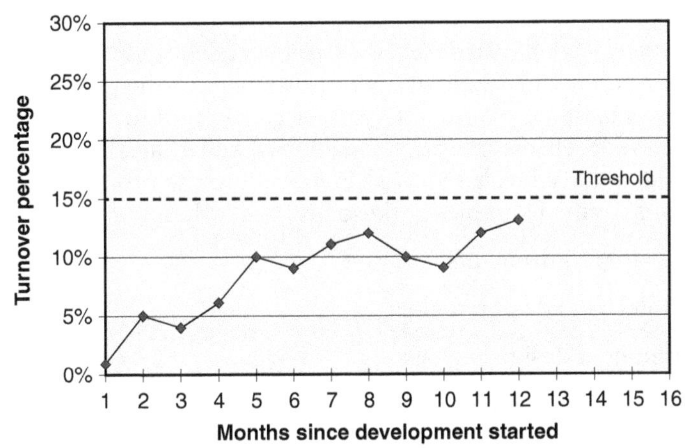

Figure 17.12 Staff turnover metrics—example.

3. PROJECT REVIEWS

> Use various types of project reviews such as phase-end, management, and retrospectives and post-project reviews to assess project performance and status, to review issues and risks, and to discover and capture lessons learned from the project. (Apply)
>
> **Body of Knowledge IV.B.3**

Project reviews bring project stakeholders together to discuss the project and exchange information. The purposes of holding project reviews are to:

- Determine if the overall status of the project work to date is acceptable
- Appraise the progress and performance of what is being accomplished on the project compared to the plans
- Provide a basis for management and engineering decisions on how to proceed
- Identify potential problems (risks) and deal with them before they negatively impact project success
- Confirm that necessary corrective action plans to control the project are put in place before unrecoverable harm is done to the project, and track those plans to closure
- Provide and receive basic project status and communications

Phase Gate Reviews

Phase gate reviews, also called *phase-end, phase transition,* or *major milestone reviews,* are event-driven reviews held at major milestones in the project. The purpose of these reviews is to act as a quality gate or checkpoint to verify that all required activities have been completed satisfactorily and that all work products have achieved the quality levels required to move into the project's next major phase, iteration, or activity. Examples of these reviews include:

- Software requirements review
- Architectural design review
- Component design reviews
- Test readiness reviews
- Ready-to-ship review

A software *requirements review* is held during the transition from the software requirement activities (phase) to the architectural design activities (phase). Examples of typical milestone-specific items reviewed at this meeting include:

- Satisfactory completion of the software requirements specification
- Results of software requirements specification peer reviews
- Traceability of software requirements to their source (e.g., system requirements, use cases, business rules, standards, or regulations)

An *architectural design review,* also called the *preliminary* or *high-level design review,* is held during the transition from the architectural design activities (phase) to the component design activities (phase). Examples of typical milestone-specific items reviewed include:

- Satisfactory completion of the software architecture specification
- Satisfactory completion of any external interface design specifications

- Traceability of software architectural design elements to software requirements
- Results of software architecture and external interface design specifications peer reviews

A *component design review,* also called the *critical* or *detailed design review,* is held during the transition from the component design activities (phase) to implementation activities (code and unit test phase). There may be multiple detailed design reviews, one for each major design element. Examples of typical milestone-specific items reviewed include:

- Satisfactory completion of the component software design
- Traceability of component design elements to software architectural design elements
- Satisfactory completion of software system (software qualification) test plans, test cases, and test procedures
- Satisfactory completion of acceptance test plans, test cases, and test procedures
- Peer-review results (component design, test plans, test cases, and test procedures)

Test readiness reviews, also called *ready to test reviews,* are held at the beginning of major test cycles (such as integration test, software system test, acceptance test, alpha test, beta test). Examples of typical milestone-specific items reviewed at an integration test readiness review include:

- Traceability of integration test cases to software design elements
- Results of peer reviews (code and integration test case and plans)
- Satisfactory implementation of the test environment, including the test data
- Availability of test personnel
- Unit test results, including execution and pass/fail status of unit test cases
- Unresolved defect reports

Examples of typical milestone-specific items reviewed at a software system test readiness review include:

- Traceability of software system test cases and test procedures to software requirements
- Integration test results, including execution and pass/fail status of integration test cases and test procedures
- Satisfactory completion of user documentation (user's manual and other user-deliverable documentation, including the installation instructions)

- Results of peer review (user documentation, software system test plans, test cases, and test procedures)
- Unresolved defect reports

A *ready-to-ship review*, also called *ready-to-release review*, is held during the transition from the final test cycle to the release (deployment) of the completed software or system. Examples of typical milestone-specific items reviewed include:

- Software or system, acceptance, and alpha or beta test results, including execution and pass/fail status of test cases and test procedures
- Satisfactory completion of the final user manual and other deliverable documentation, including the installation instructions (with any corrections to defects found in testing cycles)
- Unresolved defect reports

In addition to reviewing these items, ready-to-ship reviews confirm that the product is truly ready to be supported in the field, including verifying that:

- Training is available as needed (e.g., that training materials have been peer reviewed, that training classes have been piloted, and that training resources and personnel are available)
- Technical support personnel are trained and ready to support the product, and procedures for reporting problems and requesting enhancements are established
- Replication, ordering, and delivery procedures are established as appropriate

For Scrum projects, the sprint review meeting is a close equivalent to the ready-to-ship reviews. On the last day of the sprint, the Scrum master facilitates the sprint review meeting, where the Scrum team presents the increment of functionality developed during the sprint to the product owner and other attendees (stakeholders). Decisions made during this meeting include determining whether:

- To release the current functionality into operations
- New or updated functionality or other items (new stories) need to be added to the product backlog
- Functionality that was planned for the sprint but not delivered as expected should go back into the product backlog
- To continue the project by holding the next sprint planning meeting or to terminate the project

Project Team Reviews

Project team reviews are held periodically (daily, weekly, or monthly) throughout the project based on its size, complexity, and risk. The project manager usually runs these meetings. Attendance at these meetings typically includes:

- Project team members (or representatives of each sub-team for large projects)
- Project support personnel assigned to the project (e.g., configuration management, quality assurance, and verification and validation team members)
- Representatives from other stakeholder groups (e.g., marketing, hardware engineering, technical publications).

Depending on the project, external stakeholders (e.g., customers and suppliers) may also attend these meetings.

Project team reviews are held to monitor the ongoing, day-to-day, current status and results of the project against the project's documented estimates, plans, commitments, and requirements and to identify project issues and risks in a timely manner so that effective control actions can be taken. A sample agenda for a project team meeting might include:

- Project accomplishments for the reporting period
- Changes to the project objectives or business climate, if any
- Current project status:
  - Upcoming activities and milestones
  - Late activities—why they are late, what the impact is, what recovery plans and help are needed
  - Dependencies between groups
  - Staffing and resource availability
  - Technical status: functionality and quality
  - Cost or budget status
  - Current risk activities and new and closed risks
  - Current corrective action status
- Issues, conflicts, problems, or changes
- Review of assigned action items

For agile iterations, the agile coach or Scrum master facilitates the daily team meetings. These daily meetings help the agile team keep continuous track of the progress of the team, with each team member presenting three items:

- What was accomplished since the last meeting
- What is planned to be accomplished before the next meeting
- What are the impediments to progress, if any

These daily meetings are short, stand-up meetings (people literally stand up—no sitting in chairs allowed) of no more than 15 minutes in duration. The agile coach or Scrum master follows up outside these meetings, as needed, with the appropriate individuals to handle any identified impediments. The agile team members actively participate in these meetings. Other stakeholders or individuals may attend these meetings for informational purposes but do not participate.

Management Reviews

Management review meetings are also held periodically throughout the project based on its size, complexity, and risk, but usually less often than project team meetings (monthly or quarterly). Attendees at these meetings include senior management, project managers (several projects may be discussed in these meetings, especially if those projects are being coordinated as a program), project team representatives, management from project support groups, and management from other stakeholder groups affected by the project (e.g., marketing, hardware engineering, technical publications). Depending on the project, external stakeholders (e.g., customers and suppliers) may also attend these meetings.

Management review meetings are held to provide management awareness of, and visibility into, software project activities. A sample agenda for a management meeting might include:

- Summaries of technical (functionality and quality) performance, and cost or budget, staffing, and schedule performance (typically presented at the major feature or major milestone level)
- Changes to the project objective or business climate, if any
- Review of high-priority software project risks or risks that cannot be handled at the project level
- Issues, conflicts, problems, or changes that cannot be handled at the project level
- Review of assigned action items

Post-Project Reviews

Post-project reviews, also called *post-implementation reviews* or *post-mortems,* are meetings held at, or near, the end of the project to provide a mechanism for learning from the project experiences, both good and bad, and feed those lessons back into product and process improvements or corrective actions. These reviews examine what went well and determine the best ways to repeat these successes. These reviews also examine what went wrong and determine the best ways to keep those problems from happening again in the future. Post-project reviews should be held whether or not the project was successful. In fact, in some cases, more may be learned from a failed project than a successful one—like how not to even start one of these projects again.

For large projects, at least one representative from each major group or team involved in, or affected by, the project should be invited to participate in the post-project review. Norm Kerth (2001) expands this by recommending that everyone on the team should participate, which is what is done in agile retrospectives. Managers are typically omitted from participation because they may inhibit the candidness of the review team or the free flow of ideas. Management can hold a separate post-project management review if it is seen as value-added.

One mechanism for preparing for these meetings is to have each participant complete a post-project review form. These forms are designed to help guide participants through thinking about project successes and problems and about suggestions for future improvements. Forms can be generalized, or different forms can be customized to specific areas (e.g., project planning, software design, coding, and system testing). Forms should be tailored to meet the needs and characteristics of the specific project. Participants should focus on their personal areas of expertise and work assignments for the project being evaluated.

The review meeting atmosphere must encourage full participation and open sharing of opinions and ideas. Questions are encouraged, and different views should be expected and elicited. Having a good facilitator for the meeting can help keep it focused on the project's processes and products and not on people. The focus should also be on lessons learned and not on laying blame.

The review meeting includes the creation of two lists. The first list identifies what went right on the project, with participants presenting their ideas from the pre-prepared review forms or adding new thoughts based on actively listening to the ideas of other participants. It may be helpful to order this discussion in general life cycle order. The team can then prioritize items on the "what went right" list based on each item's benefit to the project through the use of multivoting, a prioritization matrix, or other nominal group techniques. This activity is then repeated by creating the second list of "what went wrong," prioritized by each item's negative impact on the project. Always do the positives first, because once people start talking about the negatives, it is hard to get them thinking positively again.

After these lists are created, the team selects high-priority items from the lists that need to be addressed. They develop a proposal to management on lessons learned and recommendations on how best to incorporate those lessons into future project activities. For example, additional training might be recommended to implement one lesson, or updates to the quality management system (e.g., to processes, templates, or checklists) might be recommended to implement other lessons learned.

The results and recommendations from the post-project review meeting are then presented to management. Management must commit to the recommendations for change and provide time, people, and resources to implement those changes. The lessons learned must be propagated into improvements on future projects for the post-project review meeting to be truly effective and value-added.

While these post-project reviews are one of the primary mechanisms used by waterfall-type projects to eliminate the repetition of "mistakes," identify areas for continuous improvement, and propagate good practices, post-project reviews can have the following drawbacks:

- The information from these reviews is obtained too late to help the current project
- Key team members may no longer be available to participate

- Team members may have forgotten the pain or lessons learned
- Ideas may just be documented and never actually implemented

Retrospectives

Scrum projects hold *retrospectives* after the completion of each sprint. The Scrum master facilitates the sprint retrospective meeting with the Scrum team. The intent of the sprint retrospective meeting is to identify lessons learned from the completed sprint and determine process improvement actions to improve the next sprints. During this retrospective, the Scrum team reviews what went well with the sprint so that positive lessons learned can be repeated in future sprints. They also review what did not work well on the sprint and determine process improvement actions that they need to add to the product backlog to correct those issues in future sprints.

Interim Retrospectives and Reflections

While these post-project reviews and retrospectives help future projects, industry experts (e.g., Rothman 2007 and Derby and Larson 2006) recommend that interim retrospectives be used throughout the project to feed lessons learned back into improvements as the project progresses. That way, the project benefits directly from those lessons. Improvement comes from learning about the root causes of problems (or potential ones) and finding a solution to correct (or avoid) those problems. Again, surveys can be used to gather information prior to the actual start of the retrospective meeting.

Extreme programming (XP) uses a practice called *reflections,* which asks the team members to continuously consider what they are doing and why. This provides input and feedback that allows immediate reuse of good ideas and quick elimination of problems. "Good teams don't just do their work, they think about how they are working and why they are working. Don't try to hide mistakes—expose them and learn from them. Learning is action reflected" (Beck 2005).

4. PROGRAM REVIEWS

> Define and describe various methods for reviewing and assessing programs in terms of performance, technical accomplishments, and resource utilization. (Understand)
>
> **Body of Knowledge IV.B.4**

The PMI (2013) defines a *program* as "a group of related projects managed in a coordinated way to obtain benefits and control not available from managing them individually." Tracking and controlling a program involves gathering and consolidating status information from the individual projects being managed as parts of that program.

Program review meetings are held periodically throughout the program based on its size, complexity, and risk. The program manager or a representative from the program management office typically runs these meetings. Attendees at these meetings include the project managers of the various projects that are part of the program. In addition, team representatives from various projects, management from program support groups, and management from other stakeholder groups affected by the program (e.g., marketing, hardware engineering, technical publications, suppliers, and customers) may also be in attendance. This may include acquisition project managers, supplier project managers, and other individuals from supplier-run projects as well if the program includes acquisition projects. Supplier personnel may be excluded from portions of the program review if confidential information or trade secrets are being discussed.

Program review meetings are held to track and control the coordinated program activities. These meetings examine aggregated program performance information, including cost, schedule, technical, resource, staffing, scope, risk, and supplier management information, from the individual project and nonproject activities. A sample agenda for a program review meeting might include:

- Program accomplishments for the reporting period
- Changes to the program objectives or business climate, if any
- Current program status:
    - Schedule and milestone status rollups from individual projects and non-project-related program activities
    - Dependencies between projects and groups
    - Coordination of staffing and resource availability
    - Technical status: functionality and quality
    - Rollup cost or budget status
    - Program risks and current risk-handling activities
    - Current corrective action status
- Issues, conflicts, problems, or changes that cannot be handled at the individual project-level
- Review of assigned action items

A Scrum equivalent for a programs review meeting would be *Scrum of Scrums* meetings. If the software development effort is too large for one Scrum team, one way to handle this larger scope is to create multiple Scrum teams, each working on their part of the software product. A *Scrum of Scrums team,* made up of representatives from each individual Scrum team, meets periodically to coordinate the efforts of the individual Scrum teams.

Chapter 18

Risk Management

There are many risks involved in creating high-quality software on time, within budget, with all of the required scope, and at the needed quality and performance levels while effectively and efficiently using people and resources. With ever-increasing software complexity and increasing demands for bigger and better products delivered at a faster rate, the software industry is a high-risk business. When teams don't manage risk, they leave projects and operations vulnerable to factors that can cause major rework, high cost or schedule overruns, increased product risks during operations, or complete project failure.

Adopting software risk management processes is a step that can help effectively manage software development, acquisition, and maintenance initiatives. However, for it to be worthwhile to take on these risks, the organization must be compensated with a perceived opportunity to obtain the associated rewards. The greater the risk, the greater the opportunity must be to make it worthwhile to take the chance. In software development, the possibility of rewards is high, but so is the potential for disaster. Risks exist, whether they are acknowledged or not, and unacknowledged risks can lead to unpleasant surprises when some of those risks turn into actual problems. The need for software risk management is illustrated in Gilb's risk principle: "If you don't actively attack the risks, they will actively attack you" (Gilb 1988). In order to successfully manage a software project, create high-quality products, and reap the rewards, software practitioners must learn to identify, analyze, and control these risks.

In the software industry, the future seems to be coming at us at an ever-increasing rate. Effective software managers and practitioners proactively consider all the possibilities that the future may bring, but those possibilities have uncertain outcomes. Some possibilities are called *opportunities,* or *positive risks,* if positive results are expected. For example, an opportunity exists to complete a software project and make a substantial profit, or an opportunity exists to introduce a new product into the marketplace first and capture the lion's market share. Other possibilities are called *risks,* or *negative risks,* if adverse outcomes are expected. There is also the risk of not completing that same software project successfully and losing the investment. There is also the risk of the competition beating an organization to the marketplace with a new product and losing market share. To quote Tom DeMarco, "Moving aggressively after opportunity means running toward rather than away from risk" (Hall 1998)

As illustrated in Figure 18.1, sound risk management practices are a balancing act between the risk and the opportunity. While this chapter focuses on negative

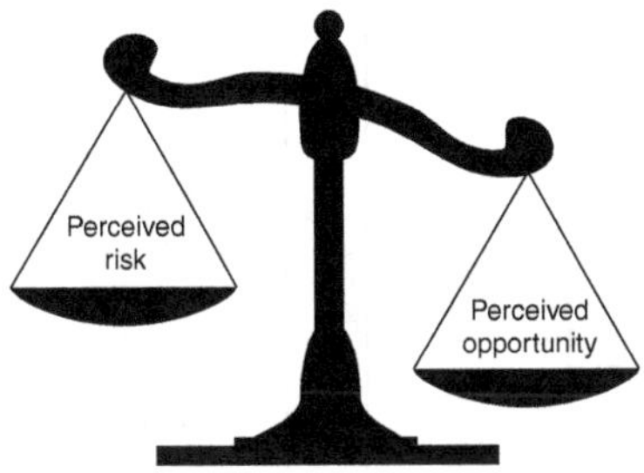

Figure 18.1 Risk/opportunity balance.

risk management, the associated opportunities must also be identified and managed. Not paying attention to opportunities and balancing opportunity management with risk management can lead to losing important opportunities.

Different people and different organizations have different *risk tolerance* levels. Risk tolerance influences a person or organization's perceived risk/opportunity balance point. *Risk-takers* are more willing to take a risk, even if the financial, economic, material, or other gains (opportunities) are less than the loss associated with the potential problem (risk). For risk-takers, the sheer pleasure or excitement of taking the risk adds weight to the opportunity side of the risk/opportunity balance. *Risk-avoiders* are averse to taking risks, and the mere presence of the risk adds weight to the risk side of the risk/opportunity balance. Risk-avoiders need additional financial, economic, material, or other incentives to take on the risk. People or organizations that are *risk-neutral* look for a balance between the risks and opportunities and have no emotional investment in either avoiding or taking the risk.

So, what are risks? A risk is simply the possibility of a problem occurring sometime in the future. "Risk, like status, is relative to a specific goal. Whereas status is a measure of progress toward a goal, risk is a measure of the probability and consequences of not achieving the goal" (Hall 1998). For example, every time a person crosses the street, that person runs the risk of being hit by a car. As illustrated in Figure 18.2, a risk starts when the commitment associated with that risk is made. The risk of getting hit by a car in the street does not exist until the person steps into the street. Of course, there is the risk of getting hit by a car while standing on the sidewalk, but that is an entirely different risk. Until the project selects Acme as its subcontractor for a component, the risks that Acme will deliver a low-reliability software component or be late in delivering it do not exist. If the project builds the software component in-house or selects a different supplier, a different set of risks will exist. A risk ends when one of two things happens:

1. Bang!! The person gets hit in the middle of the street by a car. Acme delivers a low-reliability software component or delivers it late. The problem occurs (or it has a near 100 percent probability of occurring). It is now a problem and is no longer a risk.

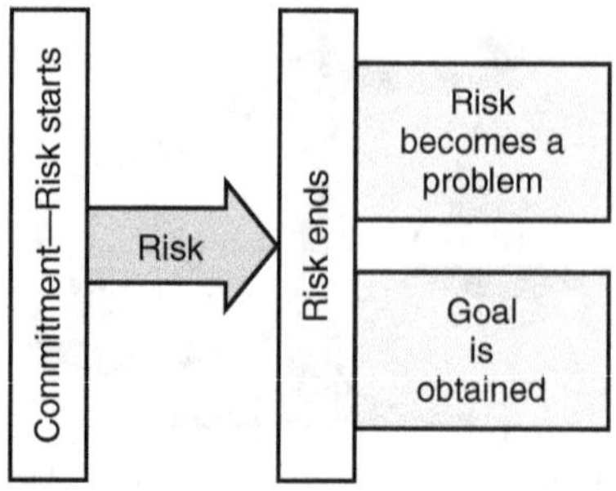

Figure 18.2 Risk duration.

2. The person safely steps onto the sidewalk on the other side of the street. Acme delivers a high-reliability software component on schedule. The risk disappears since there is no longer the possibility of a future problem because the goal has been obtained.

Acme either delivers a high-quality product component on time (goal is reached) or they don't (problem occurs). Before it becomes a problem, "a risk is just an abstraction. It's something that may affect your project, but it also may not. There is a possibility that ignoring it will not come back to bite you" (DeMarco and Lister 2003). If a person ignores the risk, however, and runs into the street without looking, there can be dire consequences. There may not be a problem every time—in fact, that person may get away with it repeatedly. But then it takes only one instance of the problem occurring for disaster to happen.

A distinction should be made between major risks and the basic day-to-day risks that are generic to almost all software projects. There will always be some risk of requirements volatility or staff turnover on every project. These types of basic risks are addressed through good project management techniques. For example, an experienced project manager always considers typical requirements volatility and staff turnover levels when estimating project scheduling, costs, and staffing. Risk management deals with the major, unique risks for the project that have the potential to impact project success. For example, risk management deals with the risk that there will be more requirements volatility or staff turnover than the project manager planned for, or that one of the critical staff members will leave the project. In good risk management practices, the "emphasis is shifted from crisis management to anticipatory management" (Down et al. 1994). Table 18.1 shows several ways in which risk management differs in perspective from project management. However, risk management is not separate from project management. According to the Project Management Institute, risk management is simply one of the process areas of good project management (PMI 2013).

Table 18.1 Project management perspective versus risk management perspective.

Project Management	Risk Management
Designed to address general or generic risks	Designed to focus on risks unique to each project
Looks at the big picture and plans for details	Looks at potential problems and plans for contingencies
Plans what should happen and looks for ways to make it happen	Evaluates what could happen and looks for ways to minimize the damage
Plans for success	Plans to manage and mitigate potential causes of failure

1. RISK MANAGEMENT METHODS

> Use risk management techniques (e.g., risk impact assessment, prevention, mitigation, and transfer) to evaluate project risks. (Evaluate)
>
> **Body of Knowledge IV.C.1**

Risk Management Process

Risks management is about:

- Identifying risks to the project, process, or products before they occur and become major problems
- Analyzing and prioritizing those risks and deciding which risks need to be mitigated
- Determining how to mitigate those risks and planning the risk mitigation activities
- Tracking and controlling all the risks to minimize disasters
- Adjusting the risk mitigation plans as new risks are identified or as new information is obtained

Risks should be taken into consideration:

- When evaluating projects for initiation and chartering
- As an integral part of initial and ongoing project and operations planning activities
- During the estimation of the project or operational effort, schedule, budget, staffing, and resource needs

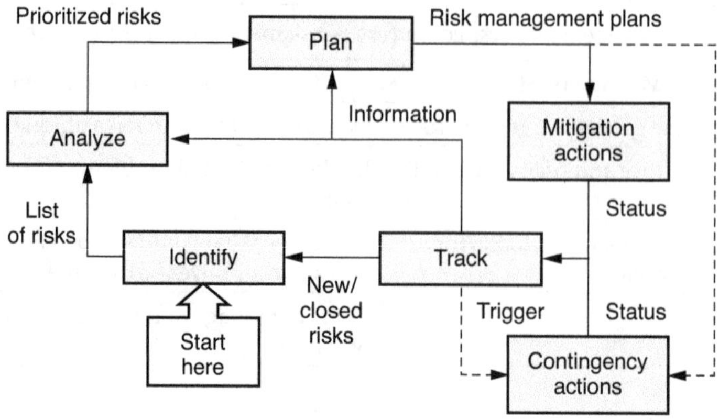

Figure 18.3 Risk management process—example.

- As an ongoing part of project, process, or product tracking and control
- As part of activity entry and exit criteria
- When making decisions about project, phase, or iteration completion and closure

Risk management creates a feedback loop, where additional information, including risk status and project or operational status, are continuously used to refine the risk list and ongoing risk management plans (as illustrated in the example of a *risk management process* in Figure 18.3):

- The risk management process starts with identifying a list of potential problems (risks).
- Each of these risks is then analyzed and prioritized.
- A risk management plan is created to identify action plans for addressing high-priority risks. These plans identify:
  - *Risk mitigation plans,* also called *risk containment plans, risk handling plans,* or *risk response strategies,* which define actions that will be taken as part of the project or operational activities to proactively reduce or eliminate either the probability that the risk will become a problem or its impact if the problem occurs
  - *Risk contingency plans,* which define actions that will be taken only if any of the associated risk triggers indicate that the risk is turning into a problem or if the problem actually occurs
- The risk mitigation plans are implemented, and the planned mitigation actions are taken as part of the ongoing project or operational execution.
- Risk tracking involves monitoring the status of known risks, as well as the results of risk management actions and other project or operational activities.

- If a trigger indicates that the risk is turning into a problem or the problem occurs, the corresponding contingency plan activities are implemented.
- As new information is obtained during execution, additional risk analysis is done and the risk management plans are updated accordingly.
- Tracking may result in the addition of newly identified risks that are added to the risk list and fed through the risk management process.
- Tracking may also result in the closure of known risks as goals are accomplished, or those risks turn into problems and move into the corrective action process.

Risk management is a systematic, consistent process of identifying, analyzing, planning, taking action to mitigate, and tracking risks. However, the specific practices of risk management implementation (the who, what, when, and how) should be adapted and tailored to the needs of the individual organization and the individual projects or operations.

Risk management is a routine part of well-managed software projects or operations. Organizations must maintain a constant vigilance and routinely identify and manage risks through all of the activities of the life cycle. Therefore, the risk management process is an ongoing part of managing the software development, acquisition, and maintenance projects and operations. In fact, DeMarco and Lister (2003) call risk management "project management for adults." Small children run into the street after the ball, but they learn to look both ways first as they mature. Risk management is designed to allow organizations to "look both ways" and avoid major catastrophes.

Risk Identification

During the first step in the software risk management process, risks are identified and added to the list of known risks. The output of this step is a list of specific risks that have the potential to compromise success, including the success of the products once those products are released. The team should be as thorough as possible on the first round of risk identification but not obsessive. It is probably impossible to identify all risks on the initial pass through the risk management process. The team does not have enough information yet. There are still technical requirements to elicit, staffing issues to decide, design trade-offs to be determined, and all kinds of commitments to make. The risk identification step will need to be revisited repeatedly throughout the life cycle, as more information is obtained from execution, tracking, and control, and as plans are progressively elaborated.

There are many techniques that can be used to identify risks. Whatever the technique selected, risk identification requires a fear-free environment where risks can be identified and discussed openly. Risk identification techniques include:

- *Interviewing or brainstorming*: With personnel, customers, users, suppliers, and other key stakeholders. Stakeholders at different levels inside and outside the organization will have different perspectives on the potential problems. Therefore, various stakeholders should be involved in the risk identification process to obtain a broader, more complete perspective of the risks. Using open-ended questions can help identify potential areas of risk. Examples include:

- What problems do you see in the future for this project or operation?
- Are there areas of this project or operation that you feel are poorly defined?
- What interface issues still need to be defined?
- What requirements exist that the team isn't sure how to implement?
- What concerns does the team have about their ability to meet the required quality levels? Performance levels? Reliability levels? Security levels? Safety levels? Or other quality attributes?
- What tools or techniques might this project or operation require that it does not currently have?
- What new or improved technologies does this project or operation require? Does the team have the expertise to implement those technologies?
- What difficulties might exist in working with this customer? Subcontractor? Partner?

- *Voluntary reporting*: Any individual who identifies a risk is encouraged to bring that risk to management's attention. This reporting requires the complete elimination of the "shoot the messenger" syndrome. Avoiding the temptation to always assign risk management actions to the person who identified the risk reduces the perception that risk identification results in being punished with additional workload. Risks can also be identified through required reporting mechanisms, like status reports or review meetings.
- *Product decomposition*: This technique can help identify additional product-specific risks. During requirements and design activities, as business requirements are decomposed into stakeholder requirements and then into product requirements, and as architectural and component design and trade-off decisions are made, opportunities exist to identify product-specific risks. As Ould (1990) states, "The most important thing about planning is writing down what you don't know because what you don't know is what you must find out."
  - Every TBD ("to be done/determined") is a potential risk.
  - Feasibility or lack of stability in the requirements or design can signal areas of risk.
  - A requirement or design element may also be risky if it requires the use of new or innovative technologies, techniques, languages, or hardware.

- Even if the technologies, techniques, languages, or hardware have been around in the industry for a while, there still may be a risk if this is the first time this organization has attempted to utilize them. For example, if this is the first time this organization has used Java on a project, there may be risks because of a lack of expertise, which might affect the quality of the end product or impact the schedule because of a learning curve.

  Remember, a risk starts when a commitment is made. As the software requirements are being defined (or allocated from the system-level requirements), the project is committing to what is going to be delivered. As the software is designed, the project is committing to choices about how the software is going to be implemented. As the team makes these commitments, they need to keep asking themselves, "What risks are associated with the commitment to meet this requirement or implement this design element?" in order to help identify the associated risks.

- *Project decomposition*: Decomposition also comes in the form of creating the work breakdown structure (WBS) during project planning, which can help identify areas of uncertainty for specific subprojects, tasks, or activities that may need to be recorded as risks. Are there any estimation, feasibility, staffing, training, or resource issues associated with each identified activity in the WBS?

- *Critical path analysis*: When performing critical path analysis, the team should remain on the alert for any associated risks. Any possibility of schedule slippage on the critical path must be considered a risk because it directly impacts the ability to meet the schedule.

- *Assumption analysis*: Assumptions are not facts. If a process, product, or planning assumption proves to be false, what potential problems might occur? In other words, what are the risks associated with each assumption? If there are not any risks associated with an assumption, then it may not be a real assumption.

- *Risk taxonomies*: If asked what will go wrong on this project, any experienced software person on the project should be able to answer with uncanny accuracy. Why? Because they know what went wrong on the last project, the one before that, and the one before that. The problems from previous projects are some of the best indicators of the risks on new or current projects. This is why risk taxonomies are such a great tool. Risk taxonomies are lists of problems that have occurred previously. They can be used as checklists to help confirm that all potential risks have been considered. Risk taxonomies can also be used during the interview or brainstorming process to help develop interview questions or focus the brainstorming. Examples of risk taxonomies include:

  - Software Engineering Institute's *Taxonomy-Based Risk Identification Report* covers 13 major risk areas with about 200 questions (SEI 1993).

- Capers Jones's book, *Assessment and Control of Software Risks,* could be viewed as one large risk taxonomy (Jones 1994).
- Steve McConnell's book, *Rapid Development: Taming Wild Software Schedules,* includes an extensive list of what he labels as "potential schedule risks" (McConnell 1996).

When an organization is just starting their risk management efforts, they can begin with taxonomies from the literature. These industry taxonomies should then be evolved and tailored over time to match the problem types encountered by that organization. One word of caution: when using risk taxonomies, a delicate balance must be maintained between making sure that known issues are handled, and focusing so much on the items from the list that new and novel risks are missed.

Once a risk is identified, it should be communicated to everyone who potentially needs to know about it. This includes communication to:

- Management
- People who could be affected if the risk became a problem
- People who will analyze the risk
- People who will take action to mitigate the risk
- Customers, users, suppliers, or other stakeholders, as needed

According to Hall (1998), "communication of identified risks is best when it is both verbal and written." Verbal communication allows discussion of the risk, which can help clarify understanding of the risk. The listener has a chance to ask questions and interact with the person communicating the risk. This two-way interaction may result in additional information about the risk, its sources, its context, and its consequences. Written communications result in historical records that can be referred to in the future. Everyone who received the written communication has identical information about the risk. Written communication also allows for easy dissemination of the risk information if the people who need the information are in multiple locations. The creation of a risk database provides a consistent and easily accessed mechanism for providing written risk information.

A written *risk statement* consists of the risk condition and its potential consequences for the project, process, or product. The *risk condition* is a brief statement of the potential problem that "describes the key circumstances, situation, and so on, causing concern, doubt, anxiety, or uncertainty" (Dorofee et al. 1996). The *risk consequence* is a brief statement that describes the immediate loss or negative outcome if the risk turns into an actual problem.

Figure 18.4 includes two example risk statements. In Example 1, a software quality engineer (SQE) reviewed the subcontractor portion of the project plan and performed an assumption analysis. The SQE noticed that the project assumed that Acme would deliver software of the required reliability and that no provisions had been made in the schedule to deal with defects found in the XYZ controller. For Example 1, the risk statement (highlighted in gray) would be: *Acme may not deliver the subcontracted XYZ controller component with the required software reliability, and, as a result, we will spend additional effort with Acme on defect resolution and regression testing.*

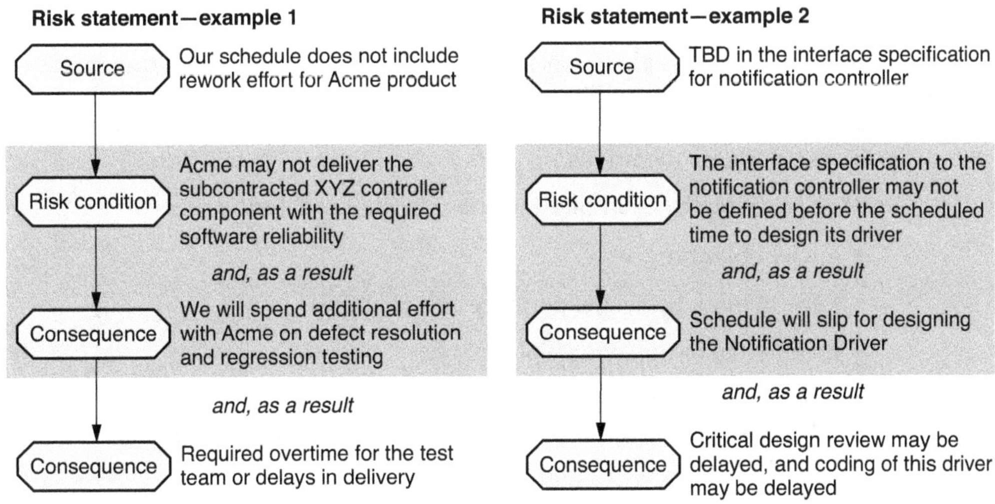

Figure 18.4 Risk statement—examples.

In Example 2, during the requirements inspection process, the inspection team noticed a TBD in the interface specification requirements for the notification controller. This controller is an external piece of equipment that the system must interface with to send delinquent notices as part of the billing system. For Example 2, the risk statement would be: *The interface specification to the notification controller may not be defined before the scheduled time to design its driver, and, as a result, the schedule will slip for designing the notification driver.* Note that in these examples, the risk statements consist only of the "if/then" components highlighted in gray in the figure. The source of the risk is not part of the risk statement and neither are any subsequent consequences to the initial consequences.

Types of Software Risks

Software development and maintenance may encounter various types of risks. Note that this list of risks by type could be used as a high-level taxonomy. Types of risks include:

- *Technical risks*: Examples include potential problems with:
  - Incomplete or ambiguous requirements or design
  - Excessive constraints
  - Large size or complexity
  - New languages, tools, or platforms
  - New or changing methods, standards, or processes
  - Dependencies on organizations outside the direct control of the team

Technical risks also include risks related to the use of the product in operations, including potential problems related to reliability, availability, functionality, safety, security, maintainability and so on. Project risk include products risks because the project is not successful if the product does not meet its requirements or its intended use in operations.

- *Management risks*: Examples include potential problems with:
  - Lack of planning
  - Lack of management experience and training
  - Incomplete or inadequate planning
  - Inadequate estimates resulting in overly compressed schedules, overly optimistic budgets, or insufficient staffing and resources
  - Communications problems
  - Organizational issues
  - Issues with customer or supplier relations
  - Lack of authority
  - Inadequate tracking or control problems
- *Financial risks*: Examples include potential problems with:
  - Cash flow
  - Capital and budgetary issues
  - Return-on-investment constraints
  - National or global economy downturns
- *Contractual and legal risks*: Examples include potential problems with:
  - Changing contractual requirements
  - Penalty clauses in the contract
  - Market-driven schedules
  - Government regulation
  - Product warranty issues
- *Personnel risks*: Examples include potential problems with:
  - Excessive staffing lags and turnover
  - Staff experience and training problems
  - Ethical and moral issues
  - Staff conflicts
  - Productivity or velocity issues

- *Resource risks*: Examples include potential problems with:
  - Unavailability or late delivery of equipment and supplies
  - Inadequate tools
  - Inadequate facilities, distributed locations
  - Unavailability of computer resources
  - Network capacity or availability issues

Risk Analysis

The primary goal of the *risk analysis* step of the risk management process is to analyze the identified list of risks and prioritize those risks for further planning and action. During the risk analysis step, each risk is assessed to determine its context, estimated probability, estimated loss or impact, and timeframe.

A *risk's context* provides the additional information that surrounds and affects that risk and helps determine its probability and potential loss. The "risk context contains the what, when, where, how, and why of the risk issue" (Hall 1998). Keywords, like the ones in Figure 18.5, can help the team explore the risk's context. Documenting the risk's context can be especially useful after time has passed and a risk is being reevaluated. Discrepancies between the original documented context and the current situation can help the team better understand how or if the risk has changed.

An estimated *risk probability* is the likelihood that the risk will turn into a problem. The risk probability can be assigned:

- *Quantitatively*: As a percent probability that the risk will turn into a problem
- *Qualitatively*:
  - As an uncertainty statement, for example, almost certain, very likely, likely, unlikely, highly unlikely

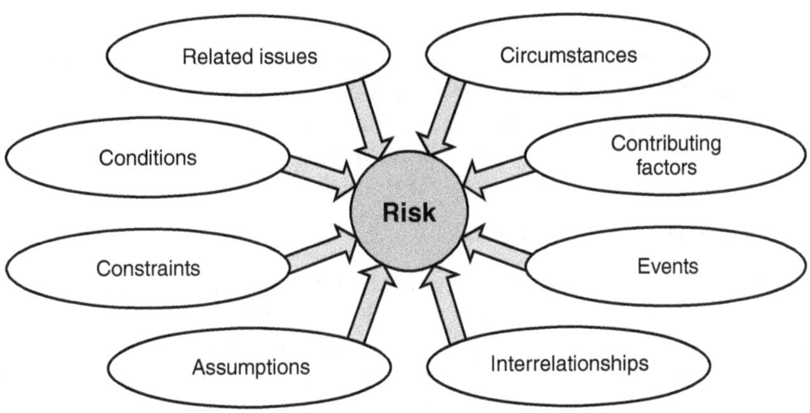

Figure 18.5 Key words for exploring a risk's context.

- As a likelihood index, for example, on a scale of 1 to 5, where 5 is almost certain

The *risk loss* is the estimated impact or consequences if the risk does turn into a problem. Losses can come in the form of:

- Additional costs (dollars or effort)
- Required changes to the schedule

Technical effects on the product being produced (e.g., its functionality, performance, or quality). Losses can also result from other types of impacts. For example, the organization could lose corporate goodwill, market share, and customer or employee satisfaction. The risk loss can be assigned:

- *Quantitatively*: As an assigned cost of the problem, if it occurs, in the amount of cash or number of effort hours
- *Qualitatively*:
  - As a loss statement, for example, catastrophic, critical, marginal, negligible
  - As a loss index, for example, on a scale of 1 to 5 where 5 is catastrophic

Not only should each risk's loss be assessed individually, but the interrelationships between risks must also be assessed to determine if compounding risk conditions exist that could magnify losses.

The level of formal risk assessment needed can range from the simple qualitative assignment of each risk to a priority (e.g., high, medium, or low) to the use of complex quantitative mathematical modeling (e.g., Monte Carlo modeling). The manager and team should use the simplest method available that allows them to make reliable risk-planning decisions. Different risks may be assessed at different levels of formality. For example, a high-impact risk with a high probability may require very formal and detailed analysis to determine the appropriate handling plans. Alternatively, knowing that a risk is unlikely to turn into a problem and that it will have very little impact if it does may be all the team needs to know about a risk to decide to accept that risk.

Risk exposure measures the impact of a risk in terms of its probabilistic expected value. Boehm (1989) defines a risk exposure equation to help quantitatively establish risk priorities, where risk exposure (RE) is defined as the probability of an undesired outcome (UO) actually occurring, multiplied by the expected loss (cost of the impact or consequences) if that UO occurs.

$$RE = \text{Probability (UO)} \times \text{Loss (UO)}$$

For example, if a risk is estimated to have a 10 percent chance of turning into a problem with an estimated impact of \$100,000, then the risk exposure for that risk is 10% × \$100,000 = \$10,000. If the risk probability and loss are analyzed qualitatively rather than quantitatively, a table like the one in Table 18.2 can be used to assign a risk exposure to each risk. The sum of all the risk exposures for a project or operation should be less than, or at least equal to, the expected value of the return on investment (opportunity), or the risk is not worth taking.

Table 18.2 Qualitative risk exposure—example.

		Likelihood			
		Probable	**Occasional**	**Remote**	**Improbable**
Impact	**Catastrophic**	High	High	High-medium	Medium
	Critical	High	High-medium	Medium	Medium-low
	Marginal	High-medium	Medium	Medium-low	Low
	Negligible	Medium	Medium-low	Low	Low

Table 18.3 Risk exposure prioritization matrix—example.

	Criteria and weights				
	Technical Exposure (.25)	**Cost Exposure (.15)**	**Schedule Exposure (.20)**	**Stakeholder Satisfaction (.40)**	**Total**
Risk 1	1	3	2	4	2.7
Risk 2	4	1	4	3	3.15
Risk 3	2	2	2	1	1.6
Risk 4	2	4	3	2	2.5

The analysis step in the risk management process includes prioritizing the list of risks. Comparing the risk exposure measurement for various risks can help identify those risks with the greatest probable negative impact to the project or operations and thus help establish which risks are candidates for further action. Risks can be prioritized using just their risk exposures or a combination of their risk exposures and timeframes. A prioritization matrix such as the one in Table 18.3 can be used when risks need to be prioritized on multiple criteria (various risk exposures). In this example, a risk exposure score of 1 to 5 (5 being the highest) is used, as illustrated in Table 18.4.

A *risk's timeframes* are significant times when the risk needs to be addressed or when the risk may turn into a problem. A risk associated with timeframes in the near future may have higher priority than other risks associated with later timeframes, even if it has a lower risk exposure.

This prioritized list of risks should be reviewed periodically. Based on changing conditions, additional information, the identification of new risk items, or simply timing, the prioritized list of risks may require periodic re-analysis and updates.

Risk Management Planning

During the planning step of the software risk management process, the appropriate risk-handling techniques are selected, and alternative risk-handling actions are evaluated. Since resource limitations rarely allow the consideration of all risks, the prioritized list of risks is used to identify the top risks for risk

Table 18.4 Risk exposure scoring for prioritization matrix—example.

Exposure Score	Technical	Schedule	Cost	Stakeholder Satisfaction
5	Unusable system	>18 months slip	>10% project budget	Will replace purchased product with competitor's product
4	Unusable function or subsystem	12 to 18 months	7% to 10% project budget	Unwilling to purchase
3	Major impact to functionality, performance, or quality	6 to 12 months slip	5% to 7% project budget	Willing to purchase for limited use
2	Minor impact to functionality, performance, or quality	3 to 6 months slip	1% to 5% project budget	Willing to purchase and use but will result in complaints
1	Minimal or no impact	<3 months slip	<1% project budget	Willing to purchase and use but may not recommend

mitigation planning and action. Some risks may require contingency plans, which will only be implemented if the risk turns into a problem. Other risks may simply have tracking mechanisms put in place to monitor them closely. At the lowest priorities, other risks are simply accepted and documented for possible future consideration. Whatever handling options are selected, any associated actions should be planned to proactively manage the risks rather than waiting for problems and reacting in a firefighting mode. The resulting risk management plans should then be incorporated into the project or operation plans, with assigned budget, schedule, staff, and resources.

Taking the prioritized risk list as input, plans are developed for the handling actions chosen for each risk. As illustrated in Figure 18.6, specific questions can be asked to help focus the type of planning required:

- *Do we know enough?* If the answer is no, plans can be made to obtain additional information through mechanisms such as prototyping, modeling, simulation, or conducting additional research. Once the additional information has been obtained, the planning step should be revisited. Based on the results of these activities and the information obtained, it may also be appropriate to repeat the analysis step for the risk, resulting in changes to its priority. Table 18.5 illustrates examples of obtaining additional information actions for the example risk statements from Figure 18.4.

- *Is it too big of a risk?* If the risk is too big to even consider trying to handle, the risk can be avoided by changing the project or operation strategies and tactics, to either choose a less risky alternative, or to decide not to do the project or operation at all (or not include the risky

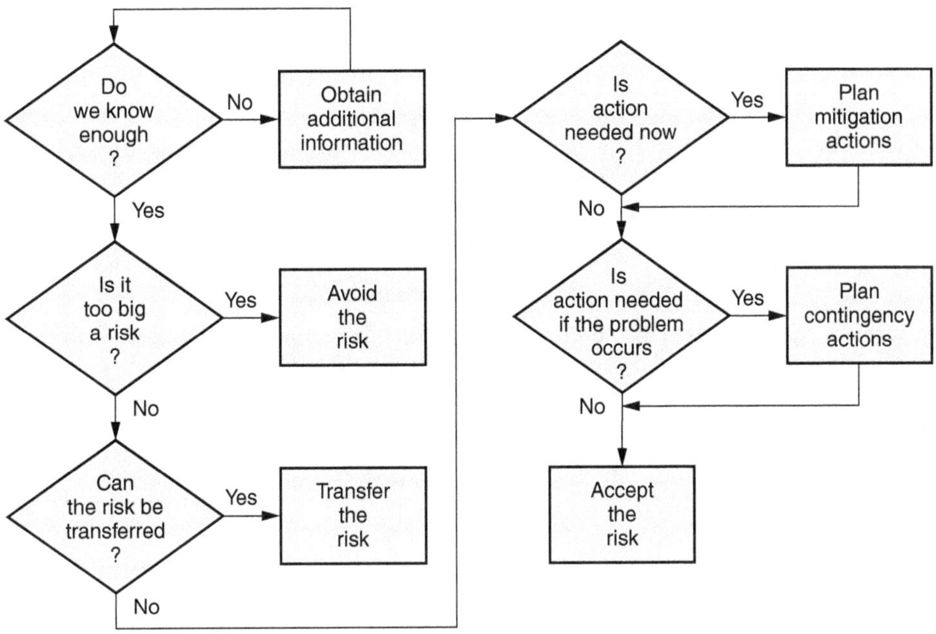

Figure 18.6 Risk-handling options.

requirements in the product). Things to remember about avoiding risks include:

- Avoiding risks may also mean avoiding opportunities
- Not all risks can or should be avoided
- Avoiding a risk in one part of the project or operation may create larger risks in other parts of the project or operation

Once the risk has been successfully avoided, it can be closed. Table 18.6 illustrates examples of *risk-avoidance* actions for the example risk statements from Figure 18.4.

- *Can the risk be transferred?* If it is not the team's risk, or if it is economically feasible to pay someone else to assume all or part of the risk, a plan can be developed to transfer the risk to another organization (e.g., buying insurance). In some cases, a risk can be closed once the risk has been successfully transferred, because it is no longer a risk. In other cases, the team may want to set up a monitoring mechanism to make sure that the people who assumed the risk are appropriately handling it. Transferring the risk to another party may also create other risks that need to be identified, analyzed, and managed. It should be remembered that transferring the risk does not eliminate it. Transferring the risk simply shifts the responsibility for the risk, and potentially changes the risk's exposure to the project. Table 18.7 illustrates examples of risk-transfer actions for the example risk statements from Figure 18.4.

Table 18.5 Obtaining additional information action—examples.

Risk	Obtain Additional Information
Acme may not deliver the subcontracted XYZ controller component with the required software reliability level, and as a result the project will spend additional effort working on defect resolution issues with Acme and doing regression testing.	• Perform a capability assessment or audit of Acme to determine if Acme even has the capability or capacity to produce software at the required quality level. • Ask for references from past customers and check on the reliability of previous Acme products.
The interface specification to the notification controller may not be defined before the scheduled time to design its driver, and as a result the schedule will slip for designing the notification driver.	• Establish communications link with device provider to obtain early design specification for the device. • Research interface specs for prior releases of the same controller or for other control devices by same provider.

Table 18.6 Risk avoidance action—examples.

Risk	Risk Avoidance Actions
Acme may not deliver the subcontracted XYZ controller component with the required software reliability level, and as a result the project will spend additional effort working on defect resolution issues with Acme and doing regression testing.	• Develop all software in-house. • Switch to a subcontractor with a proven reliability track record even though they are more expensive.
The interface specification to the notification controller may not be defined before the scheduled time to design its driver, and as a result the schedule will slip for designing the notification driver.	• Negotiate with the customer to move the implementation of this control device into a future software release. • Replace the selected control device with an older device that has a well-defined interface.

Table 18.7 Risk transfer action—examples.

Risk	Risk Transfer Actions
Acme may not deliver the subcontracted XYZ controller component with the required software reliability level, and as a result the project will spend additional effort working on defect resolution issues with Acme and doing regression testing.	• Transfer the risk to the subcontractor by building penalties into the contract for delivered software that does not have the required reliability to compensate for the potential loss.
The interface specification to the notification controller may not be defined before the scheduled time to design its driver, and as a result the schedule will slip for designing the notification driver.	• Transfer the risk to the customer by building late-delivery alternatives into the contract if the customer does not supply the specification by its due date.

Table 18.8 Risk mitigation plans—examples.

Risk	Risk Mitigation Action
Acme may not deliver the subcontracted XYZ controller component with the required software reliability level, and as a result the project will spend additional effort working on defect resolution issues with Acme and doing regression testing.	• Assign a project engineer to participate in the requirements and design inspection and to conduct alpha testing at Acme's site. • Require defect data reports from Acme on a weekly basis during integration and system test, and analyze the data for potential indications of poor quality.
The interface specification to the notification controller may not be defined before the scheduled time to design its driver, and as a result the schedule will slip for designing the notification driver.	• Assign a senior software engineer who has experience with similar control devices to the design task. • Move the design task later in the schedule and increase its effort estimate.

- *Is action needed now?* Sometimes the risk cannot be avoided or transferred, but it is still too big of a risk to merely accept. If the project team decides to proactively attack the risk directly, they typically start by creating a list of possible *risk mitigation actions*, which can be taken to reduce the risk. Two approaches to risk mitigation actions should be considered:
  - Actions that reduce the likelihood that the risk will turn into a problem
  - Actions that reduce the impact of the problem should it occur

    Table 18.8 illustrates examples of risk mitigation actions for the example risk statements from Figure 18.4.
- From the list of possible risk mitigation actions, the team selects those actions that are going to be implemented. Mitigation plans are considered effective if the risk exposure and total expected cost of the risk have been reduced to a level where the project can live with the possible impact if the risk turns into a problem.

  When considering which risk reduction activities to select, a cost/benefit analysis must be performed. Boehm (1989) defines the *risk reduction leverage (RRL)* equation to help quantitatively establish the benefit-to-cost ratio of implementing a risk reduction action. RRL measures the probabilistic benefit and cost of the alternative risk mitigation techniques, based on expected values. RRL is defined as the difference between the *risk exposure (RE)* before and after the mitigation activity divided by the cost of that activity.

  $$RRL = (RE_{Before} - RE_{After})/\text{risk reduction cost}$$

  If the RRL is less than one, it means that the cost of the risk mitigation activity outweighs the probable gain from implementing the action. Table 18.9 illustrates an example of RRL for three alternative risk reduction actions. In this example, the original risk exposure after the risk analysis was determined to be 25% × \$300K = \$75K. The project team determined that they could not accept risks over \$50K, so this

Table 18.9 Risk reduction leverage—example.

Risk #	$Probability_{Before}$	$Loss_{Before}$	RE_{Before}
143	25%	$300K	$75K

Alternative	$Probability_{After}$	$Loss_{After}$	RE_{After}	Cost	RRL
1	15%	$300K	$45K	$32K	0.94
2	25%	$180K	$45K	$15K	2.0
3	17%	$300K	$51K	$8K	3.0

risk needed a mitigation plan. Team members came up with three alternatives. The team estimates that:

- *Alternative 1*: Will reduce the risk's probability to 10 percent, so its new risk exposure after the mitigation action is 15% × $300K = $45K. The expected value of the benefit is then $75K – $45K = $30K. Since alternative 1 is estimated to cost $35K, the RRL = $30K/$32K ≈ 0.94. This alternative is then rejected because it will cost more to implement the risk mitigation than its expected benefit.
- *Alternative 2*: Will reduce the risk's cost to $180K, so its new risk exposure after the mitigation action is 25% × $180K = $45K. Therefore, alternative 2 has an RRL of ($75K – $45K)/$15K = 2. Since the RRL is greater than one, this alternative has a positive benefit to cost ratio. However, the project also has to examine the total expected cost of the risk. The new risk exposure is $45K after this alternative is implemented. The cost of implementing this alternative is $15K. Therefore the total expected cost of the risk is $45K + $15K = $60K. Since the project team determined that they could not accept risks over $50K, this alternative is rejected.
- Alternative 3: Has an RRL of ($75K – (17% × $300K)) = ($75K – $51K)/ $8K = 3. At first glance, alternative 3 has a fantastic expected benefit-to-cost ratio. However, this alternative only reduces the risk to $51K and makes the overall expected cost of the risk $51K + $8K = $59K, so this alternative is also rejected.
- Alternatives 2 and 3: Since both alternatives 2 and 3 have RRLs greater than one, the project could consider implementing both of these alternatives. Assuming that would result in a $Probability_{After}$ of 17% and a $Lose_{After}$ of $180K, then the REafter would equal 17% × $180K = $30.6K. The RRL would equal ($75K – $30.6K)/23K ≈ 1.9. The total expected cost of the risk equals ($30.6K + 15K + $8K) = $53.6K, so combining these two alternatives is also rejected.

If these are the only three alternatives, then the project team would need to go back and rethink their options. They need to either:

- Avoid this risk by not making the associated commitment by making a less risky alternative commitment.

    - Reconsider what is an acceptable risk level, since the $50K was an estimate. Could the project accept a $53.6K risk and implement alternatives 2 and 3 combined?
    - Negotiate better terms for the project in order to increase the opportunity (positive benefit of the project), allowing a risk of $53.6K to be acceptable, and implement alternatives 2 and 3 combined
    - Cancel the project.

- *Is action needed if the problem occurs?* If risk mitigation actions are not taken or those actions reduce but do not eliminate the risk, it may be appropriate to develop risk contingency plans. *Contingency plans* are plans that are implemented only if the risk turns into a problem. A *risk trigger* is a time or event in the future that acts as an early warning that the risk is turning into a problem. One or more risk triggers should be established for each risk with a contingency plan. For example, if there is a risk that outsourced software will not be delivered on schedule, the trigger could be whether the critical design review was held on schedule. A trigger can also be a relative variance or threshold metric. For example, if the risk is the availability of key personnel for the coding phase, the trigger could be a relative variance of more than 10 percent between actual and planned staffing levels. There are trade-offs in utilizing triggers in risk management. The trigger needs to be set as early as possible in order to make sure that there is plenty of time to implement risk contingency actions. It also needs to be set as late as possible, because the longer the project waits, the more information is available to make a correct decision and not implement unnecessary contingency actions. The objective is to set the triggers at the "last reasonable moment" to take actions to minimize the loss caused by the problem. Table 18.10 illustrates examples of risk contingency plans for the example risk statements from Figure 18.4.

A team is never able to remove all risk—software is a risky business. A team will, therefore, typically choose to accept many of its identified risks. The key difference is that a conscious choice has been made from a position of information and analysis rather than an unconscious choice. Even if the team accepts a risk, one or more risk triggers may be put in place to warn if the risk is turning into a problem. Risks that are assigned these triggers can then be set at a "monitor only" priority until the trigger occurs. At that time, the risk analysis step can be repeated to determine if risk reduction action is needed.

Risk management plans should be integrated back into the overall project or operation plans as appropriate. This integration should include schedules, budgets, staffing, and resource allocation, and other planning for the risk actions that will be taken. It should also include buffers to reserve time, budget, staff, and resources to handle those risks that do turn into problems. These buffers are typically established based on the sum of the risk exposures, after any risk mitigation actions are implemented, for all of the risks. Some risk-handling actions might be small enough that they can be implemented with action item lists and do not need to be incorporated into the overall project or operation plans, as activities in the WBS.

Table 18.10 Risk contingency plan—examples.

Risk	Risk Contingency Plans
Acme may not deliver the subcontracted XYZ controller component with the required software reliability level, and as a result the project will spend additional effort working on defect resolution issues withAcme and doing regression testing	• Risk assumption: this is the best subcontractor for the job, and the team will trust them to deliver reliable software • Early trigger (reassessment of risk indicated): completion of critical design review (CDR) later than June 1 • Contingency plan trigger: more than two critical and 25 major defects detected during third pass of system test • Contingency plan: assign an engineer to liaison with the subcontractor on defect resolution, and implement the regression test plan for maintenance releases from the subcontractor
The interface specification to the notification controller may not be defined before the scheduled time to design its driver, and as a result the schedule will slip for designing the notification driver	• Risk assumption: this new technology will greatly improve the usability of the system • Early trigger (reassessment of risk indicated): interface definition not received by start of preliminary design review (PDR) • Contingency plan trigger: interface definition not received by start of critical design review (CDR) • Contingency plan: hold CDR with a "to be done" and hold a second CDR for just the subsystem that uses the device

Taking Action

During the *taking action* step of risk management, the assigned individuals implement the risk management plans. Additional information is obtained, actions are taken to avoid or transfer the risks, and planned risk mitigation actions are executed. If risk triggers are activated, analysis is performed, and contingency actions are implemented as appropriate. Note that with some luck and good risk mitigation plans, many of the contingency plans may never be implemented. Contingency plans are only implemented if the associated risk turns into a problem.

Risk Tracking

The team must track the results and impacts of the risk mitigation plan implementation and any implemented contingency plans. The tracking step involves gathering data, compiling that data into information, and then reporting and analyzing that information. This step includes measuring identified risks and monitoring triggers, as well as measuring the impacts of risk-handling activities and tracking the status of the project or operation as a whole. The results of the tracking can be:

- New information that indicates that additional risk planning or analysis is needed
- Identification of new risks that need to be added to the risk list
- Validation of goal accomplishments associated with known risks so that those risks can be closed on the risk list because they are no longer a threat

- Monitoring of triggers and other risk status information that indicates the need to implement contingency plans, or the need to close risks on the risk list because they have turned into actual problems that are moved into the corrective action process

There are two primary mechanisms for tracking risks. The first is review of the items on the risk list and risk/project/operation status by staff and management. The second is through the use of metrics.

The reviews typically used to manage software projects/operations can also be used to track risks. For example, risk tracking activities should be included in project team, senior management, and phase gate review meetings discussed in Chapter 17. They should also be included in operations team and management review meetings.

At the beginning of a process, the entry criteria should be evaluated to determine if the process is truly ready to start. As part of that review, risks associated with the process and its tasks and products can also be evaluated. At the end of a process, the exit criteria should be evaluated to determine if the process is truly complete. This provides another opportunity to review the risks associated with that process and its tasks and any risks associated with the outputs (products) of that process.

Many of the software metrics typically used to manage software projects, discussed in Chapters 17, can also be used to track risks. For example, Gantt charts, earned value measures, and budget, staffing, and resource metrics can help identify and track risks involving variances between plans and actual performance or thresholds established based on project estimates. Product and process metrics discussed in Chapter 19, including requirements churn, defect arrival rates, defect backlogs, phase containment, and defect detection efficiency can also be used to track other risks, including rework risks, risks to the quality of the delivered product, and even schedule risks.

2. SOFTWARE SECURITY RISKS

> Evaluate risks specific to software security, including deliberate attacks (e.g., hacking, and sabotage), inherent defects that allow unauthorized access to data, and other security breaches. Plan appropriate responses to minimize their impact. Use threat analysis and modeling tools to identify and prevent potential security risks. (Evaluate)
>
> **Body of Knowledge IV.C.2**

Software increasingly controls or enables not just the digital infrastructure but also the physical one. Everything from transportation systems (airplanes, trains, and subways) to financial institutions, hospitals, delivery systems (postal, shipping, and pipeline), telecommunications, and e-commerce are now heavily controlled or run by software systems. Because of this increased dependence on software, the security of that software is an increasing concern. The news is full of reports of stolen credit card information, identity thefts, virus attacks, ransomware, and other software security–related issues. "As our digital infrastructure gets increasingly complex and interconnected, the difficulty of achieving application security increases exponentially" (OWASP 2013). Therefore, organizations must understand the security-related risks associated with their software and construct sufficiently robust control systems to mitigate those risks in a cost- and time-effective manner. "*Security risk management* consists of an integrated set of security-focused activities that enable cost-effective risk treatment decisions resulting in a risk level that is within the specified stakeholder risk tolerance" (NIST 2014).

Security Risks

According to McCabe, "A threat analysis is a process used to determine which components of the system need to be protected and the types of security risks (threats) they should be protected from" (McCabe 2007).

When evaluating security risks, the organization must first determine which of its assets need security protection. Who would attempt to breach the security of their systems, and what assets are they trying to attack?

Security attackers may use many different attacks following many different paths to attempt to access an organization's software and its associated data and functionality, as illustrated in Figure 18.7. Each of these paths represents a risk of a potential security breach. The potential attackers and potential paths must be identified as risks. Analysis must be performed to evaluate the risk exposure of each potential security breach, including the:

- *Probability that a security breach will occur along any given path* given the probability that:
  - A particular type of attacker will attempt a security attack on the software
  - The attacker will use a specific type of security attack
  - The attack will be along a given path with weaknesses in both the access controls and software security controls
- *Loss resulting from that security breach* including:
  - *Technical impacts*: For example, degrading of the software's safety, reliability, or availability, disclosing of sensitive data to unauthorized individuals, corrupting the software or its data, accessing functionality by unauthorized individuals
  - *Impacts to the business*: For example, damage to the organization's profits, reputation, competitive advantage, customer satisfaction, regulatory compliance, and even viability

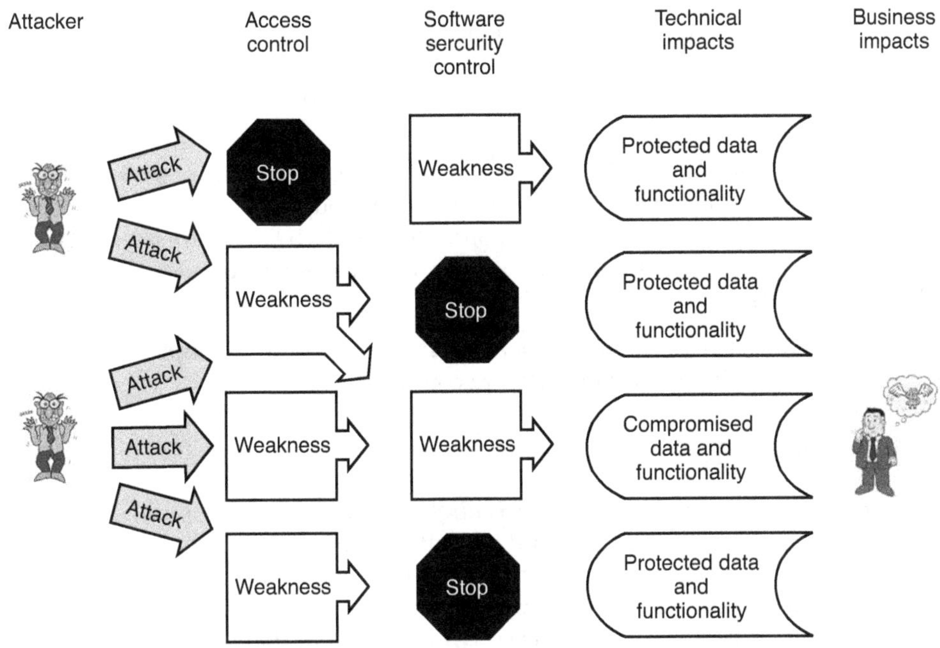

Figure 18.7 Security attackers, attacks and paths—example.

Security threats can come from outside the organization, including:

- *Hostile data injection*: Interjecting data into commands or queries in order to trick the software into executing that data, (e.g., Software Query Language [SQL], operating system [OS], or Lightweight Dictionary Access Protocol [LDAP]), in order to access data without authorization or execute unintended commands
- *Hackers*: Including worms, viruses, tunneling, spyware, or denial-of-service attacks
- *Corporate or international espionage*: Leakage of intellectual property, trade secrets, or secret or confidential information or functionality
- *Stolen identification or spoofing*: Allowing unauthorized individuals to assume the identity of authorized users
- *Cybercrime*: Including theft, fraud, and abuse
- *Cyber terrorism*: Including ransomware or taking over or sabotaging critical software infrastructure
- *Other people*: Trying to take advantage of security vulnerabilities like security holes in the software

Security threats can also come from the inside the organization, including:

- *Acts of employee sabotage and malicious code*: Including time bombs, logic bombs, or Trojan horses

- *Back doors*: Created by software practitioners to make it easy to circumvent access controls within the software during development, but which are not removed prior to release, either accidentally or intentionally
- *Security holes*: Including an unintended function, mode, or state in the software caused by design flaws or coding bugs (one or more defects) that an attacker can exploit in order to breach the software's security; the OWASP Top 10 (2013) defines and discusses prevention strategies for different types of security holes
- *Using components with known vulnerabilities*: Including libraries, widgets, frameworks, and other software components

Security Risk Analysis

Access control and *software security control* weaknesses can be found anywhere in the software. However, code that deals with the outside world and code that has anything to do with access and internal versus external privileges are naturally more vulnerable. When performing security threat analysis, particular attention needs to be paid to any part of the software or system that:

- Operates with special privileges
- Performs authentication, including verification of the digital identity of the sender (person, computer, or program) of a communication
- Provides easy access to user data or facilities (such as a network server or relational database management system)
- Includes certain language constructs in programming languages that have been identified as vulnerabilities because multiple software faults have had their root causes traced to the use of those constructs
- Stores, transmits, or backs up any confidential data or information
- Provides entry points into the system, because these are the locations where outside input is parsed, processed, or interpreted (e.g., Web forms, network sockets, and media players, as the attacker can craft bad inputs, malformed network packets, or malformed digital media in an attempt to breach security)

The first step in security risk management is to identify the assets that must be protected (data, information, functionality, or other resources of the system). Those assets must then be analyzed to determine their criticality, importance, priority, and value to determine their level of security sensitivity.

Threat/vulnerability analysis is then conducted to determine the degree and nature of any threats to those assets and the vulnerabilities to those threats being successful (see Figure 18.7) based on consideration of the following:

- *Threat*: Probability that a potential attacker would attempt to access, corrupt, use, or disrupt that asset

- *Threat*: Probability that the attacker would use a specific path of attack against that asset
- *Vulnerability*: Probability that the attacker, using that path of attack, would be successful in accessing, corrupting, using or disrupting that asset given the current/planned defenses, including vulnerabilities in physical access controls (facilities security like fences, locks, and guards), operational or business access controls (policies and procedures), and software security controls
- *Impacts*: The loss if that attacker is successful, including both technical and business impacts
- *Security risk exposure*: The level of security risk based on these probabilities and impacts

Risk exposure can then be used to prioritize security risks and select those risks that are large enough to require mitigation or contingency actions.

Security Risk Mitigation and Contingency Planning

As a result of security risk identification and analysis, "The stakeholders then can execute a balanced risk management strategy to reduce risk to an acceptable level, manage the *residual risk*, and achieve mission/business capability and trustworthiness objectives across all their risk concerns" (NIST 2014). For an organization to provide effective software security, that organization must address the following (based on Deutscher et al. 2014):

- *Confidentiality*: The data, information, functionality, or other resources of the system are accessible only to those who have either a right or a need to view it
- *Integrity*: The data, information, and other resources are accurate, valid, and reliable, which includes rules for how they can be manipulated
- *Availability*: Data, information, functionality, and other resources are available for access by authorized individuals when needed
- *Accountability*: Each transaction or action can be attributed to an accountable individual
- *Provenance*: The origin and history of each piece of data or information are known and well-defined

As with any other risks, specific risk mitigation actions are planned and implemented as part of the project's activities, as appropriate. Contingency plans may also be put in place to recover from the security breach if it occurs. In order to mitigate security risks, the organization must establish and use security policies and standardized processes as part of the organization's quality management system (QMS). This means that the organization must establish standardized, reusable security controls. Software security needs to be addressed as part of every activity in the software life cycle, including:

- *Project management*: Project plans, including schedules, budgets, staffing, and resource allocations, must include security risk management activities, as well as security risk mitigation and contingency plans.
- *Training*: Organizations should conduct a security knowledge gap analysis, provide security training to fill any identified gaps, and help educate developers, testers, and other software practitioners on software security.
- *Requirements*: The requirements must define what security means through the specification of security requirements for the software product, as discussed in Chapter 12. This includes defining requirements for data privacy, as discussed in Chapter 2 and data security, as discussed in Chapter 6. The project should consider using *misuse and abuse cases*, which are similar to the use cases discussed in Chapter 12, but with a focus on security attacks from unfriendly stakeholders, and how the software should react to repel or handle those attacks. Misuse cases for the business requirement example of *Allowing the gas station customer to pay for gas at the pump* might include the following examples:
    - Keeping credit card thieves from buying gas with stolen credit cards
    - Keeping forgers from buying gas with counterfeit credit cards
    - Keeping customers from buying gas with expired or over-the-limit credit cards
    - Keeping unauthorized individuals from changing the gas price
    - Keeping unauthorized individuals from accessing credit card information
- *Design*: The most cost-effective way to implement security is to design security into the software from the start, as discussed in Chapter 14, based on the defined security requirements. During architecture design, "The system is decomposed into security-specific architectural views and interpretations to better understand and manage the complexity associated with the engineering and assuring of the protection capability" (NIST 2014). During component design, the focus shifts to building security into the individual, security-relevant components.
- *Code*: Software should be written to include strong, usable security controls and avoid using components with known security vulnerabilities (libraries, widgets, language constructs).
- *Verification and validation*: Many security breaches are the result of the exploitation of security holes caused by defects in the design or code of the software. Therefore, in software where security is important, it is essential to incorporate security verification and validation activities

into every stage of the software development life cycle. Security threat analysis and security hole detection should be performed as part of each level of peer review and testing. For example, the software practitioners should:

- Use static code security analyzers to help identify security holes
- Include security as one of the areas of focus in requirements, design, code and test case peer reviews
- Develop appropriate security-focused and penetration test cases, based on security threat analysis for each level of testing

3. SAFETY AND HAZARD ANALYSIS

> Evaluate safety risks and hazards related to software development and implementation and determine appropriate steps to minimize their impact. (Evaluate)
>
> **Body of Knowledge IV.C.3**

Safety-critical software is software that can contribute to an accident, is intended to prevent an accident, or is intended to mitigate or recover from the results of an accident. (See Chapter 12 for a more detailed discussion of safety-critical software and accidents.) However, software by itself does not cause harm because, after all, it is just a bunch of electronic ones and zeros. Then why is there so much concern about "software safety" in the software industry? The concern results from the fact that software-enabled systems increasingly control not just our digital infrastructure but our physical one as well. Software can lead to an accident (harm) if it (Frailey 2016):

- Controls the behaviors of potentially dangerous devices, for example:
  - Transportation (airplanes, railroads, and automobiles)
  - Military (automated weaponry)
  - Energy and chemical (nuclear, electricity, oil, and gas)
  - Biomedical (dispensing of drugs, surgical instrumentation, computer-aided surgeries, implanted devices, and medical records)
  - Agriculture (food), water, and public health
  - Telecommunications (access to emergency response services)
  - Construction (dams, buildings, and energy management and building automation)
- Sends information to people who interpret or use that information in potentially dangerous ways, for example:

- Misinterpreting of medical records (allergies, dosages, or type of medication)
- Misidentifying of intruders or enemy combatants
- Misinterpreting locations of airplanes for air traffic control
- Responding incorrectly to alarms, warnings, or instructions
- Believing deceptions propagated through Internet scams or social media

Key concepts in software safety include the facts that software safety:

- *Is a risk-based activity*: It may not be cost-effective or even possible to eliminate all software-related safety hazard issues.
- *Must be considered from a system perspective*: It can only be analyzed within the context of the system it is part of, including the people who interact with that system (Frailey 2016).

According to Ericson, "system safety is the formal engineering discipline and process for identifying and controlling hazards, and the risk associated with these hazards. As systems become more complex and more hazardous, more effort is required to understand and manage system mishap risk" (Ericson 2016).

Some systems are inherently hazardous. For example, a missile guidance system is intended to direct missiles to destroy property or cause bodily harm. Fire sprinkler systems are intended to douse everything with water to put out a fire, which will always result in some property damage. For these systems, safety and hazard analysis focuses on the inadvertent or unintended occurrence of these hazards (e.g., a missile being fired inadvertently or without proper authorization or sprinkler systems going off when there is no fire).

Hazards, Accidents, and Safety Risks

A *hazard* is a potential condition existing within a system which, when actuated by a series of incidents or events, becomes an actual accident. Accidents are not just random occurrences. The right conditions have to be in place to make the accident possible. These conditions are called hazards. Hazards include:

- Hardware
- Energy
- Chemicals
- Materials

A hazard must then be followed by a series of one or more incidents or events (initiating mechanisms) to actually cause the hazardous condition to result in harm (the undesired condition). These initiating events can come from the:

- Software
- Hardware

- Humans
- Interfaces
- Functions
- Environment

An *accident*, also called a mishap, is the unplanned sequence of events that results in undesirable consequences. Accidents include:

- Injury or death of one or more people
- Damage or destruction of property
- Negative impact to the environment
- Negative impact to society at large

A *safety risk* is the risk that the hazard will be actuated and result in the undesired outcome, in other words, the possibility that the accident will occur. The safety risk starts when the hazardous condition exists. For example, spilling water on the floor creates a hazard, the potential of someone slipping and falling. The longer the water stays on the floor, the longer that hazard and its associated safety risk exists. The safety risk ends in one of two ways:

1. *The accident happens*: For example, a person walks by, that person does not notice the water on the floor, that person slips and falls, breaking their arm
2. *The hazard is eliminated*: For example, someone comes by and mops up the water, and the hazardous condition is eliminated

A *safety risk statement*, also called the hazard description, should be clear, concise, and unambiguous. According to Ericson (2016), it should include the hazard source (HS), one or more initiating mechanisms (IMs), and the target or threat outcome (TTO), which defines the accident.

Software Safety Activities

Software safety activities throughout the life cycle include:

- Establishing safety standards that must be met.
- Identifying hazards and safety risks, including ways the system can fail.
- Performing software safety and hazard analysis and determining acceptable levels of safety risks. Documenting software *safety plans* that include rigorous development processes throughout the life cycle. IEEE 1228 Standard for Software Safety Plans consists of an outline and details about what should be included in each section of this plan (IEEE 1994).
- Defining safety requirements (see Chapter 12).

- Designing and implementing safety-critical portions of the software (see Chapter 14).
- Performing verification and validation (analysis, peer reviews, testing) on safety-critical features at every step of the life cycle (see Chapters 21–24).
- Conducting independent audits of each development process to ensure implementation and effectiveness of safety-related activities (see Chapter 8).
- "Training personnel in designing, developing, testing, evaluating, or using the safety software application is critical for minimizing the consequences of software failure" (DOE 2010).
- Training users in the proper use of safety-critical systems is absolutely necessary for minimizing the consequences of a hazardous state. Improper or invalid use of the system or its software may negate the safety mitigation strategies built into the system.
- Ensuring that adequate safety records are kept.
- Identifying and mitigating security risks, because software security is directly related to software safety. If the software is not secure, it cannot be safe. Someone intending to cause harm can breach the software's security and intentionally cause that software to be unsafe.

Safety Risk Management

As with other software-related risks, safety risks must be managed (identified, analyzed, planned for and handled as appropriate, and tracked). Figure 18.8 illustrates the basic steps in the system or software hazard analysis and software *safety mitigation process.*

Safety risk analysis uses the typical risk analysis technique of determining the loss (impact or severity) if an accident does occur. The following is an example of *qualitative loss indices that are commonly* used in nuclear, aircraft, and military safety standards:

- *Catastrophic:*
  - Could result in death or permanent total disability
  - Loss exceeding $1,000,000
  - Severe environmental damage, violating law(s) or regulation(s)
- *Critical:*
  - Could result in permanent partial disability, injuries or illness affecting at least three people
  - Loss exceeding $200,000

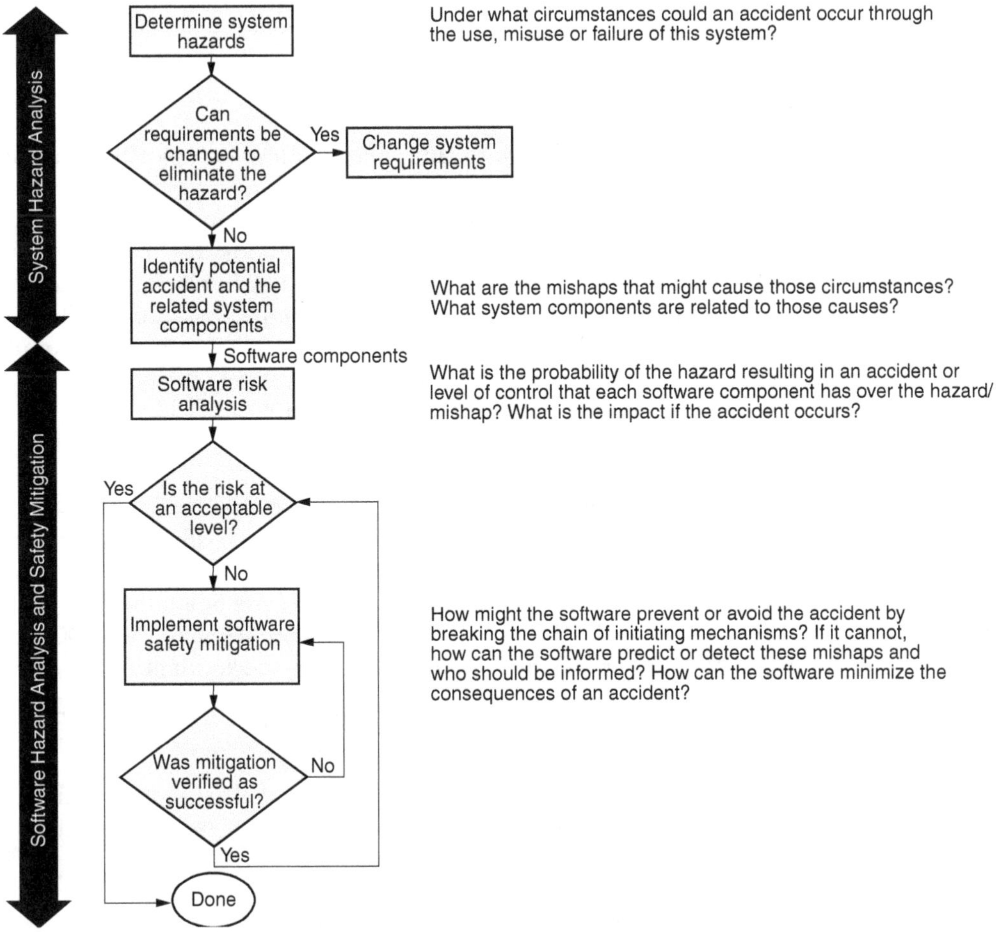

Figure 18.8 Hazard analysis and software safety mitigation process.

- Reversible damage to the environment, violating law(s) or regulation(s)

- *Marginal*:
  - Could result in injury or illness, resulting in the loss of one or more workdays
  - Loss exceeding $10,000
  - Mitigatable damage to the environment without violating any laws or regulations
- *Negligible*:
  - Could result in injury or illness, not resulting in lost workdays
  - Loss exceeding $2,000

    - Minimal damage to the environment without violating any laws or regulations

- *No impact*

Analyzing the probability of a hardware component failing is typically done by evaluating the properties of materials, the laws of physics, historic failure data, and so on. The probability of a software component failing cannot be determined using these techniques and is in fact very difficult, if not impossible, to determine. Therefore, most safety experts use *level of control* instead of *probability of failure* when performing hazard analysis on software. "The *software level of control* (LOC) is a measure of the degree to which the software is responsible for the behavior of a system, or of a specific system action, that may lead to a safety mishap" (Frailey 2016). Frailey gives the following examples of software LOC:

- A software failure may result directly in a mishap:
    - *LOC 1—Autonomous/time critical*: The software exercises autonomous control over potentially hazardous hardware systems, subsystems, or components without the possibility of intervention to preclude the occurrence of an accident. Failure of the software or a failure to prevent an event leads directly to an accident's occurrence. This LOC implies that no other failure is required to cause the accident.
- A software failure may significantly impact the margin of safety for an accident:
    - *LOC IIa—Autonomous/not time critical*: Software exercises control over potentially hazardous hardware systems, subsystems, or components and allows time for intervention by independent safety systems to mitigate the accident. This LOC implies that corrective action is possible and that a second fault or error is required for this accident to occur.
    - *LOC IIb—Information time critical*: Software provides information requiring immediate operator action to mitigate a hazard. Software failures will allow or fail to prevent the occurrence of an accident. This implies that the operator is made aware of the existence of the accident and intervention is possible.
- A software failure may have a minor impact on the margin of safety for an accident:
    - *LOC IIIa—operator controlled*: Software issues commands over potentially hazardous hardware systems, subsystems, or components requiring human action to complete the control function. There are several redundant, independent safety measures for each hazardous event.
    - *LOC IIIb—information decision*: Software generates information of a safety-critical nature used to make safety-critical decisions. There are several redundant, independent safety measures for each hazardous event.

Table 18.11 Safety risk statement—two examples.

	Accident (Mishap)	
Hazard Source (HS)	**Initiating Mechanisms (IM)**	**Target/Threat Outcome (TTO)**
The boiler contains temperature-sensitive chemicals	1. The software sends an excessive heat command to the boiler 2. The chemical mixture overheats 3. Causing a chemical reaction 4. Producing excessive pressure	Causing the boiler to explode and release noxious fumes into the surrounding city

	Accident (Mishap)	
Hazard Source (HS)	**Initiating Mechanisms (IM)**	**Target/Threat Outcome (TTO)**
Cars are massive objects that travel at high speeds	1. The car is passing a slow-moving truck. 2. The software determines that the car is getting too close to the truck and slams on the automatic breaks. 3. Another car in the passing lane does not break quickly enough.	The first car gets rear-ended, causing critical injuries and both cars to be totaled.

- A software failure will not have an impact on the margin of safety for an accident:
  - *LOC IV—not safety software*: Software does not control safety-critical hardware systems, subsystems, or components and does not provide safety-critical information.

Using a technique similar to the one used for defining qualitative risk exposure (illustrated in Table 18.2), the safety-risk analysis information can be used to determine the risk exposure of software safety risks, as illustrated in Table 18.11, where the intersecting cells identify the level of safety risk (high, medium, or not applicable) for that software component. Components can then be prioritized by that level of risk and considered for safety risk mitigation.

FMEA

One tool for performing hazard analysis is *failure mode and effects analysis (FMEA)*, also called *failure mode, effects and criticality analysis (FMECA)*. While FMEA can be used with any risks, it is a complex and rigorous process that would be considered overkill for most software risks. Therefore, FMEA is usually reserved for analyzing the most important risks, like safety-related risks. In FMEA, a component of the system or software is analyzed, as illustrated in Figure 18.9, in an attempt to reduce risk by actively analyzing the following (based on ASQ 2016):

- *Function*: What is the purpose of the system or software component; what task(s) does it perform?

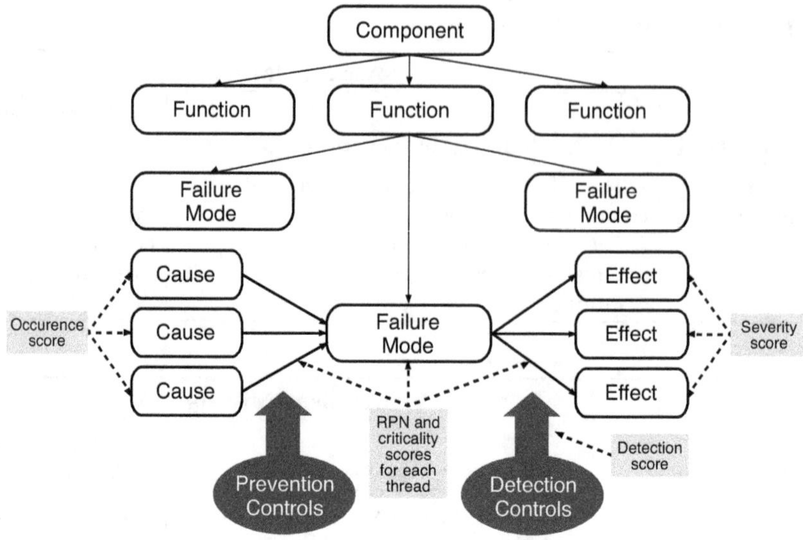

Figure 18.9 FMEA analysis of failure chain for a component—example.

- *Mode of the failure*: The way in which the system or software component may fail to deliver part or all of its function, for example:
    - Complete failure
    - Partial failure
    - Intermittent failure
    - Failure over time (deterioration over time)
    - Overperformance failure
- *Effects (consequences) of the failure*: Each failure mode can have one or more effects.
- *Severity score*: A rating of the seriousness of the impact or loss if the failure occurs—prior to mitigation actions.
- *Causes of the failure*: What are the reasons the failure might occur? Each failure mode can have one or more causes.
- *Occurrence score*: A rating of the probability of failure or, for software, the level of control—prior to mitigation actions.
- *Prevention controls*: What activities are currently in place or already planned to prevent the failure? Prevention controls are measures put in place to prevent the cause from triggering the failure mode.
- *Detection controls*: What activities are currently in place or already planned to detect the failure—prior to mitigation actions? Detection controls are measures put in place to recognize and counter the failure mode to minimize or reduce its effects.

Table 18.12 Qualitative safety risk exposure—example.

		Software Level of Control					
		LOC 1	**LOC IIb**	**LOC IIa**	**LOC IIIa**	**LOC IIIb**	**LOC IV**
Impact	**Catastrophic**	High	High	Medium	Medium	Medium	N/A
	Critical	High	High	Medium	Medium	Medium	N/A
	Marginal	Medium	Medium	Medium	Low	Low	N/A
	Negligible	Low	Low	Low	Low	Low	N/A

- *Detection score*: A rating of how likely it is that the current detection control will notice and contain the failure mode.
- *Risk priority number (RPN)*: The calculated relative risk of the failure chain, which is equal to the severity score times the occurrence score times the detection score.
- *Criticality*: The risk exposure, which is equal to the severity score times the occurrence score (see Table 18.12).

If the team performing the FMEA process determines that the RPN or risk criticality scores are high enough to indicate that mitigation is needed, the FMEA process also analyzes the following (based on ASQ 2016):

- Potential options to mitigate the failure mode:
  - Additional occurrence actions
  - Additional detection actions
- For each potential mitigation option, new scores are assessed for the failure mode's severity, occurrence, and detection, assuming the implementation of the mitigation action(s)
- For each potential mitigation option, a revised RPN is calculated:
  - Based on the new severity, occurrence, and detection scores for that mitigation
  - Used to help determine which mitigation actions should be selected to reduce the safety risk

As a result of this FMEA analysis, plans are made and actions taken to mitigate any significant safety risks. As each planned mitigation action is implemented and verified, the FMEA process may be repeated for the associated component to determine if the risk has been reduced to acceptable levels or if additional mitigation action is required.

Safety Risk Mitigation

During this step, the appropriate software risk handling techniques are selected and alternative risk mitigation actions are evaluated. Whatever mitigation options are selected, the associated actions should be planned in advance to proactively manage the safety risk rather than waiting for accidents and reacting in a firefighting mode. The resulting safety risk control plans should then be incorporated into the project plans.

There are two approaches to mitigating software safety risks, and both should be used:

- *Approach 1*: Treat it as a reliability issue by identifying components that may fail. Then implement various approaches to reduce the potential for failure by creating better software (e.g., higher-quality, more reliable, or more secure software) in order to eliminate or reduce possible failure modes. To produce better software, all of the processes, actions, methods, and techniques discussed in this handbook can be applied with more rigor.
- *Approach 2*: Treat it as a system control issue by identifying the ways the system may fail or contribute to hazardous conditions. Then define requirements or specify designs for the associated components that prohibit these conditions through additional prevention or detection controls (see Chapter 12 for a discussion of defining safety requirements and Chapter 14 for a discussion on designing for safety).
  - How might the software prevent or avoid the accident?
  - If not, how can the software predict or detect an initiating event, and who should be informed
  - How should the software detect or recover from a hazardous state?
  - How can the software minimize the consequences of an accident?

Risk management in general, and safety risk management specifically, is all about using evaluation and analysis techniques to find that optimal trade-off balance between cost, schedule, and product (including product functionality and its quality). Risk management is about making sure that the product meets the stakeholders' needs (including safety), at a price they are willing to spend within an acceptable schedule.

Part V

Software Metrics and Analysis

Chapter 19

Process and Product Metrics

Software metrics and analysis provide the data and information that allows an organization's quality management system to be based on a solid foundation of facts.

The objective is to drive continual improvement in all quality parameters through a goal-oriented measurement and analysis system.

Software metrics programs should be designed to provide the specific information necessary to manage software projects and improve software engineering processes, products, and services. The foundation of this approach is aimed at making practitioners ask not "what should I measure?" but "why am I measuring?" or "what business needs does the organization wish its measurement initiative to address?" (Goodman 1993).

Measuring is a powerful way to track progress toward project, process, product, and service goals. As Grady (1992) states, "Without such measures for managing software, it is difficult for any organization to understand whether it is successful, and it is difficult to resist frequent changes of strategy."

According to Humphrey (1989), there are four major roles (reasons) for collecting data and implementing software metrics:

- *To understand*: Metrics can be gathered to learn about software projects, processes, products, and services and their capabilities. The resulting information can be used to:
  - Establish baselines, standards, and goals
  - Derive models of the software processes
  - Examine relationships between process parameters
  - Target process, product, and service improvement efforts
  - Better estimate project effort, costs, and schedules
- *To evaluate*: Metrics can be examined and analyzed as part of the decision-making process to study projects, products, processes, or services to establish baselines, perform cost/benefit analysis, and determine if established standards, goals, and entry/exit/acceptance criteria are being met.
- *To control*: Metrics can be used to control projects, processes, products, and services by providing triggers (red flags) based on trends, variances, thresholds, control limits, standards, or performance requirements.

- *To predict*: Metrics can be used to predict the values of attributes in the future (e.g., budgets, schedules, staffing, resources, risks, quality, and reliability).

Effective software metrics provide the objective data and information necessary to help an organization, its management, its teams, and its individuals:

- Make day-to-day decisions
- Identify project, process, product, and service issues
- Correct existing problems
- Identify, analyze, and manage risks
- Evaluate performance and capability levels
- Assess the impact of changes
- Accurately estimate and track effort, costs, and schedules

The bottom line is that effective metrics help improve software projects, products, processes, and services.

1. TERMINOLOGY

> Define and describe metrics and measurement terms such as reliability, internal and external validity, explicit and derived measures, and variation. (Understand)
>
> **Body of Knowledge V.A.1**

Data, Information, and Knowledge

Once a measurement is performed, the result is one or more numbers or symbols. These are data items. *Data items* are simply "facts" that have been collected in some storable, transferable, or expressible format.

However, if the data items are going to be useful, they must be transformed into information products. Those information products can then be interpreted by people and transformed into knowledge that can be used to make better, fact-based decisions. According to the *Data Management Body of Knowledge* (DAMA 2017), "*information* is data in context." The raw material of information is data. By adding context, collected data starts to have meaning. For example, a data item stored as the number 14 does not by itself provide any usable information. If the data item's context is known, that data item is converted to information. For example, when the data item has:

- *A definition*: Such as "the number of newly detected defects"
- *A timeframe*: Such as "last week"
- *Relevance*: Such as "while system testing software product ABC"

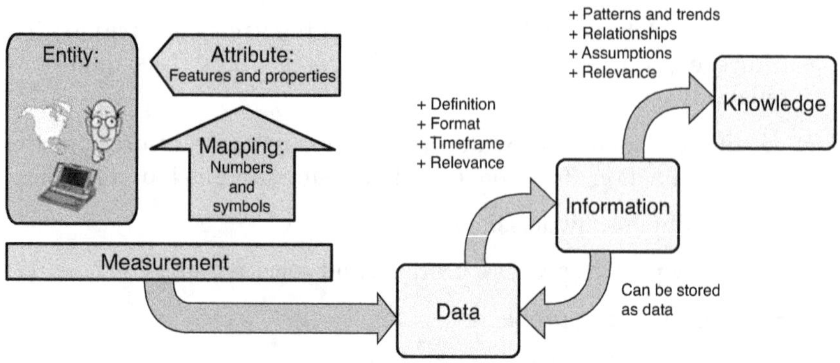

Figure 19.1 Converting measurement data into information and knowledge.

Once one or more data items are converted to information, the resulting information can also be stored as additional data items.

Information in and of itself is not useful until human intelligence is applied to convert it to *knowledge* through the identification of patterns and trends, relationships, assumptions, and relevance. Going back to the data item example, if the information regarding the 14 newly detected defects found last week during the system testing of software product ABC is put in a report that no one reads or uses, then the information is never converted to knowledge. However, if the project manager determines that this is a higher defect arrival rate (*trend*) than was found during the previous three weeks (*relationship*) and determines that corrective action is needed (*assumption*), resulting in the shifting of an additional software engineer onto problem correction (*relevance*), that information becomes knowledge. Of course, the project manager can also decide that everything is under control and that no action is necessary. If this is the case, the information has still been converted to knowledge. The right-hand side of Figure 19.1 illustrates this transformation of measurement data into information and knowledge.

The moral is that the goal should never be "To put metrics (or measurements) in place." That goal can result in an organization becoming *DRIP* (Data Rich, Information Poor). The goal should be to give people the knowledge they need to make better, fact-based decisions. Metrics are simply tools to make certain that standards exist for measuring in a consistent, repeatable manner in order to create reliable and valid data and information, (e.g., by establishing "what" is being measured through well-defined and understood entities and attributes and "how" it is being measured by standardizing the mapping system and the conditions under which measures are done). Metrics provide a standardized definition of how to turn collected data into information and how to interpret that information to create knowledge.

Reliability and Validity

Metric reliability is a function of consistency or repeatability of the measure. A *reliable metric* is one for which different people (or the same person multiple times) can use that metric over and over to measure an identical entity and obtain the same results each time. For example, if two people count the number of lines of code in a source code module, they would both have the same, consistent count (within an acceptable level of measurement error, as defined later in this chapter). Or, if one person measured the cyclomatic complexity of a detailed design element today and then measures the same design element tomorrow, that person would get the same, consistent measure (within an acceptable level of measurement error).

A *valid metric* is one that accurately measures what it is expected to measure, that is, if it captures the real value of the attribute it is intended to describe. IEEE (1998d) describes validity in terms of the following criteria:

- *Correlation*: Whether there is a sufficiently strong linear association between the attribute being measured and the metric
- *Tracking*: Whether a metric is capable of tracking changes in attributes over the life cycle
- *Consistency*: Whether the metric can accurately rank, by attribute, a set of products or processes
- *Predictability*: Whether the metric is capable of predicting an attribute with the required accuracy
- *Discriminative power*: Whether the metric is capable of separating a set of high- quality software components from a set of low-quality software components

This definition of validity actually refers to the *internal validity* of the metric or its validity in a narrow sense—does it measure the actual attribute? External validity, also called *predictive validity*, refers to the ability to generalize or transfer the metric results to other populations or conditions. For a metric to be externally valid, it must be internally valid. It can also be used as part of a prediction system, estimation process, or planning model. For example, can the mean time to fix a defect, measured for one project, be generalized to predict the mean time to fix for other projects? If so, then the mean-time-to-fix metric is considered externally valid. External validity is verified empirically through comparison of the predicted results to the subsequently observed actual results.

A metric that is reliable but not valid can be consistently measured, but that measurement does not reflect the actual attribute. An example of a metric that is reliable but not valid is using cyclomatic complexity for measuring reliability. Cyclomatic complexity can be consistently measured because it is well defined and has a consistent mapping system. However, it is internally invalid as a measure of reliability because it is a measure of complexity, not reliability.

A metric is internally valid but not reliable if it measures what it is supposed to measure but cannot be measured consistently. For example, suppose criteria are not clearly defined on how to assign numbers to the severity of a reported problem, resulting in different people assigning different severities to the same problem. If that case, the severity metric is valid but not reliable. As illustrated in Figure 19.2, the goal is to have metrics that are both reliable and internally valid. In fact, IEEE (1998d) includes reliability as one of the criteria for validity. For predictive-type metrics, external validity is also required.

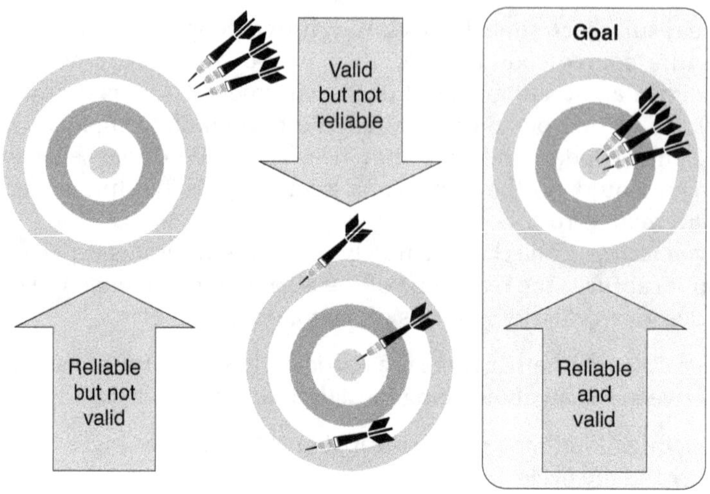

Figure 19.2 Reliability and validity.

To have reliable and valid metrics, there must be agreement on standard definitions for the entity and its attributes being measured as discussed in Chapter 6.

Explicit Measures

Data items can be either explicit or derived measures. *Explicit measures,* also called *base measures, metrics primitives,* or *direct metrics,* are measured directly. The *IEEE/ISO/IEC 24765 Systems and Software Engineering—Vocabulary* (IEEE/ISO/IEC 2017) defines a base measure as a "measure defined in terms of an attribute and the method for quantifying it." For example, explicit measures for a code inspection would include the number of:

- Lines of code inspected (loc_insp)
- Engineering hours spent preparing for the inspection (prep_hrs)
- Engineering hours spent in the inspection meeting (insp_hrs)
- Defects (defects) identified during the inspection (defects)

For explicit measures, the mapping system used to collect the data for each measure must be defined and understood, as discussed in Chapter 6. Having a clearly defined and communicated mapping system helps everyone interpret the measures the same way. The metrics mapping system and, if applicable, data needed based on the associated counting criteria define the first level of data that need to be collected in order to implement the metric.

Derived Measures

According to the *IEEE/ISO/IEC 24765 Systems and Software Engineering—Vocabulary* (IEEE/ISO/IEC 2017), a *derived measure,* also called *complex metrics,* is a "measure that is defined as a *function* of two or more values of base measures," that is, the mathematical combinations of explicit measures or other derived metrics. For a code inspection, examples of derived metrics would include:

- *Preparation rate*: The number of lines of code inspected, divided by the hours spent preparing for the inspection (loc_insp/insp_hrs)
- *Defect detection rate*: The number of defects found during the inspection, divided by the sum of the hours spent preparing for the inspections and the hours spent inspecting the work product (defects/[prep_hrs + insp_hrs])

Measurement functions should be selected and tailored to the organization's information needs, as discussed in Chapter 6.

Measurement Scales

There are four different *measurement scales* that define the type of data that is being collected. The measurement scale is important because it defines the mathematics, or kinds of statistics, that can be done using the collected data.

The *nominal scale* is the simplest form of measurement scale. In nominal scale measurements, items are assigned to a classification (one-to-one mapping), where that classification categorizes the entity's attribute. Examples of nominal scale measurements include:

- Development method (waterfall, V, spiral, other)
- Root cause (logic error, data initialization error, data definition error, other)
- Document author (Linda, Bob, Susan, Tom, other)

The categories in a nominal scale measurement must be jointly exhaustive and cover all possibilities. This means that every measurement must be assigned a classification. Many nominal scale measures include an "other" category, so everything fits somewhere. The categories must also be mutually exclusive. Each item must fit one and only one category. If an attribute is classified in one category, it cannot be classified in any other category.

The nominal scale does not make any assumptions about order or sequence. The only math that can be done on nominal scale measures is to count the number of items in each category and look at their distributions.

The *ordinal scale* classifies the attribute of the entity by order. However, there are no assumptions made about the magnitude of the differences between categories (a defect with a critical severity is not twice as bad as one with a major severity). Examples of ordinal scale measurements include:

- Defect severity (critical, major, minor)
- Requirement priority (high, medium, or low)
- CMMI® level (1 = initial, 2 = repeatable, 3 = defined,4 = managed, 5 = optimized)

- Process effectiveness (1 = very ineffective, 2 = moderately ineffective, 3 = nominally effective, 4 = moderately effective, 5 = very effective)

Since there is order, the *transitive property* is valid for ordinal scale measures. That is, if data item A > data item B and data item B > data item C, then data item A > data item C. However, mathematical operations of addition, subtraction, multiplication, and division cannot be used on ordinal scale measurement values because there is no assumption of magnitude.

For *interval scale measurements,* the exact distance between the data items is known. This knowledge allows the mathematical operations of addition and subtraction to be applied to interval scale measurement values. However, there is no absolute or nonarbitrary zero point in the interval scale, so multiplication and division do not apply. The classic example of an interval scale measure is calendar time. For example, it is valid to say, "May 1 plus ten days is May 11," but saying "May 1 times May 11" is invalid. Interval scale measurements require well-established units of measure that can be agreed upon by everyone using the measurement. The mode or the median can be used as the central tendency statistic for interval scale metrics.

In the *ratio scale,* not only is the exact distance between the scales known but there is an absolute or nonarbitrary zero point. All mathematical operations can be applied to ratio scale measurement values, including multiplication and division. Examples of ratio scale measurements include:

- Defect counts
- Defect density (defects per size)
- Minutes of outage
- Hours of effort
- Cycle times
- Rates (e.g., arrival rates and fix rates)

Note that since derived measures are mathematical combinations of two or more explicit measures or other derived measures, those explicit and derived measures must be at a minimum interval scale measurements (if only addition and subtraction are used in the measurement function), and they are typically ratio scale measurements.

Variation

To start the discussion of variation, consider an exercise from Shewhart's book, *Economic Control of Quality of Manufactured Product* (Shewhart 1980). This exercise begins by having a person write the letter "a" on a piece of paper. The person is then asked to write another "a" identical to the first one; then another just like it, then another, until the person has ten a's on the paper. When all of the letters are compared, are they all perfectly identical, or are there variations between them? The instructions were to make all of the a's identical, but as hard as the person may have tried, that person cannot make them all identical. But why can a person not write ten identical a's? Most people could probably think of multiple reasons why there are variations between the letters. Examples might include:

- The ink flow in the pen is not uniform
- The paper is not perfectly smooth
- The person's hand placement was not the same each time

People accept that there are many reasons for variation and it is impractical, and most likely impossible, to try to remove them all. If two people were asked to do this same exercise, there would probably be even more variation between one person's a's and another person's a's, because different people doing the same task adds another cause of variation.

Quality practitioners talk about two distinct sources of variation. The first of these sources is called *common causes,* also known as *random causes* or *chance causes* of variation. Common causes of variation are the normal or usual causes inherent in a consistent, repeatable process. A process is said to be in *statistical control* when it only exhibits common causes of variation. As long as the process does not change, these common causes of variation will not change. Therefore, based on historical data or a sample set of data from the current process, practitioners can quantify common causes of variation in order to predict the amount of variation that will occur in the future.

Common cause variation comes from normal, expected fluctuations in various factors, typically classified as influences from:

- People (worker) influences
- Machinery influences
- Environmental factors
- Material (input) influences
- Measurement influences
- Method influences

The only way to change or reduce the common causes of variation is to change the process (e.g., through process improvement). It should be noted that there is always some variation in every process, and that not all common causes of variation can be or should be eliminated. There are engineering trade-offs to be made. For example, eliminating common causes of variation in one part of the process may cause problems in another part of that process, or it may not be economically feasible to eliminate a source of common cause variation.

The second source of variation is *special causes,* also called *assignable causes,* which are outside the normal "noise" of the process and result in abnormal or unexpected variation. Special causes of variation include:

- *One-time perturbations*: For example, someone bumps the person's elbow while they are writing an a.
- *Things that cause trends*: For example, the person's hand starts getting tired after writing the thousandth a.

- *Shifts and jumps*: For example, making process improvement changes is a special cause that usually results in a shift in the amount of variation in the process. Another example of a shift or jump may be caused by changing the person performing the task.
- *Unexpected variation in quantity or quality of the inputs to process*: Such as raw materials, components, or subassemblies. For example, the pen's ink gets clogged or starts skipping.

If special causes of variation exist in a process, that process is not performing at its best and is considered *out of control*. In this case, the process is not sufficiently stable to use historical data or samples from the current process to predict the amount of variation that will occur in the future. That means that statistical quality control cannot be applied to processes that are out of control. Special causes of variation can usually be detected, and actions to eliminate these causes can typically be economically justified.

Statistics and Statistical Process Control

Statistics is defined as the science of the collection, organization, analysis, and interpretation of data. *Statistical quality control* refers to the use of measurement and statistical methods in the monitoring and maintenance of the quality of products and services. Faced with a large amount of data, it may seem daunting to try to turn those data into information that can be used to create knowledge. However, some basic descriptive statistics can be used to characterize a set of data items in order to provide needed information about a data set.

The *location* of the data set refers to the typical value or central tendency that is exhibited by most data sets. The location refers to the way in which data items tend to cluster around one or more values in that data set. Figure 19.3 shows examples of three data sets, with the same variance and shape but with three different locations.

Three statistics are typically used to represent the location of a data set:

- The *mean* is the arithmetic average of the numbers in the data set. The mean is "used for symmetric or near-symmetric distributions or for distributions that lack a clear, dominant single peak" (Juran 1999). The mean is calculated by summing the data values and dividing by the number of data items. For example, if the data items were 3, 5, 10, 15, 19, 21, 25, the mean would be $(3+5+10+15+19+21+25)/7=14$. The mean is probably the most used statistic in the quality field. It is used to report and predict "expected values" (e.g., average size, average yield, average percent defective, mean time to failure, and mean time to fix). However, since division is involved in calculating the mean, it can only be used on ratio scale measurement data.
- The *median* is the middle value when the numbers are arranged according to size. The median is "used for reducing the effects of extreme values or for data that can be ranked but are not economically measurable (shades of color, visual appearance, odors)" (Juran 1999). For example, if the data items were 3, 5, 10, 15, 19, 21, 102, the median would be 15. If there is an even number of items in the data set, then the

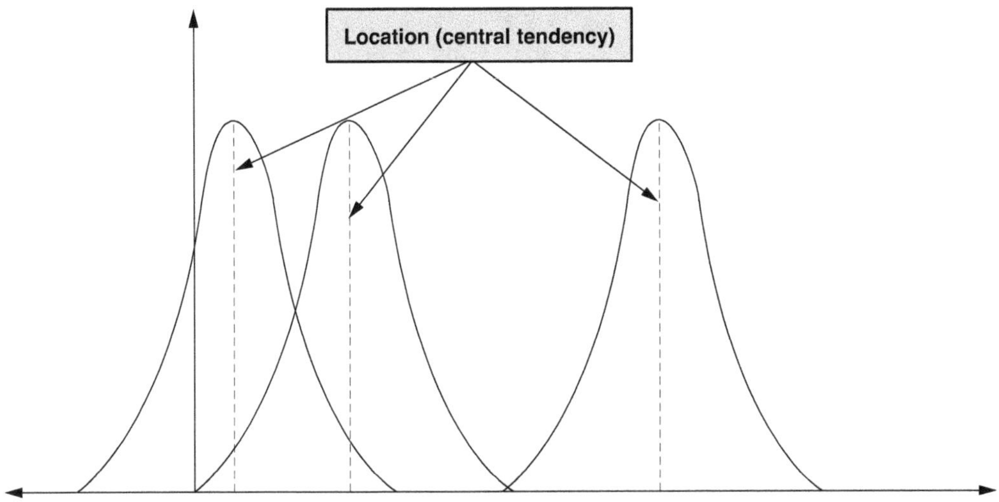

Figure 19.3 Data sets with different locations—examples.

median is calculated by adding the two middle values and dividing by two. For example, if the data items were 3, 5, 10, 15, 19, 21, 46, 102, the middle two items are 15 and 19, so the median would be calculated by (15 + 19)/2 = 17. Since the median requires the data items to be sorted in order, it cannot be used on nominal scale measurement data.

- The *mode* is the value that occurs most often in the data. The mode is "used for severely skewed distributions, describing irregular situations where two peaks are found, or for eliminating the effects of extreme values" (Juran 1999). For example, if the data items were, 1, 1, 1, 1, 1, 3, 3, 5, 10, 21, 96, the mode would be 1 (the most frequently occurring value). For the data set 1, 2, 3, 4, 5, there are five modes because all five data items have the same number of occurrences. The mode is also used for nominal scale measurement data.

In most data sets, the individual data items tend to cluster around the location and then spread or scatter out from there toward the extreme or extremes of the data set. The extent of this scattering is the *variance*, also called the *spread* or *dispersion*, of the data set. The variance is the amount by which smaller values in the data set differ from larger values. Figure 19.4 shows an example of three data sets, with the same location and shape, but with three different variances.

The simplest measure of variance is the range of the data set. The *range* is the difference between the maximum value and the minimum value in the data set. The more variance there is in the data set, the larger the value of the range will be. For example, if the data items were 3, 5, 10, 15, 19, 21, 25, the range would be equal to 25 – 3 = 22.

The most important measure of variance from the perspective of statistical quality control is standard deviation. *Standard deviation* is a measure of how much variation there is from the mean of the data set. A low value for standard deviation

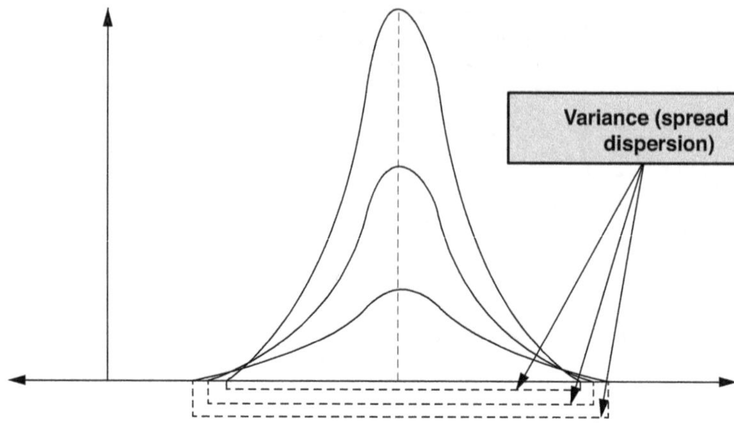

Figure 19.4 Data sets with different variances—examples.

indicates that the data values are tightly clustered around the mean. The larger the standard deviation, the more spread out the data values are from the mean. The Greek letter sigma (σ) is used to represent the standard deviation of the entire population, and the letter "s" is used to represent the standard deviation for a sample from that population.

In statistics, the *distribution* describes the shape of a set of data. More specifically, the *probability distribution* describes the range of possible values for the data in the set and the probability that any given random value selected from that population is within a measurable subset of that range. Figure 19.5 illustrates a *normal distribution,* also called a *Gaussian distribution* or *bell curve.* A *normal distribution* is a continuous distribution where there is a concentration of observations around the mean, and the shape is a symmetrical bell-like shape. In a normal distribution, the three location statistics of mean, median, and mode are all equal. Figure 19.5 also illustrates the percentage of data items under a normal distribution curve as plus and minus various standard deviations. For example, 68.26 percent of all data items in a normal distribution fall within ±1 standard deviation under the curve. At ±3 standard deviations, 99.73 percent of all data items in a normal distribution fall under the curve. Calculations for the standard deviation are dependent on the distribution of the data set. The normal distribution is the most well-known and used probability distribution. Data sets for many quality characteristics can be estimated and described using a normal distribution. As discussed in Chapter 20, normal distributions are also used as the basis for creating statistical process control charts.

A normal distribution is just one type of distribution or shape that a data set can have.

There are also other types of distributions grouped into two major classifications:

- The shape of a set of variable data is described by a *continuous distribution.* A normal distribution is one example of a continuous distribution. Uniform exponential, triangular, quadratic, cosine, and u-shaped are other common continuous distributions.

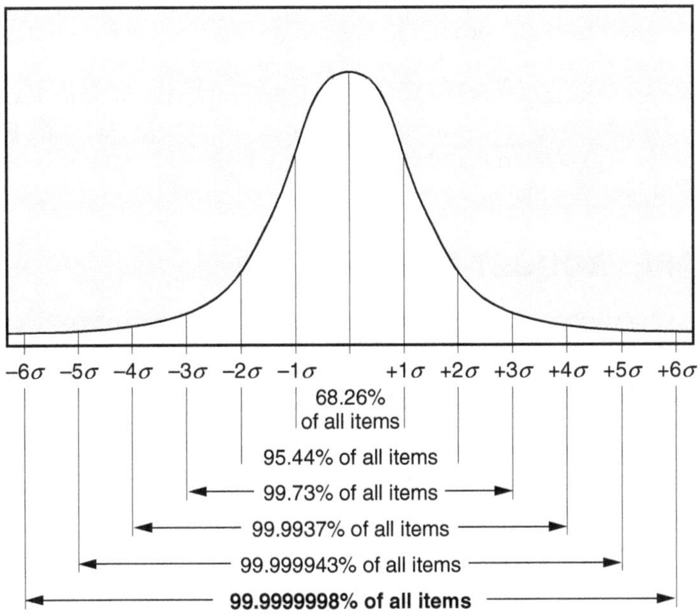

Figure 19.5 Normal distribution curve with standard deviations.

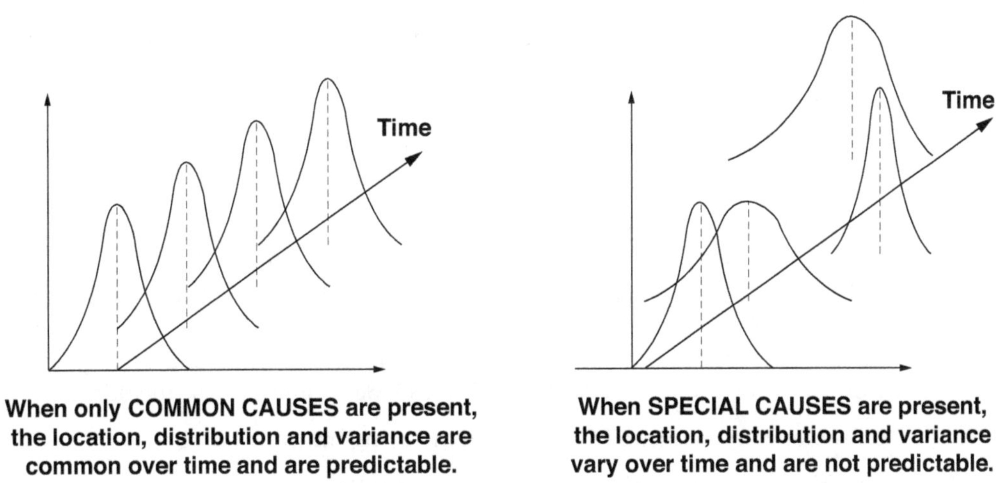

Figure 19.6 Common cause and special cause impacts on statistics—examples.

- The shape of a set of attribute data is described by a *discrete distribution*. Common discrete distributions include uniform, Poisson, binomial, and hypergeometric distributions.

When only common cause variation exists in a process, the location, distribution, and variance statistics of the data collected from that process will be stable and predictable over time, as illustrated on the left side of Figure 19.6. This means that statistics for a data set collected from the process today will be the same as for the data set collected next week, and the week after that, and the week after that, and so on.

Whenever special causes of variation are introduced into the process, the location, distribution, and variance statistics of data sets collected from that process will vary over time and will no longer be predictable, as illustrated on the right side of Figure 19.6.

2. SOFTWARE PRODUCT METRICS

> Choose appropriate metrics to assess various software attributes (e.g., size, complexity, the amount of test coverage needed, requirements volatility, and overall system performance). (Apply)
>
> **Body of Knowledge V.A.2**

With all of the possible software product and process entities and attributes, it is easy to see that a huge number of possible metrics could be implemented. The next two sections touch on only a few metrics as examples of those used in the software industry. The recommended method is to use the goal–question–metric paradigm discussed in Chapter 6, or some other mechanism, to select appropriate metrics that meet the information needs of an organization and its teams, managers, and engineers.

Size—Lines of Code

Lines of code (LOC) counts are one of the most used and most often misused of all the software metrics. Some estimation methods are based on *KLOC* (thousands of lines of code). The LOC metric may also be used in other derived metrics to normalize measures so that releases, projects, or products of different sizes can be compared (e.g., defect density or productivity). The problems, variations, and anomalies of using lines of code are well documented (Jones 1986). Some of these include:

- Problems in counting LOC for systems using multiple languages
- Difficulty in estimating LOC early in the software life cycle
- Productivity paradox—if productivity is measured in LOC per staff month, productivity appears to drop when higher-level languages are used, even though higher-level languages are inherently more productive

No industry-accepted standards exist for counting LOC. Therefore, it is critical that specific criteria for counting LOC be adopted for the organization. Considerations include:

- Are physical or logical lines of code counted?
- Are comments, data definitions, or job control language counted?

- Are macros expanded before counting? Are macros counted only once?
- How are products that include different languages counted?
- Are only new and changed lines counted, or are all lines of code counted?

The Software Engineering Institute (SEI) created a technical report specifically to present guidelines for defining, recording, and reporting software size in terms of physical and logical source statements (CMU/SEI 1992). This report includes check sheets for documenting criteria selected for inclusion or exclusion in LOC counting for physical and logical LOC.

Size—Function Points

Function points is a size metric intended to measure the size of the software without considering the language it is written in. The function point counting procedure has five steps:

Step 1: Decide what to count (total number of function points or just new or changed function points).

Step 2: Define the boundaries of the software being measured.

Step 3: Count the raw function points, which are determined by adding together the weighted counts for each of the following five function types, as illustrated in Figure 19.7:

- *External input*: The weighted count of the number of unique data or control input types that cross the external boundary of the application system and cause processing to happen within it
- *External output*: The weighted count of the number of unique data or control output types that leave the application system, crossing the boundary to the external world and going to any external application or element
- *External inquiry*: The weighted count of the number of unique input/output combinations for which an input causes and generates an immediate output

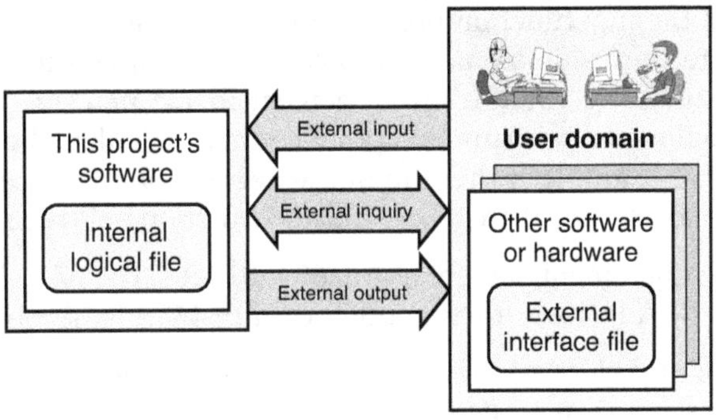

Figure 19.7 Function points.

- *Internal logical file:* The weighted count of the number of logical groupings of data and control information that are to be stored within the system
- *External interface file*: The weighted count of the number of unique files or databases that are shared among or between separate applications

Within each function type, the counts are weighted based on complexity and other contribution factors. Raw function point counts have clearly defined measurement methods, as established by the International Function Point Users Group (IFPUG).

Step 4: Assign a degree of influence to each of the 14 defined by IFPUG. These factors are used to measure the complexity of the software. Each factor is measured on a scale of zero to five (five being the highest). These adjustment factors include:

1. Data communications
2. Distributed data or processing
3. Performance objectives
4. Heavily used configuration
5. Transaction rate
6. Online data entry
7. End-user efficiency
8. Online update
9. Complex processing
10. Reusability
11. Conversion and installation ease
12. Operational ease
13. Multiple site use
14. Facilitate change

Step 5: Adjust the raw function point count. To do this, the sum of all 14 degrees of influence to the value adjustment factors is multiplied by 0.01 and added to 0.65. This results in a value from 65 percent to 135 percent. That percentage is then multiplied by the raw function point count to calculate the adjusted function point count. For example, if the sum of the degrees of influence is 42 and the raw function point count is 450, then the adjusted function point count is calculated as:

$$\text{Adjustment percentage} = 0.65 + (0.01 \times 42) = 1.07 \text{ or } 107\%$$
$$\text{Adjusted function point count} = 450 \times 107\% = 482$$

As with lines of code, function points are used as the input into many project estimation tools and may also be used in derived metrics to normalize those metrics, so that releases, projects, or products of different sizes can be compared (e.g., defect density or productivity).

Other Size Metrics

Many other metrics may be used to measure the size of different software products, including:

- In object-oriented development, size may be measured in terms of the number of objects, classes, or methods.
- Requirements size may be measured in terms of the count of unique requirements, or weights may be included in the counts (e.g., large requirements might be counted as 5, medium requirements as 3, and small requirements as 1). Other examples of weighted requirements counts include story points used in agile (the estimation of story points is discussed in Chapter 16), and use case points (based on the number and complexity of the use cases that describe the software application, and the number and type of the actors in those use cases, adjusted by factors of the technical complexity of the product and the environmental complexity of the project).
- Design size may be measured in terms of the number of design elements (configuration items, subsystems, or source code modules).
- Documentation is typically sized in terms of the number of pages or words. For graphics-rich documents, counts of the number of tables, figures, charts, and graphs may also be valuable as size metrics.
- The testing effort is often sized in terms of the number of test cases or weighted test cases (e.g., large test cases might be counted as 5, medium test cases as 3, and small test cases as 1).

Complexity—Cyclomatic Complexity

McCabe's *cyclomatic complexity* is a measure of the number of linearly independent paths through a module or detailed design element. Cyclomatic complexity can therefore be used in structural testing to determine the minimum number of path tests that must be executed for complete coverage. Empirical data indicate that source code modules with a cyclomatic complexity of 10 or greater are more defect prone and harder to maintain.

Cyclomatic complexity is calculated from a *control flow graph* by subtracting the number of nodes from the number of edges, and adding two times the number of unconnected parts of the graph (edges − nodes + 2p). In well-structured code, with one entry point and one exit point, there is only a single part to the graph, so $p = 1$. Cyclomatic complexity is a measure that is rarely calculated manually. This is a measurement where static analysis tools are extremely useful and efficient for calculating the code's cyclomatic complexity. This explanation is included here so people understand what the tool is doing when it measures cyclomatic complexity.

As illustrated in Figure 19.8:

- Straight-line code (control flow graph A) has one edge, two nodes, and one part to the graph, so its cyclomatic complexity is $1 - 2 + 2 \times 1 = 1$

- Control flow graph B is a basic if–then–else structure and has four edges and four nodes, so its cyclomatic complexity is $4 - 4 + 2 \times 1 = 2$
- Control flow graph C is a case statement structure followed by an if–then–else structure. Graph C has 12 edges and nine nodes, so its cyclomatic complexity is $12 - 9 + 2 \times 1 = 5$.

Another way of measuring the cyclomatic complexity for well-structured software (no two edges cross each other in the control flow graph, and there is one part to the graph) is to count the number of regions in the graph. For example:

- Control flow graph C divides the space into five regions, four enclosed regions and one for the outside, so its cyclomatic complexity is again 5
- Control flow graph A has a single region and a cyclomatic complexity of 1
- Control flow graph B has two regions and a cyclomatic complexity of 2
- Control flow graph D is a repeated if–then–else structure and has 13 edges and 10 nodes, so its cyclomatic complexity is $13 - 10 + 2 \times 1 = 5$ (note that it also has five regions)

Just to illustrate an example of a software component where the number of parts is greater than one, assume an old piece of poorly structured legacy code has two entry points and two exit points. For example, assume control flow graphs C and D in Figure 19.8 are both part of the same source code module. That module would have 25 edges (12 from C and 13 from D), 19 nodes (9 from C and 10 from D), and two parts (part C and part D that are not connected). The cyclomatic complexity would be $25 - 19 + (2 \times 2) = 10$. This makes sense because both control flow graphs C and D each had a cyclomatic complexity of 5, so they should have a cyclomatic complexity of 10 together.

Complexity—Structural Complexity

While cyclomatic complexity looks at the internal complexity of an individual design element or source code module, structural complexity looks at the complexity of the interactions between the modules in a calling structure (or, in

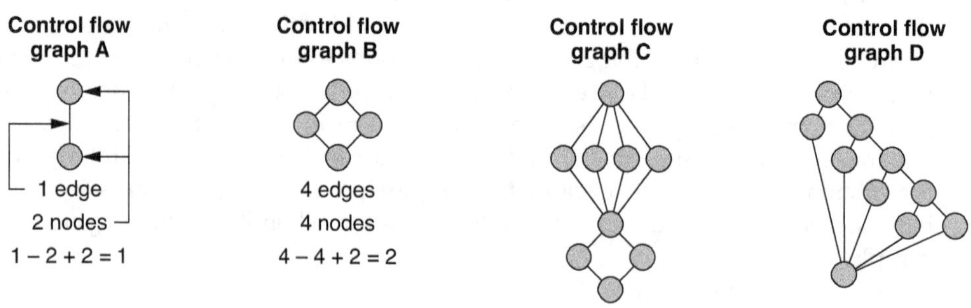

Figure 19.8 Cyclomatic complexity—examples.

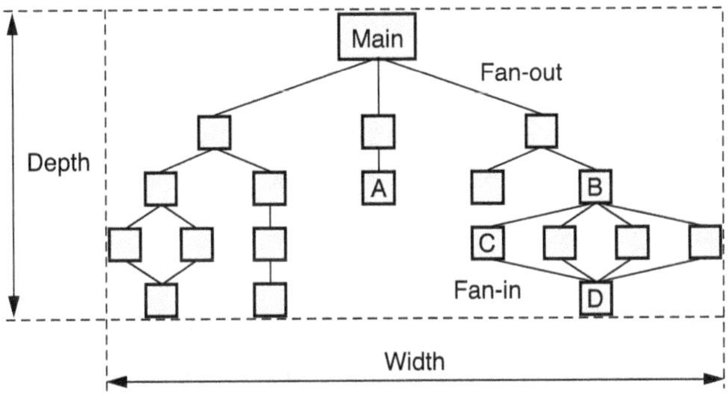

Figure 19.9 Structural complexity—examples.

the case of object-oriented development, between the classes in an inheritance tree). Figure 19.9 illustrates four structural complexity metrics. The depth and width metrics focus on the complexity of the entire structure. The fan-in and fan-out metrics focus on the individual elements within that structure.

The *depth* metric is the count of the number of levels of control in the overall structure or of an individual branch in the structure. For example, the branch in Figure 19.9 that starts with the element labeled "Main" and goes to the element labeled "A" has a depth of three. Each of the four branches that start with the element labeled "Main" and go to the element labeled "D" has a depth of five. The depth for the entire structure is measured by taking the maximum depth of the individual branches. For example, the structure in Figure 19.9 has a depth of five.

The *width* metric is the count of the span of control in the overall software system, or of an individual level within the structure. For example, the level that includes the element labeled "Main" in Figure 19.9 has a width of one. The level that includes the elements labeled "A" and "B" has a width of five. The level that includes the element labeled "C" has a width of seven. The width for the entire structure is equal to the maximum width of the individual levels. For example, the structure in Figure 19.9 has a width of seven.

Depth and width metrics can help provide information for making decisions about the integration and integration testing of the product and the amount of effort required.

Fan-out is a measure of the number of modules that are directly called by another module (or that inherit from a class). For example, the fan-out of module Main in Figure 19.9 is three. The fan-out of module A is zero, the fan-out of module B is four, and the fan-out of module C is one.

Fan-in is the count of the number of modules that directly call a module (or that the class inherits from). For example, in Figure 19.9, module A's fan-in is one, module D's fan-in is four, and module Main's is zero.

Fan-in and fan-out metrics provide information for making decisions about the integration and integration testing of the product. These metrics can also be useful in evaluating the impact of a change and the amount of regression testing needed after implementing a change.

Quality—Defect Density

Defect density is a measure of the total known defects divided by the size of the software entity being measured (Number of known defects/Size). The number of known defects is the count of total defects identified against a particular software entity during a particular time period. Examples include:

- Defects to date since the creation of the module
- Defects found in a work product during an inspection
- Defects per month since the shipment of a release to the customer

For defect density, size is used as a normalizer to allow comparisons between different software entities (e.g., source code modules, releases, or products) of different sizes. To demonstrate the calculation of defect density, Table 19.1 illustrates the size of the three major subsystems that comprise the ABC software system and the number of prerelease and post-release defects discovered in each subsystem.

Post-release defect density in defects per KLOC would be calculated as follows:

$$
\begin{aligned}
\text{Size of ABC} &= 3432 + 2478 + 6912 = 12{,}822 \text{ LOC} \\
\text{Post-release defects} &= 5 + 12 + 23 = 40 \text{ defects Defect density} \\
&= 40 \text{ defects}/12{,}822 \text{ LOC} \\
&= .00312 \text{ defects per LOC} \\
&= 3.12 \text{ defects per KLOC}
\end{aligned}
$$

Defect density compares the relative number of defects in various software components. This comparison helps identify candidates for additional inspection or testing, or possible reengineering or replacement. Identifying defect-prone components allows the concentration of limited resources into areas with the highest potential return on the investment. Typically this is done using a Pareto diagram, as illustrated in Figure 20.16.

Another use for defect density is to compare subsequent releases of a product to track the impact of defect reduction and quality improvement activities, as illustrated in Figure 19.10. Normalizing defect arrival rates by size allows releases of varying sizes to be compared. Differences between products or product lines can also be compared in this manner.

Quality—Arrival Rates

Arrival rates graph the trends of newly opened problems against a product over time. Note that this metric looks at problems, not defects. Testers, customers,

Table 19.1 Defect density inputs—example.

Subsystem	Size	Prerelease defects	Post-release defects
A	3432 LOC	64	5
B	2478 LOC	32	12
C	6912 LOC	102	23

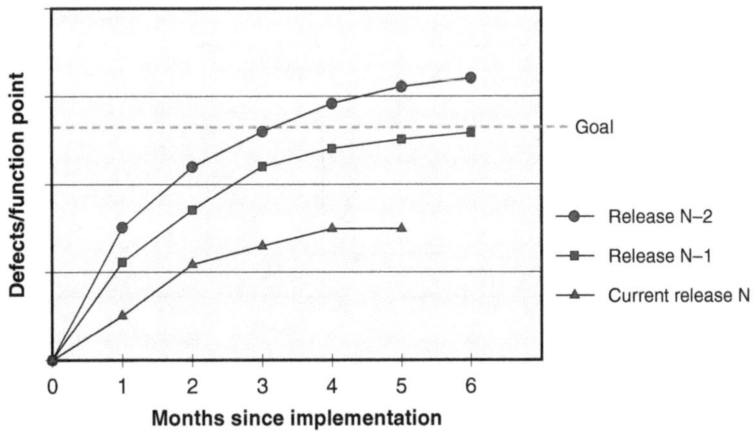

Figure 19.10 Post-release defect density—example.

technical support personnel, or other originators report problems because they think something is wrong with the software. The software developers must then debug the software to determine if there is a defect. In some cases, the problem report may be closed after this analysis as "operator error," "works as designed," "cannot duplicate," or some other non-defect-related disposition.

Prior to release, the objective of evaluating the arrival rate trends is to determine if the product is stable or moving toward stability. As illustrated in Figure 19.11, when testing is first gearing up, arrival rates may be low. During the middle of testing, arrival rates are typically higher, with some variation. However, as shown for product A, the goal is for the arrival rates to trend downward toward zero or stabilize at low levels (the time between failures should be far apart) before the completion of testing and release of the software. By stacking the arrival rates by severity, additional information can be analyzed. For example, it is one thing to have a few minor problems found near the end of testing, but if critical problems are still being found, it might be appropriate to continue testing. Product B in Figure 19.11 does not exhibit this stabilization, indicating that continued testing is appropriate. However, arrival rate trends alone are insufficient to signal the end of testing. This metric could be made to "look good" simply by slowing the effort being expended on testing. Therefore, arrival rates should be evaluated in conjunction with other metrics, including effort rates and test coverage, when evaluating test sufficiency.

Post-release arrival rate trends can also be evaluated. In this case, the goal is to determine the effectiveness of the defect detection and removal processes and process improvement initiatives. As illustrated in Figure 19.11, post-release defect arrival rates can be normalized by release size to compare them over subsequent releases. Note that this metric must wait until after debugging the problem report, because it only counts defects.

Quality—Problem Report Backlog

Arrival rates only track the number of problems being identified in the software. Those problems must also be debugged and defects corrected to increase the

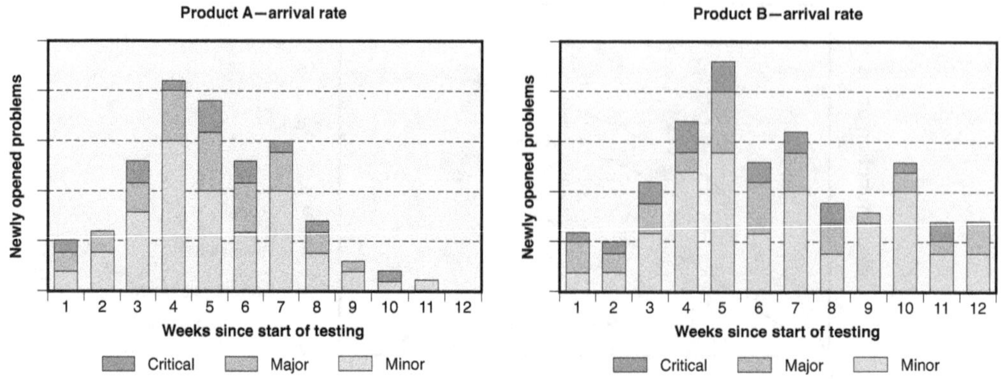

Figure 19.11 Problem report arrival rate—examples.

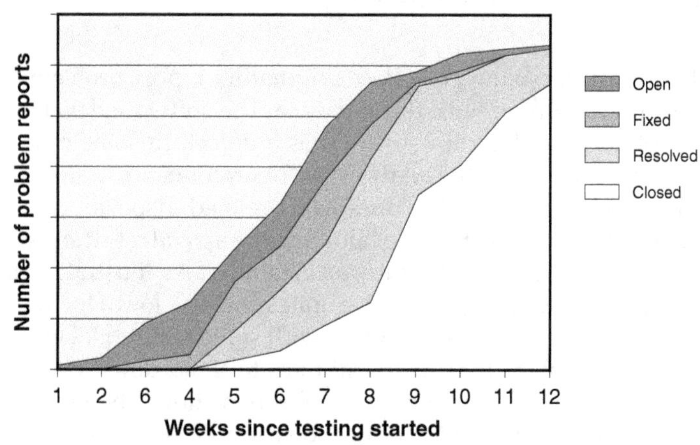

Figure 19.12 Cumulative problem reports by status—examples.

software product's quality before that software is ready for release. Therefore, tracking the *problem report backlog* over time provides additional information. The cumulative problem reports by status metric illustrated in Figure 19.12 combine arrival rate information with problem report backlog information. In this example, four problem report statuses are used:

- *Open*: The problem has not been debugged and corrected (or closed as not needing correction) by development)
- *Fixed*: Development has corrected the defect, and that correction is awaiting integration testing
- *Resolved*: The correction passed integration testing and is awaiting system testing
- *Closed*: The correction passed system testing, or the problem was closed because it was not a defect

The objectives for this metric, as testing nears completion, are for the arrival rate trend (the shape of the cumulative curve) to flatten—indicating that the software has stabilized—and for all of the known problems to have a closed status.

Quality—Data Quality

The *ISO/IEC Systems and Software Engineering—Systems and Software Quality Requirements and Evaluation (SQuaRE)—Measurement of Data Quality* standard (ISO/IEC 2015) lists the following quality measures for data quality and provides associated measures:

- *Accuracy*: A measure of the level of precision, correctness, or freedom from error in the data
- *Completeness*: A measure of the extent to which the data fully include all of the required attribute values for its associated entity
- *Consistency*: A measure of the extent to which the data are free from contradictions and inconsistencies, including the data's strict and uniform adherence to prescribed data formats, types, structures, symbols, notations, semantics, and conventions
- *Credibility*: A measure of the degree to which the users believe the data truly reflect the actual attribute values of the entities they describe
- *Currentness*: A measure of timeliness of the data—are the data up-to-date enough for its intended use
- *Accessibility*: A measure of the ease with which the data can be accessed
- *Compliance:* A measure of the data's adherence to required standards, regulations, and conventions
- *Confidentiality*: A measure of the degree to which only authorized users have access to the data
- *Efficiency*: A measure of the degree to which the data's performance characteristics can be accessed, stored, processed, and provided within the context of specific resources (including, memory, disk space, processor speed, and so on)
- *Precision*: A measure of the data's exactness and discrimination, representing the true values of the attributes for its associated entity
- *Traceability*: A measure of the degree to which an audit trail exists of any events or actions that accessed or changed the data
- *Understandability*: A measure of the degree to which the data can be correctly read and interpreted by the users
- *Availability*: A measure of the extent to which the data are available for access and use when needed

- *Portability*: A measure of the effort required to migrate the data to a different platform or environment
- *Recoverability*: A measure of the effort required to retrieve, refresh, or otherwise recover the data if they are lost, corrupted, or damaged in any way

Amount of Test Coverage Needed

There are numerous metrics that can help predict the amount of test coverage needed. For example, for unit-level white-box tests, metrics for measuring the needed amount of test coverage might include the following:

- *Number of source code modules tested*: At its simplest, coverage is measured to confirm that all source code modules are unit-tested
- *Number of lines of code*: Coverage is measured to confirm that all statements are tested
- *Cyclomatic complexity*: Coverage is measured to confirm that all logically independent paths are tested
- *Number of decisions*: Coverage is measured to confirm that all choices out of each decision are tested
- *Number of conditions*: Coverage is measured to confirm that all conditions used as a basis for all decisions are tested

For integration testing, metrics for measuring the needed amount of test coverage might include the following:

- *Fan-in and fan-out*: During integration testing, coverage is measured to confirm that all interfaces are tested
- *Integration complexity*: Coverage is measured using McCabe's integration complexity metric to confirm that independent paths through the calling tree are tested

For functional (black box) testing, metrics for measuring the needed amount of test coverage might include the number of requirements and the historic number of test cases per requirement. By multiplying these two metrics (Number of requirements × Historical number of test cases per requirement), the number of functional test cases for the project can be estimated. A forward-traceability metric can be used to track test case coverage during test design. For example, dividing the number of functional requirements that trace forward to test cases by the total number of functional requirements provides an estimate of the completeness of functional test case coverage. A graph such as the one depicted in Figure 19.13 can then be used to track the completeness of test coverage against the number of planned test cases during test execution.

Requirements Volatility

Requirements volatility, also called *requirements churn* or *scope creep*, measures the amount of change in the requirements once they are baselined. Jones (2008) reports that in the United States, requirements volatility averages range from:

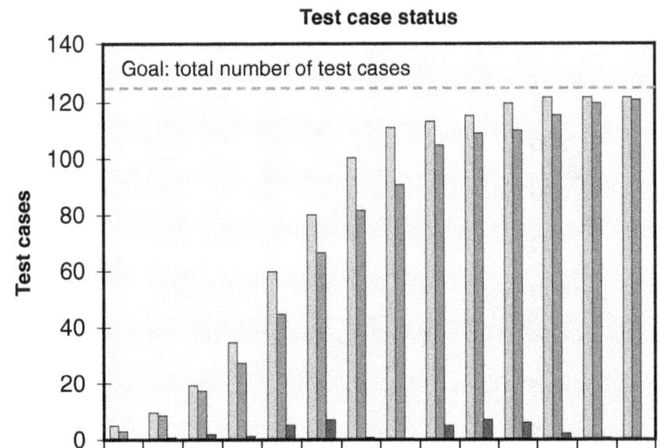

Period	1	2	3	4	5	6	7	8	9	10	11	12	13	14
Executed	5	10	20	35	60	80	100	110	113	115	120	122	122	122
Passed	4	9	18	27	45	67	82	91	105	109	110	115	120	121
Blocked		1	3	2	5	7	1	0	5	7	6	3	1	0

Figure 19.13 Completeness of test coverage—examples.

- 0.5 percent monthly for end-user software
- 1.0 percent monthly for management information systems and outsourced software
- 2.0 percent monthly for systems and military software
- 3.5 percent monthly for commercial software
- 5.0 percent monthly for Web software
- 10.0 percent monthly for agile projects, however, Jones states that the "high rate of creeping requirements for agile projects is a deliberate part of the agile method"

It is not a question of "if the requirements will change" during a project, but "how much will the requirements change?" Since requirements change is inevitable, it must be managed, and to appropriately manage requirements volatility, it must be measured. Figure 19.14 is an example of a graph that tracks requirements volatility as the number of changes to requirements over time. The line on this graph reports the current requirements size (number of requirements or weighted requirements). The data table details the number of requirements added, deleted, and modified. This detail is necessary to understand the actual requirements volatility. For example, if five requirements are modified, two new requirements are added, and two other requirements are deleted, the number of requirements remains unchanged even though significant change has occurred.

Another example of a requirements-volatility metric is illustrated in Figure 19.15. Instead of tracking the number of changes, this graph looks at the percentage of baselined requirements that have changed over time. A good project manager understands that the requirements will change and takes this into

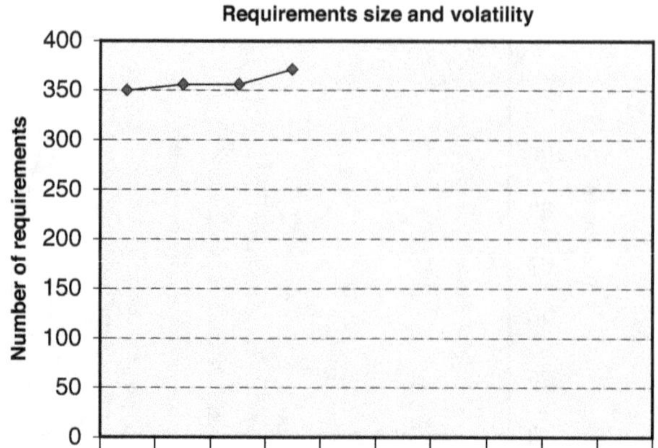

	1	2	3	4	5	6	7	8	9	10
Added	3	5	2	17						
Deleted	0	0	3	1						
Modified	6	12	18	4						
Total	352	357	356	372						

Months since initial requirements baseline

Figure 19.14 Requirements volatility: change to requirements size—example.

consideration when planning project schedules, budgets, and resource needs. The risk is not that requirements will change—that is a given. The risk is that more requirements change will occur than the project manager estimated when planning the project. In this example, the project manager estimated 10 percent requirements volatility and then tracked the actual volatility to that threshold. In this example, the number of changed requirements is the cumulative count of all requirements added, deleted, or modified after the initial requirements are baselined and before the time of data extraction. If the same requirement is changed more than once, the number of changed requirements is incremented once for each change. These metrics act as triggers for contingency plans based on this risk.

Other metrics used to measure requirements volatility include:

- *Function point churn*: The ratio of the function point changes to the total number of function points baselined
- *Use case point churn*: The ratio of the use case point changes to total number of use case points baselined

Reliability

Reliability is a quality attribute describing the extent to which the software can perform its functions without failure for a specified period of time under specified conditions. The actual reliability of a software product is typically measured in terms of the number of defects in a specific time interval (e.g., the number of failures per month) or the time between failures (mean time to failure). Software

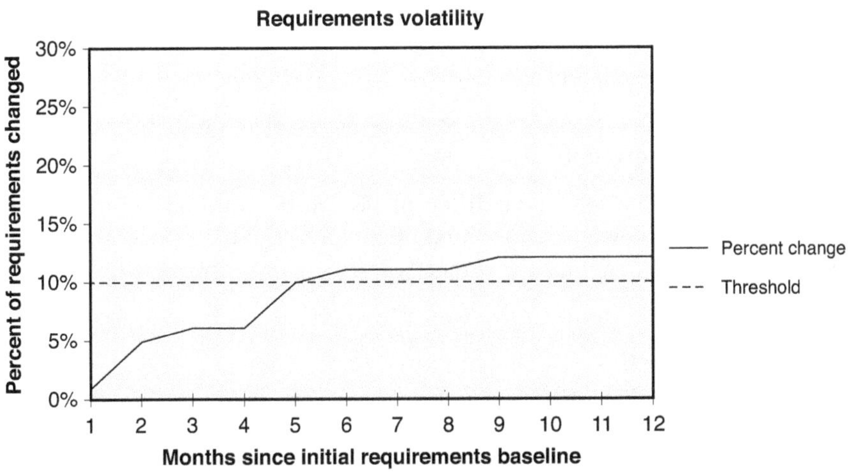

Figure 19.15 Requirements volatility: percentage requirements change—example.

reliability models are utilized to predict the software's future reliability or latent defect count. There are two types of reliability models: static and dynamic.

Static reliability models use other project or software product attributes (e.g., size, complexity, programmer capability) to predict the defect rate or number of defects. Typically, information from previous products and projects is used in these models, and the current project or product is viewed as an additional data point in the same population.

Dynamic reliability models are based on collecting multiple data points from the current product or project. Some dynamic models look at defects gathered over the entire life cycle, while others concentrate on defects found during the formal testing phases at the end of the life cycle. Examples of dynamic reliability models include:

- Rayleigh model
- Jelinski-Moranda (J-M) model
- Littlewood (LW) models
- Goel-Okumoto (G-O) imperfect debugging model
- Goel-Okumoto nonhomogeneous Poisson process (NHPP) model
- Musa-Okumoto (M-O) logarithmic Poisson execution time model
- Delayed S and inflection S models

Appropriately implementing a software reliability model requires an understanding of the assumptions underlying that model. For instance, the J-M model's five assumptions are (Kan 2003):

- There are *N* unknown software faults at the start of testing.
- Failures occur randomly, and times between failures are independent.

- All faults contribute equally to causing a failure.
- Fix time is negligible.
- A fix is perfect for each failure; there are no new faults introduced during the correction.

Various other models make different assumptions than the J-M model. When selecting a model, consideration must be given to the likelihood that the model's assumptions will be met by the project's software environment.

Availability

Availability is a quality attribute that describes the extent to which the software or a service is available for use when needed. Availability is closely related to reliability because unreliable software that fails frequently is unavailable for use because of those failures. Availability is also dependent on the ability to *restore* the software product to a working state. Therefore, availability is also closely related to maintainability in cases where a software defect caused the failure, and that defect must be corrected before the operations can continue.

In the example of an availability metric illustrated in Figure 19.16, post-release availability is calculated each month by subtracting from one the sum of all of the minutes of outage that month, divided by the sum of all the minutes of operations that month, and multiplying by 100 percent. Tracking availability helps relate software failures and reliability issues to their impact on the user community.

When defining reliability and availability metrics for software, care should be taken not to use formulas intended for calculating hardware reliability and availability to evaluate software. This is because:

- Unlike hardware, software does not wear out or fail over time because of aging

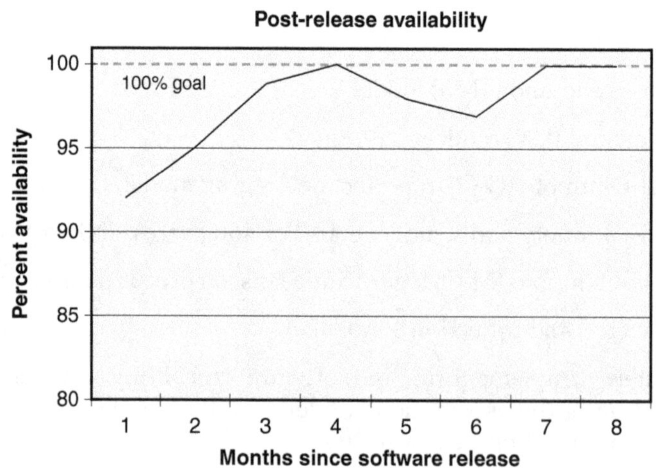

Figure 19.16 Availability—example.

- Unlike hardware, where manufacturing can interject the same error multiple times (bent pins, broken wires, solder splashed), once a defect is found and removed from the software, that defect is gone, assuming good configuration management practices do not allow the defective component to be built back into the software by mistake

System Performance

System performance metrics are used to evaluate several execution characteristics of the software. These can be measured during testing to evaluate the product's future performance, or during actual execution. Examples of performance metrics include:

- *Throughput*: A measure of the amount of work performed by a software system over a period of time, for example:
  - Mean or maximum number of transactions per time period
  - Percentage of throughputs per period that are compliant with requirements
- *Response time*: A measure of how quickly the software reacts to a given input or how quickly the software performs a function or activity, for example:
  - Mean or maximum response time
  - Percentage of response times that are compliant with requirements
- *Resource utilization*: A measure of the average or maximum amount of a given resource used by the software to perform its functions, for example:
  - Mean or maximum memory utilization
  - Mean or maximum disk space or other storage device utilization
  - Mean or maximum bandwidth utilization
  - Mean or maximum input/output (I/O) device utilization
  - Mean or maximum processor utilization
  - Percentage of the time that the utilization of a given resource is compliant with requirements
- *Accuracy*: A measure of the level of precision, correctness, or freedom from error in the calculations and outputs of the software
- *Capacity*: A measure of the maximum number of activities, actions, or events that can be concurrently handled by the software or system
  - Maximum number of users when software is in a specified mode
  - Maximum number of simultaneous transactions when software is in a specified mode
  - Maximum number of peripherals that can be interfacing with the software when that software is in a specified mode

Maintainability

Maintainability is a quality attribute describing the ease with which a software product or one of its components can be modified after it has been released. Several metrics can be used to measure maintainability. Examples include:

- *Mean time to change* the software when an enhancement request is received
- *Mean time to fix* the software when a defect is detected
- The number of function points a maintenance programmer can support in a year
- Maintainability can also be measured as a function of other metrics such as coupling, cohesion and complexity

Usability

Usability is a quality attribute describing the effort the users must expend to learn and use the software. Usability is measured in terms of five main factors:

- *Speed*: A measure of how quickly users can accomplish their tasks using the software (e.g., measured in the mean time to perform a task or the number of keystrokes or page views required to perform a task)
- *Efficiency*: A measure of how many mistakes are made by the users while performing their tasks (e.g., measured in the average number of mistakes per task, or the average number of mistakes per usage time)
- *Learnability*: A measure of how easy it is, or how much training or time it takes, to learn to use the software proficiently (e.g., measured in the average number of hours of training required to reach a specific level of proficiency, minimum speed, or error rate)
- *Memorability*: A measure of how easy it is for the users to remember how to use the software once they have learned how to use it (e.g., if a user does not use the software for a period of time, what is the average amount of time required to relearn the software once they start using it again)
- *User preference*: A measure of what the users like about the software (e.g., the number of users that reported liking the software or a software feature on a survey or during an interview)

There are typically engineering trade-offs between these various factors and between these factors and other attributes of the software product. For example, the speed of data entry might be a trade-off with efficiency. That is, the faster the user enters data, the more mistakes they make. The speed of accessing the software or its data might also require an engineering trade-off against the software's security or the data's integrity. The goal is to optimize as many of these usability factors as possible, balanced against optimizing other product attributes based on the priorities of the various software product attributes (i.e., systems thinking).

3. SOFTWARE PROCESS METRICS

> Measure the effectiveness and efficiency of software processes (e.g., functional verification tests (FVT), cost, yield, customer impact, defect detection, defect containment, total defect containment effectiveness (TDCE), defect removal efficiency (DRE), and process capability). (Apply)
>
> **Body of Knowledge V.A.3**

Software process entities can be development, maintenance, or acquisition activities, such as the entire software development process from requirements through delivery to operations, or small individual activities, such as inspecting a single document. Software process entities can also be time intervals, which may not necessarily correspond to specific activities. Examples of time intervals include the month of January or the first six months of operations after delivery.

Examples of attributes associated with process entities include cycle time, cost, the number of incidents that occurred during that process (e.g., the number of defects found, the number of pages inspected, the number of tasks completed), controllability, productivity, efficiency, effectiveness, stability, and capability.

Cost

The *cost* of a process is typically measured as either the amount of money spent or the amount of effort expended to implement an occurrence of that process, for example:

- The number of U.S. dollars (money) spent for conducting an audit
- The number of engineering hours (effort) expended preparing and conducting a peer review

Collecting measurements on the average cost of a process or cost distributions over multiple implementations of that process can help identify areas of inefficiencies and low productivity that provide opportunities for improvement to that process. After those improvements are implemented, the same cost metrics can be used to measure the impacts of the improvements on *process costs* to evaluate the success of the improvement initiatives.

Cost metrics can be used to estimate the costs of a project and project activities as part of project planning and to track actual costs against budgeted estimates as part of project tracking. The collection and evaluation of process cost metrics can help an organization understand where, when, and how they spend money and effort on their projects. This understanding can help improve the cost estimation models for predicting costs on future projects.

First-Pass Yield

First-pass yield evaluates the effectiveness of an organization's defect prevention techniques by looking at the percentage of products that do not require rework after the completion of a process. First-pass yield is calculated as a percentage of the number of items not requiring rework because of defects after the completion of the process, divided by the total number of items produced by that process. For example, if the coding process resulted in the creation and baselining of 200 source code modules, and 35 of those modules were later reworked because defects were found in those 35 modules, then the first-pass yield for the coding process is $([200-35]/200) \times 100\% = 82.5\%$. As another example, if 345 requirements are documented and baselined as part of the requirements development process, and 69 of those requirements were modified after baselining because of defects, then the first-pass yield of requirements development is $([345-69]/345) \times 100\% \approx 80\%$.

Cycle Time

Cycle time measures the amount of calendar time it takes to perform a process from start to completion. Knowing the cycle times for the processes in an organization's software development life cycle allows better estimates to be done for schedules and required resources. It also enables the organization to monitor the impacts of process improvement activities on the cycle time for those processes. Cycle time can be measured as either static or dynamic cycle time.

Static cycle time looks at the average time it takes to perform the process. Cycle time helps answer questions such as "How long, on average, does it take to code a source code module, to correct a software defect, or to execute a test case?" For example, if four source code modules were programmed this week and they took five days, ten days, seven days, and eight days, respectively, to program, the static cycle time $= (5+10+7+8)/4 = 7.5$ days per module.

Dynamic cycle time is calculated by dividing the number of items in progress (items that have only partially completed the process) by one-half of the number of new starts plus new completions during the period. For example, if 52 source code modules were started this month, 68 source code modules were completed this month, and 16 source code modules were in progress at the end of the month, the dynamic cycle time $= (16/[(52+68)/2]) \times 23$ days (the number of working days this month to convert months to days) ≈ 6.1 days per module.

Customer Impact—Customer Satisfaction

Customer satisfaction or *stakeholder satisfaction* is essential to staying in business in this modern world of global competition. An organization must satisfy and even delight its customers and other stakeholders with the value of its software products and services to gain their loyalty and repeat business. So how satisfied are an organization's customers and other stakeholders? The best way to find out is to ask them by using customer satisfaction surveys.

Several metrics can result from the data collected during these surveys. These metrics can provide management with the information they need to determine their customers' level of satisfaction with their software products and with the

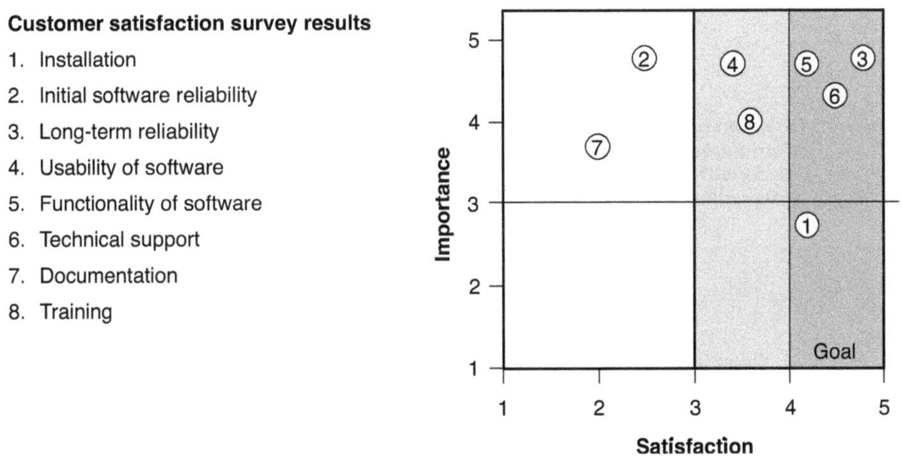

Figure 19.17 Customer satisfaction summary report—example.

services and processes associated with those products. Software engineers and other technical staff members can use these metrics to identify opportunities for ongoing process improvements and monitor the impact of those improvements. Similar metrics with different questions on the surveys can also be used to measure other types of stakeholders, (e.g., employee satisfaction surveys or supplier satisfaction surveys).

Figure 19.17 illustrates an example of a metrics report that summarizes the customer satisfaction survey results and indicates the current customer or user satisfaction level. The average satisfaction and importance values are plotted as a numbered bubble on an x–y graph for each quality attribute polled on the survey. It should be remembered that to make a calculation of average satisfaction level valid, a ratio scale measure should be used (e.g., a range of zero to five, with five being very satisfied). If an ordinal scale metric is used, the median should be used as the measure of central tendency. The darker-shaded area on this graph indicates the long-term goal of having an average satisfaction score of better than four for all quality attributes. The lighter-shaded area indicates a shorter-term goal of having an average satisfaction score better than three. From this summary report, it is possible to quickly identify "initial software reliability" (bubble 2) and "documentation" (bubble 7) as primary opportunities to improve customer satisfaction. By polling importance as well as satisfaction level in the survey, the person analyzing this metric can see that even though documentation has a poorer satisfaction level, initial software reliability is much more important to the customers and, therefore should probably be given a higher priority.

Figure 19.18 illustrates an example metrics report showing the distribution of satisfaction scores for three questions. Graphs, where the scores are tightly clustered around the mean (question A) indicate a high level of agreement among the stakeholders on their satisfaction level. Distributions that are widely spread (question B), or particularly bimodal distributions (question C), are candidates for further detailed analysis.

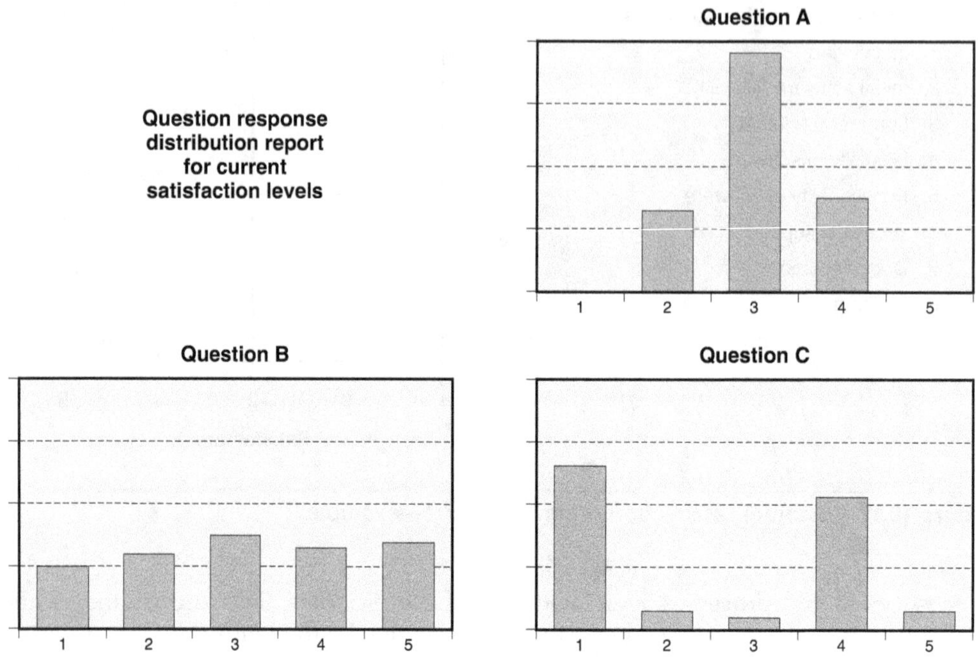

Figure 19.18 Customer satisfaction detailed report—example.

Another way to summarize the results of a satisfaction survey is to look at trends over time. Figure 19.19 illustrates an example of a metrics report that trends the initial software reliability based on quarterly surveys conducted over 18 months. Again, this graph's dark- and light-shaded areas indicate the long- and short-term satisfaction level goals. One note of caution is that to trend the results over time, the survey must remain unchanged in the area being trended. Any rewording of the questions on the survey can have a significant impact on the survey results. Therefore, historic responses from before wording changes should not be used in future trends.

The primary purpose of trend analysis is to determine if the improvements made to the projects, products, services, or processes impacted the satisfaction level. It should be remembered, however, that satisfaction is a lagging indicator because it provides information only about what has already occurred. Customers have long memories; the dismal initial quality of a software version from three releases ago may still impact their perception of the product even if the last two versions have been superior.

Customer satisfaction can be considered both a product and a *process metric.* It measures the stakeholder's satisfaction with the current product and services and the supporting projects and processes like on-time delivery, installation, and customer support. It can also be used as a process measure to measure the impacts of process improvements.

Customer satisfaction is a subjective measure. It is a measure of perception, not reality, although perception is more important than reality when it comes to happy stakeholders. One phenomenon that often occurs is that as the quality

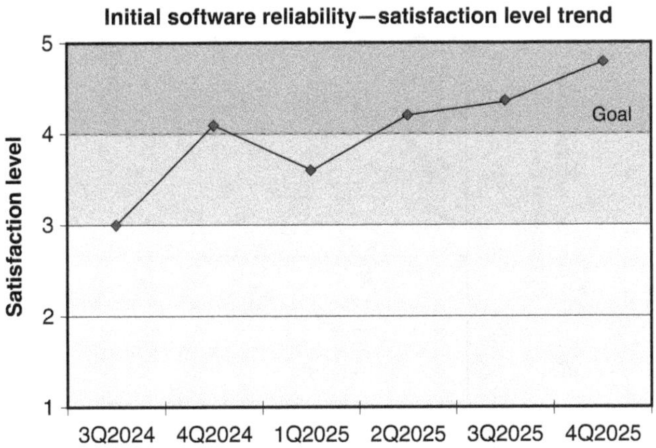

Figure 19.19 Customer satisfaction trending report—example.

of software improves, the expectations of the stakeholders also increase. Stakeholders continue to demand bigger, better, faster software. This can result in a flat trend even though the quality is continuously improving. Worse still, it can cause a declining graph if improvements to quality are not keeping up with the increases in the stakeholder's expectations. Even though this impact can be discouraging, it is valuable information that the organization needs to know in the very competitive world of software.

Customer Impact—Responsiveness to Reported Problems

When a customer or user has a problem with the software, the developer must respond quickly to resolve and close the reported problem. Service level agreements may define problem report response time goals based on the severity of the incident (e.g., critical problems within 24 hours, major problems within 30 days, and minor problems within 120 days). The graph in Figure 19.20 illustrates an example of a metric to track actual performance against these service level agreements. This graph trends the percentage of problem reports closed within the service level agreement time frames each month. This metric is a *lagging indicator,* a view of the past. Using lagging indicator metrics is like painting the front windshield of a car black and monitoring the quality of driving by counting the dead bodies that can be seen in the rearview mirror. If an organization must wait until the problem report is closed before tracking this response-time metric, it cannot take proactive action to control its responsiveness.

The graph in Figure 19.21 shows all nonclosed, major problem reports distributed by age and the number of days since they were opened. Analysis of this graph can quickly identify problem areas, including problem reports that are past their service level agreement goal and reports approaching their goal. This information allows a proactive approach to controlling responsiveness to customer- or user-reported problems. This metric is a *leading indicator* because it helps proactively identify issues. Similar graphs can also be created for minor

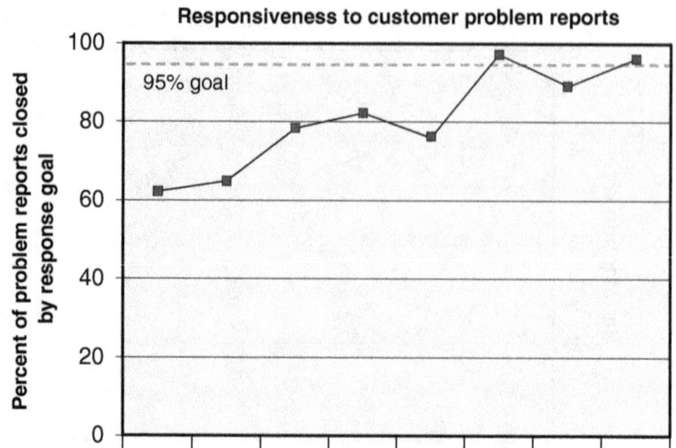

	1	2	3	4	5	6	7	8
Percent meeting goal	62	65	79	82	76	97	89	96
Number non-closed	1	3	2	1	4	2	1	0
Average age	22	34	48	30	27	38	23	0

Months since software implementation

Figure 19.20 Responsiveness to customer problems—example.

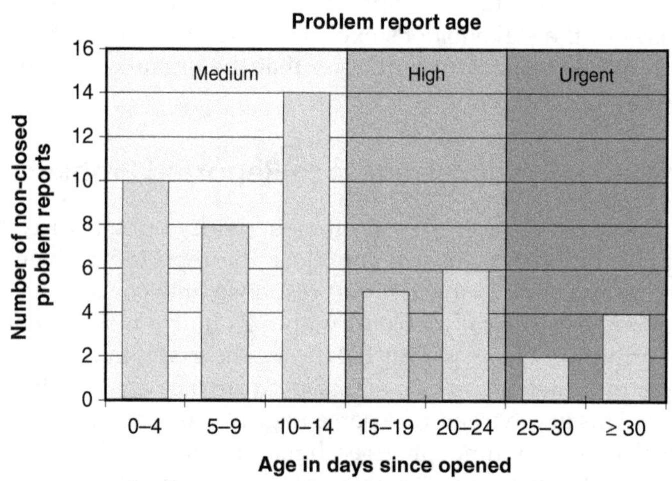

Figure 19.21 Defect backlog aging—example.

problems (since critical problems must be corrected within 24 hours, it is probably not cost-effective to track them with this metric).

Efficiency of Defect Detection—Escapes

One way to measure the effectiveness of a process is to measure the number of escapes from that process. An *escape* is any issue, defect, or nonconformance that exists during the process but is not discovered by that process. To explain the concept of escapes, let's first talk about escapes from the various defect detection processes. An analogy for these escapes is to think of each defect detection process as a screen door that catches bugs (defects). Escapes are the bugs that make it through the screen door and get farther into the house. A person cannot determine how effective the screen door is by looking at the bugs it catches. The person must go inside and count the number of bugs that got past the screen door.

In software, there is no way of counting exactly how many defects escaped a defect detection process. The number and origin of defects detected by subsequent defect detection processes must be examined to approximate the number of actual escapes. For example, as illustrated in Figure 19.22a, at the end of the first defect detection process (requirement defect detection—typically a requirement peer review), there are no known escapes.

However, as illustrated in Figure 19.22b, after analyzing the defects found by the second design defect detection process, there are not only design-type defects (seven dark gray bugs) but also requirement-type defects (three light gray bugs), so three requirement escapes have been identified.

As illustrated in Figure 19.22c, after analyzing the defects found by the third coding defect detection process, there are not only coding-type defects (seven black bugs) but also requirement-type defects (one light gray bug) and design-type defects (three dark gray bugs). So four requirement escapes and three design escapes have now been identified.

As illustrated in Figure 19.22d, everything found in the testing defect detection processes are escapes because no new defects are introduced through the testing process (new defects introduced during testing are the result of requirement, design, or code rework efforts). Analysis of defects found in testing shows requirement-type defects (two light gray bugs), design-type defects (five

Figure 19.22a Measuring escapes—requirements example.

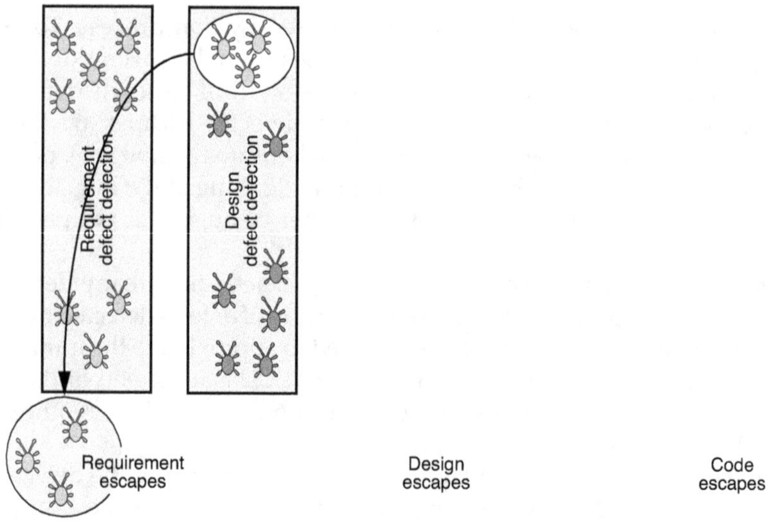

Figure 19.22b Measuring escapes—design example.

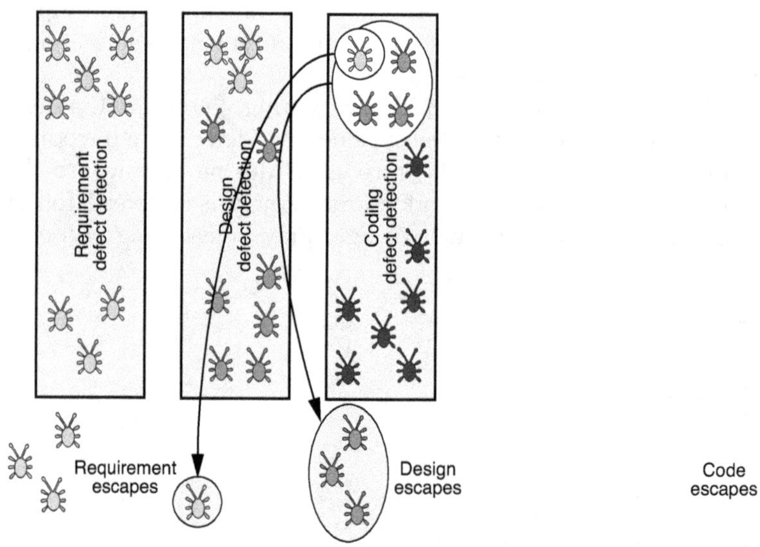

Figure 19.22c Measuring escapes—coding example.

dark gray bugs), and coding-type defects (three black bugs). There are now six requirement escapes, eight design escapes, and three coding escapes.

Finally, as illustrated in Figure 19.22e, everything found in operations is also an escape. Analysis of defects found in operations shows requirement-type defects (one light gray bug), design-type defects (three dark gray bugs), and coding-type defects (four black bugs). There are now seven requirement escapes, eleven design

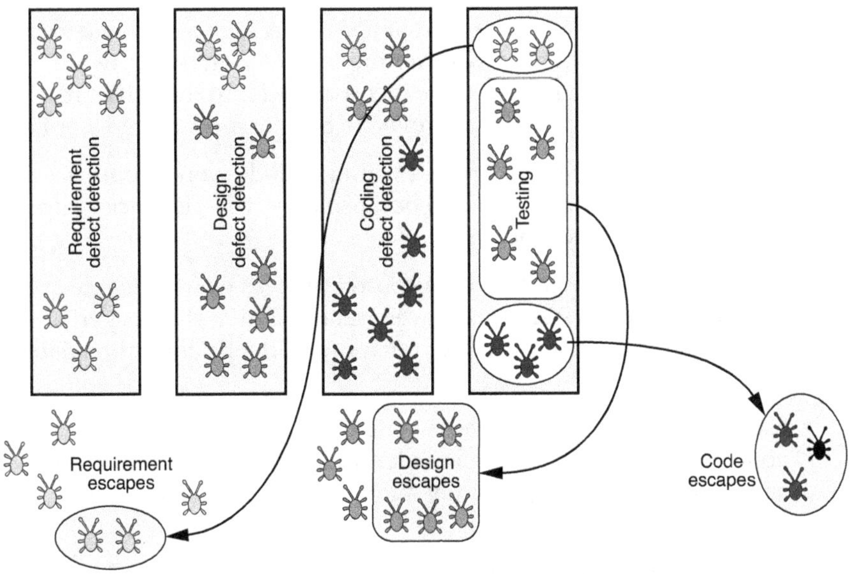

Figure 19.22d Measuring escapes—testing example.

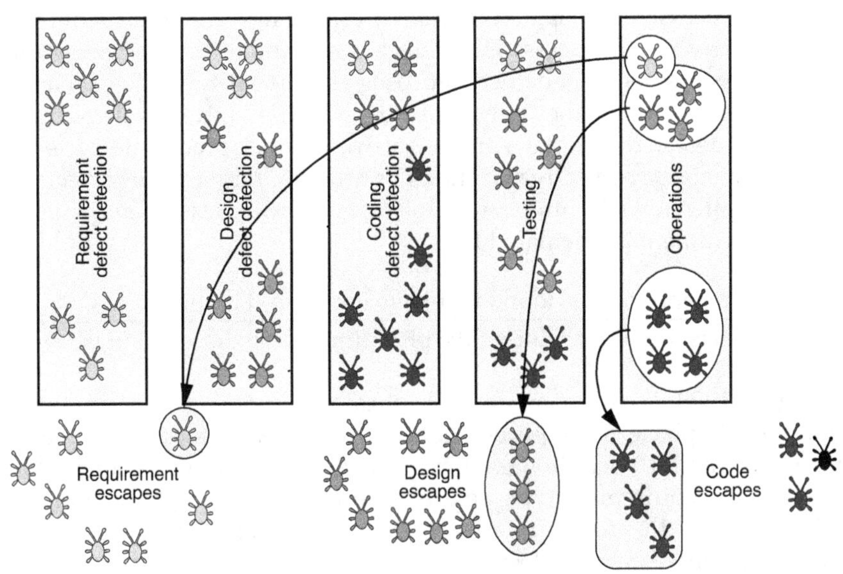

Figure 19.22e Measuring escapes—operations example.

escapes, and seven coding escapes. Of course, these counts will continue to change if additional defects are identified in operations.

However, escapes are not just limited to defect detection processes. Escapes can be examined to evaluate the effectiveness of other processes as well. For example:

- Escapes from the configuration change control process would include approved changes that were never implemented in the software, unauthorized changes that were incorporated into the software, or building the incorrect version/revision of a component into a build
- Escapes from the backup process would include work products or data items that could not be restored because they were not backed up or were corrupted after backup
- Escapes from the quality record control process would include quality records that were not retained, were retained for a shorter period of time than required, or were not retrievable, identifiable, or readable after retention
- Escapes from an audit process would include existing nonconformances that were not identified during the audit but were later discovered during another audit or that later caused issues

Defect Containment and Total Defect Containment Effectiveness (TDCE)

Defect containment effectiveness generally looks at the effectiveness of defect detection processes to keep defects from escaping into later phases or operations. *Phase containment effectiveness* is a defect containment metric used to measure the effectiveness of defect detection processes in identifying defects in the same phase (or iteration) as they were introduced. Many studies demonstrate that defects not detected until later in the software development life cycle are much more costly to correct. Organizations can better target their defect detection improvement efforts by understanding which processes are allowing escapes. Phase containment is calculated by:

$$\frac{\text{Number of defects found that were introduced in the phase}}{\text{Total defects introduced during the phase (including those found later)}} \times 100\%$$

Figure 19.23 illustrates an example of calculating phase containment. For the requirements phase, 15 requirement-type defects were found and fixed during that phase, and 10 requirement-type defects were found in later phases, for a total of 25 requirement-type defects. The requirements phase containment is 15/25 = 60%. For the design phase, 29 design-type defects were found and fixed during that phase, and 12 design-type defects were found in later phases, for a total of 41 design-type defects. The design phase containment is 29/41 ≈ 71%. To continue this example, for the code phase, 86 code-type defects were found and fixed during that phase, and 26 code-type defects were found in later phases for a total of 112 code-type defects. The code phase containment is 86/112 ≈ 77%. Since the requirements phase containment percentage is the lowest, the requirements phase would be considered the best target for process improvement.

While this example shows the phase containment metrics being calculated after the software has been in operations for some period of time, in real use, the phase containment metrics should be calculated after each phase (or iteration).

Phase detected	Phase detected defect was introduced		
	Requirements	Design	Code
Requirements	15		
Design	5	29	
Code	1	7	86
Test	3	3	19
Field	1	2	7
Total	25	41	112

Requirements: 15 / 25 = 60%
Design: 29 / 41 ≈ 71%
Code: __ / __ = __%
Prerelease: __ / __ = __%

Figure 19.23 Defect containment effectiveness—example.

At the end of the requirements phase, the requirement phase containment is 15/15 = 100%. At the end of the design phase, the requirement phase containment is 15/20 = 75%. These ongoing calculations can be compared with average values baselined from other projects to determine if corrective action is necessary at any time. For example, instead of five requirement-type defects being found in the design phase, assume that 30 requirement-type defects were found. The requirements phase containment would be calculated as 15/55 ≈ 27% in this case. This much lower value might indicate that corrective action in terms of additional requirements-type defect detection activities should be performed before proceeding into the code phase.

The *total defect containment effectiveness (TDCE)* metric shows the effectiveness of defect detection processes in identifying defects before the product is released into operations. TDCE is calculated as:

$$\frac{\text{Number of defects found prior to release}}{\text{Total defects found (including those found after release)}} \times 100\%$$

For the example in Figure 19.23, 24 requirement-type defects, 39 design-type defects, and 105 code-type defects, for a total of (24 + 39 + 105) = 168 defects, were found prior to release. Total defects found (including those found after release) is (25 + 41 + 112) = 178. The TDCE for this example is 168/178 ≈ 94%. The TDCE for this project could then be compared with the TDCE for previous projects to determine if it is at an appropriate level. If process improvements have been implemented, this comparison can be used to determine if the improvements positively impact the TDCE metric value.

Defect Removal Efficiency

Defect removal efficiency (DRE), also called *defect detection efficiency* or *defect detection effectiveness,* is a measure of the effectiveness of a defect detection and removal process. DRE is a measure of the percentage of all defects in the software that were found and removed when a detection and rework process was executed. Unlike phase containment, this metric can be calculated for each defect detection process (instead of by phase or iteration). For example, if both code reviews and unit testing were done in the coding phase, each process would have its own DRE percentage. DRE also includes all defects that process could have found, not just the ones introduced during that phase. DRE is calculated by:

$$\frac{\text{Number of defects found and removed by the process}}{\text{Total number of defects present at the start of the process}} \times 100\%$$

Figure 19.24a illustrates an example of the calculation of DRE. Note that the DRE for the requirements review process is $15/25 = 60\%$. Since this is the first defect detection and removal process and the only types of defects available to find are requirement-type defects, then the phase containment and DRE numbers are the same. However, in the design processes, not only can design-type defects be found, but the defects that escaped from the requirements phase can also be found. As illustrated in Figure 19.24a, there were five requirement-type defects and 29 design-type defects found during the design review. However, an additional five requirement-type defects and 12 design-type defects escaped the design review process and were found later. The DRE for the design review is $(5+29)/(5+29+5+12) \approx 67\%$.

As illustrated in Figure 19.24b, the DRE for the code review is $(1+3+54)/(1+3+54+4+9+58) \approx 45\%$. To complete this example, the DRE for:

- Unit testing $= (0+4+32)/(0+4+32+4+5+26) \approx 51\%$
- Integration testing $= (1+3+13)/(1+3+13+3+2+13) \approx 49\%$
- System testing $= (2+0+6)/(2+0+6+1+2+7) \approx 44\%$

Detection technique	Phase detected defect was introduced		
	Requirements	Design	Code
Requirements review	15		
Design review	5	29	
Code review	1	3	54
Unit testing	0	4	32
Integration testing	1	3	13
System testing	2	0	6
Operations	1	2	7
Total	25	41	112

Design review: (5 + 29) / (5 + 29 + 5 + 12) = 34 / 51 ≈ 67%

Figure 19.24a Defect containment effectiveness—design review example.

Detection technique	Phase detected defect was introduced		
	Requirements	Design	Code
Requirements review	15		
Design review	5	29	
Code review	1	3	54
Unit testing	0	4	32
Integration testing	1	3	13
System testing	2	0	6
Operations	1	2	7
Total	25	41	112

Code review: ((1 + 3 + 54)) / ((1 + 3 + 54) + 4 + 9 + 58) = 58 / 129 ≈ 45%

Figure 19.24b Defect containment effectiveness—code review example.

Note that of the peer reviews, the coding peer review had the lowest DRE in this example at 45 percent. Of the testing activities, system testing had the lowest DRE. Therefore, these two activities would be candidates for process improvement.

As with phase containment effectiveness, while these examples show the defect removal efficiency metrics being calculated after the software has been in operation for some period of time, in use, these metrics should be calculated after each major defect detection or removal process and compared with baseline values to identify potential issues that need corrective action.

Process Capability

When measuring process capability, the software industry offers guidance in the form of the CMMI® (2023) and the ISO/IEC 15504 (2004) set of standards. Based on these documents, each practice or process area is assigned a capability level from zero to five based on whether the implementation of that practice or process meets defined criteria for good industry practice.

Many of the metrics already discussed can also be used to understand the process capabilities of each individual process, help an organization recognize the amount of variation in its processes, and act as predictors of future process performance. These metrics include first-pass yield, phase containment, defect detection efficiency, cycle time, productivity, cost, and the product measures of the products being produced by the process. For example, the higher the defect density of requirements specification, the lower the capability of the requirements development processes.

Project and Risk Metrics

In Chapter 17, several software project metrics were discussed for tracking projects. In Chapter 18, several software risk metrics were discussed for analyzing, planning for, and tracking risks.

Agile Metrics

One of the primary practices of eXtreme programming (XP) is the *informative workspace,* which uses the workspace to communicate important, active information. For example, story cards (sorted by "done," "to be done this week," "to be done this release") or visible charts for tracking progress over time of important issues might be posted on the wall. Beck (2005) says, "An interested observer should be able to walk into the team space and get a general idea of how the project is going in 15 seconds. He should be able to get more information about real or potential problems by looking more closely."

Another agile methodology, called *crystal,* has a similar strategy called *information radiators.* Information radiators are displayed project documentation used or placed in areas where people can easily see them. Examples of information radiators include posting (Cockburn 2005):

- Flip-chart pages from facilitated sessions
- Index cards or sticky notes with documented user stories
- Butcher paper depicting project timelines and milestones
- Photographs of, or printouts from, whiteboard results from informal discussions
- Computer-generated domain models, graphs, or other outputs printed to plotters that produce large-size images

Information radiators show status such as the:

- Current iteration's work set (use cases or stories)
- Current work assignments
- Number of tests written or passed
- Number of use cases or stories delivered
- Status of key servers (up, down, or in maintenance)
- Core of the domain model
- Results of the last reflection workshop

Agile uses metrics called *burn charts,* discussed in Chapter 17. Burn charts show how the project's estimations (predictions) of the amount of work planned for the iteration or release compare to the work actually accomplished. Figure 17.9 shows a *burn-up chart* example. A burn-up chart starts with zero and tracks progress by graphing the number of items completed up to the ceiling (goal). A *burn-down chart* starts with the total number of items that need to be done and tracks progress by graphing the number of items left to do down to zero items left to do.

One of the core practices of Kanban is to visualize the workflow. To do this, a Kanban team typically uses a *cumulative flow diagram* (CFD), as illustrated in Figure 19.25. A CFD is an area graph that shows the cumulative number of work items that have gone through each activity on the Kanban board.

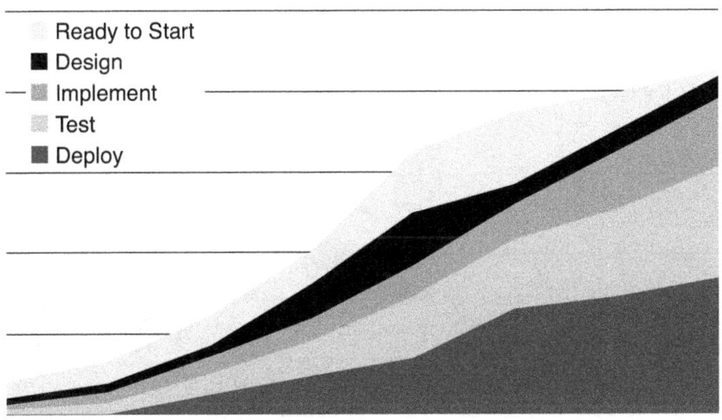

Figure 19.25 Cumulative flow diagram.

As illustrated in Figure 19.26, other Kanban metrics include:

- *Cycle time*: The elapsed time spent working on an item. This metric is measured as the average time in work from the time work starts on an item until that work is completed ("done")
- *Lead time*: The average wait time in the queue from the time the work item is defined until the time it is moved to in-work
- *New work items*: The number of newly added work items during a period of time
- *Total work in progress (WIP)*: The total number of work items that are currently being worked on. This metric is measured by adding WIP from each work activity. In the example in Figure 19.26 that would be the sum of work items in design, implement, and test.
- *Total work items remaining to be done*: The total number of work items that have been defined but not yet deployed. For the example in Figure 19.26, that would be the total work in progress plus the number of ready to start items.
- *Flow rate*: The rate at which items are moving through an activity. This metric is measured as the slope of the line at the top of that work activity. If the flow rate goes flat, it indicates a bottleneck in that work activity.

Other agile metrics include the following:

- *Rework percentage*: The effort spent on rework after the work item is "done" divided by the total effort spent on that work item
- *Release completion percentage*: The number of completed work items divided by the total number of work items planned for the release
- *Value delivered predicted vs. actual*: The product owner and or stakeholders rank each work item by value (e.g., on a scale of 1–10) before the start of the iteration and again at the demo meeting or end of the iteration

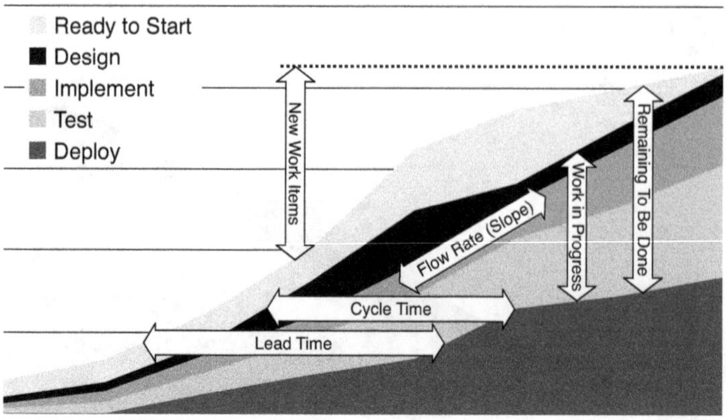

Figure 19.26 Other Kanban metrics.

Many of the other product and process metrics discussed earlier in this chapter, project tracking metrics discussed in Chapter 17, and risk management metrics discussed in Chapter 19 can also be applied to agile processes.

4. DATA INTEGRITY

> Describe the importance of data integrity from planning through collection and analysis. Apply various techniques to ensure data quality, accuracy, completeness, and timeliness. (Apply)
>
> **Body of Knowledge V.A.4**

The old saying "garbage in, garbage out" applies to software metrics. A software quality engineer (SQE) should be able to analyze the integrity of the data used as inputs when creating information products. An SQE should be able to analyze the metrics design and implementation processes and the data collection and analysis processes to evaluate the quality, accuracy, completeness, and timeliness of the data, metrics, and information products.

If the correct data items are not collected, then the objectives of the measurement program cannot be accomplished. Data analysis is pointless without good data. Therefore, establishing a data collection plan is the cornerstone of any successful metrics program. The data collection plans may be part of standardized metrics definitions at the organizational level or included or tailored in project plans or sub-plans (e.g., measurement plans, communication plans, quality assurance plans). A data collection plan should include who is responsible for the collection and validation of the data, how the data should be collected (e.g., measurement units used, and conditions under which the measurement should be collected), when the data is collected, and where and how the data are stored. Data collection plans may also include information about archiving or retiring data.

Who Should Collect the Data?

Deciding who should collect the data is an essential part of making sure that good data are collected. In most cases, the best choice is the owner of the data. The *data owner* is the person or tool with direct access to the data source and, in many cases, is actually responsible for generating the data. Table 19.2 includes a list of example data owners and the types of data that they own. For example, when a user calls into the help desk, the help desk staff has direct access to the problem identification data being reported by the user. They collect those data into a customer call report database. On the other hand, in some cases, the users themselves have access to a problem-reporting database. In that case, the users own the problem-identification data and collect those data. Whenever possible, tools should be used to collect the data and relieve people of the burden of data collection.

The benefits of having the data owner collect the data include:

- Data owners can collect the data as they are being generated, which helps increase accuracy and completeness.
- Data owners are more likely to detect anomalies in the data as they are being collected, which helps increase accuracy.
- Having data owners collect the data helps to eliminate the human error of duplicate recording (once by data recorder and again by data entry clerk), which helps increase accuracy. If a tool can collect the data that can completely eliminate the possibility of human error.

Once the people who need to gather the data are identified, they must agree to do the work. They must be convinced of the importance and usefulness of collecting the data. Management has to support the program by giving the data owners the time and resources required to perform data collection activities. To quote Watts Humphrey (1989), "The actual work of collecting data is tedious and must generally be done by software professionals who are busy with other tasks. Unless they and their immediate managers are convinced that the data is important, they either will not gather it or will not be very careful when they do."

So what can the metrics providers do to help make certain that the data owners collect good data? First, the metrics providers can design metrics that are as objective and unambiguous as possible. A data item is objective if it is collected the same way each time. Subjective data can also be valuable. For example, customer satisfaction metrics are typically subjective. However, the goal is to make the data as objective as possible. A data item is unambiguous if two different people collecting the same measure for the same item will collect the same data. This requires standardized definitions and well-defined measurement methods.

Second, the metrics provider can design the metrics and establish data collection mechanisms that are as unobtrusive and convenient as possible. Data collection must be an integral part of the software development process and not some outside step that detracts from the "real work." Data collection must be simple enough not to disrupt the working patterns of the individual collecting the

Table 19.2 Data ownership—examples.

Data Owner	Examples of Data Owned
Management	• Schedule • Budget
Software developers	• Time spent per task • Inspection data, including defects found • Root cause of defects
Testers	• Test cases planned / executed / passed • Problem identification data • Test coverage
Configuration management	• Defects corrected per build • Modules changed per build
Users	• Problem-identification data • Operational hours
Tools	• Line of code • Cyclomatic complexity

data any more than absolutely necessary. In some cases this means automating all or part of the data collection. For example, having a pull-down pick list is much more convenient than making the data owner type "critical," "major," or "minor," and it can also contribute to more accurate data by eliminating misspellings or abbreviations. As another example, do not make the data owner type the current date. The computer knows the date, so have the computer default that data item.

Third, the data owners must be trained in how to collect the data so that they understand what to do and when to do it. For simple collection mechanisms, training can be short (≤one hour). Hands-on, interactive training, where the group works with actual data collection examples, often provides the best results. Without this training, hours of support staff time can be wasted answering the same questions over and over again. An additional benefit of training is promoting a common understanding of when and how to collect the data. This reduces the risk of invalid and inconsistent data being collected.

Fourth, the metrics providers must feed the metrics information back to the data owners so that they can see that the data items are used and not just dropped into a black hole. Better yet, the metrics providers can use the data to create metrics directly useful to the data owners. There must also be people assigned to support the data collection effort so that the data collectors can get their questions answered and issues related to data, and data collection problems, are handled in a timely manner.

How Should the Data Be Collected?

The answer to the question, *how should the data be collected?* is automate, automate, automate, and automate:

- Automate data collection and validation as much as possible
- Automate the databases

- Automate data extraction
- Automate metrics reporting and delivery

There is widespread agreement that as much of the data-gathering process as possible should be automated. At a minimum, standardized forms should be used for data collection, but at some point the data from these forms must be entered into a metrics database to have any long-term usefulness. Information that stays on forms can quickly get buried in file drawers, never to see the light of day again. For data to be useful and used, easy access to the data is required. The people who need the data must be able to get to it easily. The easier the data items are to access, the easier it is to generate timely metrics reports.

Dumping raw data and hand-tallying or calculating measurements is another way to introduce human error into the measured values. Even if the data are recorded in a simple spreadsheet, automatic sorting, extracting, and calculations are available and should be used. They also increase the speed of producing the metrics and, therefore, can help increase timeliness.

Automating metrics reporting and delivery eliminates hours spent in front of copy machines. It also increases usability because the metrics are readily available on the computer. Remember, metrics are expensive. Automation can reduce costs while making the metrics available in a timely manner.

Quality Data and Measurement Error

To have high-quality data, the measurements taken need to be as free from error as possible. A *measurement error* occurs when the measured value or data item collected (the assigned number or symbol) differs from the actual value (that would be mapped to the attribute of the entity in a perfect world). For example, a person might be measured as 6 foot 2 inches tall, while in reality that person might be 6 foot 2.1385632 inches tall. The 0.1385632 inches is measurement error. According to Arthur (1985), measurement error can happen for a number of reasons, including:

- *Imperfect definition of what is being measured*: For example, if asked to measure how "big" the person is, the data owner might measure the person's height when the individual requesting the measurement actually wanted to know that person's weight.
- *Imperfect understanding of how to convert what is being measured into the measurement signal (data)*: For example, if a three-year-old is asked to measure a person's height, there might be an error because they do not understand how to even take that measurement.
- *Failure to establish the correct conditions for taking the measurement*: Should the person be measured with or without shoes? How should the data owner deal with a full head of hair piled up on the scalp?
- *Human error*: Then there is error simply because people make mistakes. For example, the data owner might have misread a 3 as a 2, or might have incorrectly converted the 75 inches they measured into 6 foot 2 inches instead of 6 foot 3 inches.

- *Defective or deficient measurement instruments*: The tool used to take the measurement might also be defective in some way. For example, a tape measure might be old and all stretched out of shape.

Data Accuracy

What if data have measurement error and are inaccurate? Can inaccurate data be accepted and used anyway? There are two reasons why the answer may be yes. First, the data may be accurate enough to meet the required objectives.

When reconstructing timesheet data, the engineer may overstate one project by a few hours one week and understate it the next. The result is still a fairly good approximation of total time spent over the life of the project. When estimating time requirements for future projects, the time card data may be accurate enough for its intended use.

The second reason for reporting inaccurate data is to make it accurate. For example, a metrics analyst created metrics reports to show the trend over time for nonclosed problems. The project managers then complained that the reports were inaccurate. They said that many of the problems were actually closed, but the database had not been updated. However, upper management utilized these reports to continue the focus on reducing the backlog of uncorrected problems. This provided the leverage needed to update the data and increase its accuracy.

If metrics providers wait for 100 percent data accuracy, they may never produce any metrics. Remember—good enough is good enough. On the other hand, metrics providers and users need to be aware of data inaccuracies and consider those inaccuracies when determining how reliable and valid the metrics are. This awareness can help determine the level of confidence that can be placed in the measurement results.

Data Completeness

What if the data are incomplete? Again, there are similar circumstances where the metrics provider can still use the data. In the timesheet example, the data might be considered incomplete because they do not include overtime. In this case, the metrics provider can change the metrics algorithm for estimating the next project to take this into consideration. If they know engineers are working about 10 hours a week unpaid overtime, they can use a 1.25 multiplication factor in the model to estimate total engineering effort for the next project.

The reporting of incomplete data can make the data more complete. When one company first started reporting code review metrics, the initial reports indicated that only a small percent of the code was being reviewed. Actually, a much higher percent was reviewed, but data were not being recorded in the database. By reporting incomplete data, emphasis was placed on the need to record accurate and complete data so that the measurement reflected the actual accomplishments. This is especially true when projects recording data use those data to demonstrate both cost and time savings to upper management.

However, as with accuracy, metrics providers and users need to be aware of data incompleteness as a consideration when determining how reliable and valid the metrics are, so they can determine the level of confidence that can be placed in the measurement results.

Data Timeliness

Data collection and reporting must be timely. For example, if a supplier is trying to make a decision on whether or not to ship a software product, having defect status data from the previous week is of little value. Two time frames need to be considered when talking about timeliness and data collection:

1. *Data collection time*: Are the data being collected in a timely manner? The longer the time period between the time an event happens and the time when the data about that event are collected, the less integrity the data probably have. For example, if an engineer records the time spent on each task (effort) as that engineer is working on those tasks, the effort data will be more accurate than they would be if the engineer waited until the end of the week to complete the data on a timesheet.
2. *Data availability time*: Are the collected data being made available in a timely manner? If there is a large gap of time between when the data are collected and when they are available in the metrics database, the integrity of the reported metrics can be impacted. For example, even if inspection data are collected on forms during the actual inspection, those forms may sit for some period of time before the data are entered into the inspection database.

There are also two time frames to consider when talking about timeliness and data reporting:

1. *Data extraction time*: Looks at the timeliness of extracting the data from the metrics database. This timing may be an issue in information warehouse environments, where data are extracted from operational systems on a periodic basis and made available for metrics reporting. For example, data extracted the previous midnight might be considered untimely for an immediate ship/no ship decision, while being perfectly timely for monthly summary reports.
2. *Data reporting time*: Looks at the timeliness of report generation and distribution of the extracted data. This timing may especially be an issue if the metrics are hand-calculated and distributed in hard copy.

The timing of data synchronization is also important. If the data for a metrics report are pulled while data are being input or refreshed, it could result in part of the data extraction being old and out of sync with another part of the extraction that reflects newer data. Data synchronization may also be a problem if data collection, availability, or extraction times vary for different data sets used in the same metrics report. If these data time frames are not synchronized for the measurements being analyzed, that analysis will be severely limited or even completely defective or erroneous.

Human Factors

A famous quote usually attributed to Mark Twain talks about the three kinds of lies—"lies, damn lies, and statistics." Marketers use statistical tricks all the time to help sell their products. When an organization puts its metrics program together, it needs to be aware of these issues and make conscious decisions on how to display the metrics for maximum usefulness and minimum "deviousness." If engineers and managers catch the metrics producer playing a marketing-type trick to make the metrics look good, they will stop believing the metrics. (See Chapter 6 for more information on human factors and metrics.)

Chapter 20

Analysis and Reporting Techniques

Software quality engineers (SQEs) need a variety of metrics reporting and analytical tools in their software quality tool belts to pull out the correct tool for the correct job.

If the SQE is defining metrics for measuring the quality of software products or software processes, the SQE must be familiar with various metric reporting tools in order to select the appropriate reporting mechanism to make it easy for stakeholders to interpret and utilize the measurement information.

The SQE must be familiar with and know how to utilize classic quality tools, including flowcharts, Pareto charts, cause-and-effect diagrams, check sheets, scatter diagrams, run charts, histograms, and control charts. The SQE must also be familiar with and know how to utilize basic problem-solving tools, including affinity diagrams, tree diagrams, matrix diagrams, interrelationship digraphs, prioritization matrices, activity network diagrams, and root cause analysis, to help teams and projects identify and solve potential quality issues, and implement improvements to software products, processes, and services.

1. METRIC REPORTING TOOLS

> Use various metric representation tools, such as dashboards and stoplight charts to report results. (Apply)
>
> **Body of Knowledge V.B.1**

As the old saying goes, "A picture is worth a thousand words." Metrics are no exception. In fact, many people will look at the picture and ignore the words. So what should the metric look like? The answers to this question is part of designing a good metrics report (see Chapter 6 for more information on designing metrics). When selecting the reporting mechanism for the metric, consideration should always be given to the "message" of the metric. The selected format must make the intended message easy to identify and interpret. Different types of charts and graphs are better at communicating different messages.

Charts and Graphs

Data tables provide a mechanism to display data in an ordered arrangement. In its simplest form, a data table is made up of rows and columns, as illustrated in Figure 20.1. The intersection of a row and column is a *cell,* which contains a data item. More complex data tables can be multidimensional.

Data tables can arrange large amounts of detailed data in a small area. Data tables are compact and well structured. However, tables show only the values for the measurements. The user of the data table must compare and evaluate the data. Data tables do not provide summary views or easily show trends. Graphs and charts are usually easier for the user to interpret than columns and rows of numbers. Consider adding a data table to the graph or chart when specific detail numbers are needed.

Pie charts, also called *circle charts,* are circular, static graphics where the circle is divided into slices. As illustrated in Figure 20.2, pie charts show the relative size of each data item as a percentage (slice) of the whole. Multiple pie charts can be used to represent multiple groups. Sometimes one or more pie slices are separated from the whole for emphasis. However, pie charts only show individual values and not trends over time. Showing more than five or six slices in a pie chart makes the chart busy and challenging to interpret (a simple bar chart might be a better choice).

Line graphs show data points plotted in the order in which they occur to show changes or trends over time. An example of a line graph is illustrated in Figure 20.3. Each line on the graph is typically independent of the other lines. Care should be taken to make certain that there are not so many lines on a single graph that the message becomes muddled. If this is the case, consider using multiple line charts as part of a dashboard, each with a clear, individual message.

	Open	Fixed	Resolved
Jan	23	13	3
Feb	27	24	11
Mar	18	26	15
Apr	12	18	27

Figure 20.1 Data table—problem report backlog example.

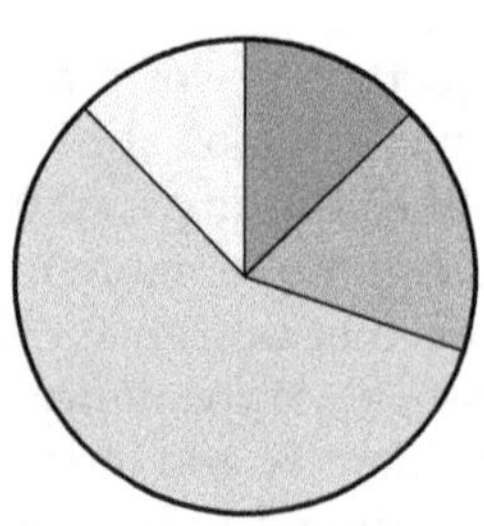

Figure 20.2 Pie chart—example.

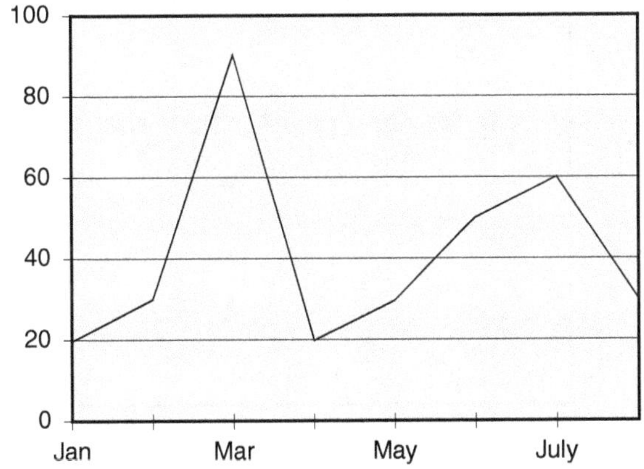

Figure 20.3 Line graph—example.

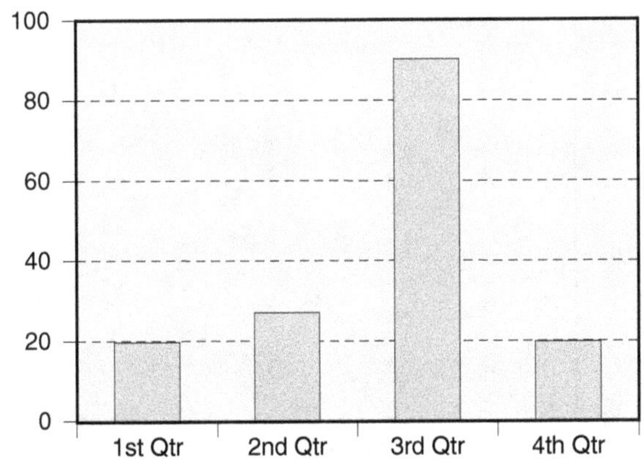

Figure 20.4 Simple bar chart—example.

Bar charts, also called *bar graphs,* represent grouped data with bars that are proportional in length to the values they represent. Bar charts are a simple way to show the relationships or comparisons between numbers, ratios, or proportions. A simple bar chart, such as the example in Figure 20.4, includes a single quantitative or discrete variable. Pareto charts and histograms are typically presented using a simple bar chart. As illustrated in Figure 20.5, bar charts can be displayed horizontally as well as vertically. Experimentation can be used to determine which type emphasizes the message of the metric more clearly.

As illustrated in Figure 20.6, *grouped bar charts* show the comparison of two or more categories, (e.g., the differences between the values of a variable for project A, and the same variable for project B, as they compare over time. Showing more than three categories typically makes the grouped bar chart busy and difficult to read.

As illustrated in Figure 20.7, *stacked bar charts,* show how a total breaks into categories, (e.g., how the total incident arrival rate for a period is distributed by incident severity). The height of the intermediate bars corresponds to the frequency

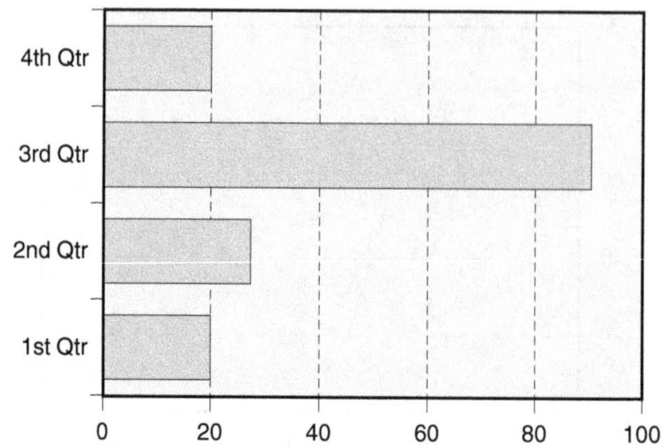

Figure 20.5 Horizontal bar chart—example.

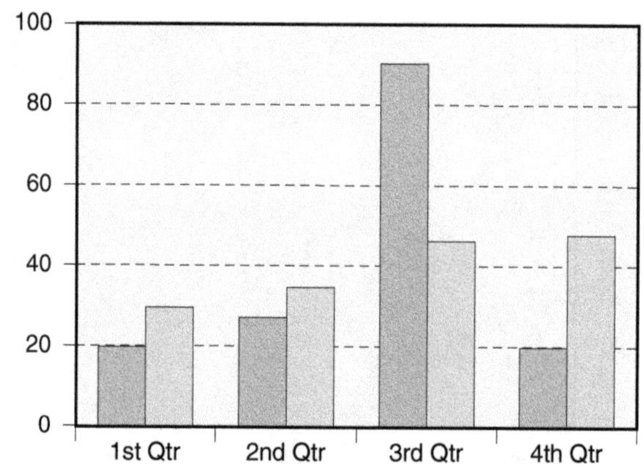

Figure 20.6 Grouped bar chart—example.

for that category. The choice between grouped or stacked bar charts is determined by the type of message that needs to be emphasized. For example, the data shown in Figures 20.6 and 20.7 are identical, but the messages shown in the graphs are very different. When the chart needs to emphasize the differences between the categories, a grouped bar chart should be chosen. When the chart needs to emphasize the differences between the totals, a stacked bar chart is easier to interpret.

As illustrated in Figure 20.8, *area graphs,* also called *area charts,* are used to show trends over time for a single attribute or for a set of related attributes. Unlike the line graph, however, if there are multiple lines in an area graph, those lines are dependent because they divide the total area into parts (similar to a stacked bar chart). An area graph is more effective than a line chart when the graph needs to emphasize both the trends and the size of or relationship between the areas.

As illustrated in Figure 20.9, *box charts,* also called *box plots* or *box-and-whisker-diagrams* compress a set of data into each box and indicate its variance. Box charts can be used to show both the differences within a group of data and between

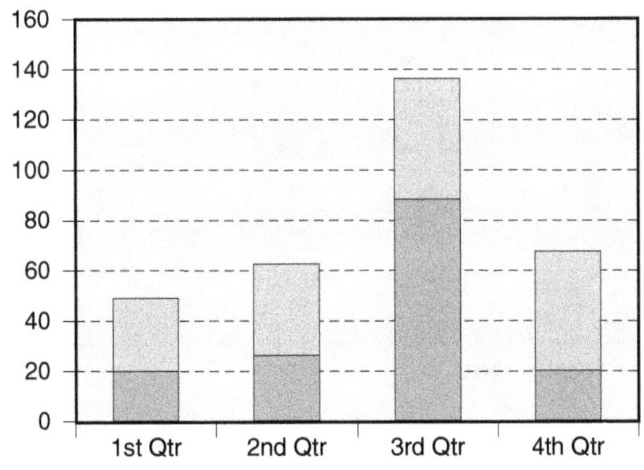

Figure 20.7 Stacked bar chart—example.

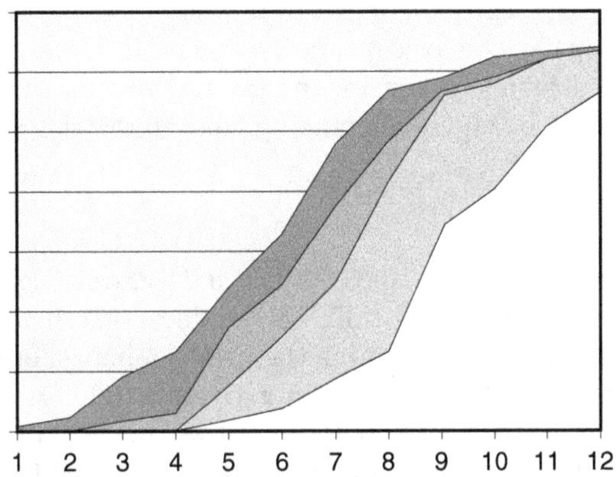

Figure 20.8 Area graph—example.

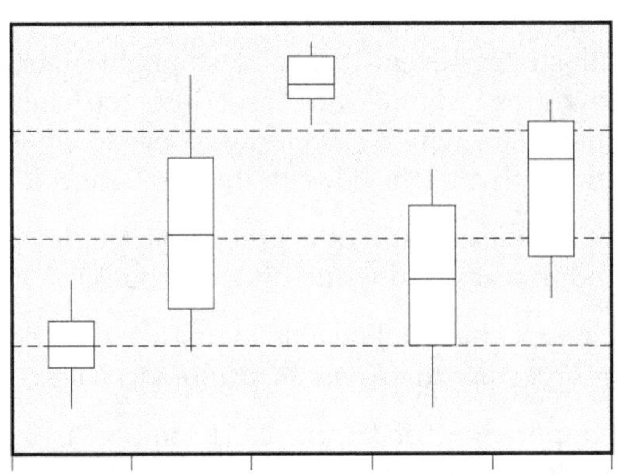

Figure 20.9 Box chart—example.

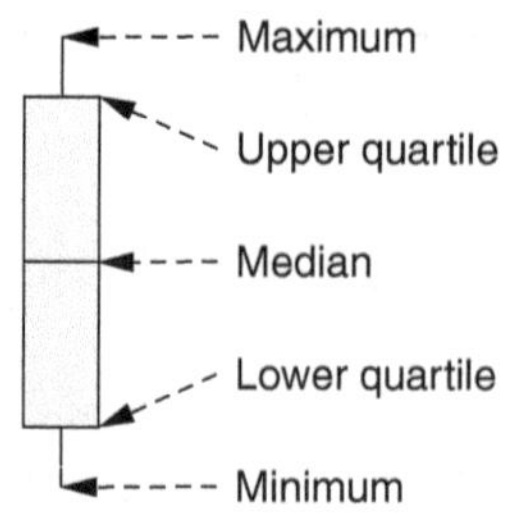

Figure 20.10 Box chart components.

groups of data. The various components of an individual box chart component are illustrated in Figure 20.10. Fifty percent of all data values are between the lower and upper quartiles. 25 percent of all data values are lower than the lower quartile, and 25 percent are higher than the upper quartile. The median is the middle data value (typically the mean or mode).

Different colors have historical or culturally bound interpretations and impacts (e.g., in the United States, green means go, red means stop, and yellow means caution). Care should be taken when using color in metrics graphs and charts.

Stoplight Charts

According to the IEEE/ISO/IEC 15939 *Systems and Software Engineering—Measurement Process* standard (IEEE/ISO/IEC/IEEE 2017b), decision criteria are the "thresholds, targets, or patterns used to determine the need for action or further investigation or to describe the level of confidence in a given result." In other words, decision criteria provide guidance that will help the metric customers interpret the measurement results. One way of doing this is to create a *stoplight chart* that provides a red/yellow/green or a red/green signal. A red signal indicates that the measurement result needs immediate action or further investigation. A yellow signal indicates that the measurement result does not need immediate attention but is in a cautionary state and should be monitored. A green signal indicates that no action or investigation is needed.

Figure 20.11 illustrates several types of stoplight charts. In the chart on the left, the red/yellow/green signals are built into the chart as colored areas in the background (one place where color is appropriate when displaying a metric). The regions on this graph are labeled with the stoplight colors because:

- It make sure people can still interpret the stoplight chart correctly even if it is printed in black and white (like this book)
- Color blind people may not be able to distinguish the colors on the graph even if they are displayed or printed in color

The table in the upper right of Figure 20.11 simply lists the metrics and their current stoplight colors. This type of summary table can include hyperlinks to the detailed metrics behind each reported stoplight status. A third example is simply a visual stoplight indicator for the metric, as illustrated in the lower right

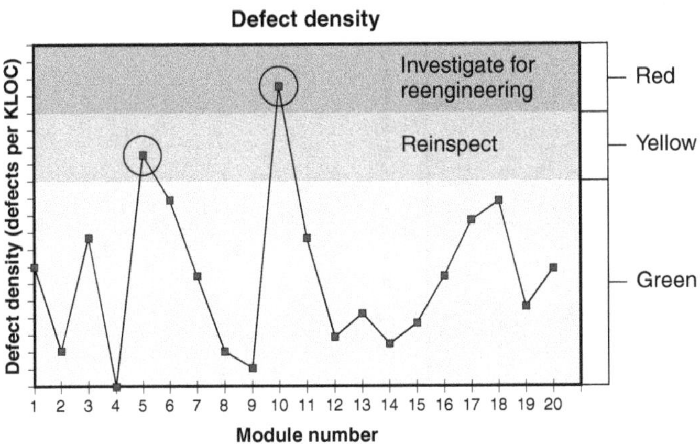

Performance metric	Status
Throughput	Green
Response times	Green
Memory utilization	Yellow
Direct access storage	Green
Capacity	Green

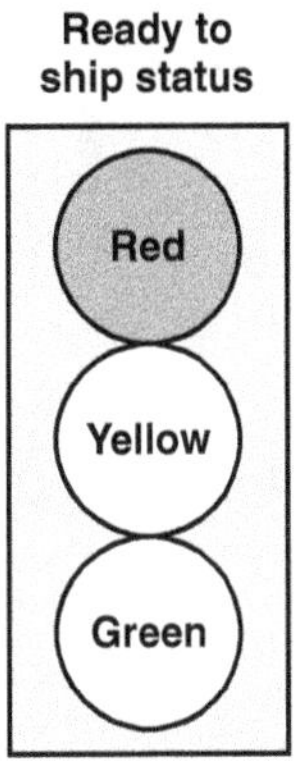

Figure 20.11 Stoplight chart—examples.

of Figure 20.11. This stoplight indicator can be placed next to a chart or graph to indicate its stoplight status.

Dashboards

The concept of a metrics dashboard is analogous to a dashboard in a car. All of the key information about the status of the car can be identified quickly by looking at the set of metrics on the dashboard, including fuel levels, speeds, engine temperature, and alarm lights for controlled items such as oil pressure and other maintenance items, doors that are ajar, or seat belts that are not fastened.

In some cases, it takes multiple metrics to get a complete picture of the status of a project, process, product, or service. A metrics *dashboard,* also called a metric *scorecard, balanced scorecard, cockpit, command center,* or *mission control,* accumulates all of the key metrics associated with a decision and allows the metrics customer to see, in a single collection, the relevant information about the subject of that dashboard. For example, a software project dashboard might include information about the schedule, budget, product functionality, product quality, and effective and efficient utilization of project staff and other resources.

Figure 20.12 shows an example of a dashboard that reports the status of the system testing process. This dashboard would be used as input into the system test sufficiency decisions. It includes graphs and charts showing:

- Test effort variances
- Test case status

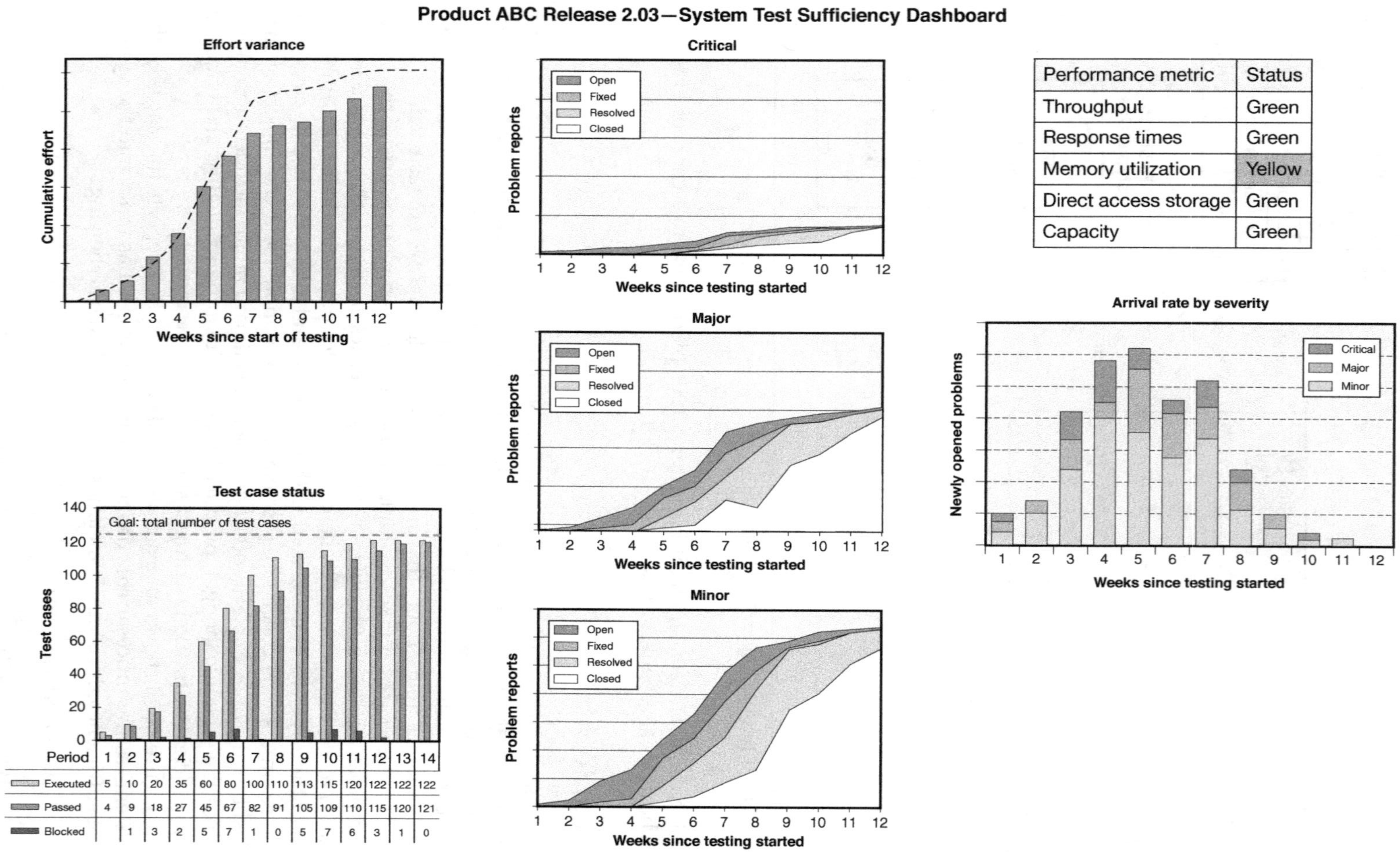

Figure 20.12 Dashboard—example.

- Cumulative defects by status (three graphs, one for each severity)
- Performance metrics stoplights
- Arrival rates by severity

Kiviat Chart

Another way to show a summary view of a set of metrics is to use a *Kiviat chart*, also called a *polar chart*, *radar chart*, or *spider chart*. In a Kiviat chart, each "spoke" represents a metric with the metric's value plotted on that spoke. The outer circle (or pattern) represents the objective or goal (ideal value). Inner circles (patterns) may be added to depict valid range areas, or alternatively the Kiviat chart can be combined with a stoplight chart by adding red, yellow, or green bands to the chart.

Figure 20.13 on the next page illustrates an example of a Kiviat chart that summarizes the customer satisfaction scores shown previously in Figure 19.20. From this chart, it is easy to identify "documentation" as the satisfaction area with the value farthest away from the outer circle goal of having a 5 as a satisfaction score. Therefore, this chart shows that documentation presents the best opportunity for process.

A Kiviat chart can also be used to compare several different items (e.g., projects, products, processes, or services) across several parameters against the ideal value by plotting additional sets of points. For example, Figure 20.14 on the next page compares customer satisfaction results for two different products.

2. CLASSIC QUALITY TOOLS

> Describe the appropriate use of classic quality tools (e.g., flowcharts, Pareto charts, cause-and-effect diagrams, control charts, and histograms) (Apply)
>
> **Body of Knowledge V.B.2**

Flowcharts

Flowcharts are used to graphically represent the inputs, actions, decision points, and outputs of a process. Historically, flowcharts were used as a software design tool to model the control flow of a source code module (component design). While flowcharts have generally been replaced with more modern design-modeling tools, they are still used extensively in process definition to map the process, as illustrated in Figure 20.15. Flowcharting a process is a good starting point to help a team understand a current process. As a quality analysis tool, flowcharts can help identify bottlenecks or problem areas (e.g., missing steps, unnecessary loops, redundant steps) in a process. Flowcharts can help a team come to a consensus on the steps in a standard process and explore where tailoring is needed. Flowcharts can also act as a training aid to help trainees visualize how the individual steps in a process interact.

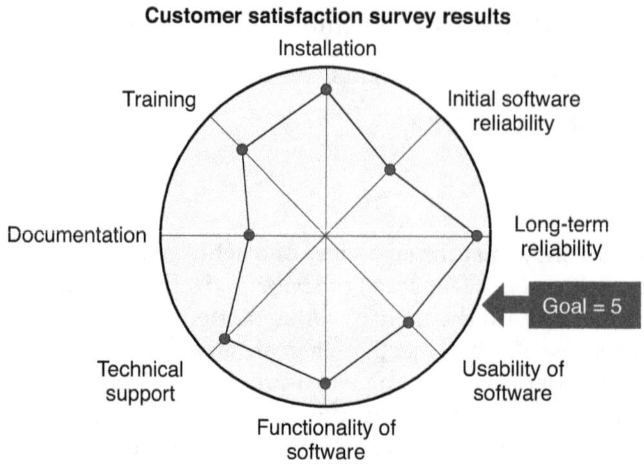

Figure 20.13 Kiviat chart—example.

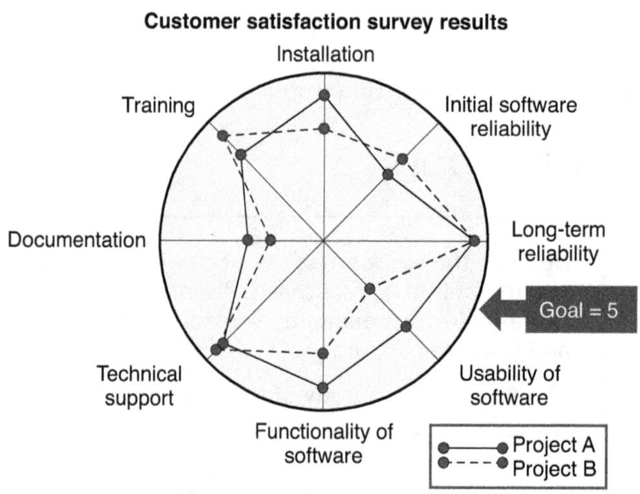

Figure 20.14 Kiviat chart comparison—example.

While there are many different flowchart symbols, Figure 20.15 on the next page illustrates a few basic symbols and an example flowchart.

Pareto Charts

Pareto analysis is the process of ranking problems or categories based on their frequency of occurrence or the size of their impact. Pareto analysis is used to determine priorities, which of many possible opportunities to pursue first because they have the greatest potential for improvement. Pareto analysis is based

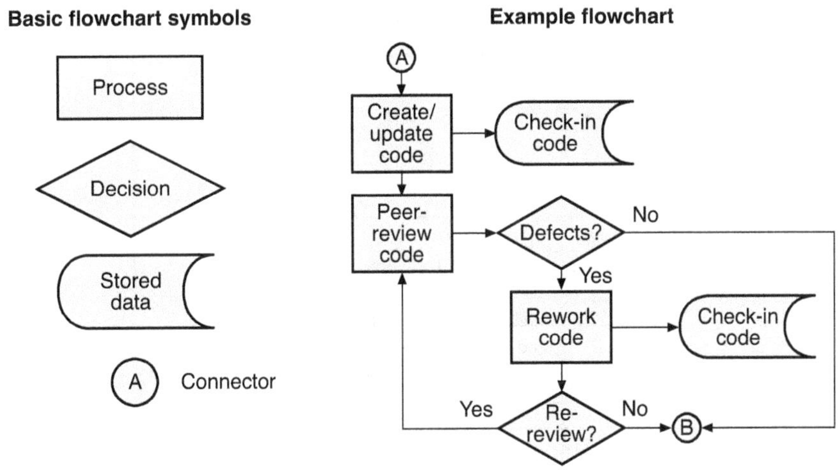

Figure 20.15 Basic flowchart symbols and flowchart—example.

on the *Pareto principle*: 80 percent of the variation in a process comes from about 20 percent of the sources. The intent is to identify the "critical few" that need the most attention among the "insignificant many."

A *Pareto chart* is a bar chart, where the height of each bar indicates the frequency or impact, of problems or categories. For example, as illustrated in Figure 20.16, a Pareto chart can identify defect-prone modules that need more testing or are candidates for reengineering. The steps to construct a Pareto chart are:

- *Step 1*: Determine the problems or categories for the chart. For example, a team might decide to examine the number of defects identified in each module or analyze problems by root cause, phase introduced, or reporting user.
- *Step 2*: Determine a unit of measure such as frequency or cost, (e.g., the number of defects in each module, the number of defects per function point in each module, or the cost (dollars or effort) of fixing defects in each module.
- *Step 3*: Select a time interval for analysis. For example, the team could look at all defects reported in the last six months.
- *Step 4*: Gather data and rank-order the problems or categories from the largest total occurrences to the smallest to create a bar chart, where the bars are arranged in descending order of height from left to right. In the example in Figure 20.16, module D has the highest number of defects and module C has the lowest.

Cause-and-Effect Diagrams

Improving quality involves taking action on the root causes of a problem or the causes of variation in a process. "With most practical applications, the number

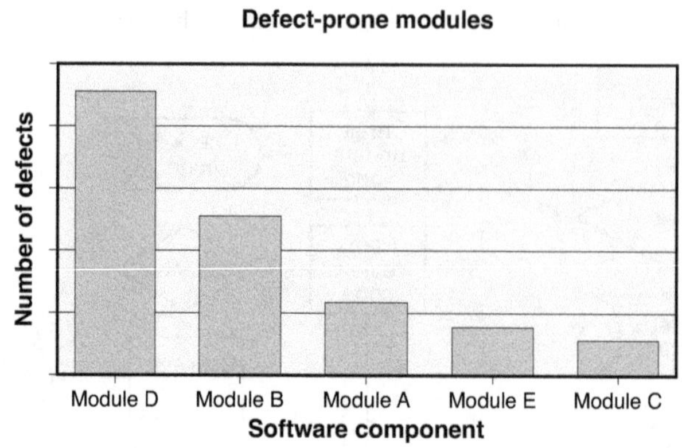

Figure 20.16 Pareto chart—example.

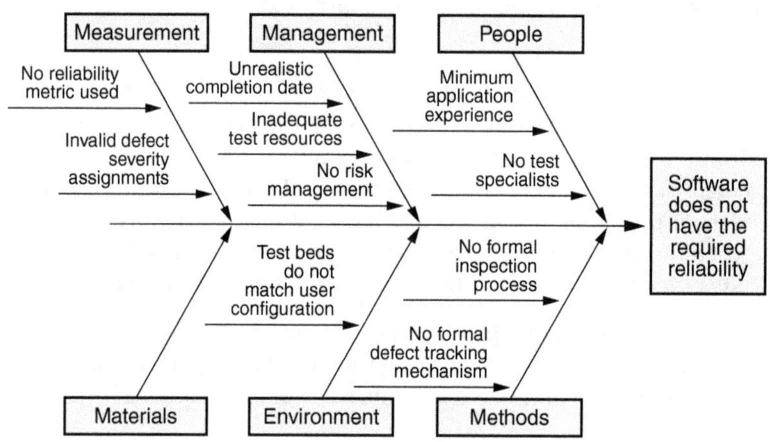

Figure 20.17 Cause-and-effect diagram—example.

of possible causes for any problem can be huge. Dr. Kaoru Ishikawa developed a simple method of graphically displaying the causes of any given quality problem" (Pyzdek 2001). In a *cause-and-effect diagram,* also referred to as an *Ishikawa diagram* or *fishbone diagram,* the problem (or effect) is put at the "head" of the fish. The major fishbones are typically major drivers such as management, people, environment, methods, measurements, and materials. From each of these major drivers, specific drivers that are causing the problem are listed. The cause-and-effect diagram example illustrated in Figure 20.17 only shows a single level of causes off the major drivers, but sublevels and sub-sublevels can also be added as additional branches off each of these branches.

A variation on the cause-and-effect diagram is the *process-type cause-and-effect diagram,* illustrated in Figure 20.18. In a process-type cause-and-effect diagram, the causes of the problem are investigated based on how they relate to each step

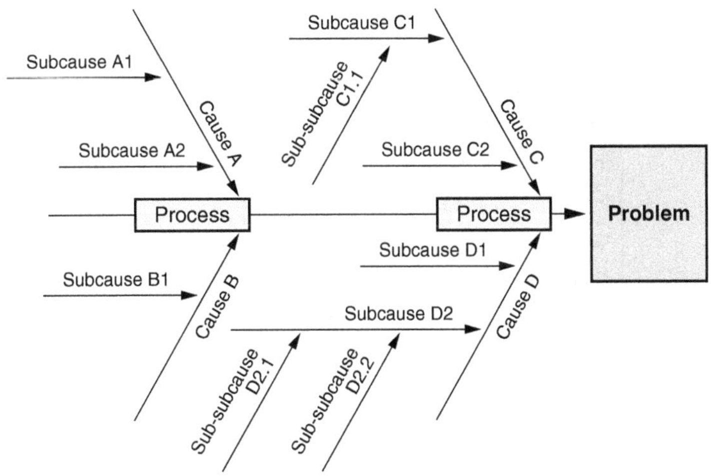

Figure 20.18 Process-type cause-and-effect diagram—example.

in a process. For example, suppose the analysis team is looking at the problem of unreliable software. In that case, they might look at the major steps in the software development process (e.g., requirements, design, code, integration test, and system test) and analyze any causes related to each process step. If the problem is that too many defects are escaping from the detailed design inspection, the team could look at the major steps in the inspection process (e.g., overview, preparation, the inspection meeting, and follow-up) and the causes related to each of those process steps.

Cause-and-effect diagrams are not just valuable for the analysis of actual problems. They can also be used to analyze potential problems and their potential causes. For example, in risk analysis or hazard analysis, the potential problem (risk or hazard) is placed at the head of the fish, and potential drivers that might cause the risk or hazard to turn into an actual problem are documented as the branches of the diagram.

Check Sheets

Check sheets are tools used in data collection. In their simplest form, check sheets make the data collection process easier by providing prewritten descriptions of categories or events being counted, either from historical data or as events occur in the future. For example, the potential causes identified in a cause-and-effect diagram could be used on a check sheet to collect the number of actual occurrences of each identified cause during the next four weeks. While check sheets can take on several forms, they typically use Xs or tally marks to count the number of occurrences. Figure 20.19 illustrates a check sheet used to collect data on the root cause of requirements-type defects. Check sheets serve as reminders that direct the data collector to items of interest and importance. The data collected on a check sheet can also be used as input into creating a Pareto chart, histogram, or other graph or data table.

Requirements defect root cause	Frequency
Missing requirement	卌 III
Ambiguous requirement	卌 卌 II
Incomplete requirement	卌 I
Incorrect requirement	卌 IIII
Contradictory requirement	III
Change to requirements not communicated	IIII

Figure 20.19 Check sheet—example.

Scatter Diagrams

A *scatter diagram* is an *x–y* plot of one variable versus another. Scatter diagrams are used to determine if there is an existing or potential correlation between the two variables. One variable, called the independent variable, is plotted on the *x*-axis. The second variable the dependent variable, is plotted on the *y*-axis. Scatter diagrams investigate whether changes to the independent variable correlate to changes in the dependent variable. Scatter diagrams can help answer questions such as:

- Does the cyclomatic complexity of a module impact the number of defects in that module?
- Does reuse cause productivity to increase?
- Is there a relationship between the number of defects found during testing and the number of defects found post-release?
- As the experience level of the programmer increases, does the number of defects in their code decrease?

As illustrated in Figure 20.20, there is a positive correlation between the independent and dependent variables if the dependent variable increases as the independent variable increases. The two variables have a negative correlation if the dependent variable decreases as the independent variable increases. The more tightly the data points are clustered, the stronger the correlation between the two variables. If the data points on the scatter diagram are randomly scattered, then no correlation exists between the two variables.

When interpreting a scatter diagram, it should be noted that correlation does not necessarily imply causation. Other interpretation considerations include:

- Making certain that enough data have been collected to indicate a trend (typically at least 20 data points)
- Making certain that the independent variable in the data set varies over a large enough range to determine correlation

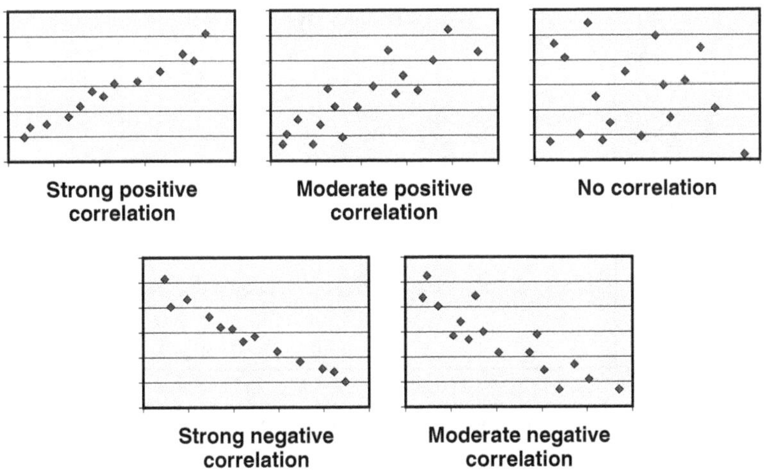

Figure 20.20 Scatter diagram—examples.

- Being careful not to use the information to predict (extrapolate) the dependent variable values using independent variable values that lie outside the study's range

Run Chart

Run charts are line graph plots of data arranged in time sequence. Over time, many factors can affect an attribute (of a project, process, product, or service). Run charts can make it easier to see what is happening to that attribute and detect trends, shifts, or patterns. Run charts can be used to monitor and detect shifts in an attribute after a process or product improvement activity has been conducted to monitor the impact of that improvement. For example, the run chart in Figure 20.21 could be used to monitor the impacts on system availability as improvements are made to network resources over time. Analysis of a run chart can also suggest areas where further investigation is needed to determine the cause before any conclusions can be reached. Patterns to watch for when analyzing run charts include:

- Repeating patterns
- Sudden shifts or stepwise jumps
- Outlying data points
- An upward or downward trend

Another type of run chart frequently used in software is the *S-curve run chart*, which tracks the cumulative progress of a parameter over time. Examples of S-curve run charts for software include:

- Completion of the design, or design reviews, of software components over time

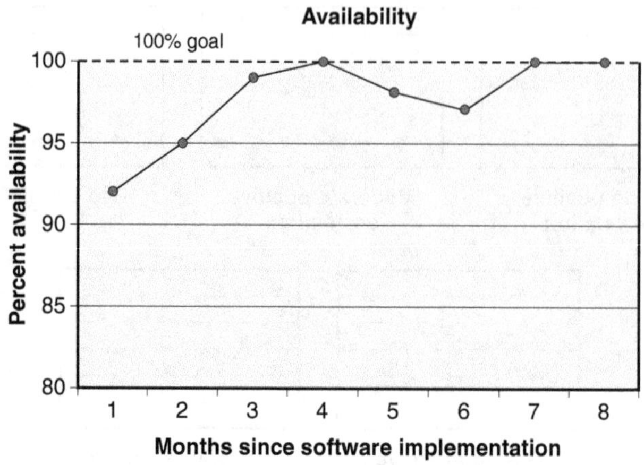

Figure 20.21 Run chart—example.

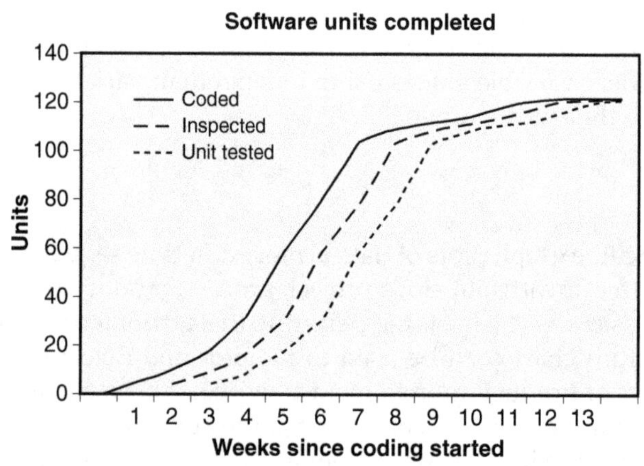

Figure 20.22 S-curve run chart—example.

- Completion of coding, inspection, or unit testing of software units, as illustrated in Figure 20.22
- Completion of writing or execution of test cases
- The arrival of problem reports during testing or post-release

These S-curve run charts are often compared to predicted values (e.g., engineering effort expended versus predicted, or dollars spent versus predicted) or against historical data (e.g., release to release defect density) so that the analysis can be placed into the proper context.

Histograms

A *histogram* is a bar chart that shows the frequency counts for a set or sample of data. The steps to create a histogram are:

- *Step 1*: Select the characteristics or ranges that will be used to sort the data and list them on one axis (typically the *x*-axis)
- *Step 2*: Count the number of occurrences in the data set for each characteristic or range
- *Step 3*: Graph a bar on the chart for each characteristic or range, where the height of the bar represents the frequency (number of elements from the data set) of that characteristic or in that range

According to Kan (2003), "In a histogram, the frequency bars are shown by order of the *X* variable, whereas in a Pareto diagram, the frequency bars are shown by order of the frequency counts." For this reason, histograms are not valid for nominal scale measures because there is no order (see Chapter 19 for more information on measurement scales). Other experts also question their validity for ordinal-scale measures, because there is no assumption about the magnitude of the differences, and therefore the "shape" of the curve may not be meaningful. Examples of software histograms include:

- Distribution of problem report arrival rates (see Figure 20.23)
- Distribution of problem report backlog by age (see Figure 20.23)
- Distribution of the staff by experience level
- Distribution of defects by severity (note that this is an ordinal-scale metric)
- Distribution of stakeholder satisfaction responses by satisfaction level (note that this may be an ordinal-scale metric depending on how the data are gathered during the survey)

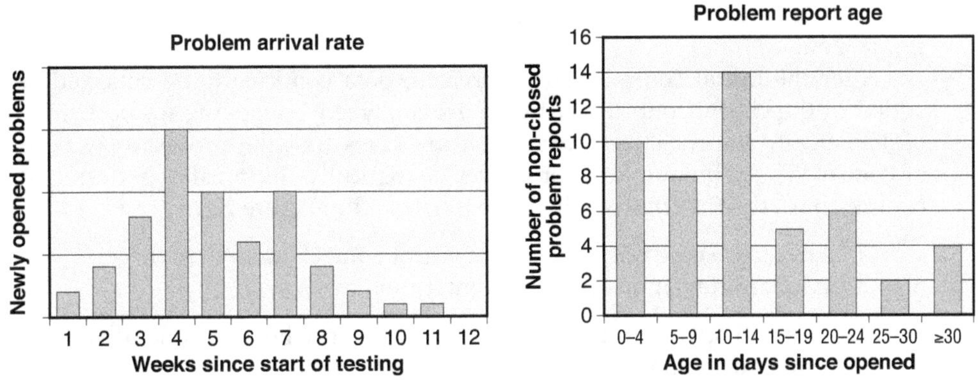

Figure 20.23 Histogram—examples.

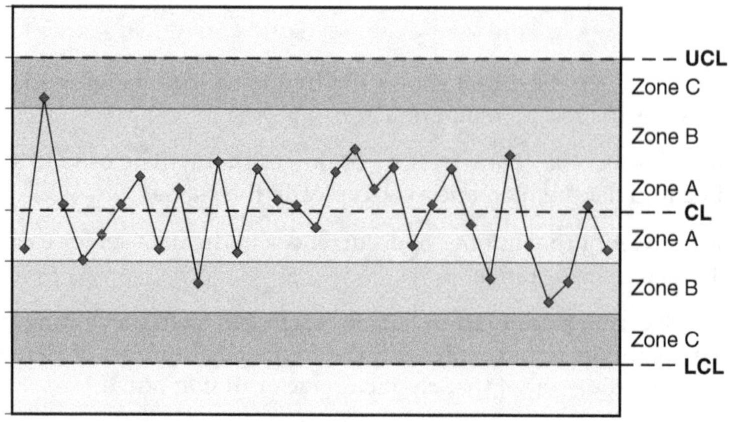

Figure 20.24 Control chart—example.

Control Charts

The purpose of a *control chart* is to control one of the attributes of a process over time. As illustrated in Figure 20.24, classic control charts have a centerline (CL), upper control limit (UCL), and lower control limit (LCL). Data are collected while the process is running and considered under control. Then the central line and control chart limits are calculated using statistical techniques. Control charts make it possible to identify improbable patterns in the process variation that may indicate that the process is out of control.

The centerline is typically the observed process average (mean), but other measures of central tendency, such as the median, can be used. The upper and lower control limits are based on the variation of the process. For classic control charts, these control limits are set at ±3 sigma (standard deviations) from the centerline. If the lower control limit falls below zero and negative numbers are not valid for the attribute, the lower control limit is set to zero. The zones on a classic control chart are areas bracketed by ±1 sigma (zone A), ±2 sigma (zone B), and ±3 sigma (zone C).

After the initial control chart is created, data continue to be collected and added to the chart. Control chart data are analyzed by looking for occurrences of statistically improbable patterns that are clues that the process may be out of control. The following are considered statistically improbable patterns on a control chart, the first five of which are illustrated in Figure 20.25:

1. Any single data point that is outside of zone C (more than three standard deviations from the centerline)
2. Any two out of three data points in a sequence beyond zone B (more than two standard deviations from the centerline)
3. Any four out of five data points in a sequence beyond zone A (more than one standard deviation from the centerline)

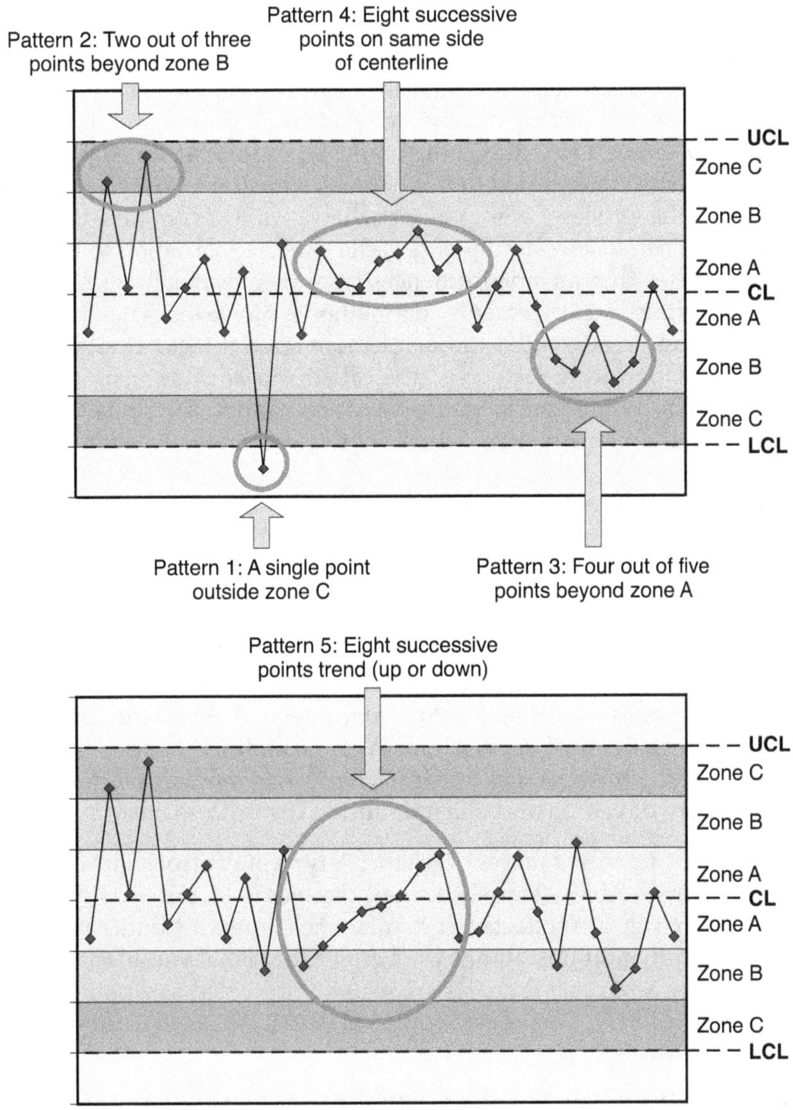

Figure 20.25 Control chart statistically improbable patterns—examples.

4. Any eight successive points on the same side of the centerline
5. Any eight successive points in a trend (all up or all down)
6. Any 14 successive points alternating up and down
7. Any 15 successive points all within zone A (less than one standard deviation from the centerline)
8. Obvious nonrandom patterns such as cycles (not illustrated in Figure 20.25)

When one of these patterns is identified, it only indicates that something improbable has happened that should be investigated. For example, observing eight successive points on the same side of the centerline is similar to flipping a coin eight times and having it come up heads each time. Can it happen? Yes, but it is unlikely enough that the coin should be examined for anomalies. If, after investigation, it is determined that the observed pattern is just part of the normal variation in the process, it is part of what is called common cause variation. *Common cause variation* is simply part of the expected variation in the process due to typical causes that include influences from people, machinery, environmental factors, materials, measurements, or methods. However, after investigation, if the pattern can be attributed to one or more special factors outside the normal expected variation in the process, it is called *special cause*, also called *assignable cause*, variation. When special cause variation occurs, the process is considered out of control, and corrective action should be implemented to eliminate the special cause(s).

There are different types of control charts. These types are based on the type of data collected (variable data versus attribute data) and how those data are collected and combined. Examples of some of the control charts that may be relevant to software processes include (Florac and Carleton 1999):

- *Average (X-bar) and standard deviation (S) charts*: A type of control chart used to monitor variable data, when large samples (usually >= 10) of varying sizes are collected at regular intervals. Standard deviation is a better estimate of variation for larger samples. X-bar and S charts are usually presented as a pair of charts, one monitoring the process standard deviation and one monitoring the process mean.

- *Average (X-bar) and ranges (R) charts*: A type of control chart used to monitor variables data when small samples (< 10, but usually 3 to 5) of constant size are collected at regular short intervals under basically the same conditions. Range is a better estimate of variation for smaller samples. X-bar and R charts are usually presented as a pair of charts, one monitoring the process standard deviation and one monitoring the process mean.

- *Individuals and moving range (XmR) charts*: A type of control chart used to monitor variables data when the sample size needs to be one (e.g., when the process is very slow, and it is a long time between samples, or when taking a sample is very expensive. XmR charts are usually presented as a pair of charts, one displaying the individual values and one monitoring the differences from one point to the next (the moving range).

- *c charts*: A type of control chart used to monitor attribute data when samples (usually >= 5) of constant size are collected for defect counts (each defect counts as a single sample, and a component can have more than one defect).

- *u charts*: A type of control chart used to monitor attribute data when samples of varying sizes are collected for defect counts (each defect counts as a single sample, and a component can have more than one defect).

3. PROBLEM-SOLVING TOOLS

> Describe the appropriate use of problem-solving tools (e.g., affinity, tree, matrix, activity network diagrams, root cause analysis and Data Flow Diagrams [DFD]). (Apply)
>
> **Body of Knowledge V.B.3**

Brainstorming

Brainstorming is a team technique for generating lots of creative ideas or suggestions in a short period of time. The basic steps in brainstorming include:

- *Step 1*: For brainstorming to be effective, the team must share a common understanding of precisely what is being brainstormed. This understanding can be accomplished through team consensus on a single statement. For example, "We are brainstorming ideas for improving our peer review process" or "We are brainstorming a list of risks for project ABC."
- *Step 2*: Team members then offer up their ideas. This can be accomplished ad hoc, with members calling out ideas as they occur to them, or it can be done in a round-robin manner, with each member taking a turn to verbalize an idea or pass. The objective during this step is to get the creative juices flowing. Team members should use the ideas of other members as a catalyst for new ideas of their own. Even the most off-the-wall suggestion may spark a brilliant idea in someone else. Therefore, during this brainstorming step, team members should refrain from analyzing or critiquing any idea. Judgment is set aside during this step to let creativity flow.
- *Step 3*: As each suggestion is made, the facilitator or recorder records that idea on a flip chart or in some other manner so that everyone in the group can read all of the ideas. This recording helps team members utilize those ideas to generate more new ideas.
- Steps 2 and 3 are iterated repeatedly until the team runs out of new ideas. The facilitator may prompt participants for more ideas if the team gets bogged down too quickly. The facilitator might pick one idea from the list and ask for more ideas similar to that idea or more ideas that expand on a similar theme. For example, if the team is brainstorming a risk list, the facilitator may pick a suggested risk that states, "We have never used Java before and do not have any real experts." The facilitator might ask, "What else are we doing on this project that we do not have experience with?"
- *Step 4*: As the last step in the brainstorming session, participants must put their analytical skills back to work and use their judgment to

discuss, clarify, combine, or remove items from the list. They can then categorize and prioritize the items on the list to make them useful.

Affinity Diagrams

An *affinity diagram* is used to organize many ideas (or data items) into categories so that they can be reviewed and analyzed. For example, an affinity diagram could be used to organize:

- Outputs of a brainstorming session into valuable categories
- Stakeholder quality requirements into quality attribute categories
- Stakeholder survey comments or stakeholder complaints into major stakeholder concern categories
- Audit observations into evidence categories to support findings
- Defect root cause data to create a root cause taxonomy
- Previously experienced problems to create a risk taxonomy

The steps in constructing an affinity diagram include:

- *Step 1*: Each idea (or data item) to be sorted is written on a piece of paper (sticky note or 3×5 card) and placed randomly on the wall or flip chart, where the entire team can access them.
- *Step 2*: Working in silence, team members move the items around, collecting similar items into related groups (categories). By doing this in silence, the team members are not influencing each other's ideas, and individual members must think for themselves. Silence also helps keep people in their right brain, enhancing creativity and pattern recognition (the left side of the brain tends to be more linear and language-oriented). If an item is being moved back and forth between groups, consider duplicating that item and putting it in both groups. Some items may stand by themselves as one-item groups.
- *Step 3*: When organizing ideas from a brainstorming session, new items can also be added to the sets of items during the generation of the affinity diagram. If, for example, a team member notices something missing once a category is taking shape, that person can create a new item and add it to that category and explain it to the team. Additional details may also be added to existing items.
- *Step 4*: After a few minutes have elapsed without additional changes, the team breaks the silence and starts a discussion of the final configuration. The team should consider whether large groups should be broken down into subgroups.
- *Step 5*: When the team has come to a consensus on the groups, the team assigns a short label (one to five words in length) to each group to describe that group.

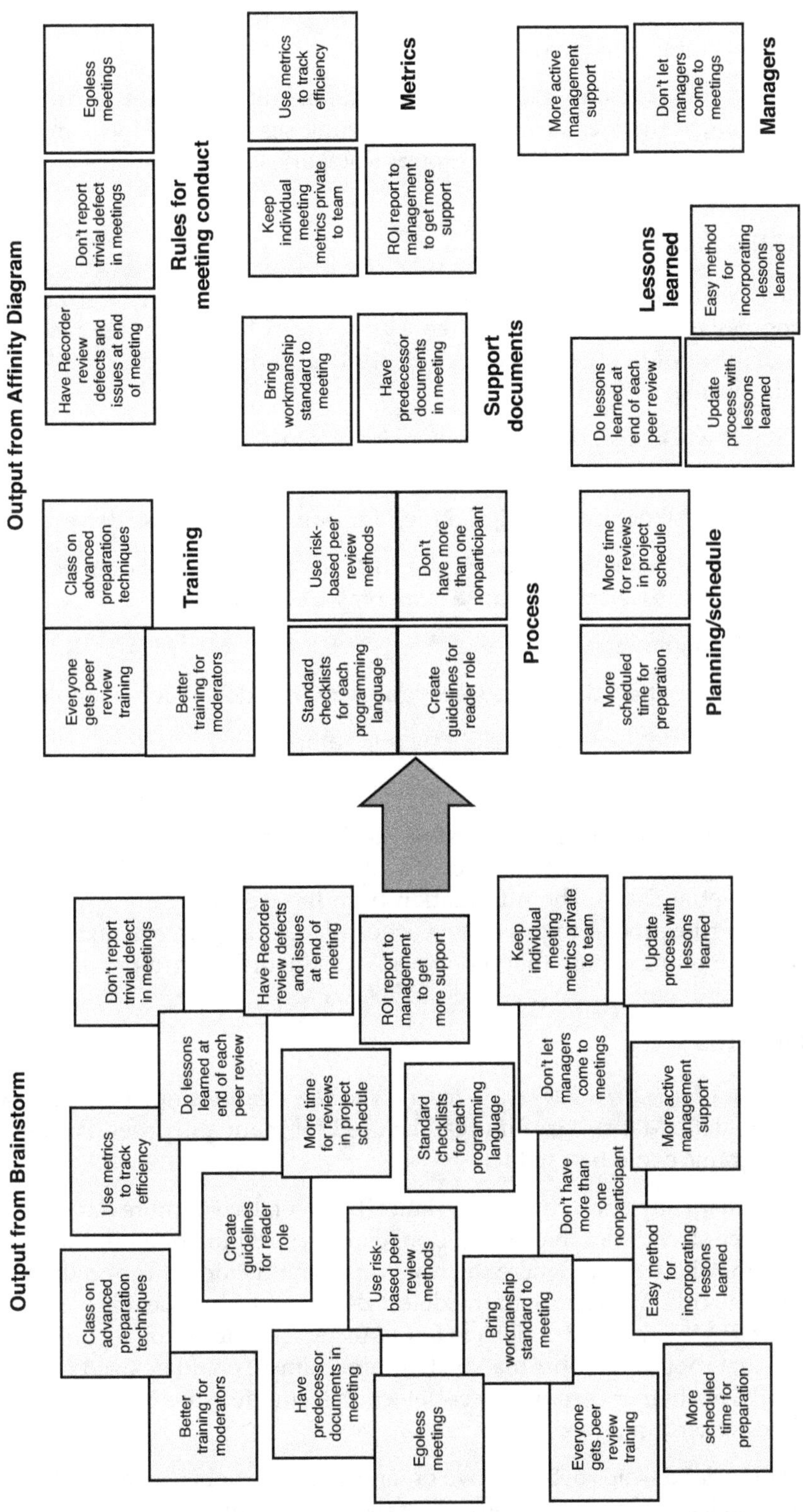

Figure 20.26 Affinity diagram—example.

- *Step 6*: The team finishes by reviewing the resulting affinity diagram and deciding what to do with the results.

Figure 20.26 illustrates an example where the affinity diagram technique was used to organize the ideas from a brainstorming session on "How can the peer review process be improved?" into groups (categories).

Tree Diagrams

A *tree diagram* is an outline or taxonomy used to break down or stratify ideas into progressively more detail. A tree diagram can help organize information and make the underlying complexities of the ideas more visible. Examples of tree diagrams include:

- Project work breakdown structures (see Figures 16.15, 16.16, and 16.17 for examples)
- Directory structures for digital files, for example, on a disk drive
- Root cause taxonomies (see Figure 20.27 for an example of one section of a software defect root cause taxonomy tree diagram)
- Risk taxonomies
- Cause-and-effect diagrams (see Figures 20.17 and 20.18 for examples)
- Decision trees (see Figure 22.13 for an example)
- Goal/question/metric diagram (see Figure 6.8 for an example)

In process improvement, tree diagrams can be "used to separate a broad goal into increasing levels of detailed actions identified to achieve the stated goal" (Christensen et al. 2007). The information from the tree diagram is then used as input to create the process improvement action plan. Using a tree diagram provides a method for making certain that action plans remain linked to improvement goals.

Matrix Diagrams

The purpose of a *matrix diagram* is to analyze the relationships between two or more sets of data. Matrix diagrams can be used for many purposes. For example, matrix diagrams could be used to:

- Systematically analyze the correlation between two or more data sets and the strength of that correlation. Figure 20.28 illustrates an example of a matrix diagram where the rows represent the defect root causes and the columns represent the defect-detection technique. Each cell contains the number of defects found using each detection technique for that root cause. This matrix diagram example could be used to analyze whether certain defect-detection techniques are better at finding certain types of defects.
- Show relationships between two or more data sets or between one data set and itself. In this case, the cells of the matrix diagrams might contain a

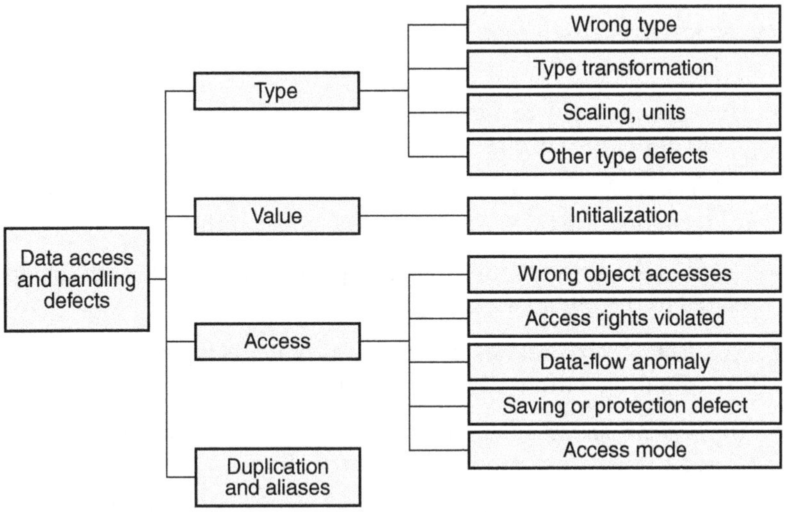

Figure 20.27 Tree diagram—example.

Root cause	Detection technique: Requirements inspection	Design inspection	Code inspection	Unit test	Integration test	System test
Requirements defects	20	5	1		2	6
Functionality implementation defects		15	5	2	2	25
Structural/control flow defects		20	15	6	5	1
Processing defects		4	18	12	4	5
Data defects		2	20	13	5	2
Internal interface defects		21	7	1	15	3
External interface defects	6	2	1	0	2	12

Figure 20.28 Matrix diagram—example.

relationship indicator (e.g., 5 = high importance to 1 = low importance, with zero indicating no relationship). A prioritization matrix is an example of this type of matrix diagram, as illustrated in Tables 20.1 and 12.4.

- Define assignments, in which case the rows might be project roles, and the columns might indicate work products or activities with the cells indicating responsibilities (e.g., P = primary responsibility, M = team member, R = reviewer, A = approval authority).

According to Christensen et al. (2007), six different shapes of matrix diagrams can be used, depending on how many data groups are being compared:

- An *L-shaped matrix* relates two groups of items to each other (or one group of items to itself)
- A *T-shaped matrix* relates three groups of items, where two of the groups are related to the third group but not to each other
- A *Y-shaped matrix* relates three groups of items, where each group is related to the other two groups
- A *C-shaped matrix* relates three groups of items together at the same time, creating a three-dimensional matrix
- An *X-shaped matrix* relates four groups of items, where each group is related to two other groups but not the third
- A *roof-shaped matrix* relates one group of items to itself (this type of matrix is typically used in combination with either an L-shaped or T-shaped matrix)

Interrelationship Digraph

An *interrelationship digraph,* also called a *relationship diagram,* organizes ideas and defines how those ideas influence each other in a cause-and-effect relationship. As in an affinity diagram, the ideas used as inputs into the interrelationship digraph can come from many sources, including brainstorming, surveys, reported defects, and so on. Using this tool, the team organizes the ideas and defines their influences by drawing arrows between the ideas, as illustrated in Figure 20.29. For example, in this interrelationship digraph, the arrow drawn from the *Inspection training ineffective* idea to the *Don't know how to prepare* idea shows a causal relationship (ineffective inspection training is a cause of reviewers not knowing how to prepare). Interrelationship digraphs can help identify the prime cause or contributing item (the item with the most arrows leaving it) and the prime outcome (the item with the most arrows arriving). This identification enables the team to focus on the most important relationships first.

Activity Network Diagrams

The purpose of an *activity network diagram* is to show the order in which activities must occur and any successor or predecessor relationships. An activity network diagram, as illustrated in Figures 16.11 and 16.12, graphically depicts what tasks in a process or project can be done in parallel and which must be done in sequence. (See Chapter 16 for a detailed discussion of activity networks.)

Root Cause Analysis

The primary objective of root cause analysis is to identify the actual cause of a problem, nonconformance, or other issue, so that the actual cause can be corrected.

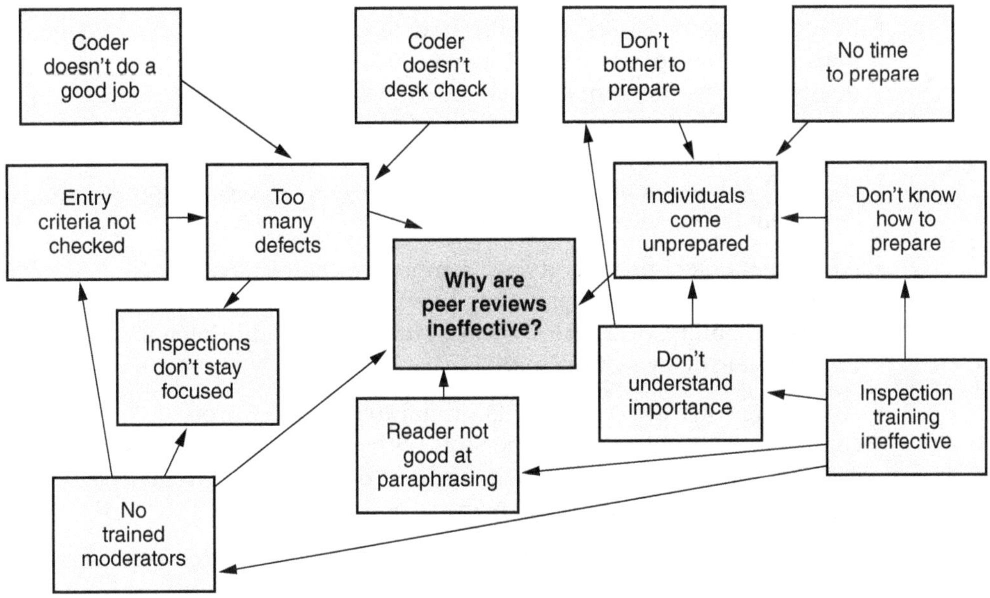

Figure 20.29 Interrelationship digraph—example.

Doing a thorough root cause analysis prevents the band-aid effect of correcting the symptoms without correcting the underlying cause. Correcting the actual root cause prevents future recurrences of the same or related issues. Root cause analysis provides opportunities to identify specific ways of improving software projects, products, processes, and services by correcting actual problems rather than their symptoms.

There are many different facets to analyzing the root cause of defects in software project, processes, products, and services:

- Analyzing when and in what processes defects are introduced into the software products can help focus corrective and preventive actions. For example, if analysis shows that 55 percent of an organization's defects were introduced during requirements elicitation, efforts should focus on those actions that will improve the elicitation process.
- Analyzing the type of defect being introduced within that phase or process (e.g., ambiguous requirement, missing requirement, overlooked stakeholder, incorrect requirement, and so on) can help further focus improvement opportunities. For example, suppose analysis shows that requirement ambiguity is a major issue, standardized requirement statement templates or additional training might help requirements analysts specify better, less ambiguous requirements.
- Analyzing defect types, combined with knowing when the defect was detected, identifies opportunities to improve the defect-detection processes.

- By identifying processes that failed to detect the defects, earlier detection rates can be improved by updating methods to look for a specific type of defect. For example, suppose analysis shows that a large percentage of ambiguous requirements are not being found during requirements peer reviews. In that case, items can be added to the requirements peer review checklist to look for ambiguity, or peer reviewers can receive training to help them better detect that type of defect.
- Analyzing defect-prone software components helps identify components where software reengineering can benefit the product's overall reliability, or where risks exist that require additional attention during peer reviews or testing.
- Analyzing software components similar to the defect-prone components can identify components with a high potential for similar defects. If a mistake was made in one part of the product, it may be repeated in a similar product component.
- Analyzing processes similar to the processes where defects were introduced, or were not detected, can identify processes with a high potential for benefiting from similar improvements.

Finally, analyzing process problems identified during assessments, audits, or lessons-learned sessions and addressing their root causes can have a long-term impact on both process effectiveness and efficiency, thereby on product quality through the prevention of defect introduction.

This book has already discussed several metrics and tools for root cause analysis:

- Defect density can be used to identify defect-prone components, and this chapter talked about using Pareto charts for the same purpose (see Figure 20.16).
- The analysis of escapes, phase containment, and defect-detection efficiency metrics can also identify ineffective detection processes.
- Check sheets (see Figure 20.19), or matrix diagrams (see Figure 20.28) are helpful in conjunction with defect taxonomies (see Figure 20.27) to analyze root causes by defect types.
- Flowcharts (see Figure 20.15) can be used to identify process inefficiencies, bottlenecks, or other problems.
- Cause-and-effect diagrams (see Figures 20.17 and 20.18) can be used to explore potential root causes.

Another tool for getting to the root cause is called the *five whys method*. In this method, the question "why?" is repeatedly asked and answered (typically, this can take five whys or maybe even more) until the root cause is identified. For example, suppose a problem exists because the interface specification to the notification controller was not defined before the scheduled time to design its driver. In that case, the individuals performing root cause analysis might ask the following five whys:

1. Why? Because the customer has not sent us the specification
2. Why? Because the customer did not have the specification
3. Why? Because the controller supplier has not provided the specification to the customer
4. Why? Because the customer has not purchased the controller
5. Why? Because the customer is waiting for a new product release that has the needed new features

The five whys method is useful in a team setting to encourage team members to identify and discuss the chain of symptoms and their causes to drill down to the initial, underlying root cause of a problem so that it can be addressed with corrective action resulting in a permanent long-term solution.

Data Flow Diagrams

A *data flow diagram* (DFD) is a graphical representation of how data flow through and are transformed by the software system, as discussed in Chapter 12 and illustrated in Figures 12.6 and 12.7. DFDs can be used to analyze requirements and can also be used as a model to represent the data flow view of the architectural or component designs (see Chapter 14).

In process definition, data flow diagrams can show how data or information flow through a process or between processes as inputs and outputs. For example, the *Project Management Body of Knowledge* (PMI 2013) uses data flow diagrams extensively to illustrate how data and information interconnect all the project management processes. In process improvement, data flow diagrams of the processes can help identify potential information bottlenecks or missing data or data flows.

Multi-voting

In the *multi-voting* method, each participant is given a specific number of votes to distribute among the ideas (or other items). For example, if there are 50 ideas, each person might be given eight votes. A participant can give one idea all eight votes, distribute votes between a few ideas, or give one vote to each of eight different ideas. The facilitator collects the votes and uses them to score and rank the ideas by priority. The group then reviews and discusses the results. A variation on this is to give each participant one small sticker for each vote to place next to the idea(s) they are voting for.

An alternative to multi-voting is the *ranked-choice method,* in which each participant is asked to select and rank their top *n* choices. For example, if there are 30 ideas, each participant is given six index cards. Each participant then writes down their top six choices along with the rank they wish to assign to each choice, from 6 (first choice) to 1 (last choice), on the index cards, one choice per card. The facilitator collects the cards and totals the scores for each idea (e.g., if an idea was ranked 6, 3, 1, and 4 by four different participants, it would get a score of 6+3+1+4=14). Ideas are prioritized based on their scores. The group then reviews and discusses the results.

Multi-voting or the ranked-choice method are used to determine whether or not there is agreement on the priorities of the ideas or items. If a few ideas get the majority of the votes, prioritization is achieved quickly. If the votes or scores are more evenly distributed across all (or many) of the ideas, it may be necessary to take a more rigorous, time-consuming approach to prioritizing the ideas. In this case, tools such as prioritization matrices and prioritization graphs can be utilized.

Prioritization Matrix

A *prioritization matrix* is used to rank ideas in order of priority based on a set of criteria. Each of these criteria can be assigned a weight based on importance. For example, in the prioritization matrix in Table 20.1, four criteria have been selected to prioritize a set of process improvement suggestions. The criteria are:

- *Bottom line*: The impact of implementing the improvement suggestion on the profits of the company (weighted at .25)
- *Easy*: How easy the improvement suggestion will be to implement (weighted at .15)
- *Staff acceptance*: How accepting the software engineering staff will be to the implementation of the improvement suggestion (weighted at .20)
- *Stakeholder satisfaction*: The impact of implementing the improvement suggestion on the satisfaction levels of the stakeholders (weighted at .40)

Each process improvement idea is then rated on a scale for each of the criteria. These scales can be an interval scale (e.g., 1 to 5, like in the example in Table 20.1, where 5 is the highest, best, or most important or significant rating, and multiple ideas can be assigned the same number). A ranking scale can also be used (e.g., if there are 6 ideas, each idea is ranked and then assigned a number from 1 to 6 based on that ranking, where 6 is assigned to the highest ranking idea and no two ideas have the same ranking). The total score for each idea is then calculated. In the example in Table 20.1, the implementation of Process Improvement 1 is expected to:

Table 20.1 Prioritization matrix—example.

Criteria and Weights					
	Bottom Line (.25)	Easy (.15)	Staff Acceptance (.20)	Stakeholder Satisfaction (.40)	Total
Process Improvement 1	1	3	2	5	3.10
Process Improvement 2	4	1	5	3	3.35
Process Improvement 3	2	2	2	1	1.60
Process Improvement 4	2	4	3	2	2.50

- Have no or insignificant impact on profitability, so the Bottom line criterion was given a score of 1
- Be about average in ease of implementation, so the Easy criterion was given a score of 3
- Have some staff resistance, so the Staff acceptance criterion was given a score of 2
- Have a major impact on the level of stakeholder satisfaction, so the Stakeholder satisfaction criterion was given a score of 5

The total score for Process Improvement $1 = (1 \times .25) + (3 \times .15) + (2 \times .20) + (5 \times .40) = 3.10$. These total scores are used to rank the improvements for implementation. Based on the ranking in this example, process improvement 2 would have the highest priority, followed by 1, then 4, and process improvement 3 would have the lowest priority.

Prioritization Graph

A *prioritization graph* is another tool that can be used to prioritize ideas when only two criteria are being used to rank items. To use this method, each item is rated on a scale (e.g., 1 to 5, where 5 is the highest priority) on each of the two criteria. The ratings for each item are then plotted on an *x–y* graph as a bubble. Figure 20.30 shows an example of a prioritization graph used to prioritize possible software inspection metrics based on two criteria, the importance of the information they provide and their ease of implementation. Items with the highest priority will be in the upper right-hand corner (both important and easy to implement). One of the advantages of this method is that it is easy to view the relationships between the items and understand the impacts of the criteria on the priority.

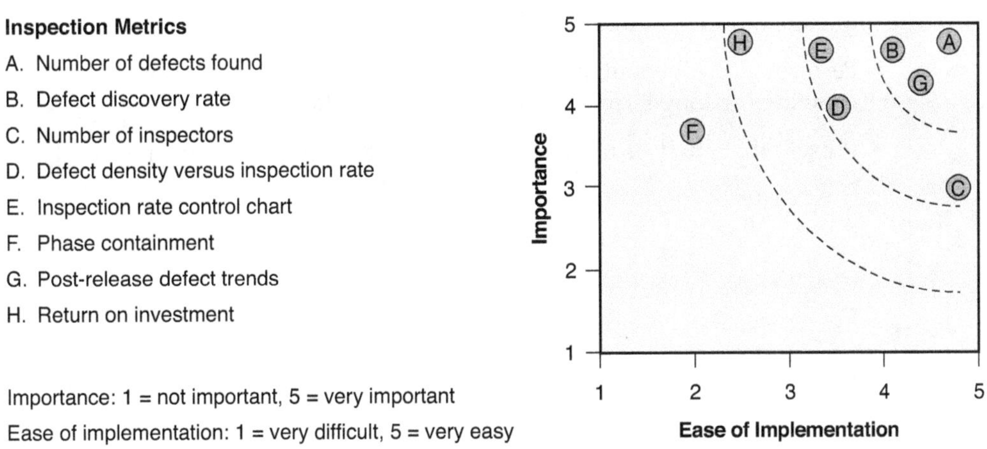

Figure 20.30 Prioritization graph—example.

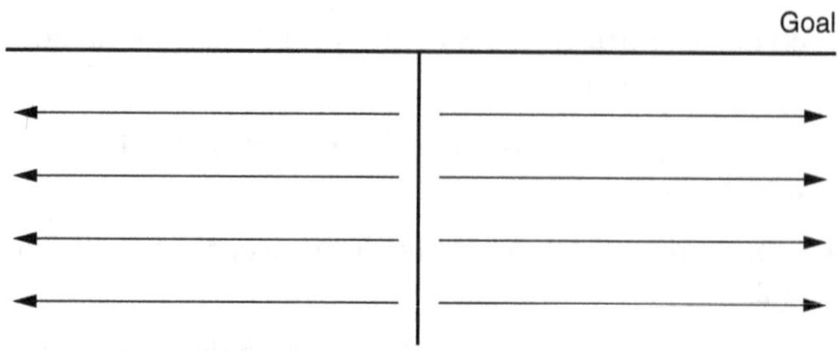

Figure 20.31 Force field analysis template—example.

Force Field Analysis

Force field analysis is a team tool used to identify factors (forces) that will aid in achieving a goal or implementing an idea and factors that will inhibit movement toward that goal. Figure 20.31 illustrates an example of a force field analysis template. Figure 4.1 in Chapter 4 shows an example of a completed force field analysis for analyzing forces for and against change.

The team can use information from the force field analysis to identify needed activities in its plans to implement the goal. Activities that enhance the factors that help move toward the goal should be identified and planned. Activities should also be identified and planned to handle or mitigate factors that inhibit or move away from the goal. Forces driving toward the goal can also be leveraged to counteract the opposing forces. For example, if:

- The goal is to "train people in peer review techniques"
- One of the driving forces toward the goal is "long-term savings in maintenance costs"
- One of the inhibiting forces is "no training budget"

The implementation plan might include actions to:

- Perform a cost/benefit analysis on peer review implementation showing the long-term return on investment (ROI)
- Present that analysis to management to get the required training funds

Part VI

Software Verification and Validation

Chapter 21

Theory (of V&V)

According to ISO/IEC/IEEE *Systems and Software Engineering—Vocabulary* (IEEE/ISO/ IEC 2017), *verification* is "the process of evaluating a system or component to determine whether the products of a given development phase satisfy the conditions imposed at the start of that phase." Verification is concerned with the process of evaluating the software to make certain that it meets its specified product requirements and adheres to the appropriate standards, practices, and conventions. Verification is implemented by looking one phase back. For example, code is verified against the detailed design, the detailed design against the architectural design, and the architecture against the specified requirements.

Validation is the "confirmation, through the provision of objective evidence, that the requirements for a specific intended use or application have been fulfilled" (IEEE/ISO/IEC/2017). Validation is concerned with the process of evaluating the software to make certain that it meets its intended use and matches the stakeholders' needs. These definitions of verification and validation are illustrated in Figure 21.1.

Verification and validation (V&V) processes are complementary and interrelated. The real strength of the V&V processes is their combined employment of various techniques to make certain that the software, its intermediate components, and its interfaces to other components of the system are high quality, comply with the specified requirements, and will be fit for use when delivered. V&V activities provide objective evaluations of work products and processes throughout the software life cycle and are an integral part of software development, acquisition, operations, and maintenance. "The purpose of V&V is to help the organization build quality into the system during the life cycle" (IEEE 2016).

Not only do V&V activities provide objective evidence that quality has been built into the software, they also identify areas where the software lacks quality, such as defects. As illustrated in Figure 21.2, every organization has processes (formal or informal) that it implements to produce its software products. These software products can include the final products that are used in operations (e.g., executable software, user manuals) and interim products used within the organization (e.g., requirements, designs, source or object code, test cases). V&V methods identify defects, or potential defects, in all of those products. If potential defects are identified, they must be analyzed to determine if actual defects exist. Identified defects are then eliminated through rework, which immediately effects the quality of the product. Feedback from V&V activities performed in parallel

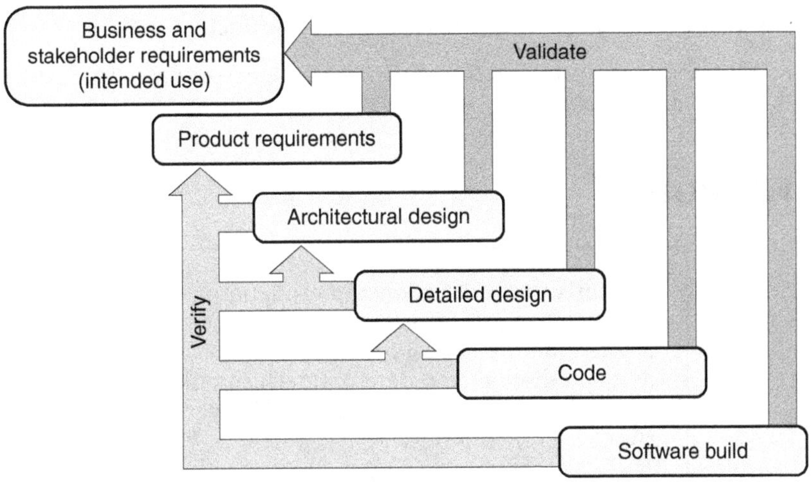

Figure 21.1 Verification and validation.

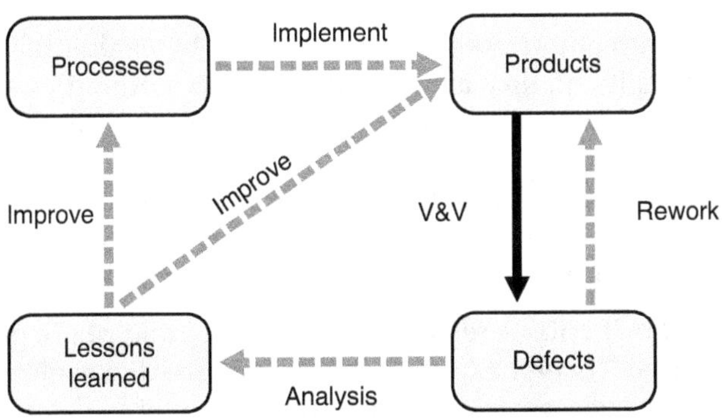

Figure 21.2 V&V techniques identify defects.

with development activities provides a cost-effective mechanism for correcting defects in a timely manner and implementing corrective actions to prevent the future recurrence of similar defects. This feedback minimizes the overall cost, schedule, and product quality impacts to the project of software defects.

The defects found with V&V methods should be analyzed to determine their root cause. The lessons learned from this analysis can be used to improve the processes used to create future products and prevent future defects, thus impacting the quality of future products produced.

The data recorded during the V&V activities can also be used for making better management decisions about issues such as:

- Whether the current quality level of the software is adequate
- What risks are involved in releasing the software in its current state

- Whether additional V&V activities are cost-effective
- What the staff, resource, schedule, and budget estimates for future V&V efforts should be

1. V&V METHODS

> Use software verification and validation methods (e.g., static analysis, structural analysis, mathematical proof, simulation, and automation) and determine which tasks should be iterated because of modifications. (Apply)
>
> **Body of Knowledge VI.A.1**

Static Analysis

Static analysis is a method used to perform V&V by evaluating a software work product to assess its quality and identify defects without executing that work product. Reviews, analysis, and mathematical proofs are all forms of static analysis.

Static analysis techniques include various *product and process reviews* to evaluate the completeness, correctness, consistency, and accuracy of the software products and the processes that created them. These reviews include:

- Entry and exit criteria reviews, quality gates, and phase gate reviews, discussed in Chapter 17. These reviews are used to evaluate whether the products of a process or major activity are of high enough quality to transition to the next activity.
- Peer reviews, including inspections, discussed in Chapter 23, can be used for engineering analysis and as defect-detection mechanisms.
- Other forms of technical reviews and pair programming, discussed in Chapter 23, can also be used to assess and evaluate software products and processes and find defects.

Static analysis can be performed using tools (e.g., compilers; requirement, design, or code analyzers; spell checkers; grammar checkers; security analyzers; and other analysis tools).

Individuals or teams can also perform analysis activities through in-depth assessments of the requirements or other software products. For example, Chapter 12 included a discussion of various models that can be useful in analyzing the software requirements for completeness, consistency, and correctness. These same models can be used to analyze the software's architecture and component designs. Other forms of analysis might include feasibility analysis, testability analysis, traceability analysis, hazard analysis, criticality analysis, security analysis, or risk analysis used to help identify and mitigate issues. Requirements allocation and traceability analysis are used to verify that the software (or system) requirements were implemented completely and accurately.

Mathematical proofs, also called *proofs of correctness,* are "a formal technique used to prove mathematically that a computer program satisfies its specified requirements" (IEEE/ISO/IEC 2017). Mathematical proofs use logic (Boolean algebra) to deduce that the design or code's logic is correct when compared to their predecessor work products.

Dynamic Analysis

Dynamic analysis methods are used to perform V&V on software components or products by executing them and comparing the actual results to the expected results. Testing and simulations, discussed in Chapters 22 and 24, are forms of dynamic analysis.

Piloting is another type of dynamic analysis technique. The piloting of a software release, also called *beta testing* or *first office verification,* rolls out the new software to just a few operational sites for evaluation before propagating it to a large number of sites to minimize the risk of full-scale implementation. For other software deliverables, such as training and help desk support, that are difficult to test, piloting can be a useful V&V technique. For example, a set of typical software issues or problem areas could be compiled during system test execution and used to create a set of "typical" problem scenarios used to pilot the help desk. The testers could call the help desk, describe one of the scenarios, and evaluate whether the help desk personnel can adequately handle the call.

Product Audits

Product audits, discussed in Chapter 8, throughout the life cycle can evaluate the product against the requirements and required standards (coding standards, naming conventions, required document formats). Product audits usually use static analysis techniques but can also use dynamic analysis techniques, for example, if the auditor selects a sampling of test cases to execute as part of the audit. Physical and functional configuration audits, discussed in Chapter 28, are also used as final checks for defects before the product is transitioned into operations.

V&V Task Iteration

Whenever a change is made to a software work product that has already undergone one or more V&V activities, an evaluation should be done to determine how many already executed V&V activities need to be re-executed (iterated). This evaluation should initially be done as part of the impact analysis performed when evaluating the change. (See Chapter 27 for a discussion of impact analysis.) Additional evaluation may be needed as the change is implemented, if the depth and breadth of that change ends up being different than expected.

The first part of the V&V task iteration decision is based on what V&V activities have already occurred. As Table 21.1 illustrates, if changes are made after the completion of just the peer reviews, then the only consideration is whether or not to re-peer review. If the change is made later in the life cycle, any completed (or partially completed) V&V activities are candidates for iteration.

The second part of the V&V task iteration decision is based on risk and the concept of V&V sufficiency. As illustrated in Figure 21.3, *V&V sufficiency analysis* weighs the risk that the software still has undiscovered defects and the potential loss associated with those defects against the cost of performing additional V&V activities and the benefits of additional V&V activities:

- Just because V&V tasks were iterated does not mean any new or additional defects will be found.
- Just because the software was changed does not mean new defects were introduced.
- Even if defects were introduced, it does not mean that those defects will cause future failures. For example, most software practitioners have experienced latent defects that have taken years to discover after they were initially introduced into a software product because the software was never before used in the exact way required to uncover the defect.

V&V task iteration risk analysis should consider the quality and integrity needs of the product or product component being evaluated. For example, software work products that have the following characteristics are typically candidates for more V&V task iteration than those that do not:

- Mission-critical
- Safety-related
- Security-related
- Highly coupled
- Defect-prone in the past
- Most used functionality
- Contains patches or other temporary fixes

Metrics can be used so that trade-off decisions can be made based on facts and information rather than "gut feelings." For example, metrics information can be used to tailor the V&V plans based on data collected on previous products or projects, on the estimated defect counts for various work products, and on an analysis of how applying the various techniques can yield the best results given the constraints at hand. As the project progresses, metrics information can also be used to refine the V&V plans using actual results to date.

Table 21.1 Verification and validation task iteration considerations.

Change(s) Made After Completion Of:	V&V Task Iteration				
	Peer Review	**Unit Test**	**Integration Test**	**Qualification Test**	**Acceptance Test**
Peer Review	Re-Peer Review Changed Work Product(s)	—	—	—	—
Unit Test	Re-Peer Review Changed Work Product(s)	Retest Changed Component(s)	—	—	—
Integration Test	Re-Peer Review Changed Work Product(s)	Retest Changed Component(s)	Retest Changed Interfaces Regression Test Impacted Interfaces Run Integration Regression Test Suite	—	—
Qualification Test	Re-Peer Review Changed Work Product(s)	Retest Changed Component(s)	Retest Changed Interfaces Regression Test Impacted Interfaces Run Integration **Regression** Test Suite	Retest Changed Functions/ Requirements/User Scenarios Regression Test Impacted Functions/ Requirements/User Scenarios Run Qualification Regression Test Suite	—
Acceptance Test	Re-Peer Review Changed Work Product(s)	Retest Changed Component(s)	Retest Changed Interfaces Regression Test Impacted Interfaces Run Integration Regression Test Suite	Retest Changed Functions/ Requirements/ User Scenarios Regression Test Impacted Functions/ Requirements/ User Scenarios Run Qualification Regression Test Suite	Retesting & Regression testing of a new release depends on customer, user & other stakeholder requirements
Release (Corrective Release or Small Enhancement)	Re-Peer Review Changed Work Product(s)	Retest Changed Component(s)	Retest Changed Interfaces Regression Test Impacted Interfaces Run Integration Regression Test Suite	Retest Changed Functions/ Requirements/User Scenarios Regression Test Impacted Functions/ Requirements/User Scenarios Run Qualification Regression Test Suite	Retesting and regression testing of a new release depends on customer, user & other stakeholder requirements
Release (New Features)	V&V Requirements Defined in Project, V&V or Test Plan				

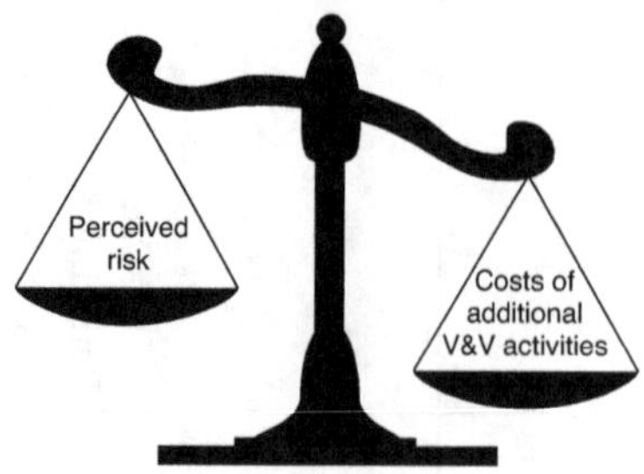

- Probability of undiscovered defects
- Loss associated with the defects

- Cost of continued V&V activities
- Benefits of additional V&V activities

Figure 21.3 Verification and validation sufficiency.

2. SOFTWARE PRODUCT EVALUATION

> Use various evaluation methods on documentation, source code, etc., to determine whether user needs and project objectives have been satisfied. (Analyze)
>
> **Body of Knowledge VI.A.2**

Evaluation Methods

Common V&V techniques used for requirements and design include the following. (Note that requirements validation is discussed in more detail in Chapter 12.)

- Analysis and reviews (e.g., interface analysis, criticality analysis, hazard analysis, testability analysis, feasibility analysis, security analysis, and risk analysis) and pair programming, peer reviews including inspections and other technical reviews to:
  - Validate that the business, stakeholder, product requirements, architectural design, and component designs meet the intended use and stakeholders' needs
  - Validate that the requirements as a set are complete, internally and externally consistent, and modifiable
  - Validate that each specified requirements is clear, unambiguous, concise, finite, measurable, feasible, testable, traceable, and value-added
  - Verify that the design completely and consistently implements the specified requirements, and that it is feasible and testable

- Validate and verify that criticality, safety, security, and risks are identified, analyzed, and mitigated in the requirements and design as appropriate
- Verify that the specified requirements and designs comply with standards, policies, regulations, laws, and business rules

- Mathematical proofs of correctness to verify the correctness of the design based on its predecessor specifications
- Functional configuration audits to verify that the completed requirements and designs conform to their requirements specification(s) in terms of completeness, performance, and functional characteristics
- Prototyping to evaluate correctness and feasibility
- Traceability matrices to verify that the requirements have been built into the software work products and that one or more test cases exist to verify and validate each requirement and its implementation
- Design of integration, system, and acceptance test cases and test procedures, and the execution of test cases and test procedures to validate and verify the correct and complete implementation of the requirements and design
- Static analysis tools including spelling and grammar checkers and tools used to search for words and phrases that are potentially problematic, for example:
  - Nonfinite words or phrases (e.g., all, always, sometimes, including but not limited to, and etc.)
  - Words ending with "ize" or "ly," (e.g., optimize, minimize, quickly, user-friendly)
  - Other ambiguous or nonmeasurable words (e.g., fast, quick, responsive, large)
  - Indications of incompleteness (e.g., "to be done," "to be determined," or "TBD")

Common V&V techniques used to evaluate source code include:

- Analysis (e.g., data flow analysis, control flow analysis, criticality, hazard, security, safety, and risk) and pair programming, peer reviews, and other types of technical reviews to:
  - Validate that the source code as written meets the intended use and stakeholders' needs
  - Verify that the source code completely and consistently implements its component design and allocated requirements
  - Verify 100 percent traceability from component design to source code

    - Verify that criticality, safety, security, and risks are mitigated in the source code as specified
    - Verify that the specified source code complies with standards, policies, regulations, laws, and business rules
- Mathematical proofs of correctness to verify the correctness of each source code module based on its component design
- Traceability matrices to verify that the requirements, architectural design, and component designs have been built into the source code and that one or more unit or integration test cases exist to verify each source code module and its interfaces
- Design and execution of unit test cases and test procedures to verify the correct and complete implementation of the code and its associated component design
- Functional configuration audits to verify that the completed source code modules conform to their requirements specification(s) in terms of completeness, performance, and functional characteristics
- Physical configuration audits to verify that the completed source code modules conform to the technical documentation that defines them
- Static analysis tools (e.g., compilers and more-sophisticated code analyzers and security analyzers) to identify defects, security holes, and other analogies

Common V&V techniques used to evaluate software builds include:

- Integration, system, alpha, beta, and acceptance test design and execution to validate and verify the correct and complete implementation of the software products, including user documentation
- Build reproducibility analysis to verify the reproducibility of the build
- Functional configuration audits to verify that the completed software products, including user documentation, conform to their requirements specification(s) in terms of completeness, performance, and functional characteristics
- Physical configuration audits to verify that the completed software products, including user documentation, conform to the technical documentation that defines them

Common V&V techniques used to evaluate test cases and test procedures include:

- Analysis, pair programming, peer reviews, and other technical reviews
    - Validate that the test cases and test procedures as written reflect the intended use and stakeholders' needs
    - Verify that the test cases and test procedures completely and consistently verify the source code, design, or requirements

  - Verify that the specified test cases and test procedures comply with standards, policies, regulations, laws, and business rules
- Traceability matrices to verify test coverage and completeness of the test cases and test procedures sets
- Manual execution of the test cases and test procedures before they are automated to identify defects and other analogies
- After automation, the test automation can be reviewed and tested before it is used to test the software to identify defects and other analogies

Common V&V methods used to evaluate documentation (e.g., training materials, user documentation) include:

- Analysis and pair programming, peer reviews, and other types of reviews
  - Validate that the documentation as written meets the intended use and stakeholders' needs
  - Verify that the documentation completely and consistently implements its allocated requirements
  - Verify 100 percent traceability from requirements to documentation
  - Verify that criticality, safety, security, and risks are mitigated in the documentation as specified or needed
  - Verify that the specified documentation complies with standards, policies, regulations, laws, and business rules
- Design and implementation of test cases and test procedures to verify the user-deliverable documentation (e.g., user manuals and installation instructions)
- Static analysis tools (e.g., spell checkers and grammar checkers) to identify defects and other analogies

Common V&V methods used to evaluate the replication and packaging process include:

- Comparing the replicated product to the original
- Delivering a complete software shipment in-house, evaluating its contents, and installing and testing or operating the included software.

Piloting is a common V&V method for evaluating training and help desk services.

Risk-Based V&V

Risk-based V&V prioritizes software items (work products, product components, or features and functions) based on their risk exposure. The *risk exposure* is calculated based on probability and impact. In risk-based V&V, *probability* is the estimated likelihood that yet undiscovered, relevant defects will exist in the software item after the completion of a V&V activity. *Impact* is the estimated cost of the result or consequence if one or more undiscovered defects escape detection during the V&V activity and cause issues in operations. The higher the risk exposure of an item, the higher the rigor and intensity needed when performing and documenting V&V activities for that item.

Multiple factors or probability indicators may contribute to a software item having a higher or lower risk probability. These probability indicators may vary from project to project and from environment to environment. Therefore, each organization or project should determine and maintain a list of probability indicators to consider when assigning risk probabilities to its software items. Figure 21.4 illustrates examples of probability indicators, including:

- The more churn there has been in the requirements allocated to the software item, the more likely it is that there are defects in that item
- Software items that have had a history of defects in the past are more likely to have additional defects
- Larger or more-complex software items are more likely to have defects than smaller or simpler software items
- The more constraints on the quality attribute (e.g., reliability, performance, safety, security, maintainability), the more likely it is that there are related defects in the software item

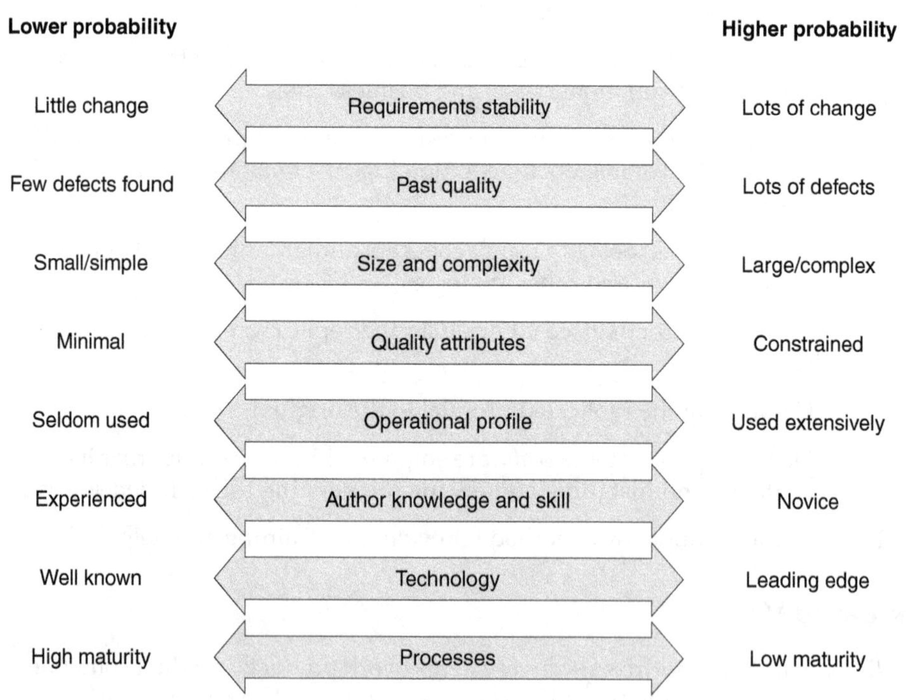

Figure 21.4 Probability indicators—examples.

- The more an item in the software is used, the more likely it is that users will encounter escaped defects in that part of the software
- Novice software developers with less knowledge and skill tend to make more mistakes than experienced developers, resulting in more defects in their work products.
- If the developers are very familiar with the programming language, tool set, and business domain, they are less likely to make mistakes than if they are working with new or leading-edge technology.
- Higher-maturity processes are more likely to help prevent defects from getting into the work products or identify them earlier in the life cycle.

Multiple and varying factors or impact indicators may also contribute to a software item having a higher or lower risk impact. Each organization or project should determine and maintain a list of impact indicators to consider when assigning risk impacts to its work products. Examples of impact indicators include:

- Schedule and effort impacts
- Development and testing costs
- Internal and external failure costs
- Corrective action costs
- High maintenance costs
- Stakeholder dissatisfaction or negative publicity resulting in lost market opportunities or sales
- Litigation, warranty costs, or penalties
- Noncompliance with laws, regulations, or policies

The IEEE *Standard for System, Software and Hardware Verification and Validation* (IEEE 2016) uses the assignment of an integrity level, varying from high to low, to determine the required set of V&V activities and their associated level of rigor and intensity for each software item (instead of risk exposure, as discussed previously). Other standards, including industry-specific standards, use similar integrity-level grading schemas. The "*integrity level* determination establishes the importance of the system to the user and acquirer based on complexity, criticality, risk, safety level, security level, desired performance, reliability or other system-unique characteristics" (IEEE 2016).

The level of integrity needed in the software varies by the context of that software. Table 21.2 shows an example of an integrity-based grading scale for V&V activities.

For example, biomedical software, airplane navigation software, and software used in military weapons are examples of software where failures may impact the health or safety of people or may damage or destroy property. Therefore, the organizations that build this type of software are extremely careful to minimize the risk of any failures during operations. The organizations building this type of software are willing to spend the time, effort, and money necessary to reduce

Table 21.2 Software V&V requirements by integrity level—example.

<table>
<tr><th rowspan="2">Software Testing Activity</th><th colspan="4">Activity Grading by Integrity Level</th></tr>
<tr><th>4</th><th>3</th><th>2</th><th>1</th></tr>
<tr><td>Requirements peer review</td><td>Required</td><td>Required</td><td>Required</td><td>Tailorable</td></tr>
<tr><td>Design peer review</td><td>Required</td><td>Required</td><td>Tailorable</td><td>Not Required</td></tr>
<tr><td>Code peer review</td><td>Required</td><td>Tailorable</td><td>Tailorable</td><td>Not Required</td></tr>
<tr><td>Software unit testing</td><td>Required</td><td>Tailorable</td><td>Tailorable</td><td rowspan="2">Combined & Tailorable</td></tr>
<tr><td>Software integration testing</td><td>Required</td><td>Required</td><td>Tailorable</td></tr>
<tr><td>Software system testing</td><td>Required</td><td>Required</td><td>Required</td><td>Tailorable</td></tr>
<tr><td>Software acceptance testing</td><td>Required</td><td>Required</td><td>Required</td><td>Tailorable</td></tr>
</table>

those risks through rigorous development and V&V practices. The customers of this type of software are also willing to pay the price necessary to purchase high-quality software.

On the other end of the spectrum, software used just for entertainment or convenience has much less risk. Examples of this type of software include gaming software, word processors, and information applications (like the one that identifies the song playing on the radio). The organizations that develop this type of software cannot afford to spend the time, effort, and money required to minimize the probability of operational failures, or their products would be so expensive that they would probably not succeed in the marketplace.

The types of testing used will also vary based on the software's context: For example:

- Banking and e-commerce software will include a focus on security testing.
- Real-time hardware control software will include a focus on performance testing.
- Software that can create safety risks will include a focus on safety testing.
- Software that will be sold internationally will include a focus on localization testing.

However, the IEEE standard leaves the determination of the actual integrity levels up to the needs of each project or program and states that the V&V plans should include (IEEE 2016):

- The characteristics used to determine the selected integrity schema
- The resulting item integrity levels
- Plans defining the minimum level of V&V activities based on each item's integrity level

V&V Plans

Each project or program defines its V&V strategies and tactics as part of its V&V plans. The IEEE *Standard for System, Software and Hardware Verification and Validation* (IEEE 2016) provides an outline and guidance for performing V&V planning. Like other planning documents, the V&V plans should include information about the project's organizational structure, schedule, budget, roles and responsibilities, resources, tools, techniques, and methods related to the V&V activities. The V&V plan also defines the integrity level schema mentioned previously.

The V&V plans define the risk-based, or integrity-based, V&V processes that are planned for each major phase or activity of the software life cycle to evaluate the work products produced or changed during those phases or activities, including:

- Acquisition
- Supply
- Project planning
- Configuration management
- Business, stakeholder, and product requirements
- Design, including architectural and component design
- Implementation
- Test (unit, integration, system, alpha, beta)
- Acceptance
- Installation
- Operations
- Maintenance
- Retirement (disposal)

The V&V plans should include plans for ongoing status reporting on identified problems and V&V activities, for V&V summary reports, and any special or other V&V reports required for the project. The V&V plans should also define V&V administrative procedures, including:

- Processes for reporting identified problems and tracking them to resolution
- Policies for V&V task iteration
- Policies for obtaining deviations or waivers
- Processes for controlling V&V-related configuration items
- V&V standards, practices, and conventions that are being used on a project or program, including those adopted or tailored from the organizational-level standards and any that are project-specific

Chapter 22

Test Planning and Design

Testing is a dynamic analysis verification and validation (V&V) method. *ISO/IEC/IEEE Systems and Software Engineering—Vocabulary* (IEEE/ISO/IEC 2017) defines *testing* as "the process of operating a system or component under specified conditions, observing or recording the results, and making an evaluation of some aspect of the system or component."

One of the testing objectives is to find yet undiscovered, relevant defects in the software. *Relevant defects* are defects that will negatively impact the stakeholders. Both software developers and software testers are responsible for the quality of the delivered software. It has to be a team effort. The job of a software developer is to prevent defects in the product and to build quality into the software products. Software developers are also responsible for V&V activities early in the life cycle, including static analysis techniques discussed in Chapter 21. However, a mental shift must occur as the software moves into testing. While software developers make the software work, testers do everything they can to find all the possible ways to "break" the software. Even if the developers are also doing testing, they are switching to the tester role and must make this shift to the tester mentality. Testers are successful every time they uncover a yet-undiscovered, relevant defect, because that means one less defect will make its way to the software users and impact the stakeholders' satisfaction with that product (assuming that any uncovered defects are corrected before release).

Given that testing, like all other development activities, has a limited amount of resources (time, money, people, and other resources), the goal of the tester is to select a set of test cases that are most likely to uncover as many different relevant defects as possible within those constraints. Therefore, the tester is part detective, hunting for clues about how the software might fail. The tester is part amateur psychologist trying to look into the mind of the programmer to figure out what kinds of mistakes they might have made when developing the software. The tester is also part psychic, trying to predict how the users might misuse or inappropriately use the software or use it in ways not accounted for by the software being tested. Testers must understand how interfacing entities (hardware, other software applications, and databases) might fail, have their communications corrupted or interrupted, or interface differently than defined in the software requirements.

As illustrated in Figure 22.1, the IEEE/ISO/IEC Test Processes Standard (IEEE/ISO/IEC 2021a) divides the software testing activities into three process groups and further into major testing processes. A test process consists of multiple activities, which can be further broken down into associated tasks. These activities

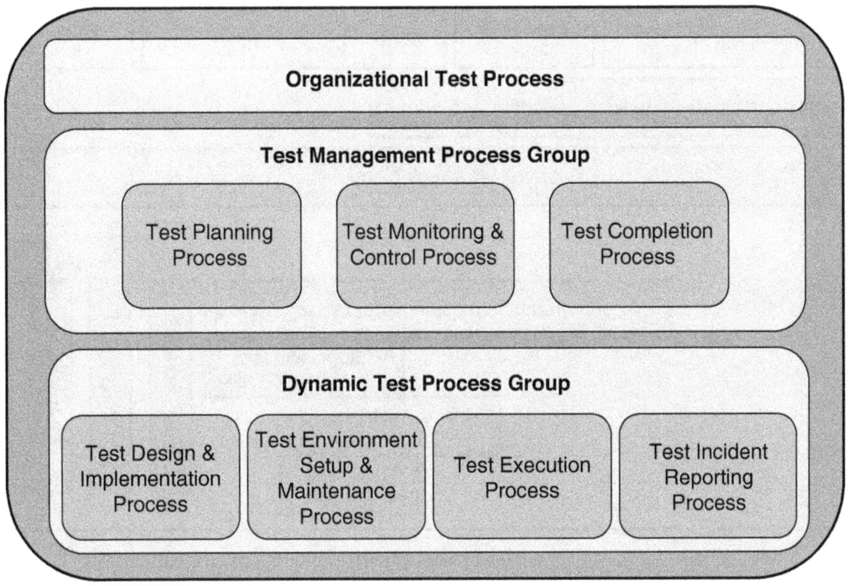

Figure 22.1 Test process groups.

and tasks will vary for projects, iterations, or releases based on the context, as we just discussed.

Although these activities will be discussed in sequential order in this book, during the actual testing processes, these activities are typically implemented iteratively. In agile development, short iterations are used to develop the software, where the software design, code, build, and test cycles happen repeatedly throughout the iteration. In sequential development, these activities may be repeated iteratively within a phase or across phases as needed. Based on their context, the testing processes should be tailored to the organization's specific needs and the individual software development, maintenance, and acquisition efforts.

The *organizational test process* defines the testing-related parts of the organization's quality management system (QMS), including test policies, processes, and work instructions, as discussed in Chapter 6. The processes of the test management process group and dynamic test process group are discussed in this chapter and in Chapter 24.

The actual act of testing requires the execution of a software system or one of its components. Therefore, test execution cannot be performed until executable software exists. With the exception of prototypes and simulations, this means that, in waterfall life cycle models, most test execution starts during or after the coding phase of the software development life cycle. However, while test execution cannot start until these later phases, test planning and design can and should be done as early as possible in the life cycle, as illustrated in Figure 22.2. Test planning and design can start as soon as the software requirements are identified. Simply designing test cases (even without executing them) can help identify defects in the software work products, and the earlier those defects are discovered, the less expensive they will be to correct. The V-model, discussed in

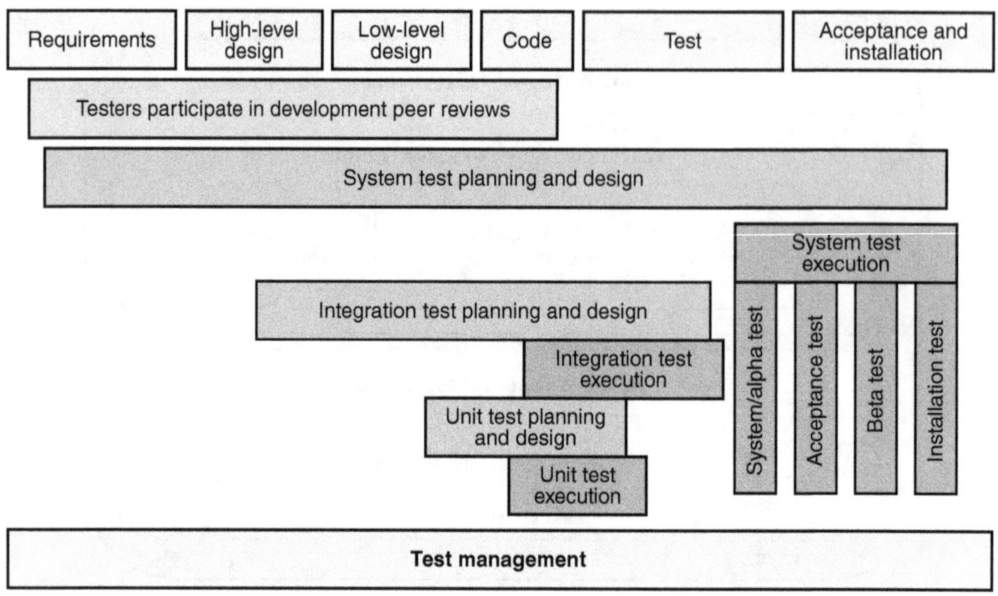

Figure 22.2 Test activities throughout waterfall life cycles.

Chapter 9 and illustrated in Figure 9.2, highlights the relationship between the testing phases and the products produced in the early life cycle phases:

- When stakeholder requirements are available, acceptance test planning and design can begin
- When product requirements are available, system test planning and design can begin
- When software architecture and component designs are available, one or more levels of integration test planning and design can begin
- When component designs and source code modules are available, unit test planning and design can begin

Testers are also excellent candidates for participating in peer reviews. This statement is especially true if the work product being reviewed will provide the basis for the test design. For example, a tester who will write test cases to validate the requirements should participate in the peer review of the requirements specification, and the integration tester should participate in the peer review of the architectural design. The testing effort must be ongoing and managed throughout the software life cycle.

In agile testing, similar test activities occur, but they are conducted iteratively throughout software development. In fact, in test-driven development, the test cases are created from the stakeholder-level requirements and replace product-level software requirements. Agile testers also participate in peer reviews, as appropriate. As source code is written, it is continuously integrated into the software executable and tested.

1. TEST STRATEGIES

> Select and analyze test strategies (e.g., test-driven design, good-enough, risk-based, time-box, top-down, bottom-up, black-box, white-box, simulation, automation,) for various situations such as Software as a Service (SaaS) and new product applications. (Analyze)
>
> **Body of Knowledge VI.B.1**

There are many different strategies that can be used when testing software. Each of these strategies has its strengths and weaknesses, so a combination of different strategies is typically the best approach for uncovering a variety of defect types.

White-Box Testing

White-box testing, also known as *structural, structure-based, clear-box,* or *glass-box testing,* is testing based on the software's internal structure and looks for issues in the framework, construction, or logic of the software. White-box testing is typically executed early in the testing cycle, starting at the module level, and may also be performed as source code modules are combined into components. White-box testing explores the internals of each source code module or component, looking at the control flow or data flow through the individual lines of source code.

White-box testing can find control or logic-flow defects that are all but invisible using a black-box testing strategy. It is also much easier to thoroughly investigate a suspicious source code module or component when the tester can see its internal structure. Early structural testing allows the tester to more easily trace defects back to their origin by isolating them to individual source code modules or paths through that code. Other strengths of white-box testing include the ability to look at:

- *Coverage*: When the testers can see into the source code module or component's internal structure, they can devise tests covering areas of the code that may not be touched using black-box testing.
- *Flow*: When the testers can see into the source code module or component's internal structure, they can determine what the software is supposed to do next as a function of its current state. The testers can use debuggers or other tools to run the software to track the sequence in which lines of code are executed and determine the values of key variables at specific points during the execution.
- *Data integrity*: When the testers know which part(s) of the software should modify a given data item, they can then determine the value a selected data item should have at a given point in the software, compare that expected value with the value the variable actually

has, and report deviations. The white-box testers can also detect data manipulation by inappropriate source code modules and components.

- *Boundaries*: The tester can see internal boundaries in the code using structural techniques that are entirely invisible to the black-box tester.
- *Algorithms*: The testers can check for common mistakes in calculation (e.g., division by zero, or the use of incorrect operators or operands) using structural techniques. Testers can also verify intermediate data values (including appropriate rounding) at each step in the algorithm or calculation.

Gray-Box Testing

When larger numbers of source code modules/components are integrated, performing pure white-box testing usually becomes unwieldly. In any complex system, there are too many possible paths through the software. As illustrated in Figure 22.3, at some point, testing typically progresses into various levels (shades) of *gray-box testing,* which is a blending of the white-box and black-box testing strategies. During development, source code modules and components are integrated into programs, programs into subsystems, and subsystems into the software system (note that the number of levels of integration can vary based on the project's needs). At the lowest level of gray-box testing, the individual source code modules or components are treated as gray boxes, where the tester peeks into the internals of each source code module or component just enough to determine how they interact and interface with each other, ignoring the rest of the internal details. At each subsequent level of integration, the individual programs or subsystems are also treated as gray boxes, where the tester peeks into their internals just enough to determine how those programs or subsystems interact and interface.

There are two basic strategies for conducting gray-box testing. The first strategy, called *top-down,* is illustrated in Figure 22.4 at the source code module to component integration level. The top-down strategy tests one or more of the

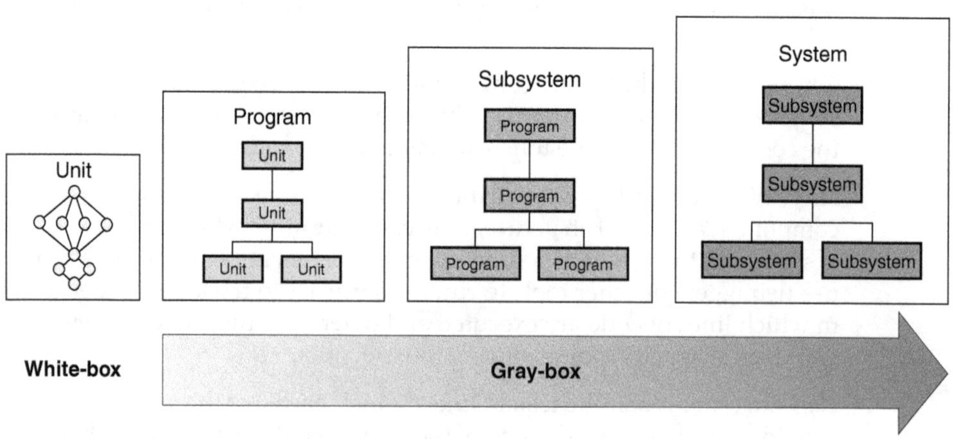

Figure 22.3 White-box testing and the transition to gray-box testing.

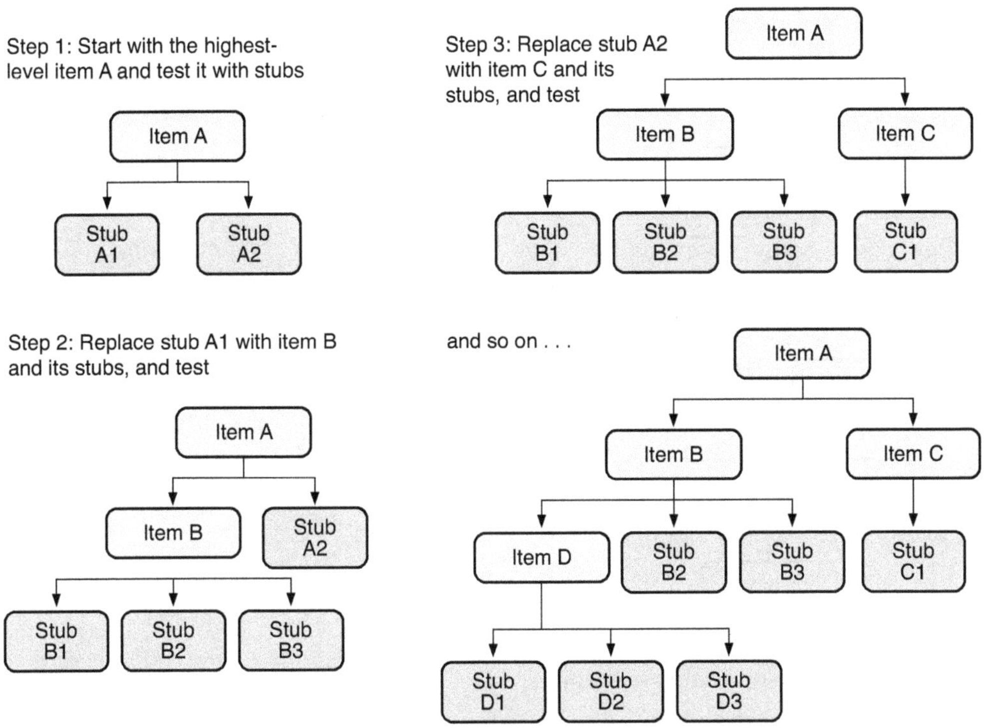

Figure 22.4 Top-down testing strategy.

highest-level software items (source code modules, components, programs, subsystems) in the tree, using stubs for lower-level software items that have not yet been integrated (or developed). These stubs simulate the required actions of those lower-level software items. The lower-level software items and their stubs are then integrated, one or more at a time, replacing these stubs.

Gray-box testing focuses on the interfaces of the newly integrated software items as they are added and their interactions with the rest of the already integrated software. The top-down strategy is typically chosen when:

- Control flow structures are critical or suspected of being defect-prone
- Critical or high-risk software items (or associated high-risk hardware items controlled by the software) are near the top of the structure since items that are integrated earlier will be tested more extensively during gray-box testing
- Top-level software items contain menus or other important external interface elements that make testing easier
- There is more fan-out than there is fan-in in the software structure being integrated
- A skeletal version would be beneficial for demonstration
- It will take less effort to create stubs than drivers

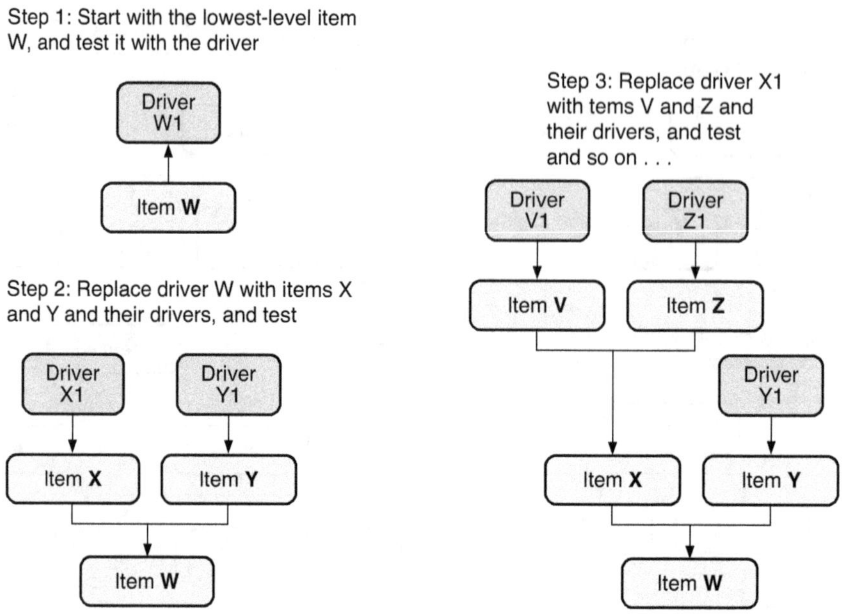

Figure 22.5 Bottom-up testing strategy.

The second gray-box testing strategy is called *bottom-up,* as illustrated in Figure 22.5. This strategy tests one or more of the lowest-level software items in the tree using drivers for higher-level calling software items that have not yet been integrated (or developed).

These drivers simulate the required actions of those higher-level software items. The higher-level software items and their drivers are then integrated one or more at a time, replacing the drivers. Testing again focuses on the interfaces to the newly integrated software items as they are added and their interactions with already integrated software. The bottom-up strategy is typically chosen when:

- Critical or high-risk software items (or associated high-risk hardware items controlled by the software) are near the bottom of the structure
- Bottom-level software items contain menus or other important external interface elements that make testing easier
- There is more fan-in than there is fan-out in the software structure being integrated
- It will take less effort to create drivers than stubs

Selection of an integration testing strategy depends on software characteristics and sometimes project schedule. Different approaches can also be used for different parts of the software. For example, some parts of the software can be integrated using a bottom-up strategy, and others using a top-down strategy. The resulting software items could then be integrated using either strategy.

It should be noted that stubs and drivers are discussed here for simplicity. Stubs and drivers can be replaced with test harnesses using modern test tools.

Black-Box Testing

Black-box testing, also known as *specification-based, data-driven, input/output-driven,* or *functional testing,* ignores the software's internal structure and tests the software's behavior from the perspective of the users. Black-box testing focuses on inputting values into the software under known conditions or states, and evaluating the resulting software outputs against expected values while treating the software itself as a black box. When the software is treated as a black box, its internal structure is not considered when designing and executing the tests. This strategy helps maintain an external focus on the software requirements and stakeholder needs.

Since black-box testing does not require an intimate knowledge of the software's internal structure, individuals other than those with knowledge of the source code can perform black-box testing. That allows testing to be performed by people who are independent of the developers and know nothing about how the software was designed. Black-box testers may not even know how to read or interpret the coding language used to write source code. Developers have a natural bias and will test the software based on the same assumptions they used to create that software. Independent testing allows for more objectivity and a different perspective that can result in more defects being identified. Developers are also focused on making the software work. It is easier for an independent tester to focus on how to "break" the software.

The assembled software system is often too large and complex to test using pure white-box testing strategies. Black-box testing can be used to test the entire system, as well as its smaller individual source code modules and components, as illustrated in Figure 22.6. Black-box testing can also find defects that are very difficult or impossible to identify using a white-box strategy. For example:

- Timing-related issues and race conditions. A *race condition* is where the software output is dependent on the sequencing of timing of uncontrolled events. A race condition becomes a defect when events do not happen in the order the programmer expects or intended. For example, when writing the source code, the programmer may have assumed that event A would always occur before event B, but somehow event B occurs first, creating an incorrect output. Another defect might occur if the programmer was unaware of the race condition and did not take it into account when programming the source code.
- Unanticipated error conditions and interoperability issues that may be initiated by another interfacing software application or hardware component
- Inconsistencies in how the users interface with the system
- Inconsistencies between different displays and reports
- Issues related to how the software tolerates real-time functioning and multitasking, including high levels or volumes of environmental load or stress

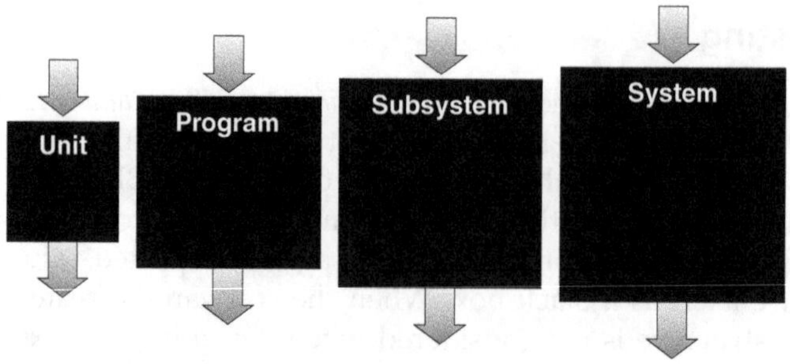

Figure 22.6 Black-box testing.

Many of the strengths of black-box testing are weaknesses of white-box testing, and many of the strengths of white-box testing are weaknesses of black-box testing. Therefore, the best strategy is usually a balancing of both approaches.

Test-Driven Design

Test-driven design (TDD), also called *test-driven development,* is an agile, iterative software development methodology. As discussed in Chapter 9, TDD implements software functionality based on writing the test cases that the code must pass. Those test cases become the product-level requirements used as the basis for implementing the code. Those test cases are then run often to verify that additional changes have not broken any existing capability (regression testing). While TDD has a testing component, it is in no way limited to being a testing strategy. TDD addresses the entire software development process.

TDD requires the maintenance of a set of automated test cases written to exhaustively test the existing code before it goes to the independent testers for their testing (and potentially into operations). As illustrated in Figure 22.7, the software is considered in the "green" state when it passes all of these automated test cases. When new functionality is identified, the first step in TDD is to write test cases for that new functionality. Since code does not yet exist to implement these test cases, they fail, and the software goes into the "red" state. The developer then writes just enough code to pass these new test cases and move the software back to the "green" state. According to Astels (2003), "that means you do the simplest thing that could possibly work." Refactoring is then done on the changed code, as necessary, to improve the quality of that new code. If changes made during the refactoring cause any automated test cases to fail, the code is corrected until the software returns to the "green" state (all test cases pass).

Risk-Based Testing, Good-Enough Software, and Time-Box Testing

According to Myers (2004), "it is impractical, often impossible, to find all the errors in a program." One reason for this is that exhaustive testing is impossible. For example, consider the black-box testing of a function that accepts a positive integer as one of its inputs. How many possible tests are there? To exhaustively test this

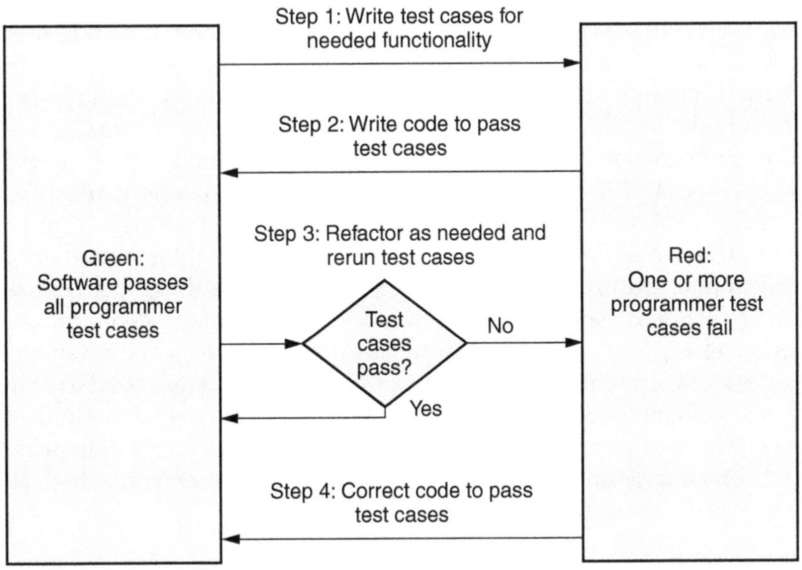

Figure 22.7 Test-driven design model—example.

single input, there are an infinite number of possible tests. First, there are an infinite number of positive integers to test. There is always the possibility that inputting the number 3,256,319 will cause some strange anomaly. Then there are also an infinite number of invalid inputs that could be tested, including negative numbers, decimal numbers, and nonnumbers (character strings including one or more alpha or other nonnumeric characters). Since exhaustive testing is impossible, all testing in the real world is based on sampling. Answering the question "How much sampling is needed?" leads to the concepts of risk-based testing, good-enough software, and time-box testing.

Risk-based testing focuses on identifying software items (e.g., work products, product components, features, functions) with the highest risk (see the discussion of risk-based V&V in Chapter 21) or that require the highest integrity levels. Limited testing resources are then proportionally distributed to spend more resources testing higher-risk or -integrity items and fewer resources testing lower-risk or -integrity items. Risk-based testing also embraces the "law of diminishing returns." At the start of testing, many problems and anomalies are discovered with little effort. As testing proceeds, discovering additional problems and anomalies requires more and more effort. At some point, the return on investment in discovering those last few defects, if they even exist, is outweighed by the cost of additional testing.

Good-enough software recognizes that not all software applications are created equal, and not all software applications, or even all components within a single software application, have an equal amount of risk. For example, word processing software does not require the same level of integrity as banking software and banking software does not require the same level of integrity as biomedical software that impacts a person's health or life. Doing good-enough software analysis has to do with making conscious, logical decisions about the trade-offs between the level of quality and integrity that the stakeholders need in the software, and the basic economic fact that increasing software quality and integrity will cost more and take longer. This fact is not meant to discount the impacts of long-term continual improvement initiatives—but right now, on today's project, with current skills and capabilities, this is reality. Hard business decisions need to be made about how much testing (and other V&V activities) a project can afford, what the stakeholders are willing to pay for it, and what the acceptable level of risk is that the developers, acquirers, and other stakeholders are willing to accept.

In project management, a time box is a fixed amount of calendar time allocated to complete a given task, like testing. In *time-box testing*, the calendar time for testing is fixed, and the scope of the testing effort must be adjusted to fit inside that time box. This method can be accomplished by prioritizing test activities and tests based on risk and benefit and then executing the activities or tests in priority order. If time runs out before all of the activities and tests are accomplished, at least the lowest-priority ones are the ones left unfinished.

Simulation

The test environment often cannot duplicate the real-world environment where the software will be executed. For example, in the test bed, the testers performing capacity testing for a telecommunications switch do not have access to 10,000 people all calling in to the switch at once. One way of solving this gap is to create a simulator that imitates the real world by mimicking those 10,000 people. Another example of when simulators might be needed is when other interfacing software applications or hardware are being developed in parallel with the software being tested and are not ready for use or unavailable during testing.

Since a simulator only mimics the people, other software, and hardware in the real-world environment, strategic decisions need to be made about when to spend testing resources creating simulators and utilizing simulation. These are, again risk-based decisions because:

- *Resources are limited*: Resources spent creating and testing the simulators are not spent testing other parts of the software.
- *Simulators are not the real environment*: If incorrect choices are made in creating the simulators, defects may still escape testing, or issues that would never have become actual defects may be reported, and unnecessary or incorrect changes could be made to the software under test.

Test Automation

Test automation uses software tools to automate the activities of test design, test execution, and the capturing and analysis of test results. Strategic decisions here involve:

- How much of the testing to automate
- Which tests or test activities to automate

- Which automation tools to purchase
- How to apply limited resources to the automation effort

Automated tests can typically be run much more quickly and, therefore, more often than manual tests, providing increased visibility into the quality of the software at any given time. This visibility can be particularly beneficial during regression testing, especially with iterative or agile development life cycles. Automation can also eliminate human error when comparing large amounts of output data with expected results.

However, not every test can or should be automated. For example, usability testing requires a human to evaluate the software. Automation may also not be cost-effective if part of the software changes so rapidly that it requires more effort to maintain the automation for that software than to test it manually.

As a trade-off, test automation requires an initial investment in tools and resources to create the automation and a long-term investment to maintain the automated test suites as the software being developed changes over time. For software items that change frequently, more time may be spent maintaining the test automation than manual testing would require. Test automation also requires a different skill set than testing itself. Therefore, additional staff or training for existing staff may be necessary to implement test automation.

Finally, test automation has limitations in that it can only compare test results with a finite set of expected results. Since automated tests only look for what they are programmed to check, they cannot identify other abnormal or unexpected behaviors that may result from software defects. For example, if the input to the test case is a number and the expected output is the square root of that number, then an automated test case would check that the square root value is correct for the selected input value. However, the automated test case would not identify that the background screen erroneously changed colors before it displayed the square root value. A human tester conducting manual testing would probably catch this type of anomaly easily.

As with other testing strategies, a mix of manual and automated testing strategies may provide the most effective approach to testing.

2. TEST PLANS

> Develop and evaluate test plans and procedures, such as system, acceptance, and validation, to determine whether project objectives are being met and risks are appropriately mitigated. (Create)
>
> **Body of Knowledge VI.B.2**

V&V Plan

Test planning starts with the overall V&V plans, as discussed in Chapter 21, where decisions are made about the levels of testing that will be done; the testing objectives, strategies, and approaches that will be used at each level; and the allocation of testing staff and other resources. V&V planning determines what V&V activities, including testing, will be used to evaluate the implementation of each requirement. The V&V plan is one of several test planning documents.

Test Plans

The test planning process includes developing, reviewing, and approving test plans. For each of the major testing levels (integration, system, acceptance, alpha, and beta testing), additional test plans may be created to refine the approach and define additional details for that level. Each test plan defines the specific strategies and tactics that will be used to make certain that project or program objectives associated with that level of testing are being met and that the requirements that are to be verified and validated at that level (as allocated by the V&V plan) are adequately tested. Each test plan may be a separate document, or the test plans can be incorporated as part of:

- A single overall test plan for the project/program
- As one or more sections in the V&V plan or software quality assurance (SQA) plan
- As one or more sections in the overall project or program plan
- As data entries in an automated test planning and management tool

Based on *IEEE Standard for Software Test Documentation* (IEEE 2008a) and IEEE/*ISO/IEC Software and Systems Engineering—Software Testing—Part 3: Test Documentation* (IEEE/ISO/IEC 2021b), test plans should include:

- *Introduction or overview*: A description of the context and structure of the plan, the plan's scope, a summary of the unique nature of this particular level of testing, and the plan's history. Other information that may be included in this section, as appropriate, includes:
  - The plan's unique identifier
  - The issuing and approving organizations
  - Assumptions and constraints
  - References
  - Glossary and acronyms
- *Project or program*: The project or program associated with the test plan.
- *Level of testing*: The level(s) of testing, or the testing subprocesses, for which this plan is being written.
- *Items to be tested*: A list of the software configuration items that will be tested under the plan, including their versions or revisions, and a list of any known defects reports associated with the items to be tested.
- *Test traceability matrix*: Include or point to the traceability matrix that provides linkages between the requirements and the test cases or procedures that will be executed as part of this testing.

- *The functions that will and will not be tested*: For example, if the project adds functions to an existing product, risk-based analysis is used to determine which existing functions are not impacted by the changes and therefore do not need to be retested. Based on risk, this section may also include the depth and level of rigor required for testing each function and establish priorities for testing.
- *The test approaches and methods that will be used for the testing level*: This includes the identification of policies, test strategies, approaches (positive and negative testing, domain testing, boundary testing, exploratory testing, *error guessing*), and manual and automated testing methods that will be used for:
    - Functional, user scenario, and operational profile testing
    - Testing quality and product attribute requirements (e.g., usability; performance; environmental load, volume, or stress; safety; security; interoperability; conversion; installability; and internationalization)
    - External interface and environmental configuration testing
    - Data and database testing
    - Testing user documentation (e.g., user manuals, operator manuals, marketing materials, and installation instructions)
    - Regression testing

  This section may include additional details in the system test matrix to define the exact approaches and methods for each requirement being tested at the stated level of testing. In addition, this section defines approaches and methods for:
    - Problem reporting and resolution
    - Issues reporting and resolution
    - Configuration management of test deliverables and testing tools
    - Test status reporting and other stakeholder communications, including reviews and metrics
- *Item pass/fail criteria*: Specific, measurable criteria for determining if the item under test has successfully passed the cycle of tests typically used as exit criteria for the testing cycle. Examples might include the:
    - Allowable minimum level of test coverage of the requirements
    - Allowable number of nonclosed problem reports by severity and requirements for workarounds
    - Allowable arrival rates of new problems
    - Allowable number of unexecuted or unpassed test cases
    - Allowable variance between planned and actual testing effort

- *Suspension criteria*: Specific, measurable criteria that, when met, indicate that the testing activities should be stopped until the resumption criteria are met. For example, if a certain number of critical defects are discovered or if a certain percentage of test cases are blocked, it may no longer be considered cost-effective to continue the testing effort until development corrects the software.
- *Resumption criteria*: Specific, measurable criteria that when met indicate that the testing activities can be restarted after suspension. For example, a new build is received from development with corrections to most of the critical defects or defects that are blocking test case execution.
- *Test deliverables*: Outputs from this cycle of testing, examples include:
    - Test plans
    - Test design specification
    - Test designs, including test cases, procedures, and scripts
    - Problem reports
    - Status reports and metrics
    - Test logs
    - Interim and final test completion reports
- *Testing tasks with assigned responsibilities (roles), resources, and schedules*: This includes the work breakdown structure, schedule, staffing and other resources allocated for this level of testing. This information may be kept as part of the project information in a project management tool and pointed to by the test plan.
- *Environmental needs*: This section defines:
    - Test bed requirements, including the physical setup, technologies, and tools
    - Mechanisms for validating the test bed setup
    - Methods for requesting updates or changes to the test bed configuration during testing
    - Allocation and schedules of test bed resources
- *Staffing plans and training needs for test personnel*: Testing personnel may include test managers, test designers, test automation specialists, testers (people to perform manual and exploratory testing), test bed coordinators, and support staff. Skill gap analysis should be performed to make sure that the testing personnel have the requisite knowledge and skills, and the training plans need to define how identified gaps will be filled.
- *Risk management plans for risks associated with this level of testing, including*:

- Product and project risks associated with this plan and any associated risk handling
- Risks that are being mitigated with this testing

Test Monitoring and Control

As illustrated in Figure 22.1, test monitoring and control is another test management process in addition to the test planning process. Test monitoring and control occur throughout the entire life cycle, from when test planning starts until the final test activities are completed. Test monitoring involves tracking actual test activity progress against the test plans, including the specified test completion criteria (exit criteria or definition of done). The monitoring may be accomplished through status reporting and meetings or through the monitoring of the test metrics specified in the test plans.

The results of ongoing test monitoring should be communicated to all relevant stakeholders. If the monitoring of the test actuals against the test plan indicates that the variation is too large, controlling actions are planned and implemented, as necessary, to make sure that the testing objectives will be met. This control may involve simple corrective action plans or necessitate revisiting and modifying the test plans.

Test Completion

As illustrated in Figure 22.1, test completion is the final test management process. Test completion is the "set of test management tasks that deal with the activities related to collecting data from completed test activities to consolidate experiences, *testware* and any other relevant information (ISTQB 2018a).

Test completion activities take place at project milestones. For example, test completion activities occur at the completion of a testing level or type, after an iteration, at product release, or if the test project is completed or canceled. Test completion activities include the following:

- Ensuring delivery of test deliverables
- Verifying that all incident reports are resolved or documented (change requests, product backlog)
- Documenting test completion and acceptance of the software
- Finalizing and archiving testware for later use, including:
  - Test environment
  - Test data
  - Test infrastructure
- Transitioning testware for later reuse, as appropriate, to:
  - Maintenance organization
  - Other project teams
  - Other stakeholders

- Analyzing lessons learned from completed activities
- Using gathered information to improve test processes

3. TEST DESIGNS

> Select and evaluate various test designs, such as fault insertion, fault-error handling, equivalence class partitioning, boundary value. (Evaluate)
>
> **Body of Knowledge VI.B.3**

As illustrated in Figure 22.1, *test design and implementation* is one of the four dynamic testing processes. According to IEEE (2021a), "The test design and implementation process is used to derive test cases and test procedures." Test design and implementation activities include performing the following:

- Analysis of the test basis and objects to determine the approach needed to evaluate the test object.
- Evaluating the test basis and test objects to verify their testability and identify any defects
- Design, documentation, and verification of the test cases and procedures
- Identification of the necessary test data
- Design of the test environment
- The optional automation of tests that will be repetitively executed
- Traceability analysis to confirm the completeness of the set of tests

In other words, test analysis determines "what to test" in terms of measurable coverage criteria (ISTQB 2018b).

Figure 22.8 outlines the steps in analyzing the test basis and objects to determine the features and feature sets to be tested and define the *test conditions* for the current level of testing being designed (unit, integration, software system testing, user acceptance testing). This activity includes risk, criticality, and hazard analysis to determine how extensively to test the test objects. This activity may optionally include definition test charters if exploratory testing is planned during testing.

The *test basis* is the "body of knowledge" used as the basis for determining what needs to be tested and the design test cases. For example, the test basis may include requirements specifications, design and implementation information, or as-built components (code, databases, interfaces).

The *test object* is the component, software, or system under test. For example, in unit testing the test object would be the unit under test. For software system

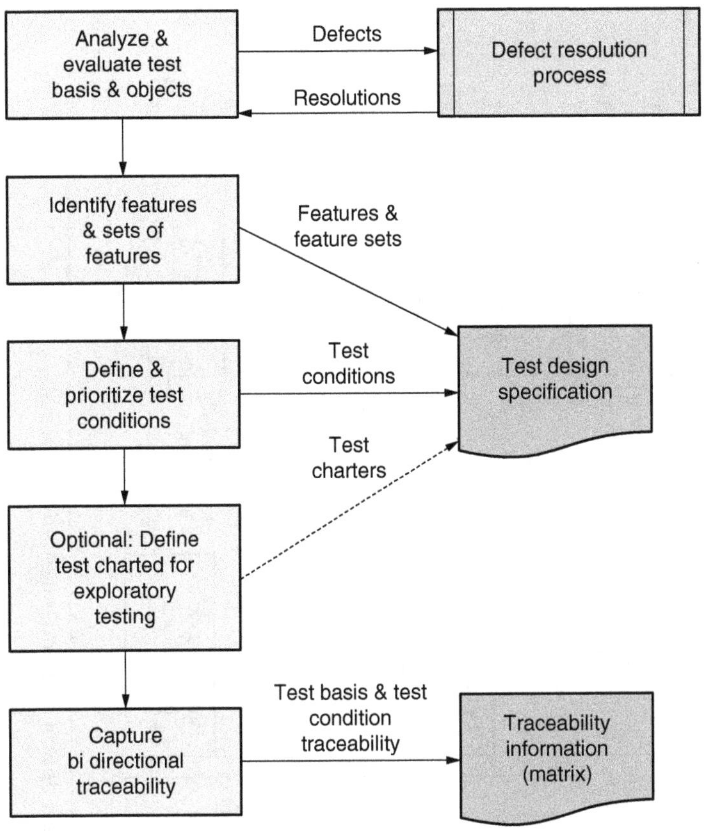

Figure 22.8 Steps in the analysis activities of the test design and implementation process.

testing, the test object(s) would be the software executable and optionally any user documentation.

The *test design specification* documents the features, feature sets, test conditions, and exploratory test charters. According to IEEE (IEEE 2008a), a test design specification is an optional document that defines the features to be tested, the approach refinements, test identification, and feature pass/fail criteria.

Test design defines "how to test" (ISTQB 2018b). As illustrated in Figure 22.9, test design includes the following:

- Turning the higher-level test conditions into test cases and sets of test cases used to evaluate individual test conditions, including prioritizing the test cases and defining each test case's inputs, steps, and expected results.
- Identifying and specifying the necessary test data to support test conditions and test cases
- Designing and specifying the test environment and required infrastructure and tools

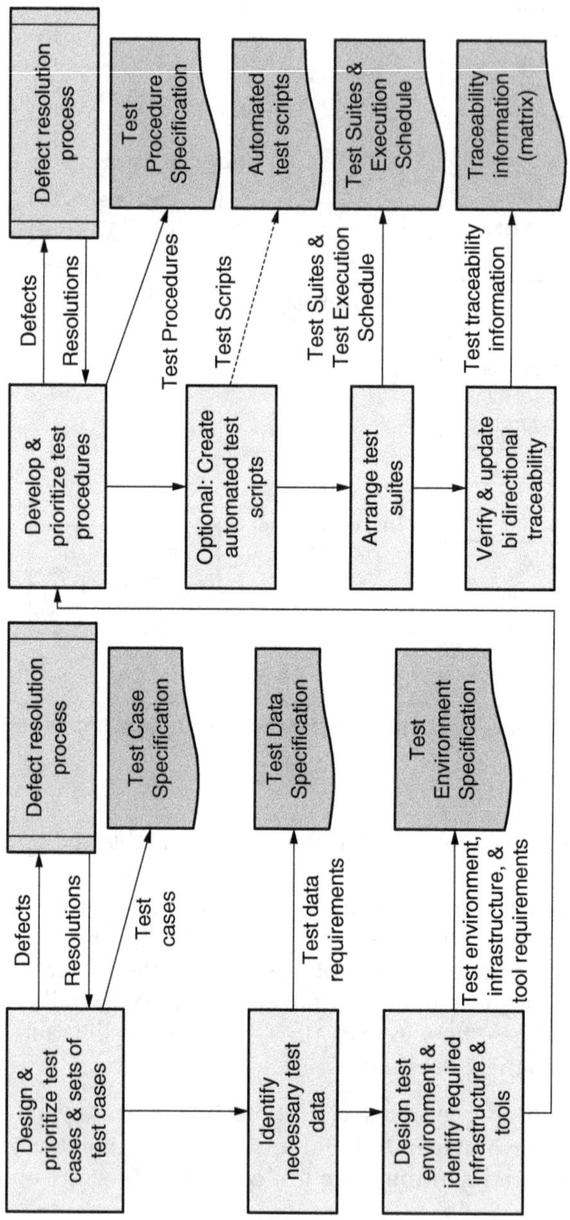

Figure 22.9 Steps in the design activities of the test design and implementation process.

- Combining test cases into test procedures to evaluate start-to-finish interactions
- Optionally automating the tests
- Creating test suites and test execution schedules
- Verifying bidirectional traceability and updating it as needed

As with test analysis, the test design activities can help identify defects in the test basis and objects. Both test analysis and test design result in the benefit of early defect detection and reducing the risk of defects propagating into other successor work products.

There are techniques the tester can use during test design to systematically minimize the number of tests while maximizing the probability of finding undiscovered relevant defects.

Equivalence Class Partitioning

Equivalence class partitioning is a test design technique that takes a set of inputs or outputs and divides them into valid and invalid data sets that are expected to be treated in the same way by the software. This technique allows the tester to sample from each equivalence class, also called a partition, during testing, using a minimum set of tests while still achieving coverage. The assumption is that if one sampled value from an equivalence class uncovers a defect, then any other sample from the same equivalence class would probably catch the same defect. If the sampled value does not catch a defect, other samples will probably not catch it. Rules of thumb when selecting equivalence classes include:

- If the input or output is a continuous range (e.g., from 1 to 100 inclusively), there is one valid input class (numbers from 1 to 100) and two invalid classes (numbers less than 1 and numbers greater than 100)
- If the input or output condition is a "must be" (e.g., the first character must be a numeric character), there is one valid input class (inputs where the first character is numeric) and one invalid class (inputs where the first character is not numeric)
- If the input or output condition specifies a set of valid input values (e.g., the jobs of programmer, analyst, tester, QA, and manager), then each valid input is an equivalence class (in this example, there are five valid classes) and there is one invalid class (everything that is not in the list of valid inputs)
- If the input or output condition is a specific number of values (e.g., one to six inspectors for a formal inspection), there is one valid equivalence class (having one to six inspectors) and two invalid classes (zero inspector and more than six inspectors)

Test cases and procedures should be written generically for each equivalency class (one test case for inputting a number from 1 to 100, another test case for inputs less than 1 and another for inputs greater than 100) so that the tester selects the sample at the time of test execution. Testers then strive to select a different sample from the equivalence class with every execution of the test case. If test execution is automated, this may require a random number generator or some other mechanism for varying the input with each execution.

Boundary Value

Boundary value analysis explores the values on or around the boundaries of the equivalence classes or for each input or output. This technique will help identify an error where the programmer uses < instead of ≤, uses > instead of ≥, or forgets or mishandles the boundary in some other way. In two-value boundary value testing, tests are created to test values at the boundaries:

- The minimum boundary value
- The maximum boundary value
- The value just below the minimum
- The value just above the maximum

For example, for the inputs in the range of 1 to 100 inclusively, boundary value tests would include the inputs of 0, 1, 100, and 101.

In three-value boundary value testing, two additional tests are created to test the values:

- Just above the minimum
- Just above the maximum

For example, for the inputs in the range of 1 to 100 inclusively discussed above, three-point boundary value tests would add the inputs of 2 and 99.

As illustrated in Table 22.1 boundary values for numeric ranges are selected to the same level of precision as the range specified in the requirements.

Another example of boundaries would be to consider an input dialog box, as illustrated in Figure 22.10. Assuming that the box displays up to 25 characters and is a scrolling text box that allows an input string of up to 255 characters, boundary value testing would test the null input, a single character, 25 characters, 26 characters, 255 characters, and 256 characters. The tester might also want to input 26 characters and then delete one character to cross the boundary in the other direction. There may also be invisible boundaries for this input that the tester might want to explore. For example, the input buffer size used by the operating system.

Other boundaries to consider when writing tests include:

- *Zero*: The boundary between positive and negative numbers (boundary value tests should consist of the inputs of –1, 0, +1 for testing around the zero boundary for numeric inputs and outputs)
- *Null*: The boundary between having an input and not having an input (boundary value tests should include the inputs of the null string and a single character string for testing around the null boundary for character string inputs and outputs)

Table 22.1 Boundary value precision—example.

Required Range	Two-Value Boundaries	Three-Value Boundaries
1–100	0, 1, 100, 101	0, 1, 2, 99, 100, 101
1.00–100.00	0.99, 1.00, 100.00, 100.01	0.99, 1.00, 1.01, 99.99, 100.00, 100.01
number < 1	0, 1	–1, 0, 1
number < 1.0	0.9, 1.0	0.8, 0.9, 1.0
all positive integers	0, 1	0, 1, 2
numbers >= 1.0	0.9, 1.0	0.9, 1.0, 1.1

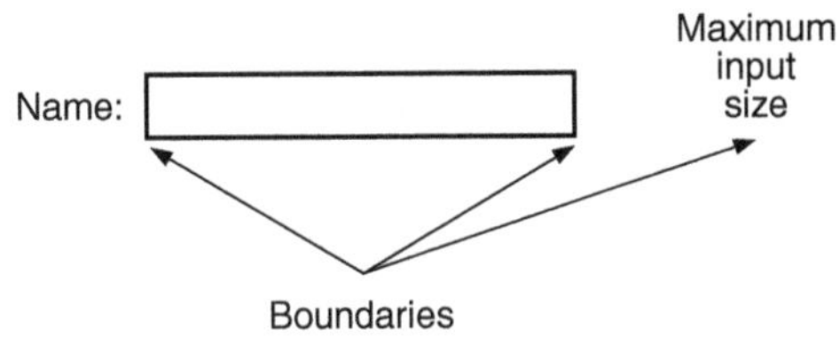

Figure 22.10 Input field boundary—example.

- *Hard copy output sizes*: Characters per line or lines per page on outputs to a printer (remember, paper sizes can vary in different countries)
- *Screen output sizes*: Screen width, length, and pixels
- *Hardware device outputs*: Communication packet sizes or block sizes used in storing data on media
- *Lists*: First item and last item in each list
- *Date and time*: The boundaries between specific times or dates, where special actions or conditions may occur that need to be tested to confirm that the software handles the boundary correctly (time zones, international date line, daylight savings time transitions, end of day, end of month, end of quarter, end of year, end of century, leap year)

Decision Table Testing

While equivalence class partitioning and boundary value analysis techniques focus on the input and output domains, *decision table testing* focuses on the handling of business logic and business rules. This focus can be especially useful when:

- Testing different combinations of input results in different actions being taken
- Testing different inputs in combination with different states results in different actions being taken

Decision tables are an excellent way to analyze the specification for completeness and ambiguities. Decision tables can also be used to represent logic relationships from which productive test cases can be identified.

The following requirements will be used to walk through an example of decision table testing:

R101 For females less than 65 years of age, the premium is $500

R102 For males less than 25 years of age, the premium is $3,000

R103 For males between the ages of 25 and 64 years of age, the premium is $1,000

R104 For anyone 65 years of age or more, the premium is $1,500

The first step of the decision table testing technique is to analyze the business logic and business rules defined by the requirements. These business logic or rules are in the form of:

IF (conditions) THEN (actions)

Business logic or rules are analyzed to identify all possible conditions, where conditions are inputs, events, or states that the system uses to makes decision. For example:

- In the foregoing set of four requirements, the conditions are the values of the two input variables Sex and Age.
- Conditions can include the various states the system is in when an input or event occurs. For example, a pay at the pump gas station system will react differently to the cancel key being pressed at the pump, depending on the state of that pump when cancel is selected.
- Conditions can also include specific events, like a plane landing or a value reaching a specified threshold (e.g., the time going to midnight or a temperature exceeding a safety threshold).

Business logic or rules are also analyzed to identify all corresponding actions that occur. Actions are taken by the system or software in response to the decisions made as a result of the combinations of one or more conditions. For example, based on the rules defined in the example requirements, the combinations of the variable Sex being input as Male and the variable Age being input with a value less than 25 would result in the action of the Premium being set to $3,000.

This analysis of the business logic or rule has a beneficial side effect of pointing out incompleteness and ambiguities in the specified requirements. For example, the example requirements do not indicate what to do if an input other than Male or Female is entered into the system for the variable Sex. The requirements also do not specify either an upper or lower limit on the variable Age. Is the company going to grant can insurance to a 12-year-old or a 99-year-old person? Because this analysis can identify potential defects in the requirements, it should be performed early in the life cycle as part of the requirements validation process discussed in Chapter 12.

For the purposes of this example, however, assume that input to both the variables Sex and Age are selected from pick lists and cannot have these invalid inputs.

The second step is to start building a decision table by generating all possible combinations of the identified conditions. Each combination creates a column in the decision table. For example, since the analysis from step 1 identified that the variable Sex has two possible conditions and the variable Age has three possible conditions, there are a total of six possible combinations (2×3) for the set of example requirement, as illustrated in Table 22.2. This table can then be reduced by removing columns that:

- Contain impossible combinations
- Contain combinations that do not affect the outcomes

In the third step of the decision table testing technique, the business logic or rules are analyzed again to determine the corresponding actions that should occur for each of the identified combinations. The actions are added under the corresponding column of the decision table.

In this example, that is the setting of the policy amount, as illustrated in Table 22.3.

This analysis can again help identify requirements defects, for example, if there is a possible combination that does not have a corresponding action identified.

In decision table testing, each decision rule (each column in the decision table which defines the relationship between a unique combination of conditions and actions) becomes a test condition. A test case can be designed for each of these test conditions. Decision table coverage is the percentage of the total number of decision rules identified that are covered by executed test cases.

Fault Insertion

Fault insertion, also called *fault seeding* or *defect salting,* estimates the number of undetected defects that escaped from a testing process. Fault insertion intentionally inserts known defects into the software (typically done by people independent of the developers or testers). The software is then tested, and the defects that are found are analyzed to determine if they are part of the intentionally inserted,

Table 22.2 Possible condition combinations for the example.

Combination	1	2	3	4	5	6
Sex	Male	Male	Male	Female	Female	Female
Age	<25	25 to 64	>= 65	<25	25 to 64	>= 65

Table 22.3 All possible condition combinations for the example.

Combination	1	2	3	4	5	6
Sex	Male	Male	Male	Female	Female	Female
Age	<25	25 to 64	>= 65	<25	25 to 64	>= 65
Policy	$3,000	$1,000	$1,500	$500	$500	$1,500

known defects or if they are new, unknown defects. The number of unfound, unknown defects is then estimated through the use of the simple ratio:

$$\frac{\text{Found unknown (unseeded) defects}}{\text{Found known (seeded) defects}} = \frac{\text{Unfound unknown (unseeded) defects}}{\text{Unfound known (seeded) defects}}$$

For example, assume 100 known defects were inserted into the software prior to system test. The software is tested, and 72 of the inserted defects were found and 165 other defects were found. That means that 28 (100 minus 72) of the inserted (known) defects were not found during system test. The number of unfound, unknown defects (UUD) can be estimated:

$$165/72 = \text{UUD}/28 \text{ or } (165/72) \times 28 = \text{UUD}$$
$$\text{Since } 165/72 \approx 2.29$$
$$2.29 \times 28 \approx \text{UUD so UUD} \approx 64 \text{ defects}$$

This technique assumes that the known defects are being found at the same rate as the unknown defects. Care must be taken to "seed" the software with a representative set of defects for this assumption to hold true. For example, if only logic errors are inserted into the code, then only logic type errors can be reasonably analyzed and predicted.

Prior to fault insertion, good configuration management is required to make certain that the software can be retrieved for use moving forward. If not, care must be taken to remove all known defects and that new defects are not introduced by mistake. Regression testing must be performed after the completion of the removal activities to verify this removal.

Fault-Error Handling

Testing techniques examining *exception handling*, also called *fault-error handling* or *negative testing*, look at how well the software handles errors, invalid conditions, invalid or out-of-sequence inputs, and invalid data.

Testing should check to confirm that exception handling is triggered whenever an exception occurs. Exception handlers should allow graceful software recovery or shutdown if necessary. They should prevent abandonment of control to the operating system (obscure crashes). Testing should also check that any reserved system resources (e.g., buffers, allocated memory, and network channels) are appropriately released by creating tests that produce multiple exceptions to see how the software handles these repeated exception conditions.

Error guessing is one technique for selecting fault-error handling tests. The main idea behind error guessing is to "attack" the software with inputs that expose potential programmer mistakes or represent the kinds of invalid inputs software would receive if user, interfacing software or hardware mistakes occurred. It is difficult to describe this technique since it is mainly intuitive and experience-based. One approach to error guessing is to brainstorm probable types of faults or errors that might occur, given the specific software. Another technique is to create and use checklists of common errors in past software products. For example:

- What mistakes do programmers make?
  - < versus ≤ or > versus ≥ (boundary value testing)
  - Not appropriately handling zero, negative numbers, or null inputs
  - Failure to initialize data or variables
  - Declaring variables as one type or size and then using them as a different type or size
  - Not performing adequate error checking on inputs
  - Allowing buffer or variable overflows
  - Not appropriately handling zero times through a loop or only once through a loop
  - Incorrect logic or control flow
  - Not releasing resources or not reinitializing variables
- What mistakes do users make?
  - Creative invalids (alpha versus numeric, decimals versus integers, alt key versus control key versus function key versus shift key, left versus right mouse clicks)
  - Too many or too few inputs (not completing required fields)
  - Invalid user entries resulting from inadequate instructions (should spaces, parentheses, and dashes be included in number sequences like phone numbers, credit card numbers, and social security numbers?)
  - Too little or too much input (not putting in an input or overflowing input fields)
  - Invalid combinations (State = Texas, but entered with a New York ZIP Code)
  - Invalid processes (attempting to delete a customer record from the database before it is added)
  - Recursive processes (a spreadsheet cell that adds that cell to itself)
  - Doing things out of sequence (trying to pump gas before selecting gas type)
  - Incorrect usage (Hint: look at problem reports from the field that were operator errors)
- What errors do machines make?
  - Defective boards (Hint: collect them as they are returned from the field)
  - Bad or corrupted data
  - Failure to communicate (not responding when expected or responding after software times out)

- Interrupted or incomplete communications
- Power failures
- Limited, full, unavailable, or busy resources
- Write protected, unavailable, or damaged/corrupted media (no media in the drive, or corrupted block or sector on the media)
- Race conditions and other timing issues

4. SOFTWARE TESTS

> Evaluate and execute various tests, including unit, functional, performance, integration, regression, usability, acceptance, certification, environmental load, stress, worst-case, perfective, exploratory, and system. (Evaluate)
>
> **Body of Knowledge VI.B.4**

Test Execution

Test execution is the process of operating a system or component under specific, known conditions, observing or recording the results, and making an evaluation of some aspect of the system or component (IEEE/ISO/IEC 2017). As illustrated in Figure 22.11, test execution activities include the following:

- Setting up and validating the testing environment, including the test data and test bed.
- Executing the tests by running predefined manual or automated test cases or procedures or performing exploratory testing.
- Documenting these test execution activities in the test log. If they are performing exploratory testing, the testers also document the test cases and procedures as they are executing them so that they are repeatable.
- Analyzing the actual test results by comparing them to the expected results. The objective of this analysis is to identify failures.
- Looking around to identify any other adverse side effects or anomalous results. The expected results typically only describe what should happen rather than what could happen. The test might, for example, enter "xyz" information, and the expected result is the display of a specific screen. It will not say, "The printer should not start spewing paper on the floor," but if the printer does start spewing paper, the tester should notice and report the anomaly.

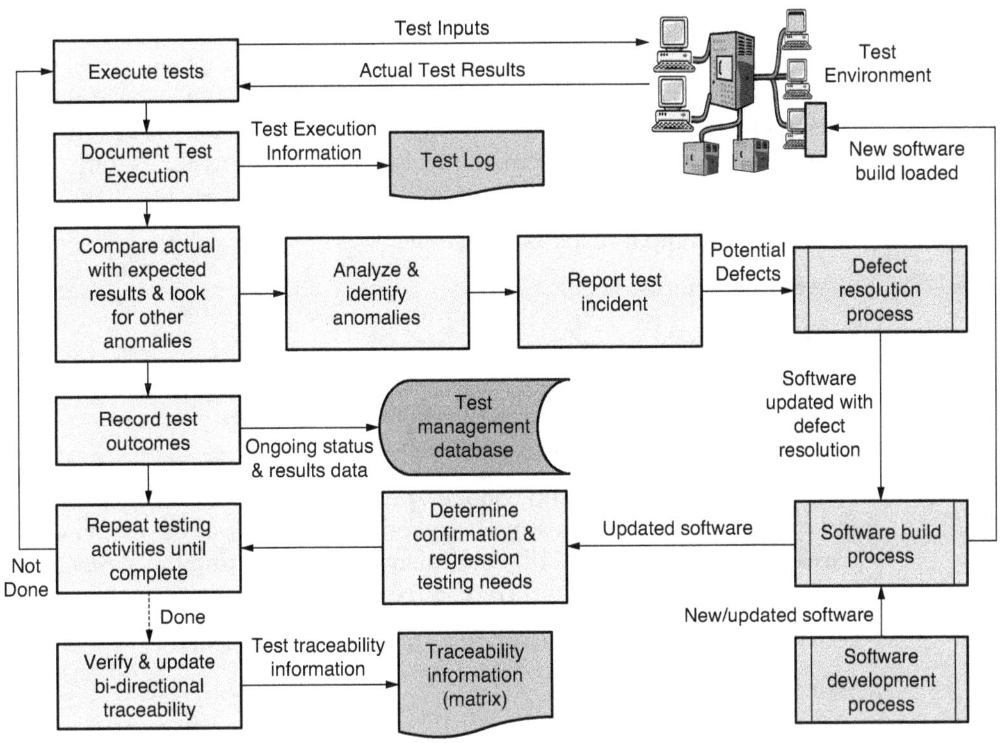

Figure 22.11 Steps in the test execution and test incident reporting processes.

- Analyze to determine root cause (potential software defect or false positive (e.g., an error in test environment setup, a defect in the test data, or a tester error)
- Attempting to repeat the steps needed to recreate the failures or other anomalous results. The problem should be reported even if the tester cannot repeat it, even if it is intermittent.
- Capturing problems, including adverse side effect information that help the authors to identify the underlying defects (e.g., screen captures, data values, database snap, memory dump shots), if failures or other anomalous results are identified. The testers report the identified failures or other anomalies in formal change requests (problem reports) if the items under test are baselined items. If the items under test are not baselined, failures and anomalies are reported directly to the item's author for resolution. The author might be able to correct it. He can always close it as "could not duplicate."
- Participating in change control processes to "champion" their problem reports and provide additional information as needed, if the problem was found in a baselined configuration item.

- Working with the author, as necessary, to isolate defects if the author cannot reproduce the failures or other anomalous results. For example, the author may not be able to reproduce the problem because the author does not have the same tools, hardware, simulators, databases, or other resources or configurations in the development environment as the testers are using in the test environment.
- Reporting ongoing test status, which includes:
  - Activities completed
  - Effort expended
  - Test cases executed, passed, failed, and blocked. A blocked test case is one that was planned for execution but could not be executed because of a problem in the software.
- Testing corrected defects and changes, performing regression analysis, and executing regression testing based on that analysis, as developers provide updated software. This may also include updating the status of the associated problem reports or change requests, as appropriate.
- Verifying and updating bidirectional traceability between:
  - Test basis
  - Test conditions
  - Test cases
  - Test procedures
  - Test results
- Tracking and controlling the ongoing test execution effort using test status reports, data, metrics, and project review meetings. Test metrics are also used to evaluate the test coverage, the completeness status of testing, and the quality of the software products.
- Writing a test report to summarize test results at the end of each major testing cycle.

Levels of Testing

Various levels of software testing can be done, as illustrated in Figure 22.12. These different levels of testing are known by different names in different organizations. Even the number of levels of testing may vary from organization to organization. The following discussion should be considered as an example of the different levels of testing and include some of the more common names for these levels.

At the lowest level, during *unit testing,* also called *module testing* or *component testing,* the individual source code modules that make up the software product are tested individually. A source code module is the smallest software element that can be separately compiled. The author of a source code module usually performs this unit testing. One of the advantages of unit testing is fault isolation. Any prob-

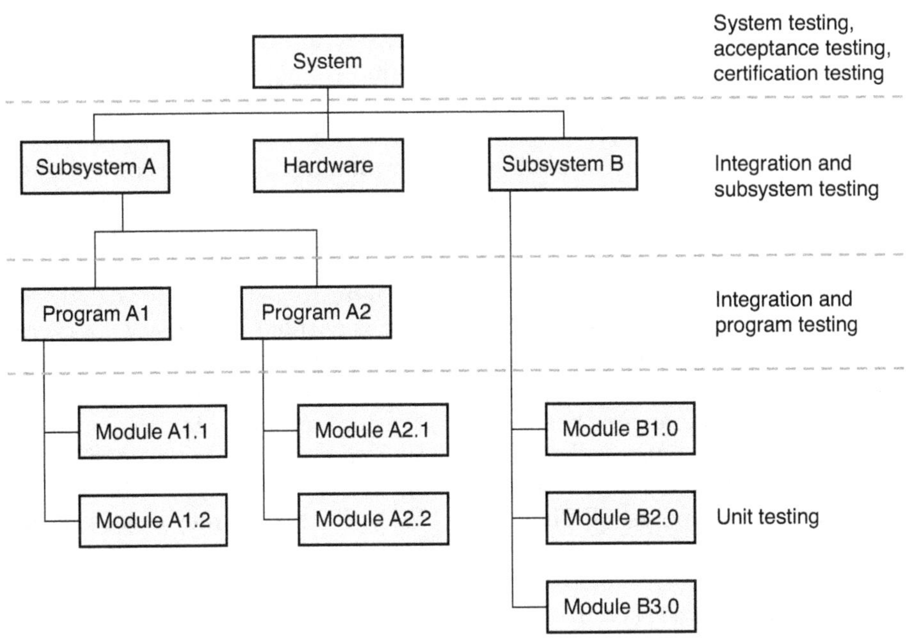

Figure 22.12 Levels of testing.

lems found during unit test execution are isolated to the individual source code module being tested. Unit testing is typically the earliest level of test execution and, therefore identifies defects earlier, when they are less expensive to fix, than if they were found during later levels of test execution. At the unit test level, it is also easier to confirm that every line of code or every branch through the source code module is tested during test execution. An analogy for unit testing is to make sure that the individual bricks are good before the workers start building the wall out of them.

During *software integration*, the aggregation two or more source code modules or other components (constituent configuration items) is used to create an integrated software component (composite configuration item). With large software development projects, *integration testing* often includes software from many developers. Integration testing may be accomplished on several levels, for example:

- Integrating source code modules into software program components and testing the interfaces and interactions between those source code modules
- Integrating software program components into software subsystem components and testing the interfaces and interactions between those software programs
- Integrating software subsystem components into the software system and testing the interfaces and interactions between those software subsystems

Depending on the organization and the size of the project, integration testing may be performed by the developers or by a separate independent test team.

Integration testing focuses on the interfaces and interactions between the integrated components and source code modules. Since the individual source code modules are already tested at the module level, integration testing can maintain this focus without the distraction of an excessive amount of module-level defects. To continue the analogy used above, integration testing focuses on making sure the mortar between the bricks is good and that the wall is solid as it is being built.

Software system testing, also called *software qualification testing,* is testing conducted on a complete, integrated software application or product to evaluate the software's compliance with its specified requirements.

During *system integration,* if the software is part of a larger system, there may be other levels of integration testing where the software, hardware, and other system components are integrated into the system and testing focuses on the interfaces and interactions between those system components.

System testing, also called *system qualification testing,* is "testing conducted on a complete, integrated system to evaluate the system's compliance with its specified requirements" (IEEE/ISO/IEC 2017). The system may be an aggregation of software, hardware, or both that is treated as a single entity for the purpose of system testing, as well as operational use after release. To finish the analogy, system testing focuses on the functionality and quality of the entire house.

If the system is part of a system of systems, yet another level of integration and testing may occur, followed by system of systems testing and so on as needed.

Acceptance testing, also called *user acceptance test (UAT),* is a particular type of testing performed by or for the acquirer of the software or system to demonstrate that the as-built product performs in accordance with its acceptance criteria. Typically, the acceptance test criteria are defined in a set of acceptance tests and agreed to as part of the contract or other early agreement between the development organization and the acquirers. Acceptance tests are designed to provide a mechanism for formal acceptance of the software as it transitions into operations.

Certification testing is also a special type of testing that is typically done by a third party (an organization other than the supplier or acquirer of the software or system). Certification test criteria and any standards the software or system must meet should be defined and agreed to in advance. Contractual agreements between the supplier and acquirer may call for certification testing in place of, or in addition to, acceptance testing. A software development organization may also voluntarily seek certification for their products for marketing or other purposes. Commercial off-the-shelf (COTS) software may also be put through certification testing as needed. For example, if the COTS software is used in or to build safety-critical or security-critical systems, certification testing may be considered a cost-effective, efficient way of performing V&V.

Functional Testing

Functional testing focuses on testing the functional requirements of the software—"what the software is supposed to do." Functional testing strategies include:

- Testing each individual function
- Testing usage scenarios
- Testing to the operational profile

To test each individual function, the tester decomposes and analyzes functional requirements and partitions the functionality into logical components (commands, actions, menu options). For each component, a list is made of detailed functions and subfunctions that need to be tested, as illustrated in the example in Figure 22.13.

According to Whittaker (2003), "functional software testing is about intelligence." First, the testers must become familiar with the environment in which each function or subfunction operates. This process involves identifying "users" of that function or subfunction from outside the software being tested. As illustrated in Figure 22.14, there are four classes of environmental users that need to be considered.

- *Humans and hardware*: Human users cannot interface directly with the software. Instead, they must communicate using hardware devices (keyboard, mouse, or button) whose inputs are processed by device drivers. Software may also interface directly with hardware without human interaction.

- *Operating system*: The operating system user provides memory and file pointers. Operating systems offer a set of functions that are needed and used by most applications (e.g., time and date functions) and provide the necessary linkages to control a computer's hardware.

- *File systems*: File system users read, write, and store data in binary or text format. Files store persistent data, instead of internally stored global or local data structures (e.g., integers, characters, arrays, strings, stacks, queues, pointers) that only exist while the program that defined them is executing. Because file contents are easily corrupted and are also easy to change from outside the confines of the applications that create and process them, the tester must be concerned with testing issues such as how the software handles:

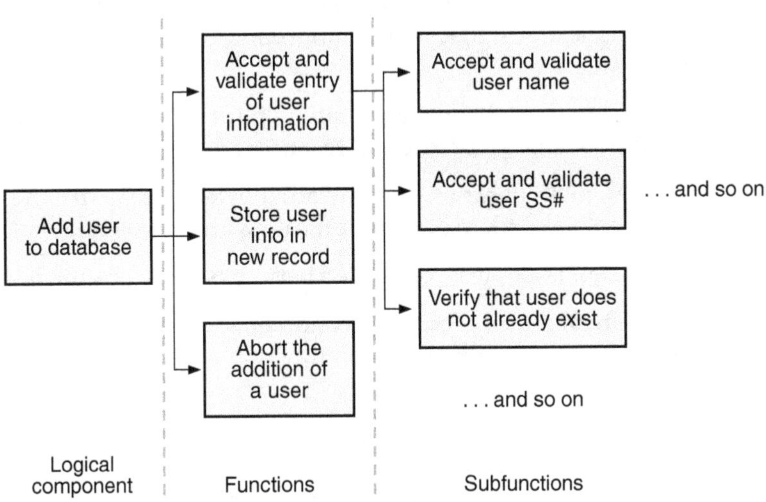

Figure 22.13 Function and subfunction list—example.

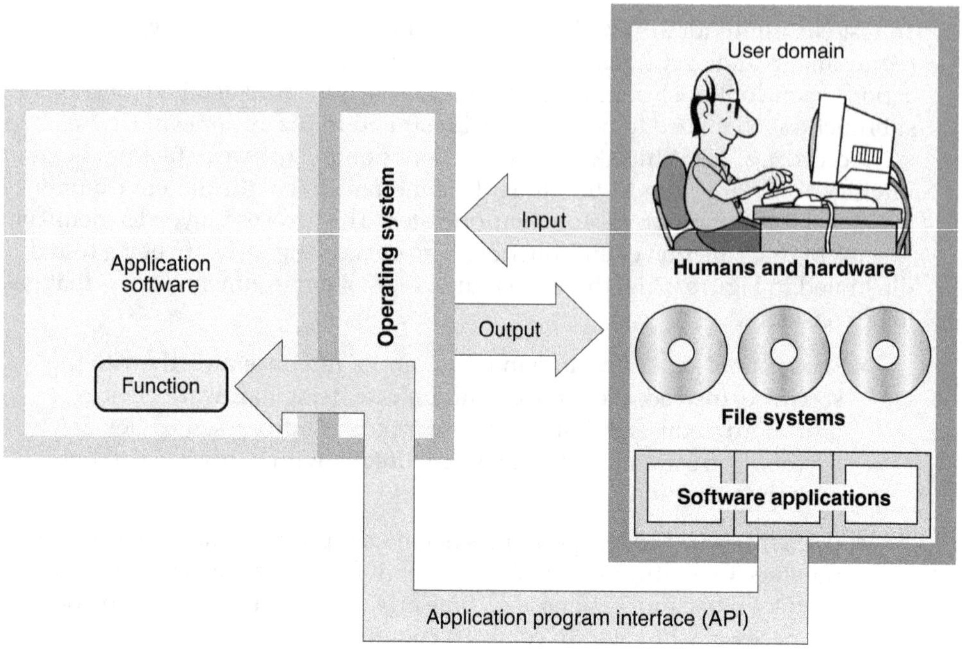

Figure 22.14 Environment in which the function operates.

- Corrupt data (e.g., wrong type, incorrectly formatted, field delimiters misplaced, fields in the wrong order, file too large for the system under test to handle)
- A privileged user changing the permissions of a file while another user is editing the file contents

- *Software applications*: Other software programs (e.g., databases, runtime libraries, applications) that supply inputs and receive outputs. Testers should consider:
  - Data that are passed to the application, return values and error codes, and failure scenarios of external resources
  - Databases, math libraries, and any other external resource the application may link to or communicate with for potential failure
  - Environment issues (e.g., congested networks, busy or slow responses)

Once the environmental users for the function or subfunction are identified, the tester must explore the capabilities of each function or subfunction as it relates to each of those users. For example:

- *What inputs, both valid and invalid, can each user provide the function/subfunction?* The tester first tests to make certain that valid input values are handled appropriately. The tester then tests to confirm that the software prevents invalid inputs from being accepted or that the software handles them correctly.

- *What outputs do the functions or subfunctions send to each user?* The tester must test to make sure that the function responds to specific conditions, inputs, and events with outputs of the correct form, fit, and function. For example, if the user hits the print command from a word processor, the appropriate document is sent to the printer (function), it is formatted correctly (form), and the right number of pages are printed (fit). The tester must also test to make certain that the function does not respond with outputs unless the proper conditions, inputs, or events occur. For example, the word processor does not send the document to the printer when the "open" function is activated or no function has been selected.
- *What data are stored or retrieved because of interactions with each user?* The tester must test to confirm that data are stored and retrieved in the proper formats and structures. Data structures should be tested to make certain that data can be appropriately added, stored, read, and deleted. Testers should test to make sure that overflowing of the data structures does not occur. Testers should also consider other data system–related issues, including data integrity and data control (e.g., security, refresh, backup, and recovery).
- *What computation(s) are done because of interactions with each user?* Testers should think about what computations occur and how they might overflow (or underflow) or interact or share data poorly with other computations or features.

To test usage scenarios, the tester chains together individual test cases, from testing of the functions and subfunctions, into test procedures (scenarios) that test start-to-finish user interactions with the system. Each feature may work independently, but they may not "play well together" to get real work accomplished for the users.

Testers perform *operational profile testing* to thoroughly evaluate the areas where defects are most likely to impact the reliability of the software. To perform operational profile testing, the tester must identify the different threads through the software. A *thread* is a sequential set of usage steps through the software that a user takes to accomplish a start-to-finish task. For example, as illustrated in the decision tree in Figure 22.13, one thread would be to enter the credit card correctly the first time, enter the PIN correctly the first time, and then select and pump regular gas. A different thread would repeat these steps except for selecting and pumping premium gas. A third thread would be to enter the card correctly, but take two tries before entering the PIN correctly and selecting and pumping regular gas. Once the threads have been identified, probabilities must be assigned to the paths out of any given state. The sum of all of the probabilities (percentages) for paths out of any given state must be 1 (100%). For example, in the decision tree in Figure 22.15:

- For the *card entry* state: Putting the card in correctly is assigned a 95% (0.95) probability, and putting the card in incorrectly is assigned a 5% (0.05) probability
- For the *await grade selection* state: Selecting regular gas is assigned a 65% (0.65) probability, selecting super gas is assigned a 25% (0.25) probability, and selecting premium gas is assigned a 10% (0.10) probability

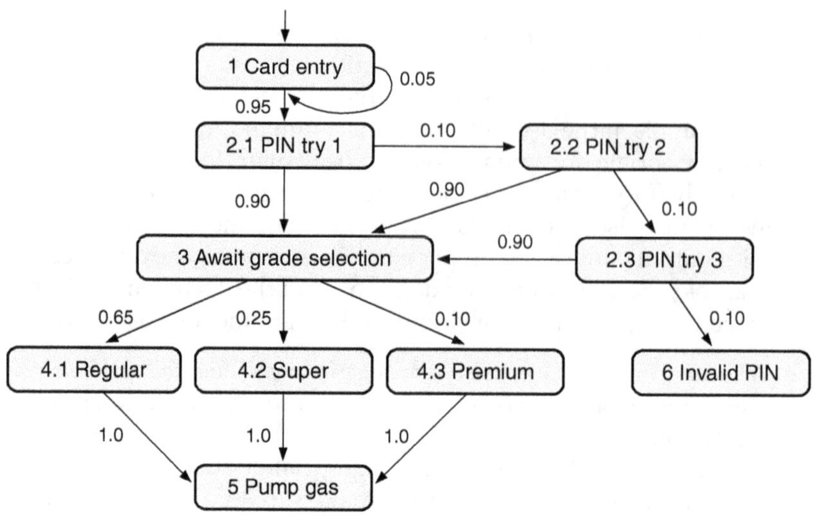

Figure 22.15 Decision tree—example.

The probability of a given thread is calculated by multiplying the assigned probabilities on that thread together. Those threads with the highest probability are the most frequently traversed threads. For example, again using the example in Figure 22.15:

- Thread 1 through states 1, 2.1, 3, 4.1, and 5 = $.95 \times .90 \times .65 \times 1 \approx 55.58\%$
- Thread 2 through states 1, 2.1, 3, 4.2, and 5 = $.95 \times .90 \times .25 \times 1 \approx 21.38\%$
- Thread 3 through states 1, 2.1, 3, 4.3, and 5 = $.95 \times .90 \times .10 \times 1 \approx 8.56\%$
- Thread 4 through states 1, 2.1, 2.2, 3, 4.3, and 5 = $.95 \times .10 \times .90 \times .10 \times 1 \approx 0.85\%$
- Thread 5 through states 1, 2.1, 2.2, 2.3, 3, 4.1, and 5 = $.95 \times .10 \times 10 \times .90 \times .65 \times 1 \approx 0.56\%$
- And so on

Using this example, if 1,000 tests are going to be executed, operational profile testing would sample 556 (1,000 × 55.58%) of those tests as variations on thread 1, for example, using different types of cards, different valid PINs and pumping different amounts of gas. A sample of 214 tests would be executed as variations on thread 2, 86 tests would be executed as variations on thread 3, and so on through each possible thread.

Performance, Environmental Load, Volume, and Stress Testing

The objective of *performance testing* is to determine if the system has any problems meeting its performance requirements, including throughput (number of transactions per time unit), response time, or capacities (e.g., the number of simultaneous users, terminals, or transactions). Performance testing is usually

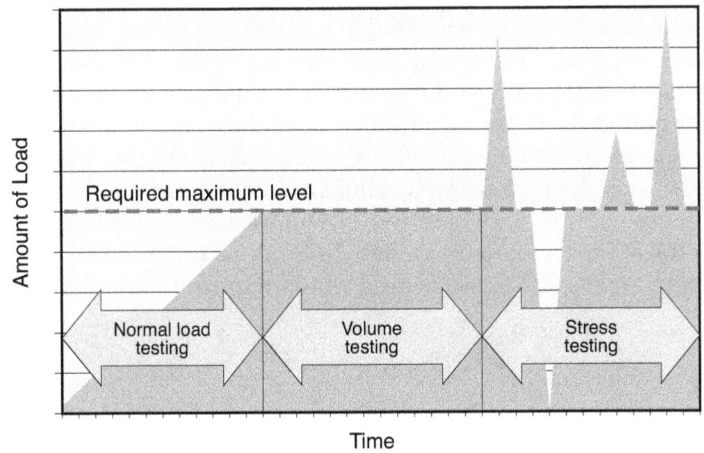

Figure 22.16 Load, stress, and volume testing.

done at the system level under full environmental load. Typically, software can perform to specification when not much is going on. However, the performance of some software applications may degrade at full volume, with multiple users interacting with the software and other applications running in the background taking up system resources. Performance tests should be performed after other testing is complete and the software is relatively problem-free.

Environmental load testing evaluates the software's performance capabilities (throughputs, response times, and capacities) under normal load conditions, as illustrated in Figure 22.16.

Volume testing, also called *endurance testing* or *soak testing,* is a special type of environmental load testing that subjects the software to heavy loads over long periods of time and evaluates the software's capability. For example, does the software have problems handling the maximum required volumes of data, transactions, users, and peripherals over several days or weeks?

Stress testing is another special type of environmental load testing that subjects the software to surges or spikes in load over short periods of time and evaluates the software's performance. For example, is the software able to appropriately deal with jumping from no load to a spike of maximum data, transactions, or users? Stress testing can also involve testing for excess load. For example, if the requirement states that the maximum capacity that a telecommunication switch can handle is 10,000, does the software appropriately handle a spike to 11,000 calls by reporting the appropriate messages ("all lines are currently busy, please call back later"), or does it result in an inappropriate failure condition?

Worst-Case Testing

One type of *worst-case testing* is an extension of boundary testing. Boundary testing investigates the boundaries of each input variable separately by exploring the minimum value, one below the minimum, the maximum value, and one above the maximum. During routine boundary testing, if multiple variables are needed as inputs to the software, each variable is boundary tested individually, while all of the other variables are input at their normal values. For example, if a software function has two inputs x with boundaries at 0 and 5 and y with boundaries at 20 and 50, boundary testing would first boundary-test variable x with y set to its typical value. Then variable x would be set to its typical value and variable y would be boundary-tested. This example would require eight tests.

Worst-case testing explores the boundaries of multiple input variables in combination at their boundary values. To continue the previous example with the two inputs x and y, worst-case testing would require 16 test cases:

1. $x = -1$ and $y = 19$
2. $x = -1$ and $y = 20$
3. $x = -1$ and $y = 50$
4. $x = -1$ and $y = 51$
5. $x = 0$ and $y = 19$
6. $x = 0$ and $y = 20$
7. $x = 0$ and $y = 50$
8. $x = 0$ and $y = 51$
9. $x = 5$ and $y = 19$
10. $x = 5$ and $y = 20$
11. $x = 5$ and $y = 50$
12. $x = 5$ and $y = 51$
13. $x = 6$ and $y = 19$
14. $x = 6$ and $y = 20$
15. $x = 6$ and $y = 50$
16. $x = 6$ and $y = 51$

Another type of worst-case testing is to execute testing under worst-case resource conditions, for example, with the minimum allowable amount of memory, bandwidth, processing speed, or network speed, or with the maximum volume of users, peripherals, or other applications running in the background. Testing performance requirements is a particular concern when testing under worst-case conditions.

Resource Utilization Testing

The objective of *resource utilization testing* is to determine if the system uses resources (e.g., memory, buffers, or disk space) at levels that exceed requirements or if the software has any problems when needed resource levels fluctuate or when resources are busy or unavailable. When performing resource utilization testing, the testers must evaluate objectives or requirements for resource availability. By testing the software with inadequate resources or by saturating resources with artificially induced overloads, the tester can force resource-related exception conditions. Other resource-related testing considerations include whether the software can appropriately handle hardware or resource failures like:

- Full disk, directory, memory area, print queue, message queue, stack
- Disk not in drive, out of service, missing or write-protected
- Printer off-line, out of paper, out of ink, jammed, or missing
- Extended memory not present or not responding

Usability Testing

The objective of *usability testing* is to make certain that the software matches the user's actual work style and determines whether the software has any areas that will be difficult or inconvenient for the users. The characteristics of usability include:

- *Accessibility*: Can users enter, navigate, and exit with relative ease?
- *Responsiveness*: Can users do what they want, when they want, quickly?
- *Efficiency*: Can users do what they want in a minimum amount of time, keystrokes, or steps?
- *Comprehensibility*: Do users understand the product structure, its help system, instructions, error messages, and documentation?
- *Aesthetics*: Are the screens, reports, and other user interfaces pleasing to the user's senses?
- *Ease of use*: How intuitive is the software to use, and how easy is it to remember how to use the software when it is used only occasionally?
- *Ease of learning*: How intuitive is it to learn how to use the software?

When designing tests for usability, the tester has to consider the different potential users of the system, including:

- Novice users
- Occasional users
- Different types of typical users (e.g., gas station managers, attendants, family car drivers, and 18-wheeler drivers)
- Power users
- Enterprise users (e.g., drivers from a taxi fleet or a trucking company that have special requirements for the pay-at-the pump software)
- Database administrators (DBAs)
- Operators or maintenance personnel
- People with disabilities (focus on accessibility)

These different types of users may interpret usability very differently. Features that make the system user-friendly to novice or occasional users may drive power users crazy.

One method that is often employed in usability testing is to have different types of actual users work with the software and observe their interactions. Usability tests can include having these users perform free-form, unplanned tasks as part of the testing process. This method allows users to work in a manner that reflects how they expect to work with the system. Alternatively, the user may be asked to perform predefined, written scripts containing step-by-step instructions for the user to follow. This method facilitates better coverage but may not reflect the user's work patterns effectively. Using a preliminary prototype or mock-up rather than the final product allows usability testing to be performed earlier in the life cycle. Conducting and observing beta testing or field trials on the software at the actual user's site or in the environment where the software will be used can also be used to perform usability testing.

Exploratory Testing

In *exploratory testing,* the testers design the test cases and procedures at the same time they are testing the software. Unlike preplanned test design, where test cases and procedures are written well in advance derived from the test basis, in exploratory testing, the testers use the information they learn about the product as they are testing it to create additional tests. Unlike ad hoc testing, exploratory testing is not just randomly wandering around the software. It systematically explores part of the software based on a test charter. The testers consciously think about what they do not know but want to discover. Also, unlike ad hoc testing, the exploratory tester should "always write down what you do and what happens when you run exploratory tests" (Kaner et al. 1999).

Exploratory testing is based on the tester's intuition, experience, and observations. For example, a tester runs a function's preplanned set of test cases, and everything matches the expected results. Normally, the tester would move on to the next set of tests. However, if things just do not seem right, the tester uses exploratory testing to create additional test cases that further investigate the feature. Another example would be when the tester does find several problems in an area of the software but believes there are still more to find, even though all the preplanned test cases are executed. Again, the tester creates additional test cases to explore the feature further, trying to confirm or refute their suspicions.

Exploratory testing can also leverage risk-based testing. For example, if a feature is expected to be low-risk, few planned test cases may be written for that feature. The exploratory testers can quickly investigate the low-risk feature to find out if they can find problems that might cause a reassessment of that feature's risk. On the other hand, for features that are expected to be high-risk, time may be reserved in the schedule to plan for additional exploratory testing of those features.

Regression Testing

Regression testing is concerned with how changes (fixes or updates) to software affect the unchanged portions of the software or hardware. Regression testing is "selective retesting of a software or software component to verify that modifications have not caused unintended effects and that the system or component still complies with its specified requirements" (IEEE/ISO/IEC 2017).

Regression analysis examines a proposed software change to determine the depth and breadth of the effects the proposed change could have on the software. It determines how extensively a change needs to be regression tested and how many already-executed test cases need to be re-executed. Regression analysis must balance the risk of releasing the software with undiscovered defects and the software's quality and integrity requirements against the cost of a more extensive testing effort.

Of course, the first testing step when software changes is to retest what changed. White-box, gray-box, and black-box testing strategies can be used to test the individual source code modules, components, interfaces, and features that changed.

Regression analysis is then used to determine how extensively other parts of the software should be retested. When deciding what other white-box and gray-box regression tests to execute, consider source code modules or components that share the same local or global variables or that directly call, or are called by, those source code modules or components that changed. More rigorous regression testing might also consider going to a second level, as illustrated in Figure 22.17. This more rigorous testing includes testing the source code modules or components that call, or are called by, the source code modules or component that directly call, or are called by, the source code modules or component that were changed. Regression analysis also includes deciding what other black-box regression tests to execute when changes are made to the software. Black-box regression analysis considers functions that:

- Perform similar functions or are closely associated with the changed functions
- Interface with the same external device(s) or user(s)
- Access the same external data or databases

The final step in regression testing is to execute the *regression test suite*. The regression test suite includes white-box, gray-box, and black-box tests that are always repeated no matter what changes in the software. Candidate tests for the regression test suite are tests for the source code modules or components and functions that:

- *Are mission-critical to the success of the software*: For example, if everything else fails, the "save" command in a word processor or the "self-destruct" in a missile system must work.
- *Are defect-prone*: Software that has been defect-prone in the past is more likely to be defect-prone in the future.
- *Are the most used (operational profile)*: Defects in areas of the software that are used most extensively will impact more users.
- *Contain patches*: Since patches are temporary fixes, it is very easy to make a mistake when adding existing patches into a new software build.

Tests included in the regression test suite are prime candidates for automation because those tests will be rerun many times (every time the software is changed).

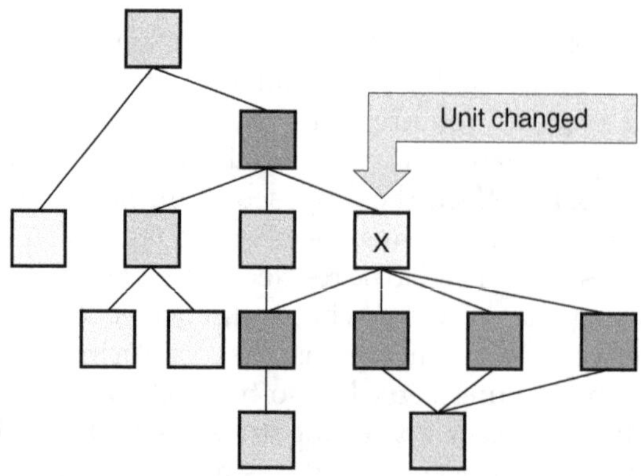

Figure 22.17 Test what might be impacted by the changes.

5. TESTS OF EXTERNAL PRODUCTS

> Determine appropriate levels of testing for integrating supplier, third-party, and subcontractor components and products. (Apply)
>
> **Body of Knowledge VI.B.5**

When an external party-provided software component or product is incorporated into a software system, it must be tested to the same level of rigor as other parts of that software system. That is, decisions about testing external software products should be based on risk analysis and the necessary level of integrity. If an externally provided component or product is used to develop, build, or test the software system, it must also be tested to confirm that it does not cause problems in the software being developed. Externally provided software components or products should be treated as an extended part of the primary development organization. The external supplier's development and test process adequacy should be prequalified and then monitored throughout the effort. Supplier processes do not necessarily have to mirror the acquirer's process, but must achieve the same quality criteria for development and testing.

Throughout the software development process, the supplier of the external product should be performing V&V activities. The amount of visibility the acquirer needs into those V&V activities may vary, depending on the acquirer's relationship with that supplier. For example, for commercial off-the-shelf (COTS) or open-source software, the acquirer may have no way of determining the extent to which the supplier tested their software. On the other hand, with custom-built software, the acquirer might actively participate in some supplier V&V activities or at least have access to the V&V results and reports.

As part of product acceptance for custom-built software, the acquirer should conduct acceptance testing against negotiated acceptance criteria, included in the supplier contract or agreement to confirm that the delivered products meet all agreed-to requirements. The acquirer may also want to conduct alpha or beta testing. If the supplier's software is being integrated into a larger product, it should be integration-tested and system-tested as part of that product. Black-box strategies are usually necessary for this testing, since access to the acquired software source code is typically unavailable.

As with other testing activities, risk should be considered when scoping any COTS testing effort. For example, for COTS software, where the mainstream functionality is being utilized, less testing may be required because this functionality has already been validated extensively through operational use. However, more extensive testing may be appropriate if more fringe functionality is being used, or if this is a newly released COTS product. If the COTS software is being integrated into the software system, it should be integration-tested and system-tested as part of that product. As with any supplier-supplied software, black-box strategies are usually necessary for this testing, since access to the software source code is typically unavailable. If the COTS software is being used to develop or test the software being developed, risk-based validation and certification testing may be used to confirm that it performs as required. Any customization or wraparounds that are done for the COTS software should also be tested.

For open-source software, risk should be considered when scoping any testing effort. If the open-source software is being integrated into the software product, it should again be integration-tested and system-tested as part of that product. If the open-source software is being used to develop or test the software being developed, risk-based validation and certification testing may be used to confirm that it performs as required as it was with COTS software. Any modifications, customization, or wraparounds that are done for the open-source software should also be tested. However, there may be legal issues associated with modifications to open-source software that may need to be considered, so legal advice should be obtained before using open-source software. Open-source software can also be problematic when it comes to keeping up with changes and performing retesting and regression testing.

For some projects, the customers may also supply software that will be integrated into the software system. If this is the case, this customer-supplied software must also be tested to the same level of rigor as other parts of the software system.

6. TEST COVERAGE SPECIFICATIONS

> Evaluate the adequacy of test specifications such as functions, states, data and time domains, interfaces, security, and configurations that include internationalization and platform variances. (Evaluate)
>
> **Body of Knowledge VI.B.6**

Test coverage looks at the completeness of the testing. Test coverage maps the tests (test cases and procedures) to some attribute of the source code modules, component, or software product being tested. These can be attributes of the specification, as discussed in this section, or attributes of the code, as discussed in the next section.

Requirements Coverage

Requirements coverage looks at the mapping of tests to each uniquely identified requirement in the specification. Traceability is one of the primary mechanisms used to make sure that all of the functional and other product requirements from the specification are covered. Every requirement should be traced forward to one or more tests that will be used to evaluate the complete and correct implementation of that requirement. Requirements coverage metrics can be used to monitor progress toward complete coverage during the testing activities. The requirements coverage metric is the percentage of the requirements that trace forward to at least one test case. If the value of this metric is less than 100 percent, additional test cases need to be written to test the uncovered requirements. However, a value of 100 percent does not guarantee thorough testing, since more than one test case may be needed to test each requirement comprehensively.

A *test matrix*, also called a *system verification matrix*, is another tool for making certain that each requirement is appropriately tested. A test matrix traces each requirement to the method that will be used to test it. Initially, a test matrix is used to plan the level of testing and test strategy for each requirement, as shown in Table 22.4. As test design progresses, additional detail can be added to the matrix to include more information about the specific approach to be taken. In this example, the test strategies include:

- *Inspection*: Evaluation of the requirements done through visual examination or static analysis of the software products (e.g., source code or documentation) rather than through dynamic execution. Examples of the types of requirements that would be inspected are:
  - Product physical characteristics (e.g., the screen will be blue)
  - Standards (e.g., adherence to specified coding, commenting, or modeling standards)
  - Specified languages or algorithms (e.g., the software will be written in C++, or the data items will be sorted using a bubble sort)
- *Black-box demonstration*: Evaluation of the requirements is done through the use of black-box strategies, where the evaluation can be accomplished through observation or visual confirmation. Examples of the types of requirements that would be demonstrated are:
  - Functional requirements
  - Performance-type requirements (e.g., throughput, speed, response times, capacity)

Table 22.4 Test matrix—example.

Number	Requirement	Test Level	Test Strategy
R103	The system responds to all user commands and data entries within three seconds	System	Black-box demonstration
R200	The software shall store all currently active transactions in memory to allow access if the storage media fails	Unit	White-box execution
R397	The source code shall be written in C++	Peer review of code	Inspection
R560	The system shall calculate sales tax at current tax rates	Unit and system	White-box and black-box analysis
R50	The system shall work with all credit cards issued by all banks	—	Not testable

- Security- or safety-type requirements
- Usability-type requirements

- *Black-box analysis or white-box analysis*: Evaluation of the requirements that necessitate the use of independent calculations or other analytical techniques used with either white-box or black-box testing strategies. Examples of the types of requirements that would be evaluated through analysis are:
    - Requirements that include calculations (e.g., calculating payroll withholding, calculating monthly sales totals)
    - Requirements that include statistical techniques (e.g., means, standard deviations, confidence intervals)
- *White-box execution*: Evaluation of the requirements is done through the use of white-box techniques. Examples of the types of requirements that would be tested in this manner are most design constraint–type requirements, including:
    - Internal communication protocol requirements
    - Internal data storage requirements
    - Output format requirements like specifics for prompt or error message wording or report layouts
- *Not testable*: Requirements that cannot be evaluated because:
    - They are not really software requirements (e.g., project constraints like delivery schedules)
    - It is not possible to test them with the current resources (e.g., requirements that are not finite or not measurable)

An advantage of creating a test matrix early in the life cycle is the identification of nontestable requirements like R50 in Table 22.4, before they propagate into other software work products. Nontestable requirements and other issues discovered during the creation of the test matrix should be reported to the requirements analyst for resolution. It should be remembered that any requirement that is not tested is a source of risk, because the implementation of that requirement may include undiscovered defects.

Functional Coverage

Of course, 100 percent requirements coverage automatically achieves 100 percent coverage of all of the functional requirements. Another way to measure functional coverage from a user perspective is to look at the percentage of threads through the program that map to test procedures.

State Coverage

State coverage looks at the mapping of tests to the various states that the software can be in, and the various transitions between those states to make certain that they are thoroughly tested. State testing is done to determine if the software:

- Switches correctly from valid state to valid state, as specified
- Performs any state transitions that are not valid (moving to a state that is not a valid transition from the previous state)
- Loses track of its current state
- Mishandles inputs or other data while it is switching states

Data Domain Coverage

Data domain coverage looks at the mapping of tests to the various data domains in the software to confirm that they are tested thoroughly. Once the equivalency classes and boundaries for the inputs and outputs have been identified, as defined previously, data domain coverage evaluates the percentage of those classes and boundaries with associated test cases.

Another data domain coverage technique creates a list of all of the data items in the software. This can be done by evaluating the data dictionary or looking at the specification to identify data items and structures. The CRUD acronym can then be used to investigate data domain coverage by making certain that tests exist to evaluate any requirements or stakeholder needs for:

- C: The creation and initialization of each data item or structure
- R: The reading, querying, or displaying of each data item or structure
- U: The updating or refreshing of each data item or structure
- D: The deleting of each data item or structure

Some organizations use the alternative CRUDL acronym, where the "L" is added for listing, or tracking each data item or structure in reports.

Date and Time Domain Coverage

Date and time domain coverage looks at the mapping of tests to the various date and time domains in the software to confirm that they are thoroughly tested. Date and time domain testing considers:

- Different times of the day, days of the week, month, or year when the software is expected to perform differently, for example, performing background activities or maintenance, refreshing data, or creating reports.
- Special dates or times when the software performs differently, for example, vacations when normal operations are shut down.
- Peak or minimum loads at certain times or dates.

Date and time domain testing also considers whether the software appropriately handles special dates and times. For example:

- Shift changes, the end of the day, week, month, quarter or year, or even the end of the century
- Holidays
- Daylight savings time changes
- Leap year
- Time zone changes or international date line changes

Interface Coverage

An *interface* is a shared boundary across which information is passed. There are interfaces internal to the software between its source code modules and components. There are also interfaces between the software and external entities (user, hardware, operating systems, file systems, and other software applications). Interface coverage looks at the mapping of tests to the various *identified interfaces* to make certain that they are thoroughly tested. Integration testing typically focuses on confirming interface coverage.

Security Testing

The objective of *security testing* is to determine if the system's security can be attacked and breached and to determine whether or not the software can handle those breaches or recover from them if security is breached. Security coverage looks at the mapping of tests to the various identified security issues and then performs penetration testing to confirm that they are thoroughly tested. As our digital infrastructure gets increasingly complex and interconnected, the difficulty of achieving application security increases exponentially, and so does the difficulty of the associated security test efforts. As with other types of testing, checklists are often used to make sure that key security areas are tested. Testers also use penetration testing techniques to attempt to break through access control and software security controls to identify weaknesses along specific paths of attack. (See Chapters 6, 12, 14, and 18 for additional discussions about software security.)

Platform Configuration Coverage

The objective of platform *configuration testing* is to determine if the software has any problems handling all the required possible hardware platforms and software configurations (e.g., various servers, browsers, operating systems, and so on). *Platform configuration coverage* looks at the mapping of tests to the various possible platforms to confirm that the software is thoroughly tested on each platform. One method used for platform configuration coverage is to use a platform configuration test matrix. In a *platform configuration test matrix,* all of the possible platforms are listed as column headers. The example in Table 22.5 includes various operating systems that the software needs to run on, but other configurations can be included as needed. The left column lists the tests to be conducted. Each cell lists the status of those tests on that platform:

- *Blank*: The test has yet to be executed on that configuration.
- *Passed*: The test passed when it was executed on that configuration.
- *Failed*: The test failed when it was executed on that configuration.
- *N/A*: The test is not applicable or does not need to be executed on that configuration. For example, N/A might indicate that the test is not expected to react differently, or have different results, on that configuration than it will have on one or more other configurations in the matrix.

This matrix only tests a single configuration entity variance per column. However, if combinations of configuration entities are expected to interact with the software in different ways, it may be necessary to test configuration combinations as well. For example, if there are multiple possible hardware platforms and multiple possible operating systems, configuration testing may require the testing of the software on each hardware platform/operating system pairing.

Internationalization Testing

The objective of *internationalization testing,* also called *localization testing,* is to determine if the software has any problems transitioning it to other geographic

Table 22.5 Platform configuration test matrix—example.

Tests	Mac OS	Windows 2007	Windows 2010	Linux	UNIX
ABC-1	Passed			N/A	N/A
ABC-2	Passed			N/A	N/A
AR-1	Passed			N/A	N/A
XYZ-10	**Passed**			**Passed**	

locations. This includes adaptation to different languages, cultures, customs, and standards. The tester must be fluent in each language used to test the system to confirm proper translation of screens, reports, error messages, help, and other items.

Internationalization configuration coverage looks at the mapping of tests to the various possible geographic locations to confirm that the software is thoroughly tested on each location. Internationalization configuration coverage can also be tracked using a configuration test matrix, similar to the one illustrated in Table 22.5. However, in the case of internationalization, all of the possible locations are listed as column headers instead of the platforms. Test considerations in internationalization testing include:

- *Character sets*: Different character sets are used in different countries. Verify that the software uses the proper set.
- *Keyboards*: Keyboards should be tested with the software to make sure that they correctly interpret key codes according to the character set.
- *Text filters*: The software may accept only certain characters in a field. Test that the software allows and displays every character in its appropriate place.
- *Loading, saving, importing, and exporting high and low ASCII*: Save and read full character sets to every file format the software supports. Display and print them to verify correctness.
- *Operating system language*: Check variances in wild card symbols, file name delimiters, and common operating system commands.
- *Hot keys*: Consider any underlined, bold, or otherwise highlighted character in a menu item (such as X in eXit). Do any hot keys from the original language have their original effect even though they do not appear in the localized menu?
- *Garbled in translation*: If the software builds messages from fragments, how does that appear in the localized version? Are there file names and data values inserted into an error message to make it more descriptive?
- *Text displays*: Different languages are oriented in different ways. For example, some languages are written right to left, and others are written left to right.
- *Expanding text*: Translated text expands and can overflow menus, dialog boxes, and internal storage, overwriting other code or data.
- *Spelling rules*: Spelling rules vary across dialects of the same language. For example, in the United States the word "organization" is spelled with a "z," whereas in England it is spelled "organisation." Does the spell-checker work correctly for the new location?
- *Hyphenation rules*: Rules are not the same across languages.
- *Sorting rules*: Character and word sorting rules vary from country to country (e.g., sorting by last name, first name). The addition of special

characters like letters with accents may not allow sorting using ASCII numeric order.

- *Case conversion*: Only when the ASCII character set is used can case conversion be done correctly by adding or subtracting 32 to or from a letter. Look for case conversion issues in any search dialog or other text pattern-matching functions.
- *Underscoring rules*: Underlining conventions differ between countries. It can be poor form to underscore punctuation, spaces, and other characters in some countries.
- *Printers*: Though most European printers are essentially the same as those in the United States, printer differences can and do occur around the world.
- *Paper size*: Paper sizes can differ from country to country. Check that the default margins are correct for each size.
- *Data format and setup*: Consider both the format and the numeric separators for time and date displays and money formats. For example, where the United States puts a comma, others put a decimal point or a space, and some subroutines that the software links into may treat commas as separators between items in a list.
- *Rulers and measurements*: Rulers, tab dialogs, grids, and every measure of length, height, volume, or weight that is displayed must be in the correct unit of measure.
- *Culture-bound graphics*: Check clip art, tool icons, screens, manuals, and even packaging for culture-bound graphics.
- *Culture-bound outputs*: Calendar formats vary, the appearance of a standard invoice varies, and address formats differ between countries. For example, in the United States, a date that reads 4/12/2016 is April 12, 2016, while in other countries, it is December 4, 2016.

7. CODE COVERAGE TECHNIQUES

> Use and identify various tools and techniques to facilitate code coverage analysis techniques such as branch coverage, condition, domain, and boundary. (Apply)
>
> **Body of Knowledge VI.B.7**

Unlike threads, which are usage- or functionality-focused, *paths* refer to control flow sequences through the internal structure of the software source code. There are typically many possible paths between the entry and exit of a typical software application. A source code module containing only straight-line code has a single path. Every Boolean decision (IF or IF THEN ELSE) doubles the number of potential paths, every case statement multiplies the number of potential paths by the number of cases, and every loop multiplies the number of potential paths by the number of different iteration values possible for the loop. For example, a source code module with a loop that can be iterated from one to 100 times has 100 possible paths. Add an if–then–else statement inside that loop and that increases the number of paths to 200. Add a case statement with four possible choices inside the loop as well, and there are 800 possible paths. Moving from the individual source code modules to the integration level, if this source code module with 800 paths is integrated with a source code module with only two sequential if–then–else statements (four paths), there are now 3,200 paths through these two source code modules in combination.

Since there are rarely enough resources to test every path through a complex software application or even a complex individual source code module, the tester uses white-box logic coverage techniques to systematically select the tests that are the most likely to help identify the yet undiscovered, relevant defects.

Statement, Decision, and Condition Coverage

To demonstrate the different statement, decision, and condition coverage techniques, the piece of nonsense code shown in Figure 22.18 will be used.

A *statement* is an instruction or a series of instructions that a computer carries out. *Statement coverage* is the extent that a given source code module or component's statements are exercised by a set of tests. Statement coverage is the least rigorous type of code coverage technique. Each statement must be executed at least once to have complete statement coverage. Table 22.6 illustrates that it only takes one test case to have statement coverage of the code in Figure 22.18. As long as input variables B and C are selected so that both decisions in this code are true, every statement is executed.

A *decision* determines the branch path that the code takes. To have *decision coverage,* also called *branch coverage,* each decision takes all possible outcomes at least once. For example, if the decision is a Boolean expression, decision coverage requires test cases for both the true and false branches. If the decision is a case statement, decision coverage requires test cases that take each case branch. Table 22.7 illustrates the test cases needed to have decision coverage of the code in Figure 22.18:

- The first test case results in the first decision being true and the second decision being false.

```
A = 300
if B > 40 and C < 100 then A = 1000
if B < 60 and C < 20 then A = 10
print A
```

Figure 22.18 Code—example.

Table 22.6 Statement coverage–example.

Test Case #	Inputs		Expected Output
	B	C	A
1	> 40 and < 60	< 20	10

Table 22.7 Decision coverage–example.

Test Case #	Inputs		Expected Output
	B	C	A
1	≥ 60	< 20	1000
2	≤ 40	< 20	10

- The second test case results in the first decision being false and the second decision being true.
- Thus, these two test cases in combination provide decision coverage because the true and false paths are taken out of each decision.

A *condition* is a state that a decision is based on. To have *condition coverage,* each condition in a decision takes all possible outcomes at least once. For the code in Figure 22.18 there are three conditions that the input variable B can have:

- $B \leq 40$
- $40 < B < 60$
- $B \geq 60$

There are also three conditions that input variable C can have:

- $C < 20$
- $20 \leq C < 100$
- $C \geq 100$

Table 22.8 illustrates one choice of test cases that combines these into condition coverage of the code in Figure 22.18. Note that condition coverage does not always imply decision coverage. For example, in the set of test cases in Table 22.8:

- Test case 1 results in the first decision being true and the second decision being false.
- Test cases 2 and 3 result in both decisions being false.
- Therefore, decision coverage has not been achieved because the true path has not been taken out of the second decision.

The next level of rigor is to have *condition and decision coverage,* where each decision takes all possible outcomes at least once, and each condition in a decision takes all possible outcomes at least once. If decision/condition coverage exists,

Table 22.8 Condition coverage–example.

Test Case #	Inputs		Expected Output
	B	C	A
1	≥ 60	< 20	1000
2	≤ 40	≥ 20 and < 100	300
3	> 40 and < 60	≥ 100	300

Table 22.9 Decision/condition coverage–example.

Test Case #	Inputs		Expected Output
	B	C	A
1	≥ 60	≥ 20 and < 100	1000
2	> 40 and < 60	< 20	10
3	≤ 40	≥ 100	300

Table 22.10 Multiple condition coverage–example.

Test Case #	Inputs		Expected Output
	B	C	A
1	≤ 40	< 20	10
2	≤ 40	≥ 20 and < 100	300
3	≤ 40	≥ 100	300
4	> 40 and < 60	< 20	10
5	> 40 and < 60	≥ 20 and < 100	1000
6	> 40 and < 60	≥ 100	300
7	≥ 60	< 20	1000
8	≥ 60	≥ 20 and < 100	1000
9	≥ 60	≥ 100	300

then condition coverage and decision coverage also both exist. Table 22.9 illustrates one choice of test cases that provides decision/condition coverage of the code in Figure 22.18.

Modified condition/decision coverage (MC/DC) expands on condition and decision coverage by adding the requirements that each condition be shown to independently affect the outcome of the decision. The requirement for independence makes certain that the effect of each condition is tested relative to the other conditions (Hayhurst et al. 2001).

To have *multiple condition coverage,* each statement is executed at least once and all possible combinations of condition outcomes in each decision occur at least once. Multiple condition coverage always results in condition, decision, and statement coverage as well. Multiple condition coverage is the most rigorous type of structural coverage testing. Table 22.10 illustrates a choice of test cases that provides multiple condition coverage of the code in Figure 22.18.

Domain and Boundary Testing

For structural testing, the conditions for the input variables define their *domains*, or *equivalence classes*. Once these domains are defined, sampling from these domains can be done when executing test cases.

Boundary value testing can also be done by sampling values on the boundaries of these domains. For example, for input variable B in the code in Figure 22.18, the two-point boundary values would be 39, 40, 60, and 61.

Boundary testing can also be performed on the loop counters when loops are included in the software source code. Boundary loop testing would have test cases for the minimum and maximum times through the loop (if there is a maximum), and one less than the minimum (which would skip the loop), and one more than the maximum. One more than the maximum is an error condition that the software should appropriately catch and handle because the loop should never be able to be executed more than its maximum number of times. If the loop can be executed more than the maximum number of times, a problem should be reported.

8. TEST ENVIRONMENTS

> Select and use simulations, test libraries, drivers, stubs, harnesses, etc. and identify parameters to establish a controlled test environment. (Analyze)
>
> **Body of Knowledge VI.B.8**

Test Beds

A *test bed* is an environment established and used for the execution of tests. The goal is to be able to create a test environment that matches the actual operating environment as closely as possible, while providing the testers with a known platform to test from and visibility into test results.

As illustrated in Figure 22.19, *test environment set-up and maintenance process* is one of the four dynamic testing processes. Before testing starts, the test environment must be set up and verified. The test data must also be prepared, loaded onto the test bed, and verified. Throughout testing, the test bed and test data must be controlled so that they do not adversely impact the test results.

The test bed may also need to be maintained and updated as needed to perform various testing activities. Multiple test environments and data may be used throughout the testing processes. There may be different test environments

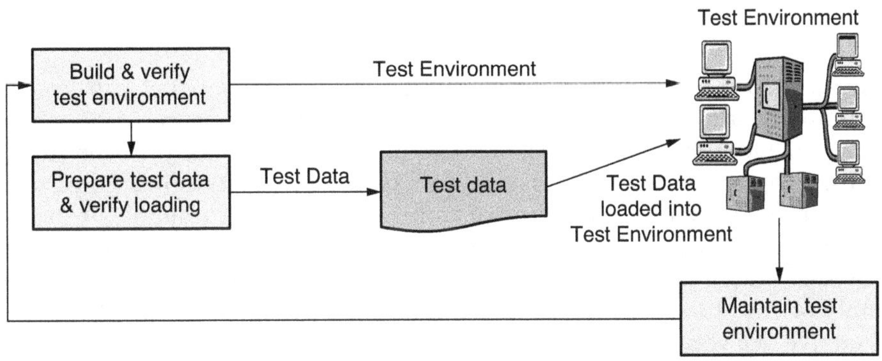

Figure 22.19 Steps in the test environment setup and maintenance process.

for performing different levels of testing (unit, integration, system) or for performing different types of testing (performance, security, safety, usability, configuration).

Requirements for the test environment may be documented in the V&V plans, test plans, test specifications, or even in the individual test cases. The test bed includes:

- *Hardware*: At a minimum, this includes the hardware required to execute the software, and it may also include additional hardware that inputs information into or accepts outputs from the software being tested.
- *Instrumentation*: This may include oscilloscopes, or other hardware or software equipment needed to probe the software or its behaviors during test execution. For example, it may include instrumentation that monitors coverage, memory, disk space, or bandwidth utilization during test execution.
- *Simulators*: As discussed earlier, simulators are used to substitute for (simulate) missing or unavailable system components (hardware, software, human) during test execution. Simulators mimic the behavior of these missing components in order to provide needed inputs to, and accept output from, the software being tested.
- *Software tools*: This may include automation tools, stubs, and drivers. Other software tools used in testing include debuggers or other tools that allow step-by-step execution or breakpoints to be inserted to "watch" code execution and examine variable and memory values. This also includes the operating system and cohabitating software (including database software or other interfacing software applications) used to run and test the software.
- *Other support elements*: For example, facilities, manuals, and other documentation.

Stubs

A *stub* is a software source code module or component that can be used to minimally simulate the actions of a called (lower-level) source code module or component. Stubs are used to simulate units that have not yet been integrated during top-down testing. Stubs can also be used in unit testing. Some required stubs need more than the minimal return response. For example, if the calling source code module or component under test expects a calculation from the invoked stub, that calculation capability may need to be programmed into the stub, or the stub can simply set the returned variable(s). Table 22.11 shows an example of a very simple stub and its associated calling routine.

A stub can also be used when the software needs to talk to an external device in order to replace that device. For example, a stub could intercept the data that would normally be sent to a hardware device and automatically analyze that data against the expected values. This analysis can often be done more quickly and accurately by a stub than it can be manually, especially if the output is large or complex. For example, humans are not good at finding a single missing or incorrect data point in a long stream of data.

Drivers

A *driver* is a source code module or component that can be used to minimally simulate the actions of a calling (higher-level) source code module or component that has not yet been integrated, during bottom-up testing. Drivers can also be used in unit testing. The driver typically establishes the values that will be passed to the source code module or component under test before calling it. Table 22.12 shows an example of a very simple driver and its associated called routine.

Table 22.11 Stub—example.

Stub	Calling (Higher-Level) Source Code Module
Procedure calculate_tax (TAXES) Return TAXES begin; return $1.32; end calculate_tax;	Procedure sale_process begin; statement 1; statement 2; TAX = calculate_tax (TAXES); statement 3; end sale_process

Table 22.12 Driver—example.

Driver	Called (Lower-Level) Source Code Module
Procedure exec_drv begin; NUM1 = 15 NUM2 = 25 mean_value(NUM1, NUM2); end exec_drv;	function mean_value (X1, X2: float) return float begin return (X1 + X2) / 2.0; end mean_value

More complex test drivers can be used to mimic large amounts of data being entered manually. For example, a driver could mimic data entry streams coming from multiple keyboards simply by reading a predefined set of inputs from a data file and using them to emulate the keyboard entries.

Harnesses

At its simplest, a *test harness* is a set of test stubs and drivers used to automate unit testing. At its most sophisticated, a test harness, also called an *automated test framework*, can be used as a scaffold for software, tests, and test data. This harness can be used to automate the testing of a source code module or component by executing a set of test cases or scripts under a variety of conditions and collecting, monitoring, and even evaluating the test results. The two primary parts of a test harness are the test execution engine and the repository of test scripts.

A *test execution engine*, also called a *test executive, test manager*, or *test sequencer*, is a test automation tool that selects, loads, and executes a test specification from the repository of test scripts. The test execution engine then monitors the execution of that test specification and reports and stores the results (pass/fail/abort status, time stamps, duration of the test, and other information) for every step in the test sequence. The test execution engine may also have an interface that allows the testers to interact with the automated tests.

A *repository of test scripts* is a collection of test specifications, also called *test sequences*, which constitute the specific test steps required to execute the tests and their associated test data.

Maintaining the Test Environments

Sometimes, the time to set up the test environment is trivial. However, according to Black (2004), "A complex test environment can take a long time and involve a lot of effort to set up and configure." Whatever the initial investment, it must be protected by managing and controlling the test environment as it is being utilized by the testers to execute their tests. The key here is to be able to restore the test environment to a known state with minimal effort because, as tests are being executed, unexpected events may occur that corrupt the data or the software being tested or cause issues with some other part of the test bed. After the software is released into operations, the test bed may also need to be reproduced to test that the software after maintenance changes are made.

Test environment control policies and processes should be established and followed to make sure that the test bed is not interjecting false positives (issues that look like software problems but are actually test bed problems) or false negatives (allowing software problems that should have been identified to be missed) into the test execution processes. The policies and procedures should include mechanisms for:

- Controlling test bed security, including control of access to the test environment.
- Controlling, backing up, and restoring test data and databases

- Controlling, restoring, "freezing," or updating the software installed on the test bed so that the testers have a steady platform to test from.
- Validating the test bed setup, the individual elements of the test bed, and their interoperability
- Requesting, approving, and implementing changes to the test bed's hardware, software, instrumentation, tool set, or other elements; change procedures should include performing impact analysis and notifying impacted stakeholders of implemented changes.
- Training testers to use the test bed elements correctly and have access to user/operator manuals, help files, or other support documentation for test bed elements.
- Providing support for the test bed's hardware, software, instrumentation, tool set, or other elements, including service level agreements and escalation processes.
- Maintaining licenses for purchased test bed elements.
- Identifying the current versions, revisions, and calibrations for the test bed elements, including the software being tested.

9. TEST TOOLS

> Identify and use test utilities, diagnostics, automation, and test management tools. (Apply)
>
> **Body of Knowledge VI.B.9**

Many different tools can be used to make software testing more efficient, consistent, and repeatable. Testing tools also reduce the tedium of repetitive work, provide objective evaluation, and provide visibility into the testing processes, test results, and test sufficiency.

Like any other software product, testing tools should go through verification and validation (V&V) prior to use. Requirements for the V&V of supplier-provided testing tools should be specified as part of acquisition plans for those tools. This may include requirements for supplier V&V activities, as well as plans for acquirer-conducted V&V (joint reviews, alpha, beta, and acceptance testing). Both supplier-provided and in-house-developed software testing tools should go through an appropriate level of rigorous V&V for the required integrity level or risk of the software they will be testing. This evaluation helps make sure that a tool does not indicate that the software passes tests that it actually failed. It also helps eliminate the waste caused when software developers spend time debugging reported software problems, which were defects in the testing tools and not in the software.

Test Planning and Management Tools

Test planning and management tools can range from simple templates in a word processor to sophisticated test management tools. Test planning and management tools include:

- *Templates*: Templates for various test documents allow the tester to focus on the content rather than the format of those documents.
- *Test management database tools*: Tools that allow test cases and procedures to be entered into test databases and allow for automated test specification generation and automated tracking of test log information. These databases can automate the traceability links between the entered test cases and procedures and the associated requirements, design elements, source code modules, change requests, and problem reports.
- *Test planning tools*: Planning tools can include project estimation and planning tools used for the testing part of the project. These tools can aid in test estimation, creating test-related work breakdown structures, test scheduling, and budgeting.
- *Test monitoring and reporting tools*: Tools that keep track of the test status, control the testing effort, log test results, and generate testing metrics and test status reporting information.
- *Test matrix tools*: Used to link requirements to test strategies and methods for testing each requirement.

Test Design and Development Tools

Test design and development tools analyze requirements and design models, software source code, and graphical user interfaces to help:

- Design and develop test cases, for example:
  - *Complexity analyzers*: Identify paths through the source code and identify the necessary input into designing unit test cases to test each path.
  - *Decision trees*: Examine the probability that a thread is executed during normal operations. This information is used in designing operational profile tests.
  - *Static code analyzers*: Tool that can be used to identify improper control flows, uninitialized variables, inconsistencies in data declarations or usage, redundant code, unreachable code, and overly complex code. This information can be used to eliminate defects before those defects need to be found using more expensive testing techniques.
  - *Cross-referencing code analyzers*: When a variable is modified in one section of code, these analyzers provide the programmers and testers with all the other the software that uses that variable and therefore

might be impacted by the change. This information can be extremely helpful when analyzing the amount of regression testing needed after a change.

- Generate test input data or equivalence classes of test inputs
- Generate expected results through automated oracles
- Select combinations of possible factors to be used in testing (Copeland 2003)

Test data preparation tools include features that provide support to (Graham et al. 2008):

- Extract data fields or records from existing real-world files or databases, and modify those data to make them anonymous (not able to be identified with real people or other entities)
- Create, generate, edit, and manipulate data for use during testing
- Sort data or records into different sequences
- Generate new records populated with pseudo-random data or data based on specific criteria (e.g., the operational profile)
- Construct large sets of data or databases populated with similar records based on a template (e.g., for use in volume testing)

Test Execution Tools

There are many *test execution* tools on the market. These include:

- *Capture-and-playback tools*: Tools that capture the keystrokes entered when performing manual testing and allow the playback of those keystrokes to re-execute that exact testing sequence
- *Programmable, integrated test tool suites*: Tools that use scripting languages to create, execute, and capture or analyze the results of automated testing
- *Test harnesses, stubs, and drivers*: As described previously
- *Simulators*: For example, tools that induce various levels of environmental load into the software for load, volume, and stress testing
- *Dynamic analysis tools*:
  - *Coverage analyzers* are used to evaluate the percentage or amount of the software that is touched during the execution of a set of tests.
  - *Resource utilization analyzers* are used for monitoring disk space, memory, bandwidth utilization, and other resource utilization during test execution
  - *Timing tools* for measuring response time and other timing parameters

- *Communication analyzers* that allow the tester to view raw data moving across the network or other communication channels
- *Debuggers* used to step through code or set breakpoints and to examine memory values during structural testing

- *Hacking tools and penetration testing tools*: Used for testing security
- *File comparison tools, screen capture and comparison tools, and other tools*: Used to capture and analyze the actual results of test execution against expected results
- *Test logging tools*: Used to log test activities and results

Test Support Tools

Support tools from other disciplines are also used in the testing effort, including:

- *Project management tools*: Used for test planning and management
- *Requirements tools and other software development tools*: Discussed in Chapter 14
- *Software configuration management (SCM) tools*: Discussed in more detail in Chapter 25
    - *SCM library tools*: Used to create and manage test libraries for controlling and tracking changes to test plans, test cases, test procedures, test specifications, test scripts, and test data. These libraries act as test repositories that can also facilitate the reuse of test products across software products or product lines.
    - *Problem reporting tools*: Used to record and track problems and other anomalies identified during testing.
    - *Configuration management status accounting tools*: Used to identify what changed (and what did not change) between a new software build and the build testers were previously testing.

Other Tools Used by Testers

There are also many simple tools used by testers to perform their work. These tools may or may not be part of the test bed discussed previously. These tools include:

- *Word processors*: Used for test plans, test reports, and other test documentation
- *Spreadsheets*: Used for test matrices, test metrics, to perform analysis calculations, to track test status, to log test results, and for other testing tasks
- *Checklists*: Lists of common defects and test heuristics so the testers remember to create tests for these items or features, or lists of attributes or other characteristics to map to tests to confirm coverage

- *Database software*: Used to create test databases as inputs into testing, to query the databases of the software being tested to analyze test results, to record and track test data, and create test metrics (e.g., a test case database or a problem-reporting database)
- *Stopwatches, oscilloscopes, or other monitoring or probe devices*: Used to analyze test results
- *Random number generators*: Used to create random samples as test inputs
- *Video cameras or other recording devices*: Used to record usability testing sessions

10. TEST DATA MANAGEMENT

> Ensure the integrity and security of test data through the use of configuration controls. (Apply)
>
> **Body of Knowledge VI.B.10**

Test Data Requirements

Once test cases, test procedures, and automated test scripts have been designed, the requirements for the specific test data items must be identified. *Test data items* include specific input data for test cases and may also include "operations-like" databases, standing data tables and global data (Graham et al. 2008).

Test data item requirements describe the data items and their characteristics and properties needed to execute the tests, including:

- The name of the test data item
- The type of the data item and, if the data item is a structure (data array, data record, or database with multiple data records), the type of each data element in that structure
- If the data item is a database, the number of records required in that database to perform the test (e.g., to execute some performance or worst-case testing, that database may need to be very small, or for other tests, it may need to be empty)
- The person or persons responsible for extracting, creating, generating, editing, and manipulating the data item or its elements
- Required values, or ranges of values, for each data item or data element contained in a data item structure, including any default values
- Units of measure for each data item or each data element contained in a data item structure, if necessary

- Requirements to sanitize any data items or data elements to make the data anonymous (not able to be identified with real people or other entities) and thus maintain the confidentiality of any data copied from any database used in operations

Testing should never directly use data items in the operational environment. If operational data are needed as the basis for test data items, that data should always be copied from the operational environment into the test environment to maintain the integrity and confidentiality of the operational data items.

Test Data Management

Test data management involves managing and controlling the test data after they are initially created. The most important methodology in test data management is utilizing a *central repository for test data items*. This repository allows test data items to be reusable so that:

- The same tests can be executed multiple times, using the same test data items with the same values each time
- Tests at different levels (unit, integration, system, acceptance) can use the same test data items, so time and effort are not wasted duplicating the same or similar data items at each level
- Test data items can be quickly refreshed if they become overwritten, corrupted, or otherwise damaged as tests are executed, rather than wasting the energy needed to recreate them

This can be accomplished through identifying test data items as configuration items and placing them into a test library (central repository for test data items) in an SCM library tool, or by creating the central repository through other means, as appropriate to the project. As test data items are needed over time for new levels of testing, they can be reused if they are already in the central repository. If they are not in the repository, they can be created and added to the repository for future reuse. The central repository also acts as a single source of content for:

- Multiple testers to find and use already existing test data items (rather than reinventing the wheel)
- Access control and other security measures
- Backing up or archiving of test data items

Test data management also includes mechanisms for:

- Defining when, and for how long each test data item is needed
- Requesting changes to test data items over time, analyzing their impacts, reviewing and approving of those changes by the appropriate authority, and implementing and verifying approved changes
- Identifying private, proprietary, or otherwise sensitive test data items and implementing security mechanisms to protect those data from

unauthorized access, unauthorized disclosure, corruption, destruction or other damage, and other breaches

- Identifying test data items that have become defective or corrupted over time, and the implementation of corrective measures, including root cause analysis and correction or refreshing of those test data items
- Determining when and under what conditions each test data item needs to be reinitialized, reset, or returned to default values
- Determining when and how test data will be archived or disposed of when testing is completed

Chapter 23

Reviews and Inspections

> Use desk checks, peer reviews, walk-throughs, inspections, etc. to identify defects. (Apply)
>
> **Body of Knowledge VI.C**

The IEEE/ISO/IEC *Systems and Software Engineering—Vocabulary* (IEEE/ISO/IEC 2017) defines a *review* as "a process or meeting during which a work product, or set of work products, is presented to project personnel, managers, users, customers, or other interested parties for comment or approval." Reviews may be conducted as part of software development, maintenance and acquisition activities.

The IEEE *Standard for Software Reviews and Audits* (IEEE 2008) defines a *management review* as "A systematic evaluation of a software product or process performed by or on behalf of management that monitors progress, determines the status of plans and schedules, confirms requirements and their system allocation, or evaluates the effectiveness of management approaches used to achieve fitness for purpose." Management reviews are used to:

- Evaluate the effectiveness and efficiency of management approaches, strategies, and plans
- Monitor project, program, and operational status and progress against plans and objectives
- Evaluate variance from plans and objectives
- Track issues, anomalies, and risks
- Provide management with visibility into the status of projects, programs, or operations
- Provide management with visibility into product and process quality
- Make decisions about corrective or preventive action
- Implement governance

Typically, management reviews evaluate the results of verification and validation (V&V) activities but are not used directly for V&V purposes. Examples of project- and program-related management reviews are discussed in Chapter 17.

However, three other major types of reviews are used directly for V&V purposes and will be discussed in this chapter, including:

- Technical reviews
- Pair programming
- Peer reviews:
  - Desk checks
  - Walk-throughs
  - Inspections

V&V Review Objectives

The primary objective of V&V reviews is to identify and remove defects in software work products as early in the software life cycle as possible. It can be very difficult for authors to find defects in their own work product. Most software practitioners have experienced situations where they hunt and hunt for that elusive defect and just cannot find it. When they ask someone else to help, the other person looks at the work product and spots the defect almost instantly. That is the power of reviews.

Another objective of V&V reviews is to provide confidence that work products meet requirements and stakeholders' needs. Reviews can be used to validate that the requirements appropriately capture the stakeholders' needs. Reviews can also be used to verify that all of the functional requirements and quality attributes have been adequately implemented in the design, code, and other software products and are being effectively evaluated by the tests.

V&V reviews are also used to check the work product for compliance with standards. For example, the design can be peer-reviewed to verify that it matches modeling standards and notations, or a source code module can be reviewed to verify that it complies with coding standards and naming conventions.

Finally, V&V reviews can be used to identify areas for improvement. This does not mean "style" issues—if it is a matter of style, the author wins. However, reviewers can identify opportunities to increase the quality of the software. For example:

- When reviewing a source code module, a reviewer might identify a more efficient sorting routine or a method of removing redundant code or even identify areas where existing code can be reused.
- During a review, a tester might identify issues with the feasibility or testability of a requirement or design.
- Reviewers might identify maintainability issues. For example, in a code review, inadequate comments, hard-coded variable values, or confusing code indentation might be identified as areas for improvement.

What to Review

Every work product created during software development can be reviewed as part of the V&V activities. However, not every work product should be. Before a review is held, practitioners should ask the question, "Will it cost more to perform this review than the benefit of holding it is worth?" Reviews, like any other process activity, should always be value-added or they should not be conducted.

Every work product delivered to a customer, user, or other external stakeholder should be considered a candidate for review. Work product deliverables include responses to requests for proposals, contracts, user manuals, requirements specifications, and, of course, the software and its subcomponents. Every deliverable is an opportunity to make a positive or negative quality impression on the organization's external stakeholders. In addition, any work products that are inputs into or have major influence on the creation of these deliverables are also candidates for review. For example, architecture and component designs, interface specifications, or test cases may never get directly delivered. Still, defects in those work products can have a significant impact on the quality of the delivered software.

So what does not get reviewed? Many work products created in the process of developing, acquiring, and maintaining software may not be candidates for reviews. For example, many quality records, such as meeting minutes, status/progress reports, test logs, and defect reports typically are not candidates for review.

Types of Reviews—Technical Reviews

The IEEE *Standard for Software Reviews and Audits* (IEEE 2008) states that the "purpose of a technical review is to evaluate a software product by a team of qualified personnel to determine its suitability for its intended use and identify discrepancies from specifications and standards." Technical reviews can also be used to evaluate alternatives to the technical solution, to provide engineering analysis, and to make engineering trade-off decisions. One of the primary differences between technical reviews in general, and peer reviews specifically, is that managers can be active participants in technical reviews while only the author's peers participate in peer reviews.

The IEEE *Standard for Software Reviews and Audits* (IEEE 2008) defines the following steps in the technical review process:

- *Management preparation*: Project management makes certain that the reviews are planned, scheduled, funded, and staffed by appropriately trained and available individuals.
- *Planning*: The review leader plans the review, which includes identifying the members of the review team, assigning roles and responsibilities, handling scheduling and logistics for the review meeting, distributing the review package, and setting the review schedule.
- *Overview*: Overviews of the review process or the software product being reviewed are presented to the review team as needed.
- *Preparation*: Each review team member reviews the work product and other inputs included in the review package before the review meeting.

These reviewers forward their comments and identified problems to the review leader, who classifies them and forwards them on to the author for disposition. The review leader also verifies that the reviewers are prepared for the meeting and takes the appropriate corrective action if not.

- *Examination*: Each software work product included in the review is evaluated to:
  - Verify the work product's completeness, suitability for its intended use, and readiness to transition to the next activity
  - Verify that any changes to the work product are appropriately implemented, including dispositions for problems identified during the preparation
  - Confirm compliance with applicable regulations, standards, specifications, and plans
  - Identify additional defects and provide other engineering inputs on the reviewed work product and other associated work products

  The results of these evaluations are documented, including a list of action items. The team leader determines if the number or criticality of any identified defects or nonconformances warrant scheduling an additional review to evaluate their resolutions.
- *Rework/follow-up*: After the evaluation, the review leader confirms that all assigned action items and identified defects are appropriately resolved.

Types of Reviews—Pair Programming

Pair programming involves two people, one of whom is constantly reviewing what the other is developing. As illustrated in Figure 23.1, pair programming offers a constant exchange of ideas about how the product is being created or

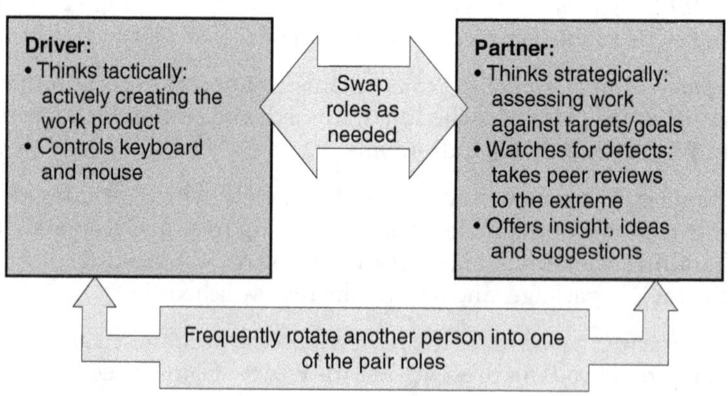

Figure 23.1 Pair programming.

modified. If one person hits a mental block, the other person can jump in and try something. Pair programming is a technique that helps people keep up the steady pace that agile methods advocate through being fully engaged in the work and keeping one another "honest" (consistent about adhering to the project's practices, standards, and so on). Martin (2003) notes that pair programming interactions can be "intense." He advocates that "over the course of the iteration, every member of the team should have worked with every other member of the team and they should have worked on just about everything that was going on in the iteration." By doing this, everyone on the team comes to understand and appreciate everyone else's work, and develops a deep understanding of the software as a whole. Pair programming is one way to avoid extraneous documentation, since everyone shares in the requirements and design knowledge of the project.

Types of Reviews—Peer Reviews

A *peer review* is a type of V&V review where one or more of the author's peers evaluate a work product to identify defects, obtain confidence that the product meets its requirements and intended use, and identify opportunities to improve that work product. The *author* of a work product is the person who either originally produced that work product or the person who is currently responsible for maintaining that work product.

Benefits of Peer Reviews

The benefits of peer reviews, especially formal inspections, are well documented in the industry. For example, more defects are typically found using peer reviews than other V&V methods. Jones (2008) reports, "The defect removal efficiency levels of formal inspections, which can top 85 percent, are about twice those of any form of testing." Well-run inspections with highly experienced inspectors can obtain 90 percent defect removal effectiveness (Wiegers 2002). "Inspections can be expected to reduce defects found in field use by one or two orders of magnitude" (Gilb 1993).

It typically takes much less time per defect to identify defects during peer reviews than it does using any of the testing techniques. For example, Kaplan reports that at IBM's Santa Teresa laboratory, it took an average of 3.5 labor hours to find a major defect using code inspection, while it took 15 to 25 hours to find a major defect during testing (Wiegers 2002). It also typically takes much less time to fix the defect because the defect is identified directly in the work product, eliminating the need for potentially time-consuming debugging activities. Peer reviews can be used early in the life cycle on work products, such as requirements and design specifications, to eliminate defects before those defects propagate into other work products and become more expensive to correct.

Peer reviews also provide opportunities for cross-training. Less-experienced practitioners can benefit from seeing what a high-quality work product looks like when they help peer review the work of more experienced practitioners. More-experienced practitioners can provide engineering analysis and improvement suggestions, which help transition knowledge when they review the work of less-experienced practitioners. Peer reviews also help spread product, project, and technical knowledge around the organization. For example, after a peer review, more than one practitioner is familiar with the reviewed work product and can potentially support it if its author is unavailable. Peer reviews of requirements and design documents aid in communications and help promote a common understanding that is beneficial in future development activities. For example, these peer reviews can help identify and clarify assumptions or ambiguities in the reviewed work products.

Peer reviews can help establish shared workmanship standards and expectations. They can build a synergistic mindset as the work products transition from individual to team ownership with the peer review.

Finally, peer reviews provide data that aid the team in assessing the quality and reliability of the work products. Peer review data can also be used to drive future defect prevention and process improvement efforts.

Selecting Peer Reviewers

Peer reviewers are selected based on the type and nature of the work product being reviewed. Reviewers should be peers of the author and possess enough technical or domain knowledge to allow for a thorough evaluation of the work product. For a peer review to be effective, the reviewers must also be available to spend the time and energy necessary to conduct the review. There are various reasons why individuals might be unavailable to participate, including time constraints, the need to focus on higher-priority tasks, or the belief that they have inadequate domain or technical knowledge. If an "unavailable" individual is considered essential to the success of the peer review, these issues need to be dealt with to the satisfaction of that individual, prior to assigning them to participate in the peer review.

Peers are the people at the same level of authority in the organization as the work product's author. The general rule for peer reviews is that managers do not participate in peer reviews—they are not peers. If managers participate in the reviews, then:

- Authors are more reluctant to have their work products peer-reviewed because the defects found might reflect badly on the author in the eyes of their management
- Reviewers are more reluctant to report defects in front of managers and make their colleagues look bad, knowing that it might be their turn next

However, it is not quite that simple. It depends a lot on the culture of the organization. In organizations where first-line managers are working managers (active team members) or where participative management styles prevail, having the immediate supervisor in the peer review may be a real benefit. They may possess knowledge and skills that are assets to the peer review team and the author. Individuals from one level above or below the author at the organizational level, as illustrated in Figure 23.2, can be considered as reviewers, but only if the author is comfortable with that selection. In other words—ask the author. If the author is comfortable having their supervisor in the review, other reviewers will probably also be comfortable.

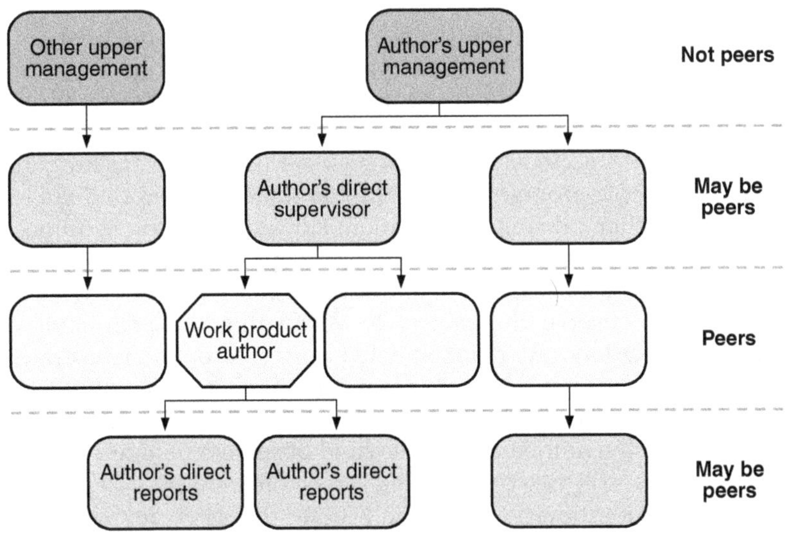

Figure 23.2 Selecting peer reviewers.

When talking about "peers" in a peer review, it does not mean "clones." For example, in a requirements peer review, where the author is a systems analyst, the peer review team should not be limited to just other systems analysts. Having reviewers with different perspectives increases the diversity of the peer review team. This diversity brings different perspectives that can increase the probability of finding additional defects and different kinds of defects. The reviewers should be selected to maximize this benefit by choosing diverse participants, including:

- The author of the work product that is the predecessor to the work product being reviewed. "The predecessor perspective is important because one . . . goal is to verify that the work product satisfies its specification" (Wiegers 2002).
- The authors of dependent (successor) work products and authors of interfacing work products, who have a strong vested interest in the quality of the work product. For example, consider inviting a designer and a tester to the requirements peer review. Not only will they help find defects, but they are excellent candidates for looking at issues such as understandability, feasibility, and testability.
- Specialists may also be called on when special expertise can add to the effectiveness of the peer review (e.g., security experts, safety experts, and human factors experts).

This diversity does not mean that others with the same job as the author are excluded from the review. They know the workmanship standards and best practices for the specific product. They are also most familiar with the common defects in that type of work product. This knowledge also adds diversity to the team.

For most peer reviews, the peers are typically other members of the same project team. However, for small projects or projects with only a single person with a specialized skill set, it may be necessary to ask people from other projects to participate in the peer reviews.

As mentioned previously, peer reviews are wonderful training grounds for teaching new people programming techniques, product and domain knowledge, processes, and other valuable information. However, too many trainees can also impact the efficiency and effectiveness of the peer review. There may also be several reasons why individuals may want to observe peer reviews. For example, an auditor may observe the review to verify that the peer review is being conducted in accordance with documented processes, or an inspection moderator trainee may be observing an experienced moderator in action. Observers can be distracting to team members, even if they are not participating in the actual review. Limiting the number of trainees and observers in a peer review meeting to no more than one or two per meeting is recommended.

Management's primary responsibility to the peer review process is to champion that process by emphasizing its value and maintaining the organization's commitment to the process. Management must understand that while the investment in peer reviews is required early in the project, their actual benefits (their return on that investment) will not be seen until the later phases of the life cycle or after the product is released. Management can champion peer reviews by:

- Providing adequate scheduling and resources for peer reviews
- Incorporating peer reviews into the project or program plans
- Making certain that training in the review techniques being used is planned and completed for all software practitioners participating in peer reviews
- Advocating and supporting the usefulness of peer reviews through communications and encouragement
- Appropriately using data and metrics from the peer reviews
- Being brave enough to stay away from the actual peer reviews unless specifically invited to participate

Informal Versus Formal Peer Reviews

Peer reviews can vary greatly in their level of formality, as illustrated in Figure 23.3. At the most informal end of the peer review spectrum, a software practitioner can ask a colleague, "Please take a look at this for me." These types of informal peer reviews are performed all the time. It is good practice to get a second pair of eyes on a work product when the practitioner has problems, needs a second opinion, or wants to verify their work. These informal reviews are done ad hoc with no formal process, preparation, quality records, or metrics. Defects are usually reported either verbally or as redlined mark-ups on a draft copy of the work product. Any rework from these informal peer reviews is up to the author's discretion.

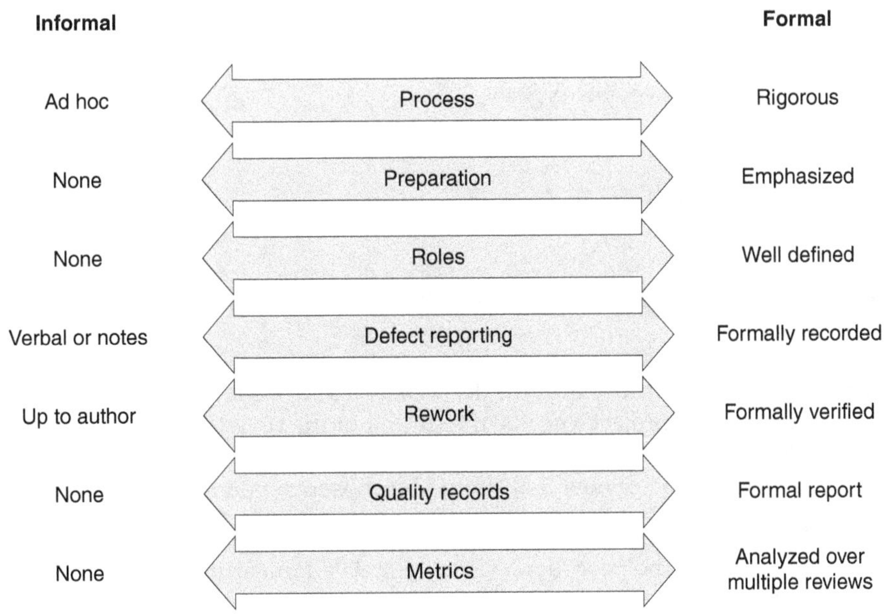

Figure 23.3 Informal versus formal peer reviews.

On the opposite end of the spectrum is the formal peer review. In formal peer reviews:

- A rigorous process is documented, followed, and continuously improved with feedback from peer reviews as the peer reviews are conducted
- Preparation before the peer review meeting is emphasized
- Peer review participants have well-defined roles and responsibilities to fulfill during the review
- Defects are formally recorded, and that list of defects, and a formal peer review report, become the quality records for the review
- The author is responsible for the rework required to correct the reported defects, and that rework is formally verified by either re-reviewing the work product or through checking done by another member of the peer review team (e.g., the inspection moderator)
- Metrics are collected and used as part of the peer review process and are also used to analyze multiple reviews over time as a mechanism for process improvement and defect prevention

Types of Peer Reviews

There are different types of peer reviews called by many different names in the software industry. Peer reviews go by names such as *team reviews, technical reviews, walk-throughs, inspections, pair reviews, pass-arounds, ad hoc reviews, desk checks,* and others. However, the author has found that most of these can be classified into one of three major peer review types:

- Desk checks
- Walk-throughs
- Inspections

Table 23.1 compares and contrasts the differences between these three major types of peer reviews

Figure 23.4 illustrates that while inspections are always very formal peer reviews, the level of formality in desk checks and walk-throughs varies greatly depending on the project's needs, the timing of the reviews, and the participants involved.

The type of peer review that should be chosen depends on several factors:

- First, inspections are focused purely on defect detection. If one of the goals of the peer review is to provide engineering analysis and improvement suggestions (e.g., reducing unnecessary complexity, suggesting alternative approaches, identifying poor methods or areas that can be made more robust), a desk check or walk-through should be used.
- The maturity of the work product being reviewed should also be considered when selecting the peer review type. Desk checks or walk-throughs can be performed very early in the life of the work product being reviewed (e.g., as soon as the source code module has a clean compile or a document has been spell-checked). In fact, walk-throughs can be used just to bounce around very early concepts before there even is a work product. However, inspections are only performed when the author thinks the work product is done and ready to transition into the next phase or activity in development.
- Staff availability and location can also be a factor. If the peer review team is geographically dispersed, it can be much easier to perform desk checks than walk-throughs or inspections. Of course, modern online meeting tools have minimized this distinction.
- Economic factors such as cost, schedule, and effort should also be considered. Team reviews tend to cost more and take longer than individuals reviewing separately. More-formal peer reviews also tend to cost more and take longer. However, the trade-off is the effectiveness of the reviews. Team peer reviews take advantage of team synergy to find more defects. More-formal reviews are typically more thorough, and therefore more effective at identifying defects.
- Risk is the final factor to consider when choosing which type of peer review to hold. Risk-based peer reviews are discussed in the next section.

Table 23.1 Compare and contrast major peer review types.

Desk Checks	Walk-Throughs	Inspections
Individual peer(s)	Team of peers	Team of peers
Evaluate product	Evaluate product	Evaluate product
First draft or clean code compile	First draft or clean code compile	Product ready for transition
Can range from informal to formal	Can range from informal to formal	Always formal
Focus on defect detection and engineering analysis	Focus on defect detection and engineering analysis	Focus only on defect detection
Find errors as early as possible	Find errors as early as possible	Find errors before transition
No preparation	Preparation less emphasized	Preparation emphasized
No one presents the work product	Author presents the work product	Separate reader role presents the work product
Checklists not required	Checklists not required	Checklists required
Might collect metrics	Might collect metrics	Emphasizes use of metrics

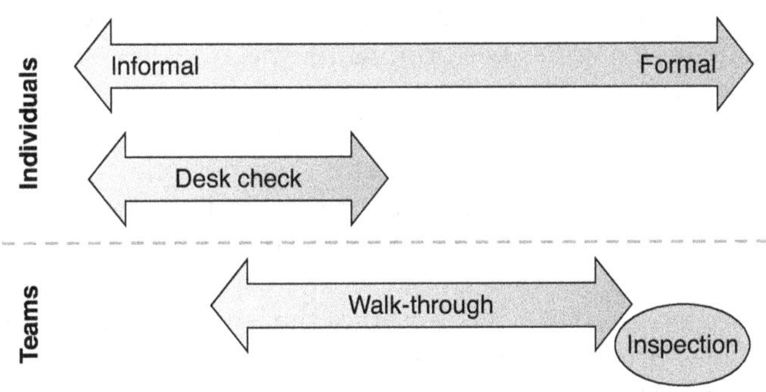

Figure 23.4 Types of peer reviews.

Risk-Based Peer Reviews

Risk-based peer reviews are simply a type of risk-based V&V activity, as discussed in Chapter 21. Risk analysis should be performed on each work product being considered for peer review to determine the probability that yet-undiscovered, relevant defects exist in that work product and the potential impact of those defects if they escape the peer review. If there is both a low probability and a low impact of undetected defects, then an informal desk check by a single peer may be appropriate, as illustrated in Figure 23.5. As the probability and impact increase, the type of peer review that is most appropriate moves to a more-formal desk check, to informal walk-through, to more-formal walk-through, and then to formal

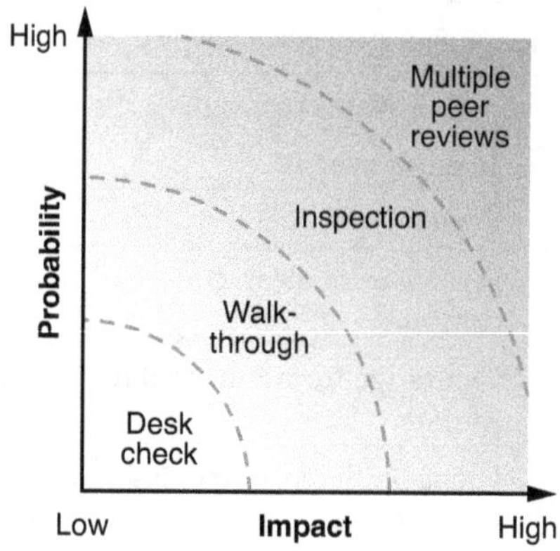

Figure 23.5 Risk-based selection of peer review types.

inspection. Having multiple peer reviews may be appropriate for a very high-risk work product. For example, for high-risk work products like requirements, each set or section of requirements may be desk-checked or have a walk-through as it is being developed. Then the entire requirements specification may be inspected late in its development, just before it is transitioned to start design.

The number of people performing the peer review may also vary based on risk. For very low-risk work products, having a single individual perform a desk check may be sufficient. For slightly higher-risk work products, it may be appropriate to have multiple people perform the desk check. For products worthy of an investment in an inspection, less risky work products may be assigned a smaller inspection team of two to four people, and higher-risk products may be assigned an inspection team of five to seven people.

Risk-based peer reviews also embrace the law of diminishing returns. When a software work product is first peer-reviewed, many of the existing defects in the product are discovered with less effort. As additional peer reviews are held, the probability of discovering any additional defects decreases. At some point, the return on investment to discover those last few defects may be outweighed by the cost of additional peer reviews.

Soft Skills for Peer Reviews

One of the challenges an author faces during a peer review is to remain "egoless" during the process (especially in the meetings) and not become defensive or argumentative. An egoless approach enables an author to step back and accept improvement suggestions and acknowledge identified defects without viewing them as personal attacks on their professional abilities or their value as an individual. One way of doing this is for the author to think of the work product as transitioning from their personal responsibility to becoming the team's product when it is submitted for review. The other peer reviewers should also take an egoless approach and not try to prove how intelligent they are by putting down the author's work, the comments of other reviewers, or the work product itself.

During a peer review, everyone should treat the work product, the other review members, and especially the author with respect.

Throughout the peer review process, the focus should be on the product, not people. One way to accomplish this is to use "I" and "it" messages, avoiding the word "you" during discussions. For example, a reviewer might say, "I did not understand this logic" or "The logic in this section of the work product has a defect" rather than saying, "You made a mistake in this logic." Another method is to word comments as neutral facts, positive critiques, or questions—not criticisms.

During the peer review meetings, team members should not interrupt each other. They should listen actively to what other team members say and build on the inputs from other reviewers to find more defects. The author should not be forced to acknowledge every defect during a peer review meeting. This requirement can beat the author down. The team should assume that the author's silence indicates their agreement that the defect exists.

Types of Peer Reviews—Desk Checks

A *desk check* is a process where one or more peers of a work product's author review that work product individually. Desk checks can be done to detect defects in the work product or to provide engineering analysis. The formality used during the desk check process can vary. Desk checks can be the most informal of the peer review processes, or more formal peer review techniques can be applied. A desk check can be a complete peer review process in and of itself or be used as part of the preparation step for a walk-through or inspection. The effectiveness of the desk check process is highly dependent on the skills of the individual reviewers and the amount of care and effort they invest in the review.

An informal desk check begins when the author of a work product asks one or more of their peers to evaluate that work product. Those peer reviewers then desk check the work product independently and feed back their comments to the author either verbally or by providing a copy of the work product that has been annotated (redlined) with their comments. This informal review is typically done on the first draft of a document or after a clean compilation of the source code module. However, desk checks can be performed at any time during the life of the work product.

Desk checks can also be done more formally. For example, when a desk check is done as part of a review and approval cycle, a defined process may be followed with specific assigned roles (software quality engineer, security specialist, or tester) with formally recorded defects. These more-formal desk checks may result in quality records being kept, including formal peer review defect logs and formal written reports. Formal desk checks can even involve reviewer follow-up by re-reviewing the updated work product to verify that defects were appropriately corrected and that engineering suggestions were incorporated.

During a desk check, the reviewers should not try to find everything in a single pass through the work product. This method defuses their focus, and fewer important defects and issues will be identified. Instead, each reviewer should make a first pass through the work product to get a general overview and understanding without looking for defects. This pass provides a context for further, more detailed review. Each reviewer then makes a second, detailed pass through the work product, documenting any identified defects. Each reviewer should then select one item from the common defect checklist, or one area of focus and make a third pass through the work product, concentrating on reviewing just that one item or area. If there is time, each reviewer can select another checklist item or focus area make a fourth pass, and so on. For example, in a requirements specification review, a reviewer could concentrate on making sure that all the requirements are finite in one pass and evaluate the feasibility of each requirement in another pass. Making multiple passes through the work product is not considered rework (repeating work already done) because each pass looks at the work product from a different focus or emphasis.

One of the desk check techniques that reviewers can use on their second, general pass through a work product is mental execution. Using the *mental execution technique,* the reviewer follows the flow of each path or thread through the work product instead of following the line-by-line order in which it is written. For example, in a design review, the reviewer might follow each logical control flow through the design. Reviewing a set of installation instructions, the reviewer might follow the flow of installation steps. A code review example might be selecting a set of input data and following the data flow through the source code module.

In the *test case review technique,* the reviewer designs a set of test cases and uses those test cases to review the work product. This technique is closely associated with the mental execution technique, since the reviewer usually does not have executable software, so they have to mentally execute their test cases. Trying to write test cases can help identify ambiguities in the work product and issues with the product's testability. By shifting from a "how does the product work" mentality to the "how can I break it" mentality of a tester, the reviewer may also identify areas that are "missing" from the work product (e.g., requirements that have not been addressed, process alternatives or exceptions that are not implemented, missing error-handling code, or areas in the code where resources are not appropriately released). The test case review technique can be used for many different types of work products. For example, the reviewer can write test cases to a requirements specification, the architecture or component design element, source code module, installation instruction, the user manual, and even test cases themselves.

One of the real advantages of having a second person look at a work product is that person can see things that the author did not consider. However, it is much harder to review what is missing from the work product than what is in the work product. Checklists and brainstorming are two techniques that can aid in identifying things missing from work products. *Checklists* remind the reviewer to look for specific item, based on historic common problems in similar work products. If checklists do not exist, the reviewer(s) can brainstorm a list of things that might be missing based on the type of work product being reviewed. For example, for a requirements specification, this list might include items such as:

- Missing requirements
- Assumptions that are not documented
- Users or other stakeholders that are not being considered

- Failure modes or error conditions that are not being handled
- Special circumstances, alternatives, or exceptions that exist but are not specified

The goal of any peer review is to find as many important defects as possible within the existing time and resource constraints. It is typically easier to find minor, insignificant defects such as typos and grammar problems than to find the major problems in the work product. If the reviewer starts finding many little defects, it is easy to get distracted by them. The reviewer should focus on finding important defects. If they notice minor defects or typos, they should document them but then return their focus to looking for the "big stuff." If the reviewers cannot resist, they should make one pass through the work product to look for the small stuff and be done with it. This distraction is why many organizations insist that all work products be spell- or grammar-checked or run through a code analysis tool (e.g., a compiler) to make sure that they are as clean as possible so reviewers can avoid wasting time finding defects that could be more efficiently found with a tool.

It is the computer age, and reviewers can take advantage of the tools computers provide. One of the tools that can be used in a desk check is the search function. If the reviewer can get an electronic copy of the work product, they can search for keywords. For example, words such as "all," "usually," "sometimes," or "every" in a requirements specification may indicate areas that need further investigation for finiteness. After a defect is identified, the search function can also help identify other occurrences of that same defect. For example, when this chapter was being written, one of the reviewers pointed out that the term "recorder" was being used in some places and the term "scribe" in others to refer to the same inspection role. Since this might add confusion, the search function was used to find all occurrences of the word "scribe" and changed them to "recorder." The moral here is *do not search manually for things the computer can find much more quickly and accurately*.

During desk checking, the work product should be compared against its predecessor for completeness and consistency. A *work product's predecessor* is the previous work product used as the basis for the creation of the work product being reviewed. Table 23.2 shows examples of work products and their predecessors.

Types of Peer Reviews—Walk-Throughs

A *walk-through* is a process where one or more peers of a work product's author meet with that author to review the work product as a team. A walk-through can be done to detect defects in the work product or to provide engineering analysis. Preparation before the walk-through meeting is less emphasized than it is in inspections. The effectiveness of the walk-through process depends on the skills of the individual reviewers, the amount of care and effort they invest in the review, and the synergy of the review team.

The formality used during the walk-through process can vary. An example of a very informal walk-through might be an author holding an impromptu "whiteboard" walk-through of an algorithm or other design element. In an informal walk-through, there may be little or no preparation.

Table 23.2 Work product predecessor—examples.

Work Product Being Peer-Reviewed	Predecessor Examples
Software requirements specification	System requirements specification, stakeholder requirements (user stories, use cases, stakeholder requirements specification), business requirements document, or marketing specification
Architectural (high-level) design	Software requirements specification
Component (detailed or low-level) design	Architectural (high-level) design
New source code module	Component (detailed or low-level) design
Modified source code module	Defect report, enhancement request, or modified component (detailed or low-level) design
Unit test cases	Source code module or component (detailed or low-level) design
System test cases	Software requirements specification
Software quality assurance (SQA) plan	Software quality assurance (SQA) standard processes, or required SQA plan template
Work instructions	Standard process documentation
Response to request for proposal	Request for proposal

As illustrated in Figure 23.6, more-formal peer review techniques can also be applied to walk-throughs. In a more formal walk-through, preparation is done before the team meeting, typically through the use of desk checking. Preparation is usually left to the discretion of the individual reviewer and may range from little or no preparation to an in-depth study of the work product under review. During the walk-through meeting, the author presents the work product one section at a time and explains each section to the reviewers. The reviewers ask questions, make suggestions (engineering analysis), or report defects found. The author answers the questions about the work product. The recorder records the discussion and any suggestions, open issues, or defects identified. After the walk-through meeting, the recorder produces the minutes from the meeting, and the author makes any required changes to the work product to incorporate suggestions, resolve issues, and correct defects.

While walk-throughs and inspections are team-oriented peer review processes, there are significant differences between the two, as illustrated in Table 23.1. However, many of the tools and techniques discussed in the inspection section can also be applied to walk-throughs depending on the amount of formality selected for the walk-through.

Types of Peer Reviews—Inspections

An *inspection* is a formal method of peer review where a team of peers, including the author, performs detailed preparation and then meets to examine a work product. The work product is inspected when the author thinks it is complete and ready for the transition to the next phase or activity. The only focus of an

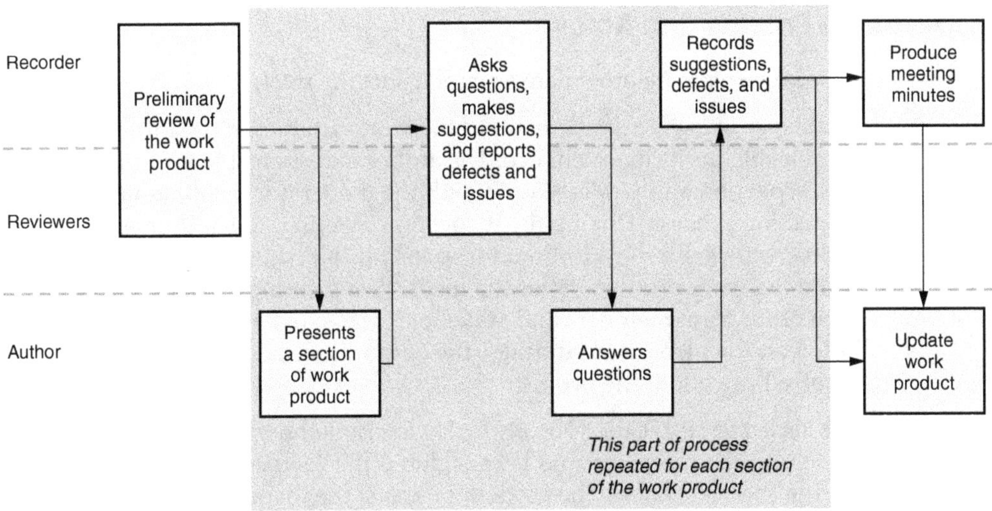

Figure 23.6 Formal walk-through process—example.

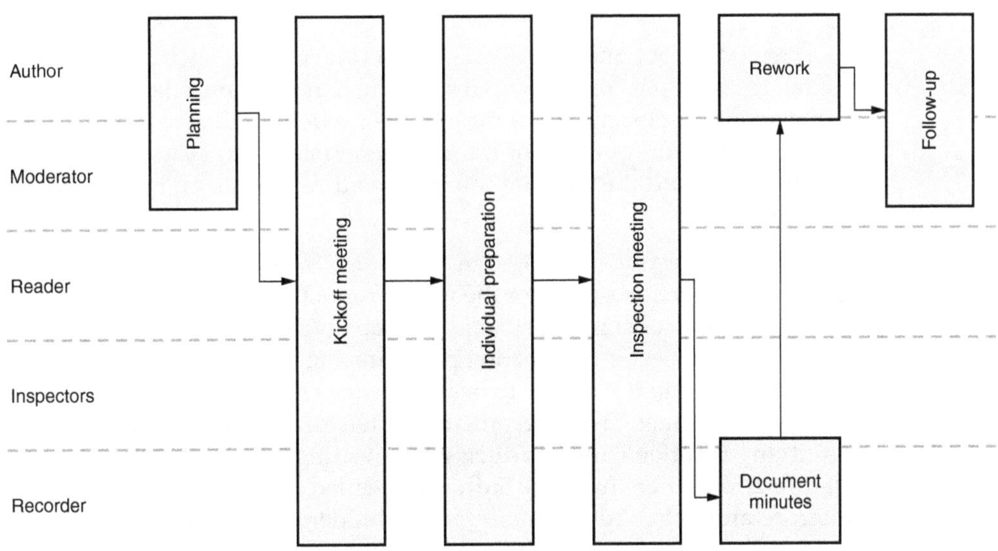

Figure 23.7 Inspection process steps.

inspection is on defect identification. Individual preparation using checklists and assigned roles is emphasized. Metrics are collected and used to determine compliance with the entry criteria for the inspection meeting, as well as for input into product or process improvement efforts. The inspection process consists of several distinct activities, as illustrated in Figure 23.7.

Inspection—Process and Roles

In an inspection, team members are assigned specific roles:

- The *author* is the original creator of the work product or the individual responsible for its maintenance. The author is responsible for initiating the inspection and works closely with the moderator throughout the inspection process. During the inspection meeting, the author acts as an inspector with the additional responsibility of answering questions about the work product. After the inspection meeting, the author works with other team members and stakeholders to close all open issues. Based on the inspection findings, the author is responsible for all required rework to the product.
- The *moderator* also called the *inspection leader,* is the coordinator and leader of the inspection, and is considered the "owner" of the process during the inspection. The moderator keeps the inspection team focused on the product and makes certain that the inspection remains "egoless." The moderator must possess strong facilitation, coordination, and meeting management skills. The moderator should receive training on how to perform their duties and on the details of the inspection process. The moderator is responsible for making sure that inspection metrics are collected and recorded in the inspection metrics database.

  The moderator handles inspection meeting logistics such as making reservations for meeting rooms and handling any special arrangements, for example, if a projector is needed for displaying sections of the work product or if the recorder needs access to a computer to record the meeting information directly into an inspection database.
- During the inspection meeting, the *reader,* also called the *presenter,* describes one section or part of the work product at a time to the other inspection team members and then opens that section up for discussion. This section-by-section paraphrasing has the added benefit of allowing the author to hear someone else's interpretation of the work product. This interpretation often leads to the author identifying additional defects during the meeting when the reader's interpretation differs from the author's intended meaning. The reader's "interpretation often reveals ambiguities, hidden assumptions, inadequate documentation, or style problems that hamper communications or are outright errors" (Wiegers 2002). The reader must have good presentation skills, product technical familiarity, and strong organizational skills.
- The *recorder,* also called the *scribe,* acts as the team's official recordkeeper. This roll includes recording the preparation and meeting metrics and all defects and open issues identified during the inspection meeting. Since the recorder is also acting as an inspector at the meeting, this person must be able to "multitask" their time and have solid writing skills.

- All of the members of the inspection team, including the author, moderator, reader, and recorder, are inspectors. The primary responsibility of the *inspectors* is to identify defects in the work product being inspected. During the preparation phase, the inspectors use desk check techniques to review the work product. They identify and note defects, questions, and other comments. During the inspection meeting, the inspectors work as a synergistic team. They actively listen to the reader paraphrase each part of the work product to determine if their personal interpretation differs from the reader's. They ask questions, discuss issues, and report defects. Inspectors should continue searching for additional defects that were not logged prior to the meeting but discovered through the synergy of the team interactions.
- The *observer* role is optional in an inspection. The observer does not take an active role in the inspection but acts as a passive observer of the inspection activities. This observation may be done to learn about the work product or the inspection process. It may also be done as part of an audit or other evaluation.

Inspection—Planning Step

When the author completes a work product and determines that it is ready for inspection, the author initiates the inspection process by identifying a moderator. There is typically a group of trained moderators available to each project from which the author can choose. The moderator works with the author to plan the inspection. The detailed process for the planning step is illustrated in Figure 23.8.

During the planning step, the moderator verifies that the inspection entry criteria, as defined in the inspection process, are met before the inspection is officially started. The moderator then verifies that the work product is ready to be inspected. This verification may include doing a desk check of a sample of the work product to make sure that it is of appropriate quality to continue. Any fundamental deficiencies should be removed from the work product before the inspection continues.

Inspection meetings are limited to no more than two hours in length. If the work product is too large to be inspected in a two-hour meeting, the author and moderator partition the product into two or more subproducts and schedule separate inspections for each partition. The same inspection team should be used for each inspection, but a single kickoff meeting may be appropriate.

The moderator and author select appropriate individuals to participate in the inspection and assign roles. The inspection teams should be kept as small as possible while still meeting the inspection objectives, including having the diversity needed to find different types of defects. A general rule of thumb is to have no more than seven inspectors per meeting.

The author and moderator determine whether or not a *kickoff meeting* is necessary. A kickoff meeting should be held, prior to the preparation step, if the inspectors need information about the important features, assumptions, background, or context of the work product to effectively prepare for the inspection

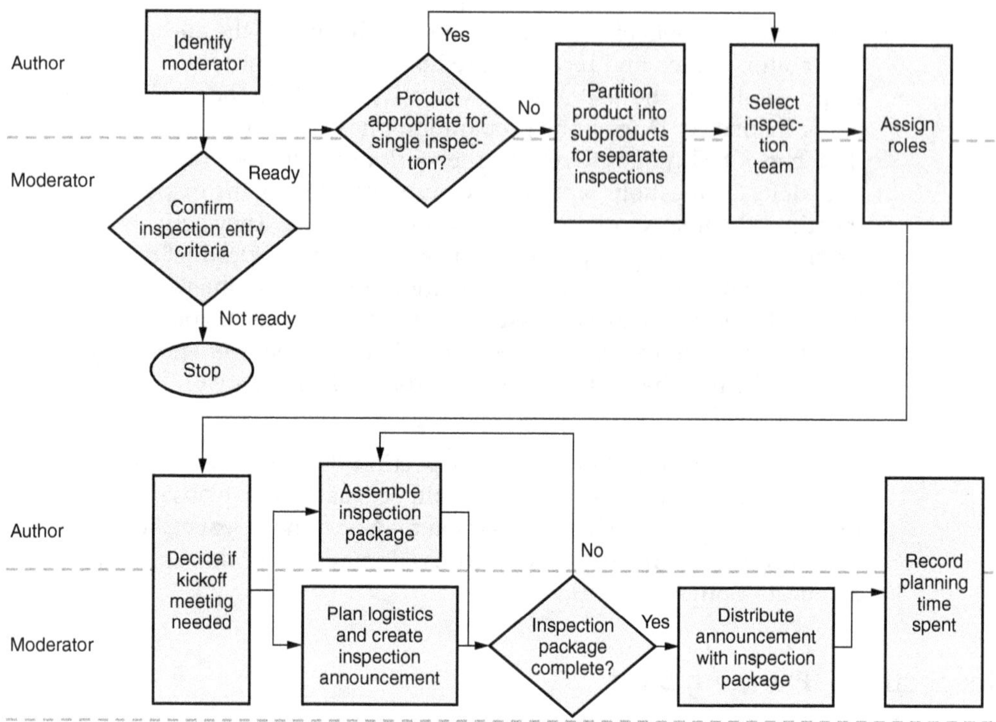

Figure 23.8 Inspection planning step process.

or if they need training on the inspection process. An alternative to holding a kickoff meeting is to include product overview information in the inspection package.

Decisions about when and where to hold the kickoff and inspection meetings should consider the participants' availability and location. No more than two inspection meetings should be scheduled per day for the same inspectors, with a long break between those two meetings. The inspection announcement should be delivered with the inspection package to all inspection team members far enough before the meeting to allow for adequate preparation (typically, a minimum of one day before the kickoff meeting and two to three days before the inspection meeting).

The primary deliverable from inspection planning is the inspection package, sent with the inspection announcement. This package, which can be sent in hard copy or electronically, includes all of the materials needed by the inspection team members to adequately prepare for the inspection, including:

- The work product being inspected
- Inspection forms (for recording defects, questions, and issues)
- Copies of or pointers to related work products, for example:
  - Predecessor work product(s)
  - Related standard(s)

    - Work product checklist(s)
    - Associated traceability matrix
- Product overview information, if a kickoff meeting is not being scheduled

Inspection—Kickoff Meeting Step

The primary objective of the kickoff meeting also called the *overview meeting,* is education. Some inspection team members may not be familiar with the work product being inspected or how it fits into the overall software product or customer requirements. First, the author gives a brief overview of the work product being inspected and answers any questions from the other team members. This product briefing section of the kickoff meeting is used to bring the team members up to speed on the scope, purpose, important functions, context, assumptions, and background of the work product being inspected.

In addition to a general evaluation of the work product from their own perspective, it may be valuable to ask individual inspectors to focus on special areas or characteristics of the work product. For example, if security is an important aspect of the work product, an inspector could be assigned to investigate that characteristic of the product. Assigning focus areas has the benefit of making sure that someone is covering that area while at the same time removing potential redundancy from the preparation process. The author is typically the best person to identify whether focus areas are appropriate for the work product under inspection. Inspectors may also be assigned to specific checklist items as areas of focus. This assignment is done to make certain that at least one inspector looks at each item on the checklist. These areas of focus and checklist items, as well as the reader and recorder roles, are assigned during the kickoff meeting.

Finally, suppose there are members of the inspection team who are not familiar with the inspection process. In this case, the kickoff meeting can also be used to provide remedial inspection process training to those individuals to make certain that all the team members understand what is expected from them. The moderator conducts an inspection process briefing to confirm that all of the inspection team members have the same understanding of the inspection process, roles, and assignments.

Inspection—Individual Preparation Step

The real work of finding defects and issues in the work product begins with the preparation step. Utilizing the information supplied in the inspection package, each inspector evaluates the work product independently using desk check techniques, with the intent of identifying and documenting as many defects, issues, and questions as possible. Checklists are used during preparation to make certain that important items are investigated. However, inspectors should look outside the checklists for other defects as well. If an inspector was assigned one or more specific areas of focus or checklist items, they should evaluate the work product from that point of view, in addition to doing their general inspection preparation. If an inspector has questions critical enough to impact that inspector's ability to prepare for the inspection, the author should be contacted. Otherwise, questions should be documented and brought to the inspection meeting.

Preparation is also essential for the reader, who must plan and lay out the schedule for conducting the meeting, including how the work product will be separated into individual, small sections for presentation and discussion. The reader must determine how to paraphrase each section of the work product, make notes to facilitate that paraphrasing during the meeting, and then practice their paraphrasing to accomplish it smoothly and quickly during the inspection meeting. Since this preparation is time-consuming, the reader is not assigned any additional focus areas.

A rule of thumb for preparation time is that the inspectors should spend from one to one and a half times as much time getting ready for the inspection meeting as the inspection meeting is expected to last. For example, if the meeting is planned to be two hours, the inspectors should spend two to three hours preparing for that meeting.

Inspection—Inspection Meeting Step

The goal of the inspection meeting is to find and classify as many important defects as possible in the time allotted. As illustrated in Figure 23.9, the inspection meeting starts with the moderator requesting the preparation time, counts of defects by severity, and counts of questions or issues from each inspector, and the recorder records these items. The moderator uses this information to determine whether or not all of the inspection team members are ready to proceed. If one or more inspectors are unprepared for the meeting, the moderator reschedules the meeting later, and the inspection process returns to the preparation step.

The moderator also uses this information to determine how to proceed with the inspection meeting. If there are a large number of major defects, questions, and issues to cover, the moderator may decide that only those major items will be brought up in the first round of discussion. If time permits, a second pass through the work product can be made to discuss minor defects. Remember, the goal is to find as many important defects as possible in the time allotted. The work product will probably require re-inspection anyway if there are numerous major defects.

After the moderator has reviewed pertinent meeting procedures, the meeting is turned over to the reader. The moderator then assumes the role of an inspector but continues to moderate the meeting to make sure that the inspection process is followed and that the inspection stays focused on the work product and not on the people. The reader begins with a call for global issues with the work product. Any issues discovered in the predecessor work products, standards, or checklists can also be reported at this time. The recorder should record these issues. After the meeting, the author should inform the appropriate responsible individuals of any identified defects in predecessor work products, standards, or checklists so they can be corrected. However, these defects are not recorded in the inspection log or counted in the inspection defect count metrics.

The reader then paraphrases one small section of the work product at a time and calls for specific issues for that part of the work product. The inspectors ask

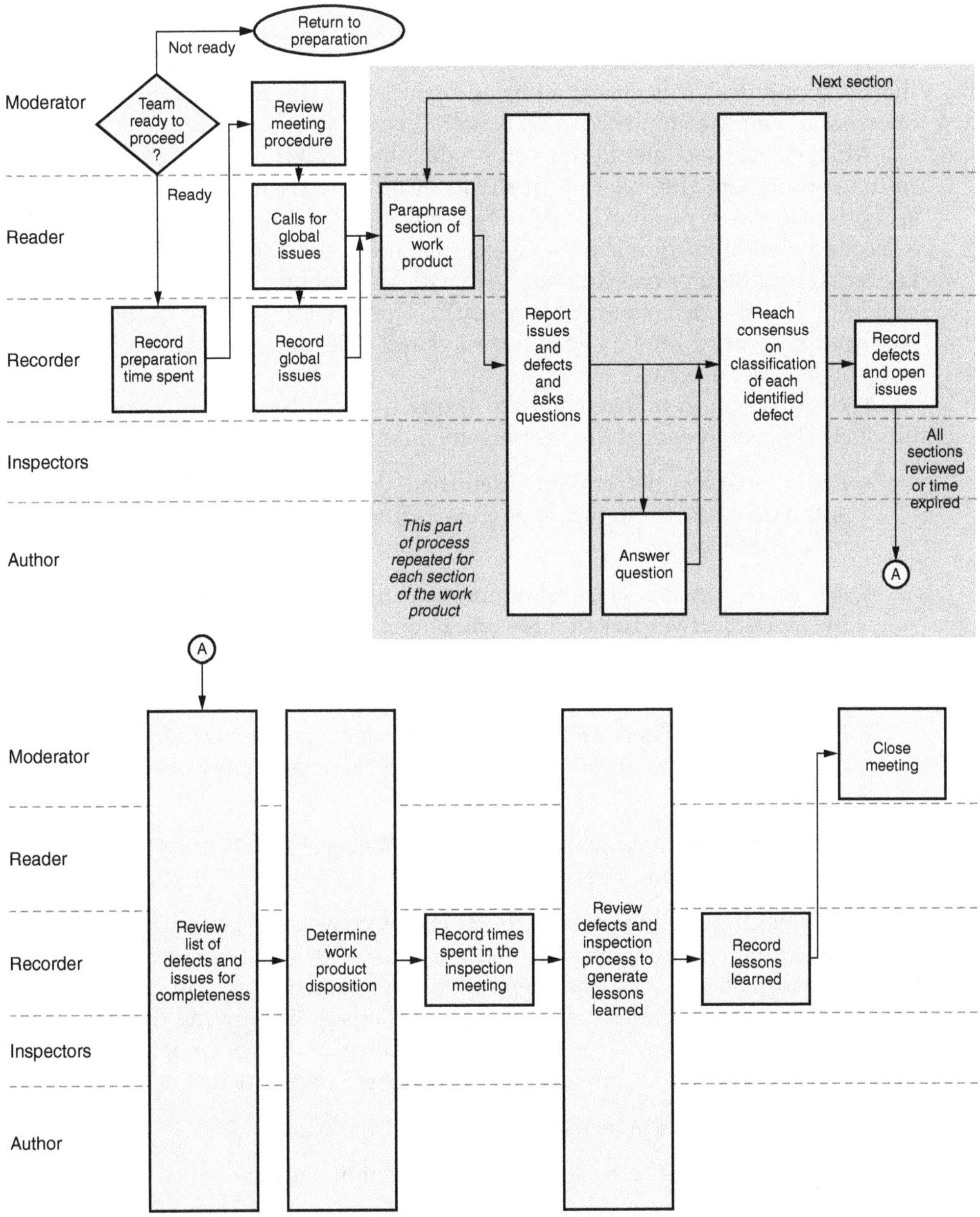

Figure 23.9 Inspection meeting step process.

questions and report issues and potential defects. The author answers questions about the work product. Team discussion on each question, issue, or potential defect should be limited to answering the questions, discussing the issue, and identifying actual defects, not discussing possible solutions. The intent of this discussion is for the team to come to a consensus on the classification (nonissue, major or minor defect, or open issue) of each question, issue, or potential defect. If consensus cannot be reached on the classification within a reasonable period (less than two minutes), it is classified as an open issue. The recorder documents all open issues and the identified defects with their severities and types.

After the last section of the work product has been discussed, the moderator resumes control of the meeting. If the meeting time has expired, but all of the sections of the work product have not been covered, the inspection team schedules a second meeting to complete the inspection. If a second continuation meeting is needed, the moderator coordinates the logistics of that meeting. The recorder then reviews all of the defects and open issues. This review is a verification step to confirm that everything has been recorded and that the description of each item is adequately documented.

The inspection team then comes to a consensus on the disposition of the work product. The work product dispositions include:

- *Pass*: If no major defects were identified, this disposition allows the author to make minor changes to correct minor issues without any required follow-up.
- *Fail—author rework*: The author performs the rework as necessary and the moderator verifies that rework. (Note: Based on the technical knowledge of the moderator, other members of the inspection team may be assigned to verify part or all of the rework.)
- *Fail—re-inspect*: There are enough major defects or open issues that the work product should be re-inspected after the rework is complete.
- *Fail—reject*: The document has so many defects or open issues that it is a candidate for reengineering.

The defect-logging portion of the meeting is now complete. The recorder notes the end time of the inspection meeting and the total effort spent (meeting duration × number of inspectors). As a final step in the inspection meeting process, the team spends the last few minutes discussing suggestions for improving the inspection process. This step can provide information that can help continuously improve the inspection process. Examples of lessons learned might include:

- Improvement to inspection forms
- Improvement to the contents of the inspection package
- Items that should be added to the checklist or standards
- Defect types that need to be added or changed
- Effective techniques used during an inspection that should be propagated to other inspections
- Ineffective techniques that need to be avoided in future inspections
- Systemic defect types found across multiple inspections that should be investigated

Inspection—Post-Meeting Steps

There are three inspection post-meeting steps:

- *Document minutes*: After the meeting, the recorder documents and distributes the inspection meeting minutes.
- *Rework*: The payback for all the time spent in the other inspection activities comes during the rework phase when the quality of the work product is improved by correcting all of the identified defects. The author is responsible for resolving all defects and documenting the corrective actions taken. The author also has primary responsibility for making certain that all open issues are handled during rework. The author, working with other inspection team member(s) and stakeholders as appropriate, either closes each open issue as resolved (if it is determined that no defect exists) or translates the open issue into one or more defect(s) and resolves those defects.
- *Follow-up*: The purpose of the follow-up step is to verify defect resolution and to confirm that other defects were not introduced during the correction process.

  Follow-up also acts as a final check to make certain that all open issues have been appropriately resolved. Depending on the work product disposition decided upon by the inspection team, the moderator may perform the follow-up step, or a re-inspection may be required.

Chapter 24

Test Execution Documentation

> Evaluate and manage test execution documents such as test results, defect reporting and tracking records, test completion metrics, trouble reports, and input/output specifications. (Evaluate)
>
> **Body of Knowledge VI.D**

Both the *IEEE Standard for Software and System Test Documentation* (IEEE 2008a) and the IEEE/*ISO/IEC Software and Systems Engineering—Software Testing—Part 3: Test Documentation* standard (IEEE/ISO/IEC 2021b) describe a set of project-level software test documents. These standards recommend the form and content for each test document, not the required set of test documents. There are a variety of different kinds of test documents, as illustrated in Figure 24.1:

- *Test planning documents* include:
  - Verification and validation (V&V) plans, discussed in Chapter 21
  - Test plans, discussed in Chapter 22
- *Test execution documentation* includes test design, case, and procedure specifications. Test design specification was discussed in Chapter 22. Planned test cases and procedures are defined and written during test design activities. Additional test cases and procedures may also be defined and written during exploratory testing done at test execution time. Test cases and procedures are utilized to run the tests during test execution activities.

Test execution documentation also includes test data and environmental requirements as discussed in Chapter 22.

- *Test execution and results documentation* are created and updated throughout the test execution process to capture the results of the testing activities. This documentation captures the test execution process and results so that people can understand and benefit from what has occurred. The results of test execution activities are documented in the following:

- Test readiness reports
- Test environment readiness reports
- Test execution logs
- Test incident reports (problem reports, issues)
- Test status reports
- Test results data and metrics
- Test completion reports

Test execution and results documentation are necessary because they record the testing process and help software testers and management know what needs to be accomplished, the status of the current testing process, what has and has not been tested, and how the testing was done. This test documentation also helps maintainers and other practitioners understand the testing performed by others and supports the repeatability of tests during debugging and regression testing. All test documentation should be an outgrowth of the testing process, not an end in itself. Testers should think carefully about the test documentation's purpose and whether it improves the final testing outcome. If not, the tester must determine if that documentation is value-added and necessary. Experience is required to strike a proper balance between too much and too little test documentation.

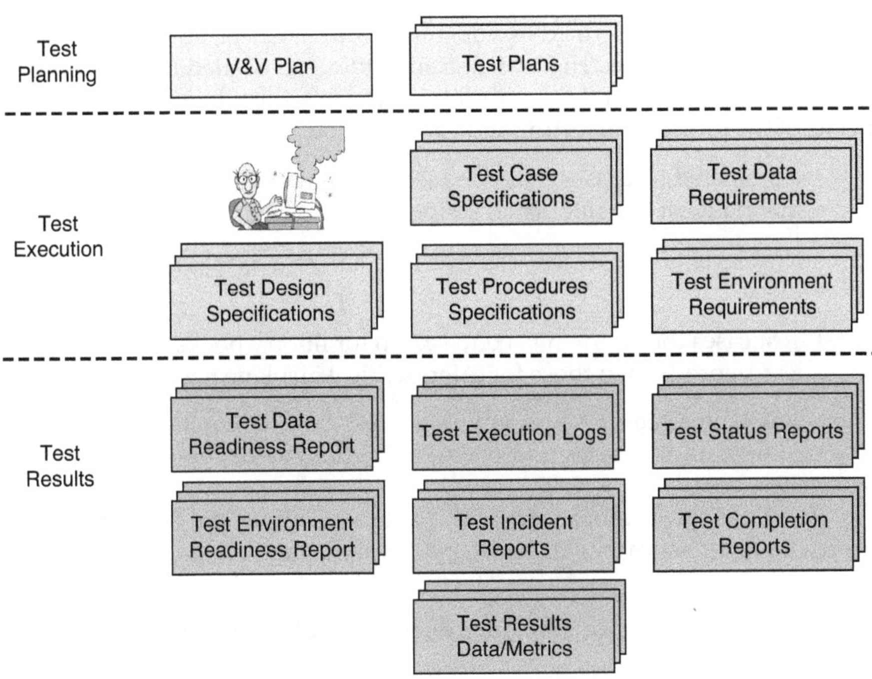

Figure 24.1 Types of testing documentation.

Test Execution Documentation—Test Case

Test cases, also called *test scripts,* are the fundamental building blocks of testing and are the smallest test unit to be directly executed during testing. Test cases are typically used to test individual paths in a source code module, interfaces, or individual functions or subfunctions of the software. Test cases define:

- The items to be tested (source code path, interface, function, or subfunction)
- Preconditions defining what must be true before the test case can be executed, including how the test environment should be set up
- What inputs are used to drive the test execution, including both test data and actions
- What the expected output is (including evaluation criteria that define successful execution and potential failures to watch for)
- The steps to execute the test

Each test case should include a description and may be assigned a priority. Test cases should always be traceable back to the requirements being tested. Depending on the level of testing, test cases may also trace back to design elements or code components. The IEEE *Standard for Software and System Test Documentation* (IEEE 2008a) states that each test case should have a unique identifier and include special procedural requirements and inter-case dependencies as part of its test case outline.

Test cases can be defined in a document, spreadsheet, automated test code, or automated testing tools. The level of rigor and detail needed in a documented test case will vary based on the risk or integrity level of the item being tested and the needs of the project and testing staff. For example:

- Automated test cases may need less documentation than those run manually, or the automated script becomes the documentation
- Unit-level test cases may need less documentation than system test cases
- Test cases for a system used to search for library books may need to be less rigorous than those for safety-critical airplane navigation systems

Figure 24.2 illustrates two examples of test case documentation: one simple example for unit test cases and one more rigorously documented example for a system test case.

The input specification part of the test case defines the requirements for the inputs into the test case. This input specification should be as generic as possible to:

- Allow the tester to sample across the possible input values for more coverage
- Minimize the number of total test cases and their associated maintenance effort over time as changes occur

Unit Test Case Example

Test case #	Inputs		Expected output
	Variable B	Variable C	Variable A
1	≥ 60	< 20	1000
2	≤ 40	≥ 20 and < 100	300
3	> 40 and < 60	≥ 100	300

System Test Case Example

Test case name	Save a file: test for invalid characters in file name	
Test equipment/environment	PC with word processing software installed	
Test setup	Run the word processing software. Create a new document with a paragraph of text in it.	
Step number	**Test execution steps and inputs**	**Expected result**
1.	Select to save file by: - Clicking on Save File icon - Selecting Save or Save As from the menu - Entering <ctrl>S - Clicking on the <X> to close	Word processor displays the Save As window
2.	In the Filename field: enter a file name consisting of < 254 alphanumeric characters with one of the following characters somewhere in the file name: • Greater than sign (>) • Less than sign (<) • Quotation mark (") • Pipe symbol (I) • Colon (:) • Semicolon (;) Left-click the <Save> button	Error message: "The filename, location, or format is not valid. Type the filename and location in the correct format"
3.	Left-click the <OK> button on the error message Press <Escape> key to exit the Save function	Return to document screen

Figure 24.2 Test case—examples.

For example, if the input is a range of values, instead of specifying a specific number such as 7 or 73, the input specification might read "input an integer from one through one hundred." In this case, the actual number chosen as input for each execution of the test case is recorded as part of the test log information to allow reproducibility of that specific test execution (and the reproducibility of any problem or anomaly identified). In other cases, when specific values are required as inputs into the test case, those specific inputs must be specified. For example, when performing boundary testing on that same range of values, those boundary values of 0, 1, 100, and 101 must be specified in the test cases.

The expected output part of the test case defines the expected results from the execution of the software with the specified inputs. This expectation might include specific outputs (e.g., reports to the printer, flashing lights, error messages, or specific numeric values). Alternatively, if inputs are selected from a range or set of possible inputs, this might include criteria for determining those outputs based on the selected input, for example:

- For numeric value outputs, the evaluation criteria might include a look-up table or an equation for determining the expected output values based on the selected inputs

- For a report output to the printer, the evaluation criteria might include the specification for judging whether the entire report was printed correctly with the correct format and that all information on the report is correct

The environmental needs (setup instructions) part of the test case defines any special test bed requirements for running the test case. For example, if it is a throughput-type requirement, there may be a need to attach a simulator that emulates multiple transaction inputs. Another example might be a requirement to run the test case while the temperature is over 140 degrees Fahrenheit for reliability testing.

Test Execution Documentation—Test Procedure

A *test procedure,* also called a *test script* or *test scenario,* is a collection of test cases intended to be executed as a set, in the specified order, for a particular purpose or objective. For example, test procedures can be used to evaluate a sequence or set of features by chaining together test cases to test a complete user scenario or usage thread within the software. For example, the "pay at the pump with a valid credit card" test procedure might chain together the following test cases:

- *Scan a readable card* test case with a valid credit card as input
- *Validate a valid credit card with the credit card clearinghouse* test case with a valid credit card and merchant information as inputs
- *Select a valid gas grade* test case with a valid gas grade as input
- *Pump gas* test case with selected gallons of gas being pumped as input
- *Complete transaction with credit card clearinghouse* test case with cost of gas pumped amount as input
- *Prompt for receipt* test case with selection of "yes" or "no" as input
- *Print receipt* test case with information on this transaction as input (this test case would only be included if "yes" was selected in the previous test case)
- *Complete transaction and store transaction information* test case with information on this transaction as input

Alternative procedures might chain many of these same test cases together with other test cases or substitute other test cases into the chain. For example, the *pay at the pump with a valid debit card* test procedure would include the addition of an *accept valid PIN* test case after the first test case and use the *validate a valid debit card with credit card clearinghouse* test case in place of the *validate a valid credit card with credit card clearinghouse* test case in the foregoing test procedure.

The documentation of a test procedure defines the purpose of the procedure, any special requirements, including environmental needs or tester skills, and the detailed instructional steps for setting up and executing that procedure. Each test procedure should have a unique identifier so that it can be easily referenced. The IEEE *Standard for Software and System Test Documentation* (IEEE 2008a) says the procedural steps should include:

- Methods and formats for logging the test execution results
- The setup actions
- The steps necessary to start, execute, and stop the test procedure (basically the chaining together of the test cases and their inputs)
- Instructions for taking test measures
- Actions needed to restore the test environment after the test procedure execution
- If a test procedure might need to be suspended, that test procedure also includes steps for shutting down and restarting the procedure
- The test procedure may also include contingency plans for dealing with risk associated with that procedure

Test procedures, like test cases, can be defined in a document, spreadsheet, automated test code, or automated testing tools. Like test cases, the level of detail needed in a documented test procedure will also vary based on the needs of the project and testing staff.

Test Execution Results Documentation—Test Log

A *test log*, also called a *test execution log*, is a chronological record that documents the execution of one or more test cases or test procedures in a testing session. The test log can be a beneficial source of information later when the authors are having trouble reproducing a reported problem or the tester is trying to replicate test results. For example, sometimes, it is not the execution of the last test case or procedure that caused the failure but the sequence of multiple test cases or procedures. The log can help repeat those sequences exactly. It may also be essential to understand the time of day or day of the week that the tests were executed, the test environment configuration, or what other applications were running in the background at the time the failed test was executed.

The test log information can be logged manually by the testers into a document or spreadsheet, directly into the test cases or procedures, or into a testing tool. If test execution is automated, some or all of the test log information may be captured as part of the outputs recorded and reported by that automation. The test log should include generic information about each test execution session, including:

- A description of the session
- The specific configuration of the test environment
- The version/revision of the software being tested
- The tester(s) and observers (if any)

As the individual tests are executed during the session, specific chronological information about the execution of each test case or procedure is also logged, including:

- Time and date of execution
- Specific input(s) selected

- Any failures or other anomalies observed
- Associated problem report identifier, if a problem report was opened
- Other observations or comments worthy of note

Test Execution Results Documentation—Problem Report

A *problem report*, also called *incident report, anomaly report, bug report, defect report, error report, failure report, issue, trouble report*, and so on, documents a failure or other anomaly observed during the execution of the tests. Problem reports may be part of the formal configuration status accounting information if the item under test is baselined, or they may be reported informally if the item is not baselined. A problem report should include information about the identified failure or anomaly with enough detail so that the appropriate change control board can analyze its impact and make informed decisions or so that the author can investigate, replicate, and resolve the problem if necessary. A description of the information recorded in a problem report type change request is discussed in Chapter 27.

Problem report descriptions should:

- Be specific enough to allow for the identification of the underlying defect(s). For example,
    - *This*: Incomplete message packets input from the XYZ controller component do not time out after two seconds and report a communication error as stated in requirements R00124.
    - *Not this*: The software hangs up.
- Be neutral and be stated in terms of observed facts, not opinions. Descriptions should treat the work product and its authors with respect. For example:
    - *This*: When the user enters invalid data into one of the fields in form ABC, the user receives the error message, "Error 397: Invalid Entry." This error message does not identify the specific fields that were incorrectly entered or provide enough guidance to the user to allow the identification and correction of mistakes.
    - *Not this*: The error messages in this software are horrible.
- Be written in a clear and easily understood manner. For example:
    - *This*: Page 12, paragraph 2 of the user manual states "that the user should respond quickly. . . ." This statement is ambiguous because the use of the term "quickly" does not indicate a specific, measurable response time.
    - *Not this*: The use of vague terminology and nonspecific references in adverbial phrasing makes the interpretation of page 12, paragraph 2 of the user manual questionable at best and may lead to misinterpretations by the individuals responsible for translating its realization into actionable responses.

- Describe a single failure or anomaly and do not combine multiple issues into a single report. However, the tester should use their judgment here. If similar failures or anomalies appear to have the same root cause, it may be appropriate to document them in the same problem report. Duplicates of the same failure or anomaly should also be recorded in the same problem report. However, if they are reported in separate reports (e.g., if they come from different testers or users), a mechanism should exist for linking duplicate reports together.

The contents and tracking of problem reports are further discussed in Chapter 27. While problem reports can be documented on paper forms, most organizations use an automated problem reporting tool or database to document these reports and track them to resolution. A project's V&V plans should specify which problem reporting and tracking mechanisms are used by that project.

Test Execution Results Documentation—Test Results Data

Test results data provide objective evidence of whether or not a specific test case or test procedure passed or failed test execution. Test results data are compared with the expected results, as defined in the test cases and evaluated to determine the outcome of the test execution.

Test Execution Results Documentation—Test Status Report

Test status reports provide ongoing information about the status of the testing part of the software projects over time. Status reports provide management visibility into the testing process, including variances between plans and actual results, the status of risks (including new or changed risks), and issues that may require corrective action. Test status reports may be written reports or may be given verbally (e.g., verbal status reports are given as part of daily meetings of an agile team).

Test Execution Results Documentation—Test Metrics

Test management and product and process quality metrics are fundamental to tracking and controlling of the project's testing activities. They also provide management with information to make better, informed decisions. Specific test measurement reports for selected test metric may be stand-alone reports (Chapter 20 included an example dashboard for a system test sufficiency dashboard) or may be included in test status reports or test completion reports.

Test management metrics report the test activities' progress against the test plans. Test management metrics provide visibility into trends in current progress, indicate accuracy in test planning estimation, flag needed control actions, can be used to forecast future progress, and provide data to indicate the completeness of the testing effort. Examples of test management metrics include:

- Test activity schedules planned versus actual
- Test effort planned versus actual
- Test costs budget versus actual

- Test resource utilization planned versus actual
- Requirements (and test case) volatility
- Testing staff turnover

Product quality metrics collected during testing can provide insight into the quality of the products being tested and the effectiveness of the testing activities. Product quality metrics also provide data to indicate the completeness of the testing effort and the product's readiness to transition to the next software life cycle activity or be released. Examples of product quality metrics include:

- Defect arrival rates
- Cumulative defects by status
- Defect density

Process quality metrics can help evaluate the effectiveness and efficiency of the testing activities and identify opportunities for future process improvement. Examples of test process quality metrics include:

- Escapes
- Defect detection efficiency
- Test cycle times
- Test productivity
- Test coverage

Test Execution Results Documentation—Test Completion Report

A *test completion report,* also called a *test report* or *test summary report,* summarizes the results and conclusions from a major cycle of testing, or a designated set of testing activities, after the completion of that testing. For example, there may be an integration test completion report for each subsystem, a system completion test report, an acceptance completion test report, or other test completion reports, as appropriate. At a minimum, every test plan should have an associated test report. A project's V&V plans should specify which test reports will be written for that project.

A test report includes a summary of the testing cycle or activities and their results, including any variances between what was planned and what was actually tested. The summary also includes an evaluation, by the testers, of the quality of the work products that were tested and their readiness (and associated risks) to transition to the next stage of development or to be released into operations. The report also includes a detailed, comprehensive assessment of:

- The testing process against the criteria specified in the test plan, including a list of any source code modules, components, interfaces, features, or functions that were or were not adequately tested
- A list of identified problems:

    - If resolved—a description of their resolution
    - If unresolved—their associated risks, potential impacts, and work-arounds
- An evaluation of each item tested based on its pass/fail criteria and test results, including an assessment of its reliability (likelihood of future failure)

Finally, the test report includes a summary of test activities and information, including test measures, testing budget, effort and staffing actuals, number of test cases and procedures executed, and other information that would be beneficial when evaluating lessons learned and planning and estimating future testing efforts.

Part VII

Software Configuration Management

INTRODUCTION TO SOFTWARE CONFIGURATION MANAGEMENT

The *IEEE Standard for Configuration Management in Systems and Software Engineering* (IEEE 2012) states that "the purpose of configuration management is to:

- Identify and document the functional and physical characteristics of any product, component, result, or service
- Control any changes to such characteristics
- Record and report each change and its implementation status
- Support the audit of products, results, services, or components to verify compliance to requirements."

Software configuration management (SCM) is the process of applying configuration management tools and techniques throughout the software product life cycle to verify the completeness, correctness, and integrity of software configuration items. SCM helps a project maintain consistency between software requirements; the resulting software design, code, documentation, and other work products; and the project's plans. SCM also protects the software intellectual capital of the organization. One way of thinking about SCM is to consider it the inventory control system for software.

Most software (or software-intensive systems) today is made up of many different components that are built using many different tools. Larger projects typically require more components and involve more people in creating, using, and sharing access to all of those components. SCM provides the mechanisms needed to identify those components, coordinate their interrelationships, manage change to those components over time, and successfully build and release a set of high-quality deliverables. As Berczuk (2003) puts it, "The way teams communicate their work products to each other is through their SCM and version control practices."

As illustrated in Figure VII.1, seven software configuration management processes create a foundation supporting software development, acquisition, and service processes. The *IEEE Standard for Configuration Management in Systems and Software Engineering* (IEEE 2012) defines the following four core processes that make up SCM:

- Software *configuration identification*: The process of partitioning the software product into formally identified configuration items, documenting each item's functional and physical characteristics, defining each item's acquisition points, assigning unique identifiers to those items, and establishing baselines
- Software *configuration control*: The process of systematically controlling changes to software items under configuration control by making sure that:
    - All proposed changes are properly identified, documented, evaluated for impact, and approved by the appropriate level of authority
    - All approved changes are incorporated and verified in all affected configuration items

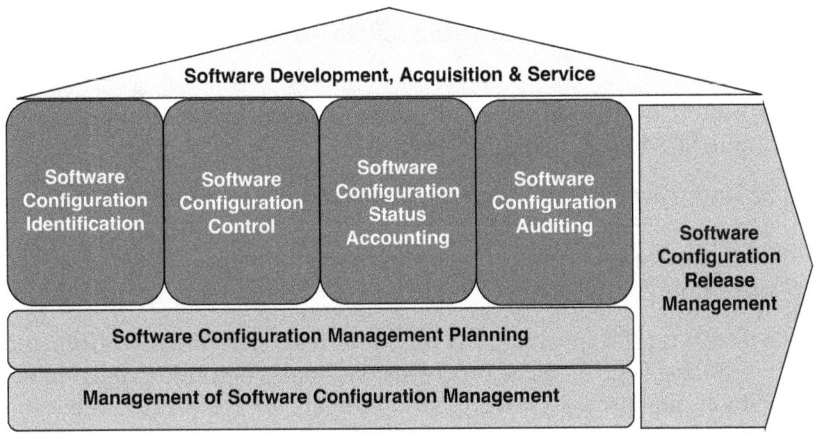

Figure VII.1 Software configuration management activities.

    - Disapproved changes and other unauthorized changes are not incorporated into any baselined configuration items

- *Software configuration status accounting*: The process of recording and reporting of information needed to manage a software configuration effectively, including the implementation status of the configuration items and the status of changes to those items
- *Software configuration auditing*: The evaluation process that provides independent, objective assurance that the software configuration items being built match their requirements, are consistent with the documentation that describes them, and, at operations, are ready to be released

In addition to these four core SCM processes, the *IEEE Standard for Configuration Management in Systems and Software Engineering* (IEEE 2012) includes three other primary processes:

- *Management of software configuration management*: The process that includes:
    - Establishing and documenting SCM policies, standards, processes, and plans for the organization or project
    - Managing SCM functions, including establishing the SCM infrastructure, assigning SCM functional responsibilities, and providing SCM training
    - Identifying, analyzing, and mitigating SCM risks
    - Monitoring the performance and effectiveness of SCM processes, the SCM group, the SCM staff, and the suppliers against requirements and plans, and making sure the implementation of corrective actions and process improvements is performed as needed

  - Retaining and protecting the software intellectual property of the organization
  - Establishing SCM metrics to monitor performance and provide input into SCM process improvement
- *Software configuration management planning*: The process of creating, communicating, controlling, and progressively elaborating SCM planning for each project, as appropriate to the needs of that project, including:
  - Defining the specific SCM roles and responsibilities, staffing and training (e.g., on SCM processes and tools), activities, deliverables, tools, resources, schedules, and budgets for performing the various project SCM activities
  - Determining how SCM will be managed for the project and planning for communicating with SCM stakeholders
  - Identifying configuration items and baselines the project will use to identify and control their software products and product components, their acquisition points and criteria, and the selected identification schemas
  - Establishing the types, levels, promotion points, owners, and mechanisms of control required for each of those configuration items and baselines
- Determining data collection and reporting requirements and selecting mechanisms for performing configuration status accounting, including:
  - Collecting and reporting on configuration management-related data
  - Defining configuration audit requirements and plans
  - Controlling supplier-produced configuration item
  - Controlling interfaces to the work products being developed by the project or its suppliers
- *Software configuration release management*: The process that supports the ability to release and distribute high-quality, stable, and reproducible software products

IEEE also identifies two special instances of these SCM processes for supplier configuration control and interface control (IEEE 2012).

The rigor and amount of documentation required for configuration management planning may vary based on the project's (program or product) needs and the associated risk. For example, for agile projects, appropriate SCM decisions and plans still need to be made and communicated to project team members. However, there may be little or no formal documentation of those plans ("just enough documentation"). Agile SCM planning is also done as needed rather than as a "big effort up front." On the other hand, large, traditional, or high-risk projects tend to formally document their SCM planning. Depending on the needs of the project, this may be done in a stand-alone document (that is, a software configuration management plan) or a section in one of the other project plans (e.g., the project management plan, project work plan, or software quality assurance plan). The *IEEE Standard for Configuration Management in Systems and Software Engineering* (IEEE 2012) includes a template for a software configuration management plan.

All seven of these SCM processes create a strong foundation that supports and provides essential services to all of the other software development, acquisition, and service processes so that they can be conducted in a stable, predictable, and efficient manner.

Examples of risks (potential problems) associated with a lack of good SCM practices include:

- Inconsistencies between various software work products (e.g., documentation that does not match the source code module or executable)
- Missing source code modules or other software work products
- Unauthorized and unwanted changes to the software
- Inability to track when and why changes were made to the software and who made them
- Inability to integrate individual "working" software modules into "working" software products
- Uncertainty about what needs to be tested
- Working software that suddenly stops working or fixed defects that reappear in subsequent builds

These risks can result in major negative consequences to software projects, including excessive rework, missed delivery deadlines, cost overruns, missing functionality or quality characteristics in deliverables, inefficient use of personnel and other resources, stakeholder dissatisfaction, and even contract or regulatory violations.

However, the level of rigor in the implementation of SCM processes should be risk-based and vary according to on the individual needs of each organization, product, or project. Too much SCM can result in costly bureaucratic overhead or delays in delivery and may even require developers to ignore the process in order to get their work done. Organizations must establish an appropriate level of SCM rigor in their processes and infrastructure. For example, agile projects focus on "just enough" process. "The great news is that both CM and agile focus heavily on responding to change." Moreira (2010) says, "Think of CM as an enabler of change for agile." SCM provides the tools and methods that efficiently facilitate the agile practices of continuous integration and builds while "melding speed with sustainability."

Chapter 25
Configuration Infrastructure

The configuration infrastructure provides the organizational structure, tools, and processes that create the discipline upon which to build a consistent and effective *software configuration management* (SCM) system. The organizational structure requires project-level SCM groups and the specific assignment of SCM roles and responsibilities at the project level. It may also include the creation of an organizational-level SCM group with defined roles and responsibilities. The selection of the appropriate SCM tools and the establishment of SCM libraries, which support the established SCM system, greatly improve the ability to build, release, and distribute high-quality software products with the least amount of effort and allow organizations to retain and protect their software *intellectual property*.

1. CONFIGURATION MANAGEMENT TEAM

> Describe the roles and responsibilities of a configuration management group. (Understand) [NOTE: The roles and responsibilities of the configuration control board (CCB) are covered in area VII.C.2.]
>
> **Body of Knowledge VII.A.1**

Trained, qualified people are needed to perform the activities required to implement an effective SCM system both at an organizational level and at a project level. The number of people needed at each level depends on the size, complexity, and objectives of the organization and the individual projects. In the case of large projects with hundreds of developers in geographically dispersed teams, with each team creating multiple software components, the project's SCM group might include several full-time specialists. For a small project, with a few co-located developers creating a limited increment of functionality, the SCM group may consist of one or more individuals performing the SCM activities as just one of their multiple project responsibilities.

Organizational-Level SCM Group Roles and Responsibilities

Many organizations establish an organizational-level SCM function to oversee the SCM system. This centralized SCM group may select industry standards and models as guidance for their SCM program to leverage good industry practices and keep from "reinventing the wheel" with each new project. The organizational-level SCM group establishes, documents, and distributes standardized SCM policies, processes, and work instructions, as well as defining and implementing standardized metrics that can be shared across multiple projects.

The organizational-level SCM group also defines, implements, and maintains an organizational-level SCM infrastructure, including:

- Standardized SCM tools
- Standardized SCM training
- Reuse libraries
- SCM databases and other historic SCM information
- Mechanisms for backup, disaster recovery, and archival

Having a standardized SCM system and infrastructure benefits the organization by making it easier for individuals to move from project to project without a steep SCM learning curve. There may also be economy-of-scale benefits derived from having a few centralized experts who act as internal consultants to projects rather than requiring expert individuals on each project. Centralizing specific SCM responsibilities provides mechanisms for software reuse across projects and product lines and for transitioning lessons learned on individual projects into an integrated SCM system to improve practices across the organization. The organizational-level SCM group also performs SCM corrective actions and process improvements as needed.

Finally, an organizational-level SCM group provides a single conduit for management visibility and oversight of the organization's SCM system. For example, the organizational-level SCM group may act as the review and approval authority as projects tailor the standardized SCM processes. This authority verifies that crucial elements are not tailored out or modified inappropriately.

Project-Level SCM Group Roles and Responsibilities

A project-level SCM group is responsible for planning, implementing, and controlling the project-specific SCM activities. Organizations that establish an organizational-level SCM function also need people at the project level who are responsible for project-specific SCM activities. This may be accomplished by having members of the organizational-level SCM group report on a dotted-line basis to the project or by having project-specific individuals assigned to SCM roles. For organizations that choose not to establish an organizational-level SCM function, the organizational-level responsibilities described previously become additional activities delegated to the project-level SCM group members as appropriate. However, this may increase the risk of inconsistent process execution and make people and other resources less interchangeable.

Each project-level SCM group performs SCM planning for their project by defining how that project intends to implement and tailor the organizational-level SCM policies, standards, processes, work instructions, and infrastructure to the needs of their specific project. Each project-level SCM group is responsible for making sure that planned SCM activities for their project are appropriately implemented and for tracking and controlling that implementation. These activities include:

- Implementing and tailoring organizational-level standardized SCM processes and work instructions or creating and implementing project-specific SCM processes and work instructions as appropriate
- Defining and implementing the project's SCM infrastructure needs, including libraries and areas, partitions, or customizations in SCM tools and databases
- Providing SCM inputs to project proposals to potential customers and requests for proposals to potential suppliers to make sure that all SCM requirements are being met
- Administering activities related to acquiring controlled configuration items
- Establishing project baselines
- Providing administrative support to the configuration control boards (CCBs)
- Verifying that only authorized changes are made to baselined software
- Verifying that all baseline changes are recorded in sufficient detail so that they can be reproduced or backed out
- Controlling commercial off-the-shelf (COTS) software; third-party software; customer-supplied products, components, or documents; and software tools as appropriate to the requirements of the project
- Supporting activities related to building the larger software composite configuration items from smaller constituent configuration items
- Participating in or conducting software configuration audits
- Making sure that configuration items are appropriately stored and protected
- Providing a mechanism for exception resolution, deviations, and waivers
- Advising the project team of SCM-related risks and problems and providing solutions

In the agile context, the project-level SCM group should be "considered as a holistic part of the agile team." The SCM activities listed previously, as well the special SCM roles and responsibilities discussed below, "are enacted by many on the team and not just one person" (based on Moreira 2010).

Special SCM Roles and Responsibilities

As part of, or in addition to, the organizational-level and project-level SCM groups, there may be individuals who perform special SCM roles. These roles include SCM managers, SCM librarians, SCM toolsmiths, builders, release managers, and software practitioners. Note that the roles and responsibilities of the CCB are discussed in Chapter 27, "Configuration Control and Status Accounting."

SCM managers are responsible for managing the SCM groups at either the organizational or project level and for the overall success of the SCM program. These responsibilities include:

- Providing the overall strategic and tactical planning for the SCM function, establishing the SCM system and group goals and objectives, and approving project-level SCM plans
- Implementing, tracking, and controlling the SCM system, including reviewing the performance of the SCM system and its practices and tools, as well as the individual performance of SCM group members
- Making sure that lessons learned are gathered, and appropriate SCM risk management and corrective actions are implemented as needed

SCM librarians are responsible for establishing, coordinating, and verifying the integrity of the controlled and static libraries for each project and for coordinating the reuse of software components between projects. SCM librarian duties may include creating branches, performing merges, and establishing baselines (e.g., through labeling). SCM librarians may also be charged with performing backup and disaster recovery functions.

SCM toolsmiths, also called *tool administrators,* are responsible for implementing and maintaining the SCM tool set. These responsibilities include:

- Making recommendations on the tool selection process
- Installing, configuring, testing, and implementing the selected tools
- Updating the SCM tools as corrective releases or new versions become available
- Customizing the SCM tools to meet organizational and project requirements
- Automating SCM activities through the use of the SCM tools or scripting routines
- Acting as subject matter experts for other project personnel with questions about or problems with the SCM tools

Builders are responsible for building source code modules into binary executables (or other constituent configuration items into composite configuration items, for example combining training chapters into completed training materials) and verifying the reproducibility of those builds. These responsibilities include making sure that:

- The correct version/revision of each constituent configuration item (source code module, library, and macro) is used to create the build
- The appropriate build platform and environment are used to create the build, including the correct version/revision of each build tool (e.g., compiler, assembler, linker, loader, scripts, and automation)
- All compile and runtime dependencies are considered
- Appropriate switches and options are set correctly

Release managers are responsible for packaging the software (or software-intensive systems) into a releasable product set for deployment into the desired environment.

Software practitioners include requirements engineers and analysts, designers, programmers, testers, suppliers, technical writers, and others who utilize the SCM processes as part of their responsibilities in creating, developing, and maintaining the software products. The SCM responsibilities of software practitioners include:

- Following the established configuration management policies and processes
- Identifying relevant configuration items
- Submitting those items for configuration control when they are baselined
- Implementing only authorized changes to baselined configuration items
- Establishing and maintaining bidirectional traceability information and other status accounting data
- Using the information produced by the status accounting system to manage the configuration items

2. CONFIGURATION MANAGEMENT TOOLS

> Describe configuration management tools as they are used for managing libraries, build systems, and defect tracking systems. (Understand)
>
> **Body of Knowledge VII.A.2**

Configuration management tools are used to automate the tedious and often error-prone work of manually performing the SCM activities, which include:

- Establishing and maintaining SCM libraries

- Establishing and maintaining bidirectional traceability
- Controlling multiple versions of work products
- Creating and releasing software builds
- Reporting and tracking defects
- Requesting and tracking change requests
- Helping with SCM audits

This automation can improve development efficiency, decrease cycle times, minimize human error, allow software work products and product information to be easily shared across large and geographically dispersed groups, and increase the integrity of product information. SCM tools are also extremely important in the agile context. In a study by Moreira (2010), agile professionals were asked, "How important are CM tools for agile projects?" Ninety-four percent of the respondents answered either "extremely important" (75 percent) or "very important" (19 percent).

SCM tools can influence how an organization works. Therefore, tools should be selected that match the organization's SCM processes. An organization should not try to force-fit its processes to its SCM tools. If SCM process improvements are needed, they should be done in conjunction with selecting or upgrading the SCM tools that support those improvements. If the tool set is cumbersome or does not match how software practitioners and others do their work, those individuals will often find ways to work around the tools or circumvent the controls provided by the tools. Other considerations when selecting SCM tools include:

- Configuration item types supported by the tool (source code, documents, executables, drawings, hardware, and so on)
- Integration with other software life cycle processes and tools (e.g., requirements tools, development tools, and testing tools)
- Naming standards and conventions and version/revision identification strategies supported by the tool
- Usability of the tool
- Reporting and query capabilities of the tool
- Security issues, including access privileges and the ability of a project to keep its libraries secure from access by other projects using the same tool
- Build methods available in the tool
- Customization and reconfigurability of the tool

SCM Library and Version Control Tools

SCM *library tools* are the tools used to establish, maintain, and manage the various SCM libraries. These tools may range from simple scripts that are run to create static archives or nightly backups to sophisticated version control tools used to establish and manage the controlled libraries. Library tools may be separate tools, or they may be integrated with version control tools.

Version control tools handle the storage of multiple versions of individual configuration items. The tool assigns a unique identifier to each version of each configuration item so that it can be referenced and retrieved as needed. Typically, the versions are stored as either forward or backward deltas to conserve space. In a backward delta system, the latest version of the item is saved along with the deltas between it and the previous versions. A forward delta system saves the original version and deltas going forward. For example, if the fifth version of a source code module were retrieved in a forward delta system, the tool would take the original file and apply four deltas to it to create the fifth version.

For configuration items like requirements or test cases, requirements management tools or test management tools may provide the mechanisms needed for version control.

Version control tools typically have a labeling mechanism that allows baselined versions of configuration items to be identified and related to other items in the same baseline. For example, when a new baseline is created, all the work products included in that baseline are labeled with that baseline's identifier. More sophisticated version control systems include automated promotion management as software development moves through the various life cycle phases.

Integrated version control tools work in conjunction with change management and build tools. These integrated tools allow change requests to be linked to the versions of the configuration items that incorporated the requested change and make certain that the correct versions of the appropriate configuration items are used when new builds are created.

Content management tools may be more appropriate than version control tools for web applications. Content management tools manage versions of the content as it evolves, including the ability to create, review or edit, publish, and view the content.

SCM Build Tools

A *software build* is a process of combining smaller software constituent configuration items into larger composite configuration items. Figure 25.1 illustrates the build process used to convert individual software source code modules into a software executable (build) that can be run on a computer. *SCM build tools* include:

- *Compilers* that convert high-level language source code into object code.
- *Assemblers* that convert assembly language source code into object code.
- *Linkers or loaders* that combine and convert object code modules into the software executable.
- *Build scripts that are* used to automate build processes that would otherwise require people to perform multiple manual operations. Build scripts can be expanded to perform tasks in addition to basic builds, including:

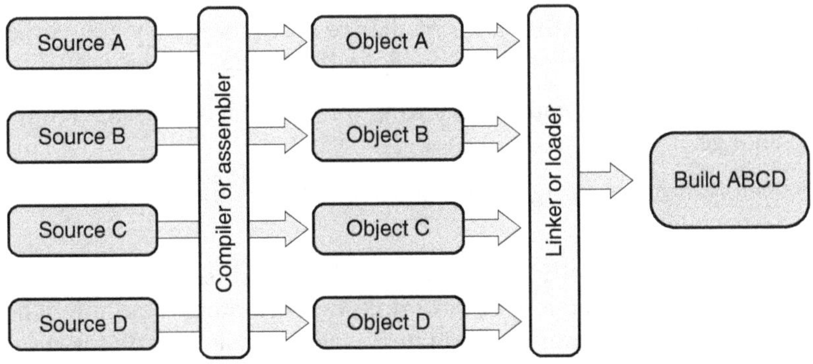

Figure 25.1 Software build process.

- Running automated test scripts
- Reporting test results
- Recording build status accounting information into the appropriate databases
- Deploying the build into the target environment

Build tools capture the details of the build and build processes so that the build can be reproduced as needed. For example, details could include what version of each configuration item was used and what switches or options were used for the compiler or linker.

SCM Change Management Tools

SCM *change management tools* automate the change control processes through which changes to the software are requested; impact analysis is performed; change requests are approved, deferred, or disapproved; and approved change requests are tracked through implementation, verification, and closure. Change management applies to both reported problems (failures, potential defects) and enhancement requests (requirements changes). Some organizations use multiple defect tracking or enhancement request tools, while others use a single integrated change management tool to track both defects and enhancement requests. Depending on their sophistication, these change management tools can be used to:

- Request changes (e.g., to request enhancements or to report problems or other anomalies), including documenting the associated information.
- Notify the appropriate individuals and disseminate information when a change is requested.
- Facilitate virtual change control board meetings and impact analysis reviews.

- Identify and track the status of change requests as they are worked to resolution, documenting each step in the change process.
- Collect and store status accounting information about each requested change.
- Determine the configuration items affected by the change (automated impact analysis).
- Automatically notify the appropriate individuals of status changes to each change request. For example, a development lead might be notified when a configuration item is updated and ready for approval, or a tester might be notified when the change is ready for testing.

Through integration with the version control tool, the change management tool can:

- Identify individuals authorized to check out controlled configuration items for change, thus helping to make sure that only authorized changes are being made to controlled items
- Support change management across parallel, concurrent, or distributed development through the automated merging of changes into items in other branches

SCM Status Accounting Tools

SCM tools can automatically provide much of the information required for good status accounting and traceability that would otherwise require a monumental manual effort to document and maintain. SCM *status accounting tools* may be separate tools or be built into the other SCM tools as data collection and reporting mechanisms. SCM status accounting tools harvest data recorded in the other SCM tools (version control tools, build tools, change management tools) to provide information for ad hoc queries and standardized reports including software measurements. These data and information support management and engineering decision-making processes. Status accounting information also supports the ability to perform SCM audits.

Other SCM Tools

Other SCM tools include:

- Virus detection and removal tools
- Back-up and archival tools
- Monitoring and auditing support tools
- Traceability tools
- Release management tools
- Release distribution tools

3. LIBRARY PROCESSES

> Describe dynamic, static, and controlled library processes and related procedures such as check-in/check-out and merge changes. (Understand)
>
> **Body of Knowledge VII.A.3**

There is a dichotomy in SCM. On one side, individual developers need the flexibility necessary to do creative work, modify code to try out what-if scenarios, and make mistakes, learn from them, and evolve better software solutions. On the other side, teams need stability to allow code to be shared with confidence, create builds and perform testing in a consistent environment, and ship high-quality products with confidence. This dichotomy requires an intricate balance to be maintained. One of the ways that SCM supports this need for both flexibility and stability is through the use of different SCM libraries. The number and kinds of libraries used in SCM will vary based on each project's size, criticality, and requirements. The contents of each library will also vary based on the types and formats of the various work products. There are four kinds of SCM libraries:

- Dynamic library
- Controlled library
- Static library
- Backup library

For each library used by a project, SCM planning should identify the name and type of the library, the information retention method (e.g., online or offline, media used, tool(s) used), the control mechanisms, and the retention policies and procedures.

Dynamic Library

A *dynamic library,* also called a *development, programmer's,* or *working library* or a *sandbox,* is used to hold products that are currently being worked on by the software practitioner (developers, testers, or other project personnel). Each practitioner typically has one or more dynamic libraries under their control that are used as personal work areas. The contents of a dynamic library are not considered to be under configuration control.

The dynamic library provides a mechanism for creating private working copies of new work products or obtaining copies of baselined work products approved for modification. Utilizing these copies, the developers can create new work products or make changes to existing products without affecting anyone else or impacting the integrity of the controlled copies. This ability gives the developer the flexibility needed to:

- Create work products
- Prototype changes
- Conduct tests and attempt to re-create problems
- Modify the code to facilitate debugging
- Make temporary changes to try out potential corrections
- Conduct trial-and-error experiments

Another example of a dynamic library is the work area testers use for test execution. These libraries allow tests to be performed that might cause failures that corrupt the software or data without impacting the work of other software practitioners. This dynamic test library is part of the test bed and is under the tester's control.

Controlled Library

A *controlled library,* also called a *master* or *system library,* is used for managing current controlled configuration items and baselines and controlling changes to those items and baselines. The controlled library contains configuration items that have been acquired and placed under formal change control. The controlled library may be made up of one or more actual libraries, directories, or repositories, depending on the project's needs. Different libraries might be used to store different types of configuration items (source code, documentation, hardware, drawings). Controlled libraries may also be implemented as areas or *codelines* inside a tool, such as a version control tool.

A controlled library is used to share work products with other members of the project team or other stakeholders. Software practitioners can freely obtain "read-only" copies of the configuration items from the controlled library and move them into their dynamic libraries. However, the appropriate authority (CCB) must authorize changes to work products in the controlled library. Work products from the controlled library are used to create official software builds from a "known" environment, such as builds used for testing and for release.

Static Library

A *static library,* also called a *software repository, archive library,* or *software product baseline library,* is used to archive important baselines, including those released into operations. Static libraries can also be used to archive ("freeze") various intermediate baselines as appropriate during the development process. Items in the static libraries can be accessed through read-only mechanisms but are not modifiable.

Backup Library

A *backup library* contains duplicates of the versions of the software and associated components, data, and documentation when the copies were made. Backups are performed periodically to make sure that the intellectual property of the organization is protected. Backup versions can be retrieved if important files are

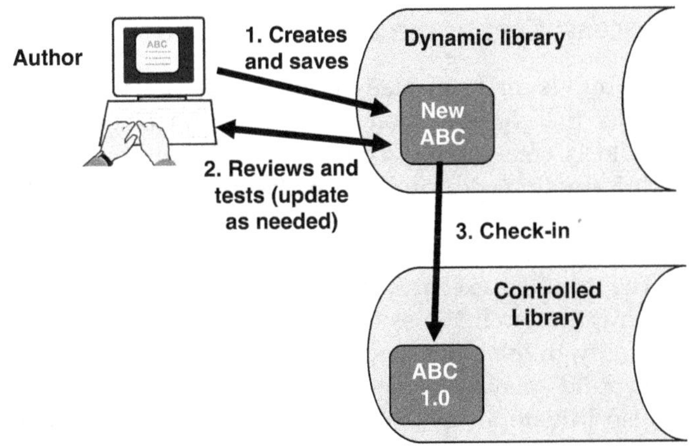

Figure 25.2 Creating a new software work product.

accidentally deleted, if a modified work product must be rolled back to a previous version, or if a hard disk crashes or other disasters occur.

SCM Library Process: Creating a New Software Work Product

A dynamic library is used when new software work products are created. For example, as illustrated in Figure 25.2, the library process steps to create a new work product include:

Step 1: The author creates and saves a new software work product ABC in a dynamic library. Examples of work products include a source code module, a document (plan, user manual, or training manual), a test case or procedure, a data file, or a build script.

Step 2: The author reviews and tests the newly created work product ABC and makes whatever updates are necessary to correct any defects found. The review process might include static analysis, peer reviews, or other types of reviews involving other software practitioners.

Step 3: When the author is satisfied that the work product is ready to share and the appropriate acquisition criteria are met, the author checks work product ABC into the controlled library, and a unique identifier is assigned to indicate the name of the work product and its version/revision (e.g., ABC 1.0).

Once a new work product has been checked into the controlled library, it is under formal change control and can be modified only through formal change control processes. The check-in process includes the capturing of information about the work product, including:

- Descriptive information about the work product, including initial creation information and subsequent change information when the work product is checked out and back in as a modified version in the future
- Check-in date and time and who performed the check-in
- File access permissions

SCM Library Process: Creating a Software Build

While software builds can be created in a dynamic library, official software builds are created from the configuration items in a controlled library. This is done so that the build is created from controlled items in a controlled environment where build information is captured to make sure that the build is reproducible. For example, as illustrated in Figure 25.3, the library process steps to creating a software build may include the following.

Step 1: The builder executes a controlled build script using controlled constituent configuration items as inputs. The builder may be a person creating the build manually, or the build may be created automatically.

Automatic builds may be done periodically, for example, every hour. Special events may also initiate an automatic build, for example, when a new version of any of a set of specified constituent configuration items is checked into the controlled library.

Step 2: The output of this execution is the software build, which is typically stored as a controlled composite configuration item in the controlled library and assigned a unique identifier.

SCM Library Process: Testing a Software Build

Assuming that the tester has already created a set of test work products (test cases, procedures, scripts, and data) and checked them into the controlled library, the library process steps for testing a software build include the following.

Step 1: A tester checks out read-only copies of the build and necessary test work products from the controlled library into the dynamic library used for testing (test bed). If testing is automated as part of an integrated build script, the script moves the build and appropriate test work products to a specified dynamic library.

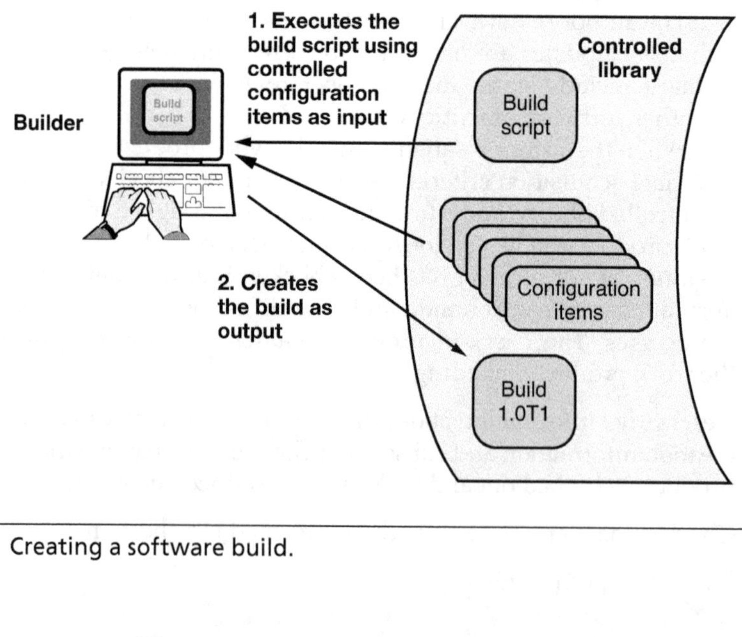

Figure 25.3 Creating a software build.

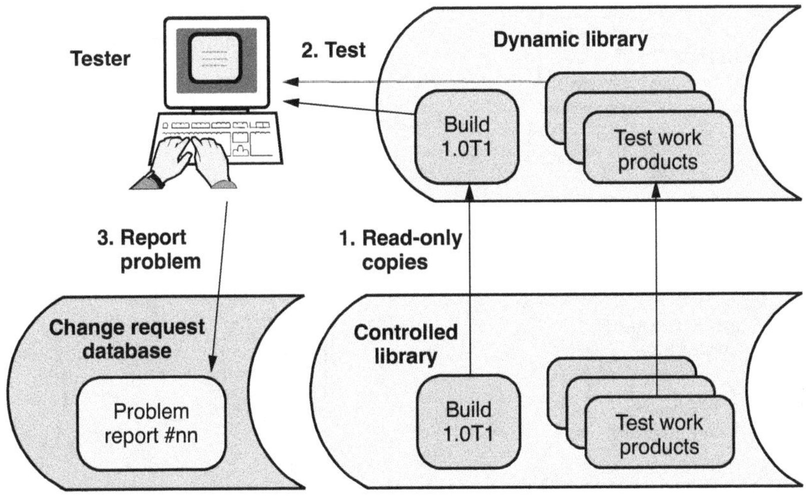

Figure 25.4 Testing a software build.

Step 2: The tester, or the automated script, executes one or more tests, and the results are either observed or recorded. An evaluation is then performed to determine if the actual results of the test matched the expected results. If the test is performed automatically, any anomalies encountered are reported to the appropriate individual(s).

Step 3: For manual testing, the tester reports any problems encountered during testing, with either the build or the test work products, into the change request (problem reporting) database. For automated testing, the person receiving the anomaly report interprets that anomaly and documents any identified problems into the change request database as appropriate.

Figure 25.4 illustrates this process for manual testing performed by a tester.

SCM Library Process: Modifying a Controlled Work Product

Once the appropriate level of authority approves a change request (problem report or enhancement request), one or more controlled work products are updated to incorporate that approved change. The library process steps to modify an existing controlled work product include the following.

Step 1: As illustrated in Figure 25.5, the assigned author checks out a read/write copy of each work product that needs to be updated from the controlled library into a dynamic library based on the impact analysis of the approved change request. The controlled library locks each work product checked out for modification so that other team members know it is being updated. This locking allows other team members to retrieve read-only copies but does not allow them to be checked out for modification. Note that some tools do not use locking, as they are sophisticated enough to allow for concurrent development. They enable developer reconciliation of multiple changes as copies of updated work products are checked back into the controlled library.

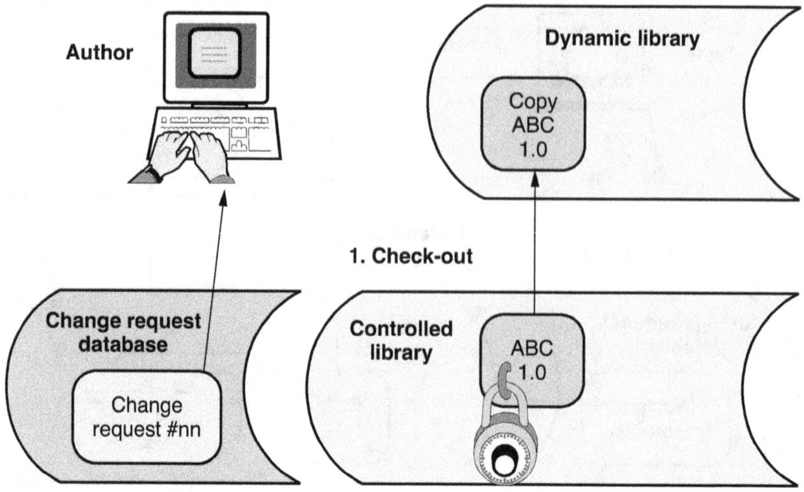

Figure 25.5 Modifying a controlled work product—check-out process.

Step 2: As illustrated in Figure 25.6, the author then makes the required changes to the copy of the checked-out work product based on the approved change request. The author saves the corrected copy in the dynamic library. The author then reviews or tests the updated work product and makes whatever additional updates are necessary to correct any defects found.

Step 3: As illustrated in Figure 25.7, when the author is satisfied that each changed work product is ready to share and the appropriate acquisition criteria are met, the author checks each updated work product back into the controlled

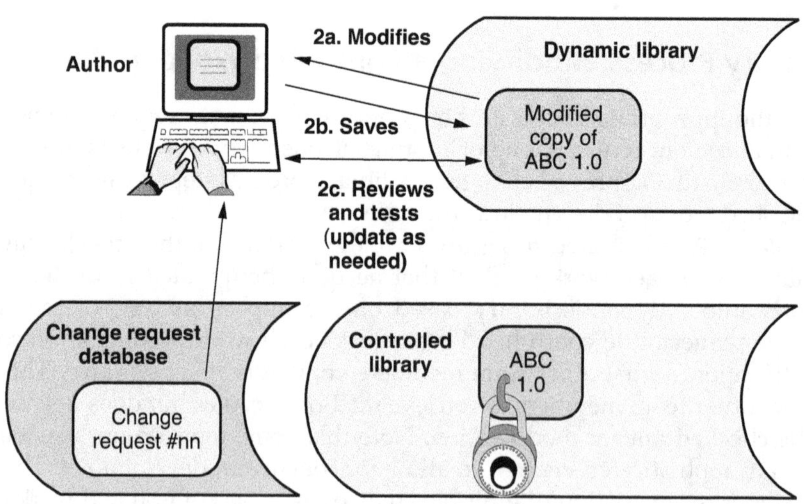

Figure 25.6 Modifying a controlled work product—modification process.

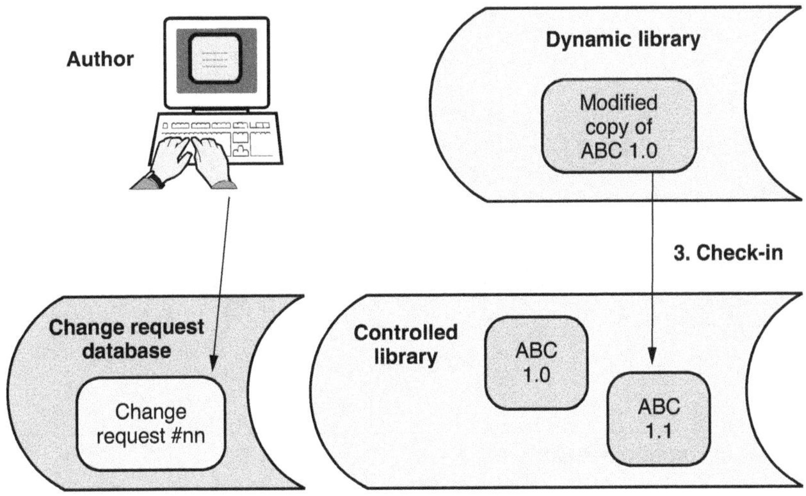

Figure 25.7 Modifying a controlled work product—check-in process.

library, and an updated version/revision number is assigned to each work product as appropriate. As it is checked in, the control library unlocks each work product so that the previous and the new versions of those work products are available for future modification. The author updates the work product library information as each product is checked in and should also update the change request with new status information.

SCM Library Process: Branching

Consider the main codeline example in Figure 25.8. Each label in this example represents the versions of a set of constituent configuration items used to create a build. Assume label 2 represents a build that has been released into operation, and label 3 represents a build that is currently under test and that contains new functionality updates to configuration items CI #1 and CI #2. A problem report comes in from operations. The label 2 build is debugged, and a defect is found in version 4 of configuration item CI #1. Version 4 cannot just be fixed and checked back in as version 8 because that would lose the new functionality or corrections added in versions 5, 6, and 7. On the other hand, the defect cannot just be fixed in version 7 to create a new build for operations (assuming that the defect even still exists in version 7 after the changes that have been made) for reasons that include the fact that the new functionality in version 7:

- May not be thoroughly tested
- May have dependencies on other updated work products
- Should not be given away for free

Branching is the software process of creating a new codeline containing one or more configuration items that are duplicated from an existing codeline in a

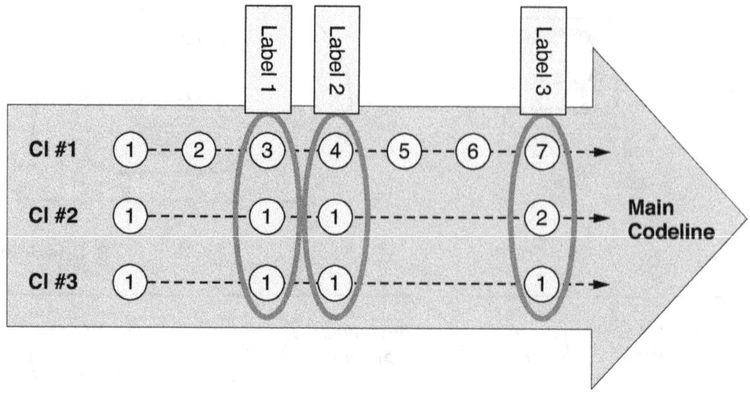

Figure 25.8 Main codeline—example.

controlled library. Branching the codeline can provide the flexibility needed to support multiple versions of the product at the same time and solve problems like the one described previously. As illustrated in Figure 25.9, a new codeline can be created from the work products in label 2. For example, version 4 of configuration item CI #1 is copied into the branched codeline, becoming version 4.1, and version 1 of configuration item CI #2 becomes version 1.1, and so on. Version 4.1 of configuration item CI #1 can then be checked out and modified to correct the defect. The updated CI #1 is then checked back in as version 4.2 into the branched codeline. The corrective build label 2.1 can now be created and released into operations containing the fix to the reported problem. At the same time, new development can continue on the main codeline.

Another example of where a branched codeline might be useful is to "freeze" a version of the product build for a testing cycle. Critical fixes that allow testing to continue can be done to the branched codeline while new development continues on the main codeline. Branching may also be used to create customized versions of work products or for concurrent, parallel development that allows multiple people to work on the same files at the same time. Branches can be taken from the original codeline, or branches can be taken from other branches. However, caution should be used when branching is implemented. Branching is the most complex of the library processes. Without a solid and disciplined branching strategy, chaos can quickly take over, with multiple copies of many configuration items needing support. A well-structured branching approach that minimizes the number of active branches at any given time reduces complexity.

One useful branching strategy is to make sure that new development is always done on the main codeline (or trunk). The advantage of this is that the branched product releases, or other branches of the main codeline, are relatively temporary in nature. Once testing is complete on a frozen branch or once a released version of the product is no longer supported in operations, its associated branch can be archived as necessary and removed to minimize the number of branches from the main codeline.

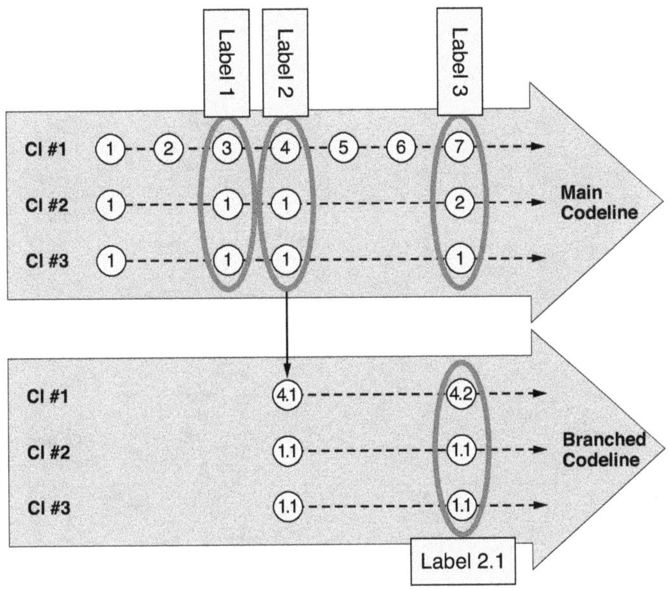

Figure 25.9 Branching—example.

SCM Library Process: Merging

Consider the defect that was corrected in the operations release label 2.1 build discussed in the last section. The current operations problem was solved with this corrective release. Suppose this defect also exists in configuration item CI 1 versions 5, 6, and 7 and it is not corrected. There is a risk that a future product release may bring back an old problem, resulting in very unhappy stakeholders. One answer to this is to check out version 7 of configuration item CI 1, edit the file to make the same correction a second time, and check it back in as version 8. Another answer is *merging*, taking two or more versions of a work product and combining them into a single new version. In the previous example, version 4.2 (with the correction) could be merged with version 7 (with the new functionality) to create a new version 8, as illustrated in Figure 25.10. Of course, merging can be done in the opposite direction, with changes being merged from the main codeline into items on a branched codeline.

Merging can also be used to combine functionality added on several concurrent, parallel development branches back into the main codeline. However, if multiple developers change the same work product, several variations can arise:

- If functionality added by the various developers does not conflict, that functionality can simply be merged. For example, one developer adds the ability to add a customized logo to a report, and another developer adds a graph to the report. These changes do not affect each other, so no conflict exists in the merge.

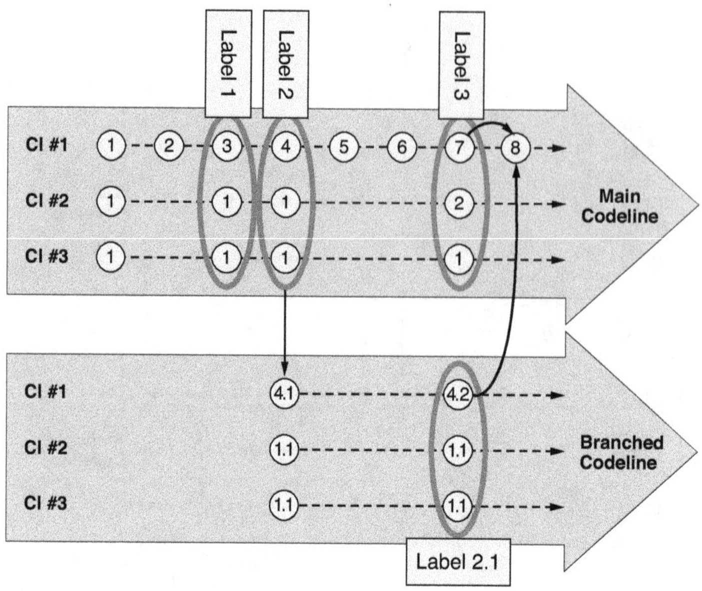

Figure 25.10 Merging—example.

- If functionality added by the various developers creates a merge conflict, that conflict may be able to be corrected through blending. In this case, one change may supersede the others. For example, one developer adds a logo to the report and another developer adds a customizable logo. Since the static logo can be added as a customized logo, the customized logo change may be chosen to supersede the other change. Changes may also be blended during a merge. In this logo example, the static logo change added by the first developer might be blended by adding it as the default to the customizable logo change.
- Suppose functionality added by the various developers creates a merge conflict and blending cannot be used to correct the conflict. In this case, decisions will have to be made to prioritize the changes and select one over the others, or additional changes will have to be made manually to incorporate the intent of all changes.

In any case, once the merge is accomplished, the resulting new version should be tested to verify that the merge did not result in new defects not identified during the merge analysis process.

Chapter 26
Configuration Identification

Software *configuration identification* is the software configuration management (SCM) activity that involves the partitioning of the software product into a hierarchy of controllable *configuration items (CIs)* and establishing baselines for those configuration items. According to Keyes (2004), "effective configuration identification is a prerequisite for the other configuration management activities (configuration control, status accounting, auditing), which all use the products of configuration identification." Configuration identification activities include:

- Determining which software work products require control as configuration items
- Identifying which documentation information is needed to describe the functional and physical characteristics for each configuration item
- Defining the acquisition points and acquisition criteria for each configuration item
- Assigning unique identifiers (naming schemes) to each configuration item and their versions and revisions
- Acquiring the identified configuration items and creating or releasing baselines for each item
- Defining the project baselines, their contents, and mechanisms used to establish those baselines
- Determining who will have access to each configuration item and at what level of access (create, read, update, delete)

Software configuration identification is the first of the four core activities of SCM to be done during a software project because it is the prerequisite for performing the other three:

- Configuration control
- Configuration status accounting
- Configuration auditing

1. CONFIGURATION ITEMS

> Describe configuration items (e.g., baselines, documentation, software code, and equipment), and identification methods (e.g., naming conventions and versioning schemes). (Understand)
>
> **Body of Knowledge VII.B.1**

Types of Work Product Control

The software development process produces many different software work products. A *software work product* is any tangible output, artifact, or specific, measurable accomplishment from the software development activity or process. Examples of software work products include source code modules, models, electronic files, documents (requirements and design specifications, plans, test documentation, user documentation, training materials, and so on), databases, services, reports, metrics, completed forms, logs, and other records. As illustrated in Figure 26.1, some software work products may be controlled, while other low-risk or temporary work products are never placed under any kind of control. Examples of temporary software work products include printed program listings with the programmer's handwritten annotations, periodic status reports, object code, or a recorder's personal notes from a meeting that are later used as input to the official meeting minutes.

Controlled software work products typically fall into two major categories. The first category is *quality records*. Work products designated as retained quality records must be controlled but are not considered to be under formal SCM. For example, for most projects, it would be considered overkill to report a correction to a set of meeting minutes by opening up a change request in the change request tool and formally approving and tracking that change to closure. In fact, no formal change control process is typically implemented for quality records.

The second category of controlled software work products is called configuration items. A *configuration item*, also called a *computer software configuration item* (CSCI), is a work product placed under SCM and treated as a single entity. "In practice, a configuration item is a stand-alone, test-alone, use-alone element of the software that requires control during development and subsequent use in the field" (Berlack 1992).

Software Work Product Partitioning

The first step in configuration identification is partitioning the software product(s) into a hierarchy of controllable items. As illustrated in Figure 26.2, *composite items* can be made up of other smaller work products called *constituent items*, which in turn can be made up of other even smaller constituent items. The entire system can be treated as a single composite item that is made up of various constituent items including hardware and software subsystems. Those software subsystems can

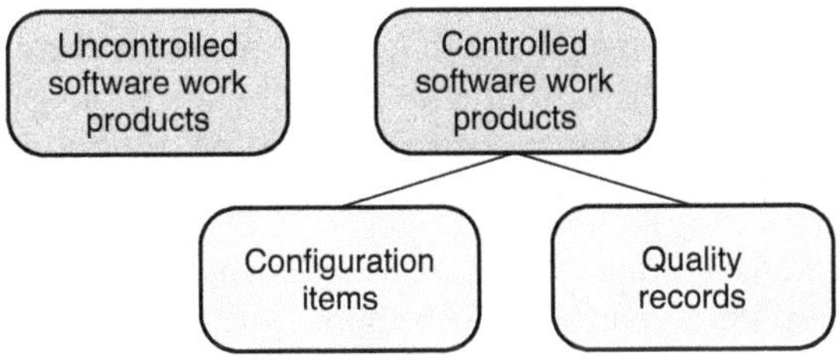

Figure 26.1 Types of software work product control.

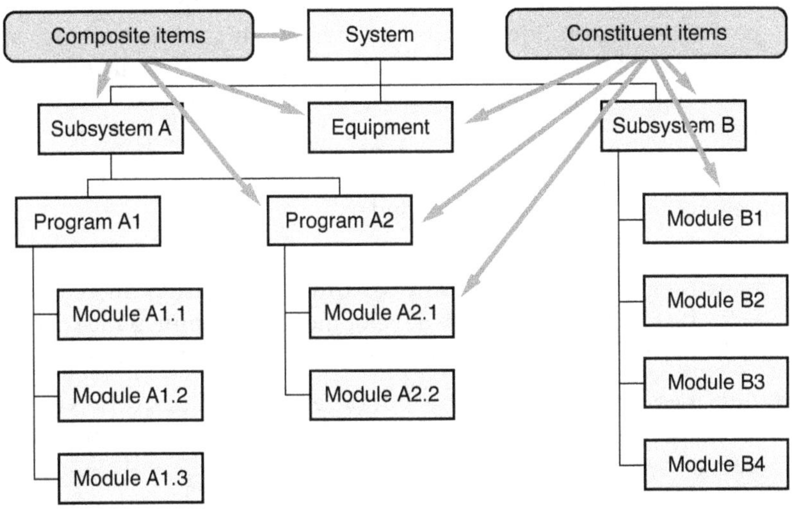

Figure 26.2 Software product hierarchy.

each be made up of smaller constituent items, including programs and individual source code modules.

Multiple technical factors should be considered when partitioning the software into items. These factors include:

- *Product architecture*: The partitioning of the software product into items typically reflects the architectural structure of the software. This means that each software product as a whole should be an item, and each sub-product that makes up that product in the architecture can be a constituent item.
- *Reuse*: Partitioning should consider whether a work product is reused by multiple subsystems or products. If part of a work product has high reuse and another part of that same work product is not being reused, it may be appropriate to break the work product into two items. If the item already exists in another product and is being reused in this product, it may also already be established as a configuration item.

- *Documentation hierarchy*: Each document in the documentation hierarchy must have a unique identifier. A decision should also be made as to whether the document should be divided into constituent items (e.g., chapters or sections).
- *Requirements allocation and traceability*: Traceability tracks the relationship between the requirements and the items to which those requirements are allocated. If more comprehensive traceability is required, a more detailed breakdown of the item hierarchy will be necessary.
- *Change and dependencies*: Another factor to consider is whether or not a work product is expected to have a high level of change over time. If part of a work product is expected to have a high rate of change and another part of that same work product is expected to have little or no change, it may be appropriate to break that work product into two items.

Multiple managerial factors should also be considered when partitioning the software into items. These factors include:

- *Scope and magnitude*: Larger and more complex software systems with more people involved typically require more rigorous and stringent SCM systems to support the development and maintenance processes. This, in turn, requires a more complex software product item hierarchy.
- *Author assignment*: If different authors are working on different parts of the same work product, it may be appropriate to break it into separate items.
- *Authority and responsibility*: The role and level of responsibility required to review and approve a software work product typically influences the product's partitioning. For example, the team lead may have the authority to approve the completion of a low-level constituent item (source code module). However, it may require a technical or project manager to approve the completion of an intermediate composite item (program or subsystem) and the director of software development to approve the completion of a high-level item (system or application). The degree of partitioning needed may also be impacted if more than one team or project will use, review, or approve the work product.
- *Planning and estimation*: The structure of the items' partitioning may also parallel the work breakdown structure used to plan and estimate the software development project and maintenance effort. The choice of an incremental or evolutionary development life cycle may also guide product partitioning.
- *Need for visibility*: Items are used to track the progress and completeness of the work. The more visibility that is needed into the status of software work products, the more detailed the hierarchy must be. The work product's criticality to the project may also need to be considered.

- *Commercial off-the-shelf (COTS) or supplier work products*: If an item is procured off-the-shelf or is produced by a supplier, it will impact the configuration item's partitioning. Only the executable software and a user manual are typically available to place under SCM and not individual source code modules.

Identifying Configuration Items

After the software work product hierarchy has been established and individual items are identified, the next step in the configuration identification is to determine which of those items should be designated as configuration items to be put under formal change control. So what items should be designated as configuration items? How much control is needed? Like many questions in software development, the answer to these questions depends on risk. As illustrated in Figure 26.3, there are many risk indicators that need to be considered when determining which software items need to be placed under configuration control. The risk indicators in this figure are just examples. Each project should perform an analysis of relevant project and product risks to guide decisions about the types of software items that should be designated as configuration items. Typically, lower-risk projects can safely select a less rigorous configuration control philosophy and designate fewer software items as configuration items. Conversely, higher-risk projects will typically require more rigorous configuration control and a larger number of controlled configuration items.

Because of the high risk of direct impact on the customers or users if issues arise, externally delivered work product items should always be designated as configuration items. Examples of externally delivered work products can include:

- Software executables (and the constituent source code modules, programs, and subsystems used to create them)

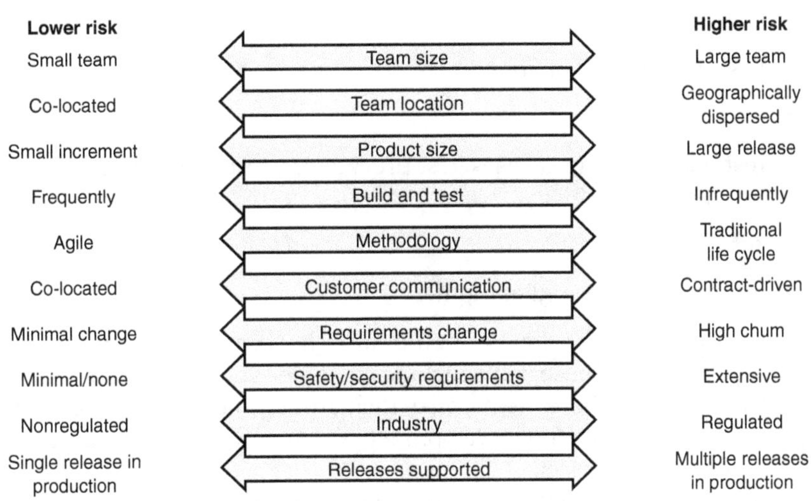

Figure 26.3 Examples of configuration management risk indicators.

- System data files and databases
- COTS or third-party software included in or delivered with the software system
- User and operational documentation (user or operations manuals, help files, installation instructions, release descriptions or notes)
- Training materials

The project may also designate other internal software items as configuration items based on the project's required rigor and level of risk. Examples of internal software items that should be considered for designation as configuration items for traditional software development include the following:

- Requirements, interface, and design specifications
- Plans (project, quality, verification and validation, test, SCM, and other plans)
- Test documents (test cases, test procedures, test data, test automation products, and other test documents)
- Project-level process and work instruction documentation
- Tools, hardware, and equipment

According to the CMMI® (2023), examples of internal software items that should be considered for designation as configuration items for agile software development include the following:

- Agile process documentation
- Product backlogs (user stories)
- Sprint backlogs
- Release plans
- Task (or Kanban) boards
- Burndown charts
- Test documents (test cases, test procedures, test data, and test automation products)
- Retrospective data
- Tools, hardware, and equipment

Externally supplied items should also be considered for designation as configuration items. These items include customer-supplied equipment, software, or documentation. These items also include supplier-related contracts, plans, or other items used in the development of the software work products or the management of the project.

"Tools and other capital assets of the project's work environment" and "other items that are used in creating and describing these work products" (CMMI 2010a) should be considered for placement under configuration control to make sure that past versions of the software work products can be re-created and maintained as necessary. Designating tools (and other assets) as configuration items may be essential when different versions of those tools were used to build multiple versions of concurrently supported software. If a tool is designated as a configuration item, both that tool's executable and associated documentation (user or operator manuals, installation instructions) should be placed under SCM. Other considerations may include keeping tool licenses current, maintaining copies of tool certifications if applicable, and maintaining a history of tool updates and customizations.

Examples of support tools and other assets that should be considered for placement under SCM include:

- Compilers, assemblers, linkers, loaders, and other build tools
- Automation scripts
- Operating systems
- Word processors and other documentation tools
- Computer-aided software engineering (CASE) tools, testing tools, and SCM tools
- Simulation and modeling tools
- Hardware and equipment

Functional and Physical Characteristics

A configuration item's functional and physical characteristics are described in or pointed to by its associated metadata. "Metadata is a database concept that means the data about the data stored in the database" (Hass 2003). In the case of a configuration item, metadata are used to define the characteristics of the configuration item, including its name, description, purpose, author, version/revision history information, status, and other information as specified (e.g., the coding language used to create the source code). For many projects, these metadata are documented inside the configuration management tools.

The physical characteristics of a configuration item are defined by maintaining references to other related configuration items that describe those characteristics. For example, the metadata for a software source code module might include references to the detailed design document that defines its internal structure or to external interface specifications. There must also be a mechanism to identify what versions or revisions of each constituent configuration item, also called *configuration components, computer software configuration components (CSCCs), configuration units,* or *computer software configuration units (CSCUs),* make up each version of each composite configuration item, also called *computer software configuration items (CSCIs).* For example, the metadata for a software user manual might include references to all of the chapters (other constituent configuration items) and their versions or revisions that are part of that manual.

As with physical characteristics, metadata define the functional characteristics of a configuration item by maintaining references to related specifications that describe those characteristics. For example, the metadata for a software build might include a reference to the software requirements (software requirement specification, use cases, user story) that describes its functional characteristics, or the metadata for a software source code module might include traceability back to the requirements that were allocated to it. These references also include the traceability of requirements to the related baselined configuration items.

Acquisition Points, Acquisition Criteria, and Baselines

It is important to define when each configuration item is acquired (placed under configuration control). Quality gates are typically used to approve acquiring a configuration item and its placement under configuration control (see Figure 26.4). Examples of quality gates include the successful completion of:

- A peer review (desk check, inspection, walk-through)
- A test activity
- A project review (phase gate review, major milestone review, iteration review)
- An independent product analysis or audit

The *acquisition point* for each configuration item, along with its associated quality gate and acquisition criteria, should be defined during software configuration identification planning. In agile software development, the definition of "done" can be used as the acquisition criteria. "Understanding the definition of 'done' is important to being able to verify and validate that the correct versions are produced and delivered from each Sprint" (CMMI 2023).

The earlier the acquisition point is in the life of a configuration item, the more rigorous the level of control and the more formal the communication about changes to that item must be, which may result in more stability. The later the acquisition point, the easier and quicker it is to make changes, resulting in more flexibility. The higher the risk that changes to a configuration item will create potential issues, the earlier the acquisition point is established for that configuration item in the life cycle. For example, consider a source code module:

- If the acquisition point is set after peer review, all defects found in unit, integration, and system tests must go through formal change control. The peer review acquisition point may be too early for most projects. However,

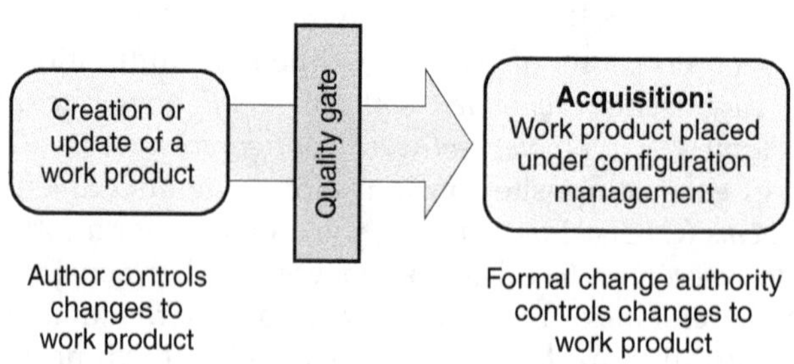

Figure 26.4 Configuration item acquisition.

if a project has an independent verification and validation (IV&V) team that does unit testing, early acquisition may provide the formality needed for the IV&V and development teams to communicate effectively.

- For many projects, an acquisition point for a source code module may be more appropriately set after unit test or integration test depending on when the handoff occurs to a testing group outside development.
- For small development teams with high levels of internal communications, it may even be appropriate for the acquisition point to be set at the point of product release so that only defects reported from operations are subject to formal change control.

For other configuration items, such as requirements or design specifications, the successful completion of the peer review, or a major phase gate or milestone review, may be the appropriate acquisition point. These points act as internal release points where the specification moves from creation by its authors to use by other members of the project team (development, test, or technical publications), so the need for more control (stability) and more formal communication about changes and their impacts may be desirable.

The *acquisition criteria* define what must be true before acquiring a configuration item. For example, if the acquisition point for a source code module is set at unit test, the acquisition criteria might include that:

- All planned unit test cases have been executed
- All test cases have passed (or any outstanding anomalies are formally reported into the change request system)

When a configuration item is acquired, it becomes a *baseline configuration item*, and is deposited into (checked into) the appropriate control library. A *baseline* is a "product or specification that has been formally reviewed and agreed upon, that thereafter serves as the basis for further development, and that can be changed only through formal change control procedures" (IEEE/ISO/IEC 2017a). Prior to baselining, the author of a configuration item can make changes to a work product quickly and informally. After baselining, changes to configuration items can only be made through formal change control (as discussed in Chapter 27). When baselined configuration items are changed, each changed item is also reacquired, creating a new baseline for that item.

Identification Methods

Another role of software configuration identification is to define an identification schema and naming conventions for assigning a *unique identifier* to each configuration item and its various versions and revisions. Several identification schemas may be used for configuration identification, including different schemas for different types of items (source code modules, builds, documents, and so on). According to Leon (2005), "a good identification scheme should:

- Facilitate the storage, retrieval, tracking, reproduction, and distribution of the configuration items

- Make it possible to understand the relationship between the configuration items from their names."

In its simplest form, the filename used to save the configuration item can be used to uniquely identify it. For example, when it was first created for edition 1, this chapter was saved in a file named *Part 7 Configuration Management—Chapter 27 Configuration Identification 01.00.doc*. The part and chapter numbers and version/revision number 01.00 was embedded in the filename to indicate that this was the first version of this chapter and that there were, as yet, no corrective revisions. As reviewing and editing occurred, the second two-digit number was incremented to create new filenames each time changes to the chapter occurred to incorporate corrections and suggestions that created new revisions. With the rewrites to this chapter for the second edition that includes new information and shifting of chapters to implement the new requirements of the updated CSQE Body of Knowledge, the chapter number was changed, and the first two-digit number indicating the chapter version was incremented to create a new filename, *Part 7 Configuration Management—Chapter 25 Configuration Identification 02.00.doc*, and so on.

Within the controlled library, configuration items may be labeled with a unique filename or other identifier. Most projects use SCM tools that automatically assign an incremented version/revision number every time a new configuration item, or a new copy of an existing item, is checked into the tool. As illustrated in Figure 26.5, for example, the three constituent configuration items have unique names CI #1, CI #2, and CI #3. They also have identifiers that indicate when they changed. CI #1 has changed seven times with version/revision identifiers 1–7, while CI #2 has changed only once with version/revision identifiers 1 and 2, and CI #3 has not changed because it is still at version/revision 1.

A project typically has multiple configuration items checked into one or more control libraries. There are times when it is desirable to identify relationships between these configuration entities (e.g., to identify the set of configuration items and their versions or revisions that make up a baseline or to identify all the constituent configuration items to be used to build a composite configuration item). These associations are accomplished using *labeling*, also called *tagging*. Most

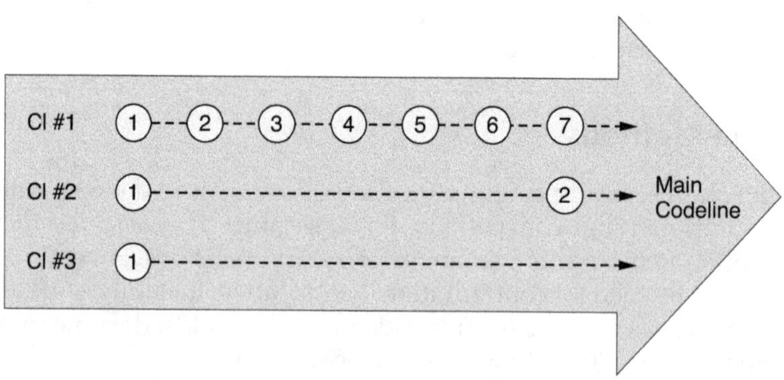

Figure 26.5 Constituent configuration item identification schema—example.

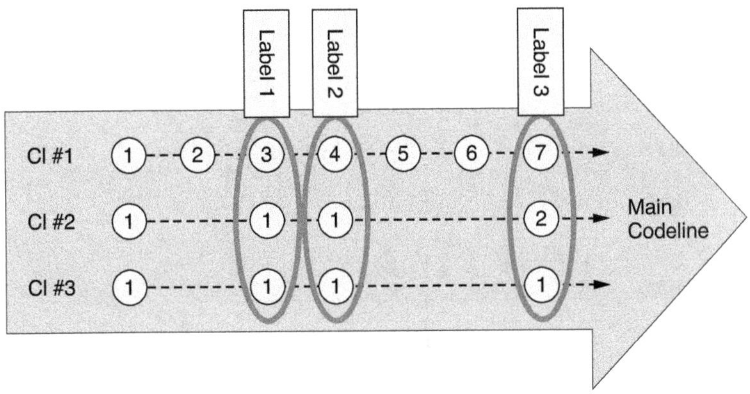

Figure 26.6 Labeling—example.

configuration management tools have the functionality to automate the labeling process.

Figure 26.6 illustrates a very simple example of labeling, where work products are labeled only in a single library. In this example, label 3 associates version/revision 7 of CI #1 with version/revision 2 of CI #2 and version/revision 1 of CI #3. Notice that the same version/revision of a work product can have more than one label. In this example, version/revision 1 of CI #3 has labels 1, 2, and 3 associated with it. More complex labeling associates work products across multiple controlled libraries and can select specified items from each library.

The labels in Figure 26.6 are version/revision-type labels, which associate the specific versions or revisions of the work products with the time the labels were created. A version/revision-type label might be useful, for example, when associating the specific versions or revisions of the source code modules that were built into a specific software build.

Another type of labeling is file labeling, where a set of files are associated without specifying their specific versions or revisions. Rules are then used to identify which versions or revisions are to be used at any given time. A simple example of these rules would be to use the latest version/revision of each file whenever the label is used. A file-type label might be useful when associating the source code modules that should be used when automatically creating builds on a periodic basis. The automation would then build the latest versions or revisions of each source code module into the new executable it creates each time it runs.

As illustrated in Figure 26.7, version/revision numbers must reflect that branching and indicate the branching relationships when branching occurs. In this identification scheme example, an extension was added to the version/revision number and label number to show the relationships. That is, CI #1 version/revision 4.1 was copied from CI 1 version/revision 4, CI #2 1.1 was copied from CI #2.1, CI #1.1 was copied from CI #3.1, and label 2.1 is an updated version/revision of label 2. Again, modern SCM tools support labeling and dynamic branching and merging functions.

The exact version/revision of anything created by a build process must also be readily identifiable. Figure 26.8 illustrates an example of a software build identification scheme. This scheme assigns:

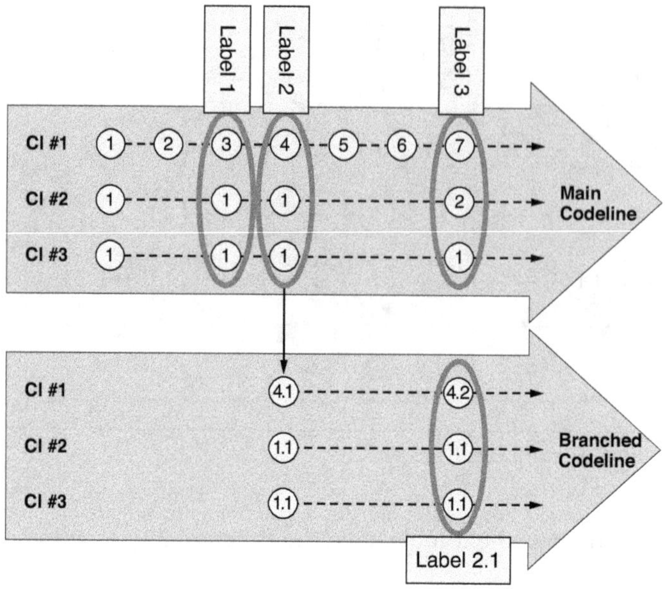

Figure 26.7 Branching identification scheme—example.

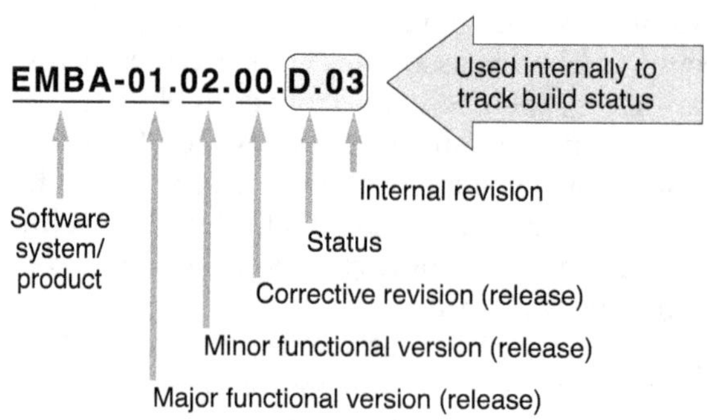

Figure 26.8 Build identification scheme—example.

- An alphanumeric name to the software system or product, in this case EMBA
- A two-digit indicator of the major functional version (release) of the build, in this case 01 (indicating that this is the first major release of this product). This major functional version indicator is incremented when a new major functional version build is created.
- A two-digit indicator of the minor functional version (release) of the build, in this case 02 (indicating that this is the second minor release of this product). This minor functional version indicator is incremented

when a new minor functional version build is created, for example, when the first build is created for the third incremental release against major release 01.

- A two-digit indicator of the corrective revision (release) of the build, in this case 00 (indicating that there are no corrective releases against this build). This corrective revision indicator is incremented when a new corrective revision build is created, such as when a service pack build, correcting defects found during operations, is created for major release 01 and minor release 02 of this product.

The final letter and two-digit number in this identification scheme are used as build status counters internally within the development organization. In this case D.03 indicates that this is the third developmental testing build of the product. Figure 26.9 illustrates an example of how this build status counter might be used during development. In this example, the letter D indicates that the build is being used for developmental testing, and the two-digit internal revision number is incremented each time a new developmental testing build is created. The final developmental testing build EMBA-01.02.00.D.nn becomes the first system testing build EMBA-01.02.00.S.01 when it passes through a quality gate (e.g., successful completion of a "ready to system test" review). The final system testing build, EMBA-01.02.00.S.nn, becomes the first beta testing build, EMBA-

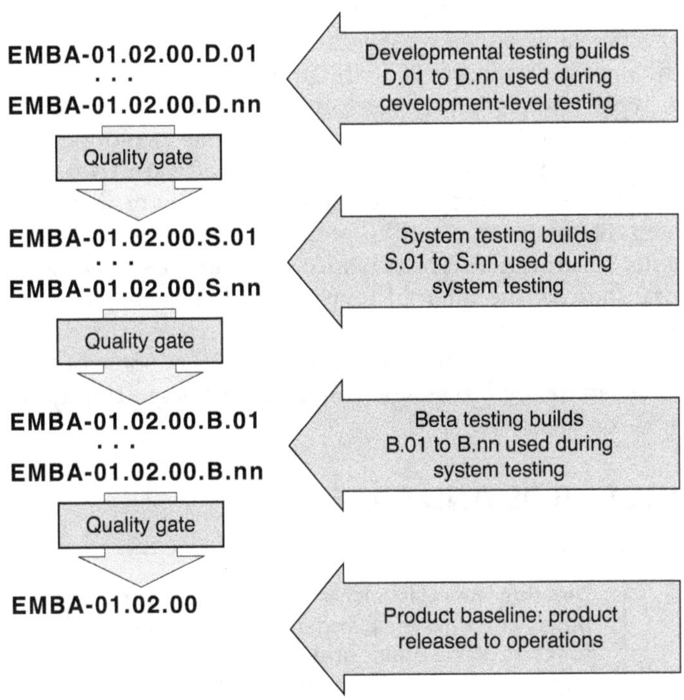

Figure 26.9 Implementing this build identification scheme across multiple builds—example.

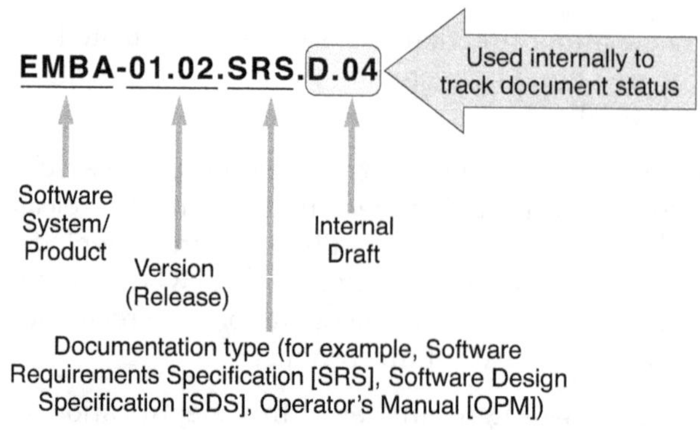

Figure 26.10 Documentation identification scheme—example.

01.02.00.B.01, when it passes through a quality gate (e.g., successful completion of a "ready to beta test" review). The final beta testing build EMBA-01.02.00.B.nn becomes the released build EMBA-01.02.00 when it passes through a quality gate (e.g., successful completion of a ready-to-ship review). Note that the internal status indicator is dropped when the build is released to operations. Many SCM tools include mechanisms for automatically assigning identifiers to individual configuration items and builds. It is recommended that the project use the identification schema from their selected tools and not reinvent a different one without specific value-added reasons.

As a final example, Figure 26.10 illustrates a document identification scheme. In this example, the alphanumeric software system or product name (EMBA), the two-digit major functional version, and the two-digit minor functional version (01.02) link the document back to the software release associated with the document. There is also a three-character acronym that indicates the type of document, in this example, SRS, indicating that the document is a software requirements specification. There is also an internal status indicator letter and number that indicate the internal revisions of the document, in this case, D.04, indicating the document's fourth developmental revision (or draft).

It should be noted that these are just examples of some of the many possible schemas that can be used to assign unique identifiers to configuration items and their versions or revisions.

2. SOFTWARE BUILDS AND BASELINES

> Describe the relationship between software builds and baselines, and describe methods for controlling builds and baselines (e.g., automation and new versions). (Understand)
>
> **Body of Knowledge VII.B.2**

Software Build

A *software build* is the process of combining constituent software configuration items into composite configuration items. If the result is an executable software work product, that executable is also called a build. According to the *IEEE/ISO/IEC Systems and Software Engineering—Vocabulary* (IEEE/ISO/IEC 2017), a build is "an operational version of a system or component that incorporates a specified subset of the capabilities that the final product will provide."

The build process includes all of the activities associated with processing source files to create one or more derived final target files (builds). According to Bays (1999), these "build activities can be broken down into several key actions:

- Dependency checking
- Source compilation
- Executable linking
- Occasionally some form of data file generation."

Guaranteeing the ability to produce high-quality, consistent, and complete builds may be the most important role of SCM, after protecting the configuration items from loss and unauthorized change.

Aiello and Sachs (2011) list the principles of build engineering, including the following:

- Builds are understood and repeatable
- Builds are fast and reliable
- Every configuration item is identifiable
- The source and compile dependencies can be easily determined
- Code should be built once and deployed anywhere
- Build anomalies are identified and managed in an acceptable way
- The cause of broken builds is quickly and easily identified (and fixed)

However, the software build process is not limited to creating the software executable. A software build can also result in a software document, set of training materials, or any other software product being created from its constituent parts. For example, a software training manual can be built from the constituent configuration items that make up its chapters. In hardware, manufacturing constitutes the build process (replicating the hardware component from its parts).

Build Automation

To make sure that the same build processes are followed consistently every time a build is performed, without the possibility of human errors, the build process should be automated as much as possible. Automation also allows builds to occur more often, providing more timely feedback as changes are made to the software and automation. Automation also eliminates the need for rigorous documentation of what files, options, and so on were used, since this information is typically documented as part of the automation. The build process is automated using build tools, or a command file called a *build script*, which specifies:

- The components of the build (source and derived)
  - Their versions
  - Their locations in the configuration libraries
- The build tools, including compilers, assemblers, linkers, loaders, and build scripts
  - Their versions
  - The required setting for options and environmental parameters
- Pointers to the appropriate macros, libraries, and other files to be included
  - Their versions
  - Their locations in the configuration libraries

For example, for UNIX these command files (build scripts) are usually shell scripts, and for C++ they are make files. These build tools or scripts may be designated as configuration items themselves and placed under configuration control to safeguard the reproducibility of the build process. The resulting builds (target files) are composite configuration items and should be considered new versions or revisions whenever the source files for the build or the build tool or script has been modified in any way.

Continuous Integration and Builds

Continuous integration is the widespread good practice from agile processes, which refers to regularly integrating, building, and testing the software solution as it is being implemented. To implement continuous integration during agile development, the automation of build activities is typically linked to automated testing, including regression testing of those builds. These automated build and test cycles may be initiated several times a day or immediately every time one or more new or updated source code modules are baselined and checked into the configuration management control libraries (which requires the automated monitoring of the controlled libraries to identify the baselining of changed components). The goal is to provide timely feedback to the developers on whether their changes adversely impacted the previously working software and to minimize the effort required to identify issues and isolate the associated defects. For example, if only one or a few code changes have occurred since the previous build when all the tests passed, then issues in the current build or test cycle are probably the direct result of those changes. In traditional development, many changes may be made in between build and test cycles, which results in more difficulty in isolating which change or changes caused the build or test issues.

Build Reproducibility

Software builds must be reproducible to release high-quality, consistent, and complete software products and to debug and correct reported problems found during testing and in operations. For a build to be reproducible, a standardized development environment with known elements is a necessity. The source files and their versions revisions used to create the build must also have a known configuration, which is why official builds are created using baselined constituent configuration items from controlled libraries. Many SCM tools make build reproducibility a trivial event if the appropriate decisions are made, and tool capabilities are appropriately implemented. However, for the build to be reproducible, the following information must be known about the tools:

- What was the configuration of the environmental platform (operating system version, hardware version, and other environmental elements that might impact the build) used to create the build?
- Which versions of which build tools (compilers, assemblers, linkers, loaders) were used to create the build, and how were the build tool options or environmental parameters set?
- Which versions of which macros, libraries, and other included files were used to create the build?

To verify that a build is reproducible, an attempt is made to reproduce that build and then compare the original build with the test build. Ideally, these two builds are created on completely separate platforms, where the test system's runtime, development, or build environments have been newly instanced using the specifications defined for those environments. The resulting second build is then compared with the original build using one or more of the following tests:

1. *Same file and directory content*: This test compares the listings of the files and directory paths in both source platforms to verify that they are identical.
2. *Identical source file content*: This test compares the contents of each pair of source files to verify that they are identical on both platforms.
3. *Same-size target files*: This test compares the target files to verify that they are identical in size.
4. *Byte-for-byte identical target files*: This test compares the contents of the target files to verify that they are identical on both platforms.
5. *Byte-for-byte identical intermediate files*: This test compares the contents of each pair of intermediate files to verify that they are identical on both platforms (Bays 1999).

These last two tests can be problematic in some systems because of complications, including:

- Compilation date and time stamps being embedded in the files
- Compilers that leave garbage in uninitialized variables

Control Points and Baselines

The *IEEE/ISO/IEC System and Software vocabulary* (IEEE/ISO/IEC 2017) has two definitions for baseline. The first definition is "A specification or product that has been formally reviewed and agreed upon, that thereafter serves as the basis for further development, and that can be changed only though formal change control procedures." The first definition of baseline refers to a configuration item baseline. This is the type of baseline that is created when a configuration item is acquired, as we discussed earlier.

Another type of baseline is a well-defined reference point that represents a snapshot of a configuration item (including all of its constituent parts) or a set of configuration items at a specific control point in the development life cycle. As illustrated in Figure 26.11, four classic types of baselines are established at traditional control points in the life cycle: functional baselines, allocated baselines, developmental baselines, and product baselines.

According to the IEEE *Standard for Configuration Management in Systems and Software Engineering* (IEEE 2012), SCM planning "shall define how baselines are established" in terms of the following:

- The events that establish a baseline
- The items that are to be controlled in the baseline
- The procedures used to establish and change the baseline
- The authority required to approve changes to the approved baselined documents

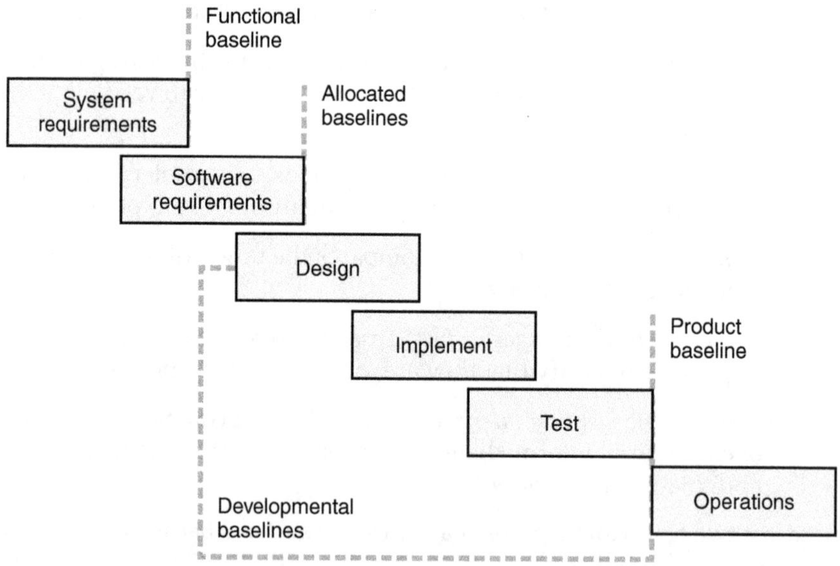

Figure 26.11 Types of control point baselines.

The *functional baseline* is typically established upon the completion and approval of the product's technical specification (system requirements). The functional baseline is an agreement between the customer(s) (or the marketing function), the organization developing the system, and other stakeholders about the needed features and functions of the system. The functional baseline becomes the basis for technical planning, designing, coding, and testing of the system and for project management planning. It also provides a basis for verification and validation planning and for system validation. Configuration items that might be included in the functional baseline are the business needs document, stakeholder-level requirements (use cases, usage scenarios, or user stories), system requirements specifications, external interface specifications, acceptance test plans and specifications, and the project charter and initial project plan. The system-level requirements are then allocated down to the highest-level constituent configuration items (hardware and software subsystems) during system design.

The *allocated baselines* are established upon the completion and approval of the technical requirements specification (hardware and software requirements) for each subsystem. For example, an allocated baseline would be established for each software application within the system when its software requirements specification is approved. There could also be one or more hardware-allocated baselines for each hardware component in the system.

The *developmental baselines* are optional, intermediate baselines established throughout the software development process. For example, an architectural design developmental baseline might be established with the approval of the software architecture. This software architectural design baseline might include the software architectural design specification, the integration test plan and specification, and the software system test cases and procedures. Another example is a source code developmental baseline established with the completion of the unit test of that source code module. This source code baseline might include the low-level design specification for that module, the source code module itself (and associated data or other files), unit test cases and procedures, and automated unit-level test scripts. There may also be developmental baselines drawn for various levels of testing.

The *product baseline,* also called the *production baseline* or *operation baseline,* "snapshots" the product as it is delivered to the customers or users for use in operations and starts the maintenance part of the life cycle. The contents of the product baseline may include configuration items promoted from the functional, allocated, and developmental baselines, as appropriate, along with the released software and other deliverables. The product baseline also includes support tool configuration items and other configuration items needed to reproduce the deliverables. The highest level of change authority typically controls the product baseline.

For agile development, since software is developed using an iterative, incremental approach rather than a traditional phased approach, these classic four baselines are not relevant The associated system requirements and their allocation are much more blurred as stories and their designs emerge over time. Control point-type baselines that might be more relevant and useful to agile development might include:

- *Product backlog baseline*: That defines the list of prioritized, known stories at the beginning of each iteration
- *Selected product backlog baseline for the iteration*: That defines the top-priority stories selected from the product backlog that will be developed during the iteration
- *Iteration baseline*: The design, code, test, and possibly documentation configuration items that were produced (created or updated) by the end of the iteration, which is analogous with a product baseline for that iteration
- *Release baseline*: The configuration items, including executables, user documentation that are delivered in the release, and the constituent configuration items used to build those deliverables, which is analogous with a product baseline for that release

Chapter 27

Configuration Control and Status Accounting

1. ITEM CHANGE AND VERSION CONTROL

> Describe processes for documentation control, item change tracking, and version control that are used to manage various configurations. Describe processes used to manage configuration item dependencies in software builds and versioning. (Understand)
>
> **Body of Knowledge VII.C.1**

Configuration Control

Software *configuration control* is "the systematic process that ensures that changes to a baseline are properly identified, documented, evaluated for impact, approved by an appropriate level of authority, incorporated into all impacted work products, and verified." (MIL-HDBK 2001). Software configuration control procedures include:

- Mechanisms for:
  - Placing configuration items under control
  - Requesting and documenting changes to controlled configuration items
  - Informing affected stakeholders of the change request and soliciting their input to the impact analysis
  - Informing affected stakeholders of the decision to accept, defer, or reject the change and for obtaining their commitment to the change if it is accepted
  - Tracking requested changes from submission through final disposition (rejection or completion of the change)
  - Verifying the implementation of approved changes
  - Obtaining deviations and waivers

- Requirements for performing impact analysis for each requested change or set of changes
- Established authorities for making decisions on accepting, deferring, or rejecting requested changes
- Controls for making sure that unauthorized change does not occur

As discussed in Chapter 25, there is a dichotomy around what level of control is necessary for software configuration management (SCM). Individual developers need flexibility to do creative work, modify source code, try out what-if scenarios, and make mistakes, learn from them, and evolve better software solutions. On the other side, teams need stability to allow software work products to be shared with confidence, create builds and perform testing in a consistent environment, and ship high-quality products with confidence. This dichotomy requires that a balanced approach to SCM, and especially to configuration control, be maintained.

The choice is not really between complete flexibility (anyone can change anything at any time) and complete stability (everything is locked down so that change only happens through rigorous control processes). As the software product is being created, reviewed, tested internally by development, independently tested, and finally released, configuration items of that product move through a continuum from complete flexibility, through various levels of more rigorous control, to a fixed configuration at release to operations. The risk-based analysis is used to decide the levels and types of control, the processes and mechanisms used to implement that control, and the number of control authority levels that are appropriate for each software configuration item. These configuration control decisions should be driven by the organization (standards and guidelines) and be "right-sized" during each project's SCM planning.

This risk-based analysis should consider the trade-off trilogy, illustrated in Figure 16.3, between cost, schedule (cycle time), and product (including product's functionality and quality). This trilogy exists for the configuration control decisions just as it does for software project management. If the rigor and formality of the configuration control process are increased to increase the integrity of the software product, then the costs (including effort and other resources) and cycle time for making changes also increase. When analyzing these trade-offs, consideration should be given to the entire useful life of the software product, including:

- Initial development costs and schedules for delivering the required product functionality, quality, reliability, safety, security, and so on
- Corrective, adaptive, preventive, and perfective maintenance costs and schedules
- The total cost of ownership (operational costs) of using and maintaining the software product in the field, including training people to use the software
- Internal and external failure costs and other impacts if the software has problems

Many software companies have focused on optimizing initial software development to the long-term detriment of the software products and processes, creating massive technical debt. For example, by overemphasizing the ship date of the initial release, shortcuts in requirements and design can result in poor-quality products that do not match their intended use, are failure-prone in operations, or are extremely difficult to modify or maintain. Of course, the reverse is also true. Overemphasizing product functionality and quality may result in analysis paralysis, function bloat, too much complexity, or missed marketing windows.

Configuration Item Attributes

For each category of identified configuration items, attributes should be specified to define how that category will be managed throughout the useful life of the software product. Different projects may use different configuration item attributes depending on the needs and context of the individual projects. Table 27.1 shows examples of typical attributes for different configuration item categories.

Descriptions of these example configuration item attributes include:

- *Configuration item attributes established as part of configuration identification*
  - *Configuration item (CI) categories*: Similar configuration items can be categorized together for configuration management if they are all controlled in the same way. For example, categories might include legacy code, new code, reused code, detailed designs, process definition documents, and so on.
  - *Acquisition point*: The point at which the configuration item is brought under configuration control. The acquisition point should be defined during configuration identification planning.
  - *Acquisition criteria*: Criteria that must be true before a configuration item can be acquired. The acquisition criteria should be defined during configuration identification planning.
  - *Location*: Where the controlled copy of the configuration items in that category are placed when acquired (where it can be retrieved from when needed).
- *Configuration item attributes established as part of configuration control:*
  - *Owner*: The owner of the configuration items in the category
  - *Access control*: Roles with create, read, update, and delete access control privileges over the configuration item.
  - *Change control type*: The types of change control (change requests through the configuration control board (CCB), review and approval of modified configuration item) for the configuration item category.
  - *Promotion points*: The point at which the level of change authority needed to approve changes to a configuration item in this category is moved to a higher level of authority.
  - *Change authority*: The person, team, or CCB

Table 27.1 Configuration item attributes—examples.

CI Category	Acquisition Point	Acquisition Criteria	Location	Owner	Access Control	Change Control Type	Promotion Points	Change Authority
Software Requirements Specification (SRS)	Creation of allocated baseline	No outstanding gating issues from Software Requirements gate review	*<project>* codeline in *<repository tool>*	Requirements engineering—assigned author(s)	Create, update, and delete: Author(s) Read: Project team members	Change requests through CCB	Acquisition	Project-level CCB
							Release	Product-level CCB
New source code file	Unit test successful completion	All unit test cases have passed (or outstanding problems recorded in CR tool)	*<project>* codeline in *<repository tool>*	Development—assigned Author(s)	Create, update and delete: Author(s) Read: Project team members	Change requests through CCB	Acquisition	Team-level CCB
							Promotion to system test	Project-level CCB
							Release	Product-level CCB
Legacy source code file	Start of project	Selected for inclusion or update in next release	*<project>* codeline in *<repository tool>*	Development—assigned Author(s)	Create, update and delete: Author(s) Read: Project team members	Change requests through CCB	Acquisition	Project-level CCB
							Release	Product-level CCB
Reused source code file	When selected for inclusion	Under configuration control in reuse library	*<reuse>* codeline in *<repository tool>*	Reuse CCB—assigned Author(s)	Create, update, and delete: Reuse SCM librarian Read: All	Change requests through CCB	Acquisition	Release-level CCB
Executable	Build successfully completed	No build errors & passes automated regression tests	*<project>* file directory	Builder(s)	Create, update, and delete: Builder(s) Read: Project team members	Change requests through CCB	Acquisition	Project-level CCB
							Release	Product-level CCB

(continued)

Table 27.1 (Continued)

CI Category	Acquisition Point	Acquisition Criteria	Location	Owner	Access Control	Change Control Type	Promotion Points	Change Authority
System test cases	System test case peer review completion	Successful completion of peer review (no outstanding defects)	*<project>* codeline in *<repository tool>*	System Test—assigned Author(s)	Create, update, and delete: Author(s) Read: Project team members	Review and approval	Acquisition	Peer-review team
User manual	User manual peer review completion	Successful completion of peer review (no outstanding defects)	*<project>* codeline in*<repository tool>*	Technical Publications—assigned Author(s)	Create, update, and delete: Author(s) Read: Project team members	Review and approval	Acquisition	Peer review
						Change requests through CCB	Release	Product level CCB
Product Baseline	Completion of last formal physical configuration audit	No outstanding gating issues	Archival of final visions of configuration items in *<archival tool>*	Release Management	Create and delete: SCM librarian Update: none Read: Project team members	Change requests through CCB	Acquisition	Product level CCB

Item Change Control Process

There are two basic types of change control. The first type, *change request through CCB-type change control,* also called simply *change control,* is the more rigorous type of configuration control providing a higher level of rigor for controlling configuration items. The change requests through CCB type of control proactively manages changes by requiring proposed or requested changes to be formally documented in change requests submitted to the CCB. The CCB then reviews each change request before it is implemented and allows only authorized changes to be made to the configuration items. Higher-risk configuration items are typically placed under this level of control. Configuration items typically controlled through this type of change control include requirements, interface and design specifications, source code modules, and executables.

The second type of change control, *review and approval of the modified configuration items,* also called *document control,* is a less rigorous level of change control. The review and approval type change control process is more reactive in managing change than the more formal change request through CCB change control process because it reviews the changed configuration items after the changes are implemented to the configuration item. While it provides a lower level of rigor for controlling configuration items, it is typically faster and less costly. It allows for more flexibility because multiple changes can be batched up and made simultaneously. All of those changes are approved together as a set when the configuration item is reviewed and approved. This type of change control also allows problems or enhancements found while making other changes to be implemented when the planned changes are being made without creating a change request or going through the CCB process first. Lower-risk configuration items can be placed under this level of change control. Configuration items typically controlled through this type of control include plans (project, software quality assurance, verification and validation, test, SCM), project-specific processes and work instructions, user documentation, version description documents (release notes), and training materials. Note that not all documents are controlled by this type of control. As mentioned previously, requirements, interface, and design specifications are typically controlled using change control.

Figure 27.1 illustrates an example of the typical steps in the change control process used for traditional (waterfall-type) software development. It shows how the two types of change control fit into this process. These steps include:

Step 1—Create work product: The first step in Figure 27.1 is not part of the change control process. Change control only applies to configuration items after they are acquired and baselined. Therefore, prior to acquisition, when one or more authors are assigned to create a new work product, they can make any changes necessary to create the work product and implement its allocated requirements. Some initial verification is also typically done before acquisition, and any defects found are considered private and can be corrected by the authors without any approval.

Step 2—Acquisition—baselined for internal use: In this step, the development of the work product has reached the acquisition point. If the acquisition criteria are met, the work product is baselined and placed under control. After this point, the configuration item can only be changed through formal change control processes.

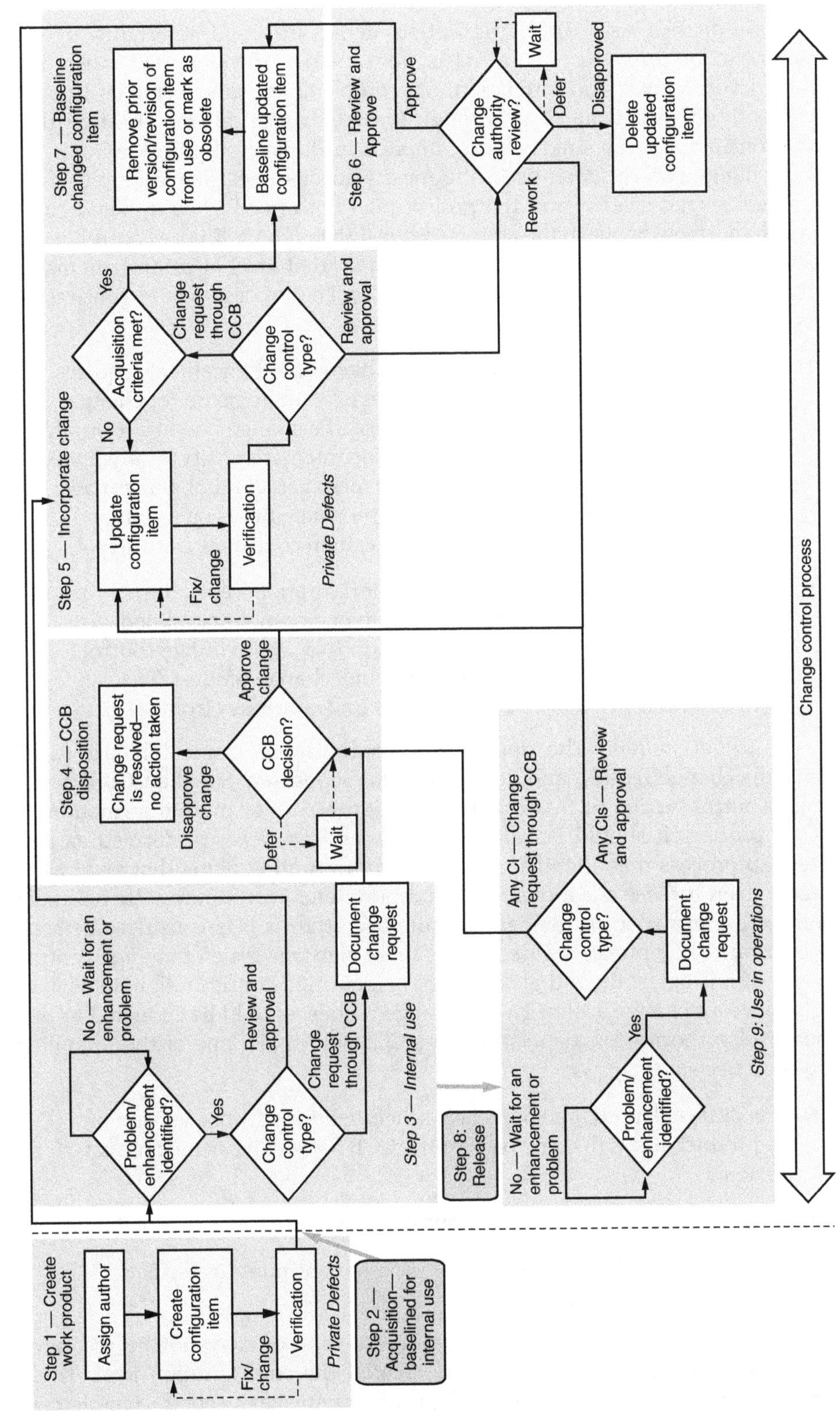

Figure 27.1 Change control process—example.

Step 3—Internal use: As configuration items are used internally by the development organization as a basis for further development, including additional verification and validation, any problems or enhancements that are subsequently identified must go through formal change control based on their change control type. If a single change impacts multiple configuration items, the level of change control defaults to the most rigorous level. For example, if the requirements specification and the project plan both need to be updated based on an enhancement request, the change request through CCB type control would be used if the requirements specification is under that level of control, no matter which type of control the project plan is under. The process would continue as follows:

- Change request through CCB: If the problem or enhancement requires change(s) to one or more baselined configuration items under change request through CCB-type change control, the change must be formally documented in a change request. This documentation is typically done by opening a change record in a change management tool, but it may be done manually by completing the appropriate fields on a change request form. The change request process then continues from step 4.
- Review and approval of the modified configuration items: If the problem or enhancement requires change(s) to one or more baselined configuration items under review and approval type change control, the change is typically not documented in a change request. The change request process then skips step 4 and continues from step 5.

Step 4—CCB disposition: The appropriate configuration control board (CCB) reviews the change request and determines its disposition. See the CCB Process: Change Control section of this chapter and Figure 27.4 for more information on the CCB process. It should be noted that the *impact analysis* performed as part of the CCB process may identify additional configuration items that need to be changed. Even if these additional configuration items would normally be under review and approval–type change control, since the CCB is considering them, they continue to be processed under this more rigorous type of change control. However, if it is determined that these newly identified configuration items have a higher level of change authority, the change request should be escalated to that higher-level authority for disposition. The CCB will make one of the following disposition decisions:

- *Defer change*: The change request is scheduled for a future version of the product and will come back to the CCB for review as part of that version.
- *Disapprove change*: The change request is resolved with no action taken.
- *Approve change*: The change request process continues to step 5.

Step 5—Incorporate change: One or more authors are assigned to update each configuration item that requires change. These may or may not be the same individuals as the original authors who created the configuration item. These authors can change their assigned configuration item(s) as needed to implement the change. Verification is also done to evaluate the updates, and any defects found are again considered private and can be corrected by the authors without any approval. However, if the configuration item being changed is under change request through CCB-type change control and any other problem or enhancement is identified that is outside the scope of the CCB-approved change, that new problem or enhancement must be documented in another change request and that new change request starts through the process at step 4. Once the change is implemented, the change control process continues based on the configuration items type of change control:

- Change request through CCB: For each changed configuration item under change request through CCB-type change control that has met its acquisition criteria, the new version/revision of the configuration item is baselined for internal use, and the process continues from step 7.
- Review and approval of the modified configuration items: For each changed configuration item under review and approval–type change control, the change has not yet been approved by the change authority, so the process continues to step 6.

Step 6—Review and approve: Each updated configuration item goes through a review cycle and must be formally approved by the appropriate level of authority before it is baselined for internal use. For review and approval–type change control, the change authority may or may not be a traditional CCB in the formal sense. Depending on the defined process, the review and approval decision might be assigned to a peer review team or even a single individual. The point is that some formal change authority approves each changed configuration item before it is baselined and used. If the change authority discovers that additional configuration items require change due to changes made to the configuration item(s) under review, one or more steps in the change control process may need to be iterated, as appropriate. The approval authority will make one of the following disposition decisions:

- *Defer change*: The changed configuration item is scheduled for a future time and will come back to the CCB for review at that time.
- *Disapprove change*: The changed configuration item is deleted (no changes are baselined).
- *Rework change*: The changed configuration item is sent back to step 5 for one or more modifications to the changes or one or more additional changes.
- *Approve change*: The change request process continues to step 7.

Step 7—Baseline changed configuration item: After each configuration item is changed and approved (either through approval of the change request by the CCB before the change(s) are made or by change authority review of the changed configuration item), the new version/revision of that configuration item is baselined and returns to step 3 for internal use. Even if the previous version of the configuration item had been released, the newly baselined configuration item would typically go through additional verification and validation (and possibly integration) before being released for external use. The previous version/revision will either be removed from use, or if it needs to continue to be used (e.g., on an older project or for an older product), it will be marked obsolete to minimize the potential for accidental use.

Step 8—Release: At some point, the software deliverables are released into operations. (See Chapter 29 for more information about release management.)

Step 9—Use in operation: As the release is used externally in operations, any problems or enhancements that are subsequently identified must go through formal change control based on their change control type (note that since these problems and enhancements are reported against released software deliverables, they are more formally tracked through change requests no matter which type of change control is being used):

- Change request through CCB: If the problem or enhancement requires change(s) to one or more baselined configuration items under change request through CCB-type change control, the change request process continues from step 4.
- Review and approval of the modified configuration items: If the problem or enhancement requires change(s) to one or more baselined configuration items under review and approval–type change control, the change request process then skips step 4 and continues from step 5.

Agile development projects are more likely to use another variation of the change control process. Problems encountered in configuration items being developed or updated during the iteration are fixed immediately and communicated to the team during the daily meetings. Other changes (enhancements and problems requiring correction after the completion of an iteration) are captured in the product backlog. In effect, the product owner and agile team act as the CCB during the iteration planning meeting by prioritizing these backlog changes and selecting them for implementation in the appropriate iteration. Final approval of the configuration items changed during any given iteration comes during the product demonstration at the end of that iteration.

Tracking Item Changes

Different mechanisms can be used to place items under configuration control, and to request and track changes to those items. At their simplest, these mechanisms can be completely manual. As illustrated in Figure 27.2, if a configuration item is being controlled manually, the change made to each version/revision of the document might be manually documented in a change history section of the configuration item (or in a separate, related change history document). The configuration item itself may be controlled using manual directories on the storage media. First, the item being changed would be moved from a manually controlled library (a controlled directory) into the author's personal dynamic library (a noncontrolled directory) while it is being changed, verified, and approved. Then the updated item would be moved into another (or the same) manually controlled library while manually marking it with the updated version/revision.

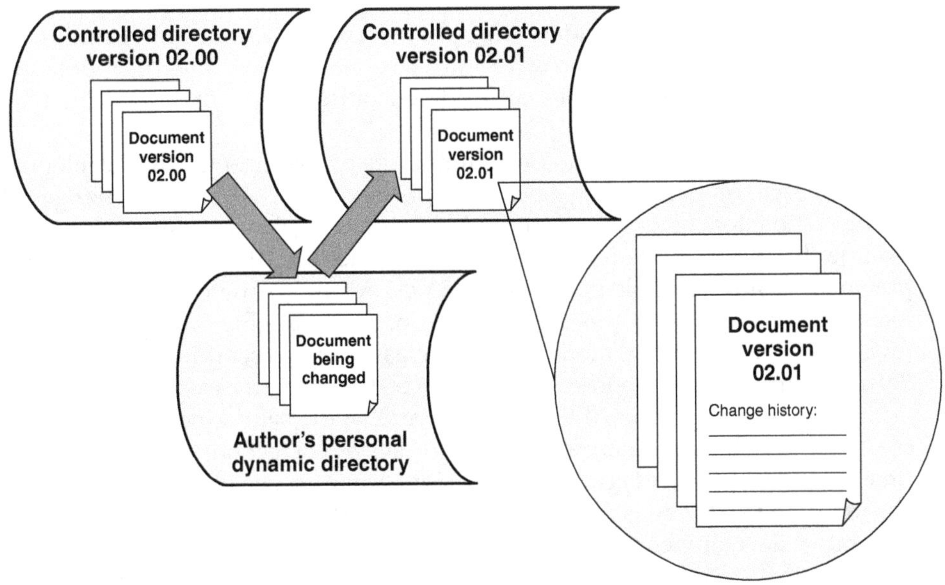

Figure 27.2 Manual tracking of item changes.

More complex mechanisms used to place configuration items under control and for requesting and tracking changes to those items can include automated version control and change management tools. Change management tools allow for the recording and reporting of the change request from origination through screening, impact analysis, CCB disposition, and the correction and verification of approved changes. Depending upon the degree of integration with the version control tool, the change management tool can either include fields that point to configuration items and their versions being updated to implement approved changes, or information and links between these tools can be shared. These tools can also provide controls to help guard against unauthorized changes. For example, the version control tool might include a required field for either the requirement identifier or change request identifier when a new version of a baselined configuration item is being checked into the tool.

Version Control

A *version* of a configuration item has a defined set of functionality. The first version of a new software product will typically include the first versions of multiple lower-level constituent configuration items. Subsequent, updated versions of existing software products will include updates of some existing versions of the lower-level constituent configuration items, and potentially add first versions of new lower-level constituent configuration items. For example, a version of a software executable is being created to add two new features to the software product. Implementing these two new features may require new versions of multiple lower-level constituent configuration items, including a new requirements specification; an updated architecture; several updated or new detailed designs and software source code modules; new or updated test cases, procedures, and scripts; an updated build script; or an updated user manual.

As these new or updated configuration items go through the development process, each of them may have multiple revisions. A *revision* changes the configuration item to correct defects without affecting functionality. For example, assume that a new version of the source code module ABC is updated from the previous version to implement part of a new feature. A defect is found during integration testing after this new version of module ABC is acquired. A new revision of this version of module ABC is created to correct this defect. A second defect is found later in integration or system testing. In that case, a second revision of this version of module ABC is created, and so on, with a new revision being created for each subsequent correction (or set of corrections done at the same time). As illustrated in Figure 26.9, revisions can also occur to a software build version as the software is rebuilt throughout the development life cycle (because of iterative development or to incorporate defect corrections). Configuration item revisions may also be created during iterative or incremental development, as functions that were always intended to be part of the versions of those items are added over time.

Configuration Item Dependencies

The real benefit of version control comes from maintaining the dependencies between versions or revisions of higher-level composite configuration items and the associated versions or revisions of lower-level constituent configuration items used to create or describe them. For example, as illustrated in Figure 27.3, each software build has dependencies with the version/revision of the:

- Specifications for requirements, architectural design, and detailed designs that describe its physical and functional characteristics
- Constituent configuration items used to build it
- Tools, macros, libraries, and the development platform used to assemble or compile those constituent configuration items and to create the build
- Test cases, procedures, and scripts used to test it
- User documentation (e.g., user manuals, operation manuals, help files, version description documents, and installation instructions) used to support it in operations
- Target platforms and environments it is intended to run on, including databases, hardware, and other software with which it interfaces

All of these dependencies should be considered when analyzing the impact of a requested change and when implementing that change. Changes to these dependencies must be documented when approved changes are implemented.

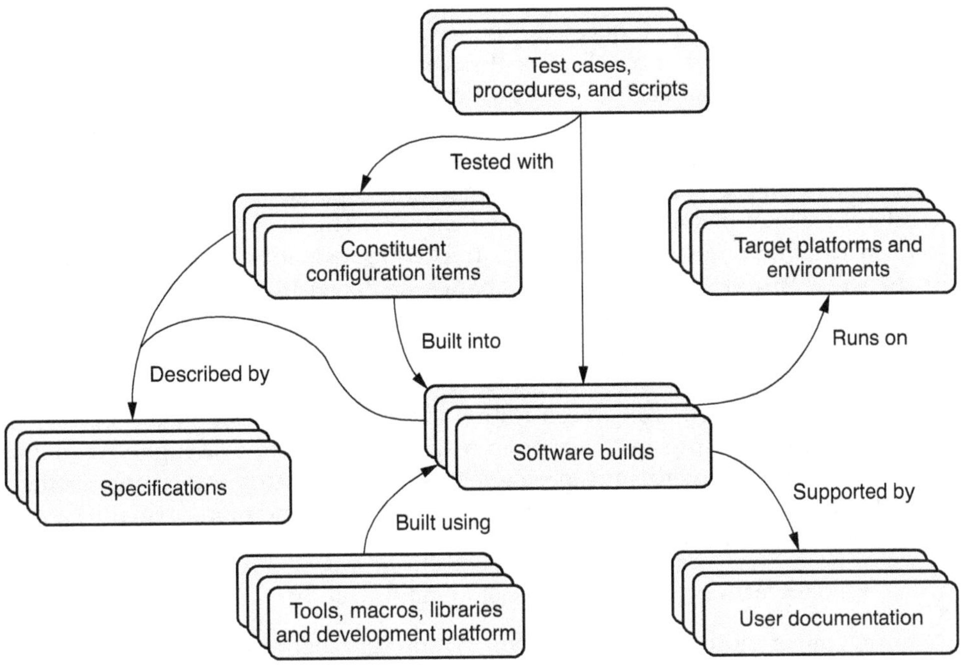

Figure 27.3 Configuration item dependencies.

2. CONFIGURATION CONTROL BOARD (CCB)

> Describe the roles, responsibilities, and processes of the Configuration Control Board (CCB). (Understand) [NOTE: The roles and responsibilities of the configuration management team are covered in area VII.A.1.]
>
> **Body of Knowledge VII.C.2**

Configuration control boards (CCBs), also called *change control boards* (CCBs), *change advisory boards* (CABs), or *engineering change boards* (ECBs), are the formal change authorities that make decisions on configuration control issues. CCBs provide the authority for approving, deferring, or disapproving requested configuration items and baseline changes. They provide visibility into the configuration control processes and facilitate communication with affected stakeholders. Multiple levels of CCBs may exist for a project. However, each configuration item or baseline has a single CCB assigned as its owner at any specific time. If a change request affects multiple configuration items or baselines with different CCB ownership, ownership is typically escalated to the highest level CCB.

As changes are requested, the CCB with ownership of the affected configuration items or baselines makes certain that the stakeholders affected by those changes are informed of the requests and can provide input into the impact analysis. Based on the impact analysis, the CCB then decides the disposition of the requested change (accept, reject, or defer) and informs the affected stakeholders of that decision. For approved changes, the owning CCB facilitates resource allocation to implement and verify the changes. The CCB also provides a vehicle to ensure that requested changes are tracked to resolution (either closed with no change required or the requested change has been implemented and verified) and that unauthorized changes do not occur. At any given CCB meeting, multiple change requests may be reviewed, analyzed for impact, dispositioned (approved, deferred, disapproved), tracked, and resolved. By performing these activities, CCBs play an integral role in controlling the software development process.

CCB membership can vary from a single individual to a highly structured and formal team with multiple members representing various organizations. CCBs with higher levels of authority tend to be larger and more formally run. The CCB membership should include:

- Policymakers, since it is a decision-making body
- One or more managers or leads to represent planning and management concerns and impacts (e.g., cost, schedule, and resource allocation)

The CCB membership may optionally include the representatives from the following groups, or these groups may be consulted by the CCB on an as-needed basis:

- Technical specialists who can provide insight into technical impact and issues associated with change requests (development, test, quality assurance, technical publications, training)
- Representatives from various groups that may be affected by the decisions made by that CCB
- Customers, users, other stakeholders, or their representatives to make sure that their needs and priorities are represented

CCB Member Roles and Responsibilities

In addition to the responsibilities of the CCB as a team, individual members of the CCB perform special CCB roles, including CCB leader, CCB members, CCB screener, and CCB recorder. Any individual CCB member can assume more than one of these roles. When a single individual acts as the entire CCB, that individual is responsible for all of these roles and responsibilities.

The change request *originator* requests the change or changes to a baselined configuration item. The originator is typically not a member of the CCB.

The *CCB leader* (chair) is responsible for planning and leading the CCB meetings. This includes handling logistics, setting priorities, making assignments, and performing other leadership activities. The CCB leader is also responsible for escalating change requests and other issues to a higher-level CCB or management as necessary. The CCB leader may delegate the actual execution of these activities to others. However, the leader is responsible for confirming the completion of those activities.

Individual *CCB members* are responsible for representing their constituency groups in the CCB meetings. This responsibility includes making certain that the groups they represent are informed of change requests and the CCB's disposition of those requests. The members act as their group's voice in impact analysis discussions and CCB decisions. Members may also be responsible for tracking approved changes assigned to their groups, reporting their status back to the CCB, and making sure that those changes are implemented and verified.

CCB screener is an optional role responsible for reviewing submitted change requests (or changed documents) and checking their correctness and completeness before CCB review. This role can help increase the efficiency and effectiveness of the CCB meetings by making certain that the CCB works from correct and adequate information. This role can also aid in the timely disposition of change requests. For example, suppose the CCB only meets weekly and rejects a change request because more information is needed from its originator; it will be another week before the request is reconsidered. Having a CCB screener who can review the requested change and obtain the additional information before the first meeting can eliminate this additional delay. The CCB screener may also have limited ability to reject a change request, such as one resulting from operator error or a duplicate of another request.

The *CCB recorder* is the record-keeper for the CCB. The recorder is responsible for recording and distributing the CCB minutes. The recorder may also record CCB information in change request records (e.g., CCB dispositions, CCB updates to change request priorities or problem severities, impact analysis information, and implementation assignments).

CCB Charter

How a CCB conducts its business can vary widely. The *CCB charter* is a document used to initiate a CCB and define the CCB's:

- *Purpose*: The purpose statement describes the CCB's objectives and its relationships with other CCBs within the organization and other decision-making bodies (e.g., the product management office and project management).
- *Scope of authority*: Define the specific scope of the decisions that the CCB can make and what decisions need to be escalated to higher-level CCBs. For example, the scope of the CCB's decision-making authority may vary from being limited to a single software component or baseline to encompassing the entire system.
- *Membership*: List the members of the CCB by name and the organization they represent. List any specific CCB roles assigned to each member.
- *Operating procedures*: Define the operating procedures for the CCB, including:
    - Frequency of regularly scheduled meetings
    - Criteria and procedures for calling additional or special meetings

- A description of how the meetings will be conducted. For example, is a quorum of members needed to conduct the business of the meeting, are guests permitted, what mechanisms are in place for recording impact analysis and CCB decisions, and what quality records are kept?
- A description of whether members can send representatives to the meeting and the roles and responsibilities of those representatives. For example, the representative must be of a high enough authority to make decisions and provide input at the same level as the person they represent.
- Processes for how impact analysis will be performed within the CCB and processes for establishing an impact analysis team to perform analysis of proposed changes when the CCB needs additional input
- Processes for calling on subject matter experts when additional technical or managerial information is needed

- *Decision-making processes*: Describe the process by which the CCB makes its decisions. For example, CCB members may have voting privileges. If so, how do members who are not in attendance at a given meeting cast their votes (e.g., through a representative or a proxy, or via e-mail)? In other cases, CCBs may be required to come to consensus on CCB decisions, or the CCB leader may be the final decision-making authority with only recommendations being input from the other CCB members.
- *Communication mechanisms*: Describe how the CCB will gather input from affected stakeholders and how CCB decisions will be communicated to those stakeholders, including higher-level or lower-level CCBs.

CCB Process: Change Control

Figure 27.4 illustrates an example of a typical process for requesting, dispositioning, and resolving a requested change to a baselined configuration item. The actual steps in the process for any given project should be defined during SCM planning and documented in (or pointed to by) the project's SCM plans, as appropriate.

Step 1—Submit change request (CR): The originator submits a formal change request for CCB review and disposition. This submittal can be accomplished through multiple techniques, completing and submitting a hard-copy change request form or the opening and submitting of a change request in the change management tool. A requested change may be submitted to correct an identified problem (defect, problem, anomaly, or issue report) or to request the implementation of one or more new or changed requirements (enhancement request).

Step 2—Initial CR review: Optionally, the CCB screener performs an initial review of the requested change.

- *Step 2 (cont.)—Rework CR*: If the change request has been inadequately completed (required fields were left blank or the description of the change does not include enough detail), the screener sends

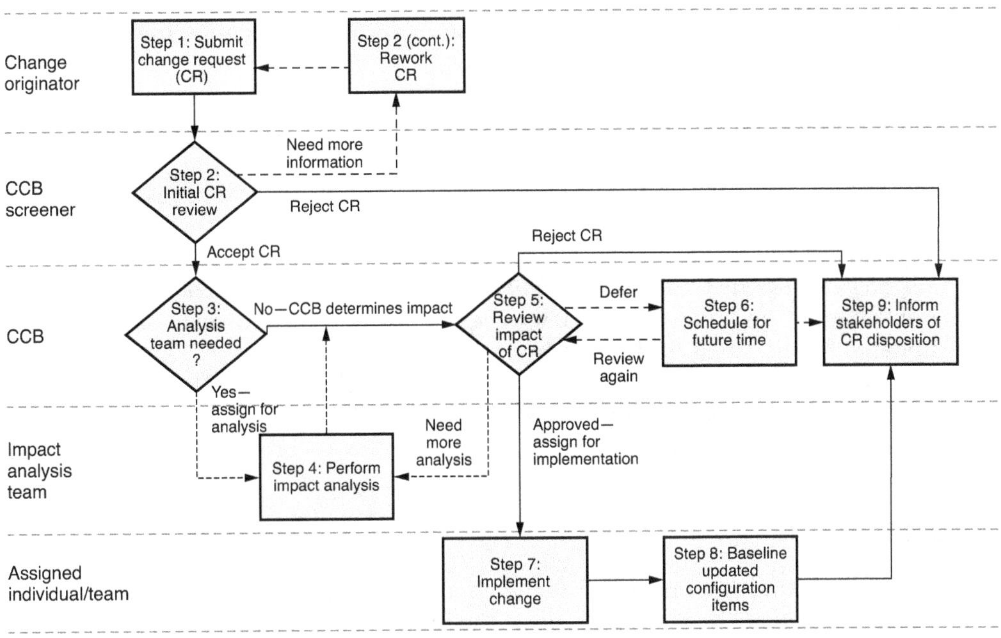

Figure 27.4 Configuration control board process for change control—example.

the change request back to its originator with a description of the additional information needed. The originator then supplies the requested information and resubmits the change request.

- The screener may also have limited authority to reject change requests, such as ones resulting from operator error or duplicates of other requests. If the change request is rejected, the appropriate stakeholders, including the originator, are informed of the reasons for rejection.
- If the change request passes screening, it is assigned to a CCB meeting. In the case of critical change requests, the screener informs the CCB leader so that a special CCB meeting can be convened.

Step 3—Analysis team needed?: During the CCB meeting, the CCB members review each requested change and determine if they have the appropriate people in the meeting and adequate information to determine the impact of the requested change.

Step 4—Perform impact analysis: If the CCB assigns an impact analysis team to the change request for further investigation, this team analyzes the change's impact and contacts affected stakeholders to elicit opinions and additional information as needed. The impact analysis team reports their findings to the CCB at a future CCB meeting.

Step 5—Review impact of CR: The CCB reviews the impact of each change request by analyzing the change request or reviewing the report from the impact analysis team. If the CCB needs more information, the change request is returned to the impact analysis team with a description of the additional information needed. Once they have adequate information, the CCB decides on the disposition of the change request.

Step 6—Schedule for future time: If the change request is deferred, it is scheduled for a future version of the product and will come back to the CCB for review as part of that version.

Step 7—Implement change: If the change request is approved, it is assigned to one or more individuals or teams to incorporate the requested change into the appropriate configuration items, and verify those changes.

Step 8—Baseline updated configuration items: After each change has been implemented and verified, and its acquisition criteria are met, baselines are updated for modified configuration items.

Step 9—Inform stakeholders of CR disposition: The appropriate stakeholders, including the originator of the change request, are then informed of the resolution of the requested change.

Impact Analysis

When analyzing the impact of a change, multiple factors should be considered, including factors beyond the scope of the immediate project. This is particularly true if the requested change could impact an item reused in multiple products. Factors to consider during impact analysis include the following.

- *Size and breadth of the change*
  - Number of configuration items touched by the change
  - Number of software products or versions of a software product that include those configuration items
  - Number of other configuration items or other software or hardware products that must be changed to stay consistent with those configuration item or product changes (data coupling or control coupling dependencies)
  - Number of new, changed, or deleted function points, lines of code, requirements, or pages of documentation
- *Complexity of the change*: The complexity of new, changed, or deleted configuration items required to implement the change
- *Severity of the change*: The impact of this change on the stakeholders. If the change is required to fix a defect, how critical is the impact of the defect on the current functioning or characteristics (e.g., safety, security, reliability) of the software or system?
- *Schedule impacts*
  - New tasks added to the schedule to implement the change and their estimated duration
  - Impacts on other tasks or dependencies
  - Overall impact on the duration of the critical path

- Availability of people and other resources to implement the change
- *Cost impacts*
  - Potential costs of implementing the change
  - Potential costs or savings to customers, users, or other stakeholders from making the change
- *Effort impacts*
  - Estimated effort to implement the change
  - Potential costs or savings to customers, users, or other stakeholders from making the change
- *Technical impacts*
  - Impact on the functionality and quality attributes (e.g., quality, performance, security, safety, portability, or efficiency) of the software
  - Future product consequences of the change (e.g., maintainability, reliability, portability, supportability, or reusability)
  - Critical computer resource impacts (e.g., memory, channel capacity, or disk space utilization)
- *Relationships to other changes*
  - Supersede or eliminate the need for other changes
  - Dependencies on other changes
  - Economies of packaging the changes with other previously approved changes to the same work products
- *Testing requirements*
  - Number of new, changed, or deleted test cases or procedures required
  - New or updated test automation required
  - The amount of retesting or regression testing that will be required
  - Special testing resource requirements (e.g., test beds or simulators)
- *Release plan impacts*: Any changes to release contents or planned release schedule or effort associated with the change
- *Contract requirements impacts:* Any changes to acquirer or supplier contracts associated with the change
- *Risks*
  - Any project risks associated with the change
  - Any product risks associated with the configuration items being changed

- *Benefits*: Special advantages to be gained from adding this change (e.g., political, marketplace advantage, or stakeholder satisfaction benefits)

When assessing the size and breadth of the change, bidirectional traceability plays a significant role. Suppose an enhancement is requested because of a change in the business or operational environment (e.g., a business objective, stakeholder requirement, or standard has changed); then if good forward traceability has been maintained, that change can be traced forward to the associated product-level requirements and all of the other configuration items that are impacted by that change. This traceability can significantly reduce the effort required to do a thorough job of impact analysis. It also reduces the risk that one of the affected work products is forgotten, resulting in an incomplete implementation of the change and a new defect being introduced.

Backward traceability is beneficial when a defect is identified in one of the work products. For example, as illustrated in Figure 27.5, if the code module X01 has a defect, the traceability matrix can be used to help determine the root cause of that defect. In this example, it is not just a code defect; it also traces back to defects in the associated design X and requirement R1302.

If design or requirements defects are discovered during impact analysis, what other work products might be affected by the defect or the correction of that defect? This leads to forward traceability to make certain that a complete analysis is performed. Figure 27.6 continues this example by illustrating the use of this forward traceability analysis. In this example, the defective requirements are traced forward to two additional design elements: Z, which has the defect, and Y, which does not. The requirements are also traced to training materials, user documentation, system test cases, and other configuration items that should be analyzed for impact. Tracing from the design elements X and Z with defects uncovers two additional source code modules, X02 and Z02, that have defects. Since design element Y did not have the defect, any source code module written based on element Y will not have to be investigated. Modules X01, X02, and Z02 that have the defect may have propagated the defect forward into their associated unit test cases (or other test documents or automation). Design elements X and Z that have the defect may have propagated the defect forward into their associated integration test cases (not shown in Figure 27.6) or other test documents or automation, and so on. The point of this example is that traceability can be used to identify all of the configuration items that a defect may impact. This can be a great benefit to the project by finding a single defect and fixing that defect in multiple places. If each of the defects identified through traceability analysis had to be found separately, there would be a much higher risk that one or more of those defects would escape detection before release.

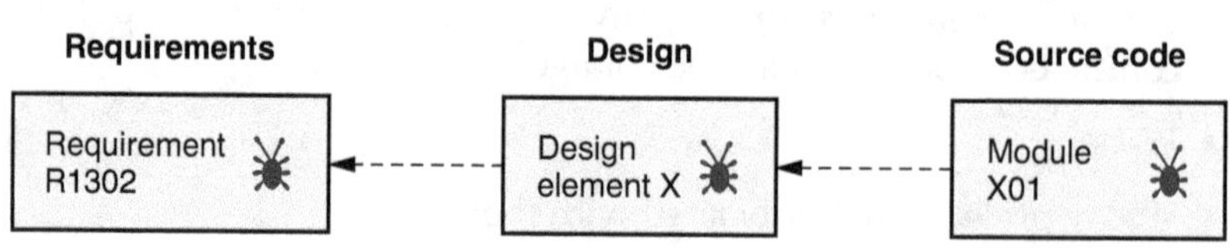

Figure 27.5 Using backward traceability for defect impact analysis—example.

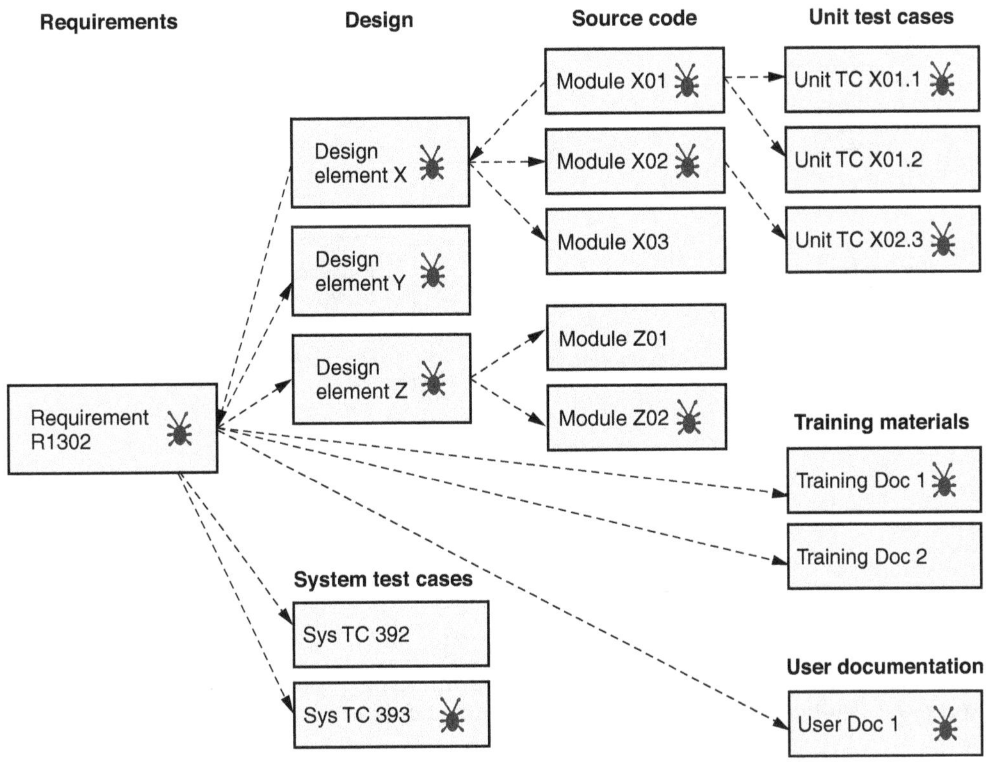

Figure 27.6 Using forward traceability for defect impact analysis—example.

Multiple Levels of CCBs

As a configuration item moves through the software development life cycle, the balance between the need for flexibility and stability typically shifts. For example, when a source code module is first created, the developer requires a high level of flexibility to experiment, create, and perfect that module. Since the module has not been acquired and no one else is using it, the need for stability is nil. As that source code module moves through the life cycle, it is integrated into larger and larger composite configuration items to create system components and eventually the system, and more and more developers, testers, and other stakeholders use it. Therefore, the need for more rigorous control and communications (stability) increases.

One mechanism for addressing this need to shift the balance from more flexible to more stable is to create multiple levels of CCBs. Having multiple levels of CCBs allows small changes of limited scope and impact to be approved by lower-level CCBs. In contrast, major changes that impact multiple stakeholders or work products can be escalated to higher-level CCBs with broader scope and more members.

In the source code module example, illustrated in Figure 27.7, the developer can change a newly created source code module as necessary. When it is initially acquired, a team-level CCB could be assigned change control authority for that

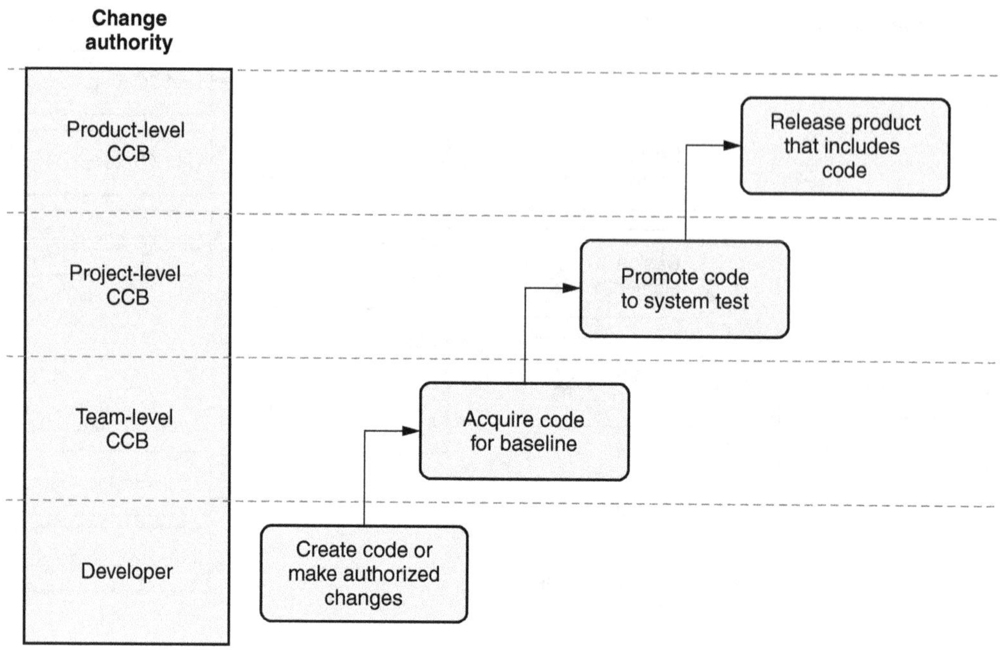

Figure 27.7 Multiple levels of CCBs—source code module example.

module because, typically, only team members need to be consulted when the code changes at that level. As illustrated in Figure 27.8, the membership of the team-level CCB might be limited to only the software architect or designer and the software engineers on the team. Since this CCB has limited members who work closely together, they can usually meet and make decisions quickly without formal meetings.

Once system testing starts, the source code module is promoted to ownership by the project-level CCB (see Figure 27.7). *Promotion* is the transition in the level of authority needed to approve changes to a baselined configuration item. Changes to the source code module at this phase of the life cycle may impact software written by other teams, the hardware, or the documentation, as well as the work of the testers, software configuration management, software quality assurance, and other specialists. Changes this late in the life cycle may also impact project schedules, costs, effort, and risks. The membership in the project-level CCB expands to include representatives from the various stakeholders that may be impacted (see Figure 27.6). In this example, the individual team architect or designer and software engineers are not members of the project-level CCB, but they may be called on to participate in CCB activities as subject matter experts if the requested change impacts their software designs or code.

Finally, when the product from this example is released into operations, the source code module is promoted to ownership by the product-level CCB, along with all of the other configuration items that become part of the product baseline (see Figure 27.7). Again, the membership of the CCB changes to include the customer or user representative and product manager who are making business decisions

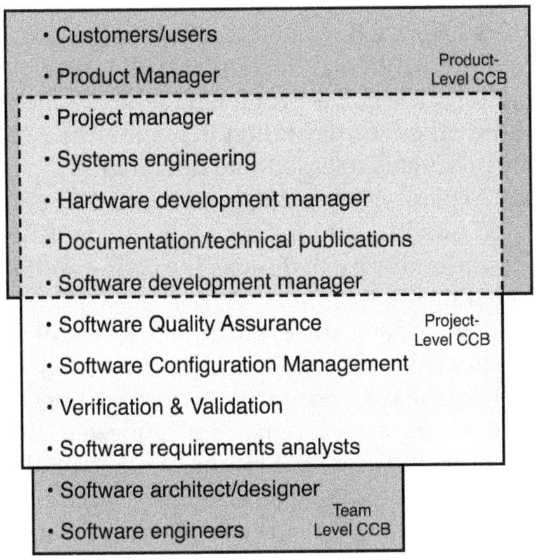

Figure 27.8 Membership of multiple levels of CCBs—example.

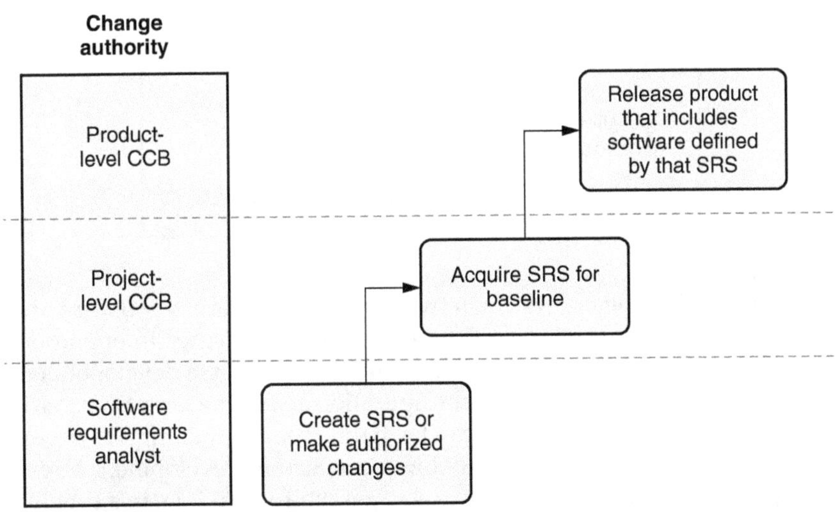

Figure 27.9 Multiple levels of CCBs—software requirements specification (SRS) example.

about the longer-term direction of the product (see Figure 27.8). However, CCB meetings at this level typically happen much less frequently. To attempt to control lower-level products early in their life cycle with this level of CCB would add an extreme time burden that would probably grind software development to a halt.

Not every configuration item needs to start at the same level of CCB ownership. Higher-risk configuration items should be assigned to higher-level CCBs when they are acquired. For example, as illustrated in Figure 27.9, the software requirements specification (SRS) for this project could go directly under the control of the project-level CCB when it is acquired because of the broader impact that changes to requirements may have across the project. When the product is released into operations, the SRS is promoted to the product-level CCB ownership, along with all of the other configuration items that become part of the product baseline, when the product is released into operations.

The project's SCM planning should define the control points where promotions occur, the associated quality gates for each type of identified configuration item, and the formal change authority that owns the configuration item at each level of acquisition or promotion.

There may also be a need for even higher levels of CCBs, at the product, product line, or even organizational level that control changes to configuration items that are reused or shared. For example, one product may "own" a database used by multiple software applications. If a requested change to that database impacts the format of one or more data records, many or all of those applications, and therefore multiple projects, may be affected. A high-level, cross-product CCB could be used to make certain that the effects to those applications and projects are considered during the impact analysis of the requested change.

3. CONCURRENT DEVELOPMENT

> Describe the use of configuration management control principles in concurrent development processes. (Understand)
>
> **Body of Knowledge VII.C.3**

Concurrent development occurs when two or more versions of a software product are in development or are being actively maintained as releases in operations simultaneously. Concurrent development may result in multiple developers needing to make changes to the same baseline configuration item at the same time. As discussed in Chapter 25, the locking process can keep this from happening. However, locking serializes changes, which can slow down concurrent development or even cause developers to work around the process to get their work done in a timely manner. To avoid this problem, sophisticated library and version control tools include nonlocked check-in/check-out processes. Nonlocking check-in/check-out processes allow multiple developers to check out the same baselined configuration items using different types of check-out. Then, after the first person performs a check-in, all subsequent check-ins include a merging process.

In addition, labeling, branching, and merging all support the configuration control of concurrent development. To illustrate the use of these processes in supporting concurrent development, let's walk through a complex example that illustrates why good tools to keep track of this complexity are essential to performing good configuration management. Figure 27.10 depicts the evolution graph for a software product currently undergoing concurrent development. Assume a decision was made to use incremental development for each major product release. Version

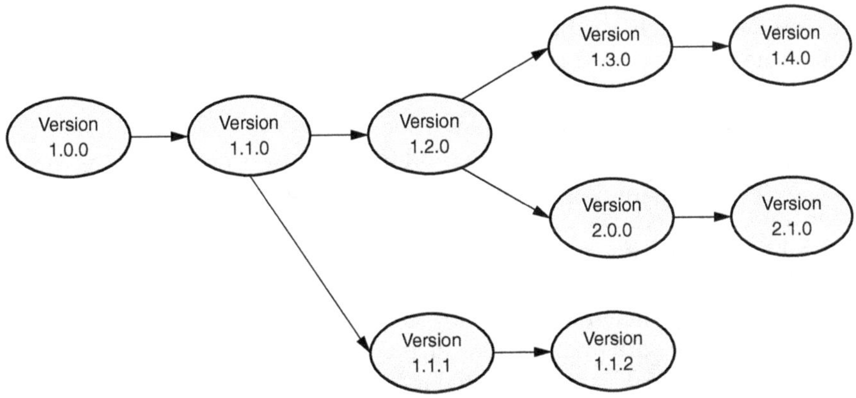

Figure 27.10 Released software product evolution graph—example.

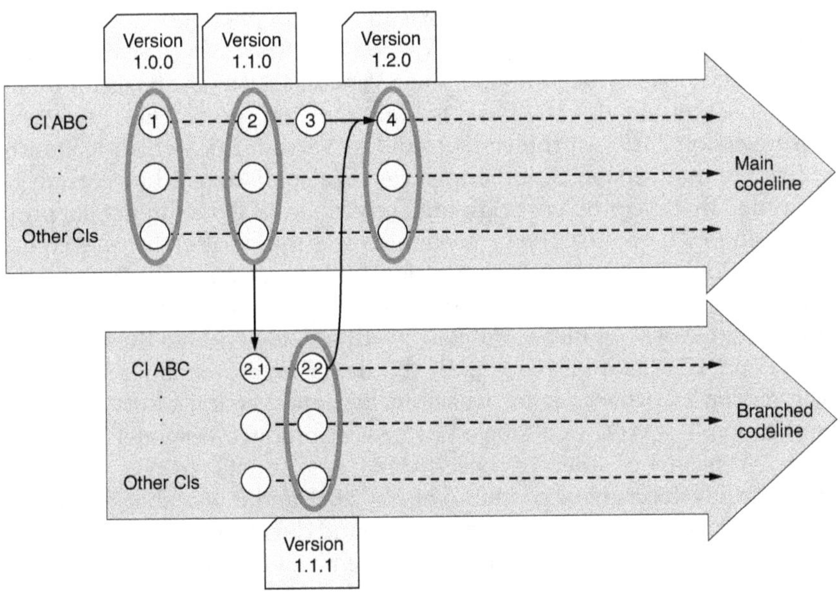

Figure 27.11 Software item versioning—example 1.

1.0.0 might represent the first incremental version of the product. As development progresses, Version 1.0.0 moves into testing, and the next increment, Version 1.1.0, starts development. When Version 1.0.0 is released, Version 1.1.0 transitions into testing and development starts on the third increment, Version 1.2.0. After the release of Version 1.1.0, a problem is discovered in that version. The problem is fixed in the corrective maintenance release Version 1.1.1, and that fix is merged into Version 1.2.0, which is still in development.

Figure 27.11 demonstrates how an individual configuration item ABC might be affected by this example. The initial controlled Version 1 of ABC was checked into the main codeline when it was acquired and later labeled as part of Version 1.0.0 of the software product when it was released. Version 1 of ABC was checked out, new functionality was added, and it was checked in as Version 2. Version 2 of ABC was checked out, and new functionality was added to it as part of the development of Version 1.2.0 of the product, and it was checked in as Version 3 of ABC. Version 2 of ABC was labeled as part of Version 1.1.0 of the software product when it was released. The problem, discovered in the released product Version 1.1.0, was debugged, and it was determined that this problem originated in Version 2 of ABC. To correct this problem, the codeline was branched, Version 2.1 of ABC was created, and the problem was fixed in Version 2.2 of ABC. After testing, Version 2.2 of ABC was labeled as part of product Version 1.1.1 when released. It was also determined that the defect also existed in Version 3 of ABC, so Version 4 of ABC was created by merging the fix from Version 2.2 of ABC into Version 3 of ABC. It should be noted that the branched codeline is created for the corrective release. This branch is expected to eventually be retired (archived or terminated) when this release is no longer supported in operations. Ongoing development continues as part of the main codeline, with branches being as temporary as possible.

To continue with the example, as Version 1.2.0 moves on to testing and becomes stable, a decision is made to not only use Version 1.2.0 as the basis for incremental Version 1.3.0 but also as the basis to start the first increment of the next major release, Version 2.0.0. A problem is found in Version 2.0.0 which was introduced and fixed in that version. Another problem is also discovered in Version 1.1.1. It is determined that the problem originated in Version 1.1.0 but did not get propagated into Version 1.2.0 because the problem area was overwritten by new code created in that version. Corrective release Version 1.1.2 is created to fix this problem.

Figure 27.12 continues the demonstration of how an individual configuration item ABC might be affected by this decision and these problems. To use Version 1.2.0 as a basis for both the incremental Version 1.3.0 and the new major Version 2.0.0, the codeline was branched, and Version 4.1 of ABC was created from Version 4 of ABC and labeled as part of product Version 1.3.0 when it was released. Version 4 of ABC was checked out, new functionality was added to it as part of the development of product Version 2.0.0, and it was checked in as Version 5 of ABC. To correct the problem that was found during Version 2.0.0 development, Version 5 of ABC was checked out, that problem was corrected, and a new Version 6 of ABC was checked in. Version 6 of ABC was later labeled as part of product Version 2.0.0 when it was released. To correct the problem that was isolated to Version 1.1.1, Version 2.2 of ABC was checked out, corrected, and checked back in as version 2.3 of ABC. Version 2.3 of ABC was labeled as part of the corrective release of product Version 1.1.2 when it was released.

As development progresses, new functionality is added to Version 1.3.0 to create the new incremental product Version 1.4.0. The latest change to the product in development was to merge the changes from product Versions 1.3.0 and 1.4.0 into Version 2.0.0 to create the next increment of the product, Version 2.1.0. Figure 27.13 demonstrates how the individual configuration item ABC might be affected by these changes. Version 4.1 of ABC was checked out, new functionality was added to it as part of the development of product Version 1.4.0, and it was checked in as Version 4.2 of ABC. Version 4.2 of ABC was labeled as part of product Version

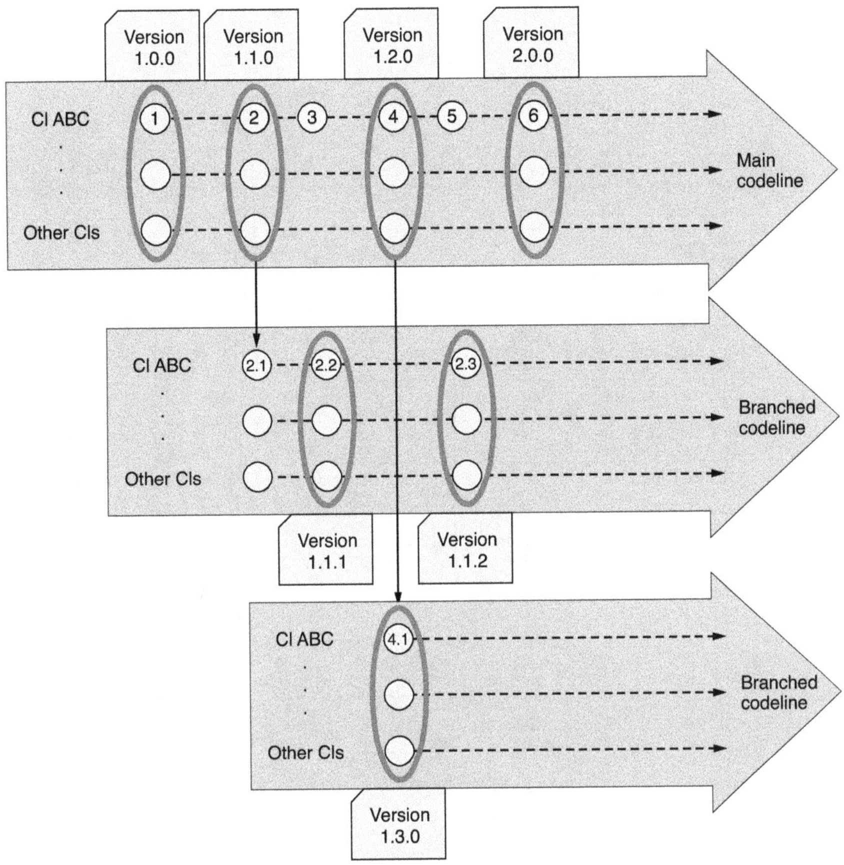

Figure 27.12 Software item versioning—example 2.

1.4.0 when that product build was done and moved into testing. Changes from Versions 4.1 and 4.2 of ABC were merged with Version 6 of ABC to create Version 7 of ABC, which was checked in. Version 7 of ABC was labeled as part of product Version 2.1.0 when that product build was done, and it moved into testing.

A significant consideration of concurrent development and supporting active maintenance on concurrent releases in operations is the need to perform more extensive impact analysis when changes are requested. For example, assume that all of the released versions of the software product from Figure 27.13 are still being used by one or more users in operations, except for Versions 1.4.0 and 2.1.0, which are still in concurrent development. The left side of Figure 27.14 illustrates the "build into" relationships between the versions of configuration item ABC and the product versions. A problem (bug) is reported from operations by a user currently using Version 1.2.0. Performing the initial debugging and impact analysis determines that this problem results from a defect in Version 4 of ABC. Since there are concurrent releases and development, impact analysis must be performed to determine which version of ABC originated the defect (in this example, it

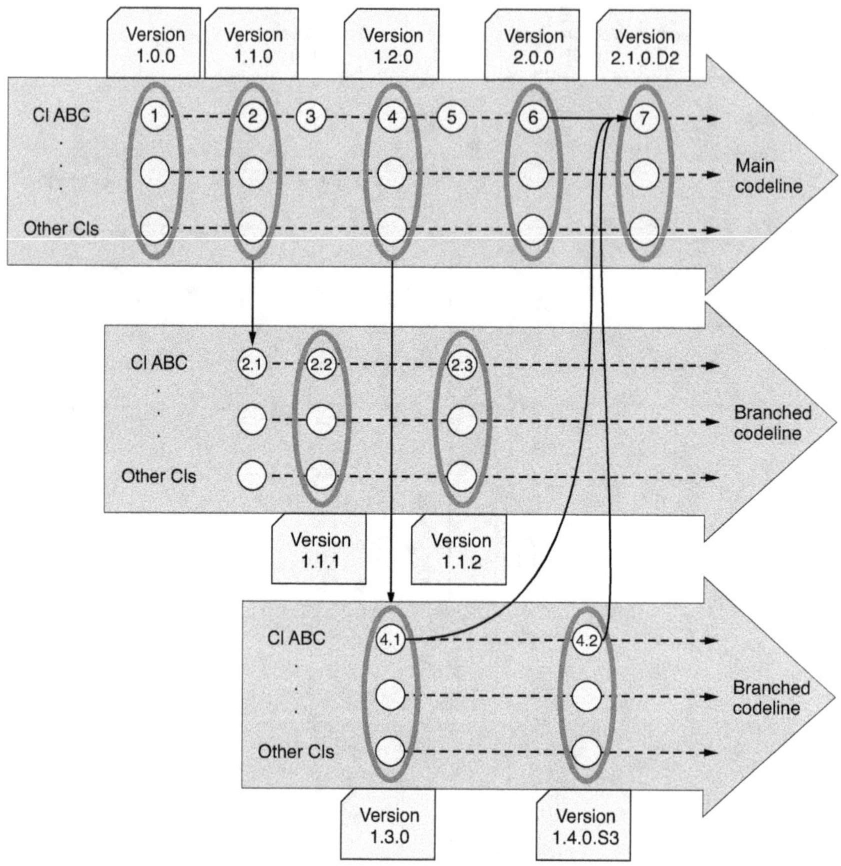

Figure 27.13 Software item versioning—example 3.

originated in Version 2 of ABC) and if it exists in any subsequent versions (in this example, it exists in all versions through Version 7 of ABC). Therefore, the defect impacts all product versions from 1.1.0 through 2.1.0. Assuming that the CCB determines that the defect is a serious enough problem to require corrective releases, a branch would be created for corrective Version 1.2.1, and Version 4 of ABC would be checked out, the defect corrected, and checked into the new branch as Version 4.0.1. (Note that the version numbering used here and elsewhere in this example will depend on the implemented version numbering schema.) As illustrated on the right side of Figure 27.14, this correction would be merged into Versions 2.3, 4.1, 4.2, 6, and 7 of ABC to create corrected Versions 2.4, 4.1.1, 4.3, 6.1, and 8 of ABC, respectively. These updated versions of ABC would then be built into corrective releases of the product in Versions 1.1.3, 1.2.1, 1.3.1, and 2.0.1. The updates would also be built into updated revisions of the product in Versions 1.4.0 and 2.1.0, currently in development. Since these two builds are currently in development, their version number was not incremented. However, the internal build revisions were updated for these two builds.

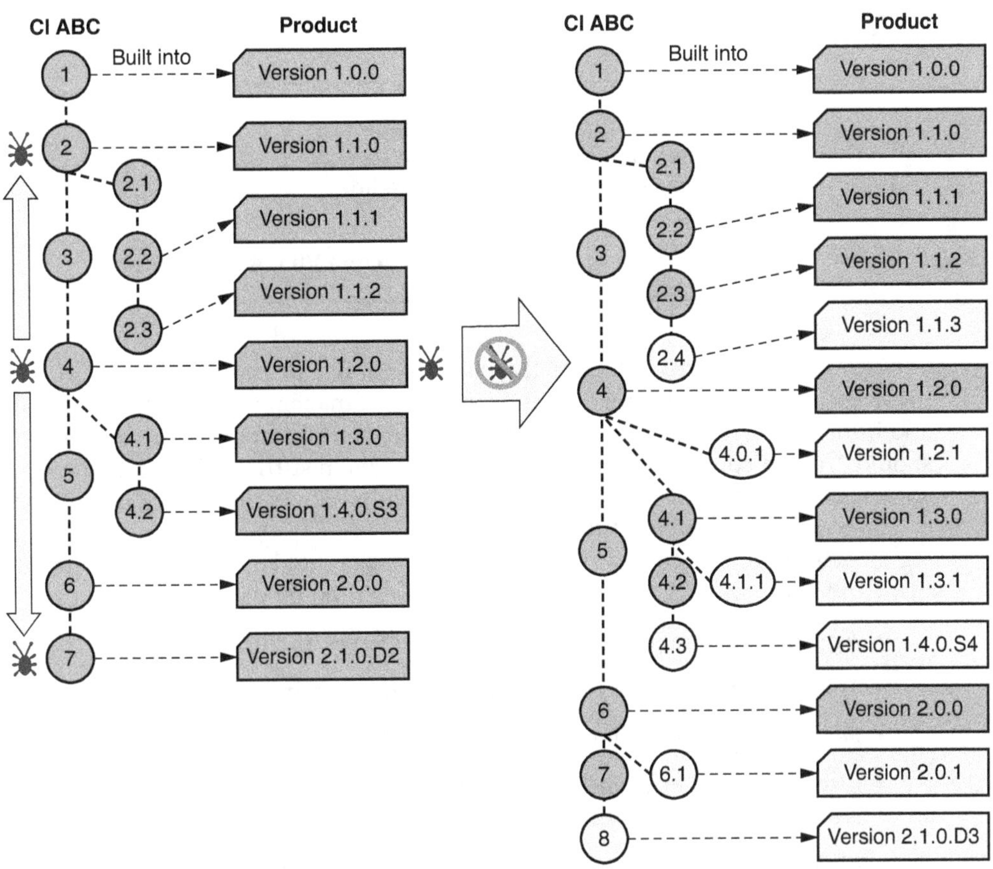

Figure 27.14 Impact analysis and concurrent development—example.

To further complicate this example, let's assume that a new version of the compiler was introduced between product Versions 1.1.0 and 1.2.0 and that a new version of the operating system is being used for major release Version 2.0.0 and its increments. This complication is why software development tools should be considered configuration items. The prior version of the compiler would be needed to build product Version 1.1.3, and the prior version of the operating system would be needed to both build and test product Versions 1.1.3, 1.2.1, 1.3.1, and 1.4.0. Again, these considerations should be included in the impact analysis.

Finally, to extend the complexity of the impact analysis even further, consider the fact that configuration item ABC might be reused in multiple products, each of which has multiple releases being supported in operations or multiple versions in concurrent development. This simple example demonstrates why most software development projects need a good SCM tool that automates the identification, versioning, branching, and merging functions.

4. STATUS ACCOUNTING

> Apply various processes for establishing, maintaining, and reporting the status of configuration items, such as baselines, builds, and tools. (Apply)
>
> **Body of Knowledge VII.C.4**

Configuration status accounting is the record-keeping activity within SCM that involves collecting and recording data, records management, and reporting data and information needed to manage the software configuration. Status accounting should "record configuration management actions in sufficient detail so that the contents and status of each configuration item is known and previous versions can be recovered" (CMMI 2010a).

Configuration status accounting provides stakeholders, including management and software practitioners, with the information they need to:

- Appropriately manage the configuration items and baselines
- Track the status of configuration items and baselines
- Perform fact-based change request impact analysis
- Effectively and efficiently track requested and approved changes to resolution and verify those changes
- Make certain that unauthorized changes are not occurring
- Analyze the software's readiness for release
- Produce version description documentation (release notes)
- Evaluate the integrity of the software products and the effectiveness of the implementation of SCM practices
- Analyze and improve the software processes, products, and services

Data Collection and Recording

The types of status accounting data collected depend on the information needs of the organization and its projects. Examples of SCM data include:

- *Configuration item metadata* (e.g., current author, current approval authority, current status, location, version/revision history, associated change requests) collected when configuration items are acquired and checked in or out of control libraries, when branches or labels are created, or when merges are performed
- *Build metadata* (e.g., build dates, versions or revisions of build contents, versions or revisions of tools, and tool options used to create the build) collected when builds are created

- *Baseline metadata* (e.g., baseline dates, versions or revisions of baseline configuration items contained in the baseline, authority that approved the baseline) collected when baselines are created or changed
- *Change request metadata* collected when problem reports or enhancement requests are created, approved, resolved, and closed
- *Configuration item dependencies and interrelationship or traceability data* (see Figures 27.3, 27.13, and 27.17) collected when baselines are created or changed
- *Release management data* (e.g., release dates and product warranty or licensing data) collected when products are released
- *Configuration management deviations and waivers data* collected when deviations and waivers are created, processed, and approved
- *Software configuration management–related project data* (e.g., effort, cost, schedule, and cycle time data) collected as part of project execution and tracking
- *Site installation history data* (e.g., what release(s) were installed on each site on what date, and site configuration history data) collected as new or updated releases are installed or as site configurations are installed or updated

The configuration status accounting data is collected and recorded while performing the other SCM, software development, testing, and release management activities. For example, the originator typically collects the following change request information when a change request is created and records this information in the change management tool or on a change request form:

- Change request identifier
- Summary of the requested change
- Originator of the change request and the originator's contact information
- Severity (impact on users or operations) and priority (to originator, to the development team)
- Creation date and time
- Product and version/revision the problem was identified in, or the enhancement is requested against
- Detailed description of the requested change. For an enhancement request, this includes one or more new or changed requirements statements. It may also include drivers for the enhancement, benefits of the proposed enhancement, or costs of not doing it. For problem reports, this includes a detailed description of the problem, including:
    - Input
    - Steps to reproduce

- Environment (including versions of operating systems, browsers, platforms, hardware configuration, and so on, as appropriate)
- Expected results
- Actual results
- Attempts to repeat and the results of those repetitions
- Anomalies (other unwanted behaviors that fall outside of the expected results)
- The phase of the life cycle in which the problem was detected
- Date and time the problem occurred (note that this may be different than the date and time the problem report was created)
- Tester or user and observer(s)

Additional information about each change request is collected and recorded as it progresses through the change process. This information may consist of the change history, including dates and times of change record modifications or status changes. This additional information may also include the following:

- As the change request is reviewed by the CCB (or other change authority), its impact analysis information, implementation priority, escalations, disposition (approve, reject, rework, or defer), and associated dates may be collected and recorded. Detailed information about the requested rework may also be documented. For deferred change requests, the deferred-to date or target release date may also be collected and recorded.
- For approved change requests, the author(s) assigned to implement the change might collect and record:
    - Its resolution description (actual versus proposed)
    - The effort expended to debug and implement the change
    - A list of (or links to) the modified versions or revisions of configuration items created or modified to implement the change request
    - Author who created or modified each of those configuration items and associated acquisition dates
    - For problem reports, as the assigned author(s) debugs a reported problem, recorded information might also include the problem's workaround and root cause information (phase of the life cycle in which the problem was introduced, root cause type of the problem, and so on)
- As the change is incorporated into one or more software builds, those build identifiers, associated build dates, and other build information (e.g., pointers to build logs or dependencies) may be collected and recorded.

- As the created or modified configuration item(s) or builds are tested, recorded information for each configuration item or build might include:
  - The effort expended to update testing documentation and to test the change at each level of testing
  - Pass/fail status for each level of testing (including the location of test data as appropriate) and associated dates

SCM planning should include the determination of the configuration management information needs of the project or the organization. This planning includes deciding on the following:

- What status accounting data needs to be collected, tracked, and reported to meet those needs
- How these data are to be collected, stored, processed, reported, and protected from loss
- How access control to these data will be implemented
- What types of status accounting reports are to be generated and at what frequency

Configuration status accounting data can be collected, recorded, and reported manually (on change request forms, in the change history of a document, in a build report, on a site installation form, or even in a spreadsheet containing a bidirectional traceability matrix). However, due to the volume of data and information required by most software development projects, configuration status accounting data collection, recording, and reporting is typically automated through various tools and databases (e.g., the library and version control tools, build tools, change management tools, or project management tools).

Status Reporting

Configuration status reporting should include both standardized reports and metrics and the ability to create ad hoc queries of the configuration status accounting data to extract additional information as needed. Depending on the requirements of the organization and project and the resulting types of configuration data being collected, a configuration status accounting system should be able to provide reports and metrics that answer questions, including:

- How many configuration items exist?
- What is the status of each version/revision of each configuration item at any specific time?
- Which versions or revisions of which constituent configuration items constitute a specific version of each composite configuration item?
- How do the versions or revisions of each configuration item differ?
- Which versions of which products are affected by a given configuration item revision?

- What documents support each version of each configuration item?
- What hardware or software dependencies exist for each configuration item?
- What are the contents of each baseline, and when was that baseline created?
- What tools and options or switches within those tools were used to create each build?
- Which releases of which products are installed at which sites, and what is the installation history of each site?
- Which configuration item versions or revisions are affected by a specific change request?
- What is the impact and status of each change request?
- How many product problems were reported or fixed in each version of each build or release?

Chapter 28
Configuration Audits

> Define and distinguish between functional and physical configuration audits and how they are used in relation to product specifications. (Understand)
>
> **Body of Knowledge VII.D**

Software configuration management (SCM) audits provide independent, objective assurance that the SCM processes are being followed, the software configuration items are being built as required, and the software products are ready to be released at delivery. By using standardized checklists tailored to the specifics of each project, these SCM audits can be more effective, efficient, and consistent, as well as aiding in continual process improvement.

Two types of SCM audits are typically performed:

- *Functional configuration audit (FCA)*: An independent evaluation of the completed software products to determine their conformance to their requirements specification(s) in terms of completeness, performance, and functional characteristics
- *Physical configuration audit (PCA)*: An independent evaluation of each configuration item to determine its conformance to the technical documentation that defines it

Functional and physical configuration audits differ from process audits of the SCM processes. Process audits, as discussed in Chapter 8, are ongoing independent evaluations conducted to provide information about compliance with processes and the effectiveness and efficiency of those processes. The primary purpose of functional and physical configuration audits is to maintain the integrity of configuration baselines. These audits verify that the baselines are complete, correct, and consistent in relation to the functional and physical specifications that were agreed to by the stakeholders. These audits verify that changes approved through configuration control were correctly implemented and that no unauthorized changes have occurred.

This Certified Software Quality Engineering Body of Knowledge section discusses configuration audits. However, for organizations that do not require a high level of rigor or independence, configuration reviews, rather than audits, can be conducted by project personnel to accomplish the same objective of maintaining the integrity of configuration baselines. For example, Moreira (2010) says, "In general, most agile teams will find the traditional CM audit approach as too heavy, even if they find value in knowing the level of integrity of their baselines." For agile development, the *configuration review* could be based on the iteration baseline and incorporated into the agile review at the end of the iteration. Combining these reviews by automating the build and configuration reporting processes could keep the process lean and remove potential redundancies from having separate baseline and end-of-iteration reviews.

FUNCTIONAL CONFIGURATION AUDIT (FCA)

According to the *IEEE/ISO/IEC Systems and Software Engineering—Vocabulary* (IEEE/ISO/IEC 2017), a *functional configuration audit* (FCA) is "an audit conducted to verify that:

- The development of a configuration item has been completed satisfactorily,
- The item has achieved the performance and functional characteristics specified in the functional or allocated configuration identification,
- The item's operational and support documents are complete and satisfactory."

A functional configuration audit provides an independent evaluation that the as-built, as-tested system or software and its deliverable documentation meet the specified functional, performance, quality attributes, and other requirements. An FCA is conducted at least once during the life cycle, typically just before the final ready-to-beta-test or ready-to-ship review, and provides input information into those reviews. However, FCAs can also be conducted throughout the life cycle at checkpoints to verify the proper transition of the requirements into the subsequent successor work products. An FCA is essentially an independent evaluation of the system's or software's verification and validation data to make sure that the deliverables meet their completion and quality requirements and that they are sufficiently mature for transition into the next phase of development or into either beta testing or operations, depending on where in the life cycle the FCA is conducted. It should be noted that it is not the role of the FCA to verify the product's conformance to the requirements. That is the role of verification and validation. The FCA's role is to make sure that all of the requirements have been completely implemented and tested through the examination of development and testing records and other objective evidence.

Table 28.1 illustrates an example of an FCA checklist and proposes possible objective evidence-gathering techniques for each example checklist item. While several suggested evidence-gathering techniques are listed for each checklist item, the level of rigor and scope for the audit will dictate which of these techniques (or other techniques) will actually be used by the auditors. The level of rigor chosen for each SCM audit should be based on a trade-off analysis of cost, schedule, product and risk. For example, when evaluating whether the code implements all and

Table 28.1 FCA checklist items and evidence-gathering techniques—example.

FCA Checklist Item	Suggestions for Evidence-Gathering Techniques
1. Do the product-level requirements implement all, and only, the documented and authorized system or software stakeholder-level requirements?	• Evaluate stakeholder-level to product-level requirements forward and backward traceability information for completeness and to verify that only authorized functionality has been implemented. • Sample a set of source code modules and compare them against their allocated requirements.
2. Does the design implement all, and only, the documented system or software requirements?	• Evaluate requirements-to-design forward and backward traceability information for completeness and to verify that only authorized functionality has been implemented. • Sample a set of requirements and, using the traceability information, review the associated design for implementation completeness and consistency. • Sample a set of approved enhancement requests and review their resolution status (or if approved for change, evaluate their associated design for implementation completeness and consistency).
3. Does the code implement all, and only, the documented system or software requirements?	• Evaluate requirements-to-source code module forward and backward traceability information for completeness and to verify that only authorized functionality has been implemented. • Sample a set of requirements and, using the traceability information, and review the associated source code modules for implementation completeness and consistency. • Sample a set of approved enhancement requests and review their resolution status (or if approved for change, evaluate their associated source code modules for implementation completeness and consistency).
4. Can each system or software requirement be traced forward into tests (test cases, procedures, scripts) that verify that requirement?	• Evaluate requirements-to-tests traceability information for completeness. • Sample a set of requirements and, using the traceability information, review the associated test documentation for verification adequacy by ensuring the appropriate level of test coverage for each requirement.
5. Is comprehensive system or software testing complete, including functional testing, interface testing, and testing required for quality attributes (performance, usability, safety, security, and so on)?	• Review approved verification and validation reports for accuracy and completeness. • Evaluate approved test documentation (e.g., test plans, defined tests) against test result data (e.g., test logs, test case status, test metrics) to confirm adequate test coverage of the requirements and system or software. • Execute a sample set of test cases to confirm that the observed results match those recorded in the test reporting documentation to evaluate the accuracy of test results.

(continued)

Table 28.1 (Continued)

FCA Checklist Item	Suggestions for Evidence-Gathering Techniques
6. Are all the problems reported during testing adequately resolved (or the appropriate waivers or deviations obtained and known defects with workarounds documented in the release notes)?	• Review a sample set of approved test problem report records for evidence of adequate resolution. • Sample a set of test problem report records and review their resolution status (or, if approved for change, evaluate their associated source code modules for implementation completeness and consistency). • Review regression test results data (e.g., test logs, test case execution outputs or status, and test metrics) to confirm adequate test coverage after defect correction.
7. Is the deliverable documentation consistent with the requirements and the as-built system/software?	• Review minutes and defect resolution information from peer reviews of deliverable documentation for evidence of consistency. • Evaluate approved test documentation (e.g., test plans, defined tests) against test results data (e.g., test logs, test case execution outputs or status, test metrics) to confirm adequate test coverage of the deliverable documentation during test execution. • Review a sample set of updates to previously delivered documents to confirm consistency with requirements and as-built system or software.
8. Are the findings from peer reviews incorporated into the software deliverables (system or software and documentation)?	• Review records from major milestone or phase gate reviews that verified the resolution of defects identified in peer reviews. • Review a sample set of peer review records for evidence of the resolution of all identified defects. • Review a sample set of minutes from peer reviews and evaluate the defect lists against the associated work products to confirm that the defects were adequately resolved.
9. Have approved corrective actions been implemented for all findings from in-process SCM audits?	• Evaluate findings from SCM audit reports against their associated corrective action status. • Re-audit against findings from in-process SCM audits to verify the implementation of corrective actions.

only the specified requirements, a less rigorous approach would be to evaluate the traceability matrix. In contrast, a more rigorous audit might examine sampled source code modules, comparing them against their allocated requirements.

PHYSICAL CONFIGURATION AUDIT (PCA)

According to the *IEEE/ISO/IEC Systems and Software Engineering—Vocabulary* (IEEE/ISO/IEC/ 2017), a physical configuration audit (PCA) is "an audit conducted to verify that each configuration item, as built, conforms to the technical documentation that defines it." A PCA verifies that:

- All items identified as being part of the configuration are present in the product baseline

- The correct version/revision of each item is included in the product baseline
- Each item corresponds to the information contained in the baseline's configuration status report

A PCA is performed to provide an independent evaluation that the software, as implemented, has been described adequately in the documentation that will be delivered with it and that the software and its documentation have been captured in the software configuration status accounting records and are ready for delivery. Finally, the PCA may also be used to evaluate adherence to legal obligations, including licensing, royalties, and export compliance requirements.

Like the FCA, the PCA is conducted at least once during the life cycle, typically just before the final ready-to-beta-test or ready-to-ship review, and provides input information into those reviews. However, PCAs can also be conducted throughout the life cycle at checkpoints to verify the proper transition of the requirements into the subsequent successor work products. The PCA is typically held in conjunction with the FCA or soon after the FCA (once any issues identified during the FCA are resolved). A PCA is essentially an evaluation of the software

Table 28.2 PCA checklist items and evidence-gathering techniques—example.

PCA Checklist Item	Suggestions for Evidence-Gathering Techniques
1. Has each nonconformance or noncompliance from the FCA been resolved?	• Review findings from the FCA audit report, associated corrective actions, follow-up and verification records to evaluate the adequacy of actions taken (or verify that appropriate approved waivers or deviations exist).
2. Have all the identified configuration items (e.g., source code modules, documentation, and so on) been baselined?	• Sample a set of configuration items that are part of the baseline according to the SCM plans and evaluate them against configuration status accounting records.
3. Has the software been built from the correct configuration items and in accordance with the specification?	• Evaluate the build records against the configuration status accounting information to confirm that each configuration item's correct version/revision was included in the build. • Evaluate any patches or temporary fixes made to the software to confirm their completeness and correctness. • Sample a set of design elements from the architectural design and trace them to their associated component design elements and source code modules. Compare those elements with the build records to evaluate them for completeness and consistency with the as-built software. • Sample and review a set of configuration items to confirm that their physical characteristics match their documented specifications.

(continued)

Table 28.2 (Continued)

PCA Checklist Item	Suggestions for Evidence-Gathering Techniques
4. Is the deliverable documentation set complete and consistent?	• Evaluate the master copy of each document against the configuration status accounting information to confirm that the correct version/revision of each document constituent configuration item (e.g., chapter, section, and figure) is included in the document. • Sample the copied documents ready for shipment and review them for completeness and quality against the master copy. • Evaluate the version description document against the build records for completeness and consistency. • Compare the current build records to the build records from the last release to identify changed configuration items. Evaluate this list of changed items against the version description document to evaluate the version description document's completeness and consistency.
5. Do the actual system deliverables conform to their specifications, and are they frozen to prevent further change? Have the deliverables been appropriately marked or labeled?	• Evaluate the items on the master media, downloadable files, or web-accessible files against the required software deliverables (executables, help files, data, and documentation) to confirm that the correct versions and revisions were included. • Rebuild a sample set of software deliverables from the SCM repository and evaluate them to confirm that the controlled configuration items match those built into the deliverables. • Sample a set of copied media, downloadable files, or web-accessible files, ready for shipment or access, and review them for completeness and quality against the master media. • Sample a set of copied media, downloadable files, or web-accessible files ready for shipment or access and review their marking or labeling against specifications.
6. Do the deliverables for shipment match the list of required deliverables?	• Evaluate the packing list against the list of documented deliverables to verify completeness. • Sample a set of ready-to-ship packages and evaluate them against the packing list to confirm that media (e.g., flash drive, disks, tape), downloadable files or web-accessible files, hard copy documentation, and other deliverables are included in each package.
7. Have all third-party licensing requirements been met?	• Evaluate the build records against configuration status accounting information to identify third-party items and license information to confirm that adequate numbers of licenses exist.
8. Have all export compliance requirements been met?	• Evaluate the build records against configuration status accounting information to identify items with export restrictions and confirmed export compliance.

configuration status accounting data to make sure that the software products and their deliverable documentation are appropriately baselined and properly built before transition into the next phase of development, release to beta testing or operations, depending on where in the life cycle the PCA is conducted.

Table 28.2 illustrates an example of a PCA checklist and proposes possible objective evidence-gathering techniques for each item.

Chapter 29

Product Release and Distribution

The development portion of the project eventually ends, and the software products and services created during development must be released into operations. While the primary focus of software quality engineering activities is to make certain that quality is being built into the software throughout development, software quality engineering must also verify quality during release and distribution.

Release is the function of configuration management that focuses on packaging the software products for promotion and distribution outside the development organization. Depending on the project, interim or preliminary releases to the customers, users, and other stakeholders may also need to be deployed, including releases into independent or beta-testing environments or other evaluation and feedback release environments.

As with any other process, defects can be introduced during the release process, impacting the quality and reliability of the product or services after release. Therefore, the software quality engineer (SQE) must be familiar with the release and distribution processes to help define these processes, evaluate their effectiveness and efficiency, help identify issues and process improvement opportunities, and perform other release and distribution quality-related activities.

1. PRODUCT RELEASE

> Assess the effectiveness of product release processes: planning, scheduling, and defining hardware and software dependencies. (Evaluate)
>
> **Body of Knowledge VII.E.1**

Types of Releases

There are two major types of product releases. The first type of release is a *corrective release,* also called a *service pack, maintenance release,* or *hot fix.* As illustrated in Figure 29.1, a corrective release is done to deliver defect corrections. This release may be done when the defects are deemed critical enough (have a large enough

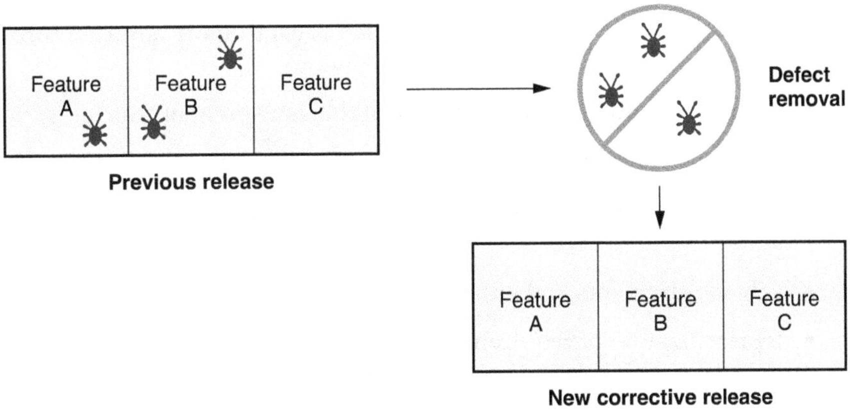

Figure 29.1 Corrective release.

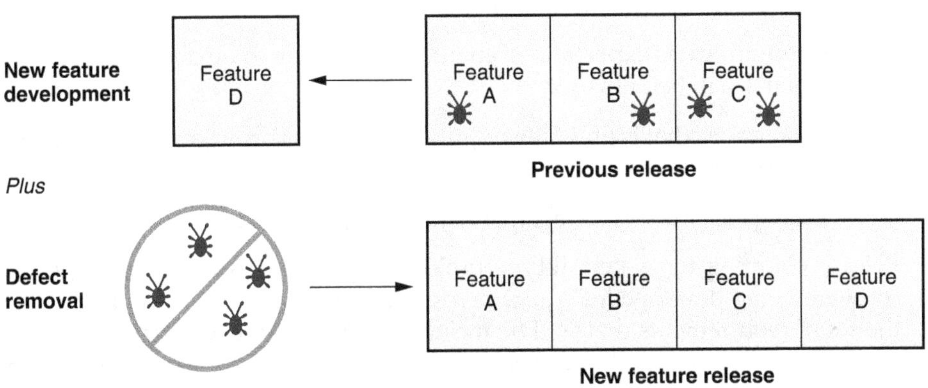

Figure 29.2 Feature release.

impact) that correction of those defects cannot wait until the next scheduled feature release of the product. Corrective releases may also occur to fix operational defects if a future feature release is not currently planned. Corrective releases may not be appropriate for some software products (e.g., products that have stringent safety or reliability requirements).

The second type of release is a *feature release*, primarily used to deliver new features or functionality to the customer or user. The first release of a new software product is always a feature release. However, as illustrated in Figure 29.2, defects may also be corrected as part of subsequent feature releases.

Release Management Planning and Scheduling

According to Pichler (2010), "release planning supports the development and launch of a successful product." *Release management planning and scheduling* define what functionality will likely be delivered in each release and when each release is scheduled to be delivered. The release plan is continually adjusted throughout the development life cycle based on the following:

- The actual productivity or velocity of the development team
- Feedback from the customers, users, and other stakeholders
- Changes in the business climate, business objectives, and competitive arena
- New requirements and enhancement requests
- Reprioritization of requirements or user stories based on factors impacting their relative value to the stakeholders

Release planning is valuable because it:

- Helps the development team and other stakeholders determine the scope of each release, which feeds into estimations for schedule, budget, resources, staffing, and so on
- Communicates expectations about what is likely to be delivered and when it will be delivered
- Makes sure that high-value requirements are appropriately prioritized in earlier releases
- Establishes milestones against which progress can be tracked

Release planning for waterfall-type projects begins by defining the scope of the software being developed. Requirements engineering practices are used to select the set of requirements that will be included in the release. Configuration control activities for waterfall-type projects updating legacy code are also used to determine which problems will be corrected in the next feature or corrective release. The release process continues throughout the design and development of the software version/revision. During implementation, changes to the scope are managed through requirements management and configuration control activities as additional enhancements are requested and new problems are discovered. Verification and validation activities are used throughout implementation to confirm that the software meets the specified requirements, stakeholder needs, and intended use.

Release planning for incremental or iterative projects starts by defining the scope of the software being developed in a manner similar to waterfall-type projects. However, for incremental or iterative projects, that scope is subdivided by prioritizing the selected requirements and problem reports and assigning them to various increments or iterations for implementation. Decisions are also made about which increments or iterations will be released. For example, several increments or iterations may be completed before enough functionality exists to warrant a release.

Agile projects can have several different release cadences. For example, some agile projects release after each feature meets its "done" criteria. Other agile projects release after every iteration or sprint or after multiple iterations or sprints. In the multiple iterations/sprints cadence, the "release plan looks forward through the release of the product, usually three to six months out at the start of a new

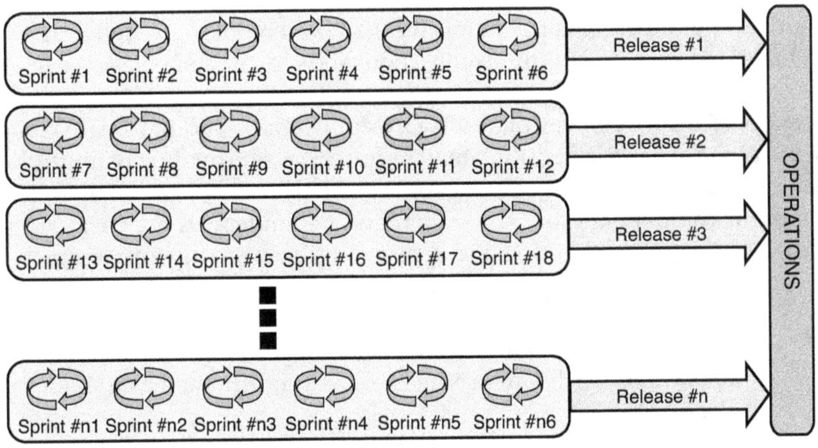

Figure 29.3 Scrum project with multiple sprints per release—example.

project. In contrast, the iteration plan looks ahead only the length of one iteration, usually two to four weeks" (Cohn 2006). Agile release planning is often date-bound instead of feature-bound, where the release date is fixed but the feature set is not. The time-box also helps fix the cost of each release. The agile team's predefined "done criteria" provides a fixed goal for software quality. For these time-box releases, the functionality included in each release varies based on the work actually accomplished during the iterations or sprints. For example, as illustrated in Figure 29.3, a Scrum project might have two-week sprints, with a release every six sprints, creating a 12-week release time-box. Using story point estimates for each item on the product backlog and historic velocity numbers for the team, release planning defines which epics or stories are targets for implementation in each future sprint and in the next release. This plan is then adjusted regularly when the product backlog is expanded and reprioritized or the team's velocity is re-estimated based on ongoing performance.

Once the software product, including its associated deliverables, has been completely implemented and is ready for release, release activities include:

- Packaging
- Creating the version description documentation (release notes)
- Performing final verification and validation activities, including final functional and physical configuration audits

A ready-to-ship or customer acceptance review is then conducted as a final quality gate before promoting the software version/revision to a release.

Release Propagation Planning

The *release propagation* planning activities are related to the release's replication, delivery, and installation. Release, distribution, and installation activities can be as trivial as installing a software package on a single computer (including users downloading the software themselves) or as complex as delivering multiple products and services to hundreds, thousands, or in some cases, even millions of sites around the globe. In complex systems or systems of systems, with dependencies between various software applications and hardware components, release planning and scheduling may require the coordination of multiple product releases from multiple projects or programs. This coordination makes certain that all system components that share changed dependencies are installed together. Release propagation planning and scheduling includes the following:

- Determining the contents of the releasable unit, for example:
  - Does the release include a single software application, or are multiple software or hardware product releases coordinated as a single releasable unit?
  - Are services like training bundled into the release unit, or are they separate release units?
- Coordinating with customers, users, and distributors on the logistics of rollout, including:
  - Which release units will be delivered
  - Quantities for each release unit
  - Delivery dates for each release unit
  - Delivery location for each release unit
  - Delivery mechanism for each release unit
- Determining release propagation schedules, budgets, staffing, and other resource requirements, communication plans, risk management plans, and so on
- Negotiating, contracting, and coordinating deliveries from suppliers, as necessary
- Handling legal issues (licensing, royalties, export regulations, and so on)
- Conducting pre-installation and installation site visits, as appropriate
- Defining installation responsibilities and performing installation and installation testing and verification
- Planning migrations from the older to the newer versions of databases if not all users are transitioned to the newer release simultaneously
- Defining and implementing back-out plans as necessary

Version Description Documentation

Each delivered release should include *version description documentation,* also called *release notes,* that describe the new version/revision of the software. This documentation should define the following:

- The environmental requirements needed to execute the software, including the operating system or browser version(s), hardware requirements (memory, processor specification, disk space, and so on)
- Hardware and software dependencies
- Installation instructions and post-installation testing instructions
- The licensing key or serial number of the product
- A list of new features, functions, or other enhancements added with this release (for feature releases)
- A list of the defects previously reported from operations that are corrected with this release
- A list of known defects, problems, or limitations in the release, including workarounds, if they are available
- Contact information for obtaining technical assistance with the release
- Instructions for reporting new problems with the release or requesting enhancements to the software

Hardware and Software Dependencies

As illustrated in Figure 29.4, *hardware dependencies* exist when:

- The current release of the software is not backward-compatible with the hardware used to run previous versions of the software. For example, a new software version is created to implement a feature set to control a new widget added to the hardware system. If that software cannot be run on older versions of the hardware that have not been upgraded to include the new widget, then a hardware dependency exists.

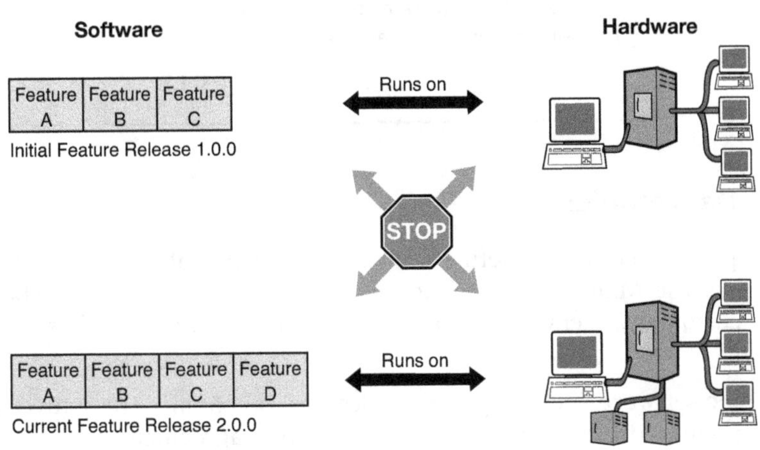

Figure 29.4 Hardware dependencies—example.

- The previous releases of the software are not forward-compatible with the current hardware.

Software dependencies exist when there are backward-compatibility or forward-compatibility issues between the current release or older releases of the software and:

- *The software platform used to run the software.* For example, the new software release uses features available only in the newest version of the operating system. It will not run on previous versions of the operating system (or older software releases are not compatible with newer versions of the operating system).
- *Other software that interface with the software.* For example, the new software release implements a new communications protocol that is also implemented in the new XYZ software, but it cannot communicate with the previous XYZ version (or older software releases cannot communicate with the new XYZ software). The current software release may also have dependencies on new applications that it interfaces with that were not needed or available to older software releases.
- *Data files that are accessed by the software.* For example, the new software implements changes to the format of the accessed data files and cannot use the older versions of the data files (or older releases of the software cannot use the new version of the data files).

2. CUSTOMER DELIVERABLES

> Assess the completeness of customer deliverables, including packaged, hosted, or downloadable products, license keys and user documentation, and marketing and training materials. (Evaluate)
>
> **Body of Knowledge VII.E.2**

Software Deliverables

Part of project planning is defining the external deliverables that will be included in each release. Many different types of software work products may be delivered to the customers, users, and other stakeholders. Examples of these deliverables include:

- The software, including executables and, potentially, the source code modules depending on contractual or other agreements
- Development documentation: depending on contractual or other agreements, documents, including requirements and design specifications; test cases, procedures, and scripts; plans; reports; and metrics

- User or operator documentation, including:
  - Installation instructions
  - Operation and maintenance instructions or manuals
  - User manuals or help files
  - System, platform, or environmental requirement specifications
  - Version description documents (release notes)
- Databases and other auxiliary data files (tables, images, audio)
- Training and training materials
- Help desk support
- Marketing materials
- Other deliverables (as required depending on contractual or other agreements)

Reviews

Peer reviews, technical reviews, and managerial reviews should be held during development to evaluate the completeness and quality of each deliverable, as appropriate.

Development Testing

Software products, including source code modules and executables, are tested by the development organization or testing teams during the various levels of testing. (See Chapter 22 for more information on the levels of testing.) During system testing, user or operator documentation and training materials are typically tested to verify that they reflect the actual performance of the as-built products. The installation of the system and its installation instructions should also be tested to verify that the system is user-friendly and efficient to install and that the resulting installation is complete and correct.

Development Audits

In-process product audits can be used throughout the software development process to evaluate the completeness and quality of software work products and their adherence to requirements and workmanship standards. Functional configuration audits can be conducted before release to verify that the development of each deliverable configuration item has been completed satisfactorily, that the item has achieved the performance and functional characteristics specified, and that its operational and support documents are complete and satisfactory. Physical configuration audits can be conducted before release to verify that each deliverable configuration item, as built, conforms to the technical documentation that defines it and that the correct versions or revisions of the constituent configuration items have been built into the delivered composite configuration items.

Release Packaging

Release packaging defines what is included on the media that is used to deliver the released software. The goal "is to create and maintain a repeatable process for packaging a release that includes a clear way to identify every component of the release" (Aiello and Sachs 2011). According to Bays (1999), there are three types of release packaging:

- *Full release*: The package contents are capable of completing a full installation of the product without requiring the prior installation of a previous version of the product
- *Partial release*: The package contents are limited to only making modifications—to add, modify, or remove portions of the full release—which requires the prior installation of a full release
- *Patch*: The package contains a temporary fix to a specific release through a modification made directly to an object or executable (which avoids the need to rebuild the product) or a temporary fix to the source itself

As illustrated in Figure 29.5, various types of packaging may be appropriate over the product's lifetime in the field. The initial release of the product must always be a full release. Subsequent corrective releases can be accomplished through partial releases and patches, since only minimal changes are made. Installing partial releases is typically much less time-consuming than installing a full release. After the initial release, major or minor feature releases may be packaged as both full and partial releases. This allows existing sites the efficiencies of partial installation. At the same time, full releases can be deployed at new sites, allowing these new sites to install a complete product without the need to install the initial full release followed by all of its subsequent partial releases.

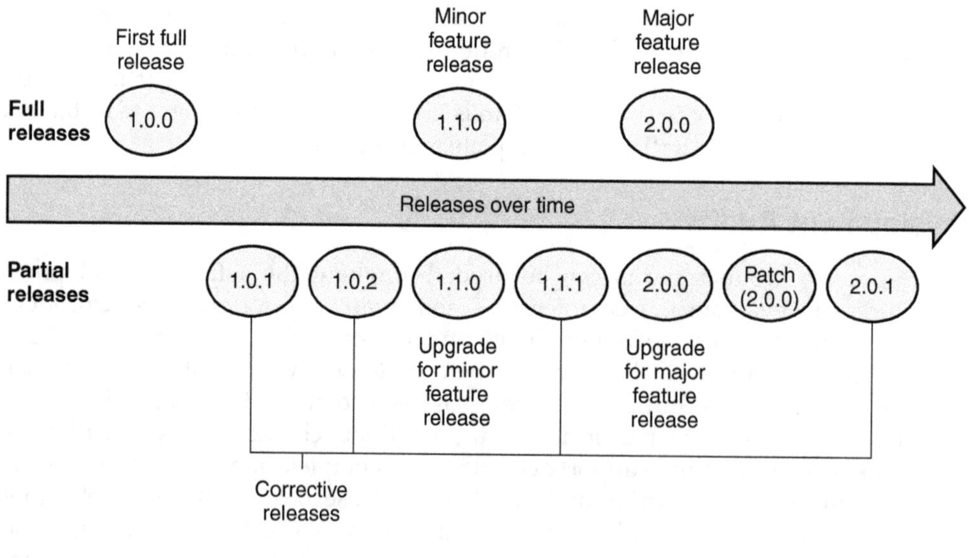

Figure 29.5 Packaging of releases over time.

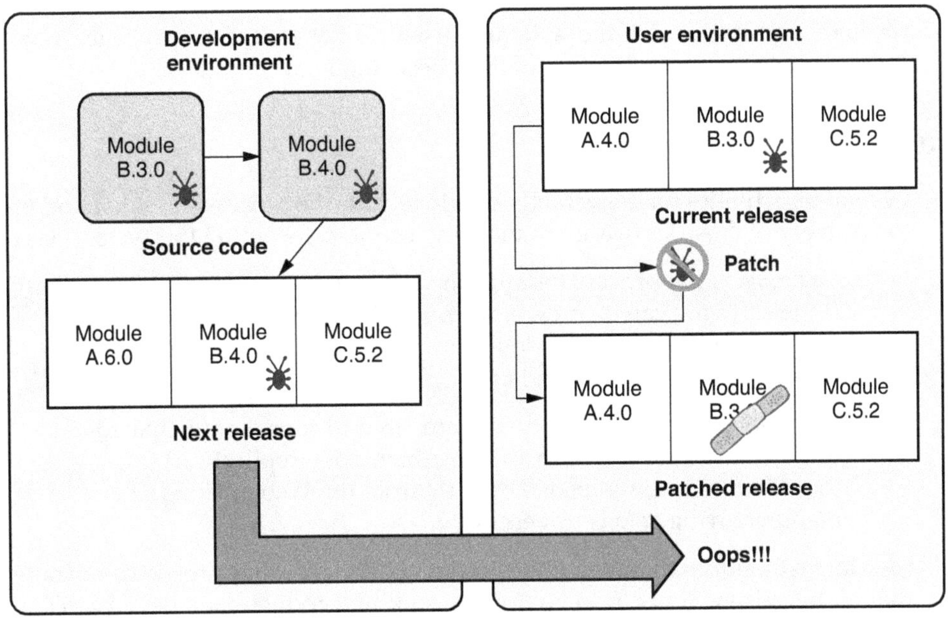

Figure 29.6 The problem with patching.

Patches are very risky because of their temporary nature and because patches to the object or executable code are often done manually. For example, as illustrated in Figure 29.6, a patch might be made to the executable code in operations to temporarily correct (put a band-aid on) a critical problem in operations. In this example, however, the defect was only fixed in the executable and not the source code module. Without good configuration management, the defective source code module might be built into a future software executable. When that executable is released into operations, the user might reencounter that same critical problem. This scenario is why retesting and regression testing should include testing all patches until they have been permanently corrected in the source code.

Localization

Localization may also impact how the software is packaged. For example, there may have to be separate packages for each different geographic location, language, or set of technical requirements for a target market.

Licensing Keys

Licensing keys, also called *product keys* or *USB flash drive keys,* are added during the manufacturing process to specifically identify the replicated software as an original, manufactured copy (and not an aftermarket, unauthorized copy). Typical license keys are a series of numbers and letters that are entered by the user during the software installation process. Some licensing keys are activated offline through verification by the software itself. Other software packages require online activation and validation of their licensing keys.

Ordering and Manufacturing

Evaluation activities for deliverables should include the process steps that lead up to the delivery of those software products and services, including the evaluations of:

- *Ordering*: Is the process for ordering the system in place and functional? If the software can be ordered through the Web or cloud, are the supporting Web pages or cloud areas in place, and has the online ordering process been validated?
- *Manufacturing*: Is manufacturing prepared to support anticipated orders, including software and documentation replication? If the software is downloaded directly from the Web or cloud, the manufacturing step is unnecessary.

Checklists or automation can be used during the ordering processes to confirm that all necessary items for a complete, working product have been considered. Checklists can also be used to validate that all ordered items have been packaged for shipment or placed on the website for downloading.

The manufacturing process for software is mostly a replication process of making copies of the software deliverables for delivery. First, verification should be done to confirm that the master source media ("gold copy") used in the replication process is free from viruses. Then, as the media is replicated, samples of the replicated product should be taken and evaluated. Copies of documentation should be sampled periodically as they are being printed, and reviewed to verify their completeness and quality. Sample copies of replicated DVDs or other replicated media could be selected, installed on test beds, and smoke-tested to verify their installability and quality. This process can also verify that the copies can be read from devices other than those they were created on. This process will catch issues with replication devices being out of calibration. Of course, the calibration of replication devices should be checked regularly.

For both manufactured and downloaded software, automated mechanisms exist to verify the replication of the software products. Most applications that perform electronic copying include verification steps that read back the information written to the media and perform a comparison to the written information. *Checksums, hash sums*, or *cyclic redundancy checks (CRCs)* can detect accidental errors that occur during transmittal (for downloadable packages) or replication. In the checksum or CRC process, an algorithm creates a calculated value for the software package. After replication or transmittal, the value is recalculated for the replicated or transmitted copy. If the original and recalculated values do not match, accidental alteration of the data has occurred.

Delivery

Delivery is the process of getting the deliverables to the customers, users, and other stakeholders. The copied products are subject to corruption in transit or storage and need appropriate protection. Again, periodic sampling can be used to verify that corruptions have not occurred during the delivery process.

Once packaged, several vehicles can be used to deliver the products into operations, including intermediate installation media, direct-access media, personal delivery, and electronic transfer.

When *intermediate installation* media is used to deliver the software product, that product is copied onto some form of intermediate media (e.g., DVD or flash drive). This intermediate media is then delivered, and the software product is installed from that media onto the target platform. These installations can be performed by the customers, users, or their representatives or by supplier personnel sent to the operational site. The intermediate media is not needed after installation, except for additional installation or reinstallation activities.

Direct-access media (e.g., EPROMS, hardcopy documentation, and in some cases DVDs or game cartridges) contain a product that is used directly from that media without the need for installation. The primary advantage of direct-access media is it does not take up space on the target computer system. The primary disadvantage is that the media have to be present to run.

Personal delivery is often used for deliverables such as training or technical support, where supplier personnel hand-deliver the product directly to the customers or users.

Today, more and more software products are being delivered via *electronic transfer* (e.g., using the Web, cloud, or network-mounted file systems). The benefits of electronic transfer as a delivery mechanism include reduced costs and immediate availability.

Installation Testing

Installation is the process of integrating the software into its operational environment or target platform by means of intermediate installation media or electronic transfer. In some cases, personal delivery may also require installation. Depending on the software (size, complexity, and criticality) and the needs of the stakeholders, it may be advisable, or even required, for the supplier of the software to send technical support personnel to the operational site(s) to verify that the installed software (or reassembled system) functions correctly. In cases where the customers or users perform the installation themselves, they may perform their version of installation testing after the installation is complete. The developer may also include automated testing in the installation scripts.

Installation testing is done by selecting a representative set of tests that will adequately verify that the software's installation (or the system's reassembly) has not degraded its functionality or performance. This is done using system or acceptance tests or with new tests specifically designed to evaluate the installation. As appropriate, installation testing should confirm that all software diagnostics are performed and results are acceptable. Software *diagnostics* are built-in tests that evaluate specific parts of the system (e.g., electronic components and printed circuit cards) and return the status to the tester. In addition, installation testing can be used to verify the license keys if they are used for the replication copies.

Another technique for verifying the entire ordering, manufacturing, delivery, and installation process is to deliver a complete release package in-house first, preferably to someone with the same level of knowledge about the system as the typical installer. This person utilizes the deliverables to install the software and verify it works in the target environment. This process helps verify that the replication process was successful, that all the needed deliverables (software, databases, documentation, and so on) were in the release package, and that the installation process can be successfully executed. The ability to back out the release, if necessary, and return the software to its previous release should also be tested.

Pilots

A *pilot* is an experimental execution (limited release) of a service to a selected audience to provide assurance that the service has the potential to succeed when released. Deliverables such as training courses can be piloted to verify the completeness, quality, and accuracy of the course content, presentation materials, teacher training, and student notebooks. Development, testing, or help desk team members might be selected to attend a pilot training class. Members of the actual customer or user community might also attend the pilot with the full knowledge that it is a pilot run and that constructive feedback is requested. Pilots can also be used effectively to verify the readiness of marketing, technical support, or other websites for rollout to support the customers or users.

The readiness of *help desk* support can also be piloted. For example, near the end of testing, the testers have gained enough knowledge of the working software to identify areas where the users may have difficulties. This knowledge could be used to create a sample set of potential user issues or questions. The testers could then pilot the help desk support by calling with these issues or questions to determine the accuracy and completeness of the assistance they receive and the overall effectiveness of the training of help desk personnel.

Customer and User Testing

In addition to the installation testing described previously, other types of testing typically performed by the customer or user include:

- *Alpha testing* is typically done by the customers or users of the software or their representatives, at the supplier's facility or on a customer or user test bed. Alpha testing may be done on an intermediate software version that does not include all of the final releasable functionality.
- *Beta testing*, also called *first office verification* or *field-testing*, is done by the customers or users in the actual operational environment, with a prerelease software product or the software that is expected to be released. Beta testing is typically done with a limited set of customers or users before the full general availability of the software. No matter how well specified and well built the simulators used during previous development and alpha testing are, there is nothing like testing the software in the actual target operational environment.

As Anderson (2007) says, "A test environment is not the operational environment. Never assume that because it works in the former, it will work in the latter." Beta testing can also be extremely useful to evaluate software applications that must function on many diverse operational environments and platforms, where it may be difficult, if not impossible, to perform configuration testing on all possible combinations of hardware, operating systems, other software background software, and so on.

- *Acceptance testing* evaluates the completed software against predefined acceptance criteria. This testing provides the customers, users, product owners, marketing, or other key stakeholders with the information they need to determine whether to allow the software to be released into operations.
- *Operational testing,* also called *in-use testing,* is performed once the software has been installed and is executed by the actual users, using normal operating processes and documentation. The objectives are to confirm that the software is operating correctly under normal operating conditions and to evaluate the effectiveness of the training the users received. Operational testing can also be performed periodically during the ongoing use of the software to verify that no problems have crept into the software or data over a period of time. For example, periodic operational testing would verify that performance has not degraded from the original baselined values, worms or viruses have not infected the software, or items in the database have not become corrupted.

Release Support

Rarely do the software supplier's responsibilities end when the release is delivered. The release must also be supported during its operational lifetime. The level and duration of this support depends upon the service level agreements between the supplier and the acquirer of the software product, which may include:

- The acquirer's ability to obtain additional copies of the release package as more operational sites are created, requiring continued replication, delivery, and installation support
- Product and process implementation and customization
- Technical support for the release as it is being used in operations, including the ability of customers or users to:
  - Get their questions answered
  - Report problems or issues and have them resolved in a timely manner
  - Request enhancements to the software

3. ARCHIVAL PROCESSES

> Assess the effectiveness of source and release archival processes: backup planning and scheduling, data retrieval, archival of build environments, retention of historical records, and offsite storage. (Evaluate)
>
> **Body of Knowledge VII.E.3**

Backups

Disasters happen. Hard disks fail or become corrupted, essential files are inadvertently deleted, buildings burn down, and Mother Nature can pack a mean punch (tornadoes, hurricanes, floods, earthquakes, and so on). When software intellectual capital assets are lost under such circumstances, they are often gone for good, or recovery or re-creation costs can be huge. Backing up the contents of the following is the single most important thing an organization can do to allow for quick recovery from these disasters:

- Dynamic, controlled, and static libraries
- Status accounting information
- Quality records
- Other intellectual capital assets

Backing up is the process of copying intellectual capital assets onto backup libraries on one or more additional storage devices to make certain that assets can be recovered if a disaster happens (the original media are lost, damaged, or corrupted or the asset is inadvertently destroyed, deleted, or modified).

Backups can be made periodically or on an event-driven basis depending on the following:

- Criticality of the work products kept in the library and the impact if they are lost
- Level of change activity
- Reliability of the system on which the electronic assets are stored

Figure 29.7 illustrates the library procedure steps for performing a backup, including:

Step 1: On a specified schedule, the system administrator creates backup copies of the contents of the dynamic, controlled, and static libraries into an on-site backup library. Each library, or type of library, can have a different backup cycle depending on the project's needs. For example, the dynamic and controlled libraries might be backed up daily, while the static library is only backed up when new releases are archived.

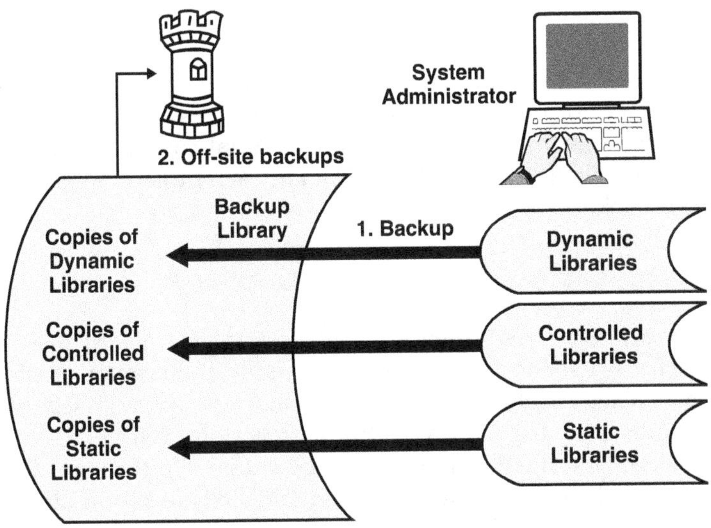

Figure 29.7 Library procedure for performing a backup.

Step 2: For disaster recovery purposes, copies of the backup library are created on a specified basis and stored in a secure, offsite repository.

Depending on these considerations, many organizations choose to back up their active (nonarchived) electronic assets on a daily basis during off hours to minimize the impact on development and operations. This process makes certain that no more than a day's worth of work is lost if a problem occurs. These backups are typically done after hours because there may be a need to freeze those assets and not allow change during the backup process. If these daily backups are incremental, a full backup is recommended at least once a week. A more modern and less-risky approach is implementing real-time, mirrored data backups to an offsite location.

Care must be taken to verify that all appropriate assets are being backed up. For example, the organization may have a policy that requires electronic assets to be stored in libraries on the shared network server and not kept on personal computers or laptops that may not be included in the official nightly backups. According to Kenefick (2003), "This has the double problem-solving ability of protecting the developer's work in between check-ins and providing access to the code should the developer be absent." Examples of events that might trigger a backup include:

- Creating one or more important assets
- Implementing changes to one or more important assets
- Creating or updating a baseline
- Attaining major milestone checkpoints
- Releasing a product

Off-Site Storage

Backup libraries can be stored on-site for quick recovery, but copies should also be stored offsite for disaster recovery. This is necessary to make certain that a single disaster does not destroy both the original copy and the backup copy of the software or intellectual capital assets. For each project, SCM planning should identify the backup and disaster-recovery plans and processes.

Archives

A *static library,* also called a *software repository, archive library,* or *software product baseline library,* is used to archive important intellectual capital assets on an event-driven basis when baselines are created, including those released into operations. Static libraries can also archive ("freeze") various intermediate baselines, as appropriate, during the development process. Items in the static libraries can be accessed through read-only mechanisms and are not modifiable. The primary goal of archiving is to make certain that intellectual capital assets from an earlier time can be recovered (e.g., a previous product release needs to be recovered). For supplier-developed software, based on contract or agreement requirements, archives may be put into escrow so that the acquirer can retrieve intellectual property assets if the supplier goes out of business or discontinues support of the software. Figure 29.8 illustrates the library procedure steps for archiving a baseline, as follows.

Step 1: The configuration librarian or release engineer archives copies of the baseline and all associated configuration items (e.g., the software build, software source code modules, documentation, tools, and data) out of the control library into the static library.

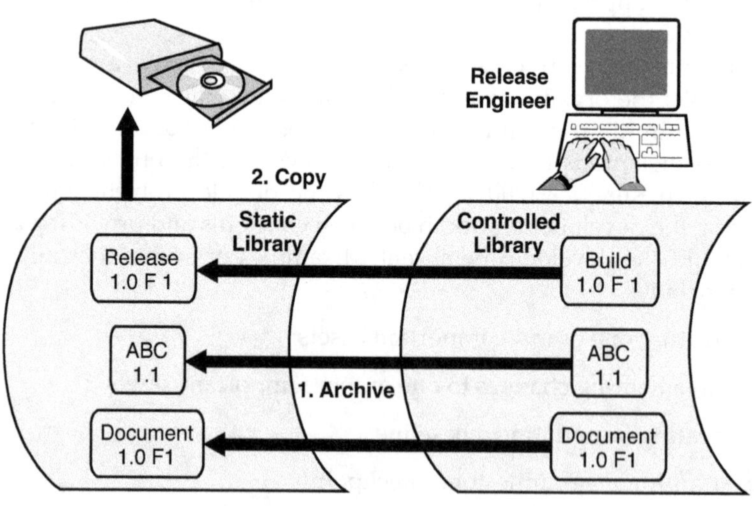

Figure 29.8 Library procedure for archiving a baseline.

Table 29.1 Differences between backups and archives.

Backups	Archives
Backups retain copies of all important current assets, which may include assets in archive libraries	Archives retain copies of baselines, including releases
Since backups are repeated regularly, they are typically kept for a short duration (e.g., a 30-day cycle, where on the 31st day, the oldest backup device is copied over, and so on)	Each archive is typically created only once and retained long-term (for a specified retention period)
Backups are used for disaster recovery	Archives are used to replicate or recover baselines or releases

Step 2: If this is a release, the archive creates a *gold copy* of the release so that it can be replicated at any time in the future. The replication process creates copies of the released software deliverables from the static library as needed for shipment to the customer, users, or stakeholders.

Table 29.1 illustrates the primary differences between backups and archives. There are also considerations concerning how and if the assets will be handled when they are backed up or archived, including:

- Dealing with open files
- Dealing with live databases (risk of data changing as it is being backed up)
- Compressing assets to save space
- Encrypting assets for security reasons
- Considering how to handle duplicate assets

Many different types of storage devices are available to use in backup or archival processes, including everything from hard copy files, magnetic media, optical media, solid-state storage, and internet-based or cloud-based remote backup services. When selecting the appropriate data storage device, concerns include geographic redundancy, size of the assets, asset security, portability, and the ability to retrieve the asset.

Archival of the Build Environment

Along with the software deliverables and their constituent configuration items, all other configuration items necessary to re-create the development, build, and test environments (including the hardware, tools, and operating system) should be archived. The goal is to be able to re-create, debug, and maintain the released software products. Other considerations may include maintaining the appropriate licenses to continue using any purchased software that is part of or used to support the released software.

Asset Retrieval

Creating a backup or an archive is not enough. The software intellectual capital assets must also be recoverable from those backups or archives. Issues with the retrieval of software intellectual capital assets include:

- *The refreshing of media*: This can be especially critical for the long-term (multiyear) archiving of electronic assets. For example, years ago, organizations used eight-inch floppy disks for backups. Those floppies started deteriorating, becoming brittle and unreadable after a year or so. This deterioration required someone to be assigned to copy archived information onto new floppies periodically to prevent the possibility of losing corporate assets. How long will CDs/DVDs last before deteriorating? Backup procedures should include mechanisms for refreshing the media used or for migrating archived assets to newer media types, depending on the media used.
- *The hardware used*: A few years ago, many organizations used Zip drives, 3.5-inch hard-sided floppy disks, or CDs as their backup media. Today those organizations may not even have the hardware necessary to read and recover information from that type of media. Of course, if the entire build environment, including hardware, is archived, this is not an issue.
- *Compressed or encrypted assets*: Confirming that these assets can be decompressed or unencrypted if they need to be recovered.
- *Single asset recovery*: If assets are backed up in sets, can a single asset be individually recovered, or must the entire set be recovered? For example, a company performs nightly backups of its source code modules. The next afternoon, a critical file is accidentally deleted. Can they retrieve just that one file, or must the entire set be retrieved? If the entire set has to be retrieved, other critical updates made that morning may be lost.

Tests should be performed when initial backups or archives are created to verify that intellectual capital assets can be retrieved. The backup, archival, and retrieval process should also be verified periodically to confirm the reliability, accuracy, and completeness of the assets being backed up or archived and the ability to retrieve those assets successfully. The full disaster recovery process should also be tested to verify that the human elements, recovery documentation, and recovery process work as a complete system.

Retention of Historical Records

While not considered configuration items under configuration management, quality records may also need to be retained. Based on risk, an organization should identify which quality records must be retained. Procedures should be established for the retention of quality records so that they are:

- Identifiable
- Appropriately collected and stored

- Protected from loss, damage, or unintended or unauthorized change
- Easily retrievable when needed

This retention is typically done through the backup and archival processes described previously. Hardcopy records may be retained through hardcopy files, or scanned and turned into electronic assets.

The retention period should be specified as part of the quality management system documentation for each type of retained quality record. Typical retention periods might be "a minimum of the duration of the project plus two years" or "a minimum of the active use of the release in operations plus one year." The retention period is specified as "at a minimum" to allow the organization to prune older quality records from active retention on a periodic basis. At the same time, this keeps auditors from reporting issues when they find quality records that have been retained for longer than an absolute specified retention period (e.g., "the completion of the project plus one year"). For supplier-developed software, the acquirer's requirements for quality record control by the supplier, including retention periods, should be specified as part of the supplier contract or agreement. The supplier should make certain that their personnel performing the archival activities know these requirements.

Procedures for controlling retained quality records should also specify what is done with those records after the retention period has expired. Are they thrown away? Do they need to be disposed of securely (e.g., through shredding)? Or are they moved into long-term archival storage, and if so, for how long? These procedures may also need to be implemented in a manner that complies with any applicable regulatory requirements or industry standards (e.g., ISO 9001) depending on the organization's needs.

In addition to quality records, other organizational process assets from the project may also be archived when the software is released. These assets might include tailored or project-specific life cycles, processes, work instructions, contracts, customer letters or e-mails, project plans, lessons learned, and other work products that were not designated as configuration items.

Acronyms

ACID—Atomic, Consistent, Isolated, Durable

AV—Actual Value

ACWP—Actual Value of Work Performed

B to B—Business to Business

B to C—Business to Consumer

B to E—Business to Employee

BCWP—Budgeted Cost of Work Performed

BCWS—Budgeted Cost of Work Scheduled

BRD—Business Requirements Document

CAB—Change Authority Board

CAR—Corrective Action Request

CCB—Configuration Control Board or Change Control Board

CDR—Critical Design Review

CEG—Cause and Effect Graph

CFD—Cumulative Flow Diagram

CI—Configuration Item

CMMI®—Capability Maturity Model Integration

COCOMO—Constructive Cost Model

COQ—Cost of Quality

COTS—Commercial-Off-the-Shelf

CPI—Cost Performance Index

CR—Change Request

CRUD—Create, Read, Update, Delete

CRUDL—Create, Read, Update, Delete, List

CSP—Cloud Service Provider

CSQE—Certified Software Quality Engineer

CTx—Critical-to-x

CV—Cost Variance

DEI—Diversity, Equity, Inclusion

DFD—Data Flow Diagram

DFSS—Design for Six Sigma

DMADV—Define, Measure, Analyze, Design, Verify

DMAIC—Define, Measure, Analyze, Improve, Control

DRE—Defect Removal Effectiveness

DRIP—Data Rich, Information Poor

ECB—Engineering Change Board

EIA—Electronic Industries Association

EPG—Engineering Process Group

ERD—Entity Relationship Diagram

ETVX—Entry criteria, Tasks, Verification steps, eXit criteria

EV—Earned Value

EVM—Earned Value Management

FaaS—Function as a Service

FCA—Functional Configuration Audit

FDD—Feature Driven Design

FMEA—Failure Modes Effects Analysis

FMECA—Failure Modes Effects Criticality Analysis

GOTS—Government-Off-the-Shelf

GROW—Goal, Reality, Options, Will Do

HLD—High Level Design

IaaS—Infrastructure as a Service

IEC—International Electrotechnical Commission

IEEE—Institute of Electrics and Electronic Engineers

IFPUG—International Function Point User Group

IM—Instant Messaging

IPT—Integrated Product Team

ISACA—Information Systems Audit and Control Association

ISO—International Organization for Standards

JAD—Joint Application Development

KLOC—Thousand (K) Lines of Code

LLD—Low Level Design

LOC—Lines of Code

LOC—Level of Control

MOTS—Modifiable-Off-the-Shelf

NGT—Nominal Group Technique

NIST—National Institute of Standards and Technology

OMG—Object Management Group

OOAD—Object Oriented Analysis and Design

PaaS—Platform as a Service

PbD—Privacy by Design

PCA—Physical Configuration Audit

PDCA—Plan, Do, Check, Act

PDR—Preliminary Design Review

PERT—Program Evaluation and Review Technique

PMBOK—Project Management Body of Knowledge

PMI—Project Management Institute

PV—Planned Value

QA—Quality Assurance

QMS—Quality Management System

RE—Risk Exposure

ROI—Return on Investment

RPN—Risk Priority Number

RRL—Risk Reduction Leverage

SaaS—Software as a Service

SAD—Structured Analysis and Design

SCM—Software Configuration Management

SCMP—Software Configuration Management Plan

SDLC—Software Development Life Cycle

SEI—Software Engineering Institute

SEPG—Software Engineering Process Group

SLA—Service Level Agreement

SLIM—Software Lifecycle Management

SMART—Specific, Measurable, Achievable, Realistic, and Time-Framed

SPI—Schedule Performance Index

SQA—Software Quality Assurance

SQAP—Software Quality Assurance Plan

SV—Schedule Variance

SwA—Software Assurance

SQA—Software Quality Assurance

SQAP—Software Quality Assurance Plan

SwA—Software Assurance

SysML—System Modeling Language

TDCE—Total Defect Containment Effectiveness

TDD—Test Driven Development

TRR—Test Readiness Review

UAT—User Acceptance Testing

VDD—Version Description Document

V&V—Verification and Validation

WBS—Work Breakdown Structure

WIP—Work in Process

XP—eXtreme Programming

YAGNI—"You Ain't Gonna Need It"

Glossary

abstraction—A design concept that emphasizes the essential attributes of the software at a specific level, while suppressing the implementation details.

acceptance—The formal acknowledgement of the customer or user's willingness to receive and use the external product deliverables.

acceptance criteria—The minimum level of quality required before a work product can pass through a quality gate including release into operations.

acceptance test—Formal black-box testing, typically conducted prior to delivery to verify that the product(s) meet predefined and agreed-to acceptance criteria. The objective is to allow the customer or users to make a determination of whether to accept the system or software.

access—The granting of permission to use the software's functionality, resources, or data (e.g., to a user or other software system).

access control—The security safeguards and mechanisms put in place to permit authorized access while detecting and denying unauthorized access.

accessibility—A quality criterion describing the ease with which the system and its functionality, resources, or data can be accessed. The degree to which the software is available to as many people as possible (the ability to access). Accessibility is often focused on people with special needs or disabilities.

accident—Any unplanned incident, event, or series of events that result in the injury or death of one or more people, the damage or destruction of property, or a negative impact to the environment or to society at large.

accountability—The ability to trace access to the software or performance of software activities to the responsible party (e.g., to a user or other software system).

accuracy—A quality attribute describing the level of precision, correctness, or freedom from error in the calculations and outputs of the software.

ACID—A data/information integrity acronym for the properties possessed by a reliability database (atomic, consistent, isolated, durable).

acquire or acquisition—The process of obtaining software through in-house development, through the purchase of commercial-off-the-shelf (COTS) software, through development by a third part vendor, or through a combination of these methods. In configuration identification, the point at which each baseline and each configuration item, component, and unit is initially brought under configuration control.

acquirer—The individuals or groups that are customers or users of the software.

acquisition criteria—The criteria that a configuration item must meet to in order to be ready to be brought under configuration control.

action item—An issue, event, action, or task that has been recorded, assigned an owner and due date for resolution, and is tracked until closure.

activity—A cohesive, individual unit of work that must be accomplished to complete a project. The lowest-level unit of work in the work breakdown structure. (Also known as a task or individual work package)

activity diagram—An analysis model that illustrates a dynamic activity-oriented view of the software product's functions.

activity network—A diagram that shows all of the activities of a project and their predecessor/successor relationships.

actor—Entities outside the scope of the system under consideration that interact with that system in a use case.

actual—Information that reports what has occurred. Examples include the actual start and end dates for an activity, the amount of effort actually expended, or the actual cost.

actual value—An earned value metric of the actual cost of work performed (ACWP).

adaptive maintenance—Modifications to an existing software product performed after delivery, to adapt that product to a new or changed external interfaces, environments, platforms, or operating systems.

additional qualifiers—The additional data items that are collected to provide the demographic data that allow the metrics to be analyzed at the various levels of this hierarchy.

adjourning—The stage of team development where the team accomplishes its mission, shuts down its efforts, and disbands.

aesthetics—A characteristic of usability that involves the extent to which screens, reports, and other user interfaces are pleasing to the senses.

affinity diagram—An analysis and problem-solving tool whose purpose is to organize a large number of ideas or items into significant categories or groups.

agile—A set of loosely affiliate software development methods and techniques including extreme programming, scrum, feature-driven development, test-driven development, crystal, lean, and others. (Also see agile coach)

Agile Alliance—A professional organization that focuses on agile methods and principles.

agile coach—Application of a specific application of coaching techniques to the agile environment. (See also Scrum master)

Agile Manifesto—A statement of the basic agile beliefs.

aging/age—A metric measuring the time since an entity (e.g., an action item, corrective action request, problem, or defect) was opened or created to the current time or to the time it was resolved/closed.

algorithm—A mathematical calculation done in the software.

allocated baseline—The baseline created when a product-level requirements specification for a system component (e.g., a software requirements specification) and its associated configuration items are brought under configuration control.

alpha testing—Testing that is typically done by the customers or users of the software or their representatives at the supplier's facility or on a user test bed.

ambiguous—Capable of being interpreted or understood in two or more senses or contexts. (See also Unambiguous)

analysis—A technique where individuals or teams perform in-depth assessments of the requirements or one or more other software work products or one or more alternatives for implementing those work products.

application layer—A layer of a TCP/IP network architecture that provides services that can be used by applications on the host to communicate data to other applications on the same or different hosts.

appraisal costs of quality—All of the costs of quality that involve finding defects in the products, processes, or services, including peer reviews, testing, audits, pilots, and quality metrics.

architectural design—For a system, the process of defining the collection of hardware, software, and manual components of a system and the interfaces between those components. For software, the process of defining the components of the software (including subsystems, programs, modules, data items, and procedural elements) and the interfaces between them. (Also called high-level design)

architectural design review—A phase gate review is a meeting held at the transition from architectural design (high-level design) to detailed component design. (Also called a preliminary design review [PDR])

archive or archival—The act of placing work products, quality records, or data into historical storage, typically for a specified period of time, so that they can be retrieved and used if needed.

area graph/chart—A type of graph used to show trends over time, where the lines divide the total area into parts (similar to a stacked bar). An area graph is typically more effective than a line chart when emphasizing both the trends and the size of or relationship between the subcomponents of the area.

arrival rate—The rate at which problem reports or corrective action requests are opened over time.

ASQ Code of Ethics—A professional code of conduct defined by ASQ to ensure that ASQ members and certified professionals "demonstrate ethical behavior in their relationships with the public, employers and clients, and peers" (Westcott 2006).

assembler—A software configuration management build tool that converts assembly language source code into object code.

assumption—An assertion, evaluation, or educated guess about how various factors or characteristics in the future will impact a project or task.

attack—An attempt to gain unauthorized access or compromise the software's integrity, availability, or confidentiality.

attacker—An individual or group that is attempting to breach (attack) the software's security.

attribute—A measurable property or characteristic of an entity.

audit—"A systematic, independent, and documented process for obtaining audit evidence and evaluating it objectively to determine the extent to which the audit criteria are fulfilled" (Coleman 2020). "An independent evaluation of a software product, process, or set of processes to assess compliance with specifications, standards, contractual agreements, or other criteria" (IEEE 2008).

audit corrective action and follow-up—The final step in the audit process, where the auditee management is responsible for creating and implementing corrective actions for any nonconformances identified during the audit. The step also included the verification follow-up by the lead auditor to make certain that those corrective actions were effectively implemented.

audit criteria—The agreed-to objective requirements, policies, standards, or processes against which conformance or compliance are evaluated during an audit. (Also called audit evaluation criteria or audit requirements)

audit execution—The step in the audit process that starts with an opening meeting, continues with the gathering of objective evidence and daily audit team and feedback meetings, and concludes with the closing meeting.

audit initiation—The first step in the audit process where the client formally initiates the audit.

audit objectives—The defined reasons (purpose) for conducting the audit.

audit plan—A document defining the plans for conducting an individual audit.

audit planning—The planning step of the audit process.

audit preparation—The step in the audit process where the audit team prepares for the audit.

audit process—The process which defines the steps for conducting an audit.

audit program—An overarching program planned as part of an organization's quality management system to ensure that required audits are performed regularly and audits of critical elements are performed more frequently.

audit purpose—The mission statement for the audit, which includes the reason for conducting the audit and the audits major objectives.

audit report—The formal documented results of an audit.

audit scope—The established boundaries of the audit that identify the exact items, groups, locations, or activities to be examined.

audit trail—The documented records of software's access and use that provide the information needed for accountability.

auditee—The organization or individual(s) being audited.

auditee management—The management of the organization being audited.

auditor or audit team—The team of one or more individuals conducting the audit, including the lead auditor.

auditor management—The management of the individuals who plan and conduct, the audit.

authentication—Security measure designed to establish the validity of a transmission, message, or originator, or a means of verifying an individual's authorization to receive specific categories of information (CNSS 2015).

author—The individual who created the work product or who is currently responsible for maintaining the work product.

availability—A quality attribute describing the extent to which the software, data, or a service is available for use when needed. Data, information, functionality, and other resources are available for access by authorized individuals when needed.

back door—An intentionally programmed mechanism in the software designed to allow access and use of the software while circumventing its security mechanisms.

backlog refinement—A Scrum meeting where items in the product backlog are prepared for the next sprint or future sprints.

backup—The act of duplicating the software or its associated components, units, data, and documentation at a specific time, typically used to protect organizational intellectual capital and assets from loss or corruption.

backup library—A software configuration management library used to contain duplicates of the versions of organizational intellectual capital and assets.

backward traceability—Traceability back to the predecessor work products (See Traceability)

bar chart/graph—A graph with rectangular bars proportional to counts, numbers, or ratios they represent.

baseline—"A specification or product that has been formally reviewed and agreed upon, that thereafter serves as the basis for further development, and that can be changed only through formal change control procedures" (IEEE/ISO/IEC 2017). A configuration item or set of configuration items formally designated and fixed at a specific time during the life cycle. An agreed-to project plan including schedule dates, effort estimates, resources, and costs that is used to compare progress (plan vs. actuals). The current or starting value of a software metric.

basic quality—The fundamental level of quality that a stakeholder expects the product to have. This basic quality comes from satisfying the requirements that are assumed by the stakeholder to be part of the product and are typically not explicitly stated or requested during requirements elicitation activities.

benchmarking—The process of identifying, understanding, adopting, and adapting outstanding practices and processes from organizations, anywhere in the world, to help an organization improve the performance of its processes, projects, products, or services.

beta test—A type of black-box testing that is typically performed by a selected subset of customers or users in an actual operational environment prior to the release of the software to general availability.

black-box testing—A form of testing that validates the external features and requirements of the software without regard to its internal structure.

bottom-up—An integration and testing strategy where testing starts with the lowest-level units or components in the calling tree being tested and tests them using drivers. The higher-level software units or components and their drivers are then integrated one or more at a time, replacing these drivers.

boundary value testing—A test design method for selecting test cases that examine the values on or around the boundaries of the equivalence classes or other boundaries in the software.

box chart—Box charts compress a set of data into each box and show its variance. Box charts can be used to show both the differences within a group of data and between groups of data.

brainstorming—An analysis and problem-solving tool whose purpose is to have a team generate lists of creative ideas in a short period of time through free association and by reserving critical judgment.

branching—The software configuration management process of creating a new codeline containing one or more configuration items that are duplicated from an existing codeline in a controlled library.

budget—A planned sequence of expenditures (e.g., dollars or engineering effort) over time with costs assigned to specific activities.

build—"An operational version of a system or component that incorporates a specified subset of the capabilities that the final product will provide" (IEEE/ISO/IEC 2017). The resulting executable software product.

builder—An individual responsible for building source code modules into binary executables (or other constituent configuration items into composite configuration items.

build reproducibility—The ability to rebuild the software product at a later time and get the same results.

burn-up and burn-down charts—Agile metrics used to track the status of work completion.

business continuity—"The capability of the organization to continue delivery of products or services at acceptable predefined levels following a disruptive incident" (ISO 2019).

business requirements—The business problems to be solved or the business opportunities to be addressed by the software product.

business rules—The specific policies, standards, practices, laws, regulations, and guidelines that define how the stakeholders do business.

calendar time—(See Duration and Schedule)

Capability Maturity Model Integrated (CMMI®)—Model for analyzing process capability.

capacity—A measure of the maximum amount of activities, actions, or events that can be concurrently handled by the software or system.

cardinality—The specification of the number of occurrences of one data item that can be related to the number of occurrences of another item.

cause-and-effect diagram—A type of tree diagram that is used to explore the multiple causes of a problem or potential causes of a risk. (Also called a fishbone diagram or an Ishikawa diagram).

causes—In failure modes and effects analysis, the reasons the failure occurred. (See also Mishap)

certification testing—A special type of testing that is typically done by a third party to certify that the product meets its certification criteria and any other standards that must be met.

champion—A senior member of management who selects and defines the team's mission, scope, and goals, setting the vision and chartering the team. (See also Team sponsor)

change—An addition, deletion, or modification to the software (e.g., to the concept, requirements, design, code, data item, or documentation). Any modification to an existing policy, system, process, product, or system. (See also Organizational change management)

change agent—An individual or team assigned the primary responsibility to plan and manage the change process.

change authority—The individuals or group who has the authority to approve, disapprove, or defer a requested change.

change control—The process by which a change to a formally controlled configuration item, component, unit, or baseline is requested, documented, evaluated, and approved or rejected by the change control board, scheduled, tracked through implementation if approved, and verified prior to its acquisition.

change request (CR)—A formally documented request to make a modification, correction, or enhancement to a controlled configuration item or baseline.

change request (CR) through CCB change control—A more rigorous type of change control where each change request must be approved by a CCB before changes are made to the associated baselined work products.

check-in—The process of placing a configuration item into a controlled library.

checklist—A type of work instruction that includes a list of yes/no questions or elements used to ensure that all items are considered during an activity (e.g., a checklist of criteria evaluated in an audit, a checklist of steps to perform in a process, or a checklist of common errors to be reviewed in an inspection).

check-out—The process of obtaining an official copy of a configuration item from the controlled library.

check sheet—An analysis and problem-solving tool used to collect data to create a frequency distribution or Pareto chart.

checksum—"Value computed on data to detect error or manipulation during transmission" (CNSS 2015).

class—In object-oriented programming, a category (classification) of objects or object types that defines all of the common properties of the different objects that belong to it.

class diagram—An analysis model that looks at the objects the system is modeling and the operations that pertain to those objects. A class diagram defines class information and the interrelationship between classes in a software product.

client—The individual who has the authority to initiate an audit and who is a primary customer of the audit results. (Also called the audit's customer or initiator)

client–server architecture—A two-tier network architecture in which each computer or process on the network is either a client or a server.

closing meeting of an audit—The last step in the execution phase of an audit where the lead auditor presents the results of the audit to the auditee management and other audit stakeholders.

cloud—"A model for enabling convenient, on-demand network access to a shared pool of configurable computing resources (e.g., networks, servers, storage, applications, and services) that can be rapidly provisioned and released with minimal management effort or service provider interaction" (NIST 2011).

coaching—A method for helping individuals and teams identify and articulate their professional challenges and goals, and support them in achieving those goals. (See also Facilitator)

COCOMO and COCOMO II models—Constructive Cost Model for project estimating.

code coverage—The amount or percentage of the code that is executed during a series of tests.

code of ethics—A set of values, principles and ethical standards to which a professional or professional organization aspire to and by which their actions are judged

codeline—A sequence of versions and revisions of a set of configuration items that does not include any branching.

coding standard—Workmanship standard that sets the specific requirements for writing software source code in a given language including style, syntax, practices, methods, and naming conventions.

cohesion—A measure of the extent to which a component performs a single task or function.

collaboration platforms—A system architecture that provides components and services to enable people to find information, communicate, and work together regardless of geographic dispersion.

commercial-off-the-shelf (COTS) software—Software that is commercially available and that is used as is.

common cause variation—When the root cause of the statistically improbable variation in a process can be attributed to normal, expected variation in the process due to typical causes including influences from people, machinery, environmental factors, materials, measurements, or methods.

communication plan—Specification of the information needs of the various project stakeholders and the communication methods and techniques that will be used to fill those needs, including descriptions, responsibilities, and schedules for project review meetings and selected metrics.

communications—Activities involving the transmitting and receiving of information between individuals.

compiler—A software configuration management build tool that converts high-level language source code into object code.

complete or completeness—A quality attribute that describes the extent to which a system or software configuration item fully implements all of its allocated requirements. A measure of the extent to which the data fully includes all of the required attribute values for its associated entity.

complexity—A quality attribute that describes the amount or degree of data or structural intricacy or interrelationships that makes the software more difficult to understand.

compliance/compliant—The degree to which a software practitioner, supplier, or organization has met an agreed-to set of requirements or other measurable criteria. A measure of the data's adherence to required standards, regulations, and conventions.

compliance requirements—Quality attribute or product attribute requirements related to the software's compliance to business rules, regulatory requirements, or safety/security needs. A measure of the data's adherence to required standards, regulations, and conventions.

component—A part or element of a larger entity (e.g., part of a system, element of the software, or part of a document).

component design—The process of refining and expanding the internal design of the components of the architectural design to contain more detailed descriptions of the processing logic, data structures, and data definitions, to the extent that the design is sufficiently complete to implement.

component design review—A phase gate review meeting held at the transition from detailed design to implementation (code and unit test). There may be multiple detailed design reviews, one for each of the major design elements. (Also called a critical design review [CDR] or a detailed design review)

composite items—A configuration item that is made up of smaller work products called constituent items.

comprehensibility—A characteristic of usability that considers factors including:

- Do users understand the product structure and its user documentation?
- Is the interface easy to remember and use even with infrequent use?
- Is the user interface intuitive to use?
- Does the system have a consistent, logical user interfaces?
- Does the software use vocabulary/terminology that is familiar to the typical user?
- Are the prompts and error messages easy to interpret?

concise, conciseness, or concision—A quality attribute describing the degree to which a function has been implemented in the minimum amount of code or documentation has been implemented in the minimum about of verbiage.

concurrent development—When two or more versions of a software product are in development or are being supported as releases in operations at the same time.

condition coverage—A white-box testing coverage technique where each statement is executed at least once and each condition in a decision takes all possible outcomes at least once.

condition and decision coverage—A white-box testing coverage technique where each statement is executed at least once, each condition in a decision takes all possible outcomes at least once, and each decision takes all possible outcomes at least once.

confidentiality—Protecting the data, information, functionality, or other resources of the system so they are accessible only to those who have either a right or a need to view it.

configuration audit—One of the four basic activities of software configuration management. An evaluation that provides independent, objective assurance that the software configuration management processes are being complied with, that the software configuration items are being built as required, and at production, that the software products are ready to be released.

configuration control—"The systematic process that ensures that changes to a baseline are properly identified, documented, evaluated for impact, approved by an appropriate level of authority, incorporated into all impacted work products, and verified" (MIL-HDBK 2001).

configuration control board (CCB)—The authority responsible for evaluating and approving, deferring, or disapproving proposed changes to a formally controlled configuration item or baselines, and for ensuring implementation and verification of approved changes. (Also called change control board [CCB], change authority board [CAB], or engineering change board [ECB])

configuration control board (CCB) charter—A document that defines how the CCB will do business.

configuration control board (CCB) process—The process by which the CCB conducts its activities.

configuration identification—One of the four basic activities of software configuration management. The partitioning of the software product into configuration items, components, and units; documenting their functional and physical characteristics; defining their acquisition; assigning unique identifiers to those items, components, and units; and establishing baselines.

configuration infrastructure—The organizational structure, tools, and processes that create the discipline upon which to build a consistent and effective software configuration management system.

configuration item (CI)—A product placed under configuration management and treated as a single entity. Configuration items are made up of configuration components, which are made up of configuration units.

configuration management—"A discipline applying technical and administrative direction and surveillance to identify and document the functional and physical characteristics of a configuration item, control changes to those characteristics, record and report change processing and implementation status, and verify compliance with specified requirements" (IEEE/ISO/IEC 2017).

configuration management plan—A planning document that describes how the organization's configuration management policies, standards, processes, work instructions, and infrastructure will be implemented and tailored to meet the needs of an individual project including specific activities, responsibilities, resource and budget allocations, tactics, tools, and methods.

configuration reviews—An evaluation that provides assurance that the software configuration management processes are being complied with, that the software configuration items are being built as required, and at production, that the soft- ware products are ready to be released.

configuration status accounting—One of the four basic activities of software configuration management. The recording and reporting on the implementation status of configuration items, components, and units and the status of changes to those items, components, or units.

configuration testing—A type of testing whose objective is to determine if the system or software has any problems handling all the required hardware and software configurations. (Also called platform configuration coverage)

conflict—A state in which the needs, values, or interests of two or more individuals, ideas, elements, or actions are in perceived or actual opposition with each other.

conflict of interest—Conflict that occurs when a person in a position of trust has competing professional or personal interests that may make it difficult for that individual to perform his or her duties in an impartial or unbiased manner.

conformance—The degree to which a product or service has met an agreed to set of requirements or other measurable criteria.

consistency—A quality attribute describing the extent to which strict and uniform adherence has been maintained to prescribed symbols, notations, terminology, and conventions. A measure of the extent to which the data are free from contradictions and inconsistencies.

constituent item—A configuration item that is part of a larger configuration item called a composite item.

constraint—A limit or condition on the project, product or process that must be met.

context diagram—A model used to identify the external entities of a system or system component and the high-level inputs to and outputs from each of those entities.

context-free questions—Questions that minimize the amount of context included in the question so that they do not limit the scope of the responses.

continuous integration—A good practice from agile processes that refers to attempting to build and deploy a new version/revision of the software product immediately after acquiring any changed constituent components for that software product.

contract—A binding, legal agreement between two parties (e.g., a supplier and an acquirer).

control chart—A graphical mechanism used to control an attribute of a process over time by identifying improbable patterns that may indicate that the process is out of control.

control flow graph—A graphical representation of the flow of control through a software component or unit.

controlled library—A software configuration management library used for managing current controlled configuration items and baselines, and controlling change to those items and baselines. (Also called the master library or system library)

control system—The system of five interrelated elements (control environment, risk assessment, control activities, information and communication, and monitoring) that helps management control processes, projects, and products.

convenient—A data collection goal where data collection must be simple enough not to disrupt the working patterns of the individual collecting the data.

conversion—A type of tort lawsuit that might be applied to software. Conversion would be involved if the software were intentionally designed to steal from the customer or destroy property. Conversion only involves tangible personal property.

copyleft—A form of licensing that allows individuals the right to disseminate copies of, or even modified versions of, a software product (or other types of works). These licenses typically stipulate that any adapted or derived versions of the original software must also be made available through the same or similar licenses.

copyright—A legal mechanism for protect original written works, such as books or software, from being coped without permission.

correction—Remedial action to correct the defect in the product or process. (Also called remedial action)

corrective action—Actions taken to eliminate the root cause of a nonconformance, noncompliance, defect, or other issue in order to prevent future recurrence.

corrective action plan—A plan that defines the steps in the planned corrective action with assigned roles, responsibilities, resources, and due dates.

corrective action request (CAR)—A formally documented request to make a modification, correction, or enhancement to a policy, standard, process, or work instruction.

corrective maintenance—"The reactive modification of a software product performed after delivery to correct discovered problems" (IEEE 2006).

corrective release—An interim release that is done to deliver defect corrections to the users in a timely manner.

cost—The total monetary amount spent to perform an activity or project, or to create a product.

cost/benefit analysis—An evaluation in which the costs for a product, process, project, or activity are matched against the anticipated value of its benefits to make a determination (e.g., whether to acquire a product, implement a process, or start or continue a project).

cost of ownership—The total cost of owning, executing, and maintaining a product in operations.

cost of quality—A measurement of the total cost of software quality that includes the sum of the costs of preventing defects, finding defects, and the costs of internal and external software failures (e.g., fixing defects, corrective maintenance, customer technical support, unhappy customers).

cost performance index (CPI)—An earned value metric defined as the earned value divided by the actual value (BCWP/ACWP).

cost variance—An earned value metric defined as the earned value minus the actual value (BCWP/ACWP).

coupling—A design concept that measures the degree to which an individual module or component is interconnected to other modules or components within the system. The more coupled a module or component is, the higher the probability that changes to the module or component will impact other parts of the system.

coverage—A software testing metric that measures the amount and completeness of testing performed by a set of tests.

credibility—A measure of the degree to which the user believes the data truly reflects the actual attribute values of the entities they describe.

critical activity—Any activity that must be completed by a certain time and has no slack time.

criticality—In failure modes and effects analysis, the calculated risk exposure, which is equal to the severity score times the occurrence score. (See also Risk exposure)

critical path—The longest path through the activities network that defines the shortest amount of time it will take to complete the project. All of the activities on the critical path must be completed on schedule for the project to be completed on schedule.

critical-to-x (CTx)—Factors or characteristics of a process or product (where x is a critical customer/stakeholder requirement including quality, cost, process, safety, delivery, and so on).

CRUD or CRUDL—Acronym (create, read, update, delete, list) used as a checklist to ensure completeness of data requirements or data domain test coverage.

crystal—An agile methodology.

culture—The beliefs, norms, and customs institutionalized into an organization.

cultural competence—The ability of individuals and organizations to work effectively with people from diverse ethnic, cultural, religious, and geographic groups.

cumulative flow diagram—An area graph that shows the cumulative number of work items that have gone through each activity on the Kanban board.

currentness—A measure of timeliness of the data—is the data up-to-date enough for its intended use?

custom-built third-party software testing—Testing of software provided by a supplier, subcontractor, vendor, or other third-party.

customer—The individuals who select, request, purchase, or pay for the product, process, or project.

customer feedback—The data from customers, users, and other external stakeholders about their experiences and perceptions when interacting with an organization or its products. (See also Customer satisfaction)

customer feedback management—Methods that provide an organization with ways to take the "voice of the customer" and turn it into:

- New and innovative products
- Enhancement and improvements to existing products
- New or improved stakeholder services
- Stakeholder interaction/communication mechanisms

customer satisfaction/customer satisfaction surveys—A measure of the extent to which the customers or other stakeholders are happy with the products, services, or results received. A requirements elicitation technique where stakeholder information is gathered through questionnaires.

cybercrime—Theft, fraud, and abuse accomplished through the breach of cyber security.

cyber security—Protection of the software from theft or damage to the hardware, the software, and the information on them, as well as from disruptions or misdirection of the services they provide.

cyber terrorism—The taking over or sabotaging of critical software infrastructure.

cycle time—A measure of the amount of calendar time it takes to go from the start to the completion of a process or activity.

cyclomatic complexity—A measure of the number of linearly independent paths through a module.

dashboard—A set of multiple metrics displayed together to provide a complete picture of the status of a product, process, or project.

data—"Facts" that have been collected in some storable, transferable, or expressible format.

data collection and storage—Designing, documenting, implementing, and improving what product and process data to collect, who collects it, how its collected, how its recorded, how its stored, and how it is made available to the people converting it into metrics and information.

data domain coverage or data domain testing—Testing analysis that looks at the mapping of tests to the various data domains in the software to ensure that they are thoroughly tested.

data dictionary—A mechanism for specification of the definition of data and information items.

data flow diagram (DFD)—A graphical representation of how data flows through and is transformed by the system or software.

data/information integrity—The extent to which the data or information derived from that data are high quality, accurate, complete, and timely.

data/information requirements—Requirements that define the specific data and information items or data structures that must be included as part of the software product.

data items—Facts that have been collected in some storable, transferable, or expressible format.

data longevity—A specification of how long data items must be retained within the system.

data management—The activities and processes involved in managing data.

data mirroring—The storing of data in multiple data stores or on multiple devices.

data owner—The person with direct access to the source of the data and who in many cases is actually responsible for generating the data.

data privacy—Issues that exist whenever personal information is collected, transmitted, or stored by the software.

data protection—Activities related to maintaining data backup and data preservation. **data quality**—Measurement of the quality of the data.

data refreshing—Mechanisms for how data from clients or other remote devices will be updated into the centralized data store.

data reporting—The conversion of data into information products and the dissemination of those information products to the metrics customers.

data security—Mechanisms that verifies that the right people can collect, access, utilize, and update data correctly, while also detecting, preventing, or recovering from inappropriate security attacks on the data.

data table—A mechanism to display large amounts of detailed data in a small area through the use of a table format. Data tables are compact and well structured. However, tables show only the values that the user must compare and evaluate.

data verification—A specification of how:

- The accuracy of input or stored data will be verified
- Inaccurate or duplicate data will be handled or reported
- Incomplete data entries will be handled or reported (e.g., from incomplete/inaccurate user entry, or because of system failures or aborts that result in partial data entry)

date and time domain coverage—Testing analysis that looks at the mapping of tests to the various date and time domains in the software to ensure that they are thoroughly tested. (Also called date and time domain testing)

debug—The process of finding and analyzing a suspected defect (fault) in a software product to identify the root cause of a failure.

decision coverage—A white-box testing coverage technique where each statement is executed at least once and each decision takes all possible outcomes at least once. (Also called branch coverage)

decision criteria—The thresholds, targets, or patterns used to determine the need for action or further investigation or to describe the level of level of confidence in a given result (IEEE/ISO/IEC 2017b).

decision tree—A tree diagram that illustrated the decisions that are made by the software.

decoupled—A lack of coupling. (See Coupling)

defect—A fault that exists in the software, which if not corrected could cause the software to fail or produce incorrect results.

defect containment effectiveness—A measure of the effectiveness of defect detection techniques to keep defects from escaping into later phases or into operations.

defect density—A measure that normalizes the number of defects by size so that comparisons can be made between software entities of different sizes.

defect detection techniques—Techniques designed to identify defects in software products (e.g., peer reviews and testing).

defect prevention—Techniques designed to prevent defects from entering software products.

defect removal effectiveness (DRE)—A measure of the effectiveness of our defect detection techniques in finding both defects introduced in the phase and in finding defects that "escaped" from previous phases. (Also called defect detection efficiency)

delegating—A situational leadership style where responsibility for decisions and implementation is turned over to the practitioners.

deliverable—A product that is delivered internally, or released and delivered to operations (e.g., to the software's users). In process definition, the tangible, physical objects or specific measurable accomplishments that are the outcomes of the tasks or verification steps.

delivery—The process of actually getting the software and its associated documentation into the operational environment for use by the user.

Delphi and wideband Delphi—Expert judgment-based estimation techniques.

denial of service—Any action or security attack that prevents authorized access to software or that prevents any part of software from functioning.

departmental audit strategy—An audit strategy that organizes the audit into manageable tasks based on the departmental or functional structure of the organization being audited, or by physical location.

dependency—An interrelationship existing between components in a system or between systems where their versions must be coordinated to avoid interoperability problems or other conflicts.

depth of a structure—The count of the maximum number of levels of control in the structure.

derived measure—Measure that is calculated using a mathematical combination of two or more explicit measures or other derived measures. (Also called a complex metric)

derived requirements—"A lower-level requirement that is determined to be necessary for a top-level requirement to be met. A requirement that is not explicitly stated in customer requirements, but is inferred from contextual requirements (such as applicable standards, laws, policies, common practices, and management decisions) or from requirements needed to specify a product or service component" (IEEE/ISO/IEC 2017).

design—Translation of the product requirements into the architecture and detailed design that define how the software will be implemented including defining the components, units, interfaces, and data internal to the software system.

design constraint—A type of nonfunctional requirement that defines a limitation on the choices that the developers can make when implementing the software or system.

Design for Six Sigma (DFSS)—The Six Sigma model used when the process or product does not exist or when evolutionary development is needed.

design verification and validation—Evaluating each design work product to ensure that it implements the requirements that were allocated to it and that it meets its intended use.

desk audit—An audit conducted at the auditor's desk.

desk check—A peer review method where one or more individuals other than the author examines a product independently and feeds back comments to the author.

detection control—In failure modes and effects analysis, the activities that are currently in place or already planned to detect the failure.

detection score—In failure modes and effects analysis, the rating of how likely it is that the current detection control will notice and contain the failure mode.

developers—Individuals and groups that are part of the organization that develops or maintains the software.

developmental baseline—A baseline created when a configuration item and its associated configuration components, units, or documentation are acquired during the development process.

deviation—A written approval from the appropriate body of authority to depart from a policy, standard, or other requirement for a specific period of time or for a specific product version.

DevOps—An expansion on traditional agile methods that brings different communities, such as the developer operations, security, and quality together to become a more efficient whole to continuously release software.

direct access media—Delivery media containing a software product that the target system can use as is, without the need for installation onto a target platform.

directing—A situational leadership style with high levels of task/directive behavior and low levels of relationship/supportive behaviors; should be used when the followers are unable to do the job and also lack the confidence to do it.

disaster recovery—A subset of business continuity that focuses on the organization's ability to recover or continue the operations of vital information technology and technical systems that support critical business functions following a disaster (natural or human induced). (See also Recovery)

distribution—In statistics, the shape of a set of data. (See also Normal distribution)

distributor—Individuals and groups that are part of the organization that distributes the software to the end users or customers.

diversity—Team diversity exists when the team members have different perspective, skills, knowledge, backgrounds, cultures, language, beliefs, and so on.

DMADV model—A Six Sigma model (define, measure, analyze, design, verify) used to define new processes and products or to perform evolutionary improvement on existing processes to achieve Six Sigma quality levels. (Also called design for Six Sigma [DFSS])

DMAIC model—A Six Sigma model (define, measure, analyze, improve, control) used to improve existing processes that are not performing at the required level through incremental improvement.

documentation plan—Specification of the list of documents to be prepared, and the tools, techniques, activities, schedules, and responsibilities for preparing, reviewing, baselining, and distributing those documents.

documentation studies—A requirements elicitation techniques where documentation is reviewed for the purpose of obtaining stakeholder information.

document identifier—A unique identifier assign to a document.

driver—A temporary piece of source code used as a replacement for a calling module when performing bottom-up integration and integration testing. A driver is used to invoke a lower-level module under test and may also provide test inputs to that module and report the results.

duration—The length of calendar time it takes to complete an activity.

dynamic analysis—Methods of performing V&V by evaluating a software component or product by executing it and comparing the actual results to expected results. Examples of dynamic analysis include testing, simulation, and piloting.

dynamic cycle time—A measure of cycle time calculated by dividing the number of items in progress (items that have only partially completed the process) by one-half of the number of new starts plus new completions during the period.

dynamic library—A software configuration management library used to hold configuration entities that are currently being worked on by software practitioners. This typically includes new products or existing product that are being modified or tested. The dynamic library is the software practitioner's work area and is under the control of that practitioner. (Also called a development, programmer's, or working library)

earned value or earned value management—A measure of the value of work performed so far. Earned value "uses the original estimates and progress to date to show whether the actual costs incurred are on budget and whether the tasks are ahead or behind the baselined plan" (Project 2000). An earned value metric of the budgeted cost of work performed (BCWP).

ease of learning—A factor of usability that describes amount of time it takes a user to become proficient in performing tasks using the software.

ease of use—A factor of usability that describes users' ability to use the software without difficulty or frustration.

effective listening—The receipt of the communications message.

effects—The consequences of a failure.

efficiency—A quality attribute describing the extent to which the software can perform its functions or the data can be accessed, stored, processed and provided while utilizing minimal amounts of computing resources (e.g., memory, bandwidth, or disk space). A usability characteristic of the extent to which the users can do what they want in a minimum amount of steps or time.

efficiency of defect detection—Measures of the ability to identify all of the defects present at the time of defect detection activity or process was executed.

effort—The amount of human work that is required to complete an activity or project.

electronic transfer—Delivery of software products via the Internet, bulletin boards, and network mounted file systems.

element method—An audit strategy that organizes the audit into manageable tasks based on the various elements of the audit requirements.

embedded system—A system where the software is embedded as part of a complete device that includes hardware or mechanical components.

emergency maintenance—Changes that must be implemented in the software and released in the software outside the normal release cycle to fix problems reported from operations.

emotional intelligence—The ability to accurately recognize, understand, and manage their emotions and those of others.

encapsulation—In object-oriented design, all that is seen of an object is its interface, namely the operations that can be performed on that object.

encryption—Set of mathematically expressed rules for rendering data unintelligible by executing a series of conversions controlled by a key.

enhancement—A change made to an existing software product to add additional functionality or capability, adapt it to a changing environment, or enhance its quality attributes.

enhancement request—A change requests reported to add an enhancement to the software.

enterprise architecture—The structure of organization-level components (e.g., processes, systems, personnel, teams, organizational sub-units, and so on), and the relationships/interfaces between them.

enterprise environmental factors—All of the elements that surround the project including items such as organizational culture, infrastructure, tools, and existing staff.

entity—In measurement, the person, place, thing, process, or time period being measured.

entity relationship diagram (ERD)—A graphical representation of how the data objects in a system relate to each other.

entry criteria—The activities that must be completed, resources that must be in place or measurable conditions that must be met prior to starting an activity or phase.

environment—The humans and hardware, operating system, files systems, and other software applications that the software must interoperate with or that must coexist with the software. There may be different environments for software development, testing and operations. (See also Test bed)

epic—A story that is too large to implement in a single iteration or sprint.

equity—The promotion of fairness and impartiality by providing equal treatment and access to opportunities, advancements, and resources.

equity theory—A theory of rewards and motivation that says that for rewards to be motivational, people have to believe that rewards (and punishments) are being equally distributed as desired.

equivalence class partitioning—A test design method where the input or output domains are divided into subsets of values that are assumed to be handled identically by the software.

ergonomics—The study and analysis of what makes things comfortable and efficient for people in their environment.

error—A human, hardware, or other software action that feeds inputs into the software that are incorrect in type, size, number, form, fit, or function. In measurement, the difference that occurs when the number or symbol (also called the measurement signal, measured value, or data item) differs from the actual value that would be mapped to the attribute of the entity in a perfect world.

error guessing—A test approach that "attacks" the software with inputs that expose potential programmer mistakes or represent the kinds of invalid inputs that the software would receive if user or hardware mistakes occurred.

error rate—A factor of usability that number of user errors per period of time.

escape—A defect that evades one or more defect detection techniques (was not identified) and that moves on to the next process or phase in the life cycle.

escort—An individual or individuals assigned to accompany the auditor(s) during objective evidence gathering activities and serve as liaison between the audit team and the auditees.

espionage—Leakage of intellectual property, trade secrets, or secret/confidential information or functionality to an unauthorized corporation or country.

estimate or estimation—The predicted future value of a software or project characteristic or attribute (e.g., size, cost, effort, schedule, quality, or reliability).

ETVX—A methodology for process definition where the entry criteria, tasks, verification, and exit criteria are defined.

event—A specific time, typically at the start or finish of an activity.

event/response table—An analysis model used to list the events that affect the software product and the software product responses to those events based on the state of the software product or one of its components.

evolutionary change—Changes that replace the existing systems, processes, or products with better ones.

evolutionary development or **evolutionary model**—A software life cycle model where the product is developed in multiple releases over time based on inputs from the utilization of that software in operations.

exciting quality—The level of quality that comes from the implementation of unrequested but innovative requirements. These are requirements that the stakeholders do not know they want but will love when they see them.

executable—A stand-alone software product that can be run directly on a computer. (See also Build)

exit criteria—The activities that must be completed, resources that must be in place, or measurable conditions that must be met before an activity or phase can be considered finished. (Also called completion criteria or acceptance criteria)

expectancy theory—A theory of rewards and motivation that says that a person will be motivated to perform an activity based on his or her belief that putting in the effort will actually lead to better results, that the extra effort will be noticed, that those better results will actually lead to personal rewards, and that those personal rewards are valuable.

expected quality—The level of quality that comes from satisfying those requirements that the stakeholder explicitly considers and requests.

expert-judgment—A set of estimation and forecasting techniques based on the judgment of the subject matter experts.

explicit knowledge—Tangible knowledge transmitted in a formal manner through various documents, formal training or education, video media, e-mail, artifacts, or books.

explicit measure—Metrics that are measured directly. (Also called a metric primitive, direct metric, or base measure)

exploratory testing—Testing where the tester designs and execute tests at the same time, based on the knowledge gained as they are testing the software. (Also called artistic testing)

expressed warranty—A warranty that is directly stated (or "expressed") verbally or in writing. (See also Warranty and Implied warranty)

external audit—An evaluation performed by a group outside the organization being audited. (Also called second-party audit or third-party audit)

external failure costs of quality—All the costs of quality that involve handling and correcting a failure that has occurred once the product is in operation.

external interface requirements—Requirements for the information flow across shared interfaces to hardware, humans, other software applications, the operating system, and file systems outside the boundaries of the software product being developed.

extranet—An intranet web architecture that has various levels of accessibility to authorized outsiders.

extreme programming (XP)—An agile method that focuses on software development techniques.

extrinsic motivation—"Satisfaction of psychological or material needs by others through incentives or rewards" (Westcott 2006).

facilitated requirements workshops—Facilitated meetings that bring together cross-functional groups of stakeholders to produce specific software requirements products.

facilitator—An individual who creates an environment in which others can direct their own learning or work.

failure—An occurrence of the software not meeting its requirements, function, or intended usage while it is being executed.

failure mode—The way in which the system/software component may fail to deliver part or all of its function (e.g., complete failure, partial failure, intermittent failure, failure over time, over-performance failure).

failure mode and effects analysis (FMEA)—Activities involving the analysis of possible software failures, the effects of those failures, and the criticality of those effects. (Also called failure mode, effect and criticality analysis [FMECA])

fan-in—In procedural language, a measure of the number of modules that directly call a module. In object-oriented development, a measure of the number of classes that a class directly inherits from.

fan-out—In procedural language, a measure of the number of modules that are directly called by another module. In object-oriented development, a measure of the number of classes that directly inherit from a class.

fault—A defect in a software product. (See also Defect)

fault-error handling—Testing techniques that look at how well the software handles errors, invalid conditions, invalid or out of sequence inputs, and invalid data. (Also called exception handling)

fault insertion—The process of intentionally adding a known number of defects to the software for the purpose of estimating the number of unknown defects remaining in the software. (Also called fault seeding)

fault tolerance—A quality attribute describing the extent to which the software can detect and handle defects, invalid inputs, or erroneous states from the environment in which it operates (e.g., hardware failures, invalid or interrupted data communications) without failure. (Also called exception handling, fault-error handling, robustness, or error tolerance)

feasible or feasibility—Something that can be implemented using available technologies, techniques, tools, resources, and personnel within the specified cost and schedule constraints.

feature—"One or more logically related system capabilities that provide value to a user and are described by a set of functional requirements" (Wiegers 2013). "A distinguishing characteristic of a system item (includes both functional and non-functional attributes such as performance and reusability)" (IEEE 2008a).

feature-driven development (FDD)—An agile development methodology for implementing software functionality that is based on breaking the requirements down into small client-valued pieces of functionality and iteratively implementing them.

feature release—A release done primarily to deliver new features or functionality to the user. Defects can also be corrected as part of a feature release.

feedback—A technique where the output or lessons learned from a process or system are loop-back into that process or system to provide information for analysis and improvement.

finding—Any nonconformance, noncompliance, observation, process improvement opportunity, best practice, or other fact worthy of reporting as a result of an audit activity.

finite—The requirement is not stated in an open-ended manner.

firmware—Software that is embedded in a piece of hardware.

first-party audit—An audit that an organization performs on itself. (Also called an internal audit)

first-pass yield—A measure of the effectiveness of defect prevention techniques.

flexibility—A quality attribute describing the ease with which the software can be modified or customized by the user.

flowchart—A graphical representation of the inputs, actions, and outputs of a process.

focus groups—Small groups of representative users (typical users) brought together to elicit information and opinions about a product and its requirements.

follow-up—A peer review activity in which someone other than the author examines the author's rework to ensure that defects and open issues were appropriately resolved. Part of the final step in the audit where the implementation of the corrective actions is verified. (See Audit corrective action and follow-up)

follow-up audit—A re-audit of one or more elements that had previous nonconformances.

force field analysis—An analysis and problem-solving tool whose purpose is to identify driving forces that help move toward reaching the goal and restraining forces inhibiting movement toward the goal.

forecasts—The predicted future value of a software or project characteristic or a period of time (e.g., staffing needs, resource needs, and project risks and opportunities).

form—A type of work instruction with a defined format where fields are completed as appropriate.

forming stage—The first stage of team development when the team is initially brought together or when new members are added to an existing team.

fraud—A type of tort lawsuit that might be applied to software. The seller of the software knowingly misrepresented the capabilities of the product.

full release—Packaging of a release that is capable of completing a full installation of the product.

function—The purpose of the system/software component or the task(s) it performs.

Functions as a Service—A cloud computing model where the cloud service provider (CSP) manages to underlying infrastructure, resources, and capacity and provides APIs to the developers who can configuration and manage those functions.

functional baseline—The baseline created when system requirements and associated configuration items are acquired.

functional configuration audit—"An audit conducted to verify that the development of a configuration item has been completed satisfactorily, the item has achieved the performance and functional characteristics specified, and its operational and support documents are complete and satisfactory" IEEE/(ISO/IEC 2017).

functional requirements—Stakeholder or product-level requirements that define the capabilities of the software (what the software must do) to satisfy the business requirements.

functional testing—Testing to evaluate the extent to which the software meets its functional requirements. (See also Black-box testing)

function points—A measurement of software size in terms of its functionality. A model-based project estimation technique where function points are used as the input.

Gantt chart—A bar chart used to show the scheduled calendar time and duration for each activity in a project and the actual calendar time and duration used.

globalization—A combination of the concepts of internationalization and localization.

goal/question/metric paradigm—A mechanism for defining a goal based measurement program through the selection of metrics that provide information to answer questions about the achievement or progress toward achieving goals.

good enough—A strategy where analysis is done to make conscious, logical decisions about the trade-offs between the level of quality and integrity that the stakeholders need in the software and the basic economic fact that increasing software quality and integrity typically costs more and takes longer.

government-off-the-shelf (GOTS) software—Software that is a government-supplied ready-to-use software product.

gray-box testing—A blending of white-box and black-box testing strategies that primarily focuses on interfaces and interactions at various levels as units are integrated into programs, programs into subsystems, and subsystems into the software system.

GROW coaching framework—One popular framework for coaching whose elements include G (goal), R(reality), O (options), and W (will do).

group consensus—An expert judgment estimation technique.

guidelines—A type of work instruction that includes suggested practice, method, or instructions that are considered good practice but are not mandatory.

hacker or hacking—An unauthorized user who attempts to or gains access into a software product, an information system, or data.

hardware dependency—A dependency that exists when the current release of the software is not backward compatible with the hardware used to run previous versions of the software.

harm—Injury or death of a human, damage or destruction of property, or negative impact to the environment or society at large.

hazard—"Anything that might go wrong that could potentially cause an accident" (Frailey 2016).

hazard analysis—The identification and analysis of potential risks that could cause harm (e.g., personal injury, property or environmental damage, negative impacts to the health, or safety of the society).

help desk—Customer technical assistance.

histogram—A bar chart that shows the distribution (shape) of a set of data.

horizontal prototype—A prototype that looks and feels like the real system from the outside perspective, but under the surface, it's all smoke and mirrors.

hostile data injection—Interjecting of data into commands or queries in order to trick the software into executing that data.

human factors and metrics—How people affect metrics and how metrics affect people.

human factors studies—Studies that consider the ways in which the human users of a software system will interact with the software.

hybrid cloud—A combination of the multitenant and private cloud platforms.

hygiene factors—Factors that cause employees to be dissatisfied at work.

identifier—A unique label associated with a configuration item, component, unit, or baseline, or one of their versions or revisions.

impact analysis—The analysis of the impacts a requested change would have on the software product, process, or project.

implementation—The software development activities that translate the design into the source code and documentation.

implied warranty—An unstated promise that comes from the understanding of the buyer or the nature of the transaction when an expressed warranty does not exist. (See also Expressed warranty and Warranty.)

inclusion—The creation of a work environment or culture that appreciates diversity. Inclusion focuses on utilizing the skills, talents, and perspectives of diverse employees to benefit the organization and those employees.

incremental change—Changes made to improve the existing systems, processes, or products to make them better.

incremental development life cycle model—A software life cycle model based on the process of constructing increasingly larger subsets of the software's requirements.

independence—An individual is free from bias and external influence.

indicator—A measure or combination of measures that provide insight into a software issue or concept (IEEE/ISO/IEC 2017).

information—"Data in context" (DAMA 2017).

information hiding—A design concept that hides a component's internal structure from its surroundings.

information radiator/informative workspaces—An agile practice where forms of (displayed) project documentation are used or placed in areas where people can easily see them.

information security—The aspect of business continuity that includes "the practice of defending information from unauthorized access, use, disclosure, disruption, modification, inspection, recording or destruction" (wikipedia .com 2016).

informative workspace—An extreme programming practice where the workspace is used to communicate important, active information.

infrastructure—The elements needed to support productive work including office space, networks, the support environment, and tools.

Infrastructure as a Service (IaaS)—A cloud computing model where the cloud service provider (CSP) owns, supplies, and manages the hardware and other infrastructure components rather than having those components on-premise at the organization.

infrastructure plan—Specification of the plan for obtaining, using, managing and controlling the infrastructure needed by a project, program, or organization.

inheritance—In object oriented programming, a relationship between classes that allows the definition of a new class based on the definition of an existing class.

input—The tangible, physical objects that are input into and utilized during the process. Data or other items that come into a software system or component from outside its boundary.

inspection—A formal method of peer review where a team of peers, including the author, meets to examine a product.

inspectors—Individuals participating in an inspection. (See also reviewers)

installability—A quality attribute describing the ease with which the software product can be installed on the target platform.

installation—The process of integrating the software into its operational environment for use by the users.

installation testing—Testing to make certain that the software was correctly installed.

integrated master schedule—A summarization of the statuses of the lower-level subprojects into a higher-level project view.

integrated product team (IPT)—A cross-functional team formed for the specific purpose of delivering an integrated product that was produced by more than one organization (e.g., partially developed in-house and partially developed by one or more suppliers).

integration—"The process of fitting together the various components of a system so that the entire system works as a whole" (Jones 1994).

integration testing—The testing of the combination of two or more software or hardware units or components in order to determine if there are any problems with their interfaces, logical communications, or other interactions.

integrity—The data, information and other resources are accurate, valid, and reliable, which includes rules for how they can be manipulated.

integrity level—A level assigned to the software to determine the amount of verification and validation rigor that is required based on the needed integrity of that software.

intellectual property—A legal area that includes inventions and ideas of the human mind, such as books, music, artwork, and software. (See also Patent, Copyright, Product license, Proprietary software license, Open source license, Copyleft trademark)

interface—A shared boundary across which information is passed.

interface coverage—The mapping of tests to the various identified interfaces to make certain that they are thoroughly tested.

interface requirements—A subset of nonfunctional requirements at the product level that define the required information flows across shared interfaces to external entities outside the boundaries of the software product being developed.

intermediate installation media—Delivery media that requires an installation of the software product onto the target system.

internal audit—An audit that an organization performs on itself. (Also called a first party audit)

internal failure costs of quality—Internal failure costs are all the costs of quality that involve handling and correcting a failure that has occurred in-house before the product has been made available to the customer.

international configuration coverage—Testing analysis that looks at the mapping of tests to the various international configurations that the software can have to ensure that they are thoroughly tested. (Also called international configuration testing or localization testing).

internationalization requirements—A subset of nonfunctional requirements that deal with developing software that can be adapted to target markets in different geographical locations around the world without the need to change the software.

internet—A web-based architecture made up of a global network connecting millions of computers.

internet layer—A TCP/IP network architecture layer that transports Internet Protocol (IP) packets from the originating host across network boundaries to the destination host.

interoperability—A quality attribute describing the degree to which the software functions properly and shares resources with other software applications or hardware operating in the same environment.

interrelationship digraph—An analysis and problem-solving tool whose purpose is to organize ideas and define the ways in which ideas influence each other.

interval scale measurement—For interval scale measurements, the exact distance between the scales is known. This allows the mathematical operations of addition and subtraction to be applied to interval scale measurement values. However, there is no zero point in the interval scale, so multiplication and division do not apply.

interview—A technique for gathering information through the process of one or more individuals (interviewers) asking questions and evaluating the answers provided by one or more other people (interviewees).

intranet—A web architecture made up of a connected network that resides behind a firewall and is only accessible to people within the organization.

intrinsic motivation—"A self-motivating process where an individual obtains reinforcement through personally valuing characteristics of the situation itself" (Westcott 2006).

ISO 27000 family of standards—Set of standards that define good practice in the area of information security, cybersecurity, and privacy protection.

iterative model—A software life cycle development process model where steps or activities are repeated multiple times.

joint application development—A facilitated team-based requirements development technique. (Also called a facilitated requirements workshop)

Kanban—An agile methodology.

Kano model—A model showing the relationship between customer satisfaction and product quality.

kick off meeting—The step in the inspection process where the author presents an overview of the work product being inspected to the inspection team. The moderator may also perform inspection training during this meeting.

kiviat chart—A circular chart where each "spoke on the wheel" represents a metric with the metric's value plotted on that spoke. (Also called a polar chart, radar chart, or spider chart)

knowledge—Information that has had human intelligence applied to it through the identification of patterns, trends, relationships, assumptions, and relevance.

knowledge management system—A system that stores and retrieves knowledge to facilitate knowledge transfer.

knowledge transfer—The action of creating, organizing, and capturing knowledge, and conveying that knowledge from one person or part of the organization to another. Knowledge transfer also involves making certain that the knowledge remains available to other stakeholders in the future.

labeling—The software configuration management process of creating a unique identifier (label) and using it to create an association between a set of configuration items in one or more controlled libraries. (Also called tagging)

lagging indicator—An information product that provides information about the past.

laws—"The system of rules which a particular country or community recognizes as regulating the actions of its members and which it may enforce by the imposition of penalties" (Oxford Dictionary).

lead auditor—The person responsible for leading the audit team and conducting the audit.

leadership—The process of influencing others to willingly work toward common, shared goals. (See also Situational leadership)

leading indicator—An information product that provides information that can be used to take proactive action to identify, control, or prevent future problems.

lead time—The average wait time in the queue from the time the work item is defined until the time it is moved to in-work.

lean—A set of agile software development techniques including eliminating waste, amplifying learning, deciding as late as possible, delivering as early as possible, empowering the team, building integrity in, and seeing the whole.

learnability—A usability measure of how easy it is, or how much training or time it takes, to learn to use the software proficiently. (See also Ease of use)

legacy software—Older or outdated software that is still used.

lessons learned—Knowledge gained from reflecting on activities, processes, or projects after they have been implemented, through performing root cause analysis of identified issues or defects, or through other empirical analysis methods that are used to prevent future problems or repeat successful actions.

level of control—"A measure of the degree to which the software is responsible for the behavior of a system, or of a specific system action, that may lead to a safety mishap" (Frailey 2016)

libraries—A configuration management mechanism for identifying, storing and labeling baselined configuration items, and for capturing and tracking the status of changes to those configuration items.

licensing key—Unique identifiers added during the manufacturing process to specifically identify the replicated software as an original, manufactured copy.

life cycle or process model—A high-level representation of the software development process, which provides a framework for the detailed definition of the process and defines the stages (phases) through which software development moves.

likeability—A factor of usability that measures the extent to which users report liking the software or one of its attributes.

limitation—Items specifically considered out of scope and will therefore be specifically excluded from the project.

line graph—Line graphs are typically used to show changes or trends over time. They are also used to report nondiscrete variables.

lines of code (LOC)—A metric for the size of a software component. A count of the number of physical or logical lines of source code in a software component.

linker or loader—A software configuration management build tool that combines and converts object code modules into the software executable.

load testing—Performance testing techniques used to determine if the system has any problems dealing with normal level, maximum level, and excess levels of capacity.

localization—The process of customizing the internationalized software to a region or location through translation or adding location-specific components.

location—In statistics, the typical value or central tendency of a data set.

maintainability—A quality attribute describing the ease with which a software system or one of its components can be modified.

maintenance—The activities involved in modifying the software after it has been released.

malicious code—An unauthorized process or function programmed or hacked into the software with the intention of adversely impacting its confidentiality, integrity, functionality, or availability.

malpractice—A type of tort lawsuit that might be applied to software. The software's author (or the program itself) provides unreasonably poor professional services.

management review—A senior management oversight mechanism for monitoring and improving the quality management system. A project review held to provide senior management awareness of and visibility into project activities.

manager/management—"A person that provides technical and administrative direction and control to those performing tasks or activities within the manager's area of responsibility. The traditional functions of a manager include planning, organizing, directing, and controlling work within an area of responsibility" (CMMI 2010a).

manufacturing perspective of quality—One of Garvin's five perspectives of quality that defines quality in terms of conformance to the specification.

mapping system—In measurement, the rules or units of measure for mapping numbers and symbols onto the attributes of entities. (Also called measurement method or counting criteria)

marketing surveys—A requirements elicitation technique where stakeholder information is gathered through questionnaires.

Maslow's Hierarchy of Needs—A model for motivation.

matrix diagram—An analysis and problem-solving tool whose purpose is to analyze the correlations between two groups of items.

mean—In statistics, the arithmetic average of the numbers in the data set.

mean-time-to-change—A maintainability metric that measures the average amount of effort it takes to make a change to the software.

mean-time-to-failure—A reliability metric that measures the average time between system/software failures.

mean-time-to-fix—A maintainability metric that measures the average amount of effort it takes to fix a defect that has been identified in the software.

measurable—The existence of specific ranges, limits, or values.

measure/measurement—"The process by which numbers or symbols are assigned to attributes of entities in the real world in such a way as to describe them according to clearly defined rules" (Fenton 1997). The number or symbol that is the result of measuring.

measurement function—The equation used to calculate a derived measure.

median—In statistics, the middle value when the numbers in a data set are arranged according to size.

meetings/meeting management—The process of preparing for, conducting, and following up on meetings.

memorability—A usability measure of how easy it is for the users to remember how to use the software once they have learned how to use it.

mentoring—The act of a more experienced individual forming a relationship with a less experienced individual in order to help that individual develop or improve his or her skill set, performance, knowledge, or other capability.

merging—The software configuration management process of taking two or more versions or revisions of a product and combining them into a single new version or revision.

messaging systems architecture—An architecture designed to accept messages from or deliver messages to other systems.

metadata—Data about the data.

method—In object-oriented programming, a piece of software code that implements an operation.

metrics—The systematic application of measurement-based techniques to software products, processes, and services to provide engineering and management information that can be used to improve those software products, processes, and services (based on Goodman 2004). The standard used for performing measurement.

metric customer—The person (or team) who will be making decisions or taking action based upon the metric. The customer is the person who needs the information supplied by the metric.

migration—Porting of the software to another environment. (See also Portability)

milestone—A significant event that marks progress in a project, usually representing the completion of a major phase of work.

mishap—Something that might cause a hazard.

mistake—Anything that a human does that could cause an issue or defect if that person does not catch their own mistake.

modality—The indication of whether or not the data item is required.

mode—In statistics, the value that occurs most often in the data set.

model—An abstract representation of an item or process from a particular point of view.

model-based estimation—A major class of estimation techniques involves models (mathematical formulas) used to calculate the estimates for the project.

moderator—The role in an inspection process that is responsible for ensuring that the inspectors correctly implement the inspection process.

modifiable—The requirements are specified in a coherent, easy-to-understand manner, they are nonredundant (that is, each requirement is stated in only one place), and each requirement can be changed without excessive impact on other requirements.

modifiable-off-the-shelf (MOTS) software—Commercially available software that may be adapted or customized to meet individual requirements.

modularity—A design concept describing the extent to which the software is partitioned into independent modules, where each module is constructed to work with the other modules without being involved with the detailed internal structure of those other modules.

motivation—The set of feelings or reasons that determines the extent to which an individual will engage in a particular activity or behavior.

motivators—Factors that drive employee work satisfaction.

multiple condition coverage—A white-box testing coverage technique where each statement is executed at least once and all possible combinations of condition outcomes in each decision occur at least once.

multicultural environments—A workforce environment that is made up of people with different cultures, norms, backgrounds, expectations, and languages.

multi-tenant cloud—The cloud computing platform where multiple cloud customers share to usage of on-demand services and resources. (Also called the public cloud)

multi-voting—A team decision-making technique where each team member casts multiple votes to aid in the selection process.

negative observation—A type of audit finding that does not raise to the level of nonconformance but is worthy of bringing to the attention of management because if left uncorrected it may cause future issues. Any potential nonconformance identified during the audit that is outside the scope of that audit.

negligence—A type of tort lawsuit that might be applied to software. The producer of the software failed to take steps that a reasonable software producer would take, and because of this failure the software injured a customer or his or her property.

negotiation—Communication designed to come to an agreement, resolve a dispute, determine a course of action, or satisfy various interests.

network interface layer—A TCP/IP network architecture layer that is responsible for packet forwarding through the network, including routing through intermediate routers, to the appropriate destination node using its unique Internet Protocol (IP) address.

nominal group techniques (NGT)—Structured, team-based analysis and problem-solving techniques used to explore issues and determine solutions.

nominal scale measurement—The simplest form of measurement is classification (one-to-one mapping) where its type categorizes the attribute of the entity. Nominal scale does not make any assumptions about order or sequence.

nonconformance or **noncompliance**—A type of audit finding that is a non-fulfillment of a specific, objective criteria.

nonfunctional requirements—Product-level requirements that define the specific characteristic that the software must possess in order to fulfill the quality attribute requirements.

normal distribution—In statistics, a continuous distribution where there is a concentration of observations around the mean, and the shape is a symmetrical bell-like shape.

norming stage—The third stage of team development when the team is coming to agreement on their methods of operations and norms for conducting their business.

n-tier architectures—N-tiered architectures break the application into two or more tiers or layers. A client–server architecture is an example of a two-tiered architecture.

object—A self-contained element in object-oriented programming that has both data associated with it and associated procedures to manipulate the data.

objective evidence—"Information, which can be proven true, based on facts obtained through observation, measurement, test, or other means" (Russell 2013).

objectivity—The absence of bias that will influence results.

object-oriented analysis and design (OOAD)—Analyzes and specifies the design in terms of the objects that the system is modeling and the operations that pertain to those objects.

observation—Any item worthy of note to management (both positive and negative) that does not rise to the level of a nonconformance.

occurrence score—In failure modes and effects analysis, the rating of the probability of failure or for software the level of control.

offshore—A type of outsourcing where the supplier is located in a different country that the acquirer.

offsite outsourcing—A type of outsourcing where the supplier is located in a different location than the acquirer.

offsite storage—The storage of archival or backup libraries in a location other than the work location.

one-way communication—Communication where the speaker sends a message but received no feedback from the receiver of that message

onsite outsourcing—A type of outsourcing where the supplier is located at the same location as the acquirer.

open-ended questions—Questions that require more than a few words to answer (e.g., who, what, when, where, why, and how type questions).

opening meeting of an audit—A meeting at the beginning of the audit execution step between the audit team and the auditee organization. The objectives of this meeting include introducing the audit team, reviewing the conduct of the audit, reviewing audit logistics, and making sure the auditee understands what to expect from the audit.

open source software—"Software with its source code made available with a license in which the copyright holder provides the rights to study, change, and distribute the software to anyone and for any purpose" (wikipedia.com 2016).

open source license—A license in which the ownership of the specific copy of the software is transferred to the end-user.

operation—In object-oriented programming, an actions that can be applied to an object to obtain a certain effect.

operational profile testing—A black-box testing method where the number of tests done on each thread is weighted based on its usage in the expected operational environment.

operational testing—Testing performed on the product during operations.

opportunity—A possibility of a positive outcome occurring during a project that will impact the success of that project. (Also called a positive risk)

ordering—The activities involved in ordering copies of the software product.

ordinal scale measurement—The ordinal scale classifies the attribute of the entity by order. However, there are no assumptions made about the magnitude of the differences between categories.

organizational change management—The management of technology transfers and other changes to the organization's methods, processes, and systems so the organization can grow and improve in order to stay competitive.

organizational process assets—Artifacts that represent organizational learning and knowledge and are considered "useful to those who are defining, implementing, and managing processes in the organization" (CMMI 2010a).

output—The tangible, physical objects that are output from a process (e.g., work products or quality records). Data or other items that come from a software system or component and are communicated outside its boundary.

outsourcing—The development or maintenance of all or part of a software product by a supplier outside the primary development organization.

packaging—Mechanisms for delivery of a release.

pair programming—An agile technique where two people work together, one of whom is constantly reviewing what the other is developing.

Pareto analysis—The process of ranking problems or categories based on their frequency of occurrence or the size of their impact in order to determine which of many possible opportunities to pursue first.

Pareto chart—A bar chart where the height of each bar indicates the frequency or impact of problems or categories.

partial release—Packaging of a release that requires a full release to already be installed.

participating—A situational leadership style with low levels of task/directive behaviors and high levels of relationship/supportive behaviors, emphasizes the sharing of ideas and responsibilities, and provides encouragement to the participants.

password—A string of characters or other identifiers used for user authentication to prove identity or access approval to gain access to a resource.

patch/patching—A modification made directly to an executable program without changing the source code (avoids reassembling or recompiling). A temporary fix to the source or object code.

patent—A legal mechanism for protecting ideas by providing "exclusive rights for novel inventions for a limited period of time" (Futrell et al. 2002).

path—A control flow-sequence through the internal structure of the software source code.

peer review—A static analysis technique for evaluating the form, structure, and content of a document, source code, or other work product using examination rather than execution.

peer reviewers/inspectors—People participating as reviewers in a peer review.

peer-to-peer architecture—A distributed applications architecture that partitions tasks or workloads between equally privileged, interconnected nodes (peers).

perfective maintenance—"Improvements in the software's performance or functionality, (e.g., in response to user suggestions or requests)" (IEEE/ISO/IEC 2017).

performance—A quality or product attribute requirements describing the levels of performance (e.g., capacity, throughput, response times) required from the software.

performance testing—A type of testing whose objective is to determine if the system has any problems meeting its performance requirements for throughput, response time, number of simultaneous users, number of terminals supported, and so on.

performing stage—The fourth stage of team development, when the team is in high performance mode and everyone knows how to work together effectively.

phase—A logical group of related activities that constitute a major step in a software project or life cycle. A major segment or component of a software life cycle.

phase containment effectiveness—A measure of the effectiveness of defect detection techniques in identifying defects in the same phase as they were introduced.

phase gate review—Reviews used to verify the successful completion of a project phase or milestone. (Also called phase transition reviews or milestone reviews)

physical configuration audit—"An audit conducted to verify that each configuration item, as built, conforms to the technical documentation that defines it" (IEEE/ISO/IEC 2017).

pie chart—The pie chart is used to illustrate how a single value is broken down into parts by category.

pilot—A verification and validation technique where the product or process is analyzed during a trial run.

plan–do–check–act model (PDCA)—A process improvement model. (Also called the Shewhart cycle or the Deming circle)

planned value (PV)—An earned value metric of the budgeted cost of work performed (BCWS).

planning poker—An agile expert judgment estimation technique where all of the members of the agile development team participate.

Platform as a Service (PaaS)—A cloud computing model where the cloud service provider (CSP) provides a cloud-based platform for software developers to develop, manage, and maintain their software applications throughout the entire life cycle.

policy—"A governing principle typically used as the basis for regulations, procedures, or standards and generally stated by the highest authority in the organization" (Humphrey 1989).

polymorphism—In object-oriented programming, a mechanism that allows the sender of a stimulus (or message) to not need to know the receiving instance's class so that receiving instance can belong to an arbitrary class.

portability—A quality attribute describing the effort required to migrate the software or data to a different platform or environment.

positive observation—An audit finding that points out something positive that is worth informing management of so that it can be propagated to other projects, processes, or areas.

post-conditions—Specific, measurable conditions that must be met before the use case is considered complete.

post-project review—A project review held at the end of a project (or at the end of one of the project's phases or activities) to evaluate what went right on the project (phase or activity) that needs to be repeated on future projects and what went wrong that could be improved on in the future. (Also called post-mortem, project retrospective, or reflection)

practitioner—The individuals performing the work including developers, testers, maintainers, support staff, and so on.

practitioner satisfaction—A measure of the extent to which the practitioners are happy with the organization, their jobs, the work they do and other factors.

precision—A measure of the data's exactness and discrimination, representing the true values of the attributes for its associated entity.

preconditions—Specific, measurable conditions that must be met before the use case can be initiated.

prevention—A quality philosophy that focuses on keeping defects out of the product, or process.

prevention control—In failure modes and effects analysis, the activities that are currently in place or already planned to prevent the failure.

prevention costs of quality—Prevention costs are all the costs of quality that involve keeping defects from getting into the software products and processes.

preventive action—Actions are taken to prevent problems that have not yet occurred through risk-based analysis of the future.

preventive maintenance—"Modification of a software product after delivery to detect and correct latent faults in the software product before they manifest as failures" (IEEE 2006).

prioritization graph—An x–y graph used to prioritize items by order of importance based on two criteria.

priority/prioritize—The rank assigned to an entity for the purpose of determining its order (e.g., the priority of competing entities within a software system determines the order in which they use system resources). The priority of a software problem determines the order in which it is debugged and corrected.

prioritization matrix—A tool used to rank items in order of priority based on a set of weighted criteria.

private cloud—The cloud computing platform where the platform us owned by and dedicated to a single user or organization.

privacy by design—Privacy that is "seamlessly integrated into product, services, and system designs by default."

problem report—A change request created to report an issue or anomaly in the software.

problem report backlog—The number of problem reports that have not been resolved/closed (are still in the works) at the point in time when the measurement was taken.

process—The step-by-step sequence of actions that must be carried out to complete an activity. A definable, repeatable, measurable sequence of tasks used to produce a quality product.

process architecture—The definition of the ordering of the individual processes, their interactions and interdependencies of product flows between the processes, and their interfaces to external processes.

process audit—An in-depth evaluation of a process or set of processes.

process capability—A measure of the extent to which the process met its purpose and objectives.

process cost—A measure of the amount of money spent or the amount of effort expended to implement an occurrence of a process.

process flow diagram—A graphical representation of the flow of control through the tasks and verification steps in a process and the roles responsible for each task or verification step.

process improvement—The planned and systematic set of all actions and activities needed to identify, plan, implement, track, and control improvements to a process.

process metric—A metric designed to report one or more measures that provide information about a process.

process stakeholder—Individuals or groups who affect or are affected by a software process and therefore have some level of influence over the requirements for that process.

product—A tangible output, artifact, or specific measurable accomplishment that is a result of a software development activity or process. Examples of these artifacts include source code, models, electronic files, documents, databases, reports, metrics, logs, and services.

product acceptance plan—Specification of the plans for obtaining the customers approval of the external product deliverables including the objective criteria for accepting each deliverable.

product attribute—Product-level requirements that define the characteristics that the software must possess to be considered a high-quality product by one or more stakeholders.

product audit—An evaluation of a product to examine its conformance to the product specification, performance, and other standards or the customer's requirements.

product backlog—In Scrum, a prioritized list of unimplemented product requirements (stories or features).

product baseline—The baseline created when the product is released into operations. (Also called the production baseline)

product decomposition—Breaking the product down from business to stakeholder to product requirements and then to the architectural and component designs.

production—Activities involving the creation of master media and the replication of that media.

productivity—A measure of the effectiveness and efficiency of executing a process typically modeled in terms of the size or amount of product produced per unit of time. A factor of usability that measures the effectiveness and efficiency of a user's ability to use the tool to perform work typically modeled in terms of tasks or actions that can be performed per unit of time.

product functional requirements—Functional and nonfunctional requirements that define what the product must do or be to meet the stakeholder level requirements.

product license—A legal mechanism for granting the license holder the right to either use or redistribute one or more copies of copyrighted products without breaking copyright law.

product limitations—A definition of items that will not be included in the product.

product metric—A metric designed to report one or more measures that provide information about a product.

product owner—The Scrum role that represents the project stakeholders, is responsible for project funding and controls and prioritizes the product backlog.

product partitioning—The partitioning of the system or software into configuration items.

product perspective of quality—One of Garvin's five perspectives of quality that ties quality to inherent characteristics or quality attributes of the product.

product scope—A definition of what will be included in the product, thus defining the boundaries of the product.

product stakeholder—Individuals or groups who affect or are affected by a software product and therefore have some level of influence over the requirements for that product.

product vision—The definition of how the new or updated software product bridges the gap between the current state and the desired future state needed to take advantage of a business opportunity or solve a business problem.

program—"A group of related projects managed in a coordinated way to obtain benefits and control not available from managing them individually" (PMI 2013).

program review—A review meeting that involves gathering and consolidating status information from the individual project being managed together as a group or program.

project—"A temporary endeavor undertaken to create a unique product, service, or result" (PMI 2013).

project audit—An audit of the processes used to plan, implement, track, control, and close a project to evaluate the processes conformance to documented instructions or standards. A project audit also looks at the effectiveness of the project management process in meeting the intended goals or objectives, and the adequacy and effectiveness of the process controls.

project charter—The document that formally authorizes the project and is created during project initiation.

project closure—The processes and activities involving the termination of a project.

project closure plan—Specification of the plans for the orderly closure of the project.

project decomposition—Using the work breakdown structure to break the project down into activities.

project execution—The processes and activities involved in implementing and deploying the project. (Also called project deployment)

project initiation—The processes and activities involved in starting a project including defining the projects charter.

project management—The planning and management of strategies and tactics of a project to direct the intent of the project, plan actions, activities, risks, resources and responsibilities, guide the ongoing project activities, and anticipate and prepare for change.

project monitoring and control—The processes and activities involved in tracking the actual results of the project's implemented actions and activities against the established plans, controlling significant deviation from the expected plans, and making any changes required to keep the project in line with its objectives. (Also called project tracking and control)

project objectives—The defined list of intended outcomes for the project.

project organization plan—A description of the internal project team organization and structure including internal interfaces (e.g., lines of authority, responsibility, and communication) within software development and supporting organizations.

project plans—The planning documentation that describes how the organization's project management policies, standards, processes, work instructions, and infrastructure will be implemented and tailored to meet the needs of an individual project, including specific activities, responsibilities, resource and budget allocations, tactics, tools, and methods. This information may be in a single project plan or separated into multiple planning documents (e.g., software quality assurance plan, verification and validation plan, and software configuration management plan).

project planning—The processes and activities involved in planning the project.

project reviews—Various types of reviews intended to track the progress of the project and identify issues.

project stakeholder—"Persons and organizations . . . that are actively involved in the project, or whose interests may be positively or negatively affected by the execution or completion of the project" (PMI 2013). (See also Stakeholders)

project team—The group of practitioners assigned to the planning, execution, tracking, and control of the project.

project team review—A project review with project team members held to monitor the current status and results of the project against the project's documented estimates, plans, commitments, and requirements and to identify project issues in a timely manner so that effective action can be taken.

promotion—The transition in the level of authority needed to approve changes to a baselined configuration item.

proof of correctness—"A formal technique used to prove mathematically that a computer program satisfies its specified requirements" (IEEE/ISO/IEC 2017). (Also called mathematical proofs)

proprietary software license—A license that grants limited rights to the user to use of one or more copies of the software, while the ownership of that software remains with the software development organization.

prototype—A partial, preliminary, or mock-up version of the software product used to elicit, analyze, and validate requirements.

provenance—The origin and history of each piece of data or information are known and well defined.

purpose—A statement of the value-added reason for performing a process.

quality—

1. The degree to which the system satisfies the stated and implied needs of its various stakeholders, and thus provides value
2. The ability of a product, service, systems, component, or process to meet customer or user needs, expectations, or requirements
3. The degree to which a set of inherent characteristics fulfills requirements (IEEE/ISO/IEC 2017)

quality attributes—Stakeholder-level requirements that define the characteristics that the software must possess to be considered a high-quality product by one or more stakeholders.

quality gate—A quality checkpoint or review that a product must pass through in order to be acquired or to transition to the next project activity.

quality goals—Specific targets established to institutionalize quality-related activities into every aspect of the organization and its key business practices.

quality management plan—Plan that specifies process improvement and quality management system activities.

quality management system (QMS)—The aggregate of the organization's quality-related organizational structure, policies, processes, work instructions, plans, supporting tools, and infrastructure.

quality objectives—Specific, measurable, achievable, realistic, and time-framed statements of what needs to be accomplished to achieve quality goals.

quality plan/quality planning—Planning documentation that defines the specifics for how a project (program or product) intends to implement and tailor the organization's quality management system to meet the needs of an individual project including specific activities, responsibilities, resource and budget allocations, tactics, tools, and methods.

quality policy—The formally documented statement of the overall intentions and direction of an organization with regard to quality, as formally expressed by top management.

quality record—An artifact that provides the objective evidence that the appropriate quality and process activities took place and that the execution of those activities met required standards, policies, procedures, or specifications. Examples of quality records include meeting minutes, logs, change requests, completed forms, completed checklists, formal sign-off, or approval pages, reports, and metrics.

random sampling—A sampling procedure where a sample of size n is drawn from the population in such a way that every possible element has the same chance of being selected.

range—In statistics, the difference between the maximum and minimum values in a data set.

ratio scale measurement—The ratio scale is an interval scale with an absolute or nonarbitrary zero point. All mathematical operations can be applied to ratio scale measurement values, including multiplication and division.

reader—The role in an inspection process that is responsible for presenting the product to the other inspectors during the meeting.

ready-to-ship review—A phase gate review held at the transition from the final test cycle to deployment of the completed system for use by the end-users. (Also called a ready-to-release review, release-to-production review, release-to operations review, and so on)

recognition and rewards—The act of acknowledging or rewarding individuals for their performance or behavior.

recorder—The person in a meeting who is responsible for keeping the record of that meeting. (Also called the scribe)

recoverability—A measure of the effort required to retrieve, refresh, or otherwise recover the data if it is lost, corrupted, or damaged in any way.

recovery—The ability to recover any changes to software products, data, or development environment made since the last backup.

reengineering—The process of rebuilding existing software products to create a product with added functionality, better performance and reliability, and improved maintainability.

refactor—The process of modifying the software code without impacting its functionality, in order to improve that software quality attributes (e.g., modularity, efficiency, maintainability, readability, and so on).

reflection—An extreme programming principle where the team continuously considers what they are doing and why. (See also Post project review).

regression testing—A type of testing whose objective is to determine if changing the software or fixing a defect causes any other problems in the system.

regulation—Rules, laws, or instructions, established by a legislative or regulatory body, which set legal standards with which organizations must comply.

reinforcement theory—A theory of rewards and motivation that says that people will be motivated to perform an activity based on their perception of a trigger (a signal to initiate the behavior) and the historic consequences of that behavior.

release—Certain promotions of the configuration item that are distributed outside development.

release management—The management of the software release process.

release manager—The individual responsible for packaging the software (or software intensive systems) into a releasable product set for deployment into the desired environment.

release plan/release planning—A specification of the release plans for releasing the product into operations.

release support—The support of the release during operations.

reliability—A quality attribute describing the extent to which the software can perform its functions without failure for a specified period of time under specified conditions.

reliable metric—A metric is reliable if it can be used by different people (or the same person multiple times) with the same result. A function of consistency or repeatability of the measure.

reorganization—The state a team that occurs when one or more team members leave the team, new members are added to the team, or major changes in work assignments occur.

replication—The process of making copies of the software and its associated documentation for shipment to the customers, users, or other stakeholders.

reproducibility—The ability to rebuild the software product at a later time and get the same results

request for proposal (RFP)—A request for a supplier to submit a proposal for building and delivering a product.

requirement—A capability, attribute, or design constraint of the software that provides value to or is needed by a stakeholder.

requirements analysis—The requirements development activity that involves evaluating and modeling the information gathered from the product's stakeholders to identify gaps, conflicts, and other issues that require additional elicitation, to organize that information into requirements, and to prioritize those requirements.

requirements churn—The level of change occurring in the baselined requirements. (Also called requirement volatility or scope creep)

requirements development—All of the activities involved in eliciting, analyzing, specifying, and validating the requirements.

requirements elicitation—The requirements development activity that includes all of the activities involved in identifying the requirement's stakeholders, selecting representatives from each stakeholder class, and collecting information to determine the needs of each class of stakeholders

requirements engineering—A disciplined, process-oriented approach to the definition, documentation, and maintenance of software requirements throughout the software development life cycle.

requirements management—The activities employed to ensure that the system and software requirements are maintained in accordance with all applicable rules, regulations, standards, procedures, and so on, and that only authorized changes are made to those requirements once they are baselined.

requirements management plan—Specification of plans, tools, techniques, schedules, and responsibilities for developing requirements and measuring, reporting, and controlling change to the product requirements.

requirements prioritization—Sorting the requirements in order of priority.

requirements review—A phase gate review held at the transition from the requirements phase into the high-level or architectural design phase.

requirements specification—The requirements development activities that involve documenting the requirements into one or more specification or other documents.

requirements test coverage—Testing analysis that looks at the mapping of tests to the various requirements for the software to ensure that they are thoroughly tested.

requirements validation—The requirements development activities that involve evaluating the requirements for completeness, correctness, consistency, feasibility, finiteness, measurability, maintainability, and other quality criteria.

residual risk—The risk that remains after risk handling actions have been implemented.

resource—The resources input into and used during a project.

resource leveling—A project management technique that involves the scheduling and allocation of resources so that they are not overloaded.

resource plans—Specification of the number of resources (other than staff) required by type, when they are needed, and source (e.g., internal transfer, purchased, or rented).

resource utilization metric—A measure of the average or maximum amount of a given resource (e.g., memory, disk space, bandwidth) used by the software to perform a function or activity.

resource utilization testing—A type of testing whose objective is to determine if the system uses resources (e.g., memory, disk space) at levels that exceed requirements.

response time/responsiveness—A measure of the average or maximum amount of calendar time required to respond to a stimulus (e.g., user command, input from another software application or hardware signal or the reporting of a problem).

restore—The ability to restore a previous state of the software products, data, or development environment from its backup.

resumption criteria—Specific, defined criteria that must be met before suspended testing activities can be restarted.

retirement—The time at which the support of a particular release of the software or the entire software product is terminated.

return on investment (ROI)—A measure of the ratio between the profits received from an activity or project and its cost.

reusability or **reuse**—A quality attribute describing the extent to which one or more components of a software product can be reused when developing other software products.

reverse engineering—The process of recreating one or more software product from successor products (e.g., recreating the software's requirements, design, or interface information from the source code).

review—"A process or meeting during which a work product, or set of work products, is presented to project personnel, managers, users, customers, or other interested parties for comment or approval" (IEEE/ISO/IEC 2017).

review and approval of the modified configuration items—A less rigorous level of change control where changes are made to a draft version of a baseline work product and change control activities are preformed through a review and approval cycle.

reviewers—Individuals participating in the review.

revision—A change to a configuration item, component, or unit that corrects one or more defects but does not incorporate any new functionality or features.

rework—Repetition of work that has already been completed because of unsatisfactory results including defects.

risk—A possibility of a major problem occurring during a project or with a product that will impact the success of that project or product. A risk starts when a commitment is made and ends when there is no longer the possibility of the problem occurring or when the problem actually does occur.

risk analysis—The activities involving exploring the context of each risk; assigning a probability, loss, and timeframe to each risk; and prioritizing the risks.

risk avoidance—A strategy for handling a risk where the risk is avoided by either not taking on the activity or project associated with the risk or selecting an alternative approach

risk-based verification and validation (testing and peer reviews)—Risk analysis used to determine the level of V&V activities needed. (Also called risk-based testing or risk-based peer reviews)

risk context—The additional information that surrounds and affects that risk and helps determine its probability and potential loss

risk contingency plans—A set of activities that are planned in advance to handle a problem and are taken only if the risk actually turns into a problem. (See also Disaster recovery plans)

risk exposure (RE)—The expected value of the loss associated with the risk calculated by multiplying the risk likelihood by the risk loss. (See also Effects and criticality)

risk identification—The activities involving the identification and communication of risks

risk impact/loss—The cost (e.g., dollars, effort hours, or schedule slippage) to the project if the risk actually turns into a problem. The impact or jeopardy to the project if the problem occurs. (Also called risk impact; see also Severity score)

risk management—The process of identifying, analyzing, planning for, taking action against, and tracking risks.

risk management plan—Specification of the methods, techniques, and tools for identifying, analyzing, and planning mitigations and tracking and controlling project-related risks.

risk mitigation plans—A set of planned activities intended to handle a risk by reducing the probability that the risk will turn into a problem or by reducing the loss to the project if the problem does occur. (Also called risk mitigation actions or risk handling activities)

risk priority number (RPN)—In failure modes and effects analysis, the calculated relative risk of the failure chain, which is equal to the severity score times the occurrence score times the detection score.

risk probability—The likelihood that the risk will turn into a major problem. (See also Occurrence score)

risk reduction leverage (RRL)—A probability-based benefit to cost ratio for a given risk handling plan.

risk statement—A documented statement of the risk condition and its potential consequences.

risk taxonomy—A categorized list of potential risks based on past problems.

risk timeframe—When the risk needs to be addressed.

risk tolerance—People perception of the risk reward balance based on their own perception of risk.

risk tracking—The activities involving tracking the risk until it is closed.

risk transfer—A risk handling action where some or all of the risk is transferred to another party or group.

role—The generic specification of or the specific individual or group responsible for one or more tasks, verification steps, or activities that are part of a process or plan.

root cause analysis—The process of assessing sets of product defects or process problems to identify their systemic cause(s).

run chart—An analytical tool that plots data arranged in time sequence.

safety—A quality attribute describing the ability to use the software without adverse impact to individuals, property, the environment, or society.

safety critical software—Software that can result in an accident that is intended to mitigate the results of an accident or that is intended to recover from the results of an accident.

safety mitigation process—The risk mitigation activities put in place to:

- Prevent or avoid a software related mishap
- Allow the software to predict, detect, or report mishaps
- Allow the software to detect or recover from a hazardous state
- Allow the software to minimize the consequences of an accident

safety plan—Specification of plans, tools, techniques, schedules, and responsibilities for performing hazard analysis and identifying, analyzing, mitigating, tracking, and controlling software related safety risks including rigorous development processes throughout the entire life cycle.

safety risk—A potential system safety problem related to software control or information provided by the software.

sampling—A mechanism for using only part of the entire population to estimate attributes of that population.

Satir change model—A model of how implementing change impacts productivity over time.

scalability—The ability of the software to be extended to accommodate more interactions, hardware connections, or users.

scatter diagram—An x–y plot of one variable versus another. One variable, called the independent variable, is typically plotted on the x-axis. The second variable, called the dependent variable, is typically plotted on the y-axis. Scatter diagrams are used to investigate whether the independent variable is causing changes in the dependent variable.

schedule or scheduling—The length of calendar time planned for the performance of a process or a project.

schedule performance index (SPI)—An earned value metric defined as the earned value divided by the planned value (BCWP/BCWS).

schedule variance (SV)—An earned value metric defined as the earned value minus the planned value (BCWP/BCWS).

SCM librarians—The role responsible for establishing, coordinating, and ensuring the integrity of the software configuration management–controlled libraries and static libraries for each project and for reuse of software components between projects

SCM management—The management activities related to software configuration management.

SCM manager—The role responsible for managing the software configuration management activities and software configuration management group.

SCM toolsmith—The role responsible for the implementation and maintenance of the software configuration management tool set.

scope—The amount of work to be accomplished or requirements to be implemented.

Scrum—A set of agile techniques that emphasize project management.

Scrum development team—A self-managed, cross-functional team that is responsible for turning each sprint backlog into a deliverable software increment.

Scrum of scrums—A team made up of representatives from each of several individual scrum teams, which meets periodically to coordinate the efforts of the individual scrum teams

Scrum master—The Scrum role that is responsible for the scrum process and fitting scrum into the organization. (Also see agile coach)

Scrum planning meeting—A meeting held to select the items from the product backlog to form a specific set (sprint backlog) of features and process improvement items for implementation in the next sprint. (Also called the sprint planning meeting)

Scrums—Daily scrum team status review meetings. (Also called daily scrums or daily stand-up meetings)

second-party audit—An audit performed by a customer (or an organization contracted by a customer) on its supplier.

secondary actors—Actors that are involved in a use case but that do not initiate the use case.

security—A quality attribute describing the probability that an attack of a specific type will be detected, repelled, or handled by the software. (See also Data security and information security)

security attack—An unauthorized attempt to access the software, its data, or its functionality

security coverage or security testing—A type of testing whose objective is to determine if the security of the system can be breached.

security hole—An unintended function, mode, or state in the software caused by design flaws or coding bugs (one or more defects) that an attacker can exploit in order to breach the software's security.

security risk—A potential software security problem.

selected product backlog items—Items from the product backlog that have been selected for implementation in the next sprint.

selling—A situational leadership style with high levels of both task/directive behaviors and relationship/supportive behaviors; decisions are explained and opportunities are provided to ask questions and clarify instructions.

sequence diagram—An analysis model that records in detail how objects interact over time to perform a task by describing the messages that pass between them.

service level agreement (SLA)—A formally negotiated agreement between two parties (typically, the maintainer and the acquirer of the software) that defines the agreed upon level of service to be provided.

service mark—An indication of the intention to create a trademark.

severity—An ordinal scale metric that measures the impact of a defect, problem or issues on the successful operation or use of the software.

severity score—In failure modes and effects analysis, the rating of the seriousness of the impact or loss if the failure occurs.

simplicity—An extreme programming (agile) value that design be kept to the most basic thing that can provide the functionality needed within the timeframe of the next delivery/iteration.

simulator/simulation—A mechanism used to imitate the real-world environment (e.g., during testing).

situational leadership—Methods where the style of leadership selected depends upon the situation or context.

Six Sigma—The Greek letter sigma (σ) is the statistical symbol for standard deviation. Therefore, Six Sigma literally means six standard deviations away from the mean. Or no more than 3.4 defects per million opportunities. As a quality strategy, Six Sigma has evolved into a comprehensive and flexible road map for both continuous (DMAIC) and evolutionary (DMADV) process improvement.

size—The amount, quantity, or volume of software and other artifacts being produced by the project.

slack—"The amount of calendar time that an activity can be delayed before it begins to impact the project's finish date" (Lewis 1995).

soft skills—Nontechnical skills needed to be effective in influencing others toward quality. Examples of "soft skills" include leadership, team building, facilitation, communication, motivation, conflict resolution, negotiation, and so on.

software architecture—The architecture composed of software-level components, interfaces between those components and the functions to be interchanged. (See also Architectural design)

Software as a Service—A cloud computing model where the cloud service provider (CSP) delivers a full application stack as a service, from underlying infrastructure to maintenance and updates to the provided application software.

software dependency—A dependency that exists when the current release of the software is not backward compatible with the software platform used to run previous versions of the software, other software that interfaced with the previous version of the software, or data files that were accessed by previous versions of the software.

software development plan—Definition of the approach to software development including the selected life cycle; selected processes and work instructions and any associated tailoring; and selected methods, tools and techniques, that will be used during the project.

software level of control (LOC)—"A measure of the degree to which the software is responsible for the behavior of a system, or of a specific system action, that may lead to a safety mishap" (Frailey 2016).

software quality assurance (SQA)—The planned and systematic set of all actions and activities needed to provide adequate confidence that:

- A product conforms to it requirements
- The organization's quality management system (or each individual process) is adequate to meet the organization's quality goals and objectives, is appropriately planned, is being followed, and is effective and efficient

software quality control—The planned and systematic set of all actions and activities needed to monitor and measure software projects, processes, and products to ensure that special causes have not introduced unwanted variation into those projects, processes and products.

software quality engineering—The study and systematic application of scientific, technological, economic, social, and practical knowledge, and empirically proven methods, to the analysis and continuous improvement of all stages of the software life cycle to maximize the quality of software processes and practices, and the products they produce. The processes and activities needed to define, plan and implement the quality management system for software related processes, projects, and products.

software quality management—The processes and activities involved in setting the organization's strategic quality goals and objectives; establishing organizational, project, and product quality planning; and providing the oversight necessary to ensure the effectiveness and efficiency of the organization's quality management system.

software security control—Logic within the software that enforces security protections.

software system testing—The testing of the software as a whole.

software verification and validation—The processes and activities used to ensure that software products meet their specified requirements and intended use in order to make sure that the "software was built right" and the "right software was built."

source code—Assembly language or higher-level language statements defining a set of instructions to be followed by the computer after translation into machine language.

special cause variation—When the root cause of the statistically improbable variation in a process can be attributed to one or more special factors outside the normal expected variation in the process.

speed—A usability measure of how quickly users can accomplish their tasks using the software

spiral model—A software life cycle model that is a risk-based model that expands on the structure of a waterfall model with details including the exploration of alternatives, prototyping, risk management, and planning.

sponsor—A senior member of management who provides ongoing funding and other necessary resources to the team.

spoofing—"Unauthorized use of legitimate identification and authentication (I&A) data, however it was obtained, to mimic a subject different from the attacker. Impersonating, masquerading, piggybacking, and mimicking are forms of spoofing" (CNSS 2015).

sprint—In scrum, a single iteration through the development cycle.

sprint backlog—In scrum, a prioritized list of product requirements (features) that are planned for implementation during the current sprint.

sprint retrospective—A meeting held after the sprint review meeting to address needed process improvements.

sprint review meeting—A scrum meeting held with stakeholders to review the output of the sprint and determine what to do next.

spyware—Unauthorized software that is intended to gather information about a person or organization without their knowledge or permission.

staff—People working on the project, process, or other activity.

staffing plans—Specification of the number of staff required by skill levels when they are needed, and source (e.g., internal transfer, new hire, or contracted).

stakeholder—Any individual or group who affects or are affected by a software product, project, or process and therefore have some level of influence over the requirements for that software product, project, or process.

stakeholder management plans—Plans define the "strategies for effectively engaging stakeholders of the project, based on their needs, interests and impact on project success" (PMI 2013).

stakeholder functional requirements—Requirements that define the needs of the various stakeholders.

standard—"A standard is a rule or basis for comparison that is used to assess size, content, value, or quality of an object or activity, typically established by common practice or designated standards body" (Humphrey 1989). "Mandatory requirements employed and enforced to prescribe a disciplined, uniform approach to software development" (IEEE/ISO/IEC 2017). "The formal requirements developed and used to prescribe consistent approaches to acquisition, development, or service" (CMMI 2010a).

standard deviation—In statistics, a measure of the variability in a data set.

state—A specific condition that an entity is in at a specific point in time.

state coverage or state testing—Testing analysis that looks at the mapping of tests to the various state transitions that can occur in the software to ensure that they are thoroughly tested.

statement coverage—White-box testing analysis that looks at the mapping of tests to the various instruction or a series of instructions that a computer carries out to ensure that they are thoroughly tested.

state transition diagram or table—A representation of the behavior of a system by depicting its states and the events that cause the system to change states.

static analysis—Methods of performing V&V by evaluating a software component or product without executing that component. Examples of static analysis techniques include peer reviews, proofs of correctness, and code analysis tools (e.g., compliers, complexity analyzers).

static cycle time—A measure of the calendar time to complete an occurrence process or the average cycle time to complete multiple occurrences of a process.

static library—A software configuration management library used to archive important baselines, including those released into operations. The static library is used for the control, preservation, and retrieval of master media. (Also called the software repository or software product baseline library).

statistical process control—The use of measurement and statistical methods in the monitoring and maintaining of the quality of products and services.

statistics—The science of the collection, organization, analysis, and interpretation of data.

stoplight chart—A chart that provides a red, yellow, and green signal to the user of the metric to aid in the interpretation of the measurement.

storming stage—The second stage of team development when the team members are aware that they are going to have to change to make the team work together and there is a period of resistance to that change.

story—A stakeholder level requirement.

storyboards—Pictorial sequences used in requirements engineering to describe the human user interfaces.

stress testing—A type of environmental load testing that subjects the software to surges or spikes in load over short periods of time, and evaluates the software's performance.

strict product liability—A type of tort lawsuit that might be applied to software. The software caused injury or property damage because it is dangerously defective.

structural complexity—The complexity of the interactions between the modules in a calling structure (or in the case of object oriented development between the classes in an inheritance tree).

structured analysis and design (SAD)—"A collection of guidelines for distinguishing between good designs and bad designs, and a collection of techniques, strategies, and heuristics that generally lead to good design" (Yourdon and Constantine 1979). (Also called structured analysis and design techniques)

stub—A temporary piece of source code used as a replacement for a called module when performing top-down integration and integration testing.

successful project—A project where the deliverable products are completed at the desired quality and performance level, within budget, on schedule, without reducing the scope of the project, while using people and other project resources effectively and efficiently.

supplier—The individuals or groups that are part of the organization that develops or maintains the software for the acquirer or are part of the organization that distribute the software to the acquirer.

supplier audit—Ongoing second party audits of a current supplier's systems, processes, projects, or products to ensure their continued capability to produce products of the required quality level. (Also called supplier surveillance audit)

supplier management and control—Ongoing activities to monitor and control the work being performed by suppliers.

supplier management plan—Specification of the selection criteria for any suppliers and a tailored management plan for managing each supplier that will contribute outsourced products to the project. (Also call the procurement plan, subcontractor management plan, or vendor management plan)

supplier qualification audit—A second party audit to evaluate the supplier's quality system to determine whether the supplier has the capability to produce products of the required quality level prior to selecting that supplier to perform work.

supportability—The ease with which technical support or a help desk can configure, monitor, support, and correct the software once it is in operation.

system—A set of interacting elements or components that work together to form a whole. (See also Quality management system)

system architecture—The highest level of design. The architecture composed of system-level components, interfaces between those components, and the functions to be interchanged.

system audits—Audits conducted on management systems that evaluate all of the policies, processes, and work instructions, supporting plans and activities, training, and other components of those systems. An in-depth evaluation of an entire system (e.g., a quality management system, environmental system, or safety system).

system integration—Integrating the various components of the system together with a focus on testing the interfaces and interactions between those components.

system of systems—A system where the interactive components are systems.

system modeling language (SysML)—A standard for defining models

system requirements—Requirements for the entire system (as opposed to software requirements).

system testing—Testing done on the entire integrated system to validate that the software meets all of its requirements as defined in the requirements specification.

tacit knowledge—Knowledge that is gained through experience rather than through formal reading or education.

tailoring—Altering or adapting the standardized policies, processes, or work instructions to meet the specific objectives, constraints, needs, or requirements of a project, program, or product.

tasks—In process definition, the individual steps or activities that must be performed to implement the process and create the resulting work product(s).

taxonomy—A categorized list of items (e.g., see risk taxonomy).

TCP/IP network architecture—A network architecture that is made up of four layers that interact to make the internet work.

team—A group of two or more people working together toward a common goal.

team development—Stages of team development include forming, storming, norming, performing, and adjourning

team leader—The person responsible for managing the team.

team member—A person on a team.

team problems—Issues that exist when the team is not functioning as it should.

technical review—Review of a software product from a technical perspective. (See also Peer reviews)

template—A type of work instruction with a predefined layout used to create a new document, page, form, or other entity of the same design, pattern, format, or style.

test/testing—A verification and validation method that involves the execution of all or part of the software in an attempt to detect the defects that exist in that software or to obtain a confidence level that the requirements are met.

"The process of operating a system or component under specified conditions, observing or recording the results, and making an evaluation of some aspect of the system or component" (IEEE/ISO/IEC 2017).

testability or testable—A quality attribute describing the effort required to perform tests on the software. There exists a reasonably cost-effective way to determine that the software satisfies the requirement.

test analysis—The activities involved in analyzing the test basis and test objects.

test automation—The use of software to automate the activities of test design, test execution, and the capturing and analysis of test results.

test basis—The body of knowledge used as the basis for test analysis and design. (ISTQB 2018c)

test bed/test environment—An environment established and used for the execution of tests including hardware, instrumentation, simulators, software tools, and other support elements.

test case, test case specification—A specific set of test data and associated procedural steps for verifying or validating a specific source code path, interface, or specific requirement.

test conditions—An aspect of the test basis that is relevant in order to achieve specific test objectives. (ISTQB 2018c)

test coverage—An evaluation to determine how completely a set of testing exercised all of the software.

test data item—Items of data that are required to perform test cases, test procedures, and automated test scripts.

test data management—The management and control of the test data items.

test design specification—A test document that is a further refinement of the test plan and is only needed if the complexity of the system is high or there is a need for additional levels of information.

test documentation—Documentation that supports test planning, design, execution, and reporting of test results.

test-driven development—An agile iterative development methodology where software functionality is implemented based on first writing the test cases that the code must pass. (Also called test-driven design)

test execution—The activities involved in actually running the tests and analyzing the results.

test execution engine—A test automation tool that selects, loads, and executes a test specification from the repository of test scripts and then monitors the execution of that test specification and reports and stores the results.

test harness—A set of test stubs and drivers or a test execution engine used to automate all or part of the testing.

test log—A test document that provides a chronological record of the execution of the tests.

test matrix—A matrix used to indicate the type of testing that will be used to validate each requirement.

test objects—The component(s) or system (s) to be tested. (ISTQB 2018c])

test plans—Management-planning documentation that addresses resources and risks, as well as establishing the methodology to be employed during the testing.

test planning and design—Activities involved in planning the testing process and designing the tests that will be run.

test procedure specification—A test document that includes the specific steps for executing the tests (including test case sequences and setting up the test environment).

test readiness review (TRR)—A phase gate review held at the beginning of a major test cycle. (Also called ready-to-test review)

test report—A report outlining the results of one or more testing cycles. (Also called a test summary report)

test resumption criteria—Specific, defined criteria that must be met before suspended testing activities can be restarted.

test suspension criteria—The criteria that must be met before a session of testing is suspended.

test tools—Tools used during the testing processes and activities.

testware—Work products produced during the test process for use in planning, designing, executing, evaluating and reporting on testing. (ISTQB 2018c and ISO/IEC/IEEE 2013)

third-party audit—An audit performed on an organization by an external auditor other than their customer.

thread—A sequential set of usage steps through the software that a user takes.

throughput—A measure of the amount of work performed by a software system over a period of time (e.g., transactions per hour, jobs per day).

time bomb—"Resident computer program that triggers an unauthorized act at a predefined time" (CNSS 2006). (See also Malicious code)

time-box testing—A testing strategy where the calendar time for testing is fixed and the scope of the testing effort must be adjusted to fit inside that time box.

timely/timeliness—A data collection goal that the data be collected and turned into information in a timely enough manner so that it is useful in making decisions.

tools—Automated or semi-automated support to the development and maintenance processes and methods. Techniques used to by team to facilitate their activities.

top-down—An integration and testing strategy where testing starts with the highest-level unit or component in the section of the calling tree being tested and tests it using stubs. The lower-level software items and their stubs are then integrated, one or more at a time, replacing the stubs.

tort lawsuits—"A wrongful act other than a breach of contract that injures another and for which the law imposes civil liabilities" (Futrell et al. 2002).

total defect containment effectiveness (TDCE)—A metric that measures the effectiveness of defect detection techniques in identifying defects before the product is released into operation.

traceability or traceable—The ability to identify the relationships between originating requirements and their resulting derived requirements, designs, code, tests, and other products. A measure of the degree to which an audit trail exists of any events or actions that accessed or changed the data.

traceability matrix—A table that defines bidirectional traceability information.

trace tags—Unique identifiers used in the successor products to identify backward traceability to the predecessor document.

tracing—An audit technique that follows the path of the chronological progress of a software module either forward or backward through its life cycle transactions.

trademark—A legal mechanism for branding products or services and distinguishing them from other similar products or services in the marketplace.

training—A mechanism for transferring knowledge or skills.

training materials—Work instructions that include reference books, user manuals, computer-based training, videos, webinars, student handbooks, tools, and other materials used during training.

training plan—Specification of the training needed to ensure that necessary skill levels needed by the project are available, and the schedules, tools, methods, and techniques for obtaining that training.

transcendental perspective of quality—One of Garvin's five perspectives of quality that says quality is something that can be recognized but not defined.

transportation layer—A layer of a TCP/IP network architecture that performs host-to-host communications.

tree diagram—An analysis and problem-solving tool whose purpose is to break down or stratify ideas into progressively more detail.

trigger—An event, variance, threshold, or pattern that indicates that a risk is turning into a problem.

tunneling—"Technology enabling one network to send its data via another network's connections. Tunneling works by encapsulating a network protocol within packets carried by the second network" (CNSS 2015).

two-way communications—Communication that includes sending messages and obtaining feedback.

unambiguous—Every requirement statement should have one and only one interpretation. A data collection goal where the data collection process is understood well enough so different people performing the same measure collect the data in the same way.

understandability—A measure of the degree to which the data can be correctly read and interpreted by the users.

unfriendly user—A user to which the software should be "unfriendly" in order to keep that user from accomplishing their goals (e.g., a hacker, thief, or spy).

unit testing—Testing that verifies that the individual software unit (component, module) meets its specification as defined in the detailed design and the requirements that are allocated to that unit.

usability—A quality attribute describing the amount of effort that the users must expend to learn and use the software.

usability testing—A type of testing whose objective is to determine if the system has any areas that will be difficult or inconvenient for the users.

usage scenario testing—The chaining together of individual test cases from testing of the functions/subfunctions, into test scenarios that test start-to-finish user interactions with the software or system.

use cases—Scenarios created to describe the threads of usage for the system or software to be implemented.

use case diagram—A graphical representation of use cases and their interactions with users and their interrelationships with each other.

user—An individual or group that utilizes the functionality of the software. A primary user executes the software directly while a secondary use utilizes one or more outputs from the software without personally executing the software.

user or operator documentation—The set of documentation delivered to the users or operators that conveys the system instructions for understanding, installing, using, and maintaining the software.

user perspective of quality—One of Garvin's five perspectives of quality that cites "fitness for use" as the appropriate measure for quality.

user preference—A usability measure of what the users like about the software. (See also Likeability)

valid metric—A metric is internally valid if it measures what we expect it to measure. A metric is externally valid (also called predictive validity) if the metric results can be generalized or transferred to other populations or conditions.

validation—The "confirmation, through the provision of objective evidence, that the requirements for a specific intended use or application have been fulfilled" (IEEE/ISO/IEC 2017). The process used "to demonstrate that a product or product component fulfills its intended use when placed in its intended environment" (CMMI 2010a).

value added—The extent to which the benefit to one or more stakeholders is greater than the cost.

value-based perspective of quality—One of Garvin's five perspectives of quality that states that quality is dependent on the amount a customer is willing to pay for it

value stream mapping—A lean technique used to trace a product from raw materials to use in order to determine areas were wastes can be eliminated.

variance—A measure of the variation in a data set equal to the square of its standard deviation.

variation—A measure of how far individual data items in a data set spread out from the mean.

velocity—A measure of the amount of work product produced per calendar time.

verification—"The process of evaluating a system or component to determine whether the products of a given development phase satisfy the conditions imposed at the start of that phase" (IEEE/ISO/IEC 2017). The process used to "ensure that selected work products meet their specified requirements" (CMMI 2010a).

verification and validation (V&V)—The combination of verification and validation techniques to help ensure the quality of the software.

verification and validation plan—A planning document that describes how the organization's verification and validation policies, standards, processes work instructions, and infrastructure will be implemented and tailored to meet the needs of an individual project including specific activities, responsibilities, resource, and budget allocations, tactics, tools, and methods.

verification and validation reviews—Reviews held with the primary purpose of identifying and removing defects and to provide confidence that the work products meet their requirements and intended use.

verification and validation sufficiency—Analysis that balances the risk that the software still has undiscovered defects and the potential loss associated with those defects against the cost of performing additional V&V activities and the benefits of additional V&V activities.

verification and validation task iteration—The repeating of one or more previously executed V&V activities.

verification step—The step(s) in the process that ensure the process was implemented correctly or the product of the process meet their requirements.

version—A configuration item, component, or unit with a defined set of functionality.

version control—The procedures and tools necessary to manage the different versions of the configuration items and software products.

version description document—A deliverable document that describes the released version of a software product including an inventory of system or component parts, new or changed features or functionality, known defects and their workarounds, and other information about that version. (Also called release notes)

virtual audit—An audit conducted remotely using a virtual environment.

virtual team—A team that is not collocated but communicates and collaborates through the use of technology.

vision—The definition of how the new or updated software product bridges the gap between the current state and the desired future state needed to take advantage of a business opportunity or solve a business problem.

V-model—A software life cycle model that is a variation on the waterfall model that highlights the relationship between the testing phases and the products produced in the early life cycle phases.

volume testing—A type of environmental load testing that subjects the software to heavy loads over long periods of time and evaluates the software's capability.

waiver—A permanent business decision can always be made to override the entry and exit criteria.

walk-throughs—A method of peer review where a team of peers meets with the author to examine a product and provide feedback.

warranty—An agreement or between the seller and buyer of a product that provides assurance that certain facts, conditions, or assertions are true. A software warranty is typically a promise by the organization developing the software, regarding the quality of that software and the remedial actions that will be taken if the software does not perform as promised. (Also see expressed warranty and implied warranty).

waste—Anything that does not add, or gets in the way of adding, value as perceived by the stakeholders.

waterfall model—A software life cycle model where each phase proceeds from start to finish before the next phase is started.

web architectures—Architectures used with web-based software applications.

white-box testing—A form of testing that exercises the internal structure of a software component in order to detect defects. (Also called structural, clear-box, or glass-box testing)

width of a structure—The count of the maximum number of the span of control in the structure.

wireless systems architecture—The architecture of a computer system that is not tethered via cabling to a network.

W-model—A software life cycle model that is a variation on the waterfall model that has two paths (or crossing Vs), each one representing the life cycle for a separate organization or team during development (the developers and the independent verification and validation team).

work breakdown structure (WBS)—"A method of subdividing the project work into smaller and smaller increments to permit accurate estimates of duration, resources, and costs" (Lewis 1995). A type of tree diagram that is a hierarchical decomposition of the project into sub-projects, tasks, and sub-tasks.

work in process (WIP)—The total number of work items that are currently being worked on.

work instructions—Specific, detailed instructions for implementing a process in a specific environment (e.g., instructions, guidelines, checklists, forms, templates and training materials).

worst-case testing—A testing technique that utilizes worst-case scenarios based on boundaries.

written communications—Communications done through the written word and or documentation.

References

Aiello, Bob, and Leslie Sachs. 2011. *Configuration Management Best Practices: Practical Methods That Work in The Real World*. Upper Saddle River, NJ: Addison-Wesley.

Ambler, Scott W., and Mark Lines. 2012. *Disciplined Agile Delivery: A Practitioner's Guide to Agile Software Delivery in the Enterprise*. Upper Saddle River, NJ: IBM Press.

ASQ. 2016. *Failure Modes Effects Analysis*. Training class taught through ASQ Learning Offerings. Milwaukee, WI: ASQ.

Anderson, Bob. 2007. "25 time tested truths in IT support." *Computerworld*. Accessed at http://www.computerworld.com/article/2553933/it-management/25-time-tested-truths-about-it-support.html.

Arter, Dennis. 1994. *Quality Audits for Improvement Performance*. 2nd ed. Milwaukee, WI: ASQ Quality Press.

Arthur, Lowell Jay. 1985. *Measuring Programmer Productivity and Software Quality*. New York, NY: John Wiley & Sons.

———. 2023a. *ASQ Code of Ethics*. Milwaukee, WI: ASQ.

———. 2016. *Failure Modes Effects Analysis*. [Training class taught through ASQ Learning Offerings]. Milwaukee, WI: ASQ.

Astels, David. 2003. *Test-Driven Development: A Practical Guide*. 2nd ed. Upper Saddle River, NJ: Prentice Hall PTR.

Astels, David, Granville Miller, and Miroslav Novak. 2002. *A Practical Guide to eXtreme Programming*. Upper Saddle River, NJ: Prentice Hall.

Bass, Len, Paul Clements, and Risk Kazman. 2013. *Software Architecture in Practice*. 3rd ed. SEI Series in Software Engineering. Upper Saddle River NJ: Addison-Wesley.

Bays, Michael E. 1999. *Software Release Methodology*. Upper Saddle River, NJ: Prentice Hall PTR.

BCI. 2018. *Good Practice Guidelines 2018 Global Edition Edited Highlights: A Guide to Global Good Practices in Business Continuity*. United Kingdom: Business Continuity Institute.

Beck, Kent (with Cynthia Andres). 2005. *Extreme Programming Explained: Embrace Change*. 2nd ed. Boston, MA: Addison-Wesley.

Berczuk, Stephan P. 2003. *Software Configuration Management Patterns: Effective Teamwork, Practical Integration*. Boston, MA: Addison-Wesley.

Berlack, H. Ronald. 1992. *Software Configuration Management*. Hoboken, NJ: John Wiley & Sons.

Bernstein, Albert J., and Sydney Craft Rozen. 1990. *Dinosaur Brains: Dealing with All Those Impossible People at Work*. New York, NY: Ballantine Books.

Black, Rex. 2004. *Critical Testing Processes: Plan, Prepare, Perform, Perfect.* Boston: Addison-Wesley Professional.

Blohowiak, Donald. 1992. *Mavericks! How to Lead Your Staff to Think Like Einstein, Create Like da Vinci, and Invent Like Edison.* Homewood, IL: Business One Irwin.

Boehm, Barry. 1981. *Software Engineering Economics.* Englewood Cliffs, NJ: Prentice Hall.

———. 1988. "A Spiral Model of Software Development and Enhancement." *IEEE Computer* (May).

———. 1989. *Tutorial: Software Risk Management.* Los Alamitos, CA: IEEE Computer Society.

Boehm, Barry, Chris Abts, A. Winsor Brown, Sunita Chulani, Bradford K. Clark, Ellis Horowitz, Ray Madachy, Donald J. Reifer, and Bert Steece. 2000. *Software Cost Estimation with COCOMO II.* Upper Saddle River, NJ: Prentice Hall PTR.

Brooks, Fredrick. 1995. *The Mythical Man-Month: Essays on Software Engineering,* Anniversary ed. Boston, MA: Addison-Wesley Professional.

Campanella, Jack. 1990. *Principles of Quality Costs: Principles, Implementation, and Use.* 2nd ed. Milwaukee, WI: ASQC Quality Press.

Christensen, Eldon H., Kathleen M. Coombes-Betz, and Marilyn S. Stein. 2007. *The Certified Quality Process Analyst Handbook.* Milwaukee, WI: ASQ Quality Press.

Clements, Paul, Felix Bachmann, Len Bass, David Garlan, James Ivers, Reed Little, Paulo Merson, Robert Nord, and Judith Stafford. 2011. *Documenting Software Architectures: Views and Beyond.* 2nd ed. Upper Saddle River, NJ: Addison-Wesley.

CMMI. 2007. *Capability Maturity Model Integration (CMMI) for Acquisition, Version 1.2, CMU/SEI-2007-TR-017, ESC-TR-2007-017.* Pittsburgh, PA: Software Engineering Institute.

———. 2010a. *Capability Maturity Model Integration (CMMI) for Development: Improving Processes for Developing Better Products and Services, Version 1.3 CMU/SEI-2010-TR-033, ESC-TR-2010-033.* Pittsburgh, PA: Carnegie Mellon University Software Engineering Institute.

———. 2010b. *Capability Maturity Model Integration (CMMI) for Service: Improving Processes for Providing Better Services, Version 1.3 CMU/SEI-2010-TR-034, ESC-TR-2010-034.* Pittsburgh, PA: Carnegie Mellon University Software Engineering Institute.

———. 2010c. *Capability Maturity Model Integration (CMMI) for Acquisition, Version 1.3 CMU/SEI-2010-TR-032, ESC-TR-2010-032.* Pittsburgh, PA: Carnegie Mellon University Software Engineering Institute.

———. 2023. *Capability Maturity Model Integration (CMMI), Version 3.0,* Schaumburg, Illinois, Information System Audit and Control Association (ISACA).

CMU/SEI. 1992, September. *Software Size Measurement: A Framework for Counting Source Statements, ESC-TR-92-020.* Technical Report, Robert E. Park, with the Size Subgroup of the Software Metrics Definition Working Group and the Software Process Measurement Project Team, Carnegie-Mellon University, Software Engineering Institute.

———. 2009. *People Capability Maturity Model (P-CMM), Version 2.0 CMU/SEI-2009-TR-003, ESC-TR-2009-003.* 2nd ed. Pittsburgh, PA: Carnegie Mellon University, Software Engineering Institute.

CNSS. 2015. "Committee on National Security Systems Instruction No. 4009." *Committee on National Security Systems (CNSS) Glossary.*

Coad, Peter, and Edward Yourdon. 1990. *Object-Oriented Analysis.* 2nd ed. Englewood Cliffs, NJ: Yourdon Press Computing Series.

Cockburn, Alistair. 2002. *Agile Software Development.* Boston, MA: Addison-Wesley.

———. 2005. *Crystal Clear: A Human-Powered Methodology for Small Teams.* Boston: Addison-Wesley.

Cohn, Mike. 2006. *Agile Estimating and Planning.* Upper Saddle River, NJ: Prentice Hall Professional Technical Reference.

Coleman, Lance B., Sr., ed. 2020. *ASQ Certified Quality Auditor.* Milwaukee, WI: ASQExcellence.

Copeland, Lee. 2003. *A Practitioner's Guide to Software Test Design.* Norwood, MA: Artech House.

Covey, Stephen R. 1998. *The 7 Habits of Highly Effective People: Powerful Lessons in Personal Change.* New York, NY: Simon & Schuster.

Crosby, Philip B. 1984. *Quality Without Tears: The Art of Hassel-Free Management.* New York, NY: Penguin Group.

DAMA. 2017. *DAMA-DMBOK: Data Management Body of Knowledge.* 2nd ed. Basking Ridge, NJ: Technics Publications.

Davies, Rachel, and Liz Sedley. 2009. *Agile Coaching.* Raleigh, NC: The Pragmatic Bookshelf.

DeMarco, Tom. 2001. *Slack.* New York, NY: Broadway Books.

DeMarco, Tom, and Timothy Lister. 1999. *Peopleware—Productive Projects and Teams.* 2nd ed. New York, NY: Dorset House.

———. 2003. *Waltzing with Bears: Managing Risk on Software Projects.* New York, NY: Dorset House.

DePree, Max. 1989. *Leadership Is an Art.* New York, NY: Dell.

Derby, Esther, and Diana Larsen. 2006. *Agile Retrospectives: Making Good Teams Great.* Raleigh, NC: The Pragmatic Bookshelf.

Deutscher, Stefan, Walter Bohmayr, William Yin, and Massimo. 2014. "Cybersecurity Meets IT Risk Management." Available online at bcgperspectives.com.

DOE. 2010. *DOE G 414.1-4, Safety Software Guide for Use with 10CFR 830 Subpart A, Quality Assurance Requirements, and DOE 414.1C, Quality Assurance.* Approved 6-17-05 Certified 11-3-10. Washington, DC: U.S. Department of Energy (DOE).

Dorofee, Audrey J., Julie A. Walker, Christopher J. Alberts, Ronald P. Higuera, Richard L. Murphy, and Ray C. Williams. 1996. *Continuous Risk Management Guidebook.* Pittsburgh, PA: Carnegie Mellon University, Software Engineering Institute.

Down, Alex, Michael Coleman, and Peter Absolon. 1994. *Risk Management for Software Projects.* London: McGraw-Hill.

Dudash, Robin. 2015. *Software Stakeholder Management—It's Not All It's Coded Up to Be.*

Dunn, Robert. 1990. *Software Quality, Concepts, and Plans.* Englewood Cliffs, NJ: Prentice Hall.

Ericson, Clifton A. 2016. *II. Hazard Analysis Techniques for System Safety.* 2nd ed. Danvers, MA: Wiley.

Fenton, Norman E., and Shari Lawrence Pfleeger. 1997. *Software Metrics: A Rigorous and Practical Approach.* 2nd ed. London: PWS Publishing.

Florac, William, and Anita Carleton. 1999. *Measuring the Software Process, Statistical Process Control for Software Process Improvement.* Reading, MA: Addison-Wesley.

Frailey, Dennis J. 2016. *Software Safety: Part 1—Background and Introduction: How Software Contributes to Safety and Why We Need to Be Concerned About It and Part 2—What We Can Do About Software Safety, System Control Analysis and Software-Specific Issues.* Course taught at the University of Texas at Dallas.

Furterer, Sandra L., and Douglas C. Woods, eds. 2021. *The ASQ Certified Manager of Quality/Organizational Excellence.* 5th ed. Milwaukee, WI: ASQExcellence.

Futrell, Robert T., Donald F. Shafer, and Linda Isabell Shafer. 2002. *Quality Software Project Management.* Upper Saddle River, NJ: Prentice Hall PTR.

Garvin, David. 1984. "What does 'product quality' really mean?" *Sloan Management Review.*

Gause, Donald, and Gerald Weinberg. 1989. *Exploring Requirements, Quality Before Design.* New York, NY: Dorset House Publishing.

GDPR. 2023. *European Union (EU) General Data Protection Regulations (GDPR).* Available at https://gdpr.eu/tag/gdpr/

GEIA. 2022. *SAE/GEIA-859B. 2022. Data Management Standard.* AE International.

Gilb, Tom. 1988. *Principles of Software Engineering Management.* Wokingham, England: Addison-Wesley.

———. 1993. *Software Inspections.* Wokingham, England: Addison-Wesley.

———. 2005. *Competitive Engineering: A Handbook for Systems Engineering, Requirements Engineering, and Software Engineering Using Planguage.* Amsterdam: Elsevier Butterworth Heinemann.

Glass, Robert. 1995. *Software Creativity.* Englewood Cliffs, NJ: Addison-Wesley.

Goodman, Paul. 1993. *Practical Implementation of Software Metrics.* London: McGraw-Hill.

———. 2004. *Software Metrics: Best Practices for Successful IT Management.* Brookfield, CT: Rothstein Associates, Inc.

Gottesdiener, Ellen. 2002. *Requirements by Collaboration,* Boston, MA: Addison-Wesley.

Grady, Robert. 1992. *Practical Software Metrics for Project Management and Process Improvement.* Englewood Cliffs, NJ: Prentice Hall PTR.

Graham, Dorothy, Erik van Veenendaal, Isabel Evans, Rex Black. 2008. *Foundations of Software Testing: ISTQB Certification.* Rev. ed. Australia: Course Technology, Cengage Learning.

GSAM. 2000. *Guidelines for Successful Acquisition and Management of Software-Intensive Systems: Weapon Systems, Command and Control Systems.* Management Information Systems (GSAM) Version 3.0. Department of the Air Force, Software Technology Support Center.

Gutierrez, Tomas. 2015. "How Do You Manage Distributed Teams? What Methodologies Do You Use? What Tools?" Available on Quora.com.

Hall, Elaine M. 1998. *Managing Risk: Methods for Software Systems Development.* Reading, MA: Addison-Wesley.

Hancock, Johnathan. 2024. *The GROW Model of Coaching and Mentoring.* MindTools. 2024 (available at https://www.mindtools.com/an0fzpz/the-grow-model-of-coaching-and-mentoring).

Harrington, H. James. 2006. *Process Management Excellence—Book 1 of the Five Pillars of Organizational Excellence.* CHICO, CA: Paton Press, LLC.

Hass, Anne Mette Jonassen. 2003. *Configuration Management Principles and Practices.* Boston, MA: Addison-Wesley.

Hayhurst, Kelly J., Dan S. Veerhusen, John J. Chileski, and Leanna K. Rierson. 2001. *A Practical Tutorial on Modified Condition/Decision Coverage, NASA/TM-2001-210876.* Hanover, MD: NASA Scientific and Technical Information (STI) Program Office.

Heller, Robert, and Tim Hindle, 1998. *Essential Manager's Manual.* New York, NY: DK Publishing.

Hersey, Paul. 1984. *The Situational Leader.* Escondido, CA: Warner Books.

Hiscock, Robb. 2023. "The 7 principles of privacy by design." *OneTrust Blog.* Available at https://www.onetrust.com/blog/principles-of-privacy-by-design/.

Humphrey, Watts S. 1989. *Managing the Software Process.* Reading, MA: Addison-Wesley.

IEEE. 1994. *IEEE 1228 Standard for Software Safety Plans.* New York, NY: The Institute of Electrical and Electronics Engineers.

———. 1998b. *IEEE Standards Software Engineering, IEEE Guide for Information Technology—System Definition—Concept of Operations (ConOps), IEEE Std. 1362-1998 (reaffirmed 2007).* New York, NY: The Institute of Electrical and Electronics Engineers.

———. 1998c. *IEEE Standards Software Engineering, IEEE Recommended Practice for Software Requirements Specifications, IEEE Std. 830-1998.* New York, NY: The Institute of Electrical and Electronics Engineers.

———. 1998d. *IEEE Standards Software Engineering, IEEE Standards for a Software Quality Metrics Methodology, IEEE Std. 1061-1998.* New York, NY: The Institute of Electrical and Electronics Engineers.

———. 2006. *IEEE Standards Software Engineering, Standard for Software Engineering—Software Life Cycle Processes—Maintenance, ISO/IEC 14764:2006.* New York, NY: The Institute of Electrical and Electronics Engineers.

———. 2008. *IEEE Standards Software Engineering, IEEE Standard for Software Reviews and Audits, IEEE Std. 1028-2008.* New York, NY: The Institute of Electrical and Electronics Engineers.

———. 2008a. *IEEE Standards Software Engineering, IEEE Standard for Software Test Documentation, IEEE Std. 829-2008.* New York, NY: The Institute of Electrical and Electronics Engineers.

———. 2012. *IEEE Standards Software Engineering, IEEE Standard for Configuration Management in Systems and Software Engineering, IEEE Std. 828-2012.* New York, NY: The Institute of Electrical and Electronics Engineers.

———. 2014. *IEEE Standards Software Engineering, IEEE Standard for Software Quality Assurance Processes, IEEE Std. 730-2014.* New York, NY: Institute of Electrical and Electronics Engineers.

———. 2016. *IEEE for System, Software and Hardware Verification and Validation, IEEE Std. 1012-2016.* New York, NY: Institute of Electrical and Electronics Engineers. IEEE/ ISO/IEC

———. 2017. *IEEE/ISO/IEC 24765 Systems and Software Engineering—Vocabulary.* Geneva, Switzerland: ISO. New York, NY: The Institute of Electrical and Electronic Engineers.

———. 2017a. *IEEE/ISO/IEC 12207 Standards Software Engineering, Systems and Software Engineering—Software Life Cycle Processes.* New York, NY: The Institute of Electrical and Electronics Engineers.

———. 2017b. *IEEE/ISO/IEC 15939. Systems and Software Engineering—Measurement Process.* New York, NY: Institute of Electrical and Electronics Engineers.

———. 2021a. *IEEE/ISO/IEC 29119-2 Systems and Software Engineering—Software Testing Part 2: Test Processes.* New York, NY: Institute of Electrical and Electronics Engineers.

———. 2021b. *IEEE/ISO/IEC 29119-3 Systems and Software Engineering—Software Testing—Part 3: Test Documentation.* New York, NY: Institute of Electrical and Electronics Engineers.

ISO. 2019. *ISO 22301:2019 Security and Resilience—Business Continuity Management System—Requirements.* Geneva, Switzerland: International Organization of Standards (ISO).

———. 2018. *ISO 19001 Guidelines for Auditing Management Systems. ISO 19011-2018.* Geneva, Switzerland: International Organization of Standards (ISO).

———. 2015. *American National Standard, Quality Management Systems—Requirements, ASQ/ANSI/ISO 9001-2015.* Milwaukee: ASQ Quality Press.

———. 2015a. *ISO. Quality Management Principles.* Geneva, Switzerland: International Organization of Standards. Available at http://www.iso.org/iso/ pub100080.pdf

"——— 2012. *ISO 22301:2012 Standard, Societal Security—Business Continuity Management System—Requirements.* Geneva, Switzerland: International Organization of Standards (ISO).

ISO/IEC. 2004. *ISO/IEC 15504-3 Information Technology—Process Assessment—Part 3: Guidance on Performing an Assessment.* Geneva, Switzerland: International Organization of Standards (ISO).

———. 2014. *ISO/IEC 25000 Systems and Software Engineering—Systems and Software Quality Requirements and Evaluation (SQuaRE)—Guide to SQuaRE,* Geneva, Switzerland: International Organization of Standards (ISO).

———. 2015. *ISO/IEC 25024 Systems and Software Engineering—Systems and Software Quality Requirements and Evaluation (SQuaRE)—Measurement of Data Quality,* Geneva, Switzerland: International Organization of Standards (ISO).

———. 2022. *ISO/IEC 27001—Information Security, Cybersecurity and Privacy Protection—Information Security Management Systems—Requirements.* Geneva, Switzerland: International Organization of Standards (ISO).

———. 2011. *ISO/IEC/IEEE 42010 Systems and Software Engineering—Architecture Description.* Geneva, Switzerland: ISO. New York, NY: The Institute of Electrical and Electronic Engineers.

———. 2017. *ISO/IEC/IEEE. 2017. ISO/IEC/IEEE 24765 Systems and Software Engineering—Vocabulary.* Geneva, Switzerland: ISO. New York, NY: The Institute of Electrical and Electronic Engineers.

ISTQB. 2018a. *Certified Tester Foundation Level Overview, Version 1.0.* International Software Testing Qualifications Board (ISTQB).

———. 2018b. *Certified Tester Foundation Level Syllabus, Version 2018.* International Software Testing Qualifications Board (ISTQB).

———. 2018c. *Standard Glossary of Terms Used in Software Testing, Version 3.1, All Terms.* International Software Testing Qualifications Board (ISTQB).

James, Michael. 2012. "Scrum Reference Card." Available at http://scrumreferencecard.com.

———. 2016. *Scrum Training Series.* Available at scrumtrainingseries.com.

Jarvis, Alka, Prakash Anand, and Johnson Jose. 2022. *Successful Management of Cloud Computing and DevOps.* Milwaukee, WI: Quality Press.

Jones, Capers. 1986. *Programming Productivity.* New York, NY: McGraw-Hill.

———. 1994. *Assessment and Control of Software Risks.* Upper Saddle River, NJ: Yourdon Press, Prentice Hall.

———. 2008. *Applied Software Measurement: Global Analysis of Productivity and Quality.* 3rd ed. New York, NY: McGraw Hill.

Juran, Joseph M. 1999. *Juran's Quality Handbook.* 5th ed. New York, NY: McGraw-Hill.

Kan, Stephen. 2003. *Metrics and Models in Software Quality Engineering.* 2nd ed. Boston: Addison-Wesley.

Kaner, Cem, Jack Faulk, and Hung Quoc Nguyen. 1999. *Testing Computer Software.*, 2nd ed. New York, NY: Wiley Computer Publishing, John Wiley & Sons.

Kappos, David. 2012. "An Examination of Software Patents." Keynote Address to the Center for American Progress, November 20, 2012. Available at http://www.uspto.gov/about-us/news-updates/examination-software-patents.

Kenefick, Sean. 2003. *Real World Software Configuration Management.* Berkeley, CA: APress.

Kerth, Norm L. 2001. *Project Retrospectives: A Handbook for Team Reviews.* New York, NY: Dorset House.

Keyes, Jessica. 2004. *Software Configuration Management.* Boca Raton, FL: Auerbach Publications.

Kim, Gene, Jez Humble, Patrick Debois, John Willis. 2023. *The DevOps Handbook, Second Edition: How to Create World-Class Agility, Reliability, & Security in Technology Organizations.* Portland, OR: IT Revolutions.

Krasner, Herb. 1998. "Using the Cost of Quality Approach for Software," *CrossTalk—The War on Bugs,* no. 11 (November).

Kruchten, Philippe. November 1995. Architectural Blueprints—The "4 + 1" View Model of Software Architecture. *IEEE Software* 12(6).

———. 2000. *The Rational Unified Process: An Introduction.* Reading, MA: Addison-Wesley.

Kubial, T. M., and Donald W. Benbow. 2009. *The Certified Six Sigma Black Belt Handbook.* 2nd ed. Milwaukee, WI: ASQ Quality Press.

Kvarme, Knut. 2013. "Distributed Teams: Looking at Distributed Teams Versus Co-located Teams." Scrum Alliance. Available at the Scrum Alliance website.

LaMalfa, Kyle. 2010. "9 ways to successfully manage customer feedback." Available at destinationcrm.com.

Lauesen, Soren. 2002. *Software Requirements: Styles and Techniques.* London: Addison-Wesley.

Leffingwell, Dean, and Don Widrig. 2000. *Managing Software Requirements: A Unified Approach.* Reading, MA: Addison-Wesley.

Leon, Alexis. 2005. *Software Configuration Management Handbook.* 2nd ed. Boston: Artech House.

Lewis, James P. 1995. *The Project Manager's Desk Reference—A Comprehensive Guide to Project Planning, Scheduling, Evaluation, Control, and Systems.* New York, NY: McGraw-Hill.

Loeb, Marshal, and Stephen Kindel. 1999. *Leadership for Dummies.* Foster City, CA: IDG Books Worldwide.

Martin, Robert C. 2003. *Agile Software Development: Principles, Patterns, and Practices.* Upper Saddle River, NJ: Prentice Hall.

McCabe, James D. 2007. *Network Analysis, Architecture, and Design*, 3rd ed. Morgan Kaufmann.

McConnell, Steve. 1996. *Rapid Development: Taming Wild Software Schedules.* Redmond, WA: Microsoft Press.

———. 2009. *Code Complete.* 2nd ed. Redmond, WA: Microsoft Press.

MIL-HDBK. 2001. *Military Handbook—Configuration Management Guidance, MIL-HDBK-61A(SE).* Washington, DC: Department of Defense.

Moreira, Mario. 2010. *Adapting Configuration Management for Agile Teams.* Chichester, West Sussex, United Kingdom: John Wiley & Sons.

Myers, Glenford J. 2004. *The Art of Software Testing.* 2nd ed. Hoboken, NJ: John Wiley & Sons.

Nekah. 2022. "What is diversity, equity, and inclusion in the workplace—defining DE&I in 2022." *Positively Nekah.* YouTube.

NIST. 2011. "FINAL Version of NIST Cloud Computing Definition." Available at https://nvlpubs.nist.gov/nistpubs/legacy/sp/nistspecialpublication800-145.pdf.

———. 2014. (Rob Ross, Janet Carrier Oren and Michael McEvilley). *Systems Security Engineering—An Integrated Approach to Building Trustworthy Resilient Systems, NIST Special Publication 800-160, Initial Public Draft.* Gaithersburg, MD: National Institute of Standards and Technology.

———. 2020. *NIST Special Publication 800-53:2020, Security and Privacy Controls for Information Systems and Organizations.* Available at the NIST website.

OMG (Object Management Group). 2015. *OMG Systems Modeling Language (OMG SysML™) Version 1.4.* Available at http://www.omg.org/spec/SysML/1.4/

Ould, Martyn A. 1990. *Strategies for Software Engineering: The Management of Risk and Quality.* Chichester, West Sussex, United Kingdom: John Wiley & Sons.

OWASP. 2009. *Some Proven Application Security Principles.* The Open Web Applications Security Project (OWASP). Available at the OWASP website.

———. 2013. *OWASP Top 10–2013: The Ten Most Critical Web Application Security Risks.* The Open Web Applications Security Project (OWASP). Available at the OWASP website.

Pande, Peter S., Robert P. Neuman, and Roland R. Cavanagh. 2000. *The Six Sigma Way.* New York, NY: McGraw-Hill.

Petchiny, Maj. Nicko. 1998. "Object Oriented Testing." [PowerPoint presentation.] Available at www.cs.queensu.ca.

Pfleeger, Shari Lawrence, and Joanne M. Atlee. 2010. *Software Engineering: Theory and Practice.* New York, NY: Pearson.

Pichler, Roman. 2010. *Agile Product Management with Scrum.* Upper Saddle River, NJ, Addison-Wesley.

PMI. 2005. PMI Global Standard, *Practice Standard for Earned Value Management.* Newton Square, PA: Project Management Institute.

———. 2013. PMI Global Standard, *Project Management Body of Knowledge (PMBOK®) Guide.* 5th ed. Newton Square, PA: Project Management Institute.

Poppendieck, Mary, and Tom Poppendieck. 2007. *Implementing Lean Software Development: From Concept to Cash.* Upper Saddle River, NJ: Addison-Wesley.

Pressman, Roger. 2005. *Software Engineering: A Practitioner's Approach.* 6th ed. New York, NY: McGraw-Hill.

Pressman, Roger Takeuchi, H., and I. Nonaka. 1986 to 1995 and Bruce R. Maxim. 2015. *Software Engineering: A Practitioner's Approach.* 8th ed. New York, NY: McGraw-Hill Project.

Project 2000. *User's Guide - Microsoft Project 2000. Microsoft Corporation. 2002.*

Putnam, Lawrence H., and Ware Myers. 1992. *Measures for Excellence—Reliable Software on Time, Within Budget.* Englewood Cliffs, NJ: Yourdon Press.

———. 2003. *Five Core Metrics: The Intelligence Behind Successful Software Management.* New York, NY: Dorset House.

Pyzdek, Thomas. 2000. *Quality Engineering Handbook.* New York, NY: Marcel Dekker.

———. 2001. *The Six Sigma Handbook.* New York, NY: McGraw-Hill, Quality Publishing Tucson.

Robertson, Suzanne, and James Robertson. 2013. *Mastering the Requirements Process: Getting Requirements Right.* Upper Saddle River, NJ: Addison-Wesley.

Rothman, Johanna. 2007. *Manage It! Your Guide to Modern, Pragmatic Project Management.* Raleigh, NC: The Pragmatic Bookshelf.

Rozanski, Nick, and Eoin Woods. 2005. *Software Systems Architecture: Working with Stakeholders Using Viewpoints and Perspectives.* Upper Saddle River, NJ: Addison-Wesley.

Russell, J. P., ed. 2013. "ASQ Quality Audit Division," in *The ASQ Auditing Handbook.* 4th ed. Milwaukee, WI: ASQ Quality Press.

Ryan, Richard, and Edward L. Deci. 2000. "Intrinsic and Extrinsic Motivations: Classic Definitions and New Directions." *Contemporary Educational Psychology.*

Scholtes, Peter R., Brian L. Joiner, Barbara J. Steibel. 2003. *The Team Handbook.* 3rd ed. Madison, WI: Oriel.

Schwaber, Ken, and Mike Beedle. 2002. *Agile Software Development with Scrum.* Upper Saddle River, NJ: Prentice Hall.

Schwaber, Ken, and Mike Cohn. 2007. From a Scrum Master training class given January 16–17, 2007 in Orlando, Florida.

SEI. 1993. Marvin J. Karr et al., "Taxonomy-Based Risk Identification," CMU/SEI-93-TR-6. Pittsburgh, PA: Software Engineering Institute, Carnegie Mellon University.

Shewhart, Walter .1980. *Economic Control of Quality of Manufactured Product.* Milwaukee, WI: ASQ Quality Press.

Sommerville, Ian. 2016. *Software Engineering.* 10th edition. Boston: Pearson.

Soni, D., R. L. Nord, and C. Hofmeister. 1995. "Software Architecture in Industry Applications," Proceedings of the 17th International Conferences on Software Engineering.

Stellman, Andrew, and Jennifer Greene. 2018. *Head First Agile: A Brain-Friendly Guide.* Beijing: O'Reilly.

Stuart, Jenny. 2017. "Kanban overview class." Construx Software Builders, Inc.

Succi, Giancarlo, and Michele Marchesi. 2001. *Extreme Programming Examined.* Boston, MA: Addison-Wesley.

———. 1995. *The Knowledge Creating Company.* New York, NY: Oxford University Press.

Thielova, Linda. 2023. "7 Steps to Comply with ISO 31700-1:2023 (Standard on Privacy by Design)." *OneTrust Blog.* https://www.onetrust.com/blog/7-steps-to-comply-with-iso-31700-12023-standard-on-privacy-by-design/.

Toastmasters International. 1990. *How to Listen Effectively.* Mission Viejo, CA: Toastmasters International.

Vienneau, Robert L., and Milton Johns. 2008. "Introduction to U.S. Intellectual Property (IP) Law," *SoftwareTech News, The Data & Acquisition Center for Software (DACS)* 11, no. 2 (August). Available at www.softwaretechnews.com.

Weinberg, Gerald. 1997. *Quality Software Management, Volume 4: Anticipating Change.* New York, NY: Dorset House Publishing.

Westcott, Russell T., ed. 2006. *ASQ Quality Management Division. The Certified Manager of Quality/Organizational Excellence Handbook.* 3rd ed. Milwaukee, WI: ASQ Quality Press.

Westfall, Linda. 2023. "12 Steps to Useful Software Metrics." Software Excellence Academy.

Whittaker, James A. 2003. *How to Break Software: A Practical Guide to Testing.* New York, NY: Addison-Wesley.

Wiegers, Karl E. 2002. *Peer Reviews in Software.* Boston, MA: Addison-Wesley

———. 2003. Wiegers, Karl E, *Software Requirements.* 2nd ed. Redmond, WA: Microsoft Press.

———. 2004. *In Search of Excellent Requirements.* Process Impact.

Wiegers, Karl, and Joy Beatty. 2013. *Software Requirements.* 3rd ed. Redmond, WA: Microsoft Press.

Withall, Stephen. 2007. *Software Requirements Patterns.* Redmond, WA: Microsoft Press.

Yourdon, Edward, and Larry Constantine. 1979. *Structured Design: Fundamentals of a Discipline of Computer Program and Systems Design.* Englewood Cliffs, NJ: Prentice Hall.

WEBSITES

Agile Alliance. Agile Alliance: http://www.agilealliance.com

ASQ. American Society for Quality: http://www.asq.org

ASQ. Software Division Software Division—American Society for Quality (asq.org): BCI The Business Continuity Institute: http://www.thebci.org

CMMI Institute: http://cmmiinstitute.com

Coursera. Online Global Learning Community, including Massive Open Online Courses (MOOCs): https://www.coursera.org/

Crossroads. http://www.cmcrossroads.com

Crosstalk. The Journal of Defense Software Engineering: https://community.apan.org/wg/crosstalk/

DAMA. Data Management Association: https://www.dama.org/

EdX. Online Global Learning Community, including Massive Open Online Courses (MOOC): https://www.edx.org/

GDPR. *European Union (EU) General Data Protection Regulations:* https://gdpr.eu/tag/gdpr/

IEEE. Institute of Electrical and Electronics Engineers: http://www.ieee.org

IEEE. Computer Society: http://www2.computer.org

IFPUG. International Function Point User's Group: http://www.ifpug.org

IIA. The Institute of Internal Auditors: http://www.theiia.org

Investopedia. https://www.investopedia.com/

ISBSG. International Software Benchmarking Standards Group: http://www.isbsg.org

ISO. International Organization for Standardization: http://www.iso.org

ITIL. IT Service Management Forum—Information Technology Infrastructure Library (ITIL): http://www.itsmfi.org/

NASA. Office of Safety and Mission Assurance-Software Assurance: http://sma.nasa.gov/sma-disciplines/software-assurance

NIST: National Institute of Standards and Technology, U.S. Department of Commerce: https://www.nist.gov/

OMG. Object Management Group: http://www.omg.org

OWASP. Open Web Application Security Project (OWASP): http://www.owasp.org

PMI. Project Management Institute: http://www.pmi.org

PMI. Project Management Magazine: http://www.projectmanagement.com/

PSM. Practical Software and System Measurement: http://www.psmsc.com

QAI. Quality Assurance Institute: http://qaiusa.com/

Scrum. Scrum Alliance: http://www.scrumalliance.org

Scrum Training Series. Free Online Scrum Master Training: http://www.scrumtrainingseries.com/

SEI. Software Engineering Institute: http://www.sei.cmu.edu

SEIR. Software Engineering Information Repository: http://seir.sei.cmu.edu/ SPMN.

Software Excellence Academy: https://www.softwareexcellenceacademy.com/

STQE. Software Testing and Quality Engineering: http://www.stickyminds.com

SWEBOK. Software Engineering Body of Knowledge. IEEE Computer Society: http://www.swebok.org

TED. TED Talks: https://www.ted.com

Westfall Team. The Westfall Team: https://www.westfallteam.com

Index

A

C

D

E

F

G

H

I

N

O

P

R

S

T

U

X

Y

www.ingramcontent.com/pod-product-compliance
Lightning Source LLC
LaVergne TN
LVHW080807220926
848485LV00011B/332